Proceedings of the Infinite Analysis 09

NEW TRENDS IN QUANTUM INTEGRABLE SYSTEMS

Proceedings of the Infinite Analysis 09

NEW TRENDS IN QUANTUM INTEGRABLE SYSTEMS

Kyoto, Japan 27–31 July 2009

Editors

Boris Feigin
Landau Institute for Theoretical Physics, Russia

Michio Jimbo
Rikkyo University, Japan

Masato Okado
Osaka University, Japan

World Scientific

NEW JERSEY • LONDON • SINGAPORE • BEIJING • SHANGHAI • HONG KONG • TAIPEI • CHENNAI

Published by

World Scientific Publishing Co. Pte. Ltd.
5 Toh Tuck Link, Singapore 596224
USA office: 27 Warren Street, Suite 401-402, Hackensack, NJ 07601
UK office: 57 Shelton Street, Covent Garden, London WC2H 9HE

British Library Cataloguing-in-Publication Data
A catalogue record for this book is available from the British Library.

NEW TRENDS IN QUANTUM INTEGRABLE SYSTEMS
Proceedings of the Infinite Analysis 09

ISBN-13 978-981-4324-36-6
ISBN-10 981-4324-36-1

Printed in Singapore by B & Jo Enterprise Pte Ltd

Tetsuji Miwa

Dedicated to Tetsuji Miwa
on the occasion of his sixtieth birthday

H.I.S.

PREFACE

The present volume is the proceedings of the international workshop "Infinite Analysis 09: New Trends in Quantum Integrable Systems ", held at Kyoto University, July 27–31, 2009. About 110 participants gathered from around the world, including Asia, Australia, Europe and US, and 21 invited talks as well as 22 posters were presented.

This workshop was planned in spirit as a continuation of the research activities at Kyoto during the last 20 years. Before 2009, the most recent was a series of workshops held in 2004 as the RIMS Research Project "Method of Algebraic Analysis in Integrable Systems". Five years have elapsed since then, and many exciting new developments have emerged during the period. We felt it was time to have an occasion for communicating some of these latest achievements.

We hope that this volume serves as a useful guide to the current advances in integrable systems and related areas in mathematics and physics.

Last but not least, we wish to dedicate this book to Tetsuji Miwa on the occasion of his 60th birthday.

Editors
Boris Feigin
Michio Jimbo
Masato Okado

WORKSHOP PROGRAM

July 27 (Mon)

10:00 – 10:30	*Registration*
10:40 – 11:40	B. M. McCoy (C.N.Yang Inst.) *Comments on Correlations*
13:30 – 14:30	J. Teschner (DESY Theory) *The Sine-Gordon model revisited*
14:40 – 15:40	T. Deguchi (Ochanomizu) *Correlation functions of the integrable spin-s XXZ spin chain via fusion method, and the form factors of exactly solvable models with the sl(2) loop algebra symmetry*
16:10 – 17:10	J. Suzuki (Shizuoka) *Quantum spin chains at finite temperatures*

July 28 (Tue)

9:30 – 10:30	V. Bazhanov (Australian National Univ.) *Fundamental mathematical structures in statistical and quantum systems*
10:40 – 11:40	F. Smirnov (CNRS, LPTHE) *Conformal Field Theory limit of fermionic structure for integrable models*
13:30 – 14:30	K. Takasaki (Kyoto) *KP and Toda tau functions in Bethe ansatz: a review*
14:40 – 15:40	A. Kuniba (Tokyo) *Spectrum in multi-species asymmetric simple exclusion process on a ring*
16:10 – 17:10	T. Nakanishi (Nagoya) *Periodicity of T-systems*

July 29 (Wed)

9:30 – 10:30	B. Feigin (Landau Inst., Indep. Univ. of Moscow) *Representations of toroidal algebras*
10:40 – 11:40	T. Miwa (Kyoto) *Fermionic and bosonic formulas*
11:40 – 12:00	*Group photo*
13:30 – 15:30	*Poster session*
16:00 – 17:00	*Concert (at Kyodai Kaikan)*
18:00 – 20:00	*Banquet (at Shiran Kaikan)*

July 30 (Thu)

9:30 – 10:30	A. Tsuchiya (IPMU) *The triplet vertex operator algebra $W(p)$ and the restricted quantum group $\bar{U}_q(sl_2)$ at $q = e^{\frac{\pi i}{p}}$*
10:40 – 11:40	E. Mukhin (Indianapolis) *Bethe algebras of the Gaudin model*
13:30 – 14:30	A. Nakayashiki (Kyusyu) *On the multivariate sigma function*
14:40 – 15:40	J. Shiraishi (Tokyo) *Hirota-Miwa equations and Macdonald operators*
16:10 – 17:10	M. Kasatani (Tokyo) *Boundary quantum Knizhnik-Zamolodchikov equation*

July 31 (Fri)

9:30 – 10:30	I. Cherednik (UNC Chapel Hill) *Affine Satake operator via DAHA*
10:40 – 11:40	P. Zinn-Justin (CNRS, LPTHE) *$O(1)$ loop model and combinatorics*
13:30 – 14:30	O. Lisovyy (Tours) *Algebraic solutions of the sixth Painleve equation*
14:40 – 15:40	R. Inoue (Suzuka Univ. of Medical Sci.) *Tropical geometry and ultradiscrete integrable systems*
16:10 – 17:10	Y. Takeyama (Tsukuba) *Differential equations compatible with boundary rational qKZ equations*

LIST OF PARTICIPANTS

Tomoyuki Arakawa
Mitsuhiro Arikawa
Susumu Ariki
Chikashi Arita
Vladimir Bazhanov
Yu-Lin Chang
Ivan Cherednik
Ed Corrigan
Etsuro Date
Tetsuo Deguchi
Patrick Dorey
Naoya Enomoto
Boris Feigin
Evgeny Feigin
Omar Foda
Frank Göhmann
Koji Hasegawa
Adrian Hemery
Jin Hong
Ayumu Hoshino
Mana Igarashi
Rei Inoue
Michio Jimbo
Eriko Kaminishi
Seok-Jin Kang
Masahiro Kasatani
Masaki Kashiwara
Akishi Kato
Shin-ichi Kato
Takahiro Kawai
Ryosuke Kodera
Kozo Koizumi
Satoshi Kondo
Yukiko Konishi
Hitoshi Konno
Atsuo Kuniba
Toshiro Kuwabara
Hyeonmi Lee
Inha Lee
Oleg Lisovyy
Giacomo Marmorini
Pierre Mathieu
Chihiro Matsui
Takuya Matsumoto
Atsushi Matsuo
Barry McCoy
Kailash Misra
Tetsuji Miwa
Evgeny Mukhin
Hajime Nagoya
Satoshi Naito
Hiraku Nakajima
Tomoki Nakanishi
Toshiki Nakashima
Toshio Nakatsu
Atsushi Nakayashiki
Katsuyuki Naoi
Hiroshi Naruse
Bernard Nienhuis
Akinori Nishino
Michitomo Nishizawa
Satoru Odake
Yasuhiro Ohta
Yousuke Ohyama
Soichi Okada
Masato Okado
Domenico Orlando
Yas-Hiro Quano
Susanne Reffert
Ingo Runkel
Yoshihisa Saito
Hidetaka Sakai
Kazumitsu Sakai
Atsumu Sasaki
Mikio Sato
Youichi Shibukawa
Mark Shimozono
Jun'ichi Shiraishi
Fedor Smirnov
Fabian Spill
Takeshi Suzuki
Junji Suzuki
Norio Suzuki
Sakie Suzuki
Taichiro Takagi
Kanihisa Takasaki
Takashi Takebe
Yoshitsugu Takei
Kouichi Takemura
Yoshihiro Takeyama
Jorg Teschner
Tetsuji Tokihiro
Zengo Tsuboi
Tadayoshi Tsuchida
Shunsuke Tsuchioka
Akihiro Tsuchiya
Teruhisa Tsuda
Benoit Vicedo
Takako Watanabe
Hidekazu Watanabe
Robert Weston
Michael Wheeler
Ralph Willox
Yasuhiko Yamada
Yuji Yamada
Daisuke Yamakawa
Seiji Yamamoto
Takao Yamazaki
Shintarou Yanagida
Paul Zinn Justin

CONTENTS

PROPERTIES OF LINEAR INTEGRAL EQUATIONS RELATED TO THE SIX-VERTEX MODEL WITH DISORDER PARAMETER

HERMANN BOOS and FRANK GÖHMANN

Fachbereich C – Physik, Bergische Universität Wuppertal,
42097 Wuppertal, Germany

One of the key steps in recent work on the correlation functions of the XXZ chain was to regularize the underlying six-vertex model by a disorder parameter α. For the regularized model it was shown that all static correlation functions are polynomials in only two functions. It was further shown that these two functions can be written as contour integrals involving the solutions of a certain type of linear and non-linear integral equations. The linear integral equations depend parametrically on α and generalize linear integral equations known from the study of the bulk thermodynamic properties of the model. In this note we consider the generalized dressed charge and a generalized magnetization density. We express the generalized dressed charge as a linear combination of two quotients of Q-functions, the solutions of Baxter's t-Q-equation. With this result we give a new proof of a lemma on the asymptotics of the generalized magnetization density as a function of the spectral parameter.

Keywords: Quantum spin chains; correlation functions.

1. Introduction

In our present understanding of the thermodynamics[9] and the finite temperature correlation functions[2,4,6,8] of the XXZ quantum spin chain certain complex valued functions defined as solutions of linear or non-linear integral equations play an important role. In first place we have to mention the so-called auxiliary function $\mathfrak{a}$, satisfying the non-linear integral equation

$$\ln(\mathfrak{a}(\lambda|\kappa)) = -2\kappa\eta - \frac{2J\,\mathrm{sh}(\eta)\mathrm{e}(\lambda)}{T} - \int_C \frac{\mathrm{d}\mu}{2\pi\mathrm{i}} K(\lambda-\mu)\ln(1+\mathfrak{a}(\mu|\kappa))\,. \tag{1}$$

Here J sets the energy scale of the spin chain, T is the temperature, and η controls the anisotropy[a]. The bare energy $\mathrm{e}(\lambda)$ and the kernel $K(\lambda)$ are defined as

$$\mathrm{e}(\lambda) = \mathrm{cth}(\lambda) - \mathrm{cth}(\lambda+\eta)\,, \quad K(\lambda) = \mathrm{cth}(\lambda-\eta) - \mathrm{cth}(\lambda+\eta)\,. \tag{2}$$

[a] The anisotropy parameter of the XXZ Hamiltonian is $\Delta = \mathrm{ch}(\eta)$ and the quantum group parameter $q = \mathrm{e}^{\eta}$. For simplicity we shall assume throughout that $\mathrm{Re}\,\eta = 0$ and $0 < \mathrm{Im}\,\eta < \pi/2$. This means to consider the XXZ chain in the critical regime.

The integration contour C encircles the real axis at a distance slightly smaller than $\gamma/2 = \mathrm{Im}\,\eta/2$. The twist parameter κ is proportional to the magnetic field h, $\kappa = h/2T\eta$.

The auxiliary function $\mathfrak{a}$ determines the free energy per lattice site,

$$f(h,T) = -\frac{h}{2} - T \int_C \frac{\mathrm{d}\lambda}{2\pi\mathrm{i}}\, \mathrm{e}(\lambda) \ln(1 + \mathfrak{a}(\lambda|\kappa))\,, \tag{3}$$

of the spin chain and, hence, all its thermodynamic properties. This explains the importance of $\mathfrak{a}$.

The magnetization, for instance, is defined as

$$m(h,T) = -\frac{\partial f(h,T)}{\partial h}\,. \tag{4}$$

It has a simple expression in terms of the logarithmic derivative of the auxiliary function,

$$\sigma(\lambda) = -T \partial_h \ln(\mathfrak{a}(\lambda|\kappa))\,, \tag{5}$$

namely,

$$m(h,T) = -\frac{1}{2} - \int_C \frac{\mathrm{d}\lambda}{2\pi\mathrm{i}} \frac{\mathrm{e}(-\lambda)\sigma(\lambda)}{1 + \mathfrak{a}(\lambda|\kappa)}\,. \tag{6}$$

The function σ satisfies the linear integral equation

$$\sigma(\lambda) = 1 + \int_C \frac{\mathrm{d}\mu}{2\pi\mathrm{i}} \frac{K(\lambda-\mu)\sigma(\mu)}{1 + \mathfrak{a}(\mu|\kappa)}\,. \tag{7}$$

Its zero temperature limit

$$\xi(\lambda) = \lim_{T\to 0+} \sigma(\lambda) \tag{8}$$

is called the dressed charge. It plays an important role in the calculation of the asymptotics of correlation functions at $T = 0$. For the lack of any better name we shall call σ, and also an α-generalization of σ to be considered below, the dressed charge as well.

Another possibility of expressing the magnetization per lattice site (6) is by means of a magnetization density G satisfying

$$G(\lambda) = \mathrm{e}(-\lambda) + \int_C \frac{\mathrm{d}\mu}{2\pi\mathrm{i}} \frac{K(\lambda-\mu)G(\mu)}{1 + \mathfrak{a}(\mu|\kappa)}\,. \tag{9}$$

Applying the 'dressed function trick' to (7) and (9) we obtain

$$m(h,T) = -\frac{1}{2} - \int_C \frac{\mathrm{d}\lambda}{2\pi\mathrm{i}} \frac{G(\lambda)}{1 + \mathfrak{a}(\lambda|\kappa)}\,. \tag{10}$$

The integrability of the XXZ chain manifests itself in the existence of commuting families of transfer matrices and Q-operators of the associated six-vertex model.[1] With an appropriate staggered choice of the horizontal spectral parameters the partition function of the six-vertex model on a rectangular lattice approximates the partition function of the XXZ chain.[9] The approximation becomes exact in the so-called Trotter limit, when the extension of the lattice in vertical direction goes to infinity. By a modification of the boundary conditions in vertical direction we can also obtain an expression for the density matrix of a finite segment of the spin chain.[5,6] The column-to-column transfer matrix in this approach is called the quantum transfer matrix. It satisfies a t-Q-equation as well, which becomes a functional equation for the eigenvalues of the involved operators due to their commutativity.

We denote the dominant eigenvalue of the quantum transfer matrix by $\Lambda(\lambda|\kappa)$. This eigenvalue alone determines the free energy in the thermodynamic limit, when the horizontal extension of the lattice tends to infinity, $f(h,T) = -T\ln\Lambda(0|\kappa)$. Let the corresponding Q-function be $Q(\lambda|\kappa)$. Then Λ and Q satisfy the t-Q-equation

$$\Lambda(\lambda|\kappa)Q(\lambda|\kappa) = q^{\kappa}a(\lambda)Q(\lambda-\eta|\kappa) + q^{-\kappa}d(\lambda)Q(\lambda+\eta|\kappa)\,, \tag{11}$$

where $a(\lambda)$ and $d(\lambda)$ are the pseudo vacuum eigenvalues of the diagonal entries of the monodromy matrix associated with the quantum transfer matrix,

$$a(\lambda) = \left(\frac{\operatorname{sh}(\lambda+\frac{\beta}{N})}{\operatorname{sh}(\lambda+\frac{\beta}{N}-\eta)}\right)^{\frac{N}{2}}, \quad d(\lambda) = \left(\frac{\operatorname{sh}(\lambda-\frac{\beta}{N})}{\operatorname{sh}(\lambda-\frac{\beta}{N}+\eta)}\right)^{\frac{N}{2}}, \tag{12}$$

and $\beta = 2J\operatorname{sh}(\eta)/T$.

Using the Q-functions corresponding to the dominant eigenvalue the auxiliary function $\mathfrak{a}$ can be expressed as

$$\mathfrak{a}(\lambda|\kappa) = \frac{q^{-2\kappa}d(\lambda)Q(\lambda+\eta|\kappa)}{a(\lambda)Q(\lambda-\eta|\kappa)}\,. \tag{13}$$

In fact, the auxiliary function $\mathfrak{a}$ is usually defined by (13), and afterwards it is shown that $\mathfrak{a}$ satisfies the non-linear integral equation (1) in the Trotter limit. To be more precise, the Q-functions, the transfer matrix eigenvalue and the vacuum expectation values depend implicitly on the Trotter number N. Hence, $\mathfrak{a}$ as defined in (13) depends on N. One can show[b] that it satisfies the non-linear integral

[b] For a recent pedagogical review on quantum spin chains within the quantum transfer matrix approach see Ref. 7, submitted to the same Festschrift volume for T. Miwa as this article.

equation

$$\begin{aligned}\ln \mathfrak{a}(\lambda|\kappa) &= -2\kappa\eta \\ &+ \ln\left[\frac{\operatorname{sh}(\lambda - \frac{\beta}{N})\operatorname{sh}(\lambda + \frac{\beta}{N} + \eta)}{\operatorname{sh}(\lambda + \frac{\beta}{N})\operatorname{sh}(\lambda - \frac{\beta}{N} + \eta)}\right]^{\frac{N}{2}} - \int_C \frac{\mathrm{d}\mu}{2\pi\mathrm{i}} K(\lambda - \mu)\ln(1 + \mathfrak{a}(\mu|\kappa)) . \qquad (14)\end{aligned}$$

Clearly this turns into (1) for $N \to \infty$. The integral equation (1) is the reason why the function $\mathfrak{a}$ is more useful for practical purposes than Q. It is hard to determine Q, and Q has no simple Trotter limit. On the other hand, (1) determines $\mathfrak{a}$ directly in the Trotter limit and can be converted into a form that can be accurately solved numerically.

Inserting (13) into (5) we obtain an expression for the dressed charge function in terms of logarithmic derivatives of Q-functions.

$$\sigma(\lambda) = 1 + \frac{1}{2\eta}\left(\frac{Q'(\lambda - \eta|\kappa)}{Q(\lambda - \eta|\kappa)} - \frac{Q'(\lambda + \eta|\kappa)}{Q(\lambda + \eta|\kappa)}\right), \qquad (15)$$

where the prime denotes the derivative with respect to κ. For the function G defined in (9) no such simple expression in terms of Q-functions is known.

Below we shall introduce generalizations of the functions σ and G that depend on additional parameters. For the generalized dressed charge we will derive a generalization of (15). This will be used in a derivation of the asymptotic behaviour of the generalized magnetization density as a function of the spectral parameter.

2. Linear integral equations

It was shown in Ref. 8 that all correlation functions of the XXZ chain regularized by a disorder parameter α can be expressed in terms of two functions, the ratio of eigenvalues

$$\rho(\lambda) = \frac{\Lambda(\lambda|\kappa + \alpha)}{\Lambda(\lambda|\kappa)} \qquad (16)$$

and a function ω with the essential part $\Psi(\lambda, \mu)$ that can be characterized in terms of solutions of certain α-dependent linear integral equations.[2] A thorough understanding of these two functions is of fundamental importance for the further study of the correlation functions of the XXZ chain and for the application of the lattice results to quantum field theory in various scaling limits.[3]

We define the 'measure'

$$\mathrm{d}m(\lambda) = \frac{\mathrm{d}\lambda}{2\pi\mathrm{i}\,\rho(\lambda)(1 + \mathfrak{a}(\lambda|\kappa))} \qquad (17)$$

and the deformed kernel

$$K_\alpha(\lambda) = q^{-\alpha}\operatorname{cth}(\lambda - \eta) - q^{\alpha}\operatorname{cth}(\lambda + \eta) . \qquad (18)$$

Then, for ν inside C the function G is, by definition, the solution of the integral equation

$$G(\lambda,\nu) = \\ q^{-\alpha}\operatorname{cth}(\lambda-\nu-\eta) - \rho(\nu)\operatorname{cth}(\lambda-\nu) + \int_C dm(\mu) K_\alpha(\lambda-\mu) G(\mu,\nu)\,. \quad (19)$$

Clearly G is a generalization of the magnetization density (9) that depends on an additional spectral parameter and on the disorder parameter α. For simplicity we keep the same notation also for the generalized function. G enters the definition of Ψ which, for ν_1,ν_2 inside C, is defined as

$$\Psi(\nu_1,\nu_2) = \int_C dm(\mu) G(\mu,\nu_2)\big(q^\alpha \operatorname{cth}(\mu-\nu_1-\eta) - \rho(\nu_1)\operatorname{cth}(\mu-\nu_1)\big)\,. \quad (20)$$

For ν or ν_1,ν_2 outside the contour, G and Ψ are given by the analytic continuations of the right and side of (19) or (20), respectively.

Lemma 2.1. *Asymptotic behaviour of G and Ψ as a functions of the spectral parameters.*

(i)

$$\lim_{\mathrm{Re}\lambda\to\infty} G(\lambda,\nu) = \lim_{\mathrm{Re}\nu\to\infty} G(\lambda,\nu) = 0\,. \quad (21)$$

(ii)

$$\lim_{\mathrm{Re}\nu_1\to\infty} \Psi(\nu_1,\nu_2) = -\frac{q^{-\alpha}-\rho(\nu_2)}{1+q^{2\kappa}}\,, \quad (22a)$$

$$\lim_{\mathrm{Re}\nu_2\to\infty} \Psi(\nu_1,\nu_2) = -\frac{q^{\alpha}-\rho(\nu_1)}{1+q^{-2\kappa}}\,. \quad (22b)$$

Proof. We may choose the contour C as the rectangular contour of hight slightly less than γ and of width $2R$ depicted in figure 1. R must be sufficiently large to include all Bethe roots. This is trivially possible for finite Trotter number, but also in Trotter limit $N\to\infty$.[7] Then the right hand side of (19) is holomorphic in λ for $|\mathrm{Im}\lambda| < \gamma/2$. It follows that

$$\lim_{\mathrm{Re}\lambda\to\infty} G(\lambda,\nu) = q^{-\alpha} - \rho(\nu) - (q^\alpha - q^{-\alpha})\int_C dm(\mu) G(\mu,\nu) = 0\,. \quad (23)$$

Here the the second equation will appear as lemma 2.3 below. We postpone the proof, because it needs some preparation.

For the calculation of the asymptotics of G for large $\mathrm{Re}\,\nu$ we have to take into account that $G(\lambda,\nu)$ as a function of λ has pole at $\lambda=\nu$ with residue $-\rho(\nu)$.

Hence, for ν outside C,

$$G(\lambda,\nu) = q^{-\alpha}\operatorname{cth}(\lambda-\nu-\eta) - \rho(\nu)\operatorname{cth}(\lambda-\nu) - \frac{K_\alpha(\lambda-\nu)}{1+\mathfrak{a}(\nu|\kappa)} + \int_C dm(\mu) K_\alpha(\lambda-\mu) G(\mu,\nu)\,. \tag{24}$$

Using that

$$\lim_{\mathrm{Re}\,\nu\to\infty} \rho(\nu) = \frac{q^{\kappa+\alpha}+q^{-\kappa-\alpha}}{q^\kappa+q^{-\kappa}}\,, \qquad \lim_{\mathrm{Re}\,\nu\to\infty} \mathfrak{a}(\nu|\kappa) = q^{-2\kappa} \tag{25}$$

and setting $g(\lambda) = \lim_{\nu\to\infty} G(\lambda,\nu)$ we obtain from (24)

$$g(\lambda) = \int_C dm(\mu) K_\alpha(\lambda-\mu) g(\mu)\,. \tag{26}$$

Then $g(\lambda)=0$, and (21) is proved.

A similar argument can be applied to prove (22b). For ν_2 outside C we have

$$\Psi(\nu_1,\nu_2) = \int_C dm(\mu) G(\mu,\nu_2)\big(q^\alpha \operatorname{cth}(\mu-\nu_1-\eta) - \rho(\nu_1)\operatorname{cth}(\mu-\nu_1)\big) - \frac{1}{1+\mathfrak{a}(\nu_2|\kappa)}\big(q^\alpha\operatorname{cth}(\nu_2-\nu_1-\eta) - \rho(\nu_1)\operatorname{cth}(\nu_2-\nu_1)\big)\,. \tag{27}$$

Using the second equation (21) and (25) we obtain (22b).

For the proof of (22a) we note that

$$\Psi(\nu_1,\nu_2) = \int_C dm(\mu) G(\mu,\nu_2)\big(q^\alpha \operatorname{cth}(\mu-\nu_1-\eta) - \rho(\nu_1)\operatorname{cth}(\mu-\nu_1)\big) - \frac{G(\nu_1,\nu_2)}{1+\mathfrak{a}(\nu_1|\kappa)} \tag{28}$$

if ν_1 is outside C. Using (23) we conclude that

$$\lim_{\nu_1\to\infty} \Psi(\nu_1,\nu_2) = -\frac{q^\kappa - q^{-\kappa}}{q^\kappa+q^{-\kappa}} \lim_{\nu_1\to\infty} G(\nu_1,\nu_2) - \frac{q^{-\alpha}-\rho(\nu_2)}{1+q^{2\kappa}}\,. \tag{29}$$

Thus, (22a) follows by means of (21). □

Here a few comments are in order. For the correlation functions of the XXZ chain,[2,8] it is actually not the function Ψ but the closely related function ω [c] which is at the heart of the theory

$$\omega(\nu_1,\nu_2) = 2\Psi(\nu_1,\nu_2)e^{\alpha(\nu_1-\nu_2)} + 4\big(\big(1+\rho(\nu_1)\rho(\nu_2)\big)g(\xi) - \rho(\nu_1)g(q^{-1}\xi) - \rho(\nu_2)g(q\xi)\big) \tag{30}$$

[c] We follow here the notation of Ref. 3. In Ref. 2 a slightly different notation for ω was used.

where $g(\xi) = \Delta_\xi^{-1}\psi(\xi)$ with $\xi = e^{\nu_1 - \nu_2}$ should be understood as in (2.10) of Ref. 3. Using the explicit form of the function $\psi(\xi) = \frac{\xi^\alpha}{2}\frac{\xi^2+1}{\xi^2-1}$ and the above lemma 2.1 one can see that

$$\lim_{\nu_1\to\pm\infty} e^{-\alpha(\nu_1-\nu_2)}\omega(\nu_1,\nu_2) = 0, \qquad \lim_{\nu_2\to\pm\infty} e^{-\alpha(\nu_1-\nu_2)}\omega(\nu_1,\nu_2) = 0 .$$

This is one of normalization conditions for ω introduced in Ref. 8.

The asymptotics of G and Ψ with respect to the first argument was derived in Ref. 2. The reasoning there was also based on the second equation (23), which was obtained rather indirectly by means of the reduction property of the density matrix and a multiple integral representation for the six-vertex model with disorder parameter. Below we shall present a more direct proof of it in lemma 2.3, based on a representation of the generalized dressed charge in terms of Q-functions.

Perhaps the simplest proof of lemma 2.3 utilizes the symmetry[2,8]

$$\Psi(\nu_1,\nu_2|\kappa,\alpha) = \Psi(\nu_2,\nu_1|-\kappa,-\alpha) . \tag{31}$$

If we combine this with $\rho(\lambda|\kappa,\alpha) = \rho(\lambda|-\kappa,-\alpha)$, then (22a) follows from (22b). But (22a) inserted into (29) implies $\lim_{\nu_1\to\infty} G(\nu_1,\nu_2) = 0$, and lemma 2.3 follows with the first equation (23).

Still, this is a little indirect and unsatisfactory. Here we are going for a more direct proof based upon the properties of the dressed charge function defined by

$$\sigma(\lambda) = 1 + \int_C dm(\mu)\sigma(\mu)K_\alpha(\mu-\lambda) . \tag{32}$$

Lemma 2.2. *Dressed charge in terms of Q-functions.*

$$\sigma(\lambda) = \frac{q^\alpha\phi(\lambda-\eta) - q^{-\alpha}\phi(\lambda+\eta)}{\phi_0(q^\alpha - q^{-\alpha})} , \tag{33}$$

where

$$\phi(\lambda) = \frac{Q(\lambda|\kappa+\alpha)}{Q(\lambda|\kappa)} , \qquad \phi_0 = \mathrm{ch}\left(\int_C \frac{d\lambda}{2\pi i}\ln\left(\frac{1+\mathfrak{a}(\lambda|\kappa+\alpha)}{1+\mathfrak{a}(\lambda|\kappa)}\right)\right) . \tag{34}$$

Proof. We recall from the appendix of Ref. 2 that

$$\phi(\lambda) = \frac{Q(\lambda|\kappa+\alpha)}{Q(\lambda|\kappa)} = \prod_{j=1}^{N/2}\frac{\mathrm{sh}(\lambda-\lambda_j(\kappa+\alpha))}{\mathrm{sh}(\lambda-\lambda_j(\kappa))} , \tag{35}$$

where the λ_j are the Bethe roots of the dominant eigenstate of the quantum transfer matrix which are located inside the contour C. Due to the t-Q-equation (11)

$$\phi(\lambda) = \frac{q^{-\alpha}\phi(\lambda+\eta)}{\rho(\lambda)} + \frac{q^\alpha\phi(\lambda-\eta) - q^{-\alpha}\phi(\lambda+\eta)}{\rho(\lambda)(1+\mathfrak{a}(\lambda|\kappa))} . \tag{36}$$

Here the first term on the right hand side is holomorphic inside C. Hence,

$$\int_C \mathrm{d}m(\mu)\big(q^{\alpha}\phi(\mu-\eta)-q^{-\alpha}\phi(\mu+\eta)\big)K_{\alpha}(\mu-\lambda)=\int_C \frac{\mathrm{d}\mu}{2\pi\mathrm{i}}\,\phi(\mu)K_{\alpha}(\mu-\lambda)\,. \quad (37)$$

The latter integral can be calculated. Note that

$$\phi(\lambda+\mathrm{i}\pi)=\phi(\lambda)\,,\quad K_{\alpha}(\lambda+\mathrm{i}\pi)=K_{\alpha}(\lambda) \quad (38)$$

and

$$\lim_{\mathrm{Re}\lambda\to\pm\infty}\phi(\lambda)=b^{\pm 1}\,,\quad b=\exp\Big(\sum_{j=1}^{N/2}\big(\lambda_j(\kappa)-\lambda_j(\kappa+\alpha)\big)\Big)\,. \quad (39)$$

With the contours sketched in figure 1 it follows that

$$\begin{aligned}\int_{\widetilde{C}+C}&\frac{\mathrm{d}\mu}{2\pi\mathrm{i}}\,\phi(\mu)K_{\alpha}(\mu-\lambda)\\ &=\int_{I_2+I_4}\frac{\mathrm{d}\mu}{2\pi\mathrm{i}}\,\phi(\mu)K_{\alpha}(\mu-\lambda)=-\frac{b+b^{-1}}{2}(q^{\alpha}-q^{-\alpha})\\ &=q^{-\alpha}\phi(\lambda+\eta)-q^{\alpha}\phi(\lambda-\eta)+\int_C\frac{\mathrm{d}\mu}{2\pi\mathrm{i}}\,\phi(\mu)K_{\alpha}(\mu-\lambda)\,. \quad (40)\end{aligned}$$

Here we have used the periodicity (38) in the first equation, the fact that the integral is independent of R and the asymptotics (39) in the second equation, and the fact that ϕ is free of poles inside $\widetilde{C}$ in the third equation. Setting $\phi_0=(b+b^{-1})/2$ and combining (37) and (40) we obtain (33).

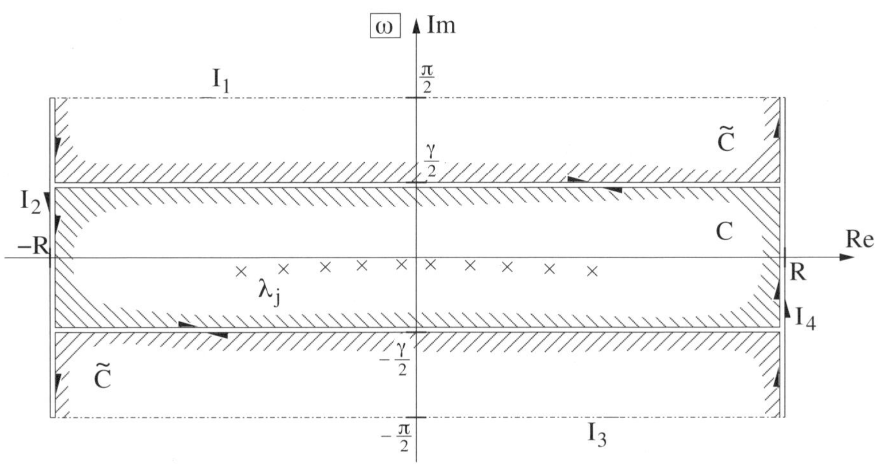

Fig. 1. Contours used in the proofs of lemma 2.2 and lemma 2.3.

It remains to show the second equation (34). It follows from

$$\int_C \frac{d\lambda}{2\pi i} \ln\bigl(1+\mathfrak{a}(\lambda|\kappa)\bigr) = -\int_C \frac{d\lambda}{2\pi i} \lambda\,\partial_\lambda \ln\bigl(1+\mathfrak{a}(\lambda|\kappa)\bigr) = -\sum_{j=1}^{N/2} \lambda_j(\kappa) - \beta/2, \tag{41}$$

and the proof is complete. □

Clearly (33) turns into (15) in the limit $\alpha \to 0$. Equipped with lemma 2.2 we can now proceed with proving

Lemma 2.3. *An identity for one-point functions.*

$$\int_C dm(\lambda) G(\lambda,\nu) = \frac{q^{-\alpha} - \rho(\nu)}{q^{\alpha} - q^{-\alpha}}. \tag{42}$$

Proof. First of all applying the dressed function trick to (19) and (32) we obtain

$$\int_C dm(\lambda) G(\lambda,\nu) = \int_C dm(\lambda)\sigma(\lambda)\bigl(q^{-\alpha}\operatorname{cth}(\lambda-\nu-\eta) - \rho(\nu)\operatorname{cth}(\lambda-\nu)\bigr). \tag{43}$$

Then we insert (33) and (36) into the right hand side. It follows that

$$\begin{aligned}
\int_C dm(\lambda) G(\lambda,\nu) &= \frac{1}{\phi_0(q^{\alpha}-q^{-\alpha})} \\
&\quad \int_C \frac{d\lambda}{2\pi i}\left(\phi(\lambda) - \frac{q^{-\alpha}\phi(\lambda+\eta)}{\rho(\lambda)}\right)\bigl(q^{-\alpha}\operatorname{cth}(\lambda-\nu-\eta) - \rho(\nu)\operatorname{cth}(\lambda-\nu)\bigr) \\
&= \frac{1}{\phi_0(q^{\alpha}-q^{-\alpha})}\Biggl(q^{-\alpha}\phi(\nu+\eta) \\
&\qquad + \int_{l_2+l_4-\tilde{C}} \frac{d\lambda}{2\pi i}\,\phi(\lambda)\bigl(q^{-\alpha}\operatorname{cth}(\lambda-\nu-\eta) - \rho(\nu)\operatorname{cth}(\lambda-\nu)\bigr)\Biggr) \\
&= \frac{q^{-\alpha}-\rho(\nu)}{q^{\alpha}-q^{-\alpha}}.
\end{aligned} \tag{44}$$

Here we have again referred to figure 1 in the second equation. □

3. Conclusions

We would like to conclude with three more remarks. First, all the above remains valid if we consider a more general six-vertex model with a more general inhomogeneous choice of parameters in vertical direction. In that case the functions $a(\lambda)$ and $d(\lambda)$ in (12) and also the driving term in (14) have to be modified as in Ref. 2.

Second, in Ref. 2 the calculation of the limit $\lim_{\nu_1\to\infty}\Psi(\nu_1,\nu_2)$ was the only point, where we had to resort to a multiple integral representation, when we showed that the functions ω as defined in Refs. 2 and 8 are identical. With the proof presented in this note the approach to the correlation functions of the XXZ chain, based on the discovery of a hidden Grassmann symmetry, as developed in Refs. 2,4,8 becomes logically independent of the multiple integral representation that was also obtained in Ref. 2.

Third, the function ϕ used above to express σ seems quite interesting and may deserve further attention. Is it possible to express the solutions of other linear integral equations in terms of ϕ? And is there a useful integral equation for ϕ itself? We hope we can come back to these questions in the future.

Acknowledgment

It a great pleasure and honour to dedicate this note to our dear friend Tetsuji Miwa on the occasion of his 60th birthday. We are grateful to Fedor Smirnov for suggesting us to prove lemma 2.3 directly. We wish to thank Alexander Seel for drawing the figure for us. Our work was generously supported by the Volkswagen foundation.

References

1. R. J. Baxter, *Exactly Solved Models in Statistical Mechanics* (Academic Press, London, 1982).
2. H. Boos and F. Göhmann, *On the physical part of the factorized correlation functions of the XXZ chain*, J. Phys. A **42** (2009) 315001.
3. H. Boos, M. Jimbo, T. Miwa and F. Smirnov, *Hidden Grassmann structure in the XXZ model IV: CFT limit*, preprint, arXiv:0911.3731 (2009).
4. H. Boos, M. Jimbo, T. Miwa, F. Smirnov and Y. Takeyama, *Hidden Grassmann structure in the XXZ model II: creation operators*, Comm. Math. Phys. **286** (2009) 875.
5. F. Göhmann, N. P. Hasenclever and A. Seel, *The finite temperature density matrix and two-point correlations in the antiferromagnetic XXZ chain*, J. Stat. Mech. (2005) P10015.
6. F. Göhmann, A. Klümper and A. Seel, *Integral representations for correlation functions of the XXZ chain at finite temperature*, J. Phys. A **37** (2004) 7625.
7. F. Göhmann and J. Suzuki, *Quantum spin chains at finite temperature*, preprint arXiv:1002.3194 (2010).
8. M. Jimbo, T. Miwa and F. Smirnov, *Hidden Grassmann structure in the XXZ model III: introducing Matsubara direction*, J. Phys. A **42** (2009) 304018.
9. A. Klümper, *Thermodynamics of the anisotropic spin-1/2 Heisenberg chain and related quantum chains*, Z. Phys. B **91** (1993) 507.

ALGEBRAIC ASPECTS OF THE CORRELATION FUNCTIONS OF THE INTEGRABLE HIGHER-SPIN XXZ SPIN CHAINS WITH ARBITRARY ENTRIES

TETSUO DEGUCHI

Department of Physics, Graduate School of Humanities and Sciences
Ochanomizu University
2-1-1 Ohtsuka, Bunkyo-ku, Tokyo 112-8610, Japan
E-mail: deguchi@phys.ocha.ac.jp

CHIHIRO MATSUI

Department of Physics, Graduate School of Science, the University of Tokyo
7-3-1 Hongo, Bunkyo-ku, Tokyo 113-0033, Japan
CREST, JST, 4-1-8 Honcho Kawaguchi, Saitama, 332-0012, Japan
E-mail: matsui@spin.phys.s.u-tokyo.ac.jp

We discuss some fundamental properties of the XXZ spin chain, which are important in the algebraic Bethe-ansatz derivation for the multiple-integral representations of the spin-s XXZ correlation function with an arbitrary product of elementary matrices.[1] For instance, we construct Hermitian conjugate vectors in the massless regime and introduce the spin-s Hermitian elementary matrices.

Keywords: Correlation functions; XXZ spin chains; algebraic Bethe ansatz; quantum groups; multiple-integral representations.

1. Introduction

The correlation functions of the spin-1/2 XXZ spin chain have attracted much interest in mathematical physics through the last two decades. One of the most fundamental results is the exact derivation of their multiple-integral representations. The multiple-integral representations of the XXZ correlation functions were derived for the first time by making use of the q-vertex operators through the affine quantum-group symmetry in the massive regime for the infinite lattice at zero temperature.[2,3] They were also derived in the massless regime by solving the q-KZ equations.[4,5] Making use of algebraic Bethe-ansatz techniques such as scalar products,[6–10] the multiple-integral representations were derived for the XXZ correlation functions under a non-zero magnetic field.[11] They were extended into those at finite temperatures,[12] and even for a large finite chain.[13] Interestingly, they are factorized in terms of single integrals.[14] Furthermore, the asymptotic expansion of a correlation function of the XXZ model has been systematically discussed.[15] Thus, the exact study of the XXZ correlation functions should play an important role not only in the mathematical physics of integrable models but also in many areas of theoretical physics.

The Hamiltonian of the spin-1/2 XXZ spin chain under the periodic boundary conditions is given by

$$\mathcal{H}_{\rm XXZ} = \frac{1}{2}\sum_{j=1}^{L}\left(\sigma_j^X\sigma_{j+1}^X + \sigma_j^Y\sigma_{j+1}^Y + \Delta\sigma_j^Z\sigma_{j+1}^Z\right) . \tag{1.1}$$

Here σ_j^a $(a = X, Y, Z)$ are the Pauli matrices defined on the jth site and Δ denotes the XXZ coupling. We define parameter q by

$$\Delta = (q + q^{-1})/2 . \tag{1.2}$$

We define η by $q = \exp\eta$. In the massive regime: $\Delta > 1$, we put $\eta = \zeta$ with $\zeta > 0$. At $\Delta = 1$ (i.e. $q = 1$) the Hamiltonian (1.1) gives the antiferromagnetic Heisenberg (XXX) chain. In the massless regime: $-1 < \Delta \le 1$, we set $\eta = i\zeta$, and we have $\Delta = \cos\zeta$ with $0 \le \zeta < \pi$ for the spin-1/2 XXZ spin chain (1.1). In the paper we consider a massless region: $0 \le \zeta < \pi/2s$ for the ground-state of the integrable spin-s XXZ spin chain.

Recently, the correlation functions and form factors of the integrable higher-spin XXX and XXZ spin chains have been derived by the algebraic Bethe-ansatz method.[1,16–18] The solvable higher-spin generalizations of the XXX and XXZ spin chains have been derived by the fusion method in several references.[19–25] In the region: $0 \le \zeta < \pi/2s$, the spin-s ground-state should be given by a set of string solutions.[26,27] Furthermore, the critical behavior should be given by the SU(2) WZWN model of level $k = 2s$ with central charge $c = 3s/(s+1)$.[24,28–40] For the integrable higher-spin XXZ spin chain correlation functions have been discussed in the massive regime by the method of q-vertex operators.[41–44]

In the present paper we discuss several important points in the algebraic Bethe-ansatz derivation of the correlation functions for the integrable spin-s XXZ spin chain where s is an arbitrary integer or a half-integer.[1] In particular, we briefly discuss a rigorous derivation of the finite-sum expression of correlation functions for the spin-s XXZ spin chain.

The content of the paper consists of the following. In section 2 we formulate the R-matrices in the homogeneous and principal gradings, respectively. They are related to each other by a similarity transformation. In section 3 we introduce the Hermitian elementary matrices and construct conjugate basis vectors for the spin-s Hilbert space in the massless regime. In section 4 we construct fusion monodromy matrices. In section 5, we first present formulas[1] for expressing the Hermitian elementary matrices in terms of global operators. Then, we review the multiple-integral representations of the spin-s XXZ correlation function for an arbitrary product of elementary matrices.[1] In section 6 we briefly sketch the derivation of the finite-sum expression of correlation functions for the spin-s XXZ spin chain, which leads to the multiple-integral representation in the thermodynamic limit. Here the spin-1/2 case corresponds to eq. (5.6) of Ref. 11.

2. Symmetric and asymmetric *R*-matrices

2.1. *R-matrix and the monodromy matrix of type* $(1, 1^{\otimes L})$

Let us now define the R-matrix of the XXZ spin chain.[7–9,11] For two-dimensional vector spaces V_1 and V_2, we define $R^{\pm}(\lambda_1 - \lambda_2)$ acting on $V_1 \otimes V_2$ by

$$R^{\pm}(\lambda_1 - \lambda_2) = \sum_{a,b,c,d=0,1} R^{\pm}(u)^{a\,b}_{c\,d}\, e^{a,\,c} \otimes e^{b,\,d} = \begin{pmatrix} 1 & 0 & 0 & 0 \\ 0 & b(u) & c^{\mp}(u) & 0 \\ 0 & c^{\pm}(u) & b(u) & 0 \\ 0 & 0 & 0 & 1 \end{pmatrix}, \tag{2.1}$$

where $u = \lambda_1 - \lambda_2$, $b(u) = \sinh u / \sinh(u + \eta)$ and $c^{\pm}(u) = \exp(\pm u) \sinh \eta / \sinh(u + \eta)$. We denote by $e^{a,\,b}$ a unit matrix that has only one nonzero element equal to 1 at entry (a, b) where $a, b = 0, 1$.

The asymmetric R-matrix (2.1), $R^{+}(u)$, is compatible with the homogeneous grading of $U_q(\widehat{sl_2})$.[3,18] We denote by $R^{(p)}(u)$ or simply by $R(u)$ the symmetric R-matrix where $c^{\pm}(u)$ of (2.1) are replaced by $c(u) = \sinh \eta / \sinh(u + \eta)$.[18] It is compatible with the affine quantum group $U_q(\widehat{sl_2})$ of the principal grading.[3,18] Hereafter, we denote them concisely by $R^{(w)}(u)$ with $w = \pm$ and p, where $w = +$ and $w = p$ in superscript show the homogeneous and the principal grading, respectively

Let s be an integer or a half-integer. We shall mainly consider the tensor product $V_1^{(2s)} \otimes \cdots \otimes V_{N_s}^{(2s)}$ of $(2s + 1)$-dimensional vector spaces $V_j^{(2s)}$ with parameters ξ_j, where $L = 2sN_s$. Here N_s denotes the lattice size of the spin-s chain. In general, we may consider the tensor product $V_0^{(2s_0)} \otimes$

$V_1^{(2s_1)} \otimes \cdots \otimes V_r^{(2s_r)}$ with $2s_1 + \cdots + 2s_r = L$, where $V_j^{(2s_j)}$ have parameters λ_j or ξ_j for $j = 1, 2, \ldots, r$. For a given set of matrix elements $A_{b,\,\beta}^{a,\,\alpha}$ for $a, b = 0, 1, \ldots, 2s_j$ and $\alpha, \beta = 0, 1, \ldots, 2s_k$, we define operator $A_{j,k}$ by

$$A_{j,k} = \sum_{a,b=1}^{\ell} \sum_{\alpha,\beta} A_{b,\,\beta}^{a,\,\alpha} I_0^{(2s_0)} \otimes I_1^{(2s_1)} \otimes \cdots \otimes I_{j-1}^{(2s_{j-1})} \otimes E_j^{a,b\,(2s_j)} \\ \otimes I_{j+1}^{(2s_{j+1})} \otimes \cdots \otimes I_{k-1}^{(2s_{k-1})} \otimes E_k^{\alpha,\beta\,(2s_k)} \otimes I_{k+1}^{(2s_{k+1})} \otimes \cdots \otimes I_r^{(2s_r)}. \tag{2.2}$$

Here $E_j^{a,b\,(2s_j)}$ denote the elementary matrices in the spin-s_j representation, each of which has nonzero matrix element only at entry (a, b).

When $s_0 = \ell/2$ and $s_1 = \cdots s_r = s$, we denote the type by $(\ell, (2s)^{\otimes N_s})$. In particular, for $s = 1/2$, we denote it by $(\ell, 1^{\otimes L})$.

2.2. *Gauge transformations*

Let us introduce operators Φ_j with arbitrary parameters ϕ_j for $j = 0, 1, \ldots, L$ as follows:

$$\Phi_j = \begin{pmatrix} 1 & 0 \\ 0 & e^{\phi_j} \end{pmatrix}_{[j]} = I^{\otimes(j)} \otimes \begin{pmatrix} 1 & 0 \\ 0 & e^{\phi_j} \end{pmatrix} \otimes I^{\otimes(L-j)}. \tag{2.3}$$

In terms of $\chi_{jk} = \Phi_j \Phi_k$, we define a similarity transformation on the R-matrix by

$$R_{jk}^{\chi} = \chi_{jk} R_{jk} \chi_{jk}^{-1} \,. \tag{2.4}$$

Explicitly, the following two matrix elements are transformed.

$$\left(R_{jk}^{\chi}\right)_{12}^{21} = c(\lambda_j, \lambda_k) e^{\phi_j - \phi_k} \,, \quad \left(R_{jk}^{\chi}\right)_{21}^{12} = c(\lambda_j, \lambda_k) e^{-\phi_j + \phi_k} \,. \tag{2.5}$$

Putting $\phi_j = \lambda_j$ for $j = 0, 1, \ldots, L$ in eq. (2.3) we have

$$R_{jk}^{\pm}(\lambda_j, \lambda_k) = (\chi_{jk})^{\pm 1} \, R_{jk}(\lambda_j, \lambda_k) \, (\chi_{jk})^{\mp 1} \quad (j, k = 0, 1, \ldots, L). \tag{2.6}$$

Thus, the asymmetric R-matrices $R_{12}^{\pm}(\lambda_1, \lambda_2)$ are derived from the symmetric one through the gauge transformation χ_{jk}.

2.3. *Monodromy matrices*

Applying definition (2.2) for matrix elements $R(u)_{cd}^{ab}$ of a given R-matrix, $R^{(w)}(u)$ for $w = \pm$ and p, we define R-matrices $R_{jk}^{(w)}(\lambda_j, \lambda_k) = R_{jk}^{(w)}(\lambda_j - \lambda_k)$

for integers j and k with $0 \le j < k \le L$. For integers j, k and ℓ with $0 \le j < k < \ell \le L$, the R-matrices satisfy the Yang-Baxter equations

$$\begin{aligned} & R_{jk}^{(w)}(\lambda_j - \lambda_k) R_{j\ell}^{(w)}(\lambda_j - \lambda_\ell) R_{k\ell}^{(w)}(\lambda_k - \lambda_\ell) \\ & = R_{k\ell}^{(w)}(\lambda_k - \lambda_\ell) R_{j\ell}^{(w)}(\lambda_j - \lambda_\ell) R_{jk}^{(w)}(\lambda_j - \lambda_k) . \end{aligned} \tag{2.7}$$

Let us introduce notation for expressing products of R-matrices.

$$\begin{aligned} R_{1,23\cdots n}^{(w)} &= R_{1n}^{(w)} \cdots R_{13}^{(w)} R_{12}^{(w)} , \\ R_{12\cdots n-1,n}^{(w)} &= R_{1n}^{(w)} R_{2n}^{(w)} \cdots R_{n-1\, n}^{(w)} . \end{aligned} \tag{2.8}$$

Here $R_{ab}^{(w)}$ denote the R-matrix $R_{ab}^{(w)} = R_{ab}^{(w)}(\lambda_a - \lambda_b)$ for $a, b = 1, 2, \ldots, n$.

We now define the monodromy matrix of type $(1, 1^{\otimes L}\, w)$, i.e. of type $(1, 1^{\otimes L})$ with grading w. Expressing the symbol $(1, 1^{\otimes L})$ briefly as (1, 1) in superscript we define it by

$$\begin{aligned} & T_{0,\, 12\cdots L}^{(1,\, 1\, w)}(\lambda_0; \{w_j\}_L) = R_{0L}^{+}(\lambda_0 - w_L) \cdots R_{02}^{+}(\lambda_0 - w_2) R_{01}^{+}(\lambda_0 - w_1) \\ & = R_{0L}^{(w)} R_{0L-1}^{(w)} \cdots R_{01}^{(w)} = R_{0,\, 12\cdots L}^{(\, w)}(\lambda_0; \{w_j\}_L) . \end{aligned} \tag{2.9}$$

Here we have put $\lambda_j = w_j$ for $j = 1, 2, \ldots, L$. They are arbitrary. We call them *inhomogeneous parameters*. We express the operator-valued matrix elements of the monodromy matrix as follows.

$$T_{0,12\cdots L}^{(1,1\,+)}(\lambda; \{w_j\}_L) = \begin{pmatrix} A_{12\cdots L}^{(1+)}(\lambda; \{w_j\}_L) & B_{12\cdots L}^{(1+)}(\lambda; \{w_j\}_L) \\ C_{12\cdots L}^{(1+)}(\lambda; \{w_j\}_L) & D_{12\cdots L}^{(1+)}(\lambda; \{w_j\}_L) \end{pmatrix} . \tag{2.10}$$

We also denote the operator-valued matrix elements by $[T_{0,12\cdots L}^{(1,1+)}(\lambda; \{w_j\}_L)]_{a,b}$ for $a, b = 0, 1$. Here $\{w_j\}_L$ denotes the inhomogeneous parameters $w_1, w_2, \ldots, w_L$. Hereafter we denote by $\{\mu_j\}_N$ the set of N numbers or parameters $\mu_1, \ldots, \mu_N$.

The monodromy matrix of principal grading, $T_{0,12\cdots L}^{(1,1\,p)}(\lambda; \{w_j\}_L)$, is related to that of homogeneous grading via similarity transformation $\chi_{01\cdots L} = \Phi_0 \Phi_1 \cdots \Phi_L$ as follows.[18]

$$\begin{aligned} & T_{0,12\cdots L}^{(1,1\,+)}(\lambda; \{w_j\}_L) = \chi_{012\cdots L} T_{0,12\cdots L}^{(1,1\,p)}(\lambda; \{w_j\}_L) \chi_{012\cdots L}^{-1} \\ & = \begin{pmatrix} \chi_{12\cdots L} A_{12\cdots L}^{(1\,p)}(\lambda; \{w_j\}_L) \chi_{12\cdots L}^{-1} & e^{-\lambda_0} \chi_{12\cdots L} B_{12\cdots L}^{(1\,p)}(\lambda; \{w_j\}_L) \chi_{12\cdots L}^{-1} \\ e^{\lambda_0} \chi_{12\cdots L} C_{12\cdots L}^{(1\,p)}(\lambda; \{w_j\}_L) \chi_{12\cdots L}^{-1} & \chi_{12\cdots L} D_{12\cdots L}^{(1\,p)}(\lambda; \{w_j\}_L) \chi_{12\cdots L}^{-1} \end{pmatrix} . \end{aligned}$$

In Ref.[18] operator $A^{(1\,+)}(\lambda)$ has been written as $A^{+}(\lambda)$.

2.4. Operator Ř: Another form of the R-matrix

Let V_1 and V_2 be $(2s+1)$-dimensional vector spaces. We define permutation operator $\Pi_{1,2}$ by

$$\Pi_{1,2}\, v_1 \otimes v_2 = v_2 \otimes v_1 \,, \quad v_1 \in V_1 \,, \, v_2 \in V_2 \,. \tag{2.11}$$

In the spin-1/2 case, we define operator $\check{R}^{(w)}_{j\, j+1}(u)$ by

$$\check{R}^{(w)}_{j,\, j+1}(u) = \Pi_{j,\, j+1}\, R^{(w)}_{j,\, j+1}(u) \,. \tag{2.12}$$

3. The quantum group invariance

3.1. Quantum group $U_q(sl_2)$

The quantum algebra $U_q(sl_2)$ is an associative algebra over **C** generated by $X^{\pm}, K^{\pm}$ with the following relations:[45–47]

$$\begin{aligned} KK^{-1} &= KK^{-1} = 1 \,, \quad KX^{\pm}K^{-1} = q^{\pm 2}X^{\pm} \,, \\ [X^+, X^-] &= \frac{K - K^{-1}}{q - q^{-1}} \,. \end{aligned} \tag{3.1}$$

The algebra $U_q(sl_2)$ is also a Hopf algebra over **C** with comultiplication

$$\begin{aligned} &\Delta(X^+) = X^+ \otimes 1 + K \otimes X^+ \,, \quad \Delta(X^-) = X^- \otimes K^{-1} + 1 \otimes X^- \,, \\ &\Delta(K) = K \otimes K \,, \end{aligned} \tag{3.2}$$

and antipode: $S(K) = K^{-1}\,, S(X^+) = -K^{-1}X^+\,, S(X^-) = -X^-K$, and coproduct: $\epsilon(X^{\pm}) = 0$ and $\epsilon(K) = 1$.

It is easy to see that the asymmetric R-matrix gives an intertwiner of the spin-1/2 representation of $U_q(sl_2)$:

$$R^+_{12}(u)\Delta(x) = \tau \circ \Delta(x) R^+_{12}(u) \quad \text{for} \quad x = X^{\pm}, K \,. \tag{3.3}$$

Here we remark that spectral parameter u is arbitrary and independent of $X^{\pm}$ or K.

3.2. Temperley-Lieb algebra

Operators $\check{R}^{\pm}_{j,j+1}(u)$ are decomposed in terms of the generators of the Temperley-Lieb algebra as follows.[48]

$$\check{R}^{\pm}_{j,j+1}(u) = I - b(u) U^{\pm}_j \,. \tag{3.4}$$

U_j^+s (U_j^-s) satisfy the defining relations of the Temperley-Lieb algebra:[48]

$$\begin{aligned} U_j^{\pm} U_{j+1}^{\pm} U_j^{\pm} &= U_j^{\pm}, \\ U_{j+1}^{\pm} U_j^{\pm} U_{j+1}^{\pm} &= U_j^{\pm}, \quad \text{for } j = 0, 1, \ldots, L-2, \\ \left(U_j^{\pm}\right)^2 &= (q + q^{-1})\, U_j^{\pm} \quad \text{for } j = 0, 1, \ldots, L-1, \\ U_j^{\pm} U_k^{\pm} &= U_k^{\pm} U_j^{\pm} \quad \text{for } |j - k| > 1 \,. \end{aligned} \tag{3.5}$$

We remark that the asymmetric R-matrices $\check{R}^{\pm}_{j,j+1}(u)$ derived from the symmetric R-matrix through the gauge transformation are related to the Jones polynomial.[49]

3.3. *Basis vectors of spin-$\ell/2$ representation of $U_q(sl_2)$*

Let us introduce the q-integer for an integer n by $[n]_q = (q^n - q^{-n})/(q - q^{-1})$. We define the q-factorial $[n]_q!$ for integers n by

$$[n]_q! = [n]_q \, [n-1]_q \, \cdots \, [1]_q \,. \tag{3.6}$$

For integers m and n satisfying $m \ge n \ge 0$ we define the q-binomial coefficients as follows

$$\begin{bmatrix} m \\ n \end{bmatrix}_q = \frac{[m]_q!}{[m-n]_q! \, [n]_q!} \,. \tag{3.7}$$

We now define the basis vectors of the $(\ell + 1)$-dimensional irreducible representation of $U_q(sl_2)$, $||\ell, n\rangle$ for $n = 0, 1, \ldots, \ell$ as follows. We define $||\ell, 0\rangle$ by

$$||\ell, 0\rangle = |0\rangle_1 \otimes |0\rangle_2 \otimes \cdots \otimes |0\rangle_\ell \,. \tag{3.8}$$

Here $|\alpha\rangle_j$ for $\alpha = 0, 1$ denote the basis vectors of the spin-1/2 representation defined on the jth position in the tensor product. We define $||\ell, n\rangle$ for $n \ge 1$ and evaluate them as follows[18] .

$$\begin{aligned} ||\ell, n\rangle &= \left(\Delta^{(\ell-1)}(X^-) \right)^n ||\ell, 0\rangle \, \frac{1}{[n]_q!} \\ &= \sum_{1 \le i_1 < \cdots < i_n \le \ell} \sigma_{i_1}^- \cdots \sigma_{i_n}^- |0\rangle \, q^{i_1 + i_2 + \cdots + i_n - n\ell + n(n-1)/2} \,. \end{aligned} \tag{3.9}$$

We define the conjugate vectors explicitly by the following:

$$\langle \ell, n|| = \begin{bmatrix} \ell \\ n \end{bmatrix}_q^{-1} q^{n(\ell-n)} \sum_{1 \le i_1 < \cdots < i_n \le \ell} \langle 0| \sigma_{i_1}^+ \cdots \sigma_{i_n}^+ \, q^{i_1 + \cdots + i_n - n\ell + n(n-1)/2} \,. \tag{3.10}$$

It is easy to show the normalization conditions:[18] $\langle \ell, n || \, || \ell, n \rangle = 1$. In the massive regime where $q = \exp \eta$ with real η, conjugate vectors $\langle \ell, n ||$ are Hermitian conjugate to vectors $|| \ell, n \rangle$.

3.4. *Conjugate vectors*

In order to construct Hermitian elementary matrices in the massless regime where $|q| = 1$, we now introduce another set of dual basis vectors. For a given nonzero integer ℓ we define $\widetilde{\langle \ell, n ||}$ for $n = 0, 1, \ldots, n$, by

$$\widetilde{\langle \ell, n ||} = \binom{\ell}{n}^{-1} \sum_{1 \le i_1 < \cdots < i_n \le \ell} \langle 0 | \sigma_{i_1}^{+} \cdots \sigma_{i_n}^{+} \, q^{-(i_1 + \cdots + i_n) + n\ell - n(n-1)/2} \, . \tag{3.11}$$

They are conjugate to $|| \ell, n \rangle$: $\widetilde{\langle \ell, m ||} \, || \ell, n \rangle = \delta_{m,n}$. Here we have denoted the binomial coefficients for integers ℓ and n with $0 \le n \le \ell$ as follows.

$$\binom{\ell}{n} = \frac{\ell!}{(\ell - n)! n!} \, . \tag{3.12}$$

We now introduce vectors $\widetilde{|| \ell, n \rangle}$ which are Hermitian conjugate to $\langle \ell, n ||$ when $|q| = 1$ for positive integers ℓ with $n = 0, 1, \ldots, \ell$. Setting the norm of $\widetilde{|| \ell, n \rangle}$ such that $\langle \ell, n || \, \widetilde{|| \ell, n \rangle} = 1$, vectors $\widetilde{|| \ell, n \rangle}$ are given by

$$\sum_{1 \le i_1 < \cdots < i_n \le \ell} \sigma_{i_1}^{-} \cdots \sigma_{i_n}^{-} | 0 \rangle q^{-(i_1 + \cdots + i_n) + n\ell - n(n-1)/2} \begin{bmatrix} \ell \\ n \end{bmatrix}_q q^{-n(\ell - n)} \binom{\ell}{n}^{-1} . \tag{3.13}$$

We have the following normalization condition:

$$\widetilde{\langle \ell, n ||} \, \widetilde{|| \ell, n \rangle} = \begin{bmatrix} \ell \\ n \end{bmatrix}_q^2 \binom{\ell}{n}^{-2} . \tag{3.14}$$

3.5. *Hermitian elementary matrices*

In the massless regime we define elementary matrices $\widetilde{E}^{m, n \, (2s+)}$ for $m, n = 0, 1, \ldots, 2s$ by

$$\widetilde{E}^{m, n \, (2s+)} = \widetilde{|| 2s, m \rangle} \, \langle 2s, n || \, . \tag{3.15}$$

In the massless regime where $|q| = 1$, matrix $|| \ell, n \rangle \widetilde{\langle \ell, n ||}$ is Hermitian: $(|| \ell, n \rangle \widetilde{\langle \ell, n ||})^{\dagger} = || \ell, n \rangle \widetilde{\langle \ell, n ||}$. However, in order to define projection operators $\tilde{P}$ such that $P \tilde{P} = P$, we have formulated vectors $\widetilde{|| \ell, n \rangle}$.

3.6. *Projection operators*

We define projection operators acting on the 1st to the ℓth tensor-product spaces by

$$P^{(\ell)}_{12\cdots\ell} = \sum_{n=0}^{\ell} ||\ell, n\rangle \, \langle \ell, n|| \, . \tag{3.16}$$

Let us now introduce another set of projection operators $\widetilde{P}^{(\ell)}_{1\cdots\ell}$ as follows.

$$\widetilde{P}^{(\ell)}_{1\cdots\ell} = \sum_{n=0}^{\ell} \widetilde{||\ell, \, n\rangle\langle\ell, \, n||} \, . \tag{3.17}$$

Projector $\widetilde{P}^{(\ell)}_{1\cdots\ell}$ is idempotent: $(\widetilde{P}^{(\ell)}_{1\cdots\ell})^2 = \widetilde{P}^{(\ell)}_{1\cdots\ell}$. In the massless regime where $|q| = 1$, it is Hermitian: $\left(\widetilde{P}^{(\ell)}_{1\cdots\ell}\right)^{\dagger} = \widetilde{P}^{(\ell)}_{1\cdots\ell}$. From (3.16) and (3.17), we show the following properties:

$$P^{(\ell)}_{12\cdots\ell} \widetilde{P}^{(\ell)}_{1\cdots\ell} = P^{(\ell)}_{12\cdots\ell} \, , \tag{3.18}$$

$$\widetilde{P}^{(\ell)}_{1\cdots\ell} P^{(\ell)}_{12\cdots\ell} = \widetilde{P}^{(\ell)}_{1\cdots\ell} \, . \tag{3.19}$$

In the tensor product of quantum spaces, $V_1^{(2s)} \otimes \cdots \otimes V_{N_s}^{(2s)}$, we define $\widetilde{P}^{(2s)}_{12\cdots L}$ by

$$\widetilde{P}^{(2s)}_{12\cdots L} = \prod_{i=1}^{N_s} \widetilde{P}^{(2s)}_{2s(i-1)+1} \, . \tag{3.20}$$

Here we recall $L = 2sN_s$.

The projection operators are also constructed by the fusion method. For $\ell > 2$ we can construct projection operators inductively with respect to ℓ as follows.[25,46]

$$P^{(\ell)}_{12\cdots\ell} = P^{(\ell-1)}_{12\cdots\ell-1} \check{R}^{+}_{\ell-1,\,\ell}((\ell-1)\eta) P^{(\ell-1)}_{12\cdots\ell-1} \, . \tag{3.21}$$

The projection operator $P^{(\ell)}_{12\cdots\ell}$ gives a q-analogue of the full symmetrizer of the Young operators for the Hecke algebra.[46]

4. Fusion construction

4.1. *Higher-spin monodromy matrix of type* $(\ell, \, (2s)^{\otimes N_s})$

We now set the inhomogeneous parameters w_j for $j = 1, 2, \ldots, L$, as N_s sets of complete $2s$-strings.[18] We define $w^{(2s)}_{(b-1)\ell+\beta}$ for $\beta = 1, \ldots, 2s$, as follows.

$$w^{(2s)}_{2s(b-1)+\beta} = \xi_b - (\beta - 1)\eta \, , \quad \text{for} \quad b = 1, 2, \ldots, N_s. \tag{4.1}$$

We shall define the monodromy matrix of type $(1, (2s)^{\otimes N_s})$ associated with homogeneous grading. We first define the massless monodromy matrix by

$$\widetilde{T}^{(1,\,2s\,+)}_{0,\,12\cdots N_s}(\lambda_0;\{\xi_b\}_{N_s}) = \widetilde{P}^{(2s)}_{12\cdots L} R^{(1,\,1\,+)}_{0,\,1\ldots L}(\lambda_0;\{w_j^{(2s)}\}_L)\widetilde{P}^{(2s)}_{12\cdots L}$$
$$= \begin{pmatrix} \widetilde{A}^{(2s+)}(\lambda;\{\xi_b\}_{N_s}) & \widetilde{B}^{(2s+)}(\lambda;\{\xi_b\}_{N_s}) \\ \widetilde{C}^{(2s+)}(\lambda;\{\xi_b\}_{N_s}) & \widetilde{D}^{(2s+)}(\lambda;\{\xi_b\}_{N_s}) \end{pmatrix} . \quad (4.2)$$

Let us introduce a set of $2s$-strings with small deviations from the set of complete $2s$-strings.

$$w^{(2s;\,\epsilon)}_{2s(b-1)+\beta} = \xi_b - (\beta-1)\eta + \epsilon r_b^{(\beta)}, \quad \text{for} \quad b = 1, 2, \cdots, N_s,$$
$$\text{and } \beta = 1, 2, \ldots, 2s. \quad (4.3)$$

Here ϵ is very small and $r_b^{(\beta)}$ are generic parameters. We express the elements of the monodromy matrix $T^{(1,1)}$ with inhomogeneous parameters given by $w_j^{(2s;\,\epsilon)}$ for $j = 1, 2, \ldots, L$ as follows.

$$T^{(1,\,1\,+)}_{0,\,12\cdots L}(\lambda;\{w_j^{(2s;\epsilon)}\}_L) = \begin{pmatrix} A^{(2s+;\,\epsilon)}_{12\cdots L}(\lambda) & B^{(2s+;\,\epsilon)}_{12\cdots L}(\lambda) \\ C^{(2s+;\,\epsilon)}_{12\cdots L}(\lambda) & D^{(2s+;\,\epsilon)}_{12\cdots L}(\lambda) \end{pmatrix} . \quad (4.4)$$

Here $A^{(2s+;\,\epsilon)}_{12\cdots L}(\lambda)$ denotes $A^{(1+)}_{12\cdots L}(\lambda;\{w_j^{(2s;\,\epsilon)}\}_L)$.

$$\widetilde{A}^{(2s+)}_{12\cdots N_s}(\lambda;\{\xi_p\}_{N_s}) = \lim_{\epsilon\to 0} \widetilde{P}^{(2s)}_{12\cdots L} A^{(2s+;\,\epsilon)}_{12\cdots L}(\lambda;\{w_j^{(2s;\,\epsilon)}\}_L)\widetilde{P}^{(2s)}_{12\cdots L} . \quad (4.5)$$

We define the massless monodromy matrix of type $(\ell, (2s)^{\otimes N_s})$ by

$$\widetilde{T}^{(\ell,\,2s\,+)}_{0,\,12\cdots N_s} = \widetilde{P}^{(\ell)}_{a_1 a_2\cdots a_\ell} \widetilde{T}^{(1,\,2s\,+)}_{a_1,\,12\cdots N_s}(\lambda_{a_1})\widetilde{T}^{(1,\,2s\,+)}_{a_2,\,12\cdots N_s}(\lambda_{a_1}-\eta)\cdots$$
$$\times\cdots\widetilde{T}^{(1,\,2s\,+)}_{a_\ell,\,12\cdots N_s}(\lambda_{a_1}-(\ell-1)\eta)\,\widetilde{P^{(\ell)}}_{a_1 a_2\cdots a_\ell} . \quad (4.6)$$

4.2. *Integrable spin-s Hamiltonians*

We define the massless transfer matrix[1] of type $(\ell, (2s)^{\otimes N_s})$ by

$$\widetilde{t}^{(\ell,\,2s\,+)}_{12\cdots N_s}(\lambda) = \mathrm{tr}_{V^{(\ell)}}\left(\widetilde{T}^{(\ell,\,2s\,+)}_{0,\,12\cdots N_s}(\lambda)\right) = \sum_{n=0}^{\ell} {}_a\langle \ell, n||\,\widetilde{T}^{(1,\,2s\,+)}_{a_1,\,12\cdots N_s}(\lambda)$$
$$\times\,\widetilde{T}^{(1,\,2s\,+)}_{a_2,\,12\cdots N_s}(\lambda-\eta)\cdots\widetilde{T}^{(1,\,2s\,+)}_{a_\ell,\,12\cdots N_s}(\lambda-(\ell-1)\eta)\,\widetilde{||\ell, n\rangle}_a . \quad (4.7)$$

It follows from the Yang-Baxter equations that the higher-spin transfer matrices commute in the tensor product space $V_1^{(2s)}\otimes\cdots\otimes V_{N_s}^{(2s)}$, which is derived by applying projection operator $P^{(2s)}_{12\cdots L}$ to $V_1^{(1)}\otimes\cdots\otimes V_L^{(1)}$.

The massless spin-s R-matrix $\widetilde{R}_{12}^{(2s,\,2s\,+)}(u)$ becomes the permutation operator at $u=0$: $\widetilde{R}_{12}^{(2s,\,2s\,+)}(0)=\Pi_{1,\,2}$.[45,50] Therefore, putting inhomogeneous parameters $\xi_p=0$ for $p=1,2,\ldots,N_s$, we show that the transfer matrix $\widetilde{t}_{12\cdots N_s}^{(2s,\,2s\,+)}(\lambda)$ becomes the shift operator at $\lambda=0$. We derive the massless spin-s XXZ Hamiltonian by the logarithmic derivative of the massless spin-s transfer matrix.

$$\mathcal{H}_{\rm XXZ}^{(2s)}=\frac{d}{d\lambda}\log\widetilde{t}_{12\cdots N_s}^{(2s,\,2s\,+)}(\lambda)\Big|_{\lambda=0,\,\xi_j=0}=\sum_{i=1}^{N_s}\frac{d}{du}\widetilde{\check{R}}_{i,i+1}^{(2s,2s)}(u)\Big|_{u=0}\,. \quad (4.8)$$

5. Spin-$\ell/2$ massless XXZ correlation functions

5.1. *Spin-s local operators in terms of global operators*

In the massless regime, we can express the Hermitian elementary matrices in terms of global operators as follows.[1] For $m\geq n$ we have

$$\begin{aligned}\widetilde{E}_i^{m,\,n\,(\ell\,+)}&=\binom{\ell}{n}\begin{bmatrix}\ell\\ m\end{bmatrix}_q\begin{bmatrix}\ell\\ n\end{bmatrix}_q^{-1}\widetilde{P}_{1\cdots L}^{(\ell)}\prod_{\alpha=1}^{(i-1)\ell}(A^{(1+)}+D^{(1+)})(w_\alpha)\\ &\times\prod_{k=1}^{n}D^{(1+)}(w_{(i-1)\ell+k})\prod_{k=n+1}^{m}B^{(1+)}(w_{(i-1)\ell+k})\prod_{k=m+1}^{\ell}A^{(1+)}(w_{(i-1)\ell+k})\\ &\times\prod_{\alpha=i\ell+1}^{\ell N_s}(A^{(1+)}+D^{(1+)})(w_\alpha)\,\widetilde{P}_{1\cdots L}^{(\ell)}\,. \end{aligned}\quad (5.1)$$

For $m\leq n$ we have

$$\begin{aligned}\widetilde{E}_i^{m,\,n\,(\ell\,+)}&=\binom{\ell}{n}\begin{bmatrix}\ell\\ m\end{bmatrix}_q\begin{bmatrix}\ell\\ n\end{bmatrix}_q^{-1}\widetilde{P}_{1\cdots L}^{(\ell)}\prod_{\alpha=1}^{(i-1)\ell}(A^{(1+)}+D^{(1+)}(w_\alpha)\\ &\times\prod_{k=1}^{m}D^{(1+)}(w_{(i-1)\ell+k})\prod_{k=m+1}^{n}C^{(1+)}(w_{(i-1)\ell+k})\prod_{k=n+1}^{\ell}A^{(1+)}(w_{(i-1)\ell+k})\\ &\times\prod_{\alpha=i\ell+1}^{\ell N_s}(A^{(1+)}+D^{(1+)})(w_\alpha)\,\widetilde{P}_{1\cdots L}^{(\ell)}\,. \end{aligned}\quad (5.2)$$

By the quantum inverse-scattering problem (QISP) of Ref. 10 the local spin operators are expressed in terms of global operators and the transfer matrices for the integrable spin-s XXX spin chain. However, it is not clear how one can derive (5.1) and (5.2) by the QISP method even for $q=1$.

5.2. Symbols for expressing sequences

Let us denote by $(a_j)_m$ a sequence of numbers a_j for $j = 1, 2, \ldots, m$, i.e. $(a_j)_m = (a_1, a_2, \cdots, a_m)$.

Definition 5.1. We say that a sequence $(b_k)_n$ is a subsequence of $(a_j)_m$ if (i) $n \le m$, (ii) $b_k \in \{a_1, \ldots, a_m\}$ for $k = 1, 2, \ldots, n$, (iii) for any pair of integers j and k satisfying $1 \le j < k \le n$, there exists a pair of integers $\ell(j)$ and $\ell(k)$ such that $a_j = b_{\ell(j)}$, $a_k = b_{\ell(k)}$ and $\ell(j) < \ell(k)$.

For a pair of sequences $(a_j)_m$ and $(b_k)_n$, we define the product $(a_j)_m \# (b_k)_n$ by a sequence $(c_\ell)_{m+n}$ such that $c_j = a_j$ for $j = 1, 2, \cdots, m$ and $c_j = b_j$ for $j = m+1, m+2, \ldots, m+n$.

5.3. Conjecture of the spin-s ground-state solution

Let us now introduce the conjecture that the ground state of the spin-s case $|\psi_g^{(2s\,+)}\rangle$ is given by $N_s/2$ sets of $2s$-strings:

$$\lambda_a^{(\alpha)} = \mu_a - (\alpha - 1/2)\eta + \epsilon_a^{(\alpha)}, \quad \text{for } a = 1, 2, \ldots, N_s/2 \text{ and } \alpha = 1, 2, \ldots, 2s. \tag{5.3}$$

Here we assume that string deviations $\epsilon_a^{(\alpha)}$ are very small when N_s is very large.[51] In terms of $\lambda_a^{(\alpha)}$, the spin-s ground state in the homogeneous grading is given by[1]

$$|\psi_g^{(2s\,+)}\rangle = \prod_{a=1}^{N_s/2} \prod_{\alpha=1}^{2s} \widetilde{B}^{(2s\,+)}(\lambda_a^{(\alpha)}; \{\xi_p\}_{N_s})|0\rangle. \tag{5.4}$$

We denote by M the number of Bethe roots: $M = 2s\, N_s/2 = sN_s$.

According to analytic and numerical studies,[24,37,39,40] we may assume the following properties of string deviations $\epsilon_a^{(\alpha)}$s. For very large N_s, the deviations are given by $\epsilon_a^{(\alpha)} = i\,\delta_a^{(\alpha)}$, where i denotes $\sqrt{-1}$ and $\delta_a^{(\alpha)}$ are real. Moreover, $\delta_a^{(\alpha)} - \delta_a^{(\alpha+1)} > 0$ for $\alpha = 1, 2, \ldots, 2s-1$, and $|\delta_a^{(\alpha)}| > |\delta_a^{(\alpha+1)}|$ for $\alpha < s$, while $|\delta_a^{(\alpha)}| < |\delta_a^{(\alpha+1)}|$ for $\alpha \ge s$.

In the limit: $N_s \to \infty$, the density of string centers, $\rho_{\rm tot}(\mu)$, is given by

$$\rho_{\rm tot}(\mu) = \frac{1}{N_s} \sum_{p=1}^{N_s} \frac{1}{2\zeta \cosh(\pi(\mu - \xi_p)/\zeta)}. \tag{5.5}$$

For the homogeneous chain where $\xi_p = 0$ for $p = 1, 2, \ldots, N_s$, we denote the density of string centers by $\rho(\lambda)$.

$$\rho(\lambda) = \frac{1}{2\zeta \cosh(\pi\lambda/\zeta)}. \tag{5.6}$$

Let us introduce useful notation of the suffix of rapidities. For rapidities $\lambda_a^{(\alpha)} = \lambda_{(a,\alpha)}$ we define integers A by $A = 2s(a-1)+\alpha$ for $a = 1, 2, \ldots, N_s/2$ and for $\alpha = 1, 2, \ldots, 2s$. We thus denote $\lambda_{(a,\alpha)}$ also by λ_A for $A = 1, 2, \ldots, sN_s$, and put $\lambda_{(a,\alpha)}$ in increasing order with respect to $A = 2s(a-1)+\alpha$ such as $\lambda_{(1,1)} = \lambda_1, \lambda_{(1,2)} = \lambda_2, \ldots, \lambda_{(N_s/2,2s)} = \lambda_{sN_s}$. In the ground state, rapidities λ_A for $A = 1, 2, \ldots, M$, are expressed by

$$\lambda_{2s(a-1)+\alpha} = \mu_a - (\alpha - 1/2)\eta + \epsilon_a^{(\alpha)} \quad (1 \le a \le N_s/2;\ 1 \le \alpha \le 2s). \tag{5.7}$$

For $A = 2s(a-1)+\alpha$ with $1 \le \alpha \le 2s$, integer a is given by $a = [(A-1)/2s]+1$, and integer α is given by $\alpha = A - 2s[(A-1)/2s]$.

For a real number x we define $[x]$ by the greatest integer less than or equal to x. We define $a(j)$ and $\alpha(j)$ for $j = 1, 2, \ldots, M$ as follows.

$$a(j) = [(j-1)/2s]+1\,, \quad \alpha(j) = j - 2s[(j-1)/2s]\,. \tag{5.8}$$

5.4. *Correlation functions of the integrable spin-s XXZ model on a long finite chain*

We define the correlation function of the integrable spin-2s XXZ spin chain for a given product of $(2s+1)\times(2s+1)$ elementary matrices such as $\widetilde{E}_1^{i_1, j_1\,(2s+)} \cdots \widetilde{E}_m^{i_m, j_m\,(2s+)}$ on the spin-s ground state, $|\psi_g^{(2s+)}\rangle$, as follows.

$$F_m^{(2s+)}(\{i_k, j_k\}) = \langle\psi_g^{(2s+)}| \prod_{k=1}^{m} \widetilde{E}_k^{i_k, j_k\,(2s+)} |\psi_g^{(2s+)}\rangle / \langle\psi_g^{(2s+)}|\psi_g^{(2s+)}\rangle\,. \tag{5.9}$$

By formulas (5.1) and (5.2) we express the mth product of $(2s+1)\times(2s+1)$ elementary matrices in terms of a $2sm$th product of 2×2 elementary matrices with entries $\{\epsilon_j, \epsilon_j^{'}\}$ as follows.

$$\prod_{b=1}^{m} \widetilde{E}_b^{i_b, j_b\,(2s+)} = C(\{i_b, j_b\})\, \widetilde{P}_{12\ldots L}^{(2s)} \cdot \prod_{k=1}^{2sm} e_k^{\epsilon_k^{'}, \epsilon_k} \cdot \widetilde{P}_{12\ldots L}^{(2s)}. \tag{5.10}$$

By making use of (5.1) and (5.2), $C(\{i_b, j_b\})$ is given by

$$C(\{i_k, j_k\}) = \prod_{b=1}^{m} \left\{ \binom{2s}{j_b} \begin{bmatrix} 2s \\ i_b \end{bmatrix}_q \begin{bmatrix} 2s \\ j_b \end{bmatrix}_q^{-1} \right\}. \tag{5.11}$$

Here $\epsilon_{2s(b-1)+\beta}$ and $\epsilon_{2s(b-1)+\beta}^{'}$ $(b = 1, \ldots, N_s;\ \beta = 1, \ldots, 2s)$ are given by

$$\epsilon_{2s(b-1)+\beta} = \begin{cases} 1 & (1 \le \beta \le j_b) \\ 0 & (j_b < \beta \le 2s) \end{cases};\quad \epsilon_{2s(b-1)+\beta}^{'} = \begin{cases} 1 & (1 \le \beta \le i_b) \\ 0 & (i_b < \beta \le 2s)\,. \end{cases} \tag{5.12}$$

We evaluate the spin-$2s$ XXZ correlation function $F_m^{(2s+)}(\{i_k, j_k\})$ by

$$F_m^{(2s+)}(\{i_k, j_k\}) = C(\{i_k, j_k\}) \langle \psi_g^{(2s+)} | \widetilde{P}_{12\ldots L}^{(2s)} \times \\ \times \prod_{j=1}^{2sm} e_j^{\epsilon_j', \epsilon_j} \cdot \widetilde{P}_{12\ldots L}^{(2s)} | \psi_g^{(2s+)} \rangle / \langle \psi_g^{(2s+)} | \psi_g^{(2s+)} \rangle . \quad (5.13)$$

Let $\boldsymbol{\alpha}^+$ be the set of j with $\epsilon_j = 0$, and $\boldsymbol{\alpha}^-$ the set of j with $\epsilon_j' = 1$:

$$\boldsymbol{\alpha}^+ = \{j;\, \epsilon_j = 0\}, \quad \boldsymbol{\alpha}^- = \{j;\, \epsilon_j' = 1\}. \quad (5.14)$$

We denote by r and r' the number of elements of the set $\boldsymbol{\alpha}^-$ and $\boldsymbol{\alpha}^+$, respectively. Due to charge conservation, we have

$$r + r' = 2sm. \quad (5.15)$$

We denote by $j_{\rm min}$ and $j_{\rm max}$ the smallest element and the largest element of $\boldsymbol{\alpha}^-$, respectively. We also denote by $j_{\rm min}'$ and $j_{\rm max}'$ the smallest element and the largest element of $\boldsymbol{\alpha}^+$, respectively.

Recall that the ground state $|\psi_g^{(2s+)}\rangle$ has M Bethe roots with $M = sN_s$. Let c_j ($j \in \boldsymbol{\alpha}^-$) and c_j' ($j \in \boldsymbol{\alpha}^+$) be integers such that $1 \le c_j \le M$ for $j \in \boldsymbol{\alpha}^-$ and $1 \le c_j' \le M + j$ for $j \in \boldsymbol{\alpha}^+$. We define sequence $(b_\ell)_{2sm}$ by

$$(b_1, b_2, \ldots, b_{2sm}) = (c_{j_{\rm max}'}', \ldots, c_{j_{\rm min}'}', c_{j_{\rm min}}, \ldots, c_{j_{\rm max}}). \quad (5.16)$$

Here sequence $(c_{j_{\rm max}'}', \ldots, c_{j_{\rm min}'}', c_{j_{\rm min}}, \ldots, c_{j_{\rm max}})$ is given by the composite sequence of c_j' s in decreasing order with respect to suffix j, and c_j s in increasing order with respect to suffix j. We introduce the following symbols:

$$\prod_{j\in\boldsymbol{\alpha}^-} \left(\sum_{c_j=1}^{M} \right) \prod_{j\in\boldsymbol{\alpha}^+} \left(\sum_{c_j'=1}^{M+j} \right) = \sum_{c_{j_{\rm min}}=1}^{M} \cdots \sum_{c_{j_{\rm max}}=1}^{M} \sum_{c_{j_{\rm min}'}'=1}^{M+j_{\rm min}'} \cdots \sum_{c_{j_{\rm max}'}'=1}^{M+j_{\rm max}'} . \quad (5.17)$$

Recall that $a(j)$ are defined in (5.8). We define $\beta(j)$ by

$$\beta(j) = j - 2s[(j-1)/2s] \quad (1 \le j \le M). \quad (5.18)$$

For $\ell, k = 1, 2, \ldots, 2sm$, we define the (ℓ, k) element of $M^{(2sm)}((b_j)_{2sm})$ by

$$\left(M^{(2sm)}((b_j)_{2sm}) \right)_{\ell,\, k} = \begin{cases} -\delta_{b_\ell - M,\, k} & (b_\ell > M) \\ \delta_{\beta(b_\ell),\, \beta(k)} \cdot \rho(\lambda_{b_\ell} - w_k^{(2s)} + \eta/2)/(N_s \rho_{\rm tot}(\mu_{a(b_\ell)})) & (b_\ell \le M) \end{cases} \quad (5.19)$$

Here, continuous variable μ, which is the argument of density $\rho_{\rm tot}(\mu)$, is evaluated at $\mu_{a(b_\ell)}$, one of the "string centers" μ_a of $2s$-strings (5.7).

We can rigorously derive a concise expression of correlation functions of the spin-s XXZ spin chain in the massless region: $0 \le \zeta < \pi/2s$ for a large finite chain. Introducing $\varphi(\lambda) = \sinh\lambda$ we have

$$
\begin{aligned}
F_m^{(2s+)}(\{i_b, j_b\}) &= C(\{i_k, j_k\}) \prod_{j\in\boldsymbol{\alpha}^-}\left(\sum_{c_j=1}^{M}\right) \prod_{j\in\boldsymbol{\alpha}^+}\left(\sum_{c'_j=1}^{M+j}\right) \det M^{(2sm)}((b_\ell)_{2sm}) \\
&\times(-1)^{r'} \frac{\prod_{j\in\boldsymbol{\alpha}^-}\left(\prod_{k=1}^{j-1}\varphi(\lambda_{c_j}-w_k^{(2s)}+\eta)\prod_{k=j+1}^{2sm}\varphi(\lambda_{c_j}-w_k^{(2s)})\right)}{\prod_{1\le k<\ell\le 2sm}\varphi(\lambda_{b_\ell}-\lambda_{b_k}+\eta)} \\
&\times \frac{\prod_{j\in\boldsymbol{\alpha}^+}\left(\prod_{k=1}^{j-1}\varphi(\lambda_{c'_j}-w_k^{(2s)}-\eta)\prod_{k=j+1}^{2sm}\varphi(\lambda_{c'_j}-w_k^{(2s)})\right)}{\prod_{1\le k<\ell\le 2sm}\varphi(w_k^{(2s)}-w_\ell^{(2s)})} + O(1/N_s)\,.
\end{aligned}
\tag{5.20}
$$

We remark that we derive (5.20) sending ϵ to zero. Before taking the limit, inhomogeneous parameters w_js are generic due to small parameter ϵ, and the sums over variables c_j in (5.20) are restricted up to M for all j.

5.5. *Multiple-integral representations of spin-s XXZ correlation function for arbitrary matrix elements*

In the thermodynamic limit: $N_s \to \infty$, rapidities λ_{b_ℓ} with b_ℓ defined in (5.16), correspond to integral variables λ_ℓ for $\ell = 1, 2, \ldots, 2sm$. For $1 \le b_\ell \le M$ they are given by the Bethe roots of $2s$-strings (5.7), while for $b_\ell > M$ they are given by complete $2s$-strings $w_k^{(2s)}$ defined by (4.1).

We define $\alpha(\lambda_j)$ by $\alpha(\lambda_j) = \gamma$ for an integer γ with $1 \le \gamma \le 2s$, if λ_j is related to integral variable μ_j by $\lambda_j = \mu_j - (\gamma - 1/2)\eta$, or if λ_j takes a value close to $w_k^{(2s)}$ with $\beta(k) = \gamma$, where $w_k^{(2s)}$ are part of complete 2s-strings (4.1). Here, variables μ_j correspond to "string centers" of variables λ_j.

We define the (j,k) element of matrix $S = S\left((\lambda_j)_{2sm}; (w_j^{(2s)})_{2sm}\right)$ by

$$
S_{j,k} = \rho(\lambda_j - w_k^{(2s)} + \eta/2)\,\delta(\alpha(\lambda_j), \beta(k))\,, \quad \text{for} \quad j,k = 1,2,\ldots,2sm\,. \tag{5.21}
$$

Here $\delta(\alpha, \beta)$ denotes the Kronecker delta, and we recall (5.18) for $\beta(k)$.

Let Γ_j be a small contour rotating counterclockwise around $\lambda = w_j^{(2s)}$.

Since $\det S$ has simple poles at $\lambda = w_j^{(2s)}$ with residue $1/2\pi i$, we have

$$\int_{-\infty+i\epsilon}^{\infty+i\epsilon} \det S((\lambda_k)_{2sm}) d\lambda_1 = \int_{-\infty-i\epsilon}^{\infty-i\epsilon} \det S((\lambda_k)_{2sm}) d\lambda_1 - \oint_{\Gamma_1} \det S((\lambda_k)_{2sm}) d\lambda_1 . \tag{5.22}$$

For sets $\boldsymbol{\alpha}^-$ and $\boldsymbol{\alpha}^+$ with relation (5.16), we define integral variables $\tilde{\lambda}_j$ for $j \in \boldsymbol{\alpha}^-$ and $\tilde{\lambda}_j^{'}$ for $j \in \boldsymbol{\alpha}^+$, respectively, by the following:

$$(\tilde{\lambda}^{'}_{j^{'}_{max}}, \ldots, \tilde{\lambda}^{'}_{j^{'}_{min}}, \tilde{\lambda}_{j_{min}}, \tilde{\lambda}_{j_{max}}) = (\lambda_1, \ldots, \lambda_{2sm}) . \tag{5.23}$$

Thus, from expression (5.20) of the correlation function in terms of a finite sum, we derive the multiple-integral representation as follows.

$$\begin{aligned} F_m^{(2s+)}(\{i_k, j_k\}) = \quad & C(\{i_b, j_b\}) \\ \times \left(\int_{-\infty+i\epsilon}^{\infty+i\epsilon} + \cdots + \int_{-\infty-i\tilde{\zeta}_s+i\epsilon}^{\infty-i\tilde{\zeta}_s+i\epsilon} \right) d\lambda_1 & \cdots \left(\int_{-\infty+i\epsilon}^{\infty+i\epsilon} + \cdots + \int_{-\infty-i\tilde{\zeta}_s+i\epsilon}^{\infty-i\tilde{\zeta}_s+i\epsilon} \right) d\lambda_{r^{'}} \\ \times \left(\int_{-\infty-i\epsilon}^{\infty-i\epsilon} + \cdots + \int_{-\infty-i\tilde{\zeta}_s-i\epsilon}^{\infty-i\tilde{\zeta}_s-i\epsilon} \right) d\lambda_{\tilde{r}} & \cdots \left(\int_{-\infty-i\epsilon}^{\infty-i\epsilon} + \cdots + \int_{-\infty-i\tilde{\zeta}_s-i\epsilon}^{\infty-i\tilde{\zeta}_s-i\epsilon} \right) d\lambda_{2sm} \\ \times Q(\{\epsilon_j, \epsilon_j^{'}\}; \lambda_1, \ldots, \lambda_{2sm}) \det S(\lambda_1, \ldots, \lambda_{2sm}) . \end{aligned} \tag{5.24}$$

Here $\tilde{\zeta}_s = (2s-1)\zeta$, $\tilde{r} = r^{'} + 1$, and $Q(\{\epsilon_j, \epsilon_j^{'}\}; \lambda_1, \ldots, \lambda_{2sm})$ is given by

$$\begin{aligned} & Q(\{\epsilon_j, \epsilon_j^{'}\}; \lambda_1, \ldots, \lambda_{2sm}) \\ & = (-1)^{r^{'}} \frac{\prod_{j \in \boldsymbol{\alpha}^-} \left(\prod_{k=1}^{j-1} \varphi(\tilde{\lambda}_j - w_k^{(2s)} + \eta) \prod_{k=j+1}^{2sm} \varphi(\tilde{\lambda}_j - w_k^{(2s)}) \right)}{\prod_{1 \le k < \ell \le 2sm} \varphi(\lambda_\ell - \lambda_k + \eta + \epsilon_{\ell,k})} \\ \times & \frac{\prod_{j \in \boldsymbol{\alpha}^+} \left(\prod_{k=1}^{j-1} \varphi(\tilde{\lambda}_j^{'} - w_k^{(2s)} - \eta) \prod_{k=j+1}^{2sm} \varphi(\tilde{\lambda}_j^{'} - w_k^{(2s)}) \right)}{\prod_{1 \le k < \ell \le 2sm} \varphi(w_k^{(2s)} - w_\ell^{(2s)})} . \end{aligned} \tag{5.25}$$

In the denominator we set $\epsilon_{k,\ell} = i\epsilon$ for $Im(\lambda_k - \lambda_\ell) > 0$ and $\epsilon_{k,\ell} = -i\epsilon$ for $Im(\lambda_k - \lambda_\ell) < 0$, where ϵ is infinitesimally small: $|\epsilon| \ll 1$. Here, $Im(a+ib) = b$ for real numbers a and b. Here, for $\boldsymbol{\alpha}^\pm$, we recall (5.14).

We evaluate $\alpha(\lambda_j)$ in (5.24), replacing paths $(-\infty - i(\gamma-1)\zeta \pm i\epsilon, \infty - i(\gamma-1)\zeta \pm i\epsilon)$ by $(-\infty - i(\gamma-1/2)\zeta, \infty - i(\gamma-1/2)\zeta)$ for $\gamma = 1, 2, \ldots, 2s$, respectively. The integrals over λ_j for $j \ge \tilde{r}$ do not change when $\epsilon \to \zeta/2$.

Thus, correlation functions (5.9) are expressed in the form of a single term of multiple integrals (5.24).

We can derive the symmetric expression for the multiple-integral representations of the spin-s correlation function $F_m^{(2s+)}(\{i_k, j_k\})$ as follows.[1]

$$
\begin{aligned}
& F_m^{(2s+)}(\{i_k, j_k\}) = C(\{i_b, j_b\}) \\
& \times \frac{1}{\prod_{1 \le \alpha < \beta \le 2s} \sinh^m(\beta - \alpha)\eta} \prod_{1 \le k < l \le m} \frac{\sinh^{2s}(\pi(\xi_k - \xi_l)/\zeta)}{\prod_{j=1}^{2s} \prod_{r=1}^{2s} \sinh(\xi_k - \xi_l + (r-j)\eta)} \\
& \times \sum_{\sigma \in \mathcal{S}_{2sm}/(\mathcal{S}_m)^{2s}} (\operatorname{sgn} \sigma) \prod_{j=1}^{r'} \left(\int_{-\infty+i\epsilon}^{\infty+i\epsilon} + \cdots + \int_{-\infty-i(2s-1)\zeta+i\epsilon}^{\infty-i(2s-1)\zeta+i\epsilon} \right) d\mu_{\sigma j} \\
& \times \prod_{j=r'+1}^{2sm} \left(\int_{-\infty-i\epsilon}^{\infty-i\epsilon} + \cdots + \int_{-\infty-i(2s-1)\zeta-i\epsilon}^{\infty-i(2s-1)\zeta-i\epsilon} \right) d\mu_{\sigma j} \\
& \times Q(\{\epsilon_j, \epsilon_j'\}; \lambda_{\sigma 1}, \ldots, \lambda_{\sigma(2sm)})) \left(\prod_{j=1}^{2sm} \frac{\prod_{b=1}^m \prod_{\beta=1}^{2s-1} \sinh(\lambda_j - \xi_b + \beta\eta)}{\prod_{b=1}^m \cosh(\pi(\mu_j - \xi_b)/\zeta)} \right) \\
& \times \frac{i^{2sm^2}}{(2i\zeta)^{2sm}} \prod_{\gamma=1}^{2s} \prod_{1 \le b < a \le m} \sinh(\pi(\mu_{2s(a-1)+\gamma} - \mu_{2s(b-1)+\gamma})/\zeta) \, .
\end{aligned}
\tag{5.26}
$$

Here λ_j are given by $\lambda_j = \mu_j - (\beta(j) - 1/2)\eta$ for $j = 1, \ldots, 2sm$.

It is straightforward to take the homogeneous limit: $\xi_k \to 0$. Here $(\operatorname{sgn} \sigma)$ denotes the sign of permutation $\sigma \in \mathcal{S}_{2sm}/(\mathcal{S}_m)^{2s}$.

6. Derivation of finite-sum expression of spin-s XXZ correlation functions with arbitrary entries

6.1. *Fundamental commutation relations*

We now discuss briefly the derivation of (5.20), which expresses the spin-s XXZ correlation functions with arbitrary entries in terms of the product of finite sums over the Bethe roots.

Let Σ_N be the set of integers $1, 2, \ldots, N$, i.e. $\Sigma_N = \{1, 2, \ldots, N\}$. Recall definition (5.14) of $\boldsymbol{\alpha}^{\pm}$ and that of integers c_j and c_j'. For a given set of c_j, c_j', we introduce $\boldsymbol{A}_j$ and $\boldsymbol{A}_j'$ by

$$
\begin{aligned}
\boldsymbol{A}_j &= \{b; 1 \le b \le M + 2sm, \, b \ne c_k, c_k' \text{ for } k < j\} \, , \\
\boldsymbol{A}_j' &= \{b; 1 \le b \le M + 2sm, \, b \ne c_k \text{ for } k \le j, b \ne c_k' \text{ for } k < j\} .
\end{aligned}
\tag{6.1}
$$

We define sets $\boldsymbol{\alpha}_j^{\pm}$ and $c(\boldsymbol{\alpha}_j^{\pm})$ as follows.

$$\boldsymbol{\alpha}_j^- = \{k;\, k < j, k \in \boldsymbol{\alpha}^-\}\,, \quad \boldsymbol{\alpha}_j^+ = \{k;\, k < j, k \in \boldsymbol{\alpha}^+\}\,, \tag{6.2}$$

$$c(\boldsymbol{\alpha}_j^-) = \{c_k;\, k \in \boldsymbol{\alpha}_j^-\}\,, \quad c(\boldsymbol{\alpha}_j^+) = \{c_k^{'};\, k \in \boldsymbol{\alpha}_j^+\}\,. \tag{6.3}$$

We have

$$\boldsymbol{A}_j = \Sigma_{M+2sm} \setminus \left(c(\boldsymbol{\alpha}_j^-) \cup c(\boldsymbol{\alpha}_j^+)\right)\,,\ \boldsymbol{A}_j^{'} = \Sigma_{M+2sm} \setminus \left(c(\boldsymbol{\alpha}_{j+1}^-) \cup c(\boldsymbol{\alpha}_j^+)\right)\,.$$

Let us denote by t the number of c_j $(j \in \boldsymbol{\alpha}^-)$ and $c_j^{'}$ $(j \in \boldsymbol{\alpha}^+)$ such that $c_j, c_j^{'} \le M$, for a given set of c_j and $c_j^{'}$. We express (5.17) as follows.

$$\sum_{t=r}^{2sm} \sum_{\{c_j, c_j^{'}\}_t} = \prod_{j \in \boldsymbol{\alpha}^-} \left(\sum_{c_j=1}^{M} \right) \prod_{j \in \boldsymbol{\alpha}^+} \left(\sum_{c_j^{'}=1}^{M+j} \right)\,. \tag{6.4}$$

Here the sum over $\{c_j, c_j^{'}\}_t$ denotes the sums over c_j and $c_j^{'}$ such that the number of $c_j^{'} \le M$ is fixed by $t - r$.

Suppose that λ_α for $\alpha = 1, 2, \ldots, M$ give a set of solutions of the Bethe ansatz equations in the spin-1/2 case with $w_j = w_j^{(2s;\,\epsilon)}$ for $j = 1, 2, \ldots, L$.[1] Here w_j are inhomogeneous parameters. We set rapidities λ_{M+j} by

$$\lambda_{M+j} = w_j = w_j^{(2s:\epsilon)}\,, \qquad j = 1, 2, \ldots, 2sm\,. \tag{6.5}$$

We can show the fundamental commutation relations as follows.[11]

$$\begin{aligned} &\langle 0| \left(\prod_{\alpha=1}^{M} C(\lambda_\alpha) \right) T_{\epsilon_1, \epsilon_1^{'}}(\lambda_{M+1}) \cdots T_{\epsilon_{2sm}, \epsilon_{2sm}^{'}}(\lambda_{M+2sm}) \\ &= \sum_{t=r}^{2sm} \sum_{\{c_j, c_j^{'}\}_t} G_{\{c_j, c_j^{'}\}}(\lambda_1, \cdots, \lambda_{M+2sm}) \langle 0| \prod_{k \in \boldsymbol{A}_{2sm+1}(\{c_j, c_j^{'}\})} C(\lambda_k)\,, \end{aligned}$$

where $d(\lambda; \{w_k^{(2s;\epsilon)}\}_L)$ and $G_{\{c_j, c_j^{'}\}}((\lambda_\alpha)_{M+2sm})$ are given by

$$\begin{aligned} d(\lambda; \{w_k^{(2s;\epsilon)}\}_L) &= \prod_{k=1}^{L} b(\lambda - w_k^{(2s;\epsilon)})\,, \\ G_{\{c_j, c_j^{'}\}}(\lambda_1, \cdots, \lambda_{M+2sm}) &= \prod_{j \in \boldsymbol{\alpha}^+} \left(\frac{\prod_{b=1; b \in \boldsymbol{A}_j^{'}}^{M+j-1} \varphi(\lambda_b - \lambda_{c_j^{'}} + \eta)}{\prod_{b=1, b \in \boldsymbol{A}_{j+1}}^{M+j} \varphi(\lambda_b - \lambda_{c_j^{'}})} \right) \\ &\quad \times \prod_{j \in \boldsymbol{\alpha}^-} \left(d(\lambda_{c_j}; \{w_k^{(2s;\,\epsilon)}\}_L) \frac{\prod_{b=1; b \in \boldsymbol{A}_j}^{M+j-1} \varphi(\lambda_{c_j} - \lambda_b + \eta)}{\prod_{b=1, b \in \boldsymbol{A}_j^{'}}^{M+j} \varphi(\lambda_{c_j} - \lambda_b)} \right)\,. \end{aligned} \tag{6.6}$$

6.2. *Finite-sum expression of correlation functions for a finite chain*

We introduce disjoint subsets of $\boldsymbol{\alpha}^{+}$, $\boldsymbol{\alpha}_J^{+}$ and $\boldsymbol{\alpha}_K^{+}$, as follows.

$$\boldsymbol{\alpha}_J^{+} = \{j;\, j \in \boldsymbol{\alpha}^{+}, 1 \le c_j^{'} \le M\}\,, \quad \boldsymbol{\alpha}_K^{+} = \{j; j \in \boldsymbol{\alpha}^{+}, c_j^{'} > M\}\,. \tag{6.7}$$

We define sets $c(\boldsymbol{\alpha}^{-})$, $c(\boldsymbol{\alpha}_J^{+})$ and $c(\boldsymbol{\alpha}_K^{+})$ as follows.

$$c(\boldsymbol{\alpha}^{-}) = \{c_k;\, k \in \boldsymbol{\alpha}^{-}\},\, c(\boldsymbol{\alpha}_J^{+}) = \{c_k;\, k \in \boldsymbol{\alpha}_J^{+}\},\, c(\boldsymbol{\alpha}_K^{+}) = \{c_k;\, k \in \boldsymbol{\alpha}_K^{+}\}\,.$$

We define a sequence $(\widetilde{b_k})_t$ by a subsequence of $(b_k)_{2sm}$ such that $\widetilde{b}_k \le M$ for $k = 1, 2, \ldots, t$. We denote sequence $(b_k)_{2sm}$ and $(\widetilde{b_k})_t$ as sets by $\boldsymbol{b}$ and $\widetilde{\boldsymbol{b}_t}$, respectively, i.e. $\boldsymbol{b} = \{b_1, b_2, \ldots, b_{2sm}\}$ and $\widetilde{\boldsymbol{b}_t} = \{\widetilde{b}_1, \ldots, \widetilde{b}_t\}$. Here we note $\widetilde{\boldsymbol{b}_t} = c(\boldsymbol{\alpha}^{-}) \cup c(\boldsymbol{\alpha}_J^{+})$. We define sequence $(b_k^{'})_{2sm-t}$ by a subsequence of $(b_k)_{2sm}$ such that $b_k^{'} > M$ for $k = 1, 2, \ldots, 2sm - t$. We denote it as a set by $\boldsymbol{b}_{2sm-t}^{'}$. Here we note $\boldsymbol{b}_{2sm-t}^{'} = c(\boldsymbol{\alpha}_K^{+})$.

We define sets Z and K by $Z = \Sigma_M \setminus \widetilde{\boldsymbol{b}_t}$ and $K = \Sigma_{2sm} \setminus \boldsymbol{b}_{2sm-t}^{'}$, respectively. We define a sequence $(z(\alpha))_{M-t}$ by putting the elements of Z in increasing order: $z(1) < z(2) < \cdots < z(M-t)$ where $Z = \{z(i); i = 1, 2, \ldots, M-t\}$, and a sequence $(\kappa_j)_t$ by putting the elements of K in increasing order: $\kappa_1 < \kappa_2 < \cdots < \kappa_t$ where $K = \{\kappa_j;\, j = 1, 2, \ldots, t\}$.

We derive the spin-s correlation functions from those of the spin-1/2 case sending ϵ to zero:

$$F_m^{(2s\,+)}(\{i_b, j_b\}; (w_j^{(2s)})_L) = C(\{i_k, j_k\}) \lim_{\epsilon \to 0} F_{2sm}^{(1\,+)}(\{\epsilon_j, \epsilon_j^{'}\}; (w_j^{(2s;\,\epsilon)})_L)\,. \tag{6.8}$$

Applying (6.6) to (5.9) (or (5.13)) we have

$$\begin{aligned}
&F_{2sm}^{(1\,+)}(\{\epsilon_j, \epsilon_j^{'}\}; (w_j^{(2s;\,\epsilon)})_L) = \sum_{t=r}^{2sm} \sum_{\{c_j, c_j^{'}\}_t} G_{\{c_j, c_j^{'}\}}(\lambda_1, \cdots, \lambda_{M+2sm}) \\
&\times \phi_{2sm}(\{\lambda_\alpha\}_M) \\
&\times \frac{\langle 0| \prod_{\alpha=1}^{M-t} C(\lambda_{z(\alpha)}) \prod_{\gamma=1}^{t} C(w_{\kappa_\gamma}^{(2s;\,\epsilon)}) \prod_{\beta=1}^{M-t} B(\lambda_{z(\beta)}) \prod_{\gamma=1}^{t} B(\lambda_{\widetilde{b_\gamma}})|0\rangle}{\langle 0| \prod_{\alpha=1}^{M-t} C(\lambda_{z(\alpha)}) \prod_{\gamma=1}^{t} C(w_{\widetilde{b_\gamma}}^{(2s;\,\epsilon)}) \prod_{\beta=1}^{M-t} B(\lambda_{z(\beta)}) \prod_{\gamma=1}^{t} B(\lambda_{\widetilde{b_\gamma}})|0\rangle} \\
&= \sum_{t=r}^{2sm} \sum_{\{c_j, c_j^{'}\}_t} \prod_{\alpha=1}^{M} \prod_{j=1}^{2sm} \frac{\varphi(\lambda_\alpha - w_j^{(2s;\,\epsilon)})}{\varphi(\lambda_\alpha - w_j^{(2s;\,\epsilon)} + \eta)} \\
&\times \prod_{j \in \boldsymbol{\alpha}^{-}} \left(\frac{\prod_{b=1, b \in A_j}^{M+j-1} \varphi(\lambda_{c_j} - \lambda_b + \eta)}{\prod_{b=1, b \in A_j^{'}}^{M+j} \varphi(\lambda_{c_j} - \lambda_b)} \right)
\end{aligned}$$

$$
\times \prod_{j \in \boldsymbol{\alpha}^+} \left(\frac{\prod_{b=1, b \in A_j'}^{M+j-1} \varphi(\lambda_b - \lambda_{c_j'} + \eta)}{\prod_{b=1, b \in A_{j+1}}^{M+j} \varphi(\lambda_b - \lambda_{c_j'})} \right) \prod_{1 \le k < \ell \le t} \frac{\varphi(\lambda_{\tilde{b}_k} - \lambda_{\tilde{b}_\ell})}{\varphi(w_{\kappa_k}^{(2s;\,\epsilon)} - w_{\kappa_\ell}^{(2s;\,\epsilon)})}
$$

$$
\times \prod_{\alpha=1}^{M-t} \prod_{\ell=1}^{t} \frac{\varphi(\lambda_{z(\alpha)} - \lambda_{\tilde{b}_\ell})}{\varphi(\lambda_{z(\alpha)} - w_{\kappa_\ell}^{(2s;\,\epsilon)})} \prod_{\alpha=1}^{M} \prod_{\ell=1}^{t} \frac{\varphi(\lambda_\alpha - w_{\kappa_\ell}^{(2s;\,\epsilon)} + \eta)}{\varphi(\lambda_\alpha - \lambda_{\tilde{b}_\ell} + \eta)}
$$

$$
\times \det \Bigg((\Phi')^{-1} \left(\lambda_{z(\alpha)})_{M-t} \# (\lambda_{\tilde{b}_\ell})_t \right)
$$

$$
\times \Psi' \left((\lambda_{z(\alpha)})_{M-t} \# (w_{\kappa_\ell}^{(2s;\,\epsilon)})_t, \, (\lambda_{z(\alpha)})_{M-t} \# (\lambda_{\tilde{b}_\ell})_t \right) \Bigg). \quad (6.9)
$$

Here, $\phi_{2sm}(\{\lambda_\alpha\}) = \prod_{j=1}^{2sm} \prod_{\alpha=1}^{M} b(\lambda_\alpha - w_j^{(2s;\,\epsilon)})$, and matrix elements $(\Psi')_{ab}$ for $a, b = 1, 2, \ldots, M$ are given by

$$
\left(\Psi'((\lambda_{z(\alpha)})_{M-t} \# (\lambda_{\tilde{b}_\ell})_t, \, (\lambda_{z(\alpha)})_{M-t} \# (w_{\kappa_\ell})_t; (w_k)_L) \right)_{a,\,b}
$$

$$
= \begin{cases} \Phi'_{a,\,b} \left((\lambda_{z(\alpha)})_{M-t} \# (\lambda_{\tilde{b}_\ell})_t \right) & \text{for } b \le M-t \\ \dfrac{\varphi(\eta)}{\varphi(\lambda_{z(a)} - w_{\kappa_k}) \varphi(\lambda_{z(a)} - w_{\kappa_k} + \eta)} & \text{for}\, b = k + M - t \ (1 \le k \le t) \end{cases}
$$

$$(6.10)$$

The matrix elements of the Gaudin matrix are given as follows.

$$
\Phi'_{a,\,b} \left((\lambda_{z(\alpha)})_{M-t} \# (\lambda_{\tilde{b}_\ell})_t; (w_k)_L \right) = \Phi'_{z(a),z(b)}((\lambda_\alpha)_M; (w_k)_L)
$$

$$
= \frac{\varphi(2\eta)}{\varphi(\lambda_a - \lambda_b + \eta) \varphi(\lambda_a - \lambda_b - \eta)} + \delta_{a,b} \left(\sum_{p=1}^{L} \frac{\varphi(\eta)}{\varphi(\lambda_a - w_p) \varphi(\lambda_a - w_p + \eta)} \right.
$$

$$
\left. - \sum_{\gamma=1}^{M} \frac{\varphi(2\eta)}{\varphi(\lambda_a - \lambda_\gamma + \eta) \varphi(\lambda_a - \lambda_\gamma - \eta)} \right). \quad (6.11)
$$

For any positive integer N_s we can rigorously calculate (6.9) as follows.[52]

Proposition 6.1.

$$
F_{2sm}^{(1\,+)}(\{\epsilon_j, \epsilon_j'\}; (w_j^{(2s;\,\epsilon)})_L) = \sum_{t=r}^{2sm} \sum_{\{c_j, c_j'\}} \left(\prod_{j,k \in \boldsymbol{\alpha}_K^+, c_j' < c_k', j<k} (-1) \right)
$$

$$
\times (-1)^{2sm-t} \prod_{j \in \boldsymbol{\alpha}_K^+} \left(\prod_{\ell \in \boldsymbol{\alpha}_J^+; \ell > j} (-1) \cdot \prod_{\kappa \in K; \kappa + M < c_j'} (-1) \right)
$$

$$
\times \det(\Phi')^{-1} \Psi'((\lambda_{z(\alpha)})_{M-t} \# (w_{\kappa_\ell}^{(2s;\,\epsilon)})_t, \, (\lambda_{z(\alpha)})_{M-t} \# (\lambda_{\tilde{b}_\ell})_t)
$$

$$\times (-1)^{r'} \frac{\prod_{j\in\boldsymbol{\alpha}^-}\left(\prod_{k=1}^{j-1}\varphi(\lambda_{c_j}-w_k^{(2s;\,\epsilon)}+\eta)\prod_{k=j+1}^{2sm}\varphi(\lambda_{c_j}-w_k^{(2s;\,\epsilon)})\right)}{\prod_{1\le k<\ell\le 2sm}\varphi(\lambda_{b_\ell}-\lambda_{b_k}+\eta)}$$

$$\times \frac{\prod_{j\in\boldsymbol{\alpha}^+}\left(\prod_{k=1}^{j-1}\varphi(\lambda_{c_j'}-w_k^{(2s;\,\epsilon)}-\eta)\prod_{k=j+1}^{2sm}\varphi(\lambda_{c_j'}-w_k^{(2s;\,\epsilon)})\right)}{\prod_{1\le k<\ell\le 2sm}\varphi(w_k^{(2s;\,\epsilon)}-w_\ell^{(2s;\,\epsilon)})}. \tag{6.12}$$

We define matrix elements (j,k) of $\phi_M^{(2sm)}((b_\ell)_{2sm})$ $(1\le j\le 2sm)$:[52]

$$\text{If}\quad b_j>M,\quad \left(\phi_M^{(2sm)}((b_\ell)_{2sm})\right)_{j,\,k}=-\delta_{b_j-M,k}\quad \text{for } k=1,2,\ldots,2sm,$$

$$\text{if}\quad b_j\le M,\quad \text{there is an integer } i \text{ such that } b_j=\tilde{b}_i$$

$$\left(\phi_M^{(2sm)}((b_\ell)_{2sm})\right)_{j,\,\kappa_k}=(\Phi')^{-1}\Psi'((\lambda_{z(\alpha)})_{M-t}\#(\lambda_{\tilde{b}_\ell})_t,$$

$$(\lambda_{z(\alpha)})_{M-t}\#(w_{\kappa_\ell}^{(2s;\,\epsilon)})_t)_{i+M-t,k+M-t}\,,\quad \text{for } k=1,2,\ldots,t,$$

$$\text{and}\quad \phi_M^{(2sm)}((b_\ell)_{2sm})_{j,b_k'}=0\quad \text{for } k=1,2,\ldots,2sm-t. \tag{6.13}$$

We can show the following proposition.[52]

Proposition 6.2.

$$\det\left((\Phi')^{-1}\Psi'((\lambda_{z(\alpha)})_{M-t}\#(w_{\kappa_\ell}^{(2s;\,\epsilon)})_t,\,(\lambda_{z(\alpha)})_{M-t}\#(\lambda_{\tilde{b}_\ell})_t)\right)$$

$$=\det\phi_M^{(2sm)}((b_\ell)_{2sm})\,(-1)^{2sm-t}\left(\prod_{j,k\in\boldsymbol{\alpha}_K^+,c_j'<c_k',j<k}(-1)\right)$$

$$\times\prod_{j\in\boldsymbol{\alpha}_K^+}\left(\prod_{\ell\in\boldsymbol{\alpha}_J^+;\ell>j}(-1)\cdot\prod_{\kappa\in K;\kappa+M<c_j'}(-1)\right). \tag{6.14}$$

When N_s is large enough, solving the integral equations for $\phi_M^{(2sm)}((b_\ell)_{2sm})$, we can show

$$\det\phi_M^{(2sm)}((b_\ell)_{2sm})=\det M^{(2sm)}((b_\ell)_{2sm})+O(1/N_s). \tag{6.15}$$

We thus obtain the finite-size spin-s XXZ correlation functions with arbitrary entries (5.20).

Acknowledgments

One of the authors (T.D.) would also like to thank K. Motegi for helpful collaboration on the spin-s R-matrices in Ref. 50, which are closely related to the present study. Furthermore, the authors would like to thank S.

Miyashita for encouragement and keen interest in this work. The authors are grateful to the organizers of the workshop "Infinite Analysis 09– New Trends in Quantum Integrable Systems –", July 27-31, 2009, Kyoto University, Japan. This work is partially supported by Grant-in-Aid for Scientific Research (C) No. 20540365.

References

1. T. Deguchi and C. Matsui, *Nucl. Phys. B* **831[FS]**, 359 (2010).
2. M. Jimbo, K. Miki, T. Miwa and A. Nakayashiki, *Phys. Lett. A* **168**, 256 (1992).
3. M. Jimbo and T. Miwa, *Algebraic Analysis of Solvable Lattice Models* (AMS, Providence, RI, 1995).
4. M. Jimbo and T. Miwa, *J. Phys. A: Math. Gen.* **29**, 2923 (1996).
5. T. Miwa and Y. Takeyama, *Contemporary Mathematics* **248**, 377 (1999).
6. N. Slavnov, *Theor. Math. Phys.* **79**, 502 (1989).
7. V. E. Korepin, N. M. Bogoliubov and A. G. Izergin, *Quantum Inverse Scattering Method and Correlation Functions* (Cambridge University Press, 1993).
8. J. M. Maillet and J. Sanchez de Santos, *Amer. Math. Soc. Transl.* **201**, 137 (2000).
9. N. Kitanine, J. M. Maillet and V. Terras, *Nucl. Phys. B* **554[FS]**, 647 (1999).
10. J. Maillet and V. Terras, *Nucl. Phys. B* **575[FS]**, 627 (2000).
11. N. Kitanine , J. M. Maillet and V. Terras, *Nucl. Phys. B* **567[FS]**, 554 (2000).
12. F. Göhmann, A. Klümper and A. Seel, *J. Phys. A: Math. Gen.* **37**, 7625 (2004).
13. J. Damerau, F. Göhmann, N. Hasenclever and A. Klümper, *J. Phys. A: Math. Theor.* **40**, 4439 (2007).
14. M. Jimbo, T. Miwa and F. Smirnov, *J. Phys. A: Math. Theor.* **42**, 304018 (2009).
15. N. Kitanine, K.K. Kozlowski, J.M. Maillet, N.A. Slavnov and V. Terras, *J. Stat. Mech.* , p. P04003 (2009).
16. N. Kitanine, *J. Phys. A: Math. Gen.* **34**, 8151 (2001).
17. O. A. Castro-Alvaredo and J. M. Maillet, *J. Phys. A: Math. Theor.* **40**, 7451 (2007).
18. T. Deguchi and C. Matsui, *Nucl. Phys. B* **814[FS]**, 405 (2009).
19. P. P. Kulish, N. Yu. Reshetikhin and E. K. Sklyanin, *Lett. Math. Phys.* **5**, 393 (1981).
20. H. M. Babujian, *Nucl. Phys. B* **215[FS7]**, 317 (1983).
21. H. M. Babujian and A. M. Tsvelick, *Nucl. Phys. B* **265** **[FS15]**, 24 (1986).
22. A. B. Zamolodchikov and V. A. Fateev, *Sov. J. Nucl. Phys.* **32**, 298 (1980).
23. K. Sogo, Y. Akutsu and T. Abe, *Prog. Theor. Phys.* **70**, 730 (1983).
24. A. N. Kirillov and N. Yu. Reshetikhin, *J. Phys. A: Math. Gen.* **20**, 1565 (1987).
25. T. Deguchi, M. Wadati and Y. Akutsu, *J. Phys. Soc. Jpn.* **57**, 1905 (1988).

26. L. A. Takhtajan, *Phys. Lett. A* **87**, 479 (1982).
27. K. Sogo, *Phys. Lett. A* **104**, 51 (1984).
28. H. Johannesson, *J. Phys. A: Math. Gen.* **21**, L611 (1988).
29. H. Johannesson, *J. Phys. A: Math. Gen.* **21**, L1157 (1988).
30. F. C. Alcaraz and M. J. Martins, *J. Phys. A: Math. Gen.* **21**, 4397 (1988).
31. I. Affleck, D. Gepner, H. J. Schultz and T. Ziman, *J. Phys. A: Math. Gen.* **22**, 511 (1989).
32. B. D. Dörfel, *J. Phys. A: Math. Gen.* **22**, L657 (1989).
33. L. V. Avdeev, *J. Phys. A: Math. Gen.* **23**, L485 (1990).
34. F. C. Alcaraz and M. J. Martins, *J. Phys. A: Math. Gen.* **22**, 1829 (1989).
35. H. Frahm, N. -C. Yu and M. Fowler, *Nucl. Phys. B* **336**, 396 (1990).
36. H. Frahm and N. C. Yu, *J. Phys. A: Math. Gen.* **23**, 2115 (1990).
37. H. J. de Vega and F. Woynarovich, *J. Phys. A: Math. Gen.* **23**, 1613 (1990).
38. A. Klümper and M. T. Batchelor, *J. Phys. A: Math. Gen.* **23**, L189 (1990).
39. A. Klümper, M. T. Batchelor and P. A. Pearce, *J. Phys. A: Math. Gen.* **24**, 3111 (1991).
40. J. Suzuki, *J. Phys. A: Math. Gen.* **32**, 2341 (1999).
41. M. Idzumi, *Thesis, University of Tokyo,* (Feb. 1993).
42. M. Idzumi, *Int. J. Mod. Phys. A* **9**, 4449 (1994).
43. A. H. Bougourzi and R. A. Weston, *Nucl. Phys. B* **417** (1994).
44. H. Konno, *Nucl. Phys. B* **432[FS]** (1994).
45. M. Jimbo, *Lett. Math. Phys.* **10**, 63 (1985).
46. M. Jimbo, *Lett. Math. Phys.* **11**, 247 (1986).
47. V. Drinfel'd, *Proc. ICM Berkeley* , 798 (1986).
48. R. J. Baxter, *Exactly Solved Models in Statistical Mechanics* (Academic Press, 1982).
49. Y. Akutsu and M. Wadati, *J. Phys. Soc. Jpn.* **56**, 3039 (1987).
50. T. Deguchi and K. Motegi, *in prep.* .
51. M. Takahashi, *Thermodynamics of One-Dimensional Solvable Models* (Cambridge University Press, 1999).
52. T. Deguchi, *in prep.* .

CHARACTERS OF COINVARIANTS IN $(1, p)$ LOGARITHMIC MODELS

BORIS L. FEIGIN

Higher School of Economics, Moscow, Russia
and Landau Institute for Theoretical Physics, Chernogolovka, 142432, Russia
E-mail: bfeigin@gmail.com

ILYA YU. TIPUNIN

Tamm Theory Division, Lebedev Physics Institute, 119991
Leninski pr., 53, Moscow, Russia
E-mail: tipunin@gmail.com

We investigate induced modules of doublet algebra in $(1, p)$ logarithmic models. We give fermionic formulas for the characters of induced modules and coinvariants with respect to different subalgebras calculated in the irreducible modules. The characters of coinvariants give multiplicities of projective modules in fusion of induced modules.

Keywords: Logarithmic conformal field theory; quantum group.

1. Introduction

Vertex–operator algebras (VOAs) in logarithmic conformal field theory are very popular last several years. The simplest family of such algebras $\mathcal{W}(p)$ was introduced in Ref. 1 and was investigated in Refs. 2,3,15 and Ref. 11. Algebras $\mathcal{W}(p)$ admit a $SL(2)$ action by symmetries. Invariants of this action is the universal enveloping of the Virasoro algebra with the central charge

$$c = 13 - 6p - \frac{6}{p}. \qquad (1)$$

In this sence $\mathcal{W}(p)$ is the "Galois extension" of the Virasoro algebra and $SL(2)$ plays a role of the Galois group. The algebra $\mathcal{W}(p)$ is called the triplet algebra[2] because it is generated by a $SL(2)$ triplet of fields. The Virasoro algebra with the central charge (1) has infinite dimensional conformal blocks but after taking the extension they become finite dimensional and the conformal field theory becomes close to a rational one.[4,5] The only difference from rationality is that the category of representations is not semisimple. (We note that the nonsemisimplisity of the representation category is equivalent to appearing of logarithms in the corellation functions.)

In the present paper we consider slightly larger "Galois extension" $\mathcal{A}(p)$ that is generated by a doublet (not a triplet like $\mathcal{W}(p)$) of fields. We prefer $\mathcal{A}(p)$ because the abelianization technique, which allows us to calculate characters of induced representations and coinvariants is easier applied to it than to $\mathcal{W}(p)$. It gives fermionic formulas for the $\mathcal{A}(p)$ characters[6] (see also Ref. 7, where fermionic formulas for $\mathcal{W}(p)$ characters were obtained). On the other hand the algebra $\mathcal{A}(p)$ is worse than $\mathcal{W}(p)$ and strictly speaking is not a vertex operator algebra because it contains half integer powers of $z-w$ in Operator Product Expansions (OPEs). However, the representation theory of $\mathcal{A}(p)$ is very similar with a representation theory of an ordinary VOA and in what follows we don't bother on nonlocality of $\mathcal{A}(p)$ strongly. In the paper, we study conformal blocks or in other language coinvariants of $\mathcal{A}(p)$. To formulate the main statement we should recall some facts and definitions about coinvariants of VOAs (that we give in Sec. 1.1) and also about relations between representation categories of VOAs and quantum group (that we give in Sec. 1.2).

1.1. *Coinvariants*

Before we start a detailed consideration of $\mathcal{A}(p)$, we should say some general words about coinvariants. General facts and references on VOAs can be found in Ref. 8. Let A be a vertex-operator algebra generated by currents $H^1(z)$, $H^2(z)$, ... Let $\mathcal{V}$ be the vacuum module of A. Let Σ be a Riemann surface. Then we have a sheaf $\mathcal{V}(\Sigma)$ on Σ with the fiber $\mathcal{V}$. A set of sections over a small punctured neibourhood of a point x generates an algebra A_x and the currents $H^1(z)$, $H^2(z)$, ... are generators of A_x, where z is a local coordinate at x. These currents have decompositions $H^i(z) = \sum_{n\in\mathbb{Z}} H^i_n z^{-n-\Delta_i}$, where Δ_i is the conformal dimension of H^i. For future convenience we introduce the vector $\boldsymbol{\Delta} = (\Delta_1, \Delta_2, \dots)$ of conformal dimensions. Let us fix n points $z_1, \dots, z_n \in \Sigma$ and n A-modules $\mathcal{V}_{z_1}, \dots, \mathcal{V}_{z_n}$. For each z_i the module $\mathcal{V}_{z_1}$ is a module over A_{z_i}. Let $S_{z_1,\dots,z_n}$ be the space of sections of $\mathcal{V}(\Sigma)$ regular outside the points $z_1, \dots, z_n$. The sections from $S_{z_1,\dots,z_n}$ act in $\mathcal{V}_{z_1} \otimes \dots \otimes \mathcal{V}_{z_n}$. The quotient

$$\mathrm{Coinv}(\mathcal{V}_1, \dots, \mathcal{V}_n) = \mathcal{V}_{z_1} \otimes \dots \otimes \mathcal{V}_{z_n} / S_{z_1,\dots,z_n} \mathcal{V}_{z_1} \otimes \dots \otimes \mathcal{V}_{z_n} \qquad (2)$$

form a bundle over the configuration space of pairiwise distinct points $z_1, \dots, z_n$ and is called coinvariants of $\mathcal{V}_1, \dots, \mathcal{V}_n$. (In the paper we consider only the cases where coinvariants are finite dimensional.)

We fix a subalgebra $A[\mathbf{u}]$ of A generated by modes of $\{H^n(z)\}$, $H^n_{-i_n-\Delta_n+m}$, $m \in \mathbb{N}$, where $\mathbf{u} = (i_1, i_2, \dots)$. Let $\mathcal{M}_\mathbf{u}$ be induced module from the trivial 1-dimensional representation of $A[\mathbf{u}]$. $\mathcal{M}_\mathbf{u}$ contains a cyclic vector $|\mathbf{u}\rangle$ annihilated by $H^n_{-i_n-\Delta_n+m}$, $m \in \mathbb{N}$. In this notation the vacuum module corresponds to $\mathbf{u} = 0$. An induced module $\mathcal{M}_\mathbf{u}$ admits a grading by an operator D commuting with A. We define first the operator d in such a way that $d|\mathbf{u}\rangle = 0$ and $[d, H^i_n] = -nH^i_n$ for generators H^i_n of A. Then, we put

$$D = L_0 - d, \tag{3}$$

where L_0 is the zero mode of the Virasoro subalgebra from A, D evidently commutes with A.

The covariant functor Coinv is representable. In particular, it means that there exists the module $\mathcal{V}_1 \dot{\otimes} \mathcal{V}_2$ (which is called the fusion of $\mathcal{V}_1$ and $\mathcal{V}_2$) such that for any module $\mathcal{V}_3$ we have

$$\mathrm{Coinv}(\mathcal{V}_1, \mathcal{V}_2, \mathcal{V}_3) = \left(Hom_{A_{z_3}}(\mathcal{V}_3, \mathcal{V}_1 \dot{\otimes} \mathcal{V}_2)\right)^*. \tag{4}$$

Using this property the module $\mathcal{V}_1 \dot{\otimes} \mathcal{V}_2$ can be constructed as an inductive system in the following way. We define a sequence $\{\mathbf{u}_m\}$ such that each component of the vector $\mathbf{u}_m$ is greater than or equal to the same component of the vector $\mathbf{u}_{m-1}$ and for any $N \in \mathbb{N}$ there exists m such that a given component of $\mathbf{u}_m$ is greater than N. This determines the sequence of subalgebras $A[\mathbf{u}_m]$; we consider the squence of modules $\mathcal{M}_{\mathbf{u}_m}$ induced from trivial 1-dimensional representations of $A[\mathbf{u}_m]$ and set

$$\mathcal{V}[\mathbf{u}]^* = \mathrm{Coinv}(\mathcal{V}_1, \mathcal{V}_2, \mathcal{M}_\mathbf{u}) = \left(Hom_{A_{z_3}}(\mathcal{M}_\mathbf{u}, \mathcal{V}_1 \dot{\otimes} \mathcal{V}_2)\right)^*. \tag{5}$$

For each vector $\mathbf{u}$, by the Frobenius duality $\mathcal{V}[\mathbf{u}] \subset \mathcal{V}_1 \dot{\otimes} \mathcal{V}_2$. This means that $\mathcal{V}_1 \dot{\otimes} \mathcal{V}_2$ is the inductive limit of the inductive system $\mathcal{V}[\mathbf{u}_m]$ enumerated by vectors $\mathbf{u}_m$.

For a wide class of VOAs and induced modules $\mathcal{M}_\mathbf{u}$ it is naturally to be expected

$$\mathcal{M}_{\mathbf{u}_1} \dot{\otimes} \mathcal{M}_{\mathbf{u}_2} \simeq \mathcal{M}_{\mathbf{u}_1 + \mathbf{u}_2}. \tag{6}$$

A typical example when (6) is satisfied is $\widehat{s\ell}(2)$ minimal models at positive integer level and a class of modules $\mathcal{M}_\mathbf{u}$.[9] For Virasoro minimal models, modules $\mathcal{M}_j$ induced from vectors $|j\rangle$ satisfying $L_{-j-2+m}|j\rangle = 0$ for $m \in$

$\mathbb{N}$ satisfy $\mathcal{M}_{j_1}\dot{\otimes}\mathcal{M}_{j_2} = \mathcal{M}_{j_1+j_2}$. This statement follows for $(2,p)$ models from Ref. 10 and is a conjecture in other cases.

For the algebra $\mathcal{A}(p)$ we also conjecture the statement similar to (6) and give in the paper some reasons why it is true.

1.2. *Representation categories*

The representation category $\mathfrak{C}$ of $\mathcal{W}(p)$ is equivalent (as quasitensor category) to the representation category of $\overline{\mathfrak{U}}_{\mathfrak{q}}s\ell(2)$ with $\mathfrak{q} = e^{i\pi/p}$. This equivalence was cojectured in Ref. 11 and was proved in Ref. 12 (see also Ref. 13). In particular $\mathcal{V}_1\dot{\otimes}\mathcal{V}_2$ corresponds to the tensor product of $\overline{\mathfrak{U}}_{\mathfrak{q}}s\ell(2)$ representations corresponding to $\mathcal{V}_1$ and $\mathcal{V}_2$. This statement can be obtained in the following way. In the representation category of Virasoro algebra a subcategory equivalent to Lusztig quantum $s\ell(2)$ can be distinguished. Then from general properties of VOAs it follows that after "Galois extension" the Lusztig quantum $s\ell(2)$ reduces to $\overline{\mathfrak{U}}_{\mathfrak{q}}s\ell(2)$. The category $\mathfrak{C}$ contains $2p$ simple objects $\mathsf{X}_r^{\pm}$, $1 \leq r \leq p$ and $2p$ projective objects $\mathsf{P}_r^{\pm}$, $1 \leq r \leq p$. Projective objects $\mathsf{P}_r^{\pm}$, $1 \leq r \leq p-1$ are not simple and $\mathsf{P}_p^{\pm} \simeq \mathsf{X}_p^{\pm}$. We call $\mathsf{X}_p^{\pm}$ Steinberg modules.

The representation category $\mathfrak{A}$ of $\mathcal{A}(p)$ is not equvalent to a representation category of a quantum group. However $\mathfrak{A}$ is obtained by a reduction of the tensor category $\mathfrak{C}$. The reduction is such that simple objects from the pairs $\mathsf{X}_r^{\pm}$ for each r becomes indistinguishable, so $\mathfrak{A}$ contains p simple (projective) objects X_r (P_r), $1 \leq r \leq p$ and $\mathsf{P}_p \simeq \mathsf{X}_p$. Many statements about $\mathcal{A}(p)$ representations can be done in terms of the category $\mathfrak{A}$. In particular dimensions of coinvariants $\mathrm{Coinv}(\mathcal{V}_1, \dots, \mathcal{V}_n)$ can be calculated in terms of the tensor category $\mathfrak{A}$. To do that one should take the tensor product of objects corresponding to $\mathcal{V}_1, \dots, \mathcal{V}_n$ and calculate the space of homomorphisms to the simple object X_1 (X_1 corresponds to trivial 1 dimensional representation). In what follows we use the same notation for $\mathcal{A}(p)$-modules and corresponding objects of $\mathfrak{A}$.

The simple objects $\mathsf{X}_r \in \mathfrak{A}$ are induced $\mathcal{A}(p)$-modules $\mathcal{M}_{\mathbf{u}_r}$ for some $\mathbf{u}_r$, which are defined in Sec. 4.1. X_1 corresponds to the vacuum $\mathcal{A}(p)$-module and $\mathsf{X}_1\dot{\otimes}\mathcal{P} = \mathcal{P}$, $\forall\mathcal{P} \in \mathfrak{A}$. We consider the module $\mathcal{M} = \mathsf{X}_2^{\dot{\otimes}n_2}\dot{\otimes}\mathsf{X}_3^{\dot{\otimes}n_3}\dot{\otimes}\cdots\dot{\otimes}\mathsf{X}_p^{\dot{\otimes}n_p}$. The module $\mathcal{M}$ as an $\mathcal{A}(p)$-module is the induced module $\mathcal{M} = \mathcal{M}_{n_2\mathbf{u}_2+n_3\mathbf{u}_3+\cdots+n_p\mathbf{u}_p}$. Each induced module admits a commuting with $\mathcal{A}(p)$ grading (3), which means that the tensor product of simple objects of category $\mathfrak{A}$ admits the grading invariant with respect to the action of the algebra. We use this grading to obtain fermionic formulas for characters of coinvariants.

1.3. *The main statement*

In this paper, we give an outline of a proof that the fusion of induced $\mathcal{A}$-modules has the same structure (6) as the fusion of induced modules of many rational VOAs. More precisely this can be formulated as follows.
Statement. *For a class of highest-weight conditions described by some vectors* $\mathbf{u}$ *(precise conditions on* $\mathbf{u}$ *are given in Sec. 4.2) two induced modules* $\mathcal{M}_{\mathbf{u}_1}$ *and* $\mathcal{M}_{\mathbf{u}_2}$ *satisfy*

$$\mathcal{M}_{\mathbf{u}_1} \dot{\otimes} \mathcal{M}_{\mathbf{u}_2} \simeq \mathcal{M}_{\mathbf{u}_1+\mathbf{u}_2}. \tag{7}$$

To give evidences for the statement, we use the fermionic formulas for the characters corresponding to the grading (3). To obtain the fermionic formulas for the characters, we use the technique of abelianization. The abelianization technique is a degeneration of the algebra to an algebra with greater number of generators but with quadratic relations. In many cases the obtained algebra is abelian whence the technique tooks its name. In our case the algebra obtained by application of the abelianization technique is not abelian but is very close to an abelian one.

The structure of the formulas for characters is similar with the structure of fermionic formulas for Kostka polynomials (see Sec. 5).

We give several examples for $p = 3$. Some low powers of "two dimensional representation" have the following decompositions

$$\mathsf{X}_2^{\dot{\otimes}2} = \mathsf{X}_1 + \mathsf{X}_3, \tag{8}$$

$$\mathsf{X}_2^{\dot{\otimes}3} = q^{-1}\mathsf{X}_2 + \mathsf{P}_2, \tag{9}$$

$$\mathsf{X}_2^{\dot{\otimes}4} = q^{-2}\mathsf{X}_1 + \mathsf{P}_1 + (q^{-2} + q^{-1} + 1)\mathsf{X}_3, \tag{10}$$

$$\mathsf{X}_2^{\dot{\otimes}5} = q^{-4}\mathsf{X}_2 + (q^{-3} + q^{-2} + q^{-1} + 1)\mathsf{P}_2 + q^{-\frac{3}{4}}(z + z^{-1})\mathsf{X}_3, \tag{11}$$

$$\begin{aligned}\mathsf{X}_2^{\dot{\otimes}6} = q^{-6}\mathsf{X}_1 &+ (q^{-4} + q^{-3} + q^{-2} + 1)\mathsf{P}_1 + q^{-\frac{5}{4}}(z + z^{-1})\mathsf{P}_2 \\ &+ (q^{-6} + q^{-5} + 2q^{-4} + q^{-3} + 2q^{-2} + q^{-1} + 1)\mathsf{X}_3,\end{aligned} \tag{12}$$

where we write characters of multiplicity spaces in the right hand sides. The variable q corresponds to the grading discussed above and z corresponds to the Cartan generator of the $s\ell(2)$ symmetry.

A structure of the proof of the statement (7) is as follows.

(1) *Show that the character of induced module* $\mathcal{M}_{\mathbf{u}}$ *is given by the Gordon formula* (69). This statement is given as Proposition 4.3 and is checked using computer for numerous examples of $\mathbf{u}$ given by (53). An outline

of the proof is given before the Proposition. Actually, we belive that the Proposition is true for wider set of vectors $\mathbf{u}$ than (53).

(2) *Show that in the decomposition* (81) *polynomials* $\hat{K}^{(p)}_{s,\mathbf{n}}(q,z)$ *are characters of coinvariants of the doublet algebra in it's irreducible modules.* This statement is checked using computer for numerous examples of vectors $\mathbf{u}$ written in Proposition 5.2. A possible way to proof the statement is outlined in Sec. 5.2 using the Felder resolution and results of Ref. 14 for lattice VOAs.

(3) *Show that in the decomposition* (81) *polynomials* $\bar{K}^{(p)}_{s,\mathbf{n}}(q,z)$ *are characters of coinvariants of the Virasoro algebra in it's irreducible modules.* This statement again is checked using computer for numerous examples, but this is the weakest point in the proof. We do not understand why it is so and the statement looks as a puzzle. We discuss this in more details in Sec. 5.1.

(4) *Three previous statements allows us to show* (7).

Unfortunately, the proof of the Statement given in the paper is partially based on calculations using computer algebra and at the moment we don't know a purely analytic proof. However, we belive that such a proof can be done and we hope to present it in a future publication.

The paper is organized as follows. In Sec. 2, we introduce notations and recall well known facts about $(1,p)$ models. In Sec. 3, we describe the representation category $\mathfrak{A}$. In Sec. 4, we investigate induced modules of $\mathcal{A}(p)$. In Sec. 5, we calculate characters of coinvariants in irreducible modules.

2. General facts about $(1,p)$ models

The $(1,p)$ models of logarithmic conformal field theory can be formulated in terms of Coulomb gas. Let φ denote the free scalar field with the OPE $\varphi(z)\,\varphi(w) = \log(z-w)$. Throughout the paper we use the standard notation

$$\alpha_+ = \sqrt{2p}\,, \qquad \alpha_- = -\sqrt{\frac{2}{p}}\,, \qquad \alpha_+\alpha_- = -2, \tag{13}$$
$$\alpha_0 = \alpha_+ + \alpha_- = \sqrt{\frac{2}{p}}(p-1),$$

where p is a positive integer grater than 1. In what follows, we drop the symbol of normal ordering in all functionals in φ. We consider the screening operator

$$F = \tfrac{1}{2\pi\mathrm{i}} \oint dz e^{\alpha_-\varphi(z)} \tag{14}$$

commuting with the Virasoro algebra corresponding to the energy–momentum tensor

$$T = \frac{1}{2}\,\partial\varphi\,\partial\varphi + \frac{\alpha_0}{2}\,\partial^2\varphi \tag{15}$$

with the central charge (1). We consider the lattice VOA $\mathcal{B}(p)$ generated by $e^{\pm\alpha_+\varphi(z)}$. The screening F acts from the vacuum module of $\mathcal{B}(p)$ to another irreducible $\mathcal{B}(p)$-module. The vacuum module of $\mathcal{W}(p)$ is the kernel of F in the vacuum module of $\mathcal{B}(p)$. So, $\mathcal{W}(p)$ is a subalgebra of $\mathcal{B}(p)$.

2.1. *The doublet algebra*

The algebra $\mathcal{A}(p)$ has a similar description. We consider the lattice VOA $\mathcal{L}(p)$ corresponding to the 1-dimensional lattice generated by the vector v, $(v,v) = p/2$. The VOA $\mathcal{L}(p)$ is generated by two vertex operators $e^{\pm\frac{\alpha_+}{2}\varphi(z)}$ (see Ref. 15). At this point we should make a remark that $\mathcal{L}(p)$ is not strictly speaking a vertex-operator algebra because whenever p is odd some OPEs contain nonlocal expressions $(z-w)^{\frac{1}{2}}$. However, one can work with $\mathcal{L}(p)$ like with a VOA. The representation category of $\mathcal{L}(p)$ is semisimple and contains p irreducible representations $\mathcal{Y}_s$ for $1 \le s \le p$. The module $\mathcal{Y}_s$ is generated from the vertex operator $V_{1,s} = e^{\frac{s-1}{2}\alpha_-\,\varphi(z)}$ and contains also the vertices

$$V_{r,s} = e^{-(\frac{r-1}{2}\alpha_+ + \frac{s-1}{2}\alpha_-)\,\varphi(z)}, \qquad r \in \mathbb{Z}. \tag{16}$$

The vacuum module of $\mathcal{L}(p)$ is $\mathcal{Y}_1$ and the screning F acts from it to $\mathcal{Y}_{p-1}$. We define the vacuum module X_1 of $\mathcal{A}(p)$ (which is equivalent to the definition of $\mathcal{A}(p)$) as a kernel of F calculated in $\mathcal{Y}_1$.

The second screning of the Virasoro algebra (15)

$$e = \frac{1}{2\pi\mathrm{i}} \oint dz e^{\alpha_+\varphi(z)} \tag{17}$$

acts in the vacuum module $\mathcal{Y}_1$ of $\mathcal{L}(p)$. The action of e can be restricted to X_1 (the vacuum module of $\mathcal{A}(p)$), where it is one of the $s\ell(2)$ algebra generators. The generator f can be constructed from F as a divided power "$F^p/[p]!$".

The algebra $\mathcal{A}(p)$ is generated by the $s\ell(2)$ doublet of fields

$$a^+(z) = e^{-\frac{\alpha_+}{2}\varphi(z)}, \qquad a^-(z) = [e, a^+(z)] = D_{p-1}(\partial\varphi(z))e^{\frac{\alpha_+}{2}\varphi(z)}, \tag{18}$$

where D_{p-1} is a degree $p-1$ differential polynomial in $\partial\varphi(z)$. The conformal dimension of these fields is $\frac{3p-2}{4}$. The fields $a^\pm(z)$ have the following OPEs

$$a^+(z)a^+(w) \sim (z-w)^{\frac{p}{2}},$$

$$a^-(z)a^-(w) \sim (z-w)^{\frac{p}{2}},$$
$$a^+(z)a^-(w) = (z-w)^{-\frac{3p-2}{2}} \sum_{n\geq 0}(z-w)^n H^n(w) \tag{19}$$

where $H^n(w)$ are fields with conformal dimension equals to n. The field H^0 is proportional to the identity field 1, $H^1 = 0$, H^2 is proportional to the energy–momentum tensor T. About other fields H^n we can say the following

$$H^{2n} = c_{2n} : T^n : + P_{2n}(T), \qquad 1 \leq n \leq p-1, \tag{20}$$
$$H^{2n+1} = c_{2n+1}\partial : T^n : + P_{2n+1}(T), \qquad 1 \leq n \leq p-2, \tag{21}$$
$$H^{2p-1} = c_{2p-1}\partial : T^{p-1} : + P_{2p-1}(T) + d_1 W^0, \tag{22}$$
$$H^{2p} = c_{2p} : T^p : + P_{2p}(T) + d_2 \partial W^0, \tag{23}$$

where $: T^n :$ is the normal ordered n-th power of the energy–momentum tensor, $P_n(T)$ is a differential polynomial in T and degree of both $P_{2n}(T)$ and $P_{2n+1}(T)$ in T is equal to $n-1$, $W^0(z) = [e, e^{-\alpha_+\varphi(z)}]$ and c_n, d_1, d_2 are some nonzero constants.

In what follows we choose the system of $p+1$ generators of $\mathcal{A}(p)$ in the form

$$a^+, a^-, H^2, H^4, \ldots, H^{2p-2}. \tag{24}$$

The corresponding vector of conformal dimensions is

$$\mathbf{\Delta} = (\frac{3p-2}{4}, \frac{3p-2}{4}, 2, 4, \ldots, 2p-2). \tag{25}$$

We defined the algebra $\mathcal{A}(p)$ as a subalgebra in $\mathcal{L}(p)$ with embedding given by (18). We note that there is another embedding $\mathcal{A}(p) \hookrightarrow \mathcal{L}(p)$ given by

$$a^+(z) = e^{\frac{\alpha_+}{2}\varphi(z)}, \qquad a^-(z) = [\bar{e}, a^+(z)], \tag{26}$$

where $\bar{e} = \frac{1}{2\pi\mathrm{i}} \oint dz e^{-\alpha_+\varphi(z)}$.

2.2. *Irreducible modules of $\mathcal{A}(p)$*

The irreducible representations of $\mathcal{A}(p)$ are described in Ref. 6. There are p irreducible representations but before we describe them we make several notations. The vertex operator algebra $\mathcal{A}(p)$ is graded (by eigenvalues of the zero mode of $\partial\varphi$)

$$\mathcal{A}(p) = \bigoplus_{\beta \in \frac{\alpha_+}{2}\mathbb{Z}} \mathcal{A}(p)^\beta \tag{27}$$

and $a^{\pm}(z) \in \mathcal{A}(p)^{\pm\frac{\alpha_+}{2}}$. We consider only the graded representations of $\mathcal{A}(p)$. For any representation $\mathcal{X} = \oplus_{t\in\mathbb{C}}\mathcal{X}^t$ we have $a^{\pm}(z) : \mathcal{X}^t \to \mathcal{X}^{t\pm\frac{\alpha_+}{2}}$ and $a^{\pm}(z)$ acting in $\mathcal{X}^t$ have the decomposition

$$a^{\pm}(z) = \sum_{n\in\pm t\frac{\alpha_+}{2}-\frac{3p-2}{4}+\mathbb{Z}} z^{-n-\frac{3p-2}{4}} a_n^{\pm}. \tag{28}$$

We note that t in fact is not arbitrary but takes the the values $t = \frac{\alpha_-}{2}n$, $n \in \mathbb{Z}$.

The irreducible $\mathcal{A}(p)$-modules can be constructed in terms of irreducible modules of lattice VOA $\mathcal{L}(p)$. Some powers of the screening operator F act between irreducible $\mathcal{L}(p)$-modules and form the Felder complex

$$\dots \xrightarrow{F^{p-s}} \mathcal{Y}_s \xrightarrow{F^s} \mathcal{Y}_{p-s} \xrightarrow{F^{p-s}} \mathcal{Y}_s \xrightarrow{F^s} \dots . \tag{29}$$

The complex is exact and the kernel of F^s in $\mathcal{Y}_s$ is irreducible $\mathcal{A}(p)$-module X_s. The irreducible representation X_s of $\mathcal{A}(p)$ is a highest-weight module generated from the vector $|s\rangle \in \mathsf{X}_s^{\frac{1-s}{2}\alpha_-}$ satisfying

$$a^{\pm}_{-\frac{3p-2s}{4}+n}|s\rangle = 0, \qquad n \in \mathbb{N},\ 1 \le s \le p. \tag{30}$$

The conformal dimension of $|s\rangle$ is $\Delta_{1,s} = \frac{s^2-1}{4p} + \frac{1-s}{2}$. The highest modes of $a^{\pm}(z)$ that generate non zero vectors from $|s\rangle$ are

$$a^{\pm}_{-\frac{3p-2s}{4}},\ 1 \le s \le p \tag{31}$$

as it shown in Fig. 1. Proceeding further we obtain the set of extremal vectors shown in Fig. 1.

We let $\mathfrak{L}_{r,s;p}$ denote the irreducible module of the Virasoro algebra with the central charge (1) and with the highest weight

$$\Delta_{r,s} = \frac{p}{4}(r^2-1) + \frac{1}{4p}(s^2-1) + \frac{1-rs}{2}, \quad 1 \le s \le p,\ r \in \mathbb{Z}. \tag{32}$$

We note that $\mathfrak{L}_{r,s;p}$ is the quotient of the Verma module by the submodule generated from one singular vector on the level rs and such modules exhaust irreducible Virasoro modules that aren't Verma modules.

The action of the $s\ell(2)$ algebra in X_s is defined similarly to the action (see (18)) in the vacuum module and X_s as a representation of $s\ell(2) \oplus \mathrm{Vir}$ decomposes as

$$\mathsf{X}_s = \oplus_{n\in\mathbb{N}} \pi_n \otimes \mathfrak{L}_{n,s;p}, \tag{33}$$

where π_n is the n-dimensional irreducible $s\ell(2)$ representation.

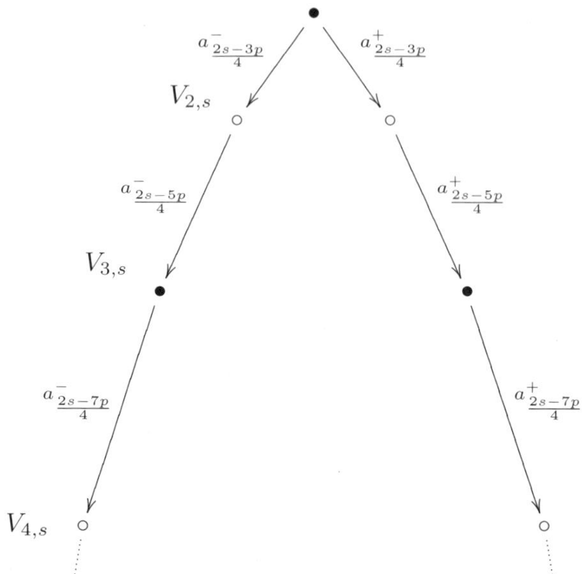

Fig. 1. The irreducible $\mathcal{A}(p)$ module X_s. The filled dot on the top is the cyclic vector $|s\rangle$. The arrows show the action of highest modes of $a^{\pm}$ that give nonzero vectors. Filled (open) dots denote vertices belonging to the triplet algebra $\mathcal{W}(p)$ representations X_s^+ (X_s^-).

The $\mathcal{L}(p)$-module $\mathcal{Y}_s$ admits two $\mathcal{A}(p)$-module structures $\mathcal{Y}_s^+$ and $\mathcal{Y}_s^-$ corresponding to the embeddings (18) and (26) respectively. We call modules $\mathcal{Y}_s^{\pm}$ the Verma modules of $\mathcal{A}(p)$ because they correspond to Verma modules in the tensor category $\mathfrak{A}$ (See detailes in Sec. 3.1). The modules $\mathcal{Y}_s^{\pm}$ are highest-weight modules generated from highest-weight vectors $|s\rangle^{\pm}$ satisfying

$$\begin{aligned} a^+_{-\frac{3p-2s}{4}+n}|s\rangle^+ &= a^-_{-\frac{3p-2s}{4}+p-s+n}|s\rangle^+ = 0,\\ a^+_{-\frac{3p-2s}{4}+p-s+n}|s\rangle^- &= a^-_{-\frac{3p-2s}{4}+n}|s\rangle^- = 0, \end{aligned} \qquad n\in\mathbb{N},\quad 1\le s\le p. \tag{34}$$

We note that $\mathcal{Y}_s^+$ is not isomorphic to $\mathcal{Y}_s^-$ excepting the case $s=p$, where two highest-weight conditions in (34) coincide.

From the Felder complex (29), we obtain 4 resolutions for the irreducible representation X_s

$$\begin{aligned} &\to \mathcal{Y}_s^+ \xrightarrow{F^s} \mathcal{Y}_{p-s}^+ \xrightarrow{F^{p-s}} \mathsf{X}_s \to 0, \qquad 0\to \mathsf{X}_s \xrightarrow{F^{p-s}} \mathcal{Y}_s^+ \xrightarrow{F^s} \mathcal{Y}_{p-s}^+ \to,\\ &\to \mathcal{Y}_s^- \xrightarrow{F^{p-s}} \mathcal{Y}_{p-s}^- \xrightarrow{F^s} \mathsf{X}_s \to 0, \qquad 0\to \mathsf{X}_s \xrightarrow{F^s} \mathcal{Y}_s^- \xrightarrow{F^{p-s}} \mathcal{Y}_{p-s}^- \to, \end{aligned} \tag{35}$$

where resolutions in each row are contragredient to each other.

3. Structure of the category $\mathfrak{A}$

3.1. *Linkage classes and indecomposable modules*

The representation category $\mathfrak{A}$ of $\mathcal{A}(p)$ is a direct sum

$$\mathfrak{A} = \bigoplus_{n=0}^{[p/2]} \mathfrak{A}_n, \tag{36}$$

where $\mathfrak{A}_n$ are full subcategories and there are no morphisms between elements from different $\mathfrak{A}_n$.

Category $\mathfrak{A}_0$ is semisimple and contains the only indecomposable object X_p. Each category $\mathfrak{A}_n$, $n > 0$ (excluded $\mathfrak{A}_{p/2}$ for even p) contains two simple objects X_s and X_{p-s}. Category $\mathfrak{A}_{p/2}$ for even p contains one simple object $\mathsf{X}_{p/2}$.

To each irreducible module X_s, the projective cover P_s corresponds. For $1 \le s \le p-1$, the projective module P_s consists of 4 subquotients (two X_s and two X_{p-s}) and $\mathsf{P}_p = \mathsf{X}_p$. Schematically the structure of P_s for $1 \le s \le p-1$ is shown in the following diagram

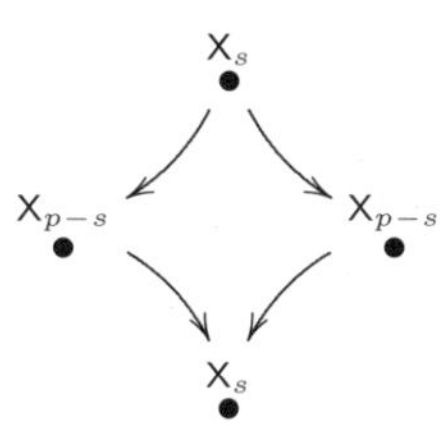

(37)

This digram corresponds to the Jordan–Holder series $0 \to \mathcal{W} \to \mathsf{P}_s \to \mathsf{X}_s \to 0$, $0 \to \mathcal{V} \to \mathcal{W} \to \mathsf{X}_{p-s} \to 0$, $0 \to \mathsf{X}_s \to \mathcal{V} \to \mathsf{X}_{p-s} \to 0$.

As it was explained in the Introduction, the quasitensor category of $\mathcal{A}(p)$ modules is equivalent to the quotient of $\overline{\mathfrak{U}}_{\mathfrak{q}} s\ell(2)$ representation category with respect to relations $\mathsf{X}_s^+ \sim \mathsf{X}_s^-$. Let $U_q(B^\pm)$ be universal enveloping of two Borel subalgebras in $\overline{\mathfrak{U}}_{\mathfrak{q}} s\ell(2)$. Then $\mathcal{Y}_s^\pm$ correspond to Verma modules induced from $U_q(B^\pm)$.

3.2. *Quasitensor structure*

The category $\mathfrak{A}$ is quasitensor category. Tensor products of irreducible modules decomposes into direct sum of irreducible and projective modules. Direct sums of irreducible and projective modules are tilting modules[16] in the category $\mathfrak{A}$. We have the following tensor products of these modules

$$\mathsf{X}_r\dot{\otimes}\mathsf{X}_s = \bigoplus_{j=|r-s|+1}^{\min(r+s-1,2p-r-s-1)}{}' \mathsf{X}_j \oplus \bigoplus_{2p-r-s+1}^{p}{}' \mathsf{P}_j, \tag{38}$$

$$\mathsf{X}_r\dot{\otimes}\mathsf{P}_s = \begin{cases} \bigoplus_{j=s-r+1}^{s+r-1}{}' \mathsf{R}_j, & r \le s,\ r+s\le p, \\ \bigoplus_{j=s-r+1}^{2p-s-r-1}{}' \mathsf{R}_j \oplus 2 \bigoplus_{j=2p-s-r+1}^{p}{}' \mathsf{R}_j, & r\le s,\ r+s>p, \\ \bigoplus_{j=r-s+1}^{2p-s-r-1}{}' \mathsf{R}_j \oplus 2 \bigoplus_{j=p+s-r+1}^{p}{}' \mathsf{R}_j, & r>s,\ r+s\le p, \\ \bigoplus_{j=r-s+1}^{2p-s-r-1}{}' \mathsf{R}_j \oplus 2 \bigoplus_{j=2p-s-r+1}^{p}{}' \mathsf{R}_j \oplus 2 \bigoplus_{j=p+s-r+1}^{p}{}' \mathsf{R}_j, & r>s,\ r+s>p, \end{cases} \tag{39}$$

and for $r \le s$

$$\mathsf{P}_r\dot{\otimes}\mathsf{P}_s = \begin{cases} 2\bigoplus_{j=s-r+1}^{s+r-1}{}' \mathsf{R}_j \oplus 2 \bigoplus_{j=p-r-s+1}^{p+r-s-1}{}' \mathsf{R}_j \oplus 4 \bigoplus_{j=p+r-s+1}^{p}{}' \mathsf{R}_j \oplus 4 \bigoplus_{j=s+r+1}^{p}{}' \mathsf{R}_j, \\ \qquad r+s\le p, \\ 2\bigoplus_{j=s-r+1}^{2p-s-r-1}{}' \mathsf{R}_j \oplus 2 \bigoplus_{j=r+s-p+1}^{p+r-s-1}{}' \mathsf{R}_j \oplus 4 \bigoplus_{j=2p-r-s+1}^{p}{}' \mathsf{R}_j \oplus 4 \bigoplus_{j=p+r-s+1}^{p}{}' \mathsf{R}_j, \\ \qquad r+s>p, \end{cases} \tag{40}$$

where we used notation $\mathsf{R}_j = \mathsf{P}_j$ for $1 \le j \le p-1$ and $\mathsf{R}_p = 2\mathsf{P}_p$ and $\bigoplus'$ is a direct sum with step 2 (for example $\bigoplus{}'^{2n}_{j=0} \mathsf{R}_j = \mathsf{R}_0 \oplus \mathsf{R}_2 \oplus \cdots \oplus \mathsf{R}_{2n}$).

For a vector $\mathbf{n}$ with nonnegative integer components $n_2, n_3, \dots, n_p$, we consider a decomposition of the tensor product

$$\mathsf{X}_2^{\dot{\otimes} n_2}\dot{\otimes}\mathsf{X}_3^{\dot{\otimes} n_3}\dot{\otimes}\dots\dot{\otimes}\mathsf{X}_p^{\dot{\otimes} n_p} = \bigoplus_{s=1}^{p-1} \mathcal{V}_s[\mathbf{n}] \boxtimes \mathsf{X}_s \bigoplus \bigoplus_{s=1}^{p} \mathcal{X}_s[\mathbf{n}] \boxtimes \mathsf{P}_s, \tag{41}$$

where $\mathcal{V}_s[\mathbf{n}]$ and $\mathcal{X}_s[\mathbf{n}]$ are vector spaces of multiplicities of the irreducible and projective modules respectively in the direct sum. The dimensions of these spaces are $\hat{N}_s[\mathbf{n}] = \dim\mathcal{V}_s[\mathbf{n}]$ and $\bar{N}_s[\mathbf{n}] = \dim\mathcal{X}_s[\mathbf{n}]$.

Remark 3.1. For example, we give decompositions for $p = 2$

$$\mathsf{X}_2^{\dot{\otimes} n} = 2^{n-2}\big(1 + (-1)^{n-1}\big)\mathsf{X}_2 + 2^{n-3}\big(1 + (-1)^n\big)\mathsf{P}_1 \tag{42}$$

and $p = 3$

$$\mathsf{X}_2^{\dot{\otimes} n} = \frac{1}{2}(1+(-1)^n)\mathsf{X}_1 + \frac{1}{2}(1-(-1)^n)\mathsf{X}_2 \tag{43}$$
$$+\frac{1}{9}(2^n+(-1)^n(3n-1))\mathsf{X}_3 + \frac{1}{216}(2^{n+3}+(-1)^n(19-48n+18n^2)-27)\mathsf{P}_1$$
$$+\frac{1}{216}(2^{n+4}+(-1)^n(11+12n-18n^2)-27)\mathsf{P}_2,$$
$$\mathsf{X}_2^{\dot{\otimes} n}\dot{\otimes}\mathsf{X}_3 = \frac{2}{3}(2^{n-1}+(-1)^n)\mathsf{X}_3$$
$$+\frac{1}{9}(2^n+(-1)^n(3n-1))\mathsf{P}_1 + \frac{1}{9}(2^{n+1}-(-1)^n(3n+2))\mathsf{P}_2,$$
$$\mathsf{X}_2^{\dot{\otimes} n}\dot{\otimes}\mathsf{X}_3^{\dot{\otimes} 2} = 2^n\mathsf{X}_3 + \frac{2}{3}(2^{n-1}+(-1)^n)\mathsf{P}_1 + \frac{2}{3}(2^n-(-1)^n)\mathsf{P}_2, \qquad n \geq 0, \tag{44}$$
$$\mathsf{X}_2^{\dot{\otimes} n}\dot{\otimes}\mathsf{X}_3^{\dot{\otimes} m} = 2^n3^{m-2}\mathsf{X}_3 + 2^n3^{m-3}\mathsf{P}_1 + 2^{n+1}3^{m-3}\mathsf{P}_2, \qquad m \geq 3 \tag{45}$$

In the previous formulas we wrote dimensions of multiplicity spaces instead of themselves.

The spaces of multiplicities $\mathcal{V}_s[\mathbf{n}]$ and $\mathcal{X}_s[\mathbf{n}]$ are $s\ell(2)$ modules; $\mathcal{V}_s[\mathbf{n}]$ is trivial module (sum of 1 dimensional modules). The $s\ell(2)$ action in the multiplicity spaces is related with the Lusztig extension of the quantum group $\overline{\mathfrak{U}}_{\mathfrak{q}}s\ell(2)$. There exists the quantum group $\overline{\mathfrak{LU}}_{\mathfrak{q}}s\ell(2)$ such that $\overline{\mathfrak{U}}_{\mathfrak{q}}s\ell(2)$ is its subgroup and the quotient is the universal enveloping of $s\ell(2)$. An irreducible representation of $\overline{\mathfrak{U}}_{\mathfrak{q}}s\ell(2)$ is the irreducible representation of $\overline{\mathfrak{LU}}_{\mathfrak{q}}s\ell(2)$ with the trivial $s\ell(2)$ action. Therefore $s\ell(2)$ acts in the multiplicity spaces. The multiplicity spaces are graded by Cartan generator h of $s\ell(2)$. For example for (45), we have for $m \geq 3$

$$\mathsf{X}_2^{\dot{\otimes} n}\dot{\otimes}\mathsf{X}_3^{\dot{\otimes} m} = (z+z^{-1})^n(z^2+1+z^{-2})^{m-2}\mathsf{X}_3$$
$$+(z+z^{-1})^n(z^2+1+z^{-2})^{m-3}\mathsf{P}_1 + +(z+z^{-1})^{n+1}(z^2+1+z^{-2})^{m-3}\mathsf{P}_2. \tag{46}$$

In the next section we investigate the additional grading given by D (3) in the spaces $\mathcal{V}_s[\mathbf{n}]$ and $\mathcal{X}_s[\mathbf{n}]$ and obtain the formulas for characters.

4. Characters of induced modules

4.1. *Induced modules of $\mathcal{A}(p)$*

We fix a $p+1$ dimensional vector $\mathbf{u}$. Components of $\mathbf{u}$ are labeled by the set of indices $\mathcal{I} = \{+,-,1,2,\ldots,p-1\}$. For a vector $\mathbf{u}$, we define the subalgebra $\mathcal{A}(p)[\mathbf{u}]^+ \subset \mathcal{A}(p)$ generated by the modes

$$a^{\pm}_{\mathbf{w}_{\pm}+m}, \qquad H^{2n}_{\mathbf{w}_n+m}, \quad 1 \le n \le p-1, \quad m \in \mathbb{N}, \tag{47}$$

where $\mathbf{w} = \mathbf{u} - \boldsymbol{\Delta}$ and $\boldsymbol{\Delta}$ is given by (25).

We define the $\mathcal{A}(p)$ module $\mathcal{M}_{\mathbf{u}}$ induced from trivial 1-dimensional $\mathcal{A}(p)[\mathbf{u}]^+$ module with the highest-weight vector $|\mathbf{u}\rangle$. The irreducible $\mathcal{A}(p)$ modules X_s are induced modules $\mathcal{M}_{\mathbf{u}_s}$ with

$$\mathbf{u}_s = (\frac{s-1}{2}, \frac{s-1}{2}, \underbrace{1, 2, \dots s-1}_{s-1}, \underbrace{s-1, \dots, s-1}_{p-s}). \tag{48}$$

This statement follows from Ref. 6.

The $\mathcal{A}(p)$ Verma modules $\mathcal{Y}^{\pm}_s$ are also induced modules.

Proposition 4.1.

$$\mathcal{Y}^{\pm}_s \simeq \mathcal{M}_{\mathbf{u}^{\pm}_s} \tag{49}$$

with

$$\mathbf{u}^{+}_s = (\frac{s-1}{2}, \frac{s-1}{2} + p - s, \underbrace{s-1, \dots, s-1}_{p-1}), \tag{50}$$

$$\mathbf{u}^{-}_s = (\frac{s-1}{2} + p - s, \frac{s-1}{2}, \underbrace{s-1, \dots, s-1}_{p-1}). \tag{51}$$

A proof of the Proposition is based on results[14] and abelianization technique. We give a sketch of the proof in Sec. 4.4.

The modules $\mathcal{M}_{\mathbf{u}_{s,r}}$ induced from the subalgebra corresponding to vectors

$$\mathbf{u}_{s,r} = (\frac{s-rp-1}{2}, \frac{(r+2)p-s-1}{2}, \underbrace{s-1, \dots, s-1}_{p-1}), \quad r \in \mathbb{Z}, \quad 1 \le s \le p \tag{52}$$

are isomorphic to $\mathcal{Y}^{+}_s$ for $r \ge 0$ and to $\mathcal{Y}^{-}_{p-s}$ for $r < 0$.

4.2. *Decompositions of induced $\mathcal{A}(p)$-modules*

We consider a set of modules induced from highest-weight conditions corresponding to vectors of the form

$$\mathbf{u} = \sum_{j=2}^{p} n_j \mathbf{u}_j + \sum_{r \in \mathbb{Z}} \sum_{s=1}^{p} n^r_s \mathbf{u}_{s,r}, \tag{53}$$

where n_j and n_s^r are nonnegative integers and only finite number of n_s^r are not equal to 0.

Proposition 4.2. *For two vectors* $\mathbf{u}$ *and* $\mathbf{u}'$ *of the form* (53) *the fusion of induced modules is induced module:*

$$\mathcal{M}_{\mathbf{u}}\dot{\otimes}\mathcal{M}_{\mathbf{u}'} = \mathcal{M}_{\mathbf{u}+\mathbf{u}'}. \tag{54}$$

For vectors

$$\mathbf{u} = n_2\mathbf{u}_2 + n_3\mathbf{u}_3 + \cdots + n_p\mathbf{u}_p, \tag{55}$$

where $n_2, n_3, \ldots, n_p$ are nonnegative integers and vectors $\mathbf{u}_s$ are given by (48) the induced module $\mathcal{M}_{\mathbf{u}}$ is tilting and therefore decomposes into a direct sum of projective and irreducible modules

$$\mathcal{M}_{\mathbf{u}} = \bigoplus_{s=1}^{p-1} \mathcal{V}_s[\mathbf{n}] \boxtimes \mathsf{X}_s \bigoplus \bigoplus_{s=1}^{p} \mathcal{X}_s[\mathbf{n}] \boxtimes \mathsf{P}_s. \tag{56}$$

The right hand side is the same as in (41) and $\mathbf{n} = (n_2, n_3, \ldots, n_p)$. We note that whenever $n_p > 0$ there is no modules X_s in the decomposition, thus $\bar{N}_s[\mathbf{n}] = 0$.

4.3. *Characters of induced modules*

The character of a module $\mathcal{M}$ is defined by

$$\bar{\chi}_{\mathcal{M}}(q) = \mathrm{Tr}_{\mathcal{M}}\, q^{L_0 - \frac{c}{24}}, \tag{57}$$

where L_0 is the zero mode of $T(z)$ (15) and c is given by (1). The characters of irreducible $\mathcal{A}(p)$ modules X_s are

$$\bar{\chi}_s(q) = \frac{1}{\eta(q)}\Big(\frac{s}{p}\left(\theta_{p-s,p}(q) + \theta_{s,p}(q)\right) + 2\left(\theta'_{p-s,p}(q) - \theta'_{s,p}(q)\right)\Big) \tag{58}$$

where the eta function is

$$\eta(q) = q^{\frac{1}{24}} \prod_{n=1}^{\infty}(1 - q^n)$$

and the theta functions

$$\theta_{s,p}(q, z) = \sum_{j \in \mathbb{Z} + \frac{s}{2p}} q^{pj^2} z^j, \quad |q| < 1,\ z \in \mathbb{C},$$

and we set $\theta_{s,p}(q) = \theta_{s,p}(q,1)$ and $\theta'_{s,p}(q) = z\frac{\partial}{\partial z}\theta_{s,p}(q,z)\big|_{z=1}$. As a q series the same characters are

$$\bar{\chi}_s(q) = \frac{q^{-\frac{1}{24}}}{\prod_{n=1}^{\infty}(1-q^n)} \sum_{n\in\mathbb{Z}} nq^{\frac{p}{4}(n-\frac{s}{p})^2}. \tag{59}$$

In what follows we work with normalized characters $\chi_{\mathcal{M}} = q^{-\Delta+\frac{c}{24}}\bar{\chi}_{\mathcal{M}}$, where Δ is the conformal dimension of the highest-weight vector in $\mathcal{M}$. We also insert in characters the dependence on additional variable z in the following way

$$\chi_{\mathcal{M}}(q) = q^{-\Delta}\mathrm{Tr}_{\mathcal{M}}\, q^{L_0} z^h, \tag{60}$$

where h is the Cartan generator of $s\ell(2)$. Then the normalized character of X_s is

$$\chi_s(q,z) = \frac{q^{-\frac{(p-s)^2}{4p}}}{\prod_{n=1}^{\infty}(1-q^n)} \sum_{n\in\mathbb{N}} \sum_{j=-\frac{n}{2}}^{\frac{n}{2}} z^{2j}\left(q^{\frac{p}{4}(n-\frac{s}{p})^2} - q^{\frac{p}{4}(n+\frac{s}{p})^2}\right). \tag{61}$$

This formula for the character immediately follows from (33). The characters of $\mathcal{W}(p)$ irreducible modules corresponds to even and odd powers of z in (61)

$$\chi_s^+(q,z) = \frac{q^{-\frac{(p-s)^2}{4p}}}{\prod_{n=1}^{\infty}(1-q^n)} \sum_{n\in\mathbb{N}_{\mathrm{even}}} \sum_{j=-\frac{n}{2}}^{\frac{n}{2}} z^{2j}\left(q^{\frac{p}{4}(n-\frac{s}{p})^2} - q^{\frac{p}{4}(n+\frac{s}{p})^2}\right), \tag{62}$$

$$\chi_s^-(q,z) = \frac{q^{-\frac{(p-s)^2}{4p}}}{\prod_{n=1}^{\infty}(1-q^n)} \sum_{n\in\mathbb{N}_{\mathrm{odd}}} \sum_{j=-\frac{n}{2}}^{\frac{n}{2}} z^{2j}\left(q^{\frac{p}{4}(n-\frac{s}{p})^2} - q^{\frac{p}{4}(n+\frac{s}{p})^2}\right), \tag{63}$$

where $\mathbb{N}_{\mathrm{even}}$ and $\mathbb{N}_{\mathrm{odd}}$ are even and odd positive integers respectively.

4.4. *Abelianization*

The Gordon-type matrix

$$\mathsf{A} = \begin{pmatrix} \frac{p}{2} & \frac{p}{2} & 1 & 2 & 3 & \dots & p-1 \\ \frac{p}{2} & \frac{p}{2} & 1 & 2 & 3 & \dots & p-1 \\ 1 & 1 & 2 & 2 & 2 & \dots & 2 \\ 2 & 2 & 2 & 4 & 4 & \dots & 4 \\ 3 & 3 & 2 & 4 & 6 & \dots & 6 \\ \dots & \dots & \dots & \dots & \dots & \dots & \dots \\ p-1 & p-1 & 2 & 4 & 6 & \dots & 2(p-1) \end{pmatrix} \tag{64}$$

determines an algebra $\bar{\mathcal{A}}(p)$ with quadratic relations that admits a realization in terms of vertex operators

$$\bar{a}^+, \bar{a}^-, \bar{H}^2, \ldots, \bar{H}^{2p-2} \tag{65}$$

with momenta

$$\bar{v}_+, \bar{v}_-, \bar{v}_1, \ldots, \bar{v}^{p-1} \tag{66}$$

with the scalar products

$$(\bar{v}_i, \bar{v}_j) = \mathsf{A}_{ij}. \tag{67}$$

We note that the matrix A is degenerate therefore for realization of $\bar{\mathcal{A}}(p)$ by vertex operators we should take the space with a dimension greater than $p+1$ with nondegenerate scalar product and construct in it the $p+1$ linearly independent vectors (66) with the scalar product (67). The algebra $\bar{\mathcal{A}}(p)$ is related to $\mathcal{A}(p)$ in the following way. The algebra $\mathcal{A}(p)$ admits such a multifiltration that the adjoint grading algebra is isomorphic to $\bar{\mathcal{A}}(p)$. In particular it means that $\mathcal{A}(p)$ can be considered a deformation of $\bar{\mathcal{A}}(p)$, i.e. there exists such a family of algebras $\bar{\mathcal{A}}_\hbar(p)$ that $\bar{\mathcal{A}}_0(p) \simeq \bar{\mathcal{A}}(p)$ and $\bar{\mathcal{A}}_\hbar(p) \simeq \mathcal{A}(p)$ whenever $\hbar \neq 0$.

Induced modules of $\bar{\mathcal{A}}(p)$ are described by the same vectors $\mathbf{u}$ and highest-weight conditions like $\mathcal{A}(p)$ ones (47). The algebra $\bar{\mathcal{A}}(p)$ admits a natural bigrading by operators $\bar{L}_0$ and h and the normalized character of induced module $\mathcal{M}$ is

$$\chi_{\mathcal{M}}(q) = \mathrm{Tr}_{\mathcal{M}}\, q^{\bar{L}_0} z^h. \tag{68}$$

Then, the normalized character of the module induced from the subalgebra described by the vector $\mathbf{u}$ is

$$\chi_{\mathbf{v}}(q,z) = \sum_{n_+,n_-,n_1,\ldots,n_{p-1}\geq 0} z^{n_+-n_-} \frac{q^{\frac{1}{2}\mathbf{n}\mathsf{A}\cdot\mathbf{n}+\mathbf{v}\cdot\mathbf{n}}}{(q)_{n_+}(q)_{n_-}(q)_{n_1}\cdots(q)_{n_{p-1}}}, \tag{69}$$

where A is given by (64) and the vector

$$\mathbf{v} = -\mathbf{u} + \mathbf{v}_1 \tag{70}$$

with

$$\mathbf{v}_1 = (\frac{p-1}{2}, \frac{p-1}{2}, 1, 2, \ldots, p-1) \tag{71}$$

and $\cdot$ is the standard scalar product. We note that z grading of $\bar{a}_n^\pm$ is ± 1 and z grading of $\bar{H}_n^i$ is zero. In Ref. 6 it was shown that the characters of $\bar{\mathcal{A}}_\hbar(p)$-modules induced from $\mathbf{u}_s$ (48) are independent of $\hbar$ and coincide with characters of $\mathcal{A}(p)$ irreducible modules (61). For irreducible characters, we have $\chi_s(q,z) = \chi_{\mathbf{v}_s}(q,z)$ with

$$\mathbf{v}_s = (\frac{p-s}{2}, \frac{p-s}{2}, \underbrace{0, \dots 0}_{s-1}, \underbrace{1, 2, \dots, p-s}_{p-s}). \tag{72}$$

Proposition 4.3. *For the vectors* $\mathbf{u}$ *of the form* (53)*, the formula* (69) *gives the character of the* $\mathcal{A}(p)$*-module induced from subalgebra described by* $\mathbf{u}$.

The characters $\psi_s(q,z)$ of projective modules P_s are

$$\psi_s(q,z) = 2\chi_s(q,z) + q^{\frac{2s-p}{4}}(z+z^{-1})\chi_{p-s}(q,z). \tag{73}$$

Proof of Prop. 4.1.. The abelianisation technique was used in Ref. 14 for lattice VOA $\mathcal{L}(p)$ with the same matrix A but for even p. We note that the results of Ref. 14 can be easily generalized for odd p. In particular, for characters of $\mathcal{Y}_s^{\pm}$ there were obtained the fermionic formula

$$\xi_s^{\pm}(q,z) = \chi_{\mathbf{v}_s^{\pm}}(q,z), \qquad \mathbf{v}_s^{\pm} = -\mathbf{u}_s^{\pm} + \mathbf{v}_1. \tag{74}$$

In the abelianization technique, we use a filtration on the algebra $\mathcal{A}(p)$. The filtration determines a filtration on the cyclic module $\mathcal{M}_{\mathbf{u}_{s,r}}$ (the highest-wight vector is choosen to be cyclic). The adjoint graded module $\bar{\mathcal{M}}_{\mathbf{u}_{s,r}}$ is a representation of $\bar{\mathcal{A}}(p)$. We consider the $\bar{\mathcal{A}}(p)$-module $\tilde{\mathcal{M}}$ induced from the same highest-weight conditions as $\mathcal{M}_{\mathbf{u}_{s,r}}$. The character of $\tilde{\mathcal{M}}$ is given by a fermionic formula and the formula coincides with (74). On the other hand the character of $\tilde{\mathcal{M}}$ is greater or equal to the character of $\mathcal{M}_{\mathbf{u}_{s,r}}$. This means that the character of induced $\mathcal{A}(p)$-module $\mathcal{M}_{\mathbf{u}_{s,r}}$ coincides with the character of the corresponding $\mathcal{Y}_s^{\pm}$. The fact that the induced module $\mathcal{M}_{\mathbf{u}_{s,r}}$ surjectively maps onto $\mathcal{Y}_s^{\pm}$ completes the proof. □

5. Characters of multiplicity spaces

For a vector $\mathbf{n} = (n_2, n_3, \dots, n_p)$ with nonnegative integer components, we introduce the vector

$$\mathbf{m} = (0, 0, n_2, n_3, \dots, n_p). \tag{75}$$

Then the vector $\mathbf{u}$ from (55) can be written as

$$\mathbf{u} = \frac{1}{2}\mathbf{m}\mathsf{A}. \tag{76}$$

We introduce polynomials $\hat{K}^{(p)}_{\ell,\mathbf{n}}(q,z)$ and $\bar{K}^{(p)}_{\ell,\mathbf{n}}(q)$, which are related to the characters of $\mathcal{X}_\ell[\mathbf{n}]$ and $\mathcal{V}_\ell[\mathbf{n}]$ respectively. These polynomials are written in terms of q-binomial coefficients

$$\begin{bmatrix} n \\ m \end{bmatrix}_q = \frac{\prod_{j=1}^{n}(1-q^j)}{\prod_{j=1}^{m}(1-q^j)\prod_{j=1}^{n-m}(1-q^j)} \tag{77}$$

for which we assume that $\left[\begin{smallmatrix} n \\ m \end{smallmatrix}\right]_q = 0$ whenever n or m is fractional or negative integer and whenever $m > n$.

For a vector $\mathbf{n} = (n_2, n_3, \ldots, n_p)$ with nonnegative integer components, we define polinomials

$$\hat{K}^{(p)}_{\ell,\mathbf{n}}(q,z) = \sum_{\mathbf{s}\in\mathbb{Z}^{p+1}} z^{s_+ - s_-}\, q^{\frac{1}{2}\mathbf{s}\mathsf{A}\cdot\mathbf{s}+\mathbf{v}_\ell\cdot\mathbf{s}} \prod_{a\in\mathcal{I}} \begin{bmatrix} \mathbf{e}_a\cdot((\frac{1}{2}\mathbf{m}-\mathbf{s})\mathsf{A}-\mathbf{v}_\ell-\mathbf{v}_1+\mathbf{s}) \\ \mathbf{e}_a\cdot\mathbf{s} \end{bmatrix}_q, \tag{78}$$

where the vector $\mathbf{m}$ is given by (75), $\mathbf{e}_a$ are the standard basis vectors and indices of each vector belong to the set $\mathcal{I}$. We also define a version of Kostka polynomials

$$\bar{K}^{(p)}_{\ell,\mathbf{n}}(q) = \begin{cases} q^{\frac{\ell-|\mathbf{n}|-1}{2}} K^{(p-2)}_{\ell-1,(n_2,n_3,\ldots,n_{p-1})}(q), & \text{for } n_p = 0, \\ 0, & \text{for } n_p > 0, \end{cases} \tag{79}$$

where standard level-restricted Kostka polynomials $K^{(k)}_{\ell,\mathbf{u}}(q)$ are given by the formula

$$K^{(k)}_{\ell,\mathbf{u}}(q) = \sum_{\substack{\mathbf{s}\in\mathbb{Z}^k_{\geq 0} \\ 2|\mathbf{s}|=|\mathbf{u}|-\ell}} q^{\mathbf{s}\bar{\mathsf{A}}\cdot\mathbf{s}+\mathbf{v}\cdot\mathbf{s}} \prod_{1\leq a\leq k} \begin{bmatrix} \mathbf{e}_a\cdot((\mathbf{u}-2\mathbf{s})\bar{\mathsf{A}}-\mathbf{v}+\mathbf{s}) \\ \mathbf{e}_a\cdot\mathbf{s} \end{bmatrix}_q, \tag{80}$$

where $\bar{\mathsf{A}}_{ij} = \min(i,j)$, the vector $\mathbf{v}$ with components $v_i = \max(i-k+\ell, 0)$ for $i = 1, 2, \ldots, k$ and $|\mathbf{u}| = \sum_{i=1}^{k} i u_i$.

Proposition 5.1. *The characters of the multiplicity spaces $\mathcal{X}_\ell[\mathbf{n}]$ and $\mathcal{V}_\ell[\mathbf{n}]$ (see* (41)*) are given by $\hat{K}^{(p)}_{\ell,\mathbf{n}}(q^{-1},z)$ and $\bar{K}^{(p)}_{\ell,\mathbf{n}}(q^{-1})$ respectively.*

The character of the induced module $\mathcal{M}_\mathbf{u}$ is given by (69) with $\mathbf{v} = -\mathbf{u}+\mathbf{v}_1$. The induced module is decomposed into a direct sum of irreducible and projective modules (56), which gives an identity for characters.

Proposition 5.2. *For given vector $\mathbf{n} = (n_2, n_3, \ldots, n_p)$ with nonnegative integer components, there is the identity*

$$\chi_{-\frac{1}{2}\mathbf{m}\mathsf{A}+\mathbf{v}_1}(q,z) = \sum_{s=1}^{p-1} \bar{K}^{(p)}_{s,\mathbf{n}}(q^{-1})\chi_s(q,z) + \sum_{s=1}^{p} \hat{K}^{(p)}_{s,\mathbf{n}}(q^{-1},z)\psi_s(q,z), \tag{81}$$

where $\chi_{\mathbf{v}}(q,z)$ *is given by* (69), $\chi_s(q,z) = \chi_{\mathbf{v}_s}(q,z)$ *with* $\mathbf{v}_s$ *given by* (72) *and* $\psi_s(q,z)$ *is given by* (73). *Multiplicities from* (41) *are* $\hat{N}_s[\mathbf{n}] = \hat{K}^{(p)}_{s,\mathbf{n}}(1,1)$ *and* $\bar{N}_s[\mathbf{n}] = \bar{K}^{(p)}_{s,\mathbf{n}}(1)$.

5.1. *Multiplicity spaces as coinvariants.*

To comment the main identity for characters (81) and fermionic formulas (78) and (79), we come back to investigation of multiplicity spaces $\mathcal{X}_s[\mathbf{n}]$ and $\mathcal{V}_s[\mathbf{n}]$. We fix a vecor $\mathbf{n} = (n_2, n_3, \dots, n_p)$ with nonnegative integer components. For the vector $\mathbf{u}$ given by (55), we define a subalgebra $\mathcal{A}(p)[\mathbf{u}]^- \subset \mathcal{A}(p)$ (compare with $\mathcal{A}(p)[\mathbf{u}]^+$ in Sec. 4.1) generated by

$$a^{\pm}_{\mathbf{w}_{\pm}-m}, \qquad H^{2n}_{\mathbf{w}_n-m}, \quad 1 \le n \le p-1, \quad m \in \mathbb{N}_0, \tag{82}$$

where $\mathbf{w} = -\mathbf{u} - \mathbf{\Delta}$.

Proposition 5.3. *The multiplicity space* $\mathcal{X}_s[\mathbf{n}]$ *can be identified with the space of coinvariants of* $\mathcal{A}(p)[\mathbf{u}]^-$ *calculated in the module* X_s, *i.e.* $\mathsf{X}_s/\mathcal{A}(p)[\mathbf{u}]^-\mathsf{X}_s$.

We note that under the identification $\mathcal{X}_s[\mathbf{n}] \simeq \mathsf{X}_s/\mathcal{A}(p)[\mathbf{u}]^-\mathsf{X}_s$ the natural gradings on these spaces differ by a sign, which leads to q^{-1} in the arguments of $\hat{K}^{(p)}_{s,\mathbf{n}}$ and $\bar{K}^{(p)}_{s,\mathbf{n}}$ in (81). The formula (78) is obtained with the abelianization procedure. The representation X_s is replaced by the representation $\bar{\mathsf{X}}_s$ (induced from the same highest-weight conditions) of the algebra $\bar{\mathcal{A}}(p)$. Then the calculation of coinvariants of $\bar{\mathcal{A}}(p)[\mathbf{u}]^-$ in $\bar{\mathsf{X}}_s$ gives the fermionic formula (78).

We consider a sequence of vectors $\mathbf{n}$ that tends to a vector $\mathbf{n}_\infty$ with at least one component equals to infinity. Then we have

for even p

$$\hat{K}^{(p)}_{s,\mathbf{n}}(q,z) \underset{\mathbf{n}\to\mathbf{n}_\infty}{\longrightarrow} \begin{cases} \chi_s(q,z), & n_2+n_3+\dots+n_p+s \quad \text{odd},\\ 0, & n_2+n_3+\dots+n_p+s \quad \text{even}, \end{cases} \tag{83}$$

for odd p

$$\hat{K}^{(p)}_{s,\mathbf{n}}(q,z) \underset{\mathbf{n}\to\mathbf{n}_\infty}{\longrightarrow} \begin{cases} \chi^-_s(q,z), & n_2+n_3+\dots+n_p+s \quad \text{odd},\\ \chi^+_s(q,z), & n_2+n_3+\dots+n_p+s \quad \text{even}, \end{cases} \tag{84}$$

where $\chi^{\pm}_s(q,z)$ are given by (62) and (63).

Thus,

$$\mathsf{X}_s = \begin{cases} \lim\limits_{\substack{\mathbf{n}\to\mathbf{n}_\infty \\ n_2+n_3+\cdots+n_p+s \text{ odd}}} \mathcal{X}_s[\mathbf{n}], & \text{for even } p, \\ \lim\limits_{\substack{\mathbf{n}\to\mathbf{n}_\infty \\ n_2+n_3+\cdots+n_p+s \text{ odd}}} \mathcal{X}_s[\mathbf{n}] \bigoplus \lim\limits_{\substack{\mathbf{n}\to\mathbf{n}_\infty \\ n_2+n_3+\cdots+n_p+s \text{ even}}} \mathcal{X}_s[\mathbf{n}], & \text{for odd } p. \end{cases} \tag{85}$$

Taking (41) into account, we obtain that sequence of induced from smaller and smaller subalgebra $\mathcal{A}(p)[\mathbf{u}]^+$ modules converges to the object in the category $\mathfrak{A}$ that is the regular $\mathcal{A}(p)$-bimodule (see a discussion in Conclusions).

Logarithmic $(1,p)$ models and $(p,p-1)$ Virasoro minimal models are in a badly understood duality. A manifestation of this duality is the fact that $\mathcal{V}_s[\mathbf{n}]$ is the space of coinvariants with respect to a subalgebra of Virasoro in an irreducible module from $(p,p-1)$ minimal model. For example, for $p=4$, characters of $\mathcal{V}_s[\mathbf{n}]$ are given by

$$\bar{K}^{(p)}_{1,(m,0,0)}(q) = \frac{1+(-1)^m}{4}\, q^{\frac{m(m-4)}{8}} \Big(\prod_{j=1}^{\frac{m}{2}}(q^{j-\frac{1}{2}}+1) + \prod_{j=1}^{\frac{m}{2}}(q^{j-\frac{1}{2}}-1)\Big), \tag{86}$$

$$\bar{K}^{(p)}_{2,(m,0,0)}(q) = \frac{1-(-1)^m}{2}\, q^{\frac{(m-1)(m-3)}{8}} \prod_{j=1}^{\frac{m-1}{2}}(q^{j}+1), \tag{87}$$

$$\bar{K}^{(p)}_{3,(m,0,0)}(q) = \frac{1+(-1)^m}{4}\, q^{\frac{(m-2)^2}{8}} \Big(\prod_{j=1}^{\frac{m}{2}}(q^{j-\frac{1}{2}}+1) - \prod_{j=1}^{\frac{m}{2}}(q^{j-\frac{1}{2}}-1)\Big). \tag{88}$$

As m tends to infinity we have

$$q^{-\frac{m(m-2)}{2}} \bar{K}^{(p)}_{1,(2m,0,0)}(q) \underset{m\to\infty}{\longrightarrow} \begin{cases} \chi_0(q), & m \text{ is even}, \\ q^{\frac{1}{2}}\chi_{\frac{1}{2}}(q), & m \text{ is odd}, \end{cases} \tag{89}$$

$$q^{-\frac{m(m-1)}{2}} \bar{K}^{(p)}_{2,(2m+1,0,0)}(q) \underset{m\to\infty}{\longrightarrow} \chi_{\frac{1}{16}}(q), \tag{90}$$

$$q^{-\frac{(m-1)^2}{2}} \bar{K}^{(p)}_{3,(2m,0,0)}(q) \underset{m\to\infty}{\longrightarrow} \begin{cases} q^{\frac{1}{2}}\chi_{\frac{1}{2}}(q), & m \text{ is even}, \\ \chi_0(q), & m \text{ is odd}, \end{cases} \tag{91}$$

where $\chi_0(q)$, $\chi_{\frac{1}{2}}(q)$ and $\chi_{\frac{1}{16}}(q)$ are characters of the Ising model irreducible representations with conformal dimensions 0, $\frac{1}{2}$ and $\frac{1}{16}$ respectively.

5.2. *Felder resolution*

The characters of the multiplicity spaces can be obtained in alternative way from the Felder resolution (35)

$$\to \mathcal{Y}_s^+ \xrightarrow{F^s} \mathcal{Y}_{p-s}^+ \xrightarrow{F^{p-s}} \mathsf{X}_s \to 0. \tag{92}$$

The multiplicity spaces $\mathcal{X}_s[\mathbf{n}]$ are spaces of coinvariants in X_s (Prop. 5.3). We let $\mathcal{Y}_s^+[\mathbf{n}]$ denote the coinvariants in $\mathcal{Y}_s^+$ with respect to the algebra $\mathcal{A}(p)[\mathbf{u}]^-$ with $\mathbf{u}$ given by (55).

Conjecture 5.1. *The Felder complex* (92) *remains exact after taking the coinvariants with respect to* $\mathcal{A}(p)[\mathbf{u}]^-$, *i.e. the complex*

$$\to \mathcal{Y}_s^+[\mathbf{n}] \xrightarrow{F^s} \mathcal{Y}_{p-s}^+[\mathbf{n}] \xrightarrow{F^{p-s}} \mathsf{X}_s[\mathbf{n}] \to 0 \tag{93}$$

is exact.

The character of $\mathcal{Y}_s^+[\mathbf{n}]$ is expressed in terms of q-supernomial coefficients, which are defined as follows. For $p-1$ dimensional vector $\mathbf{m} = (m_2, m_3, \dots, m_p)$ and a half integer number $a = j/2$, $-\mathbf{e}_p \cdot \bar{\mathsf{A}}\mathbf{m} \le j \le \mathbf{e}_p \cdot \bar{\mathsf{A}}\mathbf{m}$, we introduce q-supernomial coefficients[17]

$$\left[\!\!\left[\begin{matrix} \mathbf{m} \\ a \end{matrix} \right]\!\!\right]_q = \sum_{\substack{j_2, j_3, \dots, j_p \in \mathbb{Z} \\ j_2 + j_3 + \dots + j_p = a + \frac{1}{2}\mathbf{e}_p \cdot \bar{\mathsf{A}}\mathbf{m}}} q^{\sum_{k=2}^{p-1}(\mathbf{e}_{k+1}\cdot\bar{\mathsf{A}}\mathbf{m} - \mathbf{e}_k\cdot\bar{\mathsf{A}}\mathbf{m} - j_{k+1})j_k}$$
$$\times \begin{bmatrix} m_p \\ j_p \end{bmatrix}_q \begin{bmatrix} m_{p-1} + j_p \\ j_{p-1} \end{bmatrix}_q \dots \begin{bmatrix} m_2 + j_3 \\ j_2 \end{bmatrix}_q. \tag{94}$$

The character $\xi_s^{\pm}(q, z)$ of $\mathcal{Y}_s^{\pm}$ is given by the fermionic formula (74). The fermionic formula for the character $\xi_s^{\pm}[\mathbf{n}](q, z)$ of coinvariants $\mathcal{Y}_s^+[\mathbf{n}]$ is

$$\xi_s^{\pm}[\mathbf{m}](q, z) = q^{\frac{p-2-2|\mathbf{m}|}{4}} \sum_{r\in\mathbb{Z}} \sum_{j\in\mathbb{N}_{\mathrm{odd}}} z^r q^{\Delta_{r,s} - \Delta_{1,s} + \Delta_{j,s-pr} - \Delta_{1,-s+p(r+1)}}$$
$$\left(\left[\!\!\left[\begin{matrix} \mathbf{m} \\ \frac{-s + p(j+r) - 1}{2} \end{matrix} \right]\!\!\right]_q - \left[\!\!\left[\begin{matrix} \mathbf{m} \\ \frac{-s + p(j+r) + 1}{2} \end{matrix} \right]\!\!\right]_q \right), \tag{95}$$

where $\Delta_{r,s}$ is given by (32) and $\mathbb{N}_{\mathrm{odd}}$ denotes the odd positive integers. Whenever $m_p > 0$ formula (95) can be simplified to

$$\xi_s^{\pm}[\mathbf{m}](q, z) = \sum_{r\in\mathbb{Z}} z^r q^{\Delta_{r,s} - \Delta_{1,s}} \left[\!\!\left[\begin{matrix} \mathbf{m} - \mathbf{e}_p \\ -\frac{s}{2} + \frac{p}{2} r \end{matrix} \right]\!\!\right]_q \tag{96}$$

using identity

$$\left[\!\!\left[\begin{matrix}\mathbf{m}-\mathbf{e}_p\\ \frac{a}{2}\end{matrix}\right]\!\!\right]_q = q^{\frac{p-2-2|\mathbf{m}|}{4}} \sum_{j\in\mathbb{N}_{\mathrm{odd}}} q^{\Delta_{j,-a}-\Delta_{1,p+a}} \times\left(\left[\!\!\left[\begin{matrix}\mathbf{m}\\ \frac{a+pj-1}{2}\end{matrix}\right]\!\!\right]_q - \left[\!\!\left[\begin{matrix}\mathbf{m}\\ \frac{a+pj+1}{2}\end{matrix}\right]\!\!\right]_q\right). \quad (97)$$

The fermionic formula for coinvariants in $\mathcal{Y}_s$ with respect to some subalgebra of $\mathcal{L}(p)$ from Ref. 14 coincides with (96) up to notation.

The resolution (93) together with Conjecture 5.1 and the formula (95) for characters of $\mathcal{Y}_s^+[\mathbf{n}]$, gives an alternating sign formula for the character of $\mathsf{X}_s[\mathbf{n}]$. In the following proposition we use notation $P(z)[z^r]$, which means the coefficient of the Laurent polynomial $P(z)$ in front of z^r.

Proposition 5.4. *For $p-1$ dimensional vector $\mathbf{m}=(m_2,m_3,\ldots,m_p)$, we have*

$$\hat{K}^{(p)}_{s,\mathbf{m}}(q,z)[z^r] = q^{\frac{p-2-2|\mathbf{m}|}{4}-\Delta_{1,s}} \sum_{n,j\in\mathbb{N}_{\mathrm{odd}}} q^{\Delta_{j+r,s}+\Delta_{n,s-p(j+r)}-\Delta_{1,-s+p(j+r+1)}}$$
$$\times\left(\left[\!\!\left[\begin{matrix}\mathbf{m}\\ \frac{-s-1+p(j+n+r)}{2}\end{matrix}\right]\!\!\right]_q - \left[\!\!\left[\begin{matrix}\mathbf{m}\\ \frac{-s+1+p(j+n+r)}{2}\end{matrix}\right]\!\!\right]_q\right)$$
$$- q^{\Delta_{j+r+1,p-s}+\Delta_{n,-s-p(j+r)}-\Delta_{1,s+p(j+r+1)}}\times$$
$$\times\left(\left[\!\!\left[\begin{matrix}\mathbf{m}\\ \frac{s-1+p(j+n+r)}{2}\end{matrix}\right]\!\!\right]_q - \left[\!\!\left[\begin{matrix}\mathbf{m}\\ \frac{s+1+p(j+n+r)}{2}\end{matrix}\right]\!\!\right]_q\right). \quad (98)$$

Remark 5.1. Whenever $m_p>0$, (98) simplifies to

$$\hat{K}^{(p)}_{s,\mathbf{m}}(q,z)[z^r] = \sum_{j\in\mathbb{N}_{\mathrm{odd}}} q^{\Delta_{j+r,s}-\Delta_{1,s}} \left[\!\!\left[\begin{matrix}\mathbf{m}-\mathbf{e}_p\\ -\frac{s}{2}+p\frac{j+r}{2}\end{matrix}\right]\!\!\right]_q - q^{\Delta_{j+r+1,p-s}-\Delta_{1,s}} \left[\!\!\left[\begin{matrix}\mathbf{m}-\mathbf{e}_p\\ \frac{s}{2}+p\frac{j+r}{2}\end{matrix}\right]\!\!\right]_q, \quad (99)$$

For the Steinberg module, (98) simplifies to

$$\hat{K}^{(p)}_{p,\mathbf{m}}(q,z) = \sum_{j\in\mathbb{N}_0} q^{\Delta_{2j+1,p}-\Delta_{1,p}-\frac{|\mathbf{m}|-p+1}{2}+pj} \times\left(\left[\!\!\left[\begin{matrix}\mathbf{m}\\ pj+\frac{p-1}{2}\end{matrix}\right]\!\!\right]_q - \left[\!\!\left[\begin{matrix}\mathbf{m}\\ pj+\frac{p+1}{2}\end{matrix}\right]\!\!\right]_q\right)$$

$$+\sum_{r\in\mathbb{N}}(z^r+z^{-r})\sum_{j\in\mathbb{N}_0} q^{\Delta_{2j+r+1,p}-\Delta_{1,p}-\frac{|\mathbf{m}|-p+1}{2}+pj+\frac{p}{2}r}$$
$$\times\left(\begin{bmatrix}\mathbf{m}\\ pj+\frac{p}{2}r+\frac{p-1}{2}\end{bmatrix}_q-\begin{bmatrix}\mathbf{m}\\ pj+\frac{p}{2}r+\frac{p+1}{2}\end{bmatrix}_q\right) \tag{100}$$

and whenever $m_p > 0$ to

$$\hat{K}^{(p)}_{p,\mathbf{m}}(q,z)=\sum_{j=-\mathbf{e}_p\cdot\bar{A}\mathbf{m}}^{\mathbf{e}_p\cdot\bar{A}\mathbf{m}} z^j q^{\Delta_{j+1,p}-\Delta_{1,p}}\begin{bmatrix}\mathbf{m}-\mathbf{e}_p\\ p\frac{j}{2}\end{bmatrix}_q. \tag{101}$$

6. Conclusions

In the paper, we studed modules $\mathcal{M}_{\mathbf{u}}$ induced from smaller and smaller subalgebra $\mathcal{A}(p)[\mathbf{u}]^+$. The endomorphism algebra $\mathrm{End}(\mathcal{M}_{\mathbf{u}})$ of $\mathcal{M}_{\mathbf{u}}$ is a subquotient of $\mathcal{A}(p)$, i.e. is the quotient of a subalgebra of $\mathcal{A}(p)$ over a two-side ideal. We note that $\mathrm{End}(\mathcal{M}_{\mathbf{u}})$ is finite dimensional and can be described in quantum group terms. Indeed, $\mathcal{M}_{\mathbf{u}}$ corresponds to an object $\mathsf{M}_{\mathbf{u}}$ in the tensor category $\mathfrak{A}$ and $\mathrm{End}(\mathcal{M}_{\mathbf{u}})$ is isomorphic to endomorphisms of $\mathsf{M}_{\mathbf{u}}$ in $\mathfrak{A}$. For a sequence of vectors $\mathbf{u}$ with increasing components, algebras $\mathrm{End}(\mathcal{M}_{\mathbf{u}})$ approximate $\mathcal{A}(p)$. Thus, there is a problem to formulate the previous statement precisely, i.e. starting from a family of algebras $\mathrm{End}(\mathcal{M}_{\mathbf{u}})$ described in tensor category terms to construct the algebra $\mathcal{A}(p)$.

The algebra $\mathrm{End}(\mathcal{M}_{\mathbf{u}})$ has a complicate structure but it contains the operator corresponding to L_0 from Virasoro subalgebra in $\mathcal{A}(p)$. We calculate the action of this operator in the multiplicity spaces in tensor products. The action of L_0 in the multiplicity spaces is an additional datum to the data of quasitensor category. We would like to formulate this datum in general terms. *This additional datum allows us to reconstruct the chiral conformal field theory from tensor category.*

Another important problem is a reconstruction of a complete (chiral-antichiral) conformal field theory. The main object in the complete conformal field theory is a bimodule $\mathbb{P}$, which admits an action of two commuting copies (one depending on z and another on $\bar{z}$) of $\mathcal{A}(p)$. In terms of the category $\mathfrak{A}$ such a bimodule can be constructed in the following way. Modules $\mathcal{M}_{\mathbf{u}}$ form a projective system and the projective limit gives $\mathbb{P}$. A module $\mathcal{M}_{\mathbf{u}}$ admits the action of $\mathcal{A}(p)$ corresponding to the holomorphic sector and therefore the projective limit $\mathbb{P}$ also admits this action. The action of $\mathcal{A}(p)$ corresponding to the antiholomorphic sector and commuting with the previous one can be defined in $\mathbb{P}$ as well. We described the structure of $\mathbb{P}$ at the level of characters in (81). This $\mathbb{P}$ is the regular bimodule, i.e. it represents the identity functor in the category of $\mathcal{A}(p)$ modules.

Everything said before this line is based on the two crusial statements (7) and (81) of the paper. At the moment we do not know a proof of these statements. However, we note that to prove these statements we should only check (81). All other statements in the paper follows from (81) in more or less standard way (see Ref. 14, where all steps of a similar proof were done for lattice VOAs).

In Ref. 18 a class of lattice models was suggested. Scaling limits of these models conjecturally coincide with W-symmetric logarithmic conformal field models from Ref. 19. Strong arguments that the conjecture is true was recently obtained in Ref. 20–22. Polynomials $\hat{K}^{(p)}_{s,\mathbf{m}}(q,z)$ give some finitizations for the characters of irreducible W-modules. It would be very instructive to compare the finitizations with characters corresponding to finite lattices before taking the scaling limit.

Acknowledgments

We are grateful to E. Feigin, O.Foda, P. Pearce, J. Rasmussen and S.O. Warnaar for valuable discussions. The work of BLF was supported in part by RFBR Grant 08-01-00720, RFBR-CNRS-07-01-92214 and LSS-3472.2008.2. The work of IYuT was supported in part by LSS-1615.2008.2, the RFBR Grant 08-02-01118 and the "Dynasty" foundation.

References

1. H.G. Kausch, *Symplectic Fermions*, Nucl.Phys. B583 (2000) 513-541.
2. M.R. Gaberdiel and H.G. Kausch, *A rational logarithmic conformal field theory*, Phys. Lett. B 386 (1996) 131 [hep-th/9606050].
3. M.Flohr, *On Fusion Rules in Logarithmic Conformal Field Theories*, Int.J.Mod.Phys. A12 (1997) 1943-1958, arXiv:hep-th/9605151.
4. M.Flohr, *On Modular Invariant Partition Functions of Conformal Field Theories with Logarithmic Operators*, Int.J.Mod.Phys. A11 (1996) 4147-4172, arXiv:hep-th/9509166.
5. N.Carqueville, M.Flohr, *Nonmeromorphic operator product expansion and C_2-cofiniteness for a family of W-algebras*, J.Phys. A39 (2006) 951-966, arXiv:math-ph/0508015.
6. B.Feigin, E.Feigin, I.Tipunin, *Fermionic formulas for (1,p) logarithmic model characters in $\Phi_{2,1}$ quasiparticle realisation*, arXiv:0704.2464.
7. M.Flohr, C.Grabow, M.Koehn, *Fermionic Expressions for the Characters of c(p,1) Logarithmic Conformal Field Theories*, arXiv:hep-th/0611241.
8. A. Beilinson and V. Drinfeld, Chiral Algebras, Amer. Math. Soc. Colloq. Publ. 51, Amer. Math. Soc., Providence, 2004.

9. B.Feigin, E. Frenkel, *Coinvariants of nilpotent subalgebras of the Virasoro algebra and partition identities*, I. M. Gelfand Seminar, 139-148, Adv. Soviet Math., 16, Part 1, Amer. Math. Soc., Providence, RI, 1993.
10. B.Feigin, M. Jimbo, R. Kedem, S. Loktev, T. Miwa, *Spaces of coinvariants and fusion product. II.* $\widehat{s\ell}_2$ *character formulas in terms of Kostka polynomials*, J. Algebra 279 (2004), no. 1, 147-179.
11. B.L. Feigin, A.M. Gainutdinov, A.M. Semikhatov, and I.Yu. Tipunin, *Kazhdan–Lusztig correspondence for the representation category of the triplet W-algebra in logarithmic CFT*, math.QA/0512621
12. K. Nagatomo, A. Tsuchiya, *The Triplet Vertex Operator Algebra W(p) and the Restricted Quantum Group at Root of Unity* , arXiv:0902.4607
13. D. Adamovic, A. Milas, *Lattice construction of logarithmic modules for certain vertex algebras*, Selecta Math. New Ser. 15 (2009), 535–561.
14. B.Feigin, S.Loktev, I.Tipunin, *Coinvariants for lattice VOAs and q-supernomial coefficients*, Commun. Math. Phys. 229, 271-292 (2002).
15. J. Fuchs, S. Hwang, A.M. Semikhatov, I. Yu. Tipunin, *Nonsemisimple Fusion Algebras and the Verlinde Formula*, Commun.Math.Phys. 247 (2004) 713-742, arXiv:hep-th/0306274.
16. H.H. Andersen, *Tensor products of quantized tilting modules*, Commun. Math. Phys. 149, 149-159, (1992).
17. A. Schilling, S.O. Warnaar, *Supernomial coefficients, polynomial identities and q-series*, The Ramanujan Journal 2 (1998) 459-494, q-alg/9701007.
18. P.A. Pearce, J. Rasmussen, J.-B. Zuber, *Logarithmic Minimal Models*, J.Stat.Mech. 0611 (2006) P017, hep-th/0607232.
19. B.L. Feigin, A.M. Gainutdinov, A.M. Semikhatov, and I.Yu. Tipunin, *Logarithmic extensions of minimal models: characters and modular transformations*, Nucl.Phys. B757 (2006) 303-343, hep-th/0606196
20. P.A. Pearce, J. Rasmussen, P. Ruelle, *Integrable Boundary Conditions and W-Extended Fusion in the Logarithmic Minimal Models LM(1,p)*, arXiv:0803.0785
21. J. Rasmussen, P.A. Pearce, *W-Extended Fusion Algebra of Critical Percolation*, arXiv:0804.4335
22. J.Rasmussen, *W-Extended Logarithmic Minimal Models*, arXiv:0805.2991.

XXZ SCALAR PRODUCTS, MIWA VARIABLES AND DISCRETE KP

OMAR FODA and GUS SCHRADER

Department of Mathematics and Statistics
University of Melbourne
Parkville, Victoria 3010, Australia
E-mails: o.foda@ms.unimelb.edu.au, g.schrader@ms.unimelb.edu.au

We revisit the quantum/classical integrable model correspondence in the context of inhomogeneous finite length XXZ spin-$\frac{1}{2}$ chains with periodic boundary conditions and show that the Bethe scalar product of an arbitrary state and a Bethe eigenstate is a discrete KP τ-function. The continuous Miwa variables of discrete KP are the rapidities of the arbitrary state.

1. Introduction

Quantum models of the statistical mechanical type (the only quantum models discussed in this note) such as integrable 1-dimensional quantum spin chains, and classical models such as integrable nonlinear partial differential equations, are related in the sense that the methods used to compute in the former, particularly the quantum inverse scattering transform, also known as the algebraic Bethe Ansatz, are quantum versions of those used to compute in the latter, namely the classical inverse scattering transform. It is therefore natural to expect that the quantum integrable models have classical limits in which they reduce to classical counterparts.

What is less than natural to expect, at least to our minds, is that basic objects in quantum integrable models, such as the correlation functions, turn out to have direct interpretations in terms of objects in classical integrable models, such as solutions of integrable nonlinear partial differential and difference equations, *without taking a classical limit.* But this turns out to be the case, and it points to a direct connection between quantum and classical integrable models that is distinct from, and to our minds at least as fundamental as that obtained by taking a classical limit.

Notes on the literature. The following is far from a comprehensive survey of the relevant literature. To the best of our knowledge, a direct connection between quantum (statistical mechanical) and classical models of the type that we are interested in first appeared in Ref. 1, where Ising spin-spin correlation functions in the scaling limit were shown to satisfy Painlevé equation of the third kind, and subsequently in Refs. 2,3, where critical Ising correlation functions on the lattice were shown to satisfy the Toda lattice equation in Hirota's bilinear form. Further results, along the same lines as in Ref. 1, for the XXZ spin chain at the free fermion point, were obtained in Ref. 4, as reviewed in Ref. 5.

The fact that τ-functions (solutions of Hirota's bilinear equations) appear in the Ising model as well as in KP theory was discussed in works by the Kyoto group and reviewed in Ref. 6 where it was argued that the mathematical reason underlying this coincidence is the fact that both quantum and classical models are based on infinite dimensional Lie algebras that are realized in terms of free fermions.

Closest to the spirit of this note is the work of Krichever *et al.*,[7] reviewed in Ref. 8. The starting point of Ref. 7 is the observation that the Bethe eigenvalues satisfy Hirota's difference equation, various limits of which lead to a large number of integrable differential and difference equations.[9] We will comment on the results of Ref. 7 and how they differ from the result in this note in section **6**. More recently, studies of the ultra-discrete limit of quantum integrable spin chains revealed many classical integrable structures.[10]

Finally, while we are only interested in integrable quantum models in statistical mechanics in this note, it is important to mention bosonisation (the operator formulation of Sato's theory) as a deep and established correspondence between the quantum field theories of free fermions, which are integrable quantum models, and classical integrable hierarchies, as reviewed in Ref. 11. In this correspondence, expectation values of fermion operators have direct interpretations in terms of solutions of integrable nonlinear partial differential equations. Bosonization was further extended to connect KP theory and conformal field theories on Riemann surfaces (which are integrable quantum models) in Ref. 12.

The long term aim of our work is to develop a correspondence between integrable statistical mechanical models and classical integrable hierarchies that is as direct and detailed as that obtained by bosonisation between free fermions and classical integrable hierarchies.

Bethe scalar products and continuous KP τ-functions. Consider the inhomogeneous length-L XXZ spin-$\frac{1}{2}$ chain with periodic boundary conditions. Following Ref. 13, the Bethe scalar product $\langle \lambda_1, \cdots, \lambda_N | \mu_1, \cdots, \mu_N \rangle_\beta$ of an arbitrary state $\langle \lambda_1, \cdots, \lambda_N |$ where the auxiliary space rapidities $\{\lambda_1, \cdots, \lambda_N\}$ are free, and a Bethe eigenstate $|\mu_1, \cdots, \mu_N\rangle_\beta$ where the auxiliary space rapidities $\{\mu_1, \cdots, \mu_N\}$ obey the Bethe equations, is a polynomial τ-function of the continuous (differential) KP hierarchy. In this identification, the KP time variables $\{t_1, t_2, \cdots\}$ are power sums of the free rapidities $\{\lambda_1, \cdots, \lambda_N\}$. However, these polynomial KP τ-functions involve by construction more time variables than free rapidities. The reason is as follows.

Expanding the scalar product in terms of Schur polynomials s_λ, associated to Young diagrams $\{\lambda\}$, that are functions of the rapidities $\{\lambda_1, \cdots, \lambda_N\}$, the maximal number of rows in any Young diagram λ is N. Switching to KP time variables $\{t_1, t_2, \cdots\}$ that are powers sums in the rapidities, we obtain character polynomials χ_λ that depend on effectively as many time variables as the number of cells in (that is, the size of) λ which is larger than N. Consequently, the KP time variables $\{t_1, t_2, \cdots\}$ were *formally* considered in Ref. 13 to be independent, and the Bethe scalar product was defined as a *restricted* KP τ-function obtained by setting $\{t_1, t_2, \cdots\}$ to be power sums of a smaller number of independent variables $\{\lambda_1, \cdots, \lambda_N\}$.

Bethe scalar products and discrete KP τ-functions. In this note, we simplify the correspondence of Ref. 13 by working solely in terms of the free rapidities $\{\lambda_1, \cdots, \lambda_N\}$ which are now continuous Miwa variables and the τ-functions that we obtain are those of the discrete KP hierarchy.[14,15]

Outline of contents. In section **2**, we recall basic facts related to symmetric functions, Casoratian matrices and Casoratian determinants. In **3**, we recall basic facts related to the XXZ spin-$\frac{1}{2}$ chain, the algebraic Bethe Ansatz, the Bethe scalar product, recall Slavnov's determinant expression of the Bethe scalar product and show that it is a Casoratian determinant. In **4**, we recall basic facts related to the continuous and discrete KP hierarchies and define the Miwa variables that relate the two. In **5**, we show that Bethe scalar products in the XXZ spin-$\frac{1}{2}$ chain with periodic boundary conditions are discrete KP τ-functions. In **6**, we collect a number of remarks. Space limitations allow us to give no more than the minimal definitions necessary to fix the notation and terminology supplemented by references to relevant sources.

2. Symmetric functions and Casoratians

The canonical reference to symmetric functions is Ref. 16. Casoratian matrices and determinants are carefully discussed in Ref. 15. The definitions in Ref. 15 are more general than those used in this note.

Frequently used notation. We use $\{x\}$ for the set of finitely many variables $\{x_1, x_2, \cdots, x_N\}$, or infinitely many variables $\{x_1, x_2, \cdots\}$. The cardinality of the set should be clear from the context. We use $\{\widehat{x}_m\}$ for $\{x\}$ but with the element x_m missing. In the case of sets with a repeated variable x_i, we use the superscript (m_i) to indicate the multiplicity of x_i, as in $x_i^{(m_i)}$. For example, $\{x_1^{(3)}, x_2, x_3^{(2)}, x_4, \cdots\}$ is the same as $\{x_1, x_1, x_1, x_2, x_3, x_3, x_4, \cdots\}$ and $f\{\cdots, x_i^{(m_i)}, \cdots\}$ is equivalent to saying that f depends on m_i distinct variables all of which have the same value x_i. For simplicity, we use x_i to indicate $x_i^{(1)}$. In calculations, it is safer to think of any x_i with multiplicity $m_i > 1$ initially as distinct, that is $\{x_{i,1}, x_{i,2}, \cdots, x_{i,m_i}\}$, then set these m_i variables equal to the same value x_i at the end.

We use the bracket notation $[x] = e^x - e^{-x}$, and

$$\Delta\{x\} = \prod_{1\leq i<j\leq N} (x_i - x_j), \quad \Delta_{trig}\{\lambda\} = \prod_{1\leq i<j\leq N} [\lambda_i - \lambda_j] \tag{1}$$

for the Vandermonde determinant and its trigonometric analogue.

The elementary symmetric function $e_i\{x\}$ in N variables $\{x\}$ is the coefficient of k^i in the expansion

$$\prod_{i=1}^{N} (1 + x_i\, k) = \sum_{i=0}^{\infty} e_i\{x\}\, k^i \tag{2}$$

For example, $e_0\{x\} = 1$, $e_1(x_1, x_2, x_3) = x_1 + x_2 + x_3$, $e_2(x_1, x_2) = x_1 x_2$. $e_i\{x\} = 0$, for $i < 0$ and for $i > N$.

The complete symmetric function $h_i\{x\}$ in N variables $\{x\}$ is the coefficient of k^i in the expansion

$$\prod_{i=1}^{N} \frac{1}{1 - x_i\, k} = \sum_{i=0}^{\infty} h_i\{x\}\, k^i \tag{3}$$

For example, $h_0\{x\} = 1$, $h_1(x_1, x_2, x_3) = x_1 + x_2 + x_3$, $h_2(x_1, x_2) = x_1^2 + x_1 x_2 + x_2^2$, and $h_i\{x\} = 0$ for $i < 0$.

Useful identities for $h_i\{x\}$. From Eq. (3), it is straightforward to show that

$$h_i\{x\} = h_i\{\widehat{x}_m\} + x_m h_{i-1}\{x\} \tag{4}$$

and from that, one obtains

$$h_i\{x_1, x_2, \cdots, x_N\} = h_i\{x_1^{(2)}, x_2, \cdots, x_N\} - x_1 h_{i-1}\{x_1^{(2)}, x_2, \cdots, x_N\} \tag{5}$$

$$(x_2 - x_1) h_i\{x_1^{(2)}, x_2^{(2)}, x_3, \cdots, x_N\} = x_2 h_i\{x_1, x_2^{(2)}, \cdots, x_N\} - x_1 h_i\{x_1^{(2)}, x_2, \cdots, x_N\} \tag{6}$$

The discrete derivative $\Delta_m h_i\{x\}$ of $h_i\{x\}$ with respect to any one variable $x_m \in \{x\}$ is defined using Eq. (4) as

$$\Delta_m h_i\{x\} = \frac{h_i\{x\} - h_i\{\widehat{x}_m\}}{x_m} = h_{i-1}\{x\} \tag{7}$$

Note that the effect of applying Δ_m to $h_i\{x\}$ is a complete symmetric function $h_{i-1}\{x\}$ of degree $i - 1$ in the same set of variables $\{x\}$. The difference operator in Eq. (7) is not the most general definition of a discrete derivative, but it is sufficient for the purposes of this note. For a more general definition, see Ref. 15.

The Schur polynomial $s_\lambda\{x\}$ indexed by a Young diagram $\lambda = [\lambda_1, \ldots, \lambda_r]$ with $\lambda_i \neq 0$, for $1 \leq i \leq r$, and $\lambda_i = 0$, for $r + 1 \leq i \leq N$, is

$$s_\lambda\{x\} = \frac{\det\left(x_i^{\lambda_j - j + N}\right)_{1 \leq i,j \leq N}}{\Delta\{x\}} = \det\left(h_{\lambda_i - i + j}\{x\}\right)_{1 \leq i,j \leq N} \tag{8}$$

For example, $s_\phi\{x\} = 1$, $s_{[1]}(x_1, x_2, x_3) = x_1 + x_2 + x_3$, $s_{[1,1]}(x_1, x_2) = x_1 x_2$. The first equality in Eq. (8) is the definition of $s_\lambda\{x\}$. The second is the *Jacobi-Trudi identity* for $s_\lambda\{x\}$. $s_\lambda\{x\}$ is symmetric in the elements of $\{x\}$ and requires no more than r (the number of non-zero rows in λ) variables to be non-vanishing.

The one-row character polynomial $\chi_i\{t\}$ indexed by a one-row Young diagram of length i, is the i-th coefficient in the generating series

$$\sum_{i=0}^{\infty} \chi_i\{t\}\, k^i = \exp\left(\sum_{i=1}^{\infty} t_i\, k^i\right) \tag{9}$$

For example, $\chi_0\{t\} = 1$, $\chi_1\{t\} = t_1$, $\chi_2\{t\} = \frac{t_1^2}{2} + t_2$, $\chi_3\{t\} = \frac{t_1^3}{6} + t_1 t_2 + t_3$, and $\chi_i\{t\} = 0$ for $i < 0$. Since t_i has degree i, χ_i is not symmetric in $\{t\}$ and generally depends on as many t-variables as the row-length i.

The character polynomial $\chi_\lambda\{t\}$ indexed by a Young diagram $\lambda = [\lambda_1, \lambda_2, \ldots, \lambda_r]$ with r non-zero-length rows, $r \leq N$, is

$$\chi_\lambda\{t\} = \det\left(\chi_{\lambda_i - i + j}\{t\}\right)_{1 \leq i,j \leq n} \tag{10}$$

For example $\chi_{[1,1]}\{t\} = \frac{t_1^2}{2} - t_2$, $\chi_{[2,1]}\{t\} = \frac{t_1^3}{3} - t_3$, $\chi_{[2,2]}\{t\} = \frac{t_1^4}{12} - t_1 t_3 + t_2^2$. Notice that $\chi_\lambda\{t\}$ can depend on all t_i, for $i \leq |\lambda|$, where $|\lambda|$ is the sum of the lengths of all rows in (or area of) λ.

From character polynomials to Schur polynomials. Assuming that the t-variables are independent and that we have sufficiently many x-variables, then setting $t_m \to \frac{1}{m}\sum_{i=1}^{N} x_i^m$ sends $\chi_i\{t\} \to h_i\{x\}$. In this note, as in Ref. 13, we study Bethe scalar products that are polynomials in N variables $\{x_1, x_2, \cdots, x_N\}$. We can expand these scalar products in terms of Schur polynomials $s_\lambda\{x\}$ where $\{\lambda\}$ has at most N rows, or in terms of the corresponding character polynomials $\chi_\lambda\{t\}$ that require more t-variables (which are power sums in the x-variables) than N and therefore cannot be all independent. We choose to work in terms of the x-variables and $s_\lambda\{x\}$.

Casoratian matrices and determinants. A Casoratian matrix M of the type that appears in this note is such that the elements M_{ij} satisfy either

$$M_{i,j+1}\{x\} = \Delta_m M_{ij}\{x\}, \quad \text{or} \quad M_{i+1,j}\{x\} = \Delta_m M_{ij}\{x\} \tag{11}$$

where the discrete derivative Δ_m is taken with respect to any one variable $x_m \in \{x\}$. If M is a Casoratian matrix, then $\det M$ is a Casoratian determinant. Casoratian determinants are discrete analogues of Wronskian determinants.

3. The XXZ spin-$\frac{1}{2}$ chain and the Algebraic Bethe Ansatz

The XXZ spin-$\frac{1}{2}$ chain is discussed in detail in Ref. 17,18. A standard reference to the algebraic Bethe Ansatz, including the Bethe scalar product and Slavnov's determinant expression, is Ref. 5. We leave the definition of auxiliary and quantum spaces, auxiliary and quantum rapidities, and the precise action of the various operators to Ref. 5.

Frequently used variables. In the following, L is the number of sites in a periodic XXZ spin-$\frac{1}{2}$ chain, and N is the number of Bethe operators $B(\mu_i)$ that act on the reference state $|0\rangle$ to create an XXZ state $|\mu_1, \cdots, \mu_N\rangle$. N is also the rank of the matrix whose determinant is Slavnov's expression for the Bethe scalar product. We use the set $\{\lambda\}$ for the free auxiliary space rapidities, $\{\mu\}$ or more explicitly $\{\mu_\beta\}$ for the auxiliary space rapidities that satisfy the Bethe equations, and $\{\nu\}$ for the quantum space rapidities (the inhomogeneities). A Bethe eigenstate state whose rapidities satisfy the Bethe equations is also denoted by a subscript β, such as $|\lambda\rangle_\beta$. γ is the crossing parameter. We use the exponentiated variables $\{x_i, y_i, z_i, q\} = \{e^{\lambda_i}, e^{\mu_i}, e^{\nu_i}, e^{\gamma}\}$, but still refer to the exponentiated variables $\{x, y, z\}$ as rapidities rather than exponentiated rapidities for simplicity.

The *L*-operator of the XXZ spin-$\frac{1}{2}$ chain is

$$L_{ai}(\lambda, \nu) = \begin{pmatrix} [\lambda - \nu + \gamma] & 0 & 0 & 0 \\ 0 & [\lambda - \nu] & [\gamma] & 0 \\ 0 & [\gamma] & [\lambda - \nu] & 0 \\ 0 & 0 & 0 & [\lambda - \nu + \gamma] \end{pmatrix}_{ai} \tag{12}$$

where a is an auxiliary space index and i is a quantum space index.

The monodromy matrix of the inhomogeneous length-L XXZ spin-$\frac{1}{2}$ chain is

$$T_a(\lambda) = \begin{pmatrix} A(\lambda) & B(\lambda) \\ C(\lambda) & D(\lambda) \end{pmatrix}_a = \prod_{i=1}^{L} L_{ai}(\lambda, \nu_i) \tag{13}$$

where it is conventional to suppress the dependence on the inhomogeneous quantum space rapidities ν_i in T_a and its elements, and each of the operators

A, B, C, and D acts in the tensor product $V_1 \otimes \cdots \otimes V_L$ where V_i is a vector space isomorphic to $\mathbb{C}^2$.

The transfer matrix is the trace of the monodromy matrix over the auxiliary space,

$$\mathrm{Tr}_a T_a(\lambda) = A(\lambda) + D(\lambda) \tag{14}$$

An arbitrary state $|\mu\rangle$ is generated by the action of the $B(\mu)$ operators on the reference state $|0\rangle = \otimes^L \begin{pmatrix} 1 \\ 0 \end{pmatrix}$,

$$|\mu\rangle = B(\mu_1) \ldots B(\mu_N)|0\rangle \tag{15}$$

An arbitrary dual state $\langle\lambda|$ is generated by the action of the $C(\lambda)$ operators on the dual reference state $\langle 0| = \otimes^L \begin{pmatrix} 1 & 0 \end{pmatrix}$,

$$\langle\lambda| = \langle 0|\, C(\lambda_1) \ldots C(\lambda_N) \tag{16}$$

The scalar product of a state and a dual state is

$$\langle\lambda|\mu\rangle = \langle 0|\, C(\lambda_1) \ldots C(\lambda_N)\, B(\mu_1) \ldots B(\mu_N)|0\rangle \tag{17}$$

A Bethe eigenstate $|\mu\rangle_\beta$ is an eigenstate of the transfer matrix,

$$\Big(A(\lambda) + D(\lambda) \Big)\; |\mu\rangle_\beta = E(\lambda)|\mu\rangle_\beta \tag{18}$$

where $E(\lambda)$ is the corresponding Bethe eigenvalue. For a state $|\mu\rangle$ to be a Bethe eigenstate, its auxiliary space rapidities must satisfy a set of Bethe equations.

The Bethe equations that must be satisfied by the N auxiliary space rapidities of a state $|\mu\rangle = B(\mu_1) \cdots B(\mu_N)|0\rangle$ in order to be a Bethe eigenstate, in the specific case of the inhomogeneous length-L spin-$\frac{1}{2}$ chain, are

$$\frac{\prod_{i=1}^L [\mu - \nu_i + \gamma]}{\prod_{i=1}^L [\mu - \nu_i]} \prod_{j \neq i}^N \frac{[\mu_i - \mu_j - \gamma]}{[\mu_i - \mu_j + \gamma]} = 1 \tag{19}$$

where $\{\nu_1, \cdots, \nu_L\}$, are the quantum space rapidities, which are taken to be part of the parameters that specify the spin chain, rather than the definition of the Bethe state.

A Bethe scalar product is a scalar product of an arbitrary state $\langle \lambda |$ and a Bethe eigenstate $|\mu\rangle_\beta$,

$$\langle \lambda | \mu \rangle_\beta = \langle 0 | \, C(\lambda_1) \dots C(\lambda_N) \, B(\mu_{1,\beta}) \dots B(\mu_{N,\beta}) |0\rangle \tag{20}$$

Bethe scalar products as in Eq. (20) play a central role in computing XXZ correlation functions,[19] hence their importance.

Slavnov's determinant expression. In Ref. 20, Slavnov obtained an elegant determinant expression for the Bethe scalar product,

$$\langle \lambda | \mu \rangle_\beta = [\gamma]^N \frac{\prod_{i,j=1}^N [\lambda_i - \mu_j + \gamma]}{\Delta\{\lambda\}\Delta\{\mu\}} \prod_{k=1}^N \prod_{l=1}^L [\lambda_k - \nu_l] \, [\mu_k - \nu_l] \, \det \Omega \tag{21}$$

where the components of the $N \times N$ matrix Ω are

$$\Omega_{ij} = \frac{1}{[\lambda_i - \mu_j][\lambda_i - \mu_j + \gamma]} - \frac{1}{[\mu_j - \lambda_i][\mu_j - \lambda_i + \gamma]} \prod_{k=1}^L \frac{[\lambda_i - \nu_k + \gamma]}{[\lambda_i - \nu_k]} \prod_{l=1}^N \frac{[\lambda_i - \mu_l - \gamma]}{[\lambda_i - \mu_l + \gamma]} \tag{22}$$

Slavnov's scalar product is the main object of interest in this note. We wish to show that it is a Casoratian determinant and that the latter satisfy the bilinear identities of a discrete KP hierarchy.[15]

Re-writing Slavnov's determinant expression. In Ref. 13, it was found useful to rewrite Slavnov's determinant expression for the Bethe scalar product as follows. First, we change variables and work in terms of exponentials of the original variables as follows

$$\{e^{2\lambda_i}, e^{2\mu_i}, e^{2\nu_i}, e^{\gamma}\} \to \{x_i, y_i, z_i, q\} \tag{23}$$

but continue to call the exponentials $\{x, y, z\}$ rapidities as that is simpler and should cause no confusion. Ignoring prefactors that do not depend on $\{x\}$, it was shown in Ref. 13 that the relevant part of Slavnov's determinant expression can be re-written as

$$\det \Omega' = \frac{\det \Omega}{\Delta\{x\}}, \quad \text{where} \quad \boldsymbol{\Omega}_{ij} = \sum_{k=1}^{N+L-1} x_i^{k-1} \, \kappa_{kj}, \quad \kappa_{kj} = -\sum_{l=1}^{k} y_j^{l-k-1} \, \rho_{lj}, \tag{24}$$

and

$$\begin{aligned}\rho_{lj} = & \left(\prod_{m=1}^{L} \left(y_j q - z_m q^{-1} \right) \right) \left(\prod_{n \neq j}^{N} \left(y_j - y_n q^2 \right) \cdot e_{(L+N-l)}\{-\widehat{y}_j q^{-2}\}\{-z\} \right) \\ & - \left(\prod_{m=1}^{L} \left(y_j q - z_m q \right) \right) \left(\prod_{n \neq j}^{N} \left(y_j - y_n q^{-2} \right) \cdot e_{(L+N-l)}\{-\widehat{y}_j q^{2}\}\{-z q^{-2}\} \right)\end{aligned} \tag{25}$$

In Eq. (25), $e_k\{\widehat{y}_j\}\{z\}$ is the k-th elementary symmetric polynomial in the set of variables $\{y\} \cup \{z\}$ with the omission of y_j.

A Bethe scalar product is a Casoratian determinant. We wish to show that Slavnov's determinant expression is Casoratian in the free rapidities $\{x\}$ of the general state. Expanding $\det \Omega$, using the Cauchy-Binet identity, we obtain

$$\begin{aligned}\det \Omega &= \det \left(\sum_{k=1}^{N+L-1} x_i^{k-1} \, \kappa_{kj} \right) \\ &= \sum_{1 \leq k_1 < \cdots < k_N \leq N+L-1} \det \left(x_i^{k_j - 1} \right) \det \left(\kappa_{k_i, j} \right) \\ &= \sum_{0 \leq \lambda_N \leq \cdots \leq \lambda_1 \leq L-1} \det \left(x_i^{\lambda_{(N+1-j)} + j - 1} \right) \det \left(\kappa_{\lambda_{(N-i+1)} + i, j} \right) \\ &= \sum_{0 \leq \lambda_N \leq \cdots \leq \lambda_1 \leq L-1} \det \left(x_i^{\lambda_j + N + 1 - j - 1} \right) \det \left(\kappa_{\lambda_i + N + 1 - i, j} \right)\end{aligned} \tag{26}$$

From the definition of Schur polynomials that uses the Jacobi-Trudi identity in Eq. (8) we obtain

$$\det \Omega' = \frac{\det \Omega}{\Delta\{x\}} = \sum_{0 \le \lambda_N \le \cdots \le \lambda_1 \le L-1} \det \left(h_{\lambda_j - j + i}\{x\} \right) \det \left(\kappa_{\lambda_i + N + 1 - i, j} \right)$$

$$= \sum_{0 \le \lambda_N \le \cdots \le \lambda_1 \le L-1} \det \left(h_{\lambda_{N+1-j} - N - 1 + j + i}\{x\} \right) \det \left(\kappa_{\lambda_i + N + 1 - i, j} \right)$$

$$= \sum_{1 \le k_1 \le \cdots \le k_N \le N+L-1} \det \left(h_{k_j - N - 1 + i}\{x\} \right) \det \left(\kappa_{k_i, j} \right)$$

$$= \det \left(\sum_{k=1}^{N+L-1} h_{k-N-1+i}\{x\} \ \kappa_{kj} \right) \tag{27}$$

Hence $\det \Omega'$ is Casoratian in $\{x\}$. Next, we need to show that a Casoratian determinant is a solution of the bilinear identities of discrete KP, but this requires a number of definitions which we outline in the next section.

4. Continuous KP, Miwa variables and discrete KP

A standard introduction to the continuous KP hierarchy is Ref. 21. Miwa variables are discussed in detail in Ref. 22 where further references to their applications are provided. The discrete KP hierarchy was introduced in Ref. 9, and further studied in Refs. 14,23. In this note, we follow the treatment in Ref. 15.

Continuous KP is an infinite hierarchy of integrable partial differential equations generated in Hirota's bilinear form by expanding the bilinear identity

$$\oint_{k=k_\infty} \frac{dk}{2\pi i} \, e^{\xi(t-t',k)} \tau\{t - \epsilon(k^{-1})\} \ \tau\{t + \epsilon(k^{-1})\} = 0 \tag{28}$$

where $k \in \mathbb{P}^1$, the contour integral is around the point at infinity $k_{infinity} \in \mathbb{P}^1$, $\{t\} = \{t_1, t_2, t_3, \cdots\}$, $\xi(t,k) = \sum_{i=1}^{\infty} t_i k^i$, $\epsilon(k^{-1}) = \{\frac{1}{k}, \frac{1}{2k^2}, \frac{1}{3k^3}, \cdots\}$, $\{t \pm \epsilon(k^{-1})\} = \{t_1 \pm \frac{1}{k}, t_2 \pm \frac{1}{2k}, t_3 \pm \frac{1}{3k}, \cdots\}$. The simplest KP equation in the hierarchy is

$$\left(D_1^4 + 3D_2^2 - 4D_1 D_3 \right) \ \tau \cdot \tau = 0 \tag{29}$$

where D_i is the Hirota derivative with respect to t_i. For the precise definition of D_i and that of the notation $\tau \cdot \tau$, see Ref. 21.

Continuous and discrete Miwa variables. In Ref. 23, Miwa introduced two infinite sets of variables, the continuous variables $\{x\} = \{x_1, x_2, \cdots\}$, and the discrete (and integer valued) variables $\{m\} = \{m_1, m_2, \cdots\}$, and showed that setting

$$t_j = \sum_{i=1}^{\infty} m_i \frac{x_i^j}{j} \tag{30}$$

transforms τ-functions of continuous KP to τ-functions of a hierarchy of bilinear difference equations, namely discrete KP, studied in detail in Ref. 14. These variables are now known as continuous and discrete Miwa variables, respectively,

Multiplicities. From Eq. (30), one can see that the discrete variables $\{m\}$, where $m_i \in \mathbb{Z}$ are multiplicities of the continuous variables $\{x\}$. In other words, $m_i > 1$ is equivalent to saying that x_i occurs m_i times in $\{x\}$, or that there are m_i continuous variables that have the same value x_i.

Discrete KP is an infinite hierarchy of integrable partial *difference* equations in an infinite set of continuous Miwa variables $\{x\}$, where time evolution is obtained by changing the multiplicities $\{m\}$ of these variables. In this note, we are interested in situation where the total number of continuous Miwa variables is finite, and the sum of all multiplicities is N. In this case, the discrete KP hierarchy can be written in bilinear form as $n \times n$ determinant equations

$$\det \begin{pmatrix} 1 & x_1 & \cdots & x_1^{n-2} & x_1^{n-2}\tau_{+1}\{x\}\tau_{-1}\{x\} \\ 1 & x_2 & \cdots & x_2^{n-2} & x_2^{n-2}\tau_{+2}\{x\}\tau_{-2}\{x\} \\ \vdots & \vdots & \vdots & \vdots & \vdots \\ 1 & x_n & \cdots & x_n^{n-2} & x_n^{n-2}\tau_{+n}\{x\}\tau_{-n}\{x\} \end{pmatrix} = 0 \tag{31}$$

where $3 \leq n \leq N$, and

$$\begin{aligned} \tau_{+i}\{x\} &= \tau\{x_1^{(m_1)}, \cdots, x_i^{(m_i+1)}, \cdots, x_N^{(m_N)}\}, \\ \tau_{-i}\{x\} &= \tau\{x_1^{(m_1+1)}, \cdots, x_i^{(m_i)}, \cdots, x_N^{(m_N+1)}\} \end{aligned} \tag{32}$$

In other words, if $\tau\{x\}$ has m_i copies of the variable x_i, then $\tau_{+i}\{x\}$ has m_i+1 copies of x_i and the multiplicities of all other variables remain the same, while $\tau_{-i}\{x\}$ has one more copy of each variable except x_i. Equivalently, one can use the simpler notation

$$\tau_{+i}\{x\} = \tau\{m_1, \cdots, (m_i + 1), \cdots, m_N\},$$
$$\tau_{-i}\{x\} = \tau\{(m_1 + 1), \cdots, m_i, \cdots, (m_N + 1)\} \tag{33}$$

The simplest discrete KP bilinear difference equation, in the notation of Eq. (33), is

$$\begin{aligned} & x_i(x_j - x_k)\tau\{m_i + 1, m_j, m_k\}\tau\{m_i, m_j + 1, m_k + 1\} \\ + & x_j(x_k - x_i)\tau\{m_i, m_j + 1, m_k\}\tau\{m_i + 1, m_j, m_k + 1\} \\ + & x_k(x_i - x_j)\tau\{m_i, m_j, m_k + 1\}\tau\{m_i + 1, m_j + 1, m_k\} = 0 \end{aligned} \tag{34}$$

where $\{x_i, x_j, x_k\} \in \{x\}$ and $\{m_i, m_j, m_k\} \in \{m\}$ are any two (corresponding) triples in the sets of continuous and discrete (integral valued) Miwa variables. Eq. (34) is the discrete analogue of Eq. (29).

Discrete time evolution in discrete KP. Each continuous Miwa variable x_i corresponds to a time variable in discrete KP. Time evolution in discrete KP, in direction x_i, is given by the discrete changes in the multiplicities m_i of x_i. Notice that as a multiplicity m_i changes by ± 1, the rank of the matrix $M_{i\pm 1}$, where $\det M_{i\pm 1} = \tau_{i\pm 1}$ remains the same as the rank of M, where $\det M = \tau$.

5. Bethe scalar products are discrete KP τ-functions

In this section, we adapt the general treatment of Ref. 15 to the specific case of Slavnov's determinant expressions. We do this in detail to show explicitly that the general (and slightly abstract) identities and theorems in Ref. 15 apply to Slavnov's expressions.

Re-arranging the elements of Slavnov's determinant. Given the $N \times N$ matrix Ω' with elements

$$\omega'_{ij} = \sum_{k=1}^{N+L-1} h_{k-N-1+i}\{x\}\ \kappa_{kj} \tag{35}$$

let us consider the matrix Ω'' with elements

$$\omega_{ij} = \sum_{k=1}^{N+L-1} c_{ik}\ h_{k-j}\{x\} \tag{36}$$

which is obtained from Ω' by reordering the rows of the latter from bottom to top, changing the rows and the columns and setting $c_{ik} = \kappa_{ki}$. Notice that we use ω rather than ω'' for the elements of Ω'' to simplify the notation. Since $\det \Omega' = (-)^{N(N-1)/2} \det \Omega$, it is sufficient to show that Ω satisfies the difference bilinear identities of discrete KP.

Identities for the elements ω_{ij}. It follows from Eqs. (4–5) that the elements ω_{ij} of Ω'' satisfy analogous identities

$$\begin{aligned}\omega_{ij}\{x_1, \ldots, x_m^{(2)}, \ldots, x_N\} &= \omega_{ij}\{x_1, \ldots, x_N\} \\ &\quad + x_m \omega_{i,j+1}\{x_1, \ldots, x_m^{(2)}, \ldots, x_N\}\end{aligned} \tag{37}$$

$$\begin{aligned}&(x_r - x_s)\, \omega_{ij}\{x_1, \ldots, x_r^{(2)}, x_s^{(2)}, \ldots x_N\} = \\ &\quad x_r\, \omega_{ij}\{x_1, \ldots, x_r^{(2)}, \ldots, x_N\} - x_s\, \omega_{ij}\{x_1, \ldots, x_s^{(2)}, \ldots, x_N\}\end{aligned} \tag{38}$$

From Eq. (7), we see that

$$\Delta_m \omega_{ij}\{x_1, \ldots, x_N\} = \omega_{i,j+1}\{x_1, \ldots, x_N\} \tag{39}$$

which is equivalent to the statement that $\det \Omega$ is Casoratian.

Notation for column vectors with elements ω_{ij}. We need the column vector

$$\vec{\omega}_j = \begin{pmatrix} \omega_{1j}\{x_1^{(m_1)}, \ldots, x_N^{(m_N)}\} \\ \omega_{2j}\{x_1^{(m_1)}, \ldots, x_N^{(m_N)}\} \\ \vdots \\ \omega_{Nj}\{x_1^{(m_1)}, \ldots, x_N^{(m_N)}\} \end{pmatrix} \tag{40}$$

and write

$$\vec{\omega}_j^{[k_1, \ldots, k_n]} = \begin{pmatrix} \omega_{1j}\{x_1^{(m_1)}, \ldots, x_{k_1}^{(m_{k_1}+1)}, \ldots, x_{k_n}^{(m_{k_n}+1)}, \ldots, x_N^{(m_N)}\} \\ \omega_{2j}\{x_1^{(m_1)}, \ldots, x_{k_1}^{(m_{k_1}+1)}, \ldots, x_{k_n}^{(m_{k_n}+1)}, \ldots, x_N^{(m_N)}\} \\ \vdots \\ \omega_{Nj}\{x_1^{(m_1)}, \ldots, x_{k_1}^{(m_{k_1}+1)}, \ldots, x_{k_n}^{(m_{k_n}+1)}, \ldots, x_N^{(m_N)}\} \end{pmatrix} \tag{41}$$

for the corresponding column vector where the multiplicities of the variables $x_{k_1}, \cdots, x_{k_n}$ are increased by 1.

Notation for determinants with elements ω_{ij}. We also need the determinant

$$\tau = \det \left(\vec{\omega}_1 \, \vec{\omega}_2 \, \cdots \, \vec{\omega}_N \right) = \left| \vec{\omega}_1 \, \vec{\omega}_2 \, \cdots \, \vec{\omega}_N \right| \tag{42}$$

and the notation

$$\tau^{[k_1,\dots,k_n]} = \left| \vec{\omega}_1^{[k_1,\dots,k_n]} \, \vec{\omega}_2^{[k_1,\dots,k_n]} \, \cdots \, \vec{\omega}_N^{[k_1,\dots,k_n]} \right| \tag{43}$$

for the determinant with shifted multiplicities. Next, and closely following Ref. 15, we derive two identities involving Casoratian determinants with elements ω_{ij}.

Casoratian identity 1. The first identity that we need is

$$x_1^{n-2} \, \tau^{[1]} = \left| \vec{\omega}_1 \, \vec{\omega}_2 \, \cdots \, \vec{\omega}_{N-1} \, \vec{\omega}_{N-n+2}^{[1]} \right| \tag{44}$$

which is derived as follows. From Eq. (43), we have

$$\tau^{[1]} = \left| \vec{\omega}_1^{[1]} \, \vec{\omega}_2^{[1]} \, \cdots \, \vec{\omega}_N^{[1]} \right| \tag{45}$$

In view of Eq. (37), subtracting x_1 times column $j+1$ from column j in this determinant for $j = 1, 2, \dots, N-1$ allows us to write

$$\tau^{[1]} = \left| \vec{\omega}_1 \, \vec{\omega}_2 \, \cdots \, \vec{\omega}_{N-1} \, \vec{\omega}_N^{[1]} \right| \tag{46}$$

Multiplying column N by x_1 and adding column $N-1$ to the result, we obtain

$$x_1 \, \tau^{[1]} = \left| \vec{\omega}_1 \, \vec{\omega}_2 \, \cdots \, \vec{\omega}_{N-1} \, \vec{\omega}_{N-1}^{[1]} \right| \tag{47}$$

Similarly, multiplying column N in Eq. (47) by x_1 and subtracting column $N-2$ yields

$$x_1^2 \, \tau^{[1]} = \left| \vec{\omega}_1 \, \vec{\omega}_2 \, \cdots \, \vec{\omega}_{N-1} \, \vec{\omega}_{N-2}^{[1]} \right| \tag{48}$$

Iterating this procedure by multiplying column N by x_1 and subtracting column $N-j$, we obtain Eq. (44).

Casoratian identity 2. The second identity that we need is

$$\prod_{1\le r<s\le n}(x_r - x_s)\tau^{[1,\dots,n]} = \\ \left|\vec{\omega}_1 \ \dots \ \vec{\omega}_{N-n}\ \vec{\omega}^{[n]}_{N-n+1}\ \vec{\omega}^{[n-1]}_{N-n+1}\ \cdots\ \vec{\omega}^{[1]}_{N-n+1}\right| \tag{49}$$

which is derived as follows. From Eq. (47), it follows that

$$x_1\ \tau^{[1,2]} = \left|\vec{\omega}^{[2]}_1\ \vec{\omega}^{[2]}_2\ \cdots \vec{\omega}^{[2]}_{N-1}\ \vec{\omega}^{[1,2]}_{N-1}\right| \tag{50}$$

which we can rewrite by subtracting x_2 times column $j+1$ from column j for $j = 1, 2, \dots, N-2$ as

$$x_1\ \tau^{[1,2]} = \left|\vec{\omega}_1\ \vec{\omega}_2\ \cdots\ \vec{\omega}_{N-2}\ \vec{\omega}^{[2]}_{N-1}\ \vec{\omega}^{[1,2]}_{N-1}\right| \tag{51}$$

Multiplying column N by $(x_1 - x_2)$ and applying Eq. (38), we see that

$$(x_1 - x_2)x_1\ \tau^{[1,2]} = x_1\left|\vec{\omega}_1\ \vec{\omega}_2\ \cdots\ \vec{\omega}_{N-2}\ \vec{\omega}^{[2]}_{N-1}\ \vec{\omega}^{[1]}_{N-1}\right| \\ - x_2\left|\vec{\omega}_1\ \vec{\omega}_2\ \cdots\ \vec{\omega}_{N-2}\ \vec{\omega}^{[2]}_{N-1}\ \vec{\omega}^{[2]}_{N-1}\right| \tag{52}$$

Since the last two columns of the latter determinant are identical, we obtain

$$(x_1 - x_2)\ \tau^{[1,2]} = \left|\vec{\omega}_1\ \dots\ \vec{\omega}_{N-2}\ \vec{\omega}^{[2]}_{N-1},\ \vec{\omega}^{[1]}_{N-1}\right| \tag{53}$$

which establishes Eq. (49) for $n = 2$. Now suppose inductively that

$$\prod_{1\le r<s\le n}(x_r - x_s)\ \tau^{[1,\dots,n]} = \\ \left|\vec{\omega}_1\ \dots\ \vec{\omega}_{N-n}\ \vec{\omega}^{[n]}_{N-n+1}\ \vec{\omega}^{[n-1]}_{N-n+1}\ \cdots\ \vec{\omega}^{[1]}_{N-n+1}\right| \tag{54}$$

then analogously to Eq. (47), we have

$$\prod_{i=1}^{n} x_i \prod_{1\le r<s\le n}(x_r - x_s)\ \tau^{[1,\dots,n]} = \\ \prod_{i=1}^{n}\left|\vec{\omega}_1\ \dots\ \vec{\omega}_{N-n}\ \vec{\omega}^{[n]}_{N-n+1}\ \vec{\omega}^{[n-1]}_{N-n+1}\ \cdots\ \vec{\omega}^{[1]}_{N-n+1}\right| = \\ \left|\vec{\omega}_1\ \dots\ \vec{\omega}_{N-n}\ \vec{\omega}^{[n]}_{N-n}\ \vec{\omega}^{[n-1]}_{N-n}\ \cdots\ \vec{\omega}^{[1]}_{N-n}\right| \tag{55}$$

It follows that

$$\prod_{i=1}^{n} x_i \prod_{1\le r<s\le n} (x_r - x_s)\ \tau^{[1,\dots,n,n+1]} =$$
$$\left|\vec{\omega}_1^{[n+1]} \ \dots\ \vec{\omega}_{N-n}^{[n+1]}\ \vec{\omega}_{N-n}^{[n,n+1]}\ \vec{\omega}_{N-n}^{[n-1,n+1]}\ \cdots\ \vec{\omega}_{N-n}^{[1,n+1]}\right| =$$
$$\left|\vec{\omega}_1\ \dots\ \vec{\omega}_{N-n-1}\ \vec{\omega}_{N-n}^{[n+1]}\ \vec{\omega}_{N-n}^{[n,n+1]}\ \vec{\omega}_{N-n}^{[n-1,n+1]}\ \cdots\ \vec{\omega}_{N-n}^{[1,n+1]}\right| \tag{56}$$

Using Eq. (38) repeatedly gives

$$\prod_{1\le i\le n} (x_i - x_{n+1})\ \times$$
$$\left|\vec{\omega}_1\ \dots\ \vec{\omega}_{N-n-1}\ \vec{\omega}_{N-n}^{[n+1]}\ \vec{\omega}_{N-n}^{[n,n+1]}\ \vec{\omega}_{N-n}^{[n-1,n+1]}\ \cdots\ \vec{\omega}_{N-n}^{[1,n+1]}\right| =$$
$$\prod_{i=1}^{n} x_i \left|\vec{\omega}_1\ \dots\ \vec{\omega}_{N-n-1}\ \vec{\omega}_{N-n}^{[n+1]}\ \vec{\omega}_{N-n}^{[n]}\ \vec{\omega}_{N-n}^{[n-1]}\ \cdots\ \vec{\omega}_{N-n}^{[1]}\right| \tag{57}$$

Combining this with Eq. (56) shows that

$$\prod_{1\le r<s\le n+1} (x_r - x_s)\ \tau^{[1,\dots,n+1]} =$$
$$\left|\vec{\omega}_1\ \dots\ \vec{\omega}_{N-n-1}\ \vec{\omega}_{N-n}^{[n]}\ \vec{\omega}_{N-n}^{[n-1]}\ \cdots\ \vec{\omega}_{N-n}^{[1]}\right| \tag{58}$$

thereby completing the proof of Eq. (49). We are finally in a position to complete the proof that Slavnov's determinant expressions are discrete KP τ-functions.

Bilinear identities from Laplace expansions. Following Ref. 15, we consider the $2N\times 2N$ determinant, which is identically zero,

$$\det\begin{pmatrix} \vec{\omega}_1 \cdots \vec{\omega}_{N-1}\ \vec{\omega}_{N-n+2}^{[1]}\ 0_1 \cdots 0_{N-n+1}\ \vec{\omega}_{N-n+2}^{[n]} \cdots \vec{\omega}_{N-n+2}^{[2]} \\ 0_1 \cdots 0_{N-1}\ \vec{\omega}_{N-n+2}^{[1]}\ \vec{\omega}_1 \cdots \vec{\omega}_{N-n+1}\ \vec{\omega}_{N-n+2}^{[n]} \cdots \vec{\omega}_{N-n+2}^{[2]} \end{pmatrix} = 0 \tag{59}$$

where we have used subscripts to label the zero elements with the positions of the columns that they are in for notational clarity. Performing a Laplace expansion of the left hand side of Eq. (59) in $N\times N$ minors along the top $N\times N$ block, we obtain

$$\sum_{\nu=1}^{n}(-)^{\nu-1}\left|\vec{\omega}_1\cdots\vec{\omega}_{N-1}\vec{\omega}^{[\nu]}_{N-n+2}\right|\times$$
$$\left|\vec{\omega}_1\cdots\vec{\omega}_{N-n+1}\vec{\omega}^{[n]}_{N-n+2}\cdots\vec{\omega}^{[\nu+1]}_{N-n+2}\vec{\omega}^{[\nu-1]}_{N-n+2}\cdots\vec{\omega}^{[1]}_{N-n+2}\right|=0 \quad (60)$$

Using Eqs. (44–49), Eq. (60) becomes

$$\sum_{\nu=1}^{n}(-)^{\nu-1}x_\nu^{n-2}\,\tau^{[\nu]}\prod_{\substack{1\leq r<s\leq n\\ r,s\neq\nu}}(x_r-x_s)\tau^{[1,\ldots\hat{\nu}\ldots,n]}=0 \quad (61)$$

which we recognise as the cofactor expansion of the determinant in Eq. (31) using the last column. Hence we conclude that Slavnov's determinant expression for the XXZ Bethe scalar product is a τ-function of discrete KP.

6. Remarks

Shifted τ-functions are not Bethe scalar products. A Bethe scalar product that involves m_i rapidities x_i, for $i = 1, 2, \cdots, i_{max}$, is a Casoratian determinant of a matrix of rank $r = \sum_{i=1}^{i_{max}} m_i$. Let us denote the corresponding τ-function by $\tau = \tau\{x_1^{(m_1)}, \cdots, x_i^{(m_i)}, \cdots, x_N^{(m_N)}\}$. Now let us consider a time evolution of the latter, for example $\tau_{i+1} = \tau\{x_1^{(m_1)}, \cdots, x_i^{(m_i+1)}, \cdots, x_N^{(m_N)}\}$. Time evolution has increased the multiplicities by 1, but kept the rank of the corresponding Casoratian determinant the same, thus we cannot interpret τ_{i+1} as a Bethe scalar product and it remains unclear to us how to interpret the discrete time evolution of a Bethe scalar product in the language of the XXZ spin chain.

Fermionization remains valid. Continuous KP τ-functions can be written as expectation values of charged free fermion operators.[21] This remains the case for discrete KP τ-functions and was the starting point of the results of Refs. 14,23. In Ref. 13, the fermion expectation value version of Slavnov's determinant expression was obtained based on an earlier result.[26] It is straightforward to show that this result remains the same as the continuous KP time variables are restricted to be power sums of a finite and smaller number of continuous Miwa variables.

Relation to the work of Krichever *et al.* As mentioned earlier, our result is close in spirit to that of Krichever *et al.*[7,8] and works that followed including Refs. 24,25. The starting point of Ref. 7 is that the Bethe eigenvalues satisfy a bilinear identity that has the same structure as Hirota's bilinear difference equation and hence can be identified with τ-functions of a discrete hierarchy. From this, a large number of interesting results follow, including an identification of the fusion rules of the transfer matrices of the quantum spin chain with Hirota's difference equations, that each step in the nested Bethe Ansatz approach to the spin chain is identified with a classical Bäcklund transformation, and most interestingly that the eigenvalues of Baxter's Q operator are classical (suitably normalized) Baker-Akhiezer functions. On the other hand, our result is that it is the Bethe scalar product of a Bethe eigenstate rather than the corresponding Bethe eigenvalue that is identified with a discrete KP τ-function, and we are far from obtaining further results that are analogous to those of Ref. 7. We hope that our identification is compatible with and complements that of Ref. 7.

Relation to the work of Sato and Sato. Eq. (61) also follows from Theorem **3** of Sato and Sato.[27] We didn't know this when we obtained our proof, and the existence of more than one proof can only shed more light on the result obtained.

Acknowledgements

OF wishes to thank M Wheeler and M Zuparic for collaboration on Refs. 13, 26 and for reading the manuscript, T Shiota and K Takasaki for discussions and the anonymous referee for bringing Ref. 27 to our attention. XXZ and KP are narrow examples of the broad subjects of quantum and classical integrability that Professor T Miwa has made major contributions to. We dedicate this note to him on his 60th birthday. This work was supported by the Australian Research Council and a Pratt Foundation Scholarship.

References

1. T T Wu, B M McCoy C A Tracy and E Barouch, Phys Rev **B13** (1976) 316–374
2. J H H Perk, Phys Lett **A79** (1980) 3–5.
3. H Au-Yang and J H H Perk, Proc of Symposia in Pure Mathematics, *Theta Functions, Bowdoin 1987*, L Ehrenpreis and R C Gunning, Editors, Part I, 287–294, American Math Society, and references therein.
4. A R Its, A G Izergin, V E Korepin, and N A Slavnov, Phys Rev Lett **70** (1993) 1704–1706; *Erratum* 2357–2357
5. V E Korepin, N M Bogoliubov and A G Izergin, *Quantum inverse scattering method and correlation functions*, Cambridge University Press, (1993)

6. T Miwa, Lecture Notes in Physics **242** (1985) 96–141, Springer-Verlag
7. I Krichever, O Lipan, P Wiegmann and A Zabrodin, Comm Math Phys **188** (1997) 267–304
8. A V Zabrodin, *Bethe Ansatz and classical Hirota equations*, `hep-th/9607162`
9. R Hirota, J Phys Soc Jpn **50** (1981) 3785–3791
10. A Kuniba, R Sakamoto and Y Yamada, Nucl Phys **B786** (2007) 207–266, and references therein.
11. A Date, M Jimbo, M Kashiwara and T Miwa, in *Proceedings of RIMS Symposium on Non-Linear Integrable Systems-Classical Theory and Quantum Theory*, M Jimbo and T Miwa, Editors, World Scientific, 1983
12. N Kawamoto, Y Namikawa, A Tsuchiya and Y Yamada, Commun Math Phys **116** (1988) 247–308
13. O Foda, M Wheeler and M Zuparic, Nucl. Phys. **B820** [FS] (2009) 649–663
14. E Date, M Jimbo and T Miwa, J Phys Soc of Japan **51** (1982) 4125–4131
15. Y Ohta, R Hirota, S Tsujimoto and T Inami, J of the Phys Soc of Japan, vol **62** (1993) 1872–1886
16. I G Macdonald, *Symmetric Functions and Hall polynomials*, Oxford University Press, 2nd Edition, 1995
17. R J Baxter, *Exactly Solved Models in Statistical Mechanics*, 2nd Edition, Dover Publications, 2008
18. M Jimbo and T Miwa, *Algebraic Analysis of Solvable Lattice Models*, CBMS Regional Conference Studies in Mathematics, American Math Society, 1994
19. N Kitanine, K K Kozlowski, J M Maillet, N A Slavnov and V Terras, J Stat Mech (2009) P04003, and references therein.
20. N A Slavnov, Theor Math Phys **79** (1989) 502–508
21. T Miwa, M Jimbo and E Date, *Solitons*, Cambridge University Press, 2000
22. B Konopelchenko and L M Alonso, Phys Lett **A 258** (1999) 272–278
23. T Miwa, Proc Japan Acad. **58A** (1982) 9–12
24. A Zabrodin, Int J of Mod Phys **B 11** (1997) 3125–3158
25. O Lipan, P Wiegmann and A Zabrodin, Mod Phys Lett **A12** (1997) 1369-1378, and Comm Math Phys **193** (1998) 373-396
26. O Foda, M Wheeler and M Zuparic, J Stat Mech (2009) P03017
27. M Sato and Y Sato, in *Nonlinear Partial Differential Equations in Applied Science*, Lecture Notes in Numerical and Applied Analysis **5** (1982) 259–271.

QUANTUM SPIN CHAINS AT FINITE TEMPERATURES

FRANK GÖHMANN

Fachbereich C – Physik, Bergische Universität Wuppertal
42097 Wuppertal, Germany

JUNJI SUZUKI

Department of Physics, Faculty of Science, Shizuoka University
Ohya 836, Suruga, Shizuoka, Japan

This is a pedagogical review on recent progress in the exact evaluation of physical quantities in interacting quantum systems at finite temperatures. Quantum spin chains are discussed in detail as typical examples.

Keywords: Quantum Transfer Matrix; correlation functions at finite temperatures.

1. Introduction

The evaluation of the thermal average of physical quantities is one of the main aims in statistical mechanics. The density matrix of a system is the most fundamental quantity to achieve this aim. Its diagonalization, however, becomes exponentially difficult with growing system size L. One inevitably has to give up this procedure in the thermodynamic limit. An alternative approach for quantum systems is to diagonalize the Hamiltonian, and to sum up the contributions from each eigenstate. This means to divide the problem into two parts: (1) diagonalize, and (2) sum up. Again, both procedures become exponentially difficult with the increase of L.

In this article we re-consider this problem for integrable quantum spin chains. We will show how the integrability helps bypassing the difficulties and yields exact estimates. The first problem, the diagonalization of the Hamiltonian, can, in principle, be solved by the celebrated Bethe ansatz. The second step, however, remains as a cliff wall. A first breakthrough, the string hypothesis approach, was achieved in the early 70's.[1,2] In this approach one introduces so-called root density functions of strings and holes of various lengths for the diagonalization. The free energy becomes a functional of these density functions, which is claimed to be exact near its minimum. Therefore the variational estimate (w.r.t. density functions), with a fixed energy of the system, yields the exact free energy. The string hypothesis formulation can be regarded as a micro-canonical approach. It is supported by many consistency tests. We conclude that within the string hypothesis approach the diagonalization is achieved, but the summation is cleverly avoided.

In order to evaluate thermal expectation values of operators, it is better to deal with the canonical ensemble. We therefore consider an alternative approach based on the Quantum Transfer Matrix (QTM).[4,5] It utilizes an exact mapping between a 1D quantum system at finite temperatures and a 2D classical system. At first sight the formulation may look tautological and may seem to be suffering from the need of "summation". Yet, the main claim of the QTM formulation is that this is not the case. As in the string hypothesis approach the "summation" can be avoided. Moreover, the QTM makes the evaluation of many quantities of physical relevance straightforward.

This article is organized as follows. In Sec. 2, we present a review on the QTM formulation. The results for the bulk quantities will be summarized in Sec. 2.3. In the rest of Sec. 2, we supplement arguments to justify the formula in Sec. 2.3. The non-linear integral equation (NLIE) will be explained in Sec. 3 together with an example for the explicit evaluation of bulk quantities. The evaluation of the reduced density matrix elements (DME) will be discussed in Sec. 4.

2. The QTM formulation

2.1. *The problem*

Let $\mathcal{H}$ be the Hamiltonian of a 1D quantum system of size L and V its space of states. Our goal is to calculate the thermal expectation value of any physical quantity $\mathcal{O}$ at temperature $T(=1/\beta)$[a] in the limit $L \to \infty$,

$$\langle \mathcal{O} \rangle = \lim_{L\to\infty} \frac{\mathrm{tr}_V\, \mathcal{O}\, \mathrm{e}^{-\beta\mathcal{H}}}{Z_{1\mathrm{D}}(\beta)} \qquad Z_{1\mathrm{D}}(\beta) = \mathrm{tr}_V\, \mathrm{e}^{-\beta\mathcal{H}} = \sum_j \mathrm{e}^{-\beta E_j}. \tag{1}$$

Here E_j stands for an eigenvalue of $\mathcal{H}$.

The definition requires both diagonalization and summation. Below we shall show how we can avoid the latter within the framework of QTM.

[a]The Boltzmann constant k_B is set to be unity in this report.

2.2. *The Baxter-Lüscher formula*

To be concrete, we specify a Hamiltonian. As a prototypical integrable lattice system we choose the 1D spin $\frac{1}{2}$ XXZ model,

$$\mathcal{H} = J\sum_{j=1}^{L}\Big(\sigma_j^x\sigma_{j+1}^x + \sigma_j^y\sigma_{j+1}^y + \Delta(\sigma_j^z\sigma_{j+1}^z + 1)\Big) = \sum_{j=1}^{L}\hat{h}_{j,j+1} \qquad (2)$$

where the σ^a $(a = x, y, z)$ are the Pauli matrices. The periodic boundary conditions (PBCs) imply $\sigma_{L+1}^a = \sigma_1^a$. The anisotropy is parameterized as $\Delta = \cos\gamma$. The Hamiltonian acts on "the physical space" $V_{\text{phys}} := \bigotimes_{j=1}^{L} V_j$ where V_j denotes the jth copy of a two-dimensional vector space $c_1\mathbf{e}_+ + c_2\mathbf{e}_-$. The trace in (1) must be performed over V_{phys}. By definition the "Hamiltonian density" $\hat{h}_{j,j+1}$ is the jth summand in the first sum in (2). It acts non-trivially only on $V_j \otimes V_{j+1}$.

The above Hamiltonian is integrable in the following sense. Let $R(u,v)$ be the $U_q(\widehat{\mathfrak{sl}_2})$ R matrix,[6]

$$R(u,v) = \begin{pmatrix} [1+\frac{u-v}{2}] & & & \\ & [\frac{u-v}{2}] & q^{\frac{-u+v}{2}} & \\ & q^{\frac{u-v}{2}} & [\frac{u-v}{2}] & \\ & & & [1+\frac{u-v}{2}] \end{pmatrix} \qquad [u] := \frac{q^u - q^{-u}}{q - q^{-1}}$$

depending on the spectral parameters (or rapidities) $u, v \in \mathbb{C}$. We define E_β^α s.t. $(E_\beta^\alpha)_{i,j} = \delta_{\alpha,i}\delta_{\beta,j}$. Then the matrix elements $R_{\beta\delta}^{\alpha\gamma}$ can be read off from

$$R(u,v) = \sum_{\alpha,\beta,\gamma,\delta=1,2} R_{\beta\delta}^{\alpha\gamma}(u,v) E_\alpha^\beta \otimes E_\gamma^\delta.$$

The index 1(2) refers to $\mathbf{e}_+(\mathbf{e}_-)$. See fig. 1 for a graphic representation. We

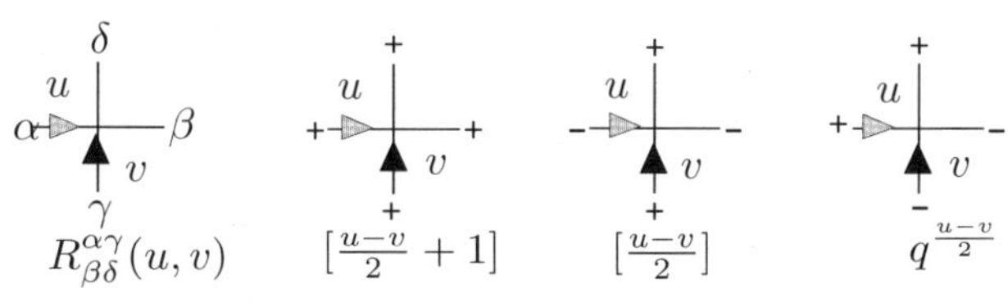

Fig. 1. A graphic representation for $R_{\beta\delta}^{\alpha\gamma}(u,v)$ and some examples

put arrows, to distinguish the R matrix from other R matrices appearing below. The reader should not confuse them with physical variables.

By $R_{j,j+1}(u,v)$ we mean the R matrix acting non-trivially only on the tensor product $V_j(u) \otimes V_{j+1}(v)$ of $U_q(\widehat{\mathfrak{sl}_2})$ modules. We also introduce the intertwiner $R^{\vee}_{j,j+1}(u,v) = P_{j,j+1}R_{j,j+1}(u,v)$, where $P : V_j(u) \otimes V_{j+1}(v) \to V_{j+1}(v) \otimes V_j(u)$. Then, with $q = \mathrm{e}^{i\gamma}$, we have the expansion

$$R^{\vee}_{j,j+1}(u,0) = 1 + \frac{\gamma}{4J\sin\gamma} u\,(\hat{h}_{j,j+1} + \hat{h}'_{j,j+1}) + O(u^2),$$

where $\hat{h}'_{j,j+1} := iJ\sin\gamma(\sigma^z_j - \sigma^z_{j+1})$. We introduce the row-to-row (RTR) transfer matrix $T_{\mathrm{RTR}}(u) \in \mathrm{End}(V_{\mathrm{phys}})$,

$$T_{\mathrm{RTR}}(u) = \mathrm{tr}_a R_{a,L}(u,0)R_{a,L-1}(u,0)\cdots R_{a,1}(u,0). \tag{3}$$

With the lattice translation operator e^{iP}, shifting the state by one site, we obtain the Baxter-Lüscher formula[3]

$$T_{\mathrm{RTR}}(u) = \mathrm{e}^{iP}\big(1 + \frac{\gamma u}{4J\sin\gamma}\mathcal{H} + O(u^2)\big). \tag{4}$$

Note that the $\hat{h}'_{j,j+1}$ terms cancel due to the PBCs. The huge symmetry $U_q(\widehat{\mathfrak{sl}_2})$ is at the bottom of the integrability of the Hamiltonian.

2.3. *A summary of results for bulk quantities*

We first present the formula for the free energy per site in the QTM formalism. A supplemental discussion will be given in subsequent sections.

We introduce the transposed R matrix[7] $R^t_{j,k}(u,v)$ by $(R^t)^{\alpha\gamma}_{\beta\delta}(u,v) = R^{\delta\alpha}_{\gamma\beta}(v,u)$. See fig. 2. The QTM does not act on V_{phys} but on a fictitious

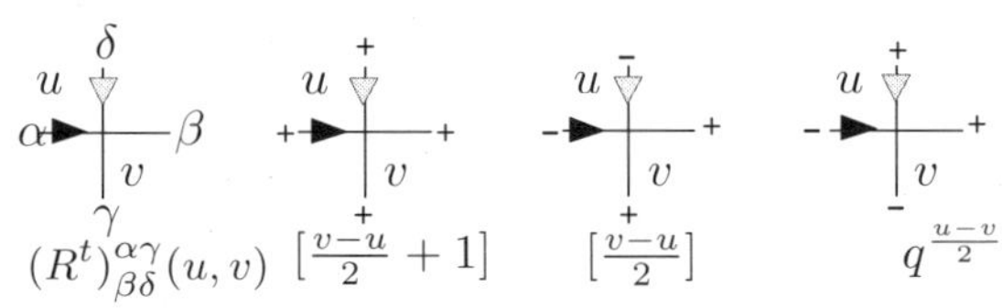

Fig. 2. A graphic representation for $(R^t)^{\alpha\gamma}_{\beta\delta}(u,v)$ and some examples

space $V_{\mathrm{Trotter}} = V_1(u) \otimes V_2(-u) \otimes \cdots V_{N-1}(u) \otimes V_N(-u)$. The fictitious system size N is often referred to as the Trotter number. The parameter u is fixed to be

$$u = \frac{4\beta J\sin\gamma}{\gamma N} = \frac{4J\sin\gamma}{\gamma NT}.$$

In its most sophisticated version, the QTM is explicitly defined by,[7]

$$T_{QTM}(x,u) = \mathrm{tr}_a R_{aN}(ix,-u)R^t_{a,N-1}(ix,u)\cdots R_{a2}(ix,-u)R^t_{a1}(ix,u). \tag{5}$$

The new parameter x will later play the role of a spectral parameter. The factor i is introduced for convenience.

We are now in a position to write down the formula for the free energy per site in the thermodynamic limit, $f = -\lim_{L\to\infty} \frac{T}{L} \ln Z_{1\mathrm{D}}(\beta)$.

Theorem 2.1. *Let Λ_0 be the largest eigenvalue of $T_{QTM}(0,u)$. Then the free energy per site is solely given by Λ_0,*

$$f = -\lim_{N\to\infty} T \ln \Lambda_0. \tag{6}$$

The limit $N \to \infty$ is referred to as the Trotter limit. As was announced earlier, eq. (6) expresses f *without recourse to any summation.* We also note that $\ln \Lambda_0$ itself is already intensive, which may reflect the size dependent interaction of the system.

The quantitative analysis of (6) is most efficiently performed by means of the NLIE. Having in mind the examples, from now on we are considering only $J = \frac{1}{4}, \gamma \to 0$ and $u = -\frac{\beta}{N}$, consequently. Let $\mathfrak{a}$ be the unique solution to the NLIE[b]

$$\ln \mathfrak{a}(x) = \beta\epsilon_0(x+i) - \int_{\mathcal{C}} \frac{2}{(x-y)^2+4} \ln \mathfrak{A}(y) \frac{dy}{\pi} \tag{7}$$

$$\epsilon_0(x) = h + \frac{2}{(x-i)(x+i)} \qquad \mathfrak{A} := 1 + \mathfrak{a}.$$

Here the contour $\mathcal{C}$ is a closed narrow loop which encircles all "Bethe roots". We added a Zeeman term $\frac{h}{2}\sum_j \sigma_j^z$ to the Hamiltonian so that $\mathrm{diag}(\exp(-\frac{\beta h}{2}), \exp(\frac{\beta h}{2}))$ is inserted in the trace in (5). Then we have the following

Theorem 2.2. *The free energy per site can be evaluated in terms of the solution to the NLIE.*

$$\beta f = \frac{\beta}{2}(1+h) - \int_{\mathcal{C}} \frac{1}{x(x+2i)} \ln \mathfrak{A}(x) \frac{dx}{\pi}. \tag{8}$$

Note that the NLIE (7) and the expression for f in (8) are independent of N. The extension to arbitrary J, γ is straightforward.

Below we shall comment on the derivation of the formula. By presenting supplementary arguments, we wish to convince the reader that the above formalism, seemingly complicated, is actually necessary and efficient for many purposes. Hereafter we set again $h = 0$ for simplicity.

---

[b]To be precise, there are, in general, several equivalent versions of NLIEs. We present one of these below.

2.4. *The 1D quantum partition function as a 2D classical partition function*

We define a rotated R matrix $\widetilde{R}(u,v)$ by $\widetilde{R}^{\alpha\gamma}_{\beta\delta}(u,v) = R^{\gamma\beta}_{\delta\alpha}(v,u)$ (fig. 3). Then we introduce a rotated transfer matrix $\widetilde{T}_{\mathrm{RTR}}(u) \in \mathrm{End}(V_{\mathrm{phys}})$ by

$$\widetilde{T}_{\mathrm{RTR}}(u) = \mathrm{tr}_a \widetilde{R}_{a,L}(-u,0)\widetilde{R}_{a,L-1}(-u,0)\cdots\widetilde{R}_{a,1}(-u,0).$$

Analogous to (4) $\widetilde{T}_{\mathrm{RTR}}(u) = \mathrm{e}^{-iP}\big(1 + u\mathcal{H} + O(u^2)\big)$. We thus obtain an

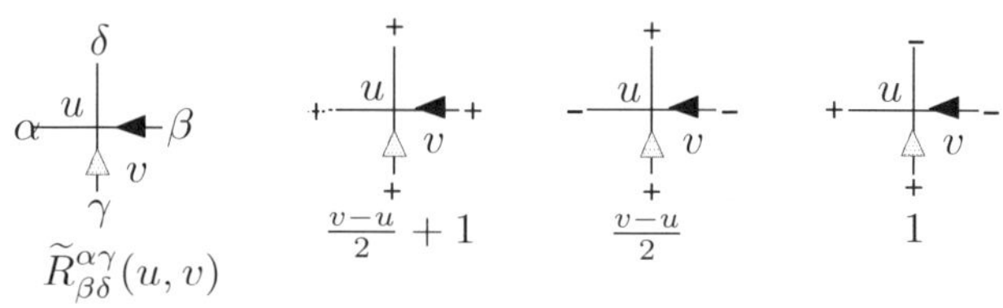

Fig. 3. A graphic representation for $\widetilde{R}^{\alpha\gamma}_{\beta\delta}(u,v)$ and some examples

important identity,

$$Z_{1\mathrm{D}}(\beta) = \mathrm{tr}_{V_{\mathrm{phys}}}\mathrm{e}^{-\beta\mathcal{H}} = \lim_{N\to\infty} \mathrm{tr}_{V_{\mathrm{phys}}}\big(T_{\mathrm{double}}(u)\big)^{\frac{N}{2}}\big|_{u\to-\frac{\beta}{N}} \tag{9}$$

where $T_{\mathrm{double}}(u) := T_{\mathrm{RTR}}(u)\widetilde{T}_{\mathrm{RTR}}(u)$. The rhs of (9) can be interpreted as a partition function of a 2D classical system defined on $N \times L$ sites (fig. 4),

$$Z_{1\mathrm{D}}(\beta) = \lim_{N\to\infty} Z_{2\mathrm{Dclassical}}(N, L, u = -\frac{\beta}{N}).$$

This equivalence lies in the heart of the QTM formalism. The expression (9) itself, however, is of no direct use for the actual evaluation of physical quantities for the following reason. Let the eigenvalue spectrum of $T_{\mathrm{RTR}}(u)$ be $\lambda_0(x) > \lambda_1(x) \geq \lambda_2(x) \geq \cdots$. We introduced $x = i^{-1}(u+1)$ for technical reasons. It is easy to see that $\widetilde{T}_{\mathrm{RTR}}(u)$ has the same spectrum. Thus,

$$\mathrm{tr}_{V_{\mathrm{phys}}}\big(T_{\mathrm{double}}(u)\big)^{\frac{N}{2}} = \big(\lambda_0(x)\big)^N\Big(1 + \big(\frac{\lambda_1(x)}{\lambda_0(x)}\big)^N + \big(\frac{\lambda_2(x)}{\lambda_0(x)}\big)^N + \cdots\Big). \tag{10}$$

The eigenvalue $\lambda_j(x)$ is characterized by its zeros $\pm\theta_a$ $(a = 1, 2, \dots)$ on the real axis (holes). We know numerically that for low excitations, $\theta_a \sim \ln L$ and also that $\lambda_j(x)$ is analytic and nonzero in the strip $|\Im\mathrm{m} x| \leq 1$ except at $\pm\theta_a$. Let us introduce an analytic and nonzero function near the real axis, $\lambda_j^\sharp$, by $\lambda_j(x) = \prod_a \mathrm{th}\,\frac{\pi}{4}(x-\theta_a)\,\mathrm{th}\,\frac{\pi}{4}(x+\theta_a)\lambda_j^\sharp(x)$. It approximately satisfies the inversion relation for $L \gg 1$,

$$\lambda_j^\sharp(x-i)\lambda_j^\sharp(x+i) = \phi(x), \tag{11}$$

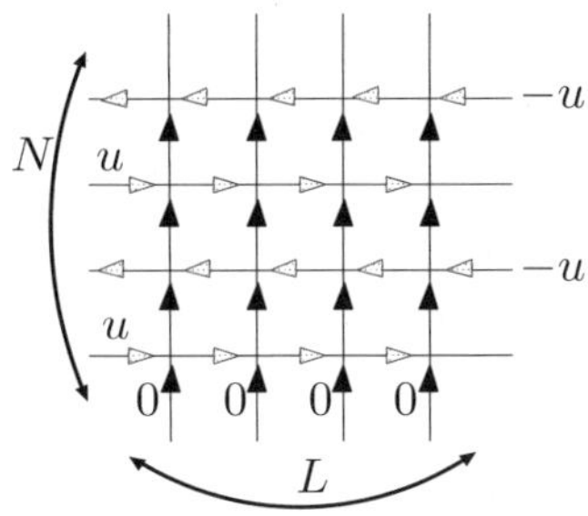

Fig. 4. Fictitious two-dimensional system

where $\phi(x)$ is a known function common to any j. Thus, we simply have

$$\left|\frac{\lambda_j(x)}{\lambda_0(x)}\right| = \left|\prod_a \operatorname{th}\frac{\pi}{4}(x-\theta_a)\operatorname{th}\frac{\pi}{4}(x+\theta_a)\right|.$$

For very low excitations, we take a single pair of holes, substitute $\theta_a \sim \frac{2}{\pi}\ln\frac{2\pi L}{\Delta_j}$ and take the large L limit. Then we arrive at the estimate ($u \sim 0$)

$$\left|\frac{\lambda_j(x)}{\lambda_0(x)}\right| \sim \mathrm{e}^{-\frac{|u|\Delta_j}{L}} \quad \text{thus} \quad \left|\frac{\lambda_j(x)}{\lambda_0(x)}\right|^N \sim \mathrm{e}^{-\Delta_j\frac{N}{L}|u|}. \tag{12}$$

For a usual 2D classical system we can consider an infinitely long cylinder and take $\frac{N}{L} \gg 1$. We thus have to take into account only the first term on the rhs in (10). By contrast, the spectral parameter depends on the fictitious system size $u = -\frac{\beta}{N}$ in the present case. Therefore, as long as $T \neq 0$, we have

$$\left|\frac{\lambda_j(x)}{\lambda_0(x)}\right|^N \sim \mathrm{e}^{-\Delta_j\frac{\beta}{L}} = O(1) \quad \text{for } L \gg 1.$$

Fig. 5 presents numerical evidence for the above argument. The left figure shows the histogram of the distribution of $|\lambda_j/\lambda_0|$ for $q = 1, L = 10, u = -0.01$ in the sector with vanishing magnetization. One clearly sees that the maximum of the distribution lies near $|\lambda_j/\lambda_0| \sim 1$. The right figure magnifies the region near $|\lambda_j/\lambda_0| \sim 1$. The maximum is located around $|\lambda_j/\lambda_0| \sim 0.96$. We believe that, with increasing L, the peak moves towards $|\lambda_j/\lambda_0| \sim 1$. These findings are consistent with (12). Hence, we conclude that infinitely many terms of the sum in the rhs of (10) contribute non-trivially, and eq. (9) is of no practical use.

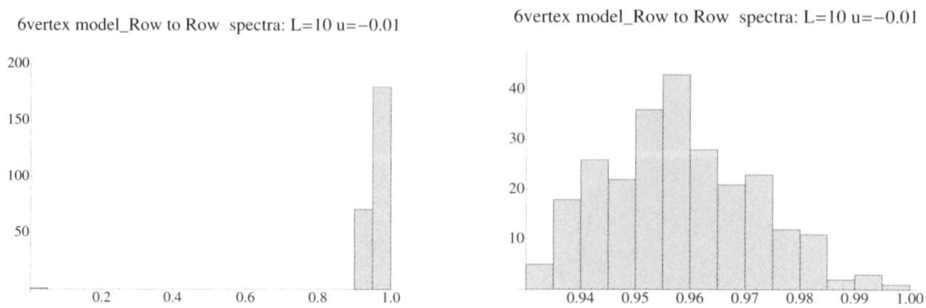

Fig. 5. The distribution of eigenvalues. The horizontal axis is the absolute value of the eigenvalues normalized by the largest one. The left figures ranges over [0,1] in the horizontal direction, and the right one is zoomed into the range [0.93,1].

2.5. *Commuting QTM*

A crucial observation was made in Ref. 4. We start from the same two-dimensional classical model in fig. 4. We consider, however, the transfer matrix propagating in horizontal direction, that is, $T'_{\text{QTM}}(u)$. Equivalently, one can rotate the system by 90°. Then we define a transfer matrix propagating in vertical direction, $T_{\text{QTM}}(u)$ (see fig. 2.5). The latter is more convenient for our formulation.

The partition function is then given by,

$$Z_{1\text{D}}(\beta) = \lim_{N\to\infty} \text{tr}_{V_{\text{Trotter}}} \left(T_{QTM}(u)\right)^L \Big|_{u=-\frac{\beta}{N}}. \tag{13}$$

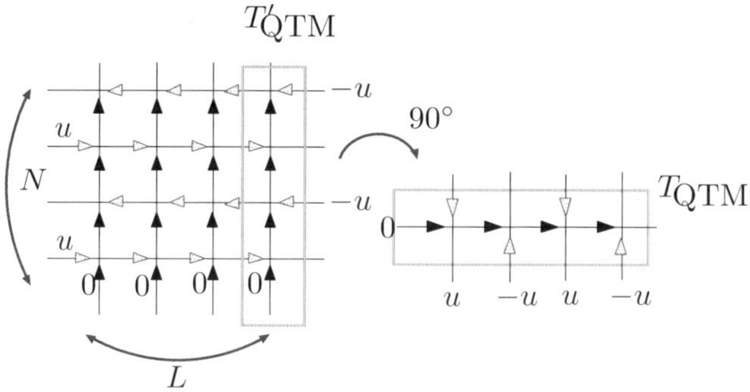

Fig. 6. The graphical definition of T_{QTM}.

Let the eigenvalue spectrum of $T_{\text{QTM}}(u)$ be $\Lambda_0(u) > \Lambda_1(u) \geq \Lambda_2(u) \geq$

$\cdots$. Then we have an expansion similar to (10)

$$\mathrm{tr}_{V_{\mathrm{Trotter}}}\big(T_{QTM}(u)\big)^L = \big(\Lambda_0(u)\big)^L\Big(1+\big(\frac{\Lambda_1(u)}{\Lambda_0(u)}\big)^L+\big(\frac{\Lambda_2(u)}{\Lambda_0(u)}\big)^L+\cdots\Big). \quad (14)$$

Our physical interest is in the free energy per site f in the thermodynamic limit $L \to \infty$.

$$f = -\frac{1}{\beta}\lim_{L\to\infty}\lim_{N\to\infty}\Big\{\ln\Lambda_0(u) + \frac{1}{L}\ln\Big(1+\Big(\frac{\Lambda_1(u)}{\Lambda_0(u)}\Big)^L+\Big(\frac{\Lambda_2(u)}{\Lambda_0(u)}\Big)^L+\cdots\Big)\Big\}\Big|_{u=-\frac{\beta}{N}}. \quad (15)$$

Proposition 2.1. *The two limits in (15) are exchangeable.*

We supplement an argument which claims that the second term in the second line in (15) is negligible for $L \to \infty$. The previous argument, using the inversion relation (11) can not be applied directly as the spectral parameter u is already fixed as $-\frac{\beta}{N}$ in the present problem.

We introduce a slight generalization, a commuting QTM $T_{QTM}(x,u)$, by assigning the parameter ix in "horizontal" direction.[10] The substitution $x = 0$ recovers the previous results. The precise definition is shown in (5). Hereafter we drop the u dependence as it is always $-\frac{\beta}{N}$. Let $\mathcal{T}_{\mathrm{QTM}}(x)$ be the corresponding monodromy matrix. Then it is easy to see that monodromy matrices are intertwined by the same R matrix as in the RTR case,

$$R(x,x')\mathcal{T}_{\mathrm{QTM}}(x)\otimes\mathcal{T}_{\mathrm{QTM}}(x') = \mathcal{T}_{\mathrm{QTM}}(x')\otimes\mathcal{T}_{\mathrm{QTM}}(x)R(x,x'). \quad (16)$$

This immediately proves the commutativity of $T_{\mathrm{QTM}}(x)$ with different x's.

The most important consequence of introducing x is that we have the inversion relation in this new "coordinate",

$$\Lambda_j^\sharp(x-i)\Lambda_j^\sharp(x+i) = \psi(x,u) \quad (17)$$

where we set again $\Lambda_j(x) = \prod_a \mathrm{th}\,\frac{\pi}{4}(x-\theta_a)\,\mathrm{th}\,\frac{\pi}{4}(x+\theta_a)\Lambda_j^\sharp(x)$. Note that $\Lambda_j^\sharp(x)$ also depends on the "old" spectral parameter u, which is set to be $-\frac{\beta}{N}$ on both sides. The known function ψ is again independent of j. The analysis of the Bethe ansatz equation associated to the QTM implies that $\theta_a \sim \frac{2}{\pi}\ln\frac{4\beta}{\Delta_j}$ for large β. Then, proceeding as before, we obtain,

$$\left|\frac{\Lambda_j(x)}{\Lambda_0(x)}\right| \sim \mathrm{e}^{-\frac{\Delta_j}{\beta}\mathrm{ch}\,\frac{\pi}{2}x} \qquad \text{thus} \qquad \left|\frac{\Lambda_j(x)}{\Lambda_0(x)}\right|^L \sim \mathrm{e}^{-\frac{\Delta_j L}{\beta}\mathrm{ch}\,\frac{\pi}{2}x}.$$

The diagonalization for fixed N clearly shows the gap between the eigenvalues, which is consistent with the above argument. Thus, at any finite temperature, the second term in (15) is negligible for $L \to \infty$. We then conclude that the formula (6) is valid.

Although we made use of the integrability of the model in the above argument, the validity of the formula is actually independent of it. See the proof in Ref. 4.

3. Diagonalization and NLIE

3.1. *Bethe roots*

Thanks to (16), one can apply the machinery of the quantum inverse scattering method, devised originally for the diagonalization of $T_{\rm RTR}$, to the diagonalization of $T_{\rm QTM}$. We skip the derivation and present only results relevant for our subsequent discussion[c]. We fix N for a while. Then the eigenvalue of $T_{\rm QTM}$ is given by

$$\Lambda^{(N)}(x) = a(x)\frac{Q(x-2i)}{Q(x)} + d(x)\frac{Q(x+2i)}{Q(x)} \tag{18}$$

$$a(x) := \phi_+(x+2i)\phi_-(x) \qquad\qquad d(x) := \phi_-(x-2i)\phi_+(x)$$

$$Q(x) := \prod_{j=1}^{m}(x-x_j) \qquad\qquad \phi_\pm(x) := \left(\frac{x \pm iu}{\pm 2i}\right)^{\frac{N}{2}}.$$

The different sets of Bethe roots $\{x_j\}$ correspond to the different eigenvalues. They satisfy the Bethe ansatz equation (BAE),

$$\frac{a(x_j)}{d(x_j)} = -\frac{Q(x_j+2i)}{Q(x_j-2i)} \qquad 1 \le j \le m. \tag{19}$$

For the largest eigenvalue the number of roots m equals $\frac{N}{2}$.

To evaluate f via (6) we need the largest $\Lambda^{(N)}$ for $N \to \infty$. This means we must deal with infinitely many roots in the limit, which resembles the situation encountered in the evaluation of the free energy in the thermodynamic limit of a classical 2D model by means of the RTR transfer matrix. Still, we would like to comment on the qualitative difference in the root distribution between such "standard" case and the problem under discussion.

Fig. 7 shows the distribution of the positive half of BAE roots for the largest eigenvalue of $T_{\rm RTR}$ (left) and $T_{\rm QTM}$ (right) for various system sizes. The distribution of RTR roots behaves smoothly for large system size. The limiting shape of the distribution (the root density function) is a smooth function satisfying a linear integral equation. For $T_{\rm QTM}$, on the other hand,

[c] A technical remark: the vacuum is conveniently chosen $(+,-,+,-,\cdots)$.

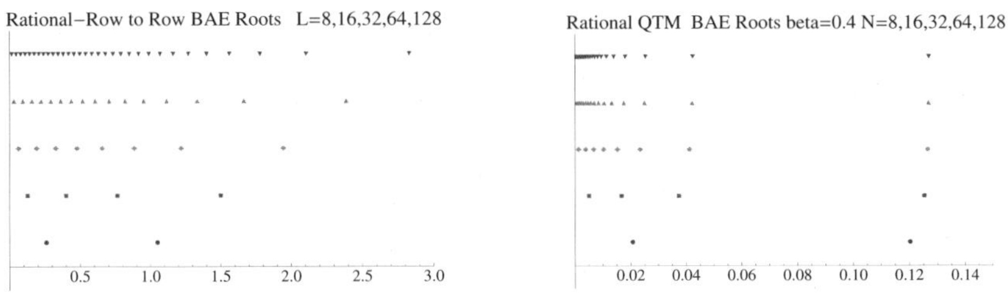

Fig. 7. The positive half of BAE roots for RTR (left) and for QTM (right) with system size, from 8 (bottom) to 128 (top).

a few large roots remain isolated at almost the same positions as N increases, while close to the origin more and more Bethe roots cluster.

Let us describe this in detail. Using the NLIE technique, we can derive an approximate BAE equation (see the discussion after (25)),

$$\frac{N}{2}\ln\Big(-\frac{\mathrm{th}\,\frac{\pi}{4}(x_j-\frac{\beta}{N}i)}{\mathrm{th}\,\frac{\pi}{4}(x_j+\frac{\beta}{N}i)}\Big)\sim(2I_j+1)\pi i. \tag{20}$$

The hole θ_a corresponds to the branch cut integer $I_{\max}=\frac{N}{4}-\frac{1}{2}$, and this implies $\theta_a\sim\frac{1}{\pi}\ln\beta$ for $\beta\gg1$, which was used in the last subsection.

Near the origin, we set $x_j=\frac{\beta}{N}\widehat{x}_j$, and obtain the approximate distribution of $\widehat{x}_j$ as an algebraic function,

$$\rho(\widehat{x})=\lim_{N\to\infty}\frac{I_{j+1}-I_j}{N(\widehat{x}_{j+1}-\widehat{x}_j)}\sim\frac{1}{2\pi(\widehat{x}^2+1)}.$$

This differs from the usual root density function which decays exponentially as $|x|\to\infty$. In the original variable, if we take the Trotter limit naively,

$$\rho(x)\sim\lim_{|u|\to0}\frac{|u|}{2\pi(x^2+u^2)}\sim\frac{1}{2}\delta(x).$$

Namely the distribution of the BAE roots for T_{QTM} is singular in the Trotter limit. We thus conclude that the usual root density method may not be applicable, and we have to devise a different tool.

Let us stress again that the cancellation of the order of N many terms in $\ln\Lambda^{(N)}$ is a unique property of the QTM. In the RTR case $O(L)$ many terms can survive, and we obtain intensive quantities (e.g., the free energy per site) only after dividing by L. On the other hand, $\ln\Lambda^{(N)}$ is already an intensive quantity, as remarked after Theorem 2.1. The cancellation is thus vital. The $O(e^N)$ terms, $a(x), d(x)$ must be canceled by the denominator $Q(x)$, resulting in $O(1)$ quantities. According to this point of view, (18) is (again) not practical, as the ratios of $O(e^N)$ terms are still present. Thus, we understand (18) still as a starting point, not as the goal.

One must intrinsically deal with a finite size system with coupling constant u depending on the Trotter size, and then take the Trotter limit. Such an attempt was executed first numerically[8] by extrapolation in N. The analytic low temperature expansion was performed[9] based on the Wiener-Hopf method. Below we shall present the most sophisticated approach which utilizes the commuting QTM in a most efficient manner.[7]

3.2. *Non-linear Integral Equation (NLIE)*

The introduction of the new spectral parameter x plays a fundamental role. Instead of dealing with the BAE roots directly, we make use of the analyticity of specially chosen auxiliary functions in the complex x plane.

There are various approaches. One of them is to introduce the fusion hierarchy of the QTM, which contains the original $T_{\rm QTM}$ as T_1. In place of the BAE one uses the functional relations among fusion transfer matrices,

$$T_m(x-i)T_m(x+i) = \psi_m(x) + T_{m-1}(x)T_{m+1}(x). \tag{21}$$

Here ψ_1 is nothing but ψ in the inversion relation (17), where the small term T_0T_2/ψ_1 was neglected. As $\{T_m\}$ constitutes a commuting family, the same relation holds among the eigenvalues. We thus use the same symbol T_m for the eigenvalue. After a change of variables, $y_m(x) = T_{m-1}(x)T_{m+1}(x)/\psi_m(x)$, one can transform the algebraic equations into integral equations under certain assumptions on the analyticity of y_m. The resultant equations coincide with the Thermodynamic Bethe Ansatz equations.[10,11] The string hypothesis is thus replaced by an assumption on the analyticity of y_m. The coupled set of equations may fix the values of $T_1(x)$. Then $\Lambda_0 = T_1(0)$ yields the free energy per site f. A technical problem in this approach is that we must deal with an infinite number of y_m functions, which requires a truncation of the equations in an approximate manner.

Below we shall discuss another approach originally devised in the context of the evaluation of finite size corrections.[12] We define the auxiliary function $\mathfrak{a}_N(x)$ by the ratio of the two terms in $\Lambda^{(N)}(x)$ (18),

$$\mathfrak{a}_N(x) = \frac{d(x)}{a(x)}\frac{Q(x+2i)}{Q(x-2i)} \qquad \mathfrak{A}_N(x) = 1 + \mathfrak{a}_N(x).$$

The suffix N is introduced to recall that we are fixing N finite here. The BAE (19) is equivalent to the condition

$$\mathfrak{a}_N(x_j) = -1 \qquad \text{or} \qquad \ln \mathfrak{a}_N(x_j) = (2I_j+1)\pi i. \tag{22}$$

We also note that $\lim_{|x|\to\infty} \mathfrak{a}_N(x) = 1$ by construction.

We then adopt the following assumptions for the analytic properties of $\mathfrak{A}_N(x)$ corresponding to the largest eigenvalue. They are supported by numerical calculations.

(1) There are $\frac{N}{2}$ simple zeros of $\mathfrak{A}_N(x)$ on the real axis. They coincide with the BAE roots. There are additional zeros, sufficiently far away from the real axis, so that $\mathcal{C}$ does not include them inside.
(2) The only pole of $\mathfrak{A}_N(x)$ in $\Im x \in [-1,1]$ is located at $x = iu$ and is of order $\frac{N}{2}$.

Once these assumptions are granted, one immediately derives the following NLIE,

$$\ln \mathfrak{a}_N(x) = \ln \frac{\phi_-(x+2i)\phi_+(x)}{\phi_+(x+2i)\phi_-(x)} - \int_{\mathcal{C}} \frac{2}{(x-y)^2+4} \ln \mathfrak{A}_N(y) \frac{dy}{\pi}. \tag{23}$$

The largest eigenvalue Λ can be similarly represented by

$$\ln\Lambda^{(N)}(x) = \ln\big(\phi_+(x+2i)\phi_-(x-2i)\big) + \int_{\mathcal{C}} \frac{\ln \mathfrak{A}_N(y)}{(x-y)(x-y-2i)} \frac{dy}{\pi}. \tag{24}$$

Note that only the driving term in (23) depends on N. We can thus take the Trotter limit easily, with $\mathfrak{a} := \lim_{N\to\infty} \mathfrak{a}_N$, and obtain the NLIE in (7) (for $h = 0$). To evaluate the free energy one has to first set $x = 0$, then take the Trotter limit. Or otherwise one meets a spurious divergence. Then we obtain the expression for the free energy in Theorem 2.2.

One still needs to make an effort to achieve a high numerical accuracy, especially at very low temperatures. The introduction of another pair of auxiliary functions solves this problem. We define $\bar{\mathfrak{a}}_N, \bar{\mathfrak{A}}_N$ by $\bar{\mathfrak{a}}_N(x) = (\mathfrak{a}_N(x))^{-1}$, $\bar{\mathfrak{A}}_N(x) = 1 + \bar{\mathfrak{a}}_N(x)$.

Numerically one finds that $|\mathfrak{a}_N| \lessgtr 1$ for $\Im x \gtrless 0$. Thus, we use $\mathfrak{a}_N, \mathfrak{A}_N$ in the upper half plane and $\bar{\mathfrak{a}}_N, \bar{\mathfrak{A}}_N$ in the lower half plane. It is straightforward to rewrite (23) in the coupled form,

$$\ln\mathfrak{a}_N(x) = D_+^{(N)}(x) + \int_{C_+} F(x-y) \ln\mathfrak{A}_N(y) \frac{dy}{2\pi} - \int_{C_-} F(x-y) \ln\bar{\mathfrak{A}}_N(y) \frac{dy}{2\pi} \tag{25}$$

$$\ln\bar{\mathfrak{a}}_N(x) = D_-^{(N)}(x) + \int_{C_-} F(x-y) \ln\bar{\mathfrak{A}}_N(y) \frac{dy}{2\pi} - \int_{C_+} F(x-y) \ln\mathfrak{A}_N(y) \frac{dy}{2\pi}$$

$$D_\pm^{(N)} = \frac{N}{2} \ln\Big(\frac{\mathrm{th}\frac{\pi}{4}(x+iu)}{\mathrm{th}\frac{\pi}{4}(x-iu)}\Big), \quad F(x) = \int_{-\infty}^{\infty} \frac{\mathrm{e}^{-ikx}}{1+\mathrm{e}^{2|k|}} dk,$$

where $C_+(C_-)$ is a straight contour slightly above (below) the real axis. In the first (second) equation we understand that $x \in C_+(C_-)$. Note that the convolution terms bring only minor contributions as they are defined on those contours where the auxiliary functions are small. Therefore the main contributions come from the known functions. This enables us to perform numerics with high accuracy. We can drop the convolution terms for the lowest order approximation. Thanks to eq. (22) this leads to eq. (20).

Similarly, for the largest eigenvalue we have

$$\ln \Lambda^{(N)}(x) = \varepsilon^{(N)}(x) + \int_{C_+} K_+(x-x')\ln \mathfrak{A}_N(x')\frac{dx'}{2\pi} + \int_{C_-} K_-(x-x')\ln \bar{\mathfrak{A}}_N(x')\frac{dx'}{2\pi},$$

$$K_\pm(x) = K(x \pm i), \qquad K(x) = \frac{\pi}{2\,\mathrm{ch}\,\pi x/2},$$

$$\varepsilon^{(N)} = \ln \phi_+(x+2i)\phi_-(x-2i) - \frac{N}{2}\int \mathrm{e}^{-|k|-ikx}\frac{\mathrm{sh}\,uk}{k\,\mathrm{ch}\,k}dk.$$

We obtain the NLIE and the eigenvalue in the Trotter limit by replacing $\mathfrak{a}_N \to \mathfrak{a}$ etc. and

$$D_\pm^{(N)} \to -\frac{\pi i\beta}{2\,\mathrm{sh}\,\frac{\pi}{2}x} \quad \varepsilon^{(N)} \to -\frac{\beta}{2}\left(1 - \int \frac{1}{1+\mathrm{e}^{2|k|}}dk\right).$$

For the actual calculation, it is even better to deal with $\mathfrak{b}(x) := \mathfrak{a}(x+i)$ and $\bar{\mathfrak{b}}(x) := \bar{\mathfrak{a}}(x-i)$ so that the singularities of $\ln(1+\mathfrak{b}), \ln(1+\bar{\mathfrak{b}})$ are away from the integration contours. We omit, however, the details.

As a concrete example for the evaluation of bulk quantities we plot the susceptibility, $\chi = \partial_h^2 f$, in fig. 8 (left). Note that at low temperatures the physical result in the Trotter limit (solid line) deviates from its finite Trotter number approximation. The above approach has been successfully

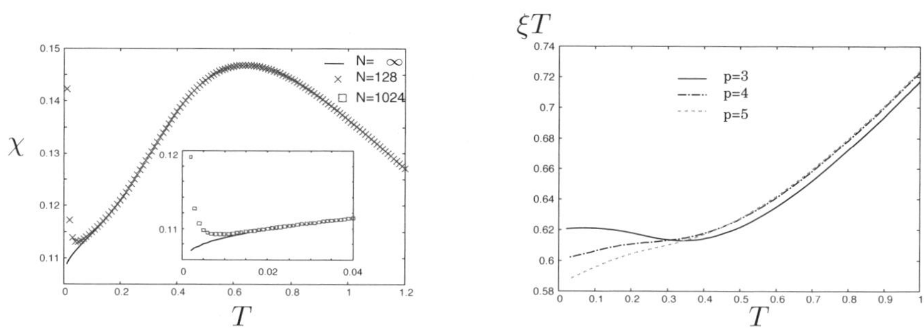

Fig. 8. Left: the susceptibility of the $s = \frac{1}{2}$ XXX model, in the Trotter limit (solid line) and for fixed Trotter number (crosses: $N = 128$, squares: $N = 1024$). Right: a plot of ξT against temperature for the XXZ model.

applied to many models of physical relevance.[13–19]

The correlation length ξ characterizes the decay of correlation functions at large distance, e.g.,

$$\langle \sigma_x^+ \sigma_y^- \rangle \sim \mathrm{e}^{-\frac{|x-y|}{\xi}} \qquad |x-y| \gg 1. \tag{26}$$

It is evaluated from the ratio of the largest and the second largest eigenvalues of the QTM.[9,20–22] For the second largest eigenvalue state, our assumption (1) on $\mathfrak{A}$ is no longer valid: a pair of holes θ_a lies on the real axis, and they are zeros of $\mathfrak{A}$ other than Bethe roots. Nevertheless, a small modification leads to a set of equations that fixes the second largest eigenvalue. The resultant NLIE has a form similar to (25) containing, however, additional inhomogeneous terms. Fig. 8 (right) shows a plot of ξT against temperature for the XXZ model with $q = \mathrm{e}^{i\frac{\pi}{p}}$ for $p = 3, 4, 5$.[21]

When $h \neq 0$, we have to replace $a(x), d(x)$ in (18) by $\mathrm{e}^{-\beta h/2}a(x)$, $\mathrm{e}^{\beta h/2}d(x)$. Then we add βh $(-\frac{\beta h}{2})$ to the rhs of (23) ((24)). Also $D_{\pm}^{(N)}$ must be replaced by $D_{\pm}^{(N)} \pm \frac{\beta h}{2}$.

Before closing this section, we would like to mention another formulation of thermodynamics also based on the QTM.[23] It is described by a NLIE for Λ_j directly and allows one to efficiently calculate high temperature expansions. Good numerical accuracy in the low temperature region is, however, hard to achieve. Moreover, we point out that the equation is the same for any eigenvalue. Thus, one should know a priori good initial values in order to select the convergence to the desired eigenvalue.

4. DME (density matrix elements) at finite temperatures

The deep understanding of a model requires ample knowledge of its correlation functions. We would therefore like to go beyond their asymptotic characterization by the correlation length ξ (26).

The evaluation of correlation functions has been defying many challenges in the past. Considerable progress was made only recently for the $T = 0$ correlations, based on vertex operators,[24] on the qKZ equation[25] and on QISM.[26] The third approach is the most relevant for our purpose. For $T = 0$ it first requires the solution of the "inverse problem", that is, one has to represent the spin operators in terms of the QISM operators $A(u), B(u), C(u), D(u)$. Then, by algebraic manipulations, one obtains the correlation functions as combinatorial sums of expectation values of QISM operators, which are finally converted into (multiple) integrals.

At first glance, the case $T > 0$ seems far more difficult, as one expects that a summation of the contributions from all excited states is necessary. We argue here that, as above, the QTM helps us to avoid this summation and that, moreover, one does not have to solve the "inverse problem" within the QTM framework.[27] The combinatorics, on the other hand, can be done in parallel to $T=0$, because the QISM algebra is the same in both cases.

Let us explain why we can bypass the inverse problem in the QTM formalism. This can be most quickly done in a graphical manner. To be specific, we need to evaluate DME,

$$D^{\alpha_1\cdots\alpha_m}_{\beta_1\cdots\beta_m} := \langle E^{\alpha_1}_{\beta_1}\cdots E^{\alpha_m}_{\beta_m}\rangle = \frac{\mathrm{tr}_{V_{\rm phys}} \mathrm{e}^{-\beta\mathcal{H}} E^{\alpha_1}_{\beta_1}\cdots E^{\alpha_m}_{\beta_m}}{\mathrm{tr}_{V_{\rm phys}} \mathrm{e}^{-\beta\mathcal{H}}}.$$

Using the logic of section 2, we can represent $\mathrm{e}^{-\beta\mathcal{H}}$ by a "2D partition function". Therefore $D^{\alpha_1\cdots\alpha_n}_{\beta_1\cdots\beta_n}$ can be represented by a modified 2D partition function: Start from the $N\times L$ classical system (fig. 4) with periodic boundaries in both directions. Then cut n successive vertical bonds at the bottom row, and fix the variables at both sides of the cut. As we are adopting PBCs in the vertical direction this is equivalent to fixing the configuration of n successive bonds at the top and at the bottom. See fig. 9 (left). As

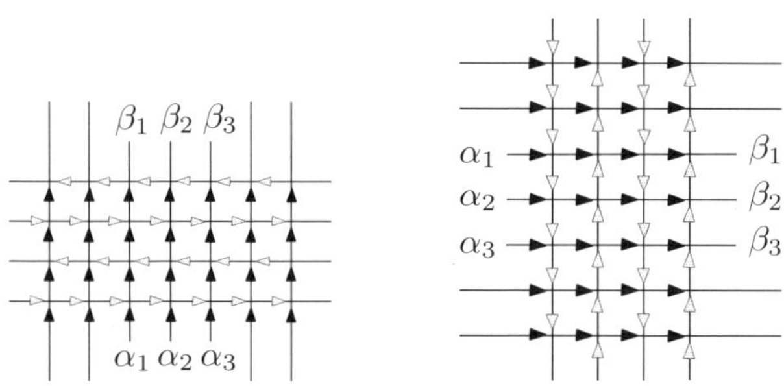

Fig. 9. Left: A graphically representation of $D^{\alpha_1,\alpha_2,\alpha_3}_{\beta_1,\beta_2,\beta_3}$. Right: The same figure rotated by 90°.

previously, we rotate the lattice by 90°. See fig. 9 (right). Obviously we can write $D^{\alpha_1\cdots\alpha_n}_{\beta_1\cdots\beta_n}$ in terms of elements of the monodromy matrix $\mathcal{T}_{\rm QTM}(x)$. By introducing independent spectral parameters ξ_i we obtain

$$\Big(D\Big)^{\alpha_1,\cdots,\alpha_n}_{\beta_1,\cdots,\beta_n}(\xi_1,\cdots,\xi_n) = \frac{\langle\Phi_0|(\mathcal{T}_{\rm QTM})^{\alpha_1}_{\beta_1}(\xi_1)\cdots(\mathcal{T}_{\rm QTM})^{\alpha_n}_{\beta_n}(\xi_n)|\Phi_0\rangle}{\langle\Phi_0|\mathcal{T}_{\rm QTM}(\xi_1)\cdots\mathcal{T}_{\rm QTM}(\xi_n)|\Phi_0\rangle}.$$

Here Φ_0 denotes the largest eigenvalue state of the QTM, which is given by acting with B operators on the vacuum. As any $(\mathcal{T}_{\rm QTM})^{\alpha_i}_{\beta_i}(\xi_i)$ is represented by a QISM operator, we reach an expression for DME purely in terms of QISM operators without solving the inverse problem.

At the same time, the problem for $T > 0$ is not so simple in view of the analyticity. We consider D^{++}_{++} as a concrete example. After employing the standard QISM algebra, one obtains

$$D^{++}_{++}(\xi_1,\xi_2) \times \mathfrak{A}(\xi_1)\mathfrak{A}(\xi_2) = \Big(\sum_{j,k} \tfrac{(x_k-\xi_2)(x_j-\xi_1-2i)}{\xi_{2,1}(x_j-x_k-2i)} \begin{vmatrix} w_{j,1} & w_{k,1} \\ w_{j,2} & w_{k,2} \end{vmatrix}$$
$$- \tfrac{\xi_{1,2}+2i}{\xi_{1,2}} \sum_j \tfrac{(x_j-\xi_2)}{(x_j-\xi_2+2i)} w_{j,1} - \tfrac{\xi_{2,1}+2i}{\xi_{2,1}} \sum_j \tfrac{(x_j-\xi_1)}{(x_j-\xi_1+2i)} w_{j,2} + 1 \Big).$$

Here x_j denotes a BAE root, $\xi_{i,j} = \xi_i - \xi_j$ and $\mathfrak{A}$ is the auxiliary function. We introduced $w_{j,k}$ in order to deal with the ratio of inner products of wave functions. $w_{j,k}$ is characterized by a simple algebraic relation. Note that the above algebraic expression is formally identical for $T = 0$ and $T > 0$: one only has to replace x_j and $w_{j,k}$ for $T = 0$ by those for $T > 0$.

In the case $T = 0$ there are several simplifications. First, the auxiliary function is by construction trivial, $\mathfrak{A} = 1$. Second, we can introduce the root density function in the thermodynamic limit. Then the algebraic relation for $w_{j,k} \to g(x_j, \xi_k)$ is solved with the explicit result $g(x,\xi) = \frac{1}{4\,\mathrm{ch}\,\frac{\pi}{2}(x-\xi+i)}$.

$$D^{++}_{++}(\xi_1,\xi_2) = \Big(\int dx dx' \tfrac{(x'-\xi_2)(x-\xi_1-2i)}{\xi_{2,1}(x-x'-2i)} \begin{vmatrix} g(x,\xi_1) & g(x',\xi_1) \\ g(x,\xi_2) & g(x',\xi_2) \end{vmatrix}$$
$$- \tfrac{\xi_{1,2}+2i}{\xi_{1,2}} \int dx \tfrac{(x-\xi_2)}{(x-\xi_2+2i)} g(x,\xi_1) - \tfrac{\xi_{2,1}+2i}{\xi_{2,1}} \int dx \tfrac{(x-\xi_1)}{(x-\xi_1+2i)} g(x,\xi_2) + 1 \Big).$$

Third, we can freely move the integration contours. Every time it passes the singularity of $g(x)$, it brings extra contributions and they cancel the "tails" (the 2nd to the 4th terms above). We finally obtain

$$D^{++}_{++}(\xi_1,\xi_2) = \int_{-\infty}^{\infty} dx dx' \frac{(x'-\xi_2+i)(x-\xi_1-i)}{\xi_{2,1}(x-x'-2i)} \begin{vmatrix} g(x+i,\xi_1) & g(x'+i,\xi_1) \\ g(x+i,\xi_2) & g(x'+i,\xi_2) \end{vmatrix}. \qquad (27)$$

Without such a compact expression, it is hard to proceed further.

On the other hand, $\mathfrak{A}$ is quite non-trivial for $T > 0$. As noted previously, we can not resort to the root density function. The explicit form of $w_{j,k}$ in the Trotter limit is thus unknown. The most significant difference is that the integration contour is already fixed for $T > 0$. Thus, we cannot apply the above trick to swallow tails into the ground state.

Nevertheless, with an appropriate choice of a further auxiliary function $G(x,\xi)$, it was shown that a compact multiple integral representation, similar to (27) is also possible for $T > 0$,[27,28]

$$D^{++}_{++}(\xi_1,\xi_2) = \int_C \frac{dx}{\mathfrak{A}(x)} \int_C \frac{dx'}{\mathfrak{A}(x')} \frac{(x-\xi_1-2i)(x'-\xi_2)}{4\pi^2 \xi_{1,2}(x-x'-2i)} \begin{vmatrix} G(x,\xi_1) & G(x',\xi_1) \\ G(x,\xi_2) & G(x',\xi_2) \end{vmatrix}.$$

The formula for any other DME is similarly known.

It is a big progress to obtain the multiple integral representation for DMEs. The representation is, however, not yet optimal. Although one can use it for the numerical analysis at sufficiently high temperatures, it suffers from numerical inaccuracy at low temperatures.[29,30] We thus would like to reduce it to (a sum of) products of single integrals.

The factorization of DME at $T = 0$ has been performed by brute force, with the extensive use of the shift of contour technique.[31,32] Based on studies of the qKZ equation, a hidden Grassmannian structure behind DME has been found.[33] It naturally explains the factorization of the multiple integral formula through the nilpotency of operators. The explicit form of DME consists of two pieces, the algebraic part, evaluated from a matrix element of q-oscillators, and the transcendental part, related to the spinon-S matrix.

On the other hand, we still do not have a finite temperature analogue of the qKZ equation. We are nevertheless able to factorize the multiple integrals for small segments.[34] The explicit results also consist of two parts. Surprisingly the algebraic part remains identical to the $T = 0$ case, while the transcendental part can be interpreted as a proper finite temperature analogue to the spinon-S matrix. This finally enables us to perform an accurate numerical analysis of the correlation functions.[35] We show plots of $\langle \sigma_1^z \sigma_4^z \rangle$ for $\Delta = \frac{1}{\sqrt{2}}$ with various magnetic fields in fig. 10. The results from a brute force calculation are also plotted, which supports the validity of our formula. Such high accuracy calculation can clarify the quantitative

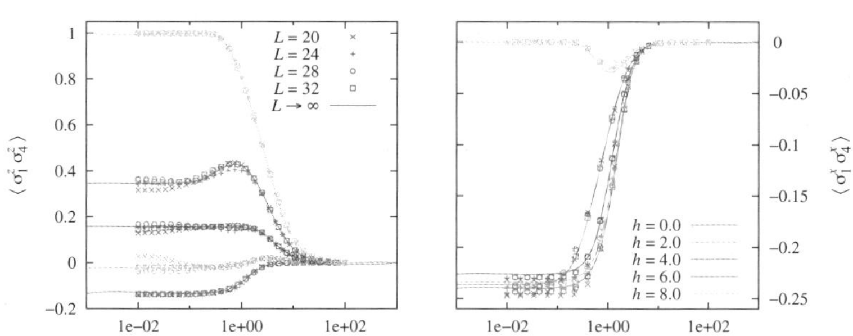

Fig. 10. The plots of $\langle \sigma_1^z \sigma_4^z \rangle$ (left) and $\langle \sigma_1^x \sigma_4^x \rangle$ (right) for $\Delta = \frac{1}{\sqrt{2}}$ with various magnetic fields by NLIEs (continuous line).

nature of interesting phenomena such as the quantum-classical crossover.[36]

Recently a proof of the existence of factorization of the DMEs for $T > 0$ was obtained, again by using the Grassmannian structure.[37] See also the further development[38] in this direction based on NLIEs.

5. Summary and discussion

We presented a brief review on the recent progress with the exact thermodynamics of 1D quantum systems. The QTM is found to be an efficient tool, and it offers a framework to evaluate quantities of physical interest, including DME. The NLIE combines into the framework nicely, yielding high accuracy numerical results.

The factorization of the multiple integral formula at $T > 0$ is yet to be further explored. It seems e.g. quite plausible that the qKZ equation could be extended to finite temperatures. This might be an important next step.

There are certainly many interesting questions left open. For example, can we have the QTM formulation starting from a continuum system? What is the generalization of the multiple integral formula to models with higher spin? The study of such questions is underway.

Acknowledgments

The authors take pleasure in dedicating this review to Professor Tetsuji Miwa on the occasion of his sixtieth birthday. They thank the organizers of "Infinite Analysis 09" for their warm hospitality.

References

1. M. Gaudin, Phys. Rev. Lett. **26** (1971) 1301.
2. M. Takahashi and M. Suzuki, Prog. Theor. Phys. **48** (1972) 2187-2209.
3. R. J. Baxter, Ann. Phys. **76** (1973) 1, ibid, 25, ibid 48.
4. M. Suzuki, Phys. Rev. B **31** (1985) 2957-2965.
5. For earlier reviews on the QTM and finite size correction method, see e.g. J. Suzuki, T. Nagao and M. Wadati, Int. J. Mod. Phys. B **6** (1992) 1119-1180. A. Klümper, Lecture Notes in Phys. **645** (2004) 349-379, or the review chapters in the book by F. H. L. Essler, H. Frahm, F. Göhmann, A. Klümper and V. E. Korepin, "The One-Dimensional Hubbard Model" (Cambridge Press 2005).
6. M. Jimbo, Lett. Math. Phys. **10** (1985) 63-69.
7. A. Klümper, Z. Phys. B **91** (1993) 507-519.
8. T. Koma, Prog. Theor. Phys. **78** (1987) 1213-1218.
9. J. Suzuki, Y. Akutsu and M. Wadati, J. Phys. Soc. Jpn. **59** (1990) 2667-2680.
10. A. Klümper, Ann. Physik (Lpz.) **1** (1992) 540.
11. A. Kuniba, K. Sakai and J. Suzuki, Nucl. Phys. **B 525** [FS] (1998) 597-626.

12. A. Klümper, M. T. Batchelor and P. A. Pearce, J. Phys. A **24** (1991) 3111-3133.
13. G. Jüttner and A. Klümper, Euro. Phys. Lett. **37** (1997) 335-340.
14. G. Jüttner, A. Klümper and J. Suzuki, Nucl. Phys. B **486** (1997) 650-574, J. Phys. A **30** (1997) 1881-1886, Nucl. Phys. B **512** (1998) 581-600, Nucl. Phys. B **522** (1998) 471-502.
15. J. Suzuki, Nucl. Phys. B **528** (1998) 683-700.
16. A. Fujii and A. Klümper, Nucl. Phys. B **546** (1999) 751-764.
17. J. Suzuki, J. Phys. A **32** (1999) 2341-2359.
18. M. Bortz and A. Klümper, Eur. Phys. J. B **40** (2004) 25-42.
19. J. Damerau and A. Klümper, JSTAT (2006) P12014.
20. M. Suzuki and M. Inoue, Prog. Theor. Phys. **78** (1987) 787.
21. K. Sakai, M. Shiroishi, J. Suzuki and Y. Umeno, Phys. Rev. B **60** (1999) 5186-5201.
22. A. Klümper, J. R. Reyes Martinez, C. Scheeren, M. Shiroishi, J. Stat. Phys. **102** (2001) 937-951.
23. M. Shirosihi and M. Takahashi, Phys. Rev. Lett. **89** (2002) 117201(4pp). Z. Tsuboi, J. Phys. A **37** (2004) 1747-1758.
24. M. Jimbo, K. Miki, T. Miwa, A. Nakayashiki, Phys. Lett. A **16** (1992) 256-263.
25. M. Jimbo and T. Miwa, J. Phys. A **29** (1996) 2923-2958.
26. N. Kitanine, J. M. Maillet and V. Terras, Nucl. Phys. B **567** (2000) 554-582. N. Kitanine, J. M. Maillet, N. A. Slavnov, V. Terras, Nucl. Phys. B **641** (2002) 487-518.
27. F. Göhmann, A. Klümper and A. Seel, J. Phys. A **37** (2004) 7625-7651.
28. F. Göhmann, N. Hasenclever and A. Seel, JSTAT (2005) P10015.
29. M. Bortz and F. Göhmann, Eur. Phys. J. B **46** (2005) 399-408.
30. Z. Tsuboi, Physica A **377** (2007) 95-101.
31. H. E. Boos and V. E. Korepin, J. Phys. A **34** (2001) 5311-5316. H. E. Boos, V. E. Korepin and F. A. Smirnov, Nucl. Phys. B **658** (2003) 417-439, J. Phys. A **37** (2004) 323-336.
32. K. Sakai, M. Shiroishi, Y. Nishiyama and M. Takahashi, Phys. Rev. E **67** (2003) 06510(R) (4pp). G. Kato, M. Shiroishi, M. Takahashi, K. Sakai, J. Phys. A **36** (2003) L337-L344. J. Sato, M. Shiroishi, M. Takahashi JSTAT (2006) P017.
33. H. Boos, M. Jimbo, T. Miwa, F. Smirnov, Y. Takeyama, Comm. Math. Phys. **261** (2006) 245-276, Annales Henri-Poincare **7** (2006) 1395-1428, Comm. Math. Phys **272** (2007) 263-281, Comm. Math. Phys. **286** (2009) 875-932.
34. H. E. Boos, F. Göhmann, A. Klümper and J. Suzuki, JSTAT 0604 (2006) P001, J. Phys. A **40** (2007) 10699-10728.
35. H. E. Boos, J. Damerau, F. Göhmann, A. Klümper and J. Suzuki and A. Weiße, JSTAT 0808 (2008) P08010.
36. K. Fabricius and B. M. McCoy, Phys. Rev. B **60** (1999) 381 386. K. Fabricius, A. Klümper and B. M. McCoy, Phys. Rev. Lett. **82** (1999) 5365-5368.
37. M. Jimbo, T. Miwa and F. Smirnov, J. Phys. A **42** (2009) 304018 (31pp).
38. H. Boos and F. Göhmann, J. Phys. A**42** (2009) 315001 (27pp).

TROPICAL SPECTRAL CURVES, FAY'S TRISECANT IDENTITY, AND GENERALIZED ULTRADISCRETE TODA LATTICE

REI INOUE

Faculty of Pharmaceutical Sciences, Suzuka University of Medical Science
3500-3 Minami-tamagaki, Suzuka, Mie, 513-8670, Japan
E-mail: reiiy@suzuka-u.ac.jp

SHINSUKE IWAO

Graduate School of Mathematical Sciences, The University of Tokyo
3-8-1 Komaba Meguro-ku, Tokyo 153-8914, Japan
E-mail: iwao@ms.u-tokyo.ac.jp

We study the generalized ultradiscrete periodic Toda lattice $\mathcal{T}(M,N)$ which has tropical spectral curve. We introduce a tropical analogue of Fay's trisecant identity, and apply it to construct a general solution to $\mathcal{T}(M,N)$.

Keywords: Tropical geometry; Riemann's theta function; Toda lattice.

1. Introduction

The *ultradiscrete* periodic Toda lattice is an integrable system described by a piecewise-linear map.[9] Recently, its algebro geometrical aspect is clarified[2–4] by applying the tropical geometry, a combinatorial algebraic geometry rapidly developed during this decade.[1,5,13] This system has tropical spectral curves, and what proved are that its general isolevel set is isomorphic to the tropical Jacobian of the tropical hyperelliptic curve, and that its general solution is written in terms of the tropical Riemann's theta function. The key to the solution is the tropical analogue of Fay's trisecant identity for a special family of hyperelliptic curves.[3]

On the other hand, there exists a generalization of *discrete* periodic Toda lattice $T(M,N)$, where M (resp. N) is a positive integer which denotes the level of generalization (resp. the periodicity) of the system. The $M=1$ case, $T(1,N)$, is the original discrete Toda lattice of N-periodicity. When $\gcd(M,N)=1$, $T(M,N)$ reduces to a special case of the integrable multidiagonal Lax matrix,[15] and the general solution to $T(M, N)$ is recently constructed.[8]

The aim of this paper is twofold: the first one is to introduce the tropical analogue of Fay's trisecant identity not only for hyperelliptic but also for more general tropical curves. The second one is to study the generalization of *ultradiscrete* periodic Toda lattice $\mathcal{T}(M, N)$ by applying the tropical Fay's trisecant identity, as a continuation of the study on $\mathcal{T}(1, N)$.[2–4]

This paper is organized as follows: in §2 we review some notion of tropical geometry,[6,11,12] and introduce the tropical analogue of Fay's trisecant identity (Theorem 2.2) by applying the correspondence of integrations over complex and tropical curves.[6] In §3 we introduce the generalization of the discrete periodic Toda lattice $T(M, N)$ and its ultradiscretization $\mathcal{T}(M, N)$. We reconsider the integrability of $T(M, N)$ (Proposition 3.1). In §4 we demonstrate the general solution to $\mathcal{T}(3, 2)$, and give conjectures on $\mathcal{T}(M, N)$ (Conjectures 4.1 and 4.2).

In closing the introduction, we make a brief remark on the interesting close relation between the ultradiscrete periodic Toda lattice and the *periodic box and ball system* (pBBS),[9] which is generalized to that between $\mathcal{T}(M, N)$ and the pBBS of M kinds of balls.[7,14] When $M = 1$, the relation is explained at the level of tropical Jacobian.[2] We expect that our conjectures on $\mathcal{T}(M, N)$ also account for the tropical geometrical aspects of the recent results[10] on the pBBS of M kinds of balls.

2. Tropical curves and Riemann's theta function

2.1. *Good tropicalization of algebraic curves*

Let K be a subfield of $\mathbb{C}$ and K_ε be the field of convergent Puiseux series in $\mathbf{e} := \mathrm{e}^{-1/\varepsilon}$ over K. Let val : $K_\varepsilon \to \mathbb{Q} \cup \{\infty\}$ be the natural valuation with respected to $\mathbf{e}$. Any polynomial f_ε in $K_\varepsilon[x, y]$ is expressed uniquely as

$$f_\varepsilon = \textstyle\sum_{w=(w_1,w_2)\in\mathbb{Z}^2} a_w(\varepsilon) x^{w_1} y^{w_2}, \qquad a_w(\varepsilon) \in K_\varepsilon.$$

Define the tropical polynomial $\mathrm{Val}(X, Y; f_\varepsilon)$ associated with f_ε by the formula: $\mathrm{Val}(X, Y; f_\varepsilon) := \min_{w\in\mathbb{Z}^2} [\mathrm{val}(a_w) + w_1 X + w_2 Y]$. We call $\mathrm{Val}(X, Y; f_\varepsilon)$ the *tropicalization* of f_ε.

We take $f_\varepsilon \in K_\varepsilon[x, y]$. Let $C^0(f_\varepsilon)$ be the affine algebraic curve over K_ε defined by $f_\varepsilon = 0$. We write $C(f_\varepsilon)$ for the complete curve over K_ε such that $C(f_\varepsilon)$ contains $C^0(f_\varepsilon)$ as a dense open subset, and that $C(f_\varepsilon) \setminus C^0(f_\varepsilon)$ consists of non-singular points. The *tropical curve* $TV(f_\varepsilon)$ is a subset of $\mathbb{R}^2$

defined by:

$$TV(f_\varepsilon) = \left\{ (X,Y) \in \mathbb{R}^2 \;\middle|\; \begin{array}{c} \text{the function } \mathrm{Val}(X,Y;f_\varepsilon) \\ \text{is not smooth at } (X,Y) \end{array} \right\}.$$

Denote $\Lambda(X,Y;f_\varepsilon) := \{w \in \mathbb{Z}^2 \mid \mathrm{Val}(X,Y;f_\varepsilon) = \mathrm{val}(a_w) + w_1 X + w_2 Y\}$. The definition of the tropical curve can be put into:

$$TV(f_\varepsilon) = \{(X,Y) \in \mathbb{R}^2 \mid \sharp\Lambda(X,Y;f_\varepsilon) \geq 2\}.$$

For $P = (X,Y) \in \mathbb{R}^2$, we define $f_\varepsilon^P := \sum_{w \in \Lambda(X,Y;f_\varepsilon)} a_w x^{w_1} y^{w_2}$.

To make use of the results of tropical geometry for real/complex analysis, we introduce the following condition as a criterion of genericness of tropical curves.

Definition 2.1. We say $C(f_\varepsilon)$ has a *good tropicalization* if:

(1) $C(f_\varepsilon)$ is an irreducible reduced non-singular curve over K_ε,
(2) $f_\varepsilon^P = 0$ defines an affine reduced non-singular curve in $(K_\varepsilon^\times)^2$ for all $P \in TV(f_\varepsilon)$ (maybe reducible).

Remark 2.1. The notion of a *good tropicalization* was first introduced in [6, Section 4.3]. The above definition gives essentially the same notion.

2.2. *Smoothness of tropical curves*

For the tropical curve $\Gamma := TV(f_\varepsilon)$, we define the set of vertices $V(\Gamma)$:

$$V(\Gamma) = \{(X,Y) \in \Gamma \mid \sharp\Lambda(X,Y;f_\varepsilon) \geq 3\}.$$

We call each disjointed element of $\Gamma \setminus V(\Gamma)$ an edge of Γ. For an edge e, we have the *primitive tangent vector* $\xi_e = (n,m) \in \mathbb{Z}^2$ as $\gcd(n,m) = 1$. Note that the vector ξ_e is uniquely determined up to sign.

Definition 2.2. [11, §2.5] The tropical curve Γ is *smooth* if:

(1) All the vertices are trivalent, *i.e.* $\sharp\Lambda(X,Y;f_\varepsilon) = 3$ for all $(X,Y) \in V(\Gamma)$.
(2) For each trivalent vertex $v \in V(\Gamma)$, let ξ_1, ξ_2 and ξ_3 be the primitive tangent vectors of the three outgoing edges from v. Then we have $\xi_1 + \xi_2 + \xi_3 = 0$ and $|\xi_i \wedge \xi_j| = 1$ for $i \neq j \in \{1,2,3\}$.

When Γ is smooth, the genus of Γ is $\dim H_1(\Gamma, \mathbb{Z})$.

2.3. *Tropical Riemann's theta function*

For an integer $g \in \mathbb{Z}_{>0}$, a positive definite symmetric matrix $B \in M_g(\mathbb{R})$ and $\beta \in \mathbb{R}^g$ we define a function on $\mathbb{R}^g$ as

$$q_\beta(\mathbf{m}, \mathbf{Z}) = \frac{1}{2}\mathbf{m}B\mathbf{m}^\perp + \mathbf{m}(\mathbf{Z} + \beta B)^\perp \qquad (\mathbf{Z} \in \mathbb{R}^g,\ \mathbf{m} \in \mathbb{Z}^g).$$

The tropical Riemann's theta function $\Theta(\mathbf{Z}; B)$ and its generalization $\Theta[\beta](\mathbf{Z}; B)$ are given by[3,12]

$$\Theta(\mathbf{Z}; B) = \min_{\mathbf{m}\in\mathbb{Z}^g} q_0(\mathbf{m}, \mathbf{Z}),$$
$$\Theta[\beta](\mathbf{Z}; B) = \frac{1}{2}\beta B\beta^\perp + \beta\mathbf{Z}^\perp + \min_{\mathbf{m}\in\mathbb{Z}^g} q_\beta(\mathbf{m}, \mathbf{Z}).$$

Note that $\Theta[0](\mathbf{Z}; B) = \Theta(\mathbf{Z}; B)$. The function $\Theta[\beta](\mathbf{Z}; B)$ satisfies the quasi-periodicity:

$$\Theta[\beta](\mathbf{Z} + K\mathbf{l}) = -\frac{1}{2}\mathbf{l}K\mathbf{l}^\perp - \mathbf{l}\mathbf{Z}^\perp + \Theta[\beta](\mathbf{Z}) \qquad (\mathbf{l} \in \mathbb{Z}^g).$$

We also write $\Theta(\mathbf{Z})$ and $\Theta[\beta](\mathbf{Z})$ for $\Theta(\mathbf{Z}; B)$ and $\Theta[\beta](\mathbf{Z}; B)$ without confusion. We write $\mathbf{n} = \arg_{\mathbf{m}\in\mathbb{Z}^g} q_\beta(\mathbf{m}, \mathbf{Z})$ when $\min_{\mathbf{m}\in\mathbb{Z}^g} q_\beta(\mathbf{m}, \mathbf{Z}) = q_\beta(\mathbf{n}, \mathbf{Z})$.

2.4. *Tropical analogue of Fay's trisecant identity*

For each $\bar{\varepsilon} \in \mathbb{R}_{>0}$, we write $C(f_{\bar{\varepsilon}})$ for the base change of $C(f_\varepsilon)$ to $\mathbb{C}$ via a map $\iota:\ K_\varepsilon \to \mathbb{C}$ given by $\varepsilon \mapsto \bar{\varepsilon}$.

Theorem 2.1. *[6, Theorem 4.3.1] Assume $C(f_\varepsilon)$ has a good tropicalization and $C(f_{\bar{\varepsilon}})$ is non-singular. Let $B_{\bar{\varepsilon}}$ and B_T be the period matrices for $C(f_{\bar{\varepsilon}})$ and $TV(f_\varepsilon)$ respectively. Then we have the relation*

$$\frac{2\pi\bar{\varepsilon}}{\sqrt{-1}} B_{\bar{\varepsilon}} \sim B_T \quad (\bar{\varepsilon} \to 0).$$

(It follows from the assumption that the genus of $C(f_\varepsilon)$ and $C(f_{\bar{\varepsilon}})$ coincide.)

A nice application of this theorem is to give the tropical analogue of Fay's trisecant identity. For the algebraic curve $C(f_\varepsilon)$ of Theorem 2.1, we have the following:

Theorem 2.2. *We continue the hypothesis and notation in Theorem 2.1, and assume $TV(f_\varepsilon)$ is smooth. Let g be the genus of $C(f_\varepsilon)$ and $(\alpha, \beta) \in \frac{1}{2}\mathbb{Z}^{2g}$ be a non-singular odd theta characteristic for* $\mathrm{Jac}(C(f_\varepsilon))$.

For P_1, P_2, P_3, P_4 on the universal covering space of $TV(f_\varepsilon)$, we define the sign $s_i \in \{\pm 1\}$ $(i = 1, 2, 3)$ as $s_i = (-1)^{k_i}$, where

$$k_1 = 2\alpha \cdot \left(\arg_{\mathbf{m}\in\mathbb{Z}^g} q_\beta(\mathbf{m}, \int_{P_3}^{P_2}) + \arg_{\mathbf{m}\in\mathbb{Z}^g} q_\beta(\mathbf{m}, \int_{P_1}^{P_4}) \right),$$

$$k_2 = 2\alpha \cdot \left(\arg_{\mathbf{m}\in\mathbb{Z}^g} q_\beta(\mathbf{m}, \int_{P_3}^{P_1}) + \arg_{\mathbf{m}\in\mathbb{Z}^g} q_\beta(\mathbf{m}, \int_{P_4}^{P_2}) \right),$$

$$k_3 = 1 + 2\alpha \cdot \left(\arg_{\mathbf{m}\in\mathbb{Z}^g} q_\beta(\mathbf{m}, \int_{P_3}^{P_4}) + \arg_{\mathbf{m}\in\mathbb{Z}^g} q_\beta(\mathbf{m}, \int_{P_1}^{P_2}) \right).$$

Set the functions $F_1(\mathbf{Z}), F_2(\mathbf{Z}), F_3(\mathbf{Z})$ of $\mathbf{Z} \in \mathbb{R}^g$ as

$$F_1(\mathbf{Z}) = \Theta(\mathbf{Z} + \int_{P_1}^{P_3}) + \Theta(\mathbf{Z} + \int_{P_2}^{P_4}) + \Theta[\beta](\int_{P_3}^{P_2}) + \Theta[\beta](\int_{P_1}^{P_4}),$$

$$F_2(\mathbf{Z}) = \Theta(\mathbf{Z} + \int_{P_2}^{P_3}) + \Theta(\mathbf{Z} + \int_{P_1}^{P_4}) + \Theta[\beta](\int_{P_3}^{P_1}) + \Theta[\beta](\int_{P_4}^{P_2}),$$

$$F_3(\mathbf{Z}) = \Theta(\mathbf{Z} + \int_{P_1+P_2}^{P_3+P_4}) + \Theta(\mathbf{Z}) + \Theta[\beta](\int_{P_4}^{P_3}) + \Theta[\beta](\int_{P_1}^{P_2}).$$

Then, the formula

$$F_i(\mathbf{Z}) = \min[F_{i+1}(\mathbf{Z}), F_{i+2}(\mathbf{Z})] \tag{1}$$

holds if $s_i = \pm 1, s_{i+1} = s_{i+2} = \mp 1$ for $i \in \mathbb{Z}/3\mathbb{Z}$.

This theorem generalizes [2, Theorem 2.4], where $C(f_\varepsilon)$ is a special hyperelliptic curve. We introduce the following lemma for later convenience:

Lemma 2.1. *[2, Proposition 2.1] Let C be a hyperelliptic curve of genus g and take $\beta = (\beta_j)_j \in \frac{1}{2}\mathbb{Z}^g$. Set $\alpha = (\alpha_j)_j \in \frac{1}{2}\mathbb{Z}^g$ as*

$$\alpha_j = -\frac{1}{2}\delta_{j,i-1} + \frac{1}{2}\delta_{j,i},$$

where i is defined by the condition $\beta_j = 0$ $(1 \le j \le i-1)$ and $\beta_i \neq 0$ mod $\mathbb{Z}$. Then (α, β) is a non-singular odd theta characteristic for $\mathrm{Jac}(C)$.

2.5. *Tropical Jacobian*

When the positive definite symmetric matrix $B \in M_g(\mathbb{R})$ is the period matrix of a smooth tropical curve Γ, the g-dimensional real torus $J(\Gamma)$ defined by

$$J(\Gamma) := \mathbb{R}^g / \mathbb{Z}^g B$$

is called the tropical Jacobian[12] of Γ.

3. Discrete and ultradiscrete generalized Toda lattice

3.1. *Generalized discrete periodic Toda lattice* $T(M,N)$

Fix $M, N \in \mathbb{Z}_{>0}$. Let $T(M,N)$ be the generalization of discrete periodic Toda lattice defined by the difference equations[8,14]

$$\begin{aligned} I_n^{t+1} + V_{n-1}^{t+\frac{1}{M}} &= I_n^t + V_n^t, \\ V_n^{t+\frac{1}{M}} I_n^{t+1} &= I_{n+1}^t V_n^t, \end{aligned} \qquad (n \in \mathbb{Z}/N\mathbb{Z},\ t \in \mathbb{Z}/M), \tag{2}$$

on the phase space T:

$$\begin{aligned} \Big\{ & (I_n^t, I_n^{t+\frac{1}{M}}, \ldots, I_n^{t+\frac{M-1}{M}}, V_n^t)_{n=1,\ldots,N} \in \mathbb{C}^{(M+1)N} \\ & \Big| \prod_{n=1}^N V_n^t,\ \prod_{n=1}^N I_n^{t+\frac{k}{M}}\ (k = 0, \ldots, M-1) \text{ are distinct} \Big\}. \end{aligned} \tag{3}$$

Eq. (2) is written in the Lax form given by

$$L^{t+1}(y) M^t(y) = M^t(y) L^t(y),$$

where

$$L^t(y) = M^t(y) R^{t+\frac{M-1}{M}}(y) \cdots R^{t+\frac{1}{M}}(y) R^t(y), \tag{4}$$

$$R^t(y) = \begin{pmatrix} I_2^t & 1 & & & \\ & I_3^t & 1 & & \\ & & \ddots & \ddots & \\ & & & I_N^t & 1 \\ y & & & & I_1^t \end{pmatrix}, \quad M^t(y) = \begin{pmatrix} 1 & & & & \frac{V_1^t}{y} \\ V_2^t & 1 & & & \\ & V_3^t & 1 & & \\ & & \ddots & \ddots & \\ & & & V_N^t & 1 \end{pmatrix}.$$

The Lax form ensures that the characteristic polynomial $\mathrm{Det}(L^t(y) - x\mathbb{I}_N)$ of the Lax matrix $L^t(y)$ is independent of t, namely, the coefficients of $\mathrm{Det}(L^t(y) - x\mathbb{I}_N)$ are integrals of motion of $T(M,N)$.

Assume $\gcd(M,N) = 1$ and set $d_j = [\frac{(M+1-j)N}{M}]$ $(j = 1, \ldots, M)$. We consider three spaces for $T(M,N)$: the phase space T (3), the coordinate space L for the Lax matrix $L^t(y)$ (4), and the space F of the spectral curves. The two spaces L and F are given by

$$L = \{(a_{i,j}^t, b_i^t)_{i=1,\ldots,M,\ j=1,\ldots,N} \in \mathbb{C}^{(M+1)N}\}, \tag{5}$$

$$\begin{aligned} F = \Big\{ & y^{M+1} + f_M(x) y^M + \cdots + f_1(x) y + f_0 \in \mathbb{C}[x,y] \ \Big| \\ & \deg_x f_j(x) \le d_j\ (j = 1, \ldots, M),\ -f_1(x) \text{ is monic} \Big\}, \end{aligned} \tag{6}$$

where each element in L corresponds to the matrix:

$$L^t(y) = \begin{pmatrix} a^t_{1,1} & a^t_{2,2} & \cdots & a^t_{M,M} & 1 & & & \frac{b^t_N}{y} \\ b^t_1 & a^t_{1,2} & a^t_{2,3} & \cdots & a^t_{M,M+1} & 1 & & \\ & \ddots & \ddots & \ddots & & \ddots & \ddots & \\ & & \ddots & & & & a^t_{M,N-1} & 1 \\ y & & & \ddots & & & a^t_{M-1,N-1} & a^t_{M,N} \\ ya^t_{M,1} & y & & & & & & a^t_{M-1,N} \\ \vdots & \ddots & \ddots & & & \ddots & \ddots & \vdots \\ ya^t_{2,1} & \cdots & ya^t_{M,M-1} & y & & & b^t_{N-1} & a^t_{1,N} \end{pmatrix}.$$

Define two maps $\psi : T \to L$ and $\phi : L \to F$ given by

$$\psi((I^t_n, I^{t+\frac{1}{M}}_n, \ldots, I^{t+\frac{M-1}{M}}_n, V^t_n)_{n=1,\ldots,N}) = L^t(y)$$
$$\phi(L^t(y)) = (-1)^{N+1} y \operatorname{Det}(L^t(y) - x\mathbb{I}_N),$$

Via the map ψ (resp. $\phi \circ \psi$), we can regard F as a set of polynomial functions on L (resp. T). We write n_F for the number of the polynomial functions in F, which is $n_F = \frac{1}{2}(M+1)(N+1)$.

Proposition 3.1. *The n_F functions in F are functionally independent in $\mathbb{C}[T]$.*

To prove this proposition we use the following:

Lemma 3.1. *Define*

$$I^{t+\frac{k}{M}} = \prod_{n=1}^{N} I^{t+\frac{k}{M}}_n, \quad V^t = \prod_{n=1}^{N} V^t_n \quad (t \in \mathbb{Z},\ k = 0, \ldots, M-1).$$

The Jacobian of ψ does not vanish iff $I^{t+\frac{k}{M}} \neq I^{t+\frac{j}{M}}$ for $0 \le k < j \le M-1$ and $I^{t+\frac{k}{M}} \neq V^t$ for $0 \le k \le M-1$.

Proof. Since the dimensions of T and L are same, the Jacobian matrix of ψ becomes an $(M+1)N$ by $(M+1)N$ matrix. By using elementary transformation, one sees that the Jacobian matrix is block diagonalized into $M+1$ matrices of N by N, and the Jacobian is factorized as

$$\pm \operatorname{Det} B \cdot \prod_{k=1}^{M-1} \operatorname{Det} A^{(k)},$$

where $A^{(k)}$ and B are

$$A^{(k)} = P(I^{t+\frac{k}{M}}, I^{t+\frac{k-1}{M}})P(I^{t+\frac{k+1}{M}}, I^{t+\frac{k-1}{M}}) \cdots P(I^{t+\frac{M-1}{M}}, I^{t+\frac{k-1}{M}}) \quad (k = 1, \dots, M-1),$$

$$B = P(I^{t+\frac{M-1}{M}}, V^t)P(I^{t+\frac{M-2}{M}}, V^t) \cdots P(I^{t+\frac{1}{M}}, V^t)P(I^t, V^t),$$

$$P(J, K) = \begin{pmatrix} J_1 & & & & -K_N \\ -K_1 & J_2 & & & \\ & \ddots & \ddots & & \\ & & \ddots & \ddots & \\ & & & -K_{N-1} & J_N \end{pmatrix} \in M_N(\mathbb{C}).$$

Since $\mathrm{Det}\, P(J, K) = \prod_{n=1}^N J_n - \prod_{n=1}^N K_n$, we obtain

$$\mathrm{Det}\, B \cdot \prod_{k=1}^{M-1} \mathrm{Det}\, A^{(k)} = \prod_{0 \le k \le M-1} (I^{t+\frac{k}{M}} - V^t) \cdot \prod_{0 \le j < k \le M-1} (I^{t+\frac{k}{M}} - I^{t+\frac{j}{M}}).$$

Thus the claim follows. □

Remark 3.1. The above lemma is true for $\gcd(M, N) > 1$, too.

Proof. (Proposition 3.1)
Take a generic $f \in F$ such that the algebraic curve C_f given by $f = 0$ is smooth. The genus g of C_f is $\frac{1}{2}(N-1)(M+1)$, and we have $\dim L = n_F + g$. Due to the result by Mumford and van Moerbeke [15, Theorem 1], the isolevel set $\phi^{-1}(f)$ is isomorphic to the affine part of the Jacobian variety $\mathrm{Jac}(C_f)$ of C_f, which denotes $\dim_{\mathbb{C}} \phi^{-1}(f) = g$. Thus F has to be a set of n_F functionally independent polynomials in $\mathbb{C}[L]$. Then the claim follows from Lemma 3.1. □

3.2. *Generalized ultradiscrete periodic Toda lattice $\mathcal{T}(M, N)$*

We consider the difference equation (2) on the phase space T_ε:

$$\Big\{ (I_n^t, I_n^{t+\frac{1}{M}}, \dots, I_n^{t+\frac{M-1}{M}}, V_n^t)_{n=1,\dots,N} \in K_\varepsilon^{N(M+1)} \;\Big|\; \mathrm{val}\Big(\prod_{n=1}^N I_n^{t+\frac{k}{M}}\Big) < \mathrm{val}\Big(\prod_{n=1}^N V_n^t\Big) \ (k = 0, \dots, M-1),$$
$$\mathrm{val}\Big(\prod_{n=1}^N I_n^{t+\frac{k}{M}}\Big) \ (k = 0, \dots, M-1) \text{ are distinct} \Big\}.$$

We assume $\gcd(M,N)=1$. Let $F_\varepsilon \subset K_\varepsilon[x,y]$ be the set of polynomials over K_ε defined by the similar formula to (6), *i.e.*

$$F_\varepsilon = \Big\{ y^{M+1} + \sum_{j=0}^{M} f_j(x) y^j \in K_\varepsilon[x,y] \;\Big|\; \deg_x f_j \le d_j, \;\; -f_1(x) \text{ is monic} \Big\}.$$

The *tropicalization* of the above system becomes the generalized ultra-discrete periodic Toda lattice $\mathcal{T}(M,N)$, which is the piecewise-linear map:

$$\begin{aligned} Q_n^{t+1} &= \min[W_n^t, Q_n^t - X_n^t], \\ W_n^{t+\frac{1}{M}} &= Q_{n+1}^t + W_n^t - Q_n^{t+1}, \end{aligned} \qquad (n \in \mathbb{Z}/N\mathbb{Z},\ t \in \mathbb{Z}/M), \tag{7}$$

$$\text{where } X_n^t = \min_{k=0,\dots,N-1} [\sum_{j=1}^{k} (W_{n-j}^t - Q_{n-j}^t)],$$

on the phase space $\mathcal{T}$:

$$\mathcal{T} = \Big\{ (Q_n^t, Q_n^{t+\frac{1}{M}}, \dots, Q_n^{t+\frac{M-1}{M}}, W_n^t)_{n=1,\dots,N} \in \mathbb{R}^{(M+1)N} \;\Big|\; \sum_n Q_n^{t+\frac{k}{M}} < \sum_n W_n^t \ (k=0,\dots,M-1), \ \sum_n Q_n^{t+\frac{k}{M}} \ (k=0,\dots,M-1) \text{ are distinct} \Big\}.$$

Here we set $\mathrm{val}(I_n^t) = Q_n^t$ and $\mathrm{val}(V_n^t) = W_n^t$. The tropicalization of F_ε becomes the space of tropical polynomials on $\mathcal{T}$:

$$\begin{aligned} \mathcal{F} = \Big\{ \min\Big[(M+1)Y, \min_{j=1,\dots,M} \big[jY + \min[d_j X + F_{j,d_j}, \dots, \\ X + F_{j,1}, F_{j,0}] \big], \ F_0 \Big] \;\Big|\; F_{1,d_1} = 0, \ F_{j,i}, F_0 \in \mathbb{R} \Big\}. \end{aligned} \tag{8}$$

We write Φ for the map $\mathcal{T} \to \mathcal{F}$.

3.3. *Spectral curves for $T(M,N)$ and good tropicalization*

We continuously assume $\gcd(M,N)=1$.

Proposition 3.2. *For a generic point $\tau \in T_\varepsilon$, that is a point in a certain Zariski open subset of T_ε, the spectral curve $\phi \circ \psi(\tau)$ has a good tropicalization.*

To show this proposition, we use the following lemma:

Lemma 3.2. *Fix $l \in K_\varepsilon[x,y]$ and set $h_t = y^{M+1} - x^N y + tl$, where $t \in K_\varepsilon$. Then $C^0(h_t)$ is non-singular in $(K_\varepsilon^\times)^2$ except for finitely many t.*

Proof. Fix $a, b \in \mathbb{Z}$ as $Ma - Nb = 1$. (It is always possible since $\gcd(M, N) = 1$.) Then the map $\nu : (K_\varepsilon^\times)^2 \to (K_\varepsilon^\times)^2$; $(x, y) \mapsto (u, v) = (x^N/y^M, x^a/y^b)$ is holomorphic. The push forward of h_t/y^{M+1} by ν becomes

$$\tilde{h} := (h_t/y^{M+1})_* = (1 - u + t\tilde{l}) \quad (\tilde{l} \in K_\varepsilon[u, v, u^{-1}, v^{-1}]).$$

By using the following claim, we see that $C^0(\tilde{h})$ is non-singular.

Claim 3.1. *Fix $f, g \in K_\varepsilon[u, v]$ such that $C^0(f)$ is non-singular, and f and g are coprime to each other. Define*

$$U = \{t \in K_\varepsilon \mid C^0(f + tg) \text{ is singular } \} \subset K_\varepsilon.$$

Then U is a finite algebraic set.

Then the lemma follows. □

Proof. (Proposition 3.2)
Recall the definition of good tropicalization (Definition 2.1).

The part (1) follows from Proposition 3.1 immediately.

The part (2): For any $f_\varepsilon \in F_\varepsilon$, it can be easily checked that if two points $P_1, P_2 \in TV(f_\varepsilon)$ exist on a same edge of the tropical curve, then $f_\varepsilon^{P_1} = f_\varepsilon^{P_2}$. This fact implies that the set $\{f_\varepsilon^P \mid P \in TV(f_\varepsilon)\}$ is finite. Therefore, the set

$$\triangle = \{f_\varepsilon \in F_\varepsilon \mid C^0(f_\varepsilon^P) \text{ is non-reduced or singular in } (K_\varepsilon^\times)^2 \text{ for some } P \in TV(f_\varepsilon)\}$$

is a union of finitely many non-trivial algebraic subsets of $F_\varepsilon \simeq K_\varepsilon^{n_F}$. Using Proposition 3.1 (with the map ι with any $\bar{\varepsilon} \in \mathbb{R}_{>0}$) and Lemma 3.2, we conclude that $(\phi \circ \psi)^{-1}(\triangle) \subset T_\varepsilon$ is an analytic subset with positive codimension. (We need Lemma 3.2 when f_ε^P includes $y^{M+1} - x^N y$.) □

4. General solutions to $\mathcal{T}(M, N)$

4.1. *Bilinear equation*

In the following we use a notation $[t] = t \bmod 1$ for $t \in \frac{\mathbb{Z}}{M}$. The following proposition gives the bilinear form for $\mathcal{T}(M, N)$:

Proposition 4.1. *Let $\{T_n^t\}_{n \in \mathbb{Z}; t \in \frac{\mathbb{Z}}{M}}$ be a set of functions with a quasi-periodicity, $T_{n+N}^t = T_n^t + (an + bt + c)$ for some $a, b, c \in \mathbb{R}$. Fix $\delta^{[t]}, \theta^{[t]} \in \mathbb{R}$ such that*

(*a*) *$\delta^{[t]} + \theta^{[t]}$ does not depend on t,* (*b*) *$2b - a < N\theta^{[t]}$ for $t \in \mathbb{Z}/M$.*

Assume T_n^t satisfies

$$T_n^t + T_n^{t+1+\frac{1}{M}} = \min[T_n^{t+1} + T_n^{t+\frac{1}{M}}, T_{n-1}^{t+1+\frac{1}{M}} + T_{n+1}^t + \theta^{[t]}]. \tag{9}$$

Then T_n^t gives a solution to (7) via the transformation:

$$\begin{aligned} Q_n^t &= T_{n-1}^t + T_n^{t+\frac{1}{M}} - T_{n-1}^{t+\frac{1}{M}} - T_n^t + \delta^{[t]}, \\ W_n^t &= T_{n-1}^{t+1} + T_{n+1}^t - T_n^t - T_n^{t+1} + \delta^{[t]} + \theta^{[t]}. \end{aligned} \tag{10}$$

We omit the proof since it is essentially same as that of $M = 1$ case in [3, §3].

Remark 4.1. Via (10), (7) is directly related to

$$\begin{aligned} T_n^t + T_n^{t+1+\frac{1}{M}} &= T_n^{t+1} + T_n^{t+\frac{1}{M}} + X_{n+1}^t, \\ X_n^t &= \min_{j=0,\dots,M-1} \Big[j\theta^{[t]} + T_{n-j-1}^{t+\frac{1}{M}} + T_{n-j-1}^{t+1} + T_n^t + T_{n-1}^t \\ &\qquad - (T_{n-1}^{t+1} + T_{n-1}^{t+\frac{1}{M}} + T_{n-j}^t + T_{n-j-1}^t) \Big]. \end{aligned}$$

This is shown to be equivalent to (9) under the quasi-periodicity of T_n^t. See [3, Propositions 3.3 and 3.4] for the proof.

4.2. *Example:* $\mathcal{T}(3,2)$

We demonstrate a general solution to $\mathcal{T}(3,2)$. Take a generic point $\tau \in T_\varepsilon$, and the spectral curve $C(f_\varepsilon)$ for $T(3,2)$ on K_ε is given by the zero of $f_\varepsilon = \phi \circ \psi(\tau) \in F_\varepsilon$:

$$f_\varepsilon = y^4 + y^3 f_{30} + y^2(x f_{21} + f_{20}) + y(-x^2 + x f_{11} + f_{10}) + f_0.$$

Due to Proposition 3.2, $C(f_\varepsilon)$ has a good tropicalization. The tropical curve $\Gamma := TV(f_\varepsilon)$ in $\mathbb{R}^2$ is the indifferentiable points of $\xi := \mathrm{Val}(X, Y; f_\varepsilon)$:

$$\min\big[4Y, 3Y + F_{30}, 2Y + \min[X + F_{21}, F_{20}], Y + \min[2X, X + F_{11}, F_{10}], F_0\big].$$

We assume that Γ is smooth, then its genus is $g = 2$. See Figure 1 for Γ, where we set the basis γ_1, γ_2 of $\pi_1(\Gamma)$. The period matrix B for Γ becomes

$$B = \begin{pmatrix} 2F_0 - 7F_{11} + F_{20} & F_{11} - F_{20} \\ F_{11} - F_{20} & F_{11} + F_{20} \end{pmatrix},$$

and the tropical Jacobi variety $J(\Gamma)$ of Γ is

$$J(\Gamma) = \mathbb{R}^2 \,/\, \mathbb{Z}^2 B.$$

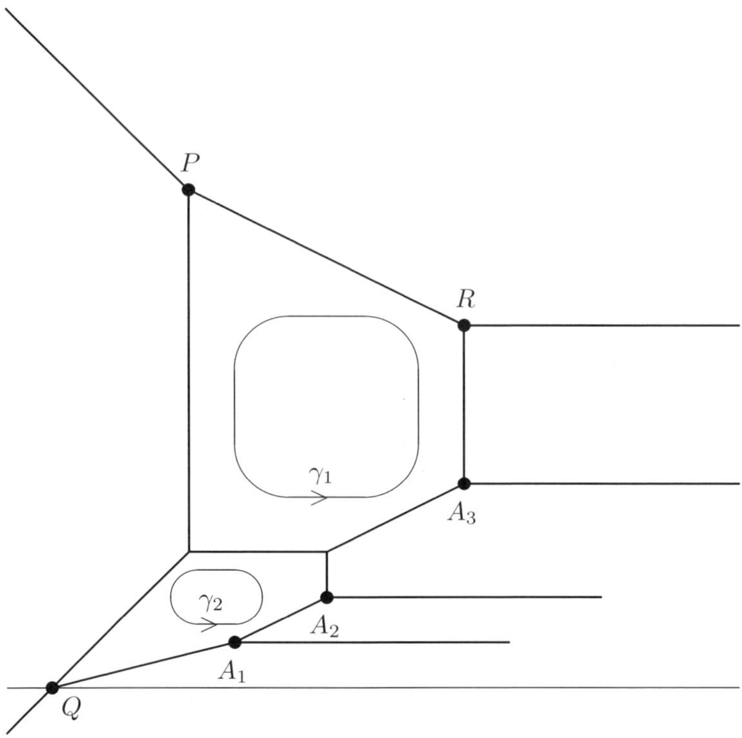

Fig. 1. Tropical spectral curve Γ for $\mathcal{T}(3,2)$

We fix 6 points on the universal covering space of Γ as follows:

$$\vec{L} = \int_P^Q = (F_0 - 3F_{11}, F_{11}),$$
$$\vec{\lambda}_1 = \int_Q^{A_3} = (F_{10} - 2F_{11}, -F_{20}), \quad \vec{\lambda}_2 = \int_Q^{A_2} = (0, F_{20} - F_{30}),$$
$$\vec{\lambda}_3 = \int_Q^{A_1} = (0, F_{30}), \quad \vec{\lambda} = \int_R^P = (F_{10} - 2F_{11}, 0).$$

Here the path $\gamma_{Q\to A_3}$ from Q to A_3 is chosen as $\gamma_{Q\to A_3} \cap \gamma_1 \cap \gamma_2 \neq \emptyset$. Remark that $\vec{\lambda} = \vec{\lambda}_1 + \vec{\lambda}_2 + \vec{\lambda}_3$ holds.

Proposition 4.2. *Fix $\mathbf{Z}_0 \in \mathbb{R}^2$. The tropical theta function $\Theta(\mathbf{Z}; B)$ satis-*

fies the following identities:

$$\begin{aligned}&\Theta(\mathbf{Z}_0) + \Theta(\mathbf{Z}_0 + \vec{\lambda} + \vec{\lambda}_i)\\ &\quad = \min[\Theta(\mathbf{Z}_0 + \vec{\lambda}) + \Theta(\mathbf{Z}_0 + \vec{\lambda}_i), \Theta(\mathbf{Z}_0 - \vec{L}) + \Theta(\mathbf{Z}_0 + \vec{L} + \vec{\lambda} + \vec{\lambda}_i) + \theta_i],\end{aligned} \tag{11}$$

for $i = 1, 2, 3$, *where* $\theta_1 = F_0 - 3F_{11}$ *and* $\theta_2 = \theta_3 = F_0 - 2F_{11}$.

Proof. Since the curve $C(f_\varepsilon)$ is hyperelliptic, we fix a non-singular odd theta characteristic as $(\alpha, \beta) = ((\frac{1}{2}, 0), (\frac{1}{2}, \frac{1}{2}))$ following Lemma 2.1. By setting $(P_1, P_2, P_3, P_4) = (R, Q, P, A_{4-i})$ in Theorem 2.2 for $i = 1, 2, 3$, we obtain (11). □

Now it is easy to show the following:

Proposition 4.3. *(i) Fix* $\mathbf{Z}_0 \in \mathbb{R}^2$ *and* $\{i, j\} \subset \{1, 2, 3\}$, *and define* T_n^t *by*

$$\begin{aligned}T_n^t &= \Theta(\mathbf{Z}_0 - \vec{L}n + \vec{\lambda}t),\\ T_n^{t+\frac{1}{3}} &= \Theta(\mathbf{Z}_0 - \vec{L}n + \vec{\lambda}t + \vec{\lambda}_i), \qquad\qquad (t \in \mathbb{Z}).\\ T_n^{t+\frac{2}{3}} &= \Theta(\mathbf{Z}_0 - \vec{L}n + \vec{\lambda}t + \vec{\lambda}_i + \vec{\lambda}_j),\end{aligned}$$

Then they satisfy the bilinear equation (9) *with* $\theta^{[0]} = \theta_i$, $\theta^{[\frac{1}{3}]} = \theta_j$ *and* $\theta^{[\frac{2}{3}]} = \theta_k$, *where* $\{k\} = \{1, 2, 3\} \setminus \{i, j\}$.
(ii) With (i) and $\delta^{[\frac{k}{3}]} = F_0 - 2F_{11} - \theta^{[\frac{k}{3}]}$ $(k = 0, 1, 2)$, *we obtain a general solution to* $\mathcal{T}(3, 2)$.

Remark 4.2. Depending on a choice of $\{i, j\} \subset \{1, 2, 3\}$, we have $3! = 6$ types of solutions. This suggests a claim for the isolevel set $\Phi^{-1}(\xi)$:

$$\Phi^{-1}(\xi) \simeq J(\Gamma)^{\oplus 6}.$$

4.3. *Conjectures on $\mathcal{T}(M, N)$*

We assume $\gcd(M, N) = 1$ again. Let Γ be the smooth tropical curve given by the indifferentiable points of a tropical polynomial $\xi \in \mathcal{F}$ (8). We fix the basis of $\pi_1(\Gamma)$ by using $\gamma_{i,j}$ $(i = 1, \ldots, M,\ j = 1, \ldots, d_i)$ as Figure 2. The genus $g = \frac{1}{2}(N-1)(M+1)$ of Γ can be obtained by summing up d_j from $j = 1$ to $\max_{j=1,\ldots,M}\{j \mid d_j \geq 1\}$.

Fix three points P, Q, R on the universal covering space $\tilde{\Gamma}$ of Γ as Figure 2, and define

$$\vec{L} = \int_P^Q, \qquad \vec{\lambda} = \int_R^P.$$

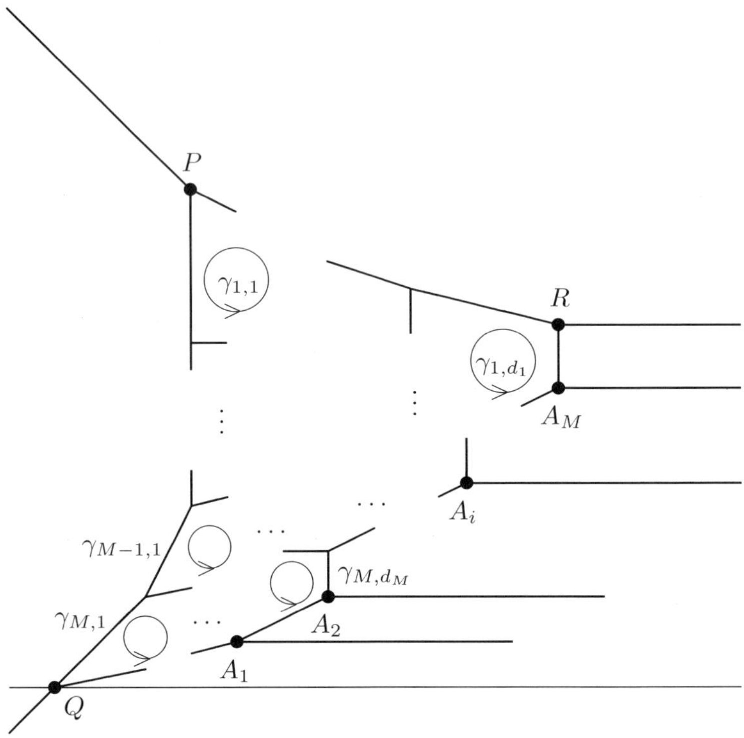

Fig. 2. Tropical spectral curve Γ for $\mathcal{T}(M,N)$

Fix A_i $(i=1,\ldots,M)$ on $\tilde{\Gamma}$ as Figure 2, such that

$$\vec{\lambda}_i = \int_Q^{A_{M+1-i}} \quad (i=1,\ldots,M)$$

satisfy $\vec{\lambda} = \sum_{i=1}^M \vec{\lambda}_i$.

We expect that the bilinear form (9) is obtained as a consequence of the tropical Fay's identity (1), by setting $(P_1,P_2,P_3,P_4)=(R,Q,P,A_i)$ in Theorem 2.2. The followings are our conjectures:

Conjecture 4.1. *Let $\mathcal{S}_M$ be the symmetric group of order M. Fix $\mathbf{Z}_0 \in \mathbb{R}^g$ and $\sigma \in \mathcal{S}_M$, and set*

$$T_n^{t+\frac{k}{M}} = \Theta(\mathbf{Z}_0 - \vec{L}n + \vec{\lambda}t + \sum_{i=1}^{k} \lambda_{\sigma(i)})$$

for $k=0,\ldots,M-1$. Then the followings are satisfied:

(i) T_n^t *satisfy* (9) *with some* $\theta^{[t]}$.
(ii) T_n^t *gives a general solution to* $\mathcal{T}(M,N)$ *via* (10).

Conjecture 4.2. *The above solution induces the isomorphism map from* $J(\Gamma)^{\oplus M!}$ *to the isolevel set* $\Phi^{-1}(\xi)$.

Remark 4.3. In the case of $\mathcal{T}(1,g+1)$ and $\mathcal{T}(2g-1,2)$, the smooth tropical spectral curve Γ is hyperelliptic and has genus g. For $\mathcal{T}(1,g+1)$, Conjectures 4.1 and 4.2 are completely proved.[3,4] For $\mathcal{T}(3,2)$, Conjecture 4.1 is shown in §4.2.

Acknowledgements

R. I. thanks the organizers of the international conference "Infinite Analysis 09 — New Trends in Quantum Integrable Systems" held in Kyoto University in July 2009, for giving her an opportunity to give a talk. She also thanks Takao Yamazaki for advice on the manuscript.

R. I. is partially supported by Grant-in-Aid for Young Scientists (B) (19740231).

S. I. is supported by KAKENHI (21-7090).

References

1. M. Einsiedler, M. Kapranov and D. Lind, *Non-Archimedean amoebas and tropical varieties*, J. Reine Angew. Math. **601**, 139-157 (2006).
2. R. Inoue and T. Takenawa, *Tropical spectral curves and integrable cellular automata*, Int. Math. Res. Not. IMRN **27** Art ID. rnn019, 27 pp (2008)
3. R. Inoue and T. Takenawa, *A tropical analogue of Fay's trisecant identity and an ultra-discrete periodic Toda lattice*, Comm. Math. Phys., **289**, pp 995-1021 (2009).
4. R. Inoue and T. Takenawa, *Tropical Jacobian and the generic fiber of the ultra-discrete periodic Toda lattice are isomorphic*, RIMS Kôkyûroku Bessatsu, B13, pp 175-190 (2009).
5. I. Itenberg, G. Mikhalkin and E. Shustin, *Tropical algebraic geometry*, Oberwolfach Seminars, 35 (Birkhüser Verlag, Basel, 2007).
6. S. Iwao, *Integration over tropical plane curves and ultradiscretization*, Int. Math. Res. Not. IMRN **2010** (1) 112–148 (2009).
7. S. Iwao, *The periodic box-ball system and tropical curves*, RIMS Kôkyûroku Bessatsu, B13, pp 157-174 (2009).
8. S. Iwao, *Solution of the generalized periodic discrete Toda equation II; Theta function solution*, arXiv:0912.2213.
9. T. Kimijima and T. Tokihiro, *Initial-value problem of the discrete periodic Toda equations and its ultradiscretization*, Inverse Problems **18**, 1705-1732 (2002).

10. A. Kuniba and T. Takagi, *Bethe Ansatz, Inverse Scattering Transform and Tropical Riemann Theta Function in a Periodic Soliton Cellular Automaton for* $A_n^{(1)}$, arXiv:0909.3759.
11. G. Mikhalkin, *Enumerative tropical algebraic geometry in* $\mathbb{R}^2$, J. Amer. Math. Soc. **18**, 313-377 (2005).
12. G. Mikhalkin and I. Zharkov, *Tropical curves, their Jacobians and theta functions*, Contemp. Math., **465**, 203–230 (2008).
13. D. Speyer and B. Sturmfels, *Tropical Mathematics*, Math. Mag., **82**, no. 3, 163–173 (2009).
14. T. Tokihiro, A. Nagai and J. Satsuma, *Proof of solitonical nature of box and ball system by means of inverse ultra-discretization*, Inverse Problems, **15**, 1639–1662 (1999).
15. P. van Moerbeke and D. Mumford, *The spectrum of difference operators and algebraic curves*, Acta Math. **143**, 93-154 (1979).

ON ONE-POINT FUNCTIONS OF DESCENDANTS IN SINE-GORDON MODEL

MICHIO JIMBO

Department of Mathematics, Rikkyo University, Toshima-ku, Tokyo 171-8501, Japan
E-mail: jimbomm@rikkyo.ac.jp

TETSUJI MIWA

Department of Mathematics, Graduate School of Science, Kyoto University, Kyoto 606-8502, Japan
E-mail: tmiwa@math.kyoto-u.ac.jp

FEDOR SMIRNOV*

Laboratoire de Physique Théorique et Hautes Energies, Université Pierre et Marie Curie, Tour 16 1er étage, 4 Place Jussieu 75252 Paris Cedex 05, France
E-mail: smirnov@lpthe.jussieu.fr

We apply the fermionic description of CFT developed in our previous work to the computation of the one-point functions of the descendent fields in the sine-Gordon model.

Keywords: Conformal field theory; sine-Gordon model.

1. Introduction

The sine-Gordon (sG) model is the most famous example of two-dimensional integrable Quantum Field Theory (QFT). The sG model is defined in two-dimensional Minkowski space with coordinate $\mathbf{x} = (x_0, x_1)$ by the action

$$\mathcal{A}_{\mathrm{sG}} = \int \left\{ \frac{1}{16\pi} (\partial_\mu \boldsymbol{\varphi}(\mathbf{x}))^2 + \frac{2\boldsymbol{\mu}^2}{\sin(\pi\beta^2)} \cos(\beta\boldsymbol{\varphi}(\mathbf{x})) \right\} d^2\mathbf{x} \,. \tag{1}$$

The normalisation of the dimensional coupling constant in front of $\cos(\beta\boldsymbol{\varphi}(\mathbf{x}))$ is chosen for future convenience. This model has been a subject of intensive study during the last 30 years. First, by semi-classical methods

*Membre du CNRS

the exact spectrum was computed, the factorisation of scattering was predicted, and the exact S-matrix was found for certain values of the coupling constant (in the absence of reflection of solitons).[2,3]

The most significant further results were found by the bootstrap method. In Refs. 4,5 the exact S-matrix was found, and in Refs. 6,7 the exact form factors were computed for the energy-momentum tensor, topological current and the operators $e^{\pm i\beta\varphi(\mathbf{x})}$, $e^{\pm\frac{i\beta}{2}\varphi(\mathbf{x})}$. Then the latter result was generalised to the operator $e^{ia\varphi(\mathbf{x})}$ with an arbitrary a,[8] using the methods which go back to the study of the closely related XXZ spin chain.[9,10]

It should be said that many important technical and conceptual methods of the modern theory of quantum integrable models originate in the quantum inverse scattering method.[11,12] It provides a clear mathematical interpretation of the work by R. Baxter.[13] In particular, in Ref. 12 the scattering matrix of the sG solitons was reproduced using the Bethe Ansatz.

The knowledge of form factors allows us to write a series representation for the two-point function

$$\langle e^{ia_1\varphi(\mathbf{x})}e^{ia_2\varphi(0)}\rangle_{\mathrm{sG}} \,. \tag{2}$$

In this paper we shall consider only the space-like region $\mathbf{x}^2 < 0$. We shall use a lattice regularisation which breaks the Lorentz invariance. So we shall take $\mathbf{x} = (0, x)$, and use the notation $\varphi(x) = \varphi(0, x)$. In this case the integrals over the form factors are rapidly convergent. It is rather hard to give a mathematically rigorous proof of the convergence of the series, but nobody doubts that they converge. Actually the convergence was proved[14] for a certain particular reduction of the sG model, known as the scaling Lee-Yang theory. So, the situation looks completely satisfactory. However, the series over the form factors converge slowly in the ultraviolet region (for small values of $-x^2$). To give an efficient description to its ultraviolet behaviour remained a problem largely open for some time. Before describing the way of solving this problem let us explain how it is solved in a particular case.

It is known that at $\beta^2 = 1/2$ the sine-Gordon model is equivalent to the free theory of a Dirac fermion. The correlation functions of (2) are non-trivial. The sG field $\varphi(x)$ is bilinear in fermions, and one has to compute the correlation function of two exponentials of bilinear forms. The result is obtained in Ref. 15, generalising the seminal work[16] on the scaling Ising model. Namely, it is shown that the correlation function satisfies an equivalent of a Painlevé III equation. The form factors are very simple in this case, and the form factor series coincides with the Fredholm determinant representation of this solution. Still the problem of describing the ultra-violet behaviour is non-trivial. It amounts to finding the connection coefficients for the Painlevé equation, which is done by studying the Riemann-Hilbert problem.[17] Through this analysis, one draws an important conclusion. Since CFT completely describes the ultra-violet limit, one might naïvely expect that the asymptotic behaviour of the two-point function may be obtained via the perturbation theory. Such an assumption would imply that the dependence on the mass scale is analytic in $\boldsymbol{\mu}^2$ which has the dimension $(\text{mass})^{2(1-\beta^2)}$. However this is not the case even for the Painlevé solution.

Correct understanding of the conformal perturbation theory is one of the most important problems in the theory of quantum integrable systems. This problem was studied in Ref. 18. In fact, the naïve perturbation theory suffers from both ultraviolet and infrared divergencies. The idea of Ref. 18 is to absorb all these divergencies into non-perturbative data: one-point functions of primary fields and their descendants. Once it is done, the remaining task is a convergent version of conformal perturbation theory. So the problem is divided into two steps. The first one requires some non-perturbative information. The second one is of genuinely CFT origin: actually, it is reduced to the computation of some Dotsenko-Fateev Coulomb gas integrals with screenings.[19]

In principle, the procedure described in Ref. 18 provides an asymptotic series in the ultra-violet domain which agrees with the structure expected from the Painlevé case: it is not just a power series in $\boldsymbol{\mu}^2$, but includes non-analytic contributions with fractional powers of $\boldsymbol{\mu}^2$.

So the main problem is to compute the one-point functions. The first important result in this direction was achieved in Ref. 20, where an exact formula for the one-point functions of the primary fields was conjectured. Then by several ingenious tricks (such as going form sine-Gordon to sinh-Gordon and back) a procedure was described in Refs. 21–23, which must in principle allow us to compute the correlation functions of the descendants. Unfortunately, this procedure involves certain matrix Riemann-Hilbert problem which has not been solved in general up to now. At the same time, this way of computation looks very indirect, and involves steps which are hard to justify. Still, the predicted result for the first not-trivial descendant[23] is quite remarkable. Even though it was obtained by a complicated and non-rigorous procedure, it was checked against many particular cases. So we do not doubt in its validity. It will be used to check the result of the present paper.

Let us mention here a deep relation between the sG model and the $\Phi_{1,3}$-

perturbation of $c < 1$ models of CFT.[24] On the formal level the relation is simple. In the action (1) one can split $\cos(\beta\boldsymbol{\varphi}(\mathbf{x})) = \frac{1}{2}(e^{-i\beta\boldsymbol{\varphi}(\mathbf{x})} + e^{i\beta\boldsymbol{\varphi}(\mathbf{x})})$ and consider the first term as a part of the Liouville action and the second as a perturbation. The Liouville model with an imaginary exponent is nothing but CFT with the central charge

$$c = 1 - 6(\beta - 1/\beta)^2\,,$$

and $e^{i\beta\boldsymbol{\varphi}(\mathbf{x})}$ is the field $\Phi_{1,3}(\mathbf{x})$. So, formally there is no difference between the sG model and the $\Phi_{1,3}$-perturbation of $c < 1$ CFT. The situation becomes interesting in the case of rational β^2. It was shown in Refs. 25,26 that in the computation of the correlation functions of $e^{\frac{in\beta}{2}\varphi(x)}$ $(n = 1, 2, 3, \cdots)$, a restriction of degrees of freedom takes place for the intermediate states of solitons. The mechanism of reduction is similar to the RSOS restriction for vertex models. This phenomenon is known as the restriction of the sG model. For example, if $\beta^2 = 3/4$ the solitons reduce to Majorana fermions, and the restricted model is nothing but the scaling Ising model.

In the series of papers, Refs. 27,28,29,1 which we will refer to as I,II,III,IV, respectively, we studied the hidden fermionic structure of the XXZ spin chain. In particular, in IV the relation between our fermions in the scaling limit and $c < 1$ CFT was established. The long distance behaviour of the XXZ model and the short distance behaviour of the sG model are described by the same CFT: free bosons with the compactification radius β^2. For the XXZ model, we use the coupling parameter ν related to β^2 via

$$\beta^2 = 1 - \nu\,.$$

This identification is used when relating the results of IV to those of Refs. 30,31, which were important for us. The relevant CFT has the central charge

$$c = 1 - 6\frac{\nu^2}{1-\nu}\,.$$

Here we do not consider the peculiarity of a rational ν, so using the $c < 1$ CFT means just the usual modification of the energy-momentum tensor. In the ultraviolet limit, the sG model is described by two chiral copies of CFT. We use the notation

$$\Phi_\alpha(x) = e^{\frac{\nu}{1-\nu}\alpha\left\{\frac{i\beta}{2}\varphi(x)\right\}}\,.$$

The field $i\beta\boldsymbol{\varphi}(x)$ splits into two chiral fields $2\varphi(x) + 2\bar{\varphi}(x)$. (We follow the normalisation of the fields $\varphi(x), \bar{\varphi}(x)$ given in Refs. 30,31 and in IV.) Our goal is to compute the one point functions

$$\frac{\langle P(\{\mathbf{l}_{-m}\})\bar{P}(\{\bar{\mathbf{l}}_{-m}\})\Phi_\alpha(0)\rangle_{sG}}{\langle\Phi_\alpha(0)\rangle_{sG}},$$

where $P(\{\mathbf{l}_{-m}\})$, $\bar{P}(\{\bar{\mathbf{l}}_{-m}\})$ are polynomials in the generators of two chiral copies of the Virasoro algebra with the central charge c.

The universal enveloping algebra of the Virasoro algebra contains the local integrals of motion $\mathbf{i}_{2k-1}$ $(\mathbf{i}_1 = \mathbf{l}_{-1})$[32] which survive the $\Phi_{1,3}$-perturbation. Clearly, the one-point functions of descendants created by them vanish. We assume that it is possible to write any element of the chiral Verma module generated by $\Phi_\alpha(0)$ as $P_1(\{\mathbf{i}_{2k-1}\})P_2(\{\mathbf{l}_{-2k}\})\Phi_\alpha(0)$. So, actually we are interested in computing

$$\frac{\langle P(\{\mathbf{l}_{-2m}\})\bar{P}(\{\bar{\mathbf{l}}_{-2m}\})\Phi_\alpha(0)\rangle_{sG}}{\langle\Phi_\alpha(0)\rangle_{sG}}.$$

For the simplest non-trivial case $\mathbf{l}_{-2}\bar{\mathbf{l}}_{-2}\Phi_\alpha(0)$ the problem was solved in Ref. 23. If one wants to consider the descendants created by the Heisenberg algebra it is easy to do using the formulae

$$(1-\nu)T(z) =: \varphi'(z)^2 : +\nu\varphi(z), \quad (1-\nu)\bar{T}(\bar{z}) =: \bar{\varphi}'(\bar{z})^2 : +\nu\bar{\varphi}(\bar{z}).$$

In this paper we shall consider only the domain $0 < \alpha < 2$, but the final results allow analytical continuation for all values of α.

The whole idea of IV is that the usual basis of the Verma module is not suitable for the perturbation, and we have to introduce another one. For the moment we consider only one chirality. Working modulo action of the local integrals of motion the new basis is provided by uncharged products of two fermions $\boldsymbol{\beta}^{\mathrm{CFT}*}_{2k-1}$ and $\boldsymbol{\gamma}^{\mathrm{CFT}*}_{2k-1}$. The fermions respect the Virasoro grading:

$$[\mathbf{l}_0, \boldsymbol{\beta}^{\mathrm{CFT}*}_{2j-1}\boldsymbol{\gamma}^{\mathrm{CFT}*}_{2k-1}] = (2j+2k-2)\boldsymbol{\beta}^{\mathrm{CFT}*}_{2j-1}\boldsymbol{\gamma}^{\mathrm{CFT}*}_{2k-1}.$$

We have

$$\boldsymbol{\beta}^{\mathrm{CFT}*}_{I^+}\boldsymbol{\gamma}^{\mathrm{CFT}*}_{I^-}\Phi_\alpha(0) = \Big\{P^{\mathrm{even}}_{I^+,I^-}(\{\mathbf{l}_{-2k}\}) + d_\alpha P^{\mathrm{odd}}_{I^+,I^-}(\{\mathbf{l}_{-2k}\})\Big\}\Phi_\alpha(0), \quad (3)$$

where I^+ and I^- are ordered multi-indices: for $I = (2r_1-1, \cdots 2r_n-1)$ with $r_1 < \cdots < r_n$, we set

$$\boldsymbol{\beta}^{\mathrm{CFT}*}_{I} = \boldsymbol{\beta}^{\mathrm{CFT}*}_{2r_1-1}\cdots\boldsymbol{\beta}^{\mathrm{CFT}*}_{2r_n-1}, \quad \boldsymbol{\gamma}^{\mathrm{CFT}*}_{I} = \boldsymbol{\gamma}^{\mathrm{CFT}*}_{2r_n-1}\cdots\boldsymbol{\gamma}^{\mathrm{CFT}*}_{2r_1-1},$$

and similarly for $\bar{\boldsymbol{\beta}}_I$, $\bar{\boldsymbol{\gamma}}_I$. We require $\#(I^+) = \#(I^-)$. In the right hand side of (3), $P^{\mathrm{even}}_{I^+,I^-}(\{\mathbf{l}_{-2k}\})$ and $P^{\mathrm{odd}}_{I^+,I^-}(\{\mathbf{l}_{-2k}\})$ are homogeneous polynomials, the constant d_α is given by

$$d_\alpha = \frac{\nu(\nu-2)}{\nu-1}(\alpha-1) = \tfrac{1}{6}\sqrt{(25-c)(24\Delta_\alpha+1-c)},$$

and the separation into even and odd parts is determined by the reflection $\boldsymbol{\beta}^{\rm CFT*}_{I^+}\boldsymbol{\gamma}^{\rm CFT*}_{I^-} \leftrightarrow \boldsymbol{\beta}^{\rm CFT*}_{I^-}\boldsymbol{\gamma}^{\rm CFT*}_{I^+}$.

The coefficients of the polynomials $P^{\rm even}_{I^+,I^-}$ and $P^{\rm odd}_{I^+,I^-}$ are rational functions of c and

$$\Delta_\alpha = \frac{\alpha(\alpha-2)\nu^2}{4(1-\nu)}$$

only. The denominators factorise into multipliers $\Delta_\alpha + 2k$, $k = 0, 1, 2, \cdots$. Exact formulae up to the level 6 can be found in IV, Eq. (12.4). In particular, on the level 2 we have $P^{\rm even}_{1,1} = 1$, $P^{\rm odd}_{1,1} = 0$. The transformation (3) is invertible.

The operators $\boldsymbol{\beta}^{\rm CFT*}_{2k-1}$ and $\boldsymbol{\gamma}^{\rm CFT*}_{2k-1}$ were found as the scaling limit of the fermions which create the quasi-local fields for the XXZ spin chain. This became possible after the computation of the expectation values of the quasi-local fields on the cylinder (see III). Actually, two different scaling limits are possible, and the second one provides the fermionic operators for the second chirality: $\bar{\boldsymbol{\beta}}^{\rm CFT*}_{2k-1}$ and $\bar{\boldsymbol{\gamma}}^{\rm CFT*}_{2k-1}$. The same story repeats for these operators, in particular, we have

$$\bar{\boldsymbol{\beta}}^{\rm CFT*}_{\bar{I}^+}\bar{\boldsymbol{\gamma}}^{\rm CFT*}_{\bar{I}^-}\Phi_\alpha(0) = \left\{P^{\rm even}_{\bar{I}^+,\bar{I}^-}(\{\bar{\mathbf{l}}_{-2k}\}) - d_\alpha P^{\rm odd}_{\bar{I}^+,\bar{I}^-}(\{\bar{\mathbf{l}}_{-2k}\})\right\}\Phi_\alpha(0)\,. \quad (4)$$

The main statement of this paper is that in the fermionic basis the sG one-point functions are simple:

$$\frac{\langle \bar{\boldsymbol{\beta}}^{\rm CFT*}_{\bar{I}^+}\bar{\boldsymbol{\gamma}}^{\rm CFT*}_{\bar{I}^-}\boldsymbol{\beta}^{\rm CFT*}_{I^+}\boldsymbol{\gamma}^{\rm CFT*}_{I^-}\Phi_\alpha(0)\rangle_{\rm sG}}{\langle\Phi_\alpha(0)\rangle_{\rm sG}} = (-1)^{\#(I^+)}\delta_{\bar{I}^-,I^+}\delta_{\bar{I}^+,I^-} \quad (5)$$

$$\times\left(\frac{M\,\sqrt{\pi}\,\Gamma(\frac{1}{2\nu})}{2\sqrt{1-\nu}\,\Gamma(\frac{1-\nu}{2\nu})}\right)^{2|I^+|+2|I^-|}\prod_{2n-1\in I^+}G_n(\alpha)\prod_{2n-1\in I^-}G_n(2-\alpha)\,,$$

Here

$$G_n(\alpha) = (-1)^{n-1}((n-1)!)^2\frac{\Gamma\left(\frac{\alpha}{2}+\frac{1-\nu}{2\nu}(2n-1)\right)\Gamma\left(1-\frac{\alpha}{2}-\frac{1}{2\nu}(2n-1)\right)}{\Gamma\left(1-\frac{\alpha}{2}-\frac{1-\nu}{2\nu}(2n-1)\right)\Gamma\left(\frac{\alpha}{2}+\frac{1}{2\nu}(2n-1)\right)}\,,$$

$|I|$ stands for the sum of elements of I, and M is the mass of soliton. We recall that $\#(I^+) = \#(I^-)$ is required in order to stay in the same Verma module, and $\#(\bar{I}^+) = \#(\bar{I}^-)$ follows.

Using (5) we can find all the one-point functions of descendants. Up to the level 6 the results of IV can be used. In particular, at the level 2 we find a perfect agreement with the formula (1.8) of Ref. 23 after the identification: $\eta = \alpha - 1$, $\xi = \frac{1-\nu}{\nu}$. To proceed to levels higher than 6 one should perform further computations in the spirit of IV.

Let us explain how we proceed in justification of the main formula (5). For CFT we use the lattice regularisation by the six vertex model (equivalently XXZ spin chain). For this model we use the fermionic description of the space of quasi-local operators found in I,II. On the lattice we have creation operators $\mathbf{b}^*(\zeta), \mathbf{c}^*(\zeta)$ and annihilation operators $\mathbf{b}(\zeta), \mathbf{c}(\zeta)$. The most honest way to proceed to the sG model would be to regularise it via the eight vertex model, and then to consider the scaling limit. On this way we would meet two problems. The first is the $U(1)$ symmetry which is broken in the eight vertex model. It is hard to introduce the lattice analogue of Φ_α with arbitrary α. This difficulty may be overcome by going to the SOS model. The second problem is conceptually more difficult. For the elliptic R-matrices we do not have an analogue of the q-oscillators which is crucial for the construction of our fermionic operators. For the moment we do not know how to attack this problem. Let us notice, however, that in the case $\alpha = 0$ an analogue of bilinear combinations of the annihilation operators $\mathbf{b}(\zeta)\mathbf{c}(\xi)$ exists. It is defined in the papers Refs. 33,34.

So, having problems with the eight vertex model we are forced to take another approach. In the paper Ref. 35 the sG or the massive Thirring models was obtained as a limit of the inhomogeneous six vertex model with the inhomogeneity ζ_0 (see section 3 below). Notice that, for this limit to make sense, one has first to consider the finite lattice on $\mathbf{n}$ sites, and then take the limit $\zeta_0 \to \infty$, $\mathbf{n} \to \infty$ in a concerted way in order that the finite mass scale appears. But exactly this kind of procedure became very natural for us after we had computed in III the expectation values of quasi-local operators on the cylinder. The compact direction on the cylinder is called the Matsubara direction, and its size $\mathbf{n}$ is what is needed for considering the limit in the spirit of Ref. 35.

The paper is organised as follows. In Section 2 we review our previous paper IV, and explain how to obtain the fermionic description for two chiral CFT models from the XXZ spin chain (six vertex model). In Section 3 we introduce the inhomogeneous six vertex model and consider the continuous limit which produces the sG model according to Ref. 35. We derive the one-point functions using the fermionic description of ultra-violet CFT.

2. Two scaling limits of the XXZ model and two chiralities

Our study of the XXZ model is based on the fermionic operators defined in I,II. This definition allowed us to compute in III the following expectation value. Consider a homogeneous six vertex model on an infinite cylinder. Let $T_{\mathrm{S,M}}$ be the monodromy matrix, where S refers to the infinite direction (called the space direction), and M refers to the compact circular direction (called the Matsubara direction). We use $\mathbf{n}$ to denote the length of the latter. We follow the notations in IV. For a quasi-local operator $q^{2\alpha S(0)}\mathcal{O}$ on the spacial lattice, we consider

$$Z_{\mathbf{n}}^{\kappa}\Big\{q^{2\alpha S(0)}\mathcal{O}\Big\} = \frac{\mathrm{Tr}_{\mathrm{S}}\mathrm{Tr}_{\mathrm{M}}\Big(T_{\mathrm{S,M}}q^{2\kappa S+2\alpha S(0)}\mathcal{O}\Big)}{\mathrm{Tr}_{\mathrm{S}}\mathrm{Tr}_{\mathrm{M}}\Big(T_{\mathrm{S,M}}q^{2\kappa S+2\alpha S(0)}\Big)}. \tag{6}$$

The generalisation of this functional $Z_{\mathbf{n}}^{\kappa,s}$ was introduced in IV. In the scaling limit, the introduction of s in this functional amounts to changing (screening) the background charge at $x=-\infty$ by $-2s\frac{1-\nu}{\nu}$. It enables us to deal with the special case of the functional for which the effective action of local integrals of motion becomes trivial.

The quasi-local operators are created from the primary field $q^{2\alpha S(0)}$ by action of the creation operators $\mathbf{t}^*(\zeta), \mathbf{b}^*(\zeta), \mathbf{c}^*(\zeta)$. Actually they act on the space

$$\mathcal{W}^{(\alpha)} = \bigoplus_{s=-\infty}^{\infty} \mathcal{W}_{\alpha-s,s}\,,$$

where $\mathcal{W}_{\alpha-s,s}$ denotes the space of quasi-local operators of spin s with tail $\alpha-s$.

In this paper we shall consider the subspace $\mathcal{W}_{\mathrm{ferm}}^{(\alpha)}$ of the space $\mathcal{W}^{(\alpha)}$ which are created from the primary fields only by fermions $\mathbf{b}_p^*$, $\mathbf{c}_p^*$ (see (12) below). On $\mathcal{W}_{\mathrm{ferm}}^{(\alpha)}$ the functional $Z_{\mathbf{n}}^{\kappa,s}$ allows the determinant form which is convenient to summarise as

$$Z_{\mathbf{n}}^{\kappa,s}\big\{q^{2\alpha S(0)}\mathcal{O}\big\} = \frac{\mathrm{Tr}_{\mathrm{S}}\left(e^{\Omega_{\mathbf{n}}}\left(q^{2\alpha S(0)}\mathcal{O}\right)\right)}{\mathrm{Tr}_{\mathrm{S}}\left(q^{2\alpha S(0)}\right)}\,,$$

where

$$\Omega_{\mathbf{n}} = \frac{1}{(2\pi i)^2}\int_{\Gamma}\int_{\Gamma} \omega_{\mathrm{rat},\mathbf{n}}(\zeta,\xi)\mathbf{c}(\xi)\mathbf{b}(\zeta)\frac{d\zeta^2}{\zeta^2}\frac{d\xi^2}{\xi^2}\,,$$

where the contour Γ goes around $\zeta^2=1$.

The function $\omega_{\mathrm{rat},\mathbf{n}}(\zeta,\xi)$ is defined by the Matsubara data (see III, Ref. 37, IV). Besides the length of the Matsubara chain $\mathbf{n}$ it depends on the parameters κ, α, s and on possible inhomogenieties in the Matsubara chain. However, we mark explicitly only the dependence on $\mathbf{n}$ which is the most important for us here.

We have ignored the descendants created by $\mathbf{t}^*(\zeta)$. Actually, they do not give any non-trivial contributions in the limit $\mathbf{n}\to\infty$, which we shall be interested in. Then Z^{κ}_{∞} is automatically reduced to the the quotient space:

$$\mathcal{W}^{(\alpha)}_{\mathrm{quo}}=\mathcal{W}^{(\alpha)}/(\mathbf{t}^*(\zeta)-2)\mathcal{W}^{(\alpha)}\,,$$

which is obviously isomorphic to $\mathcal{W}^{(\alpha)}_{\mathrm{ferm}}$ as a linear space.

We shall not repeat the definition of $\omega_{\mathrm{rat},\mathbf{n}}(\zeta,\xi)$ because in the present paper we shall use only very limited information about it. Let us explain, however, the suffix "rat": the function $(\xi/\zeta)^{\alpha}\omega_{\mathrm{rat},\mathbf{n}}(\zeta,\xi)$ is a rational function of ζ^2 and ξ^2. The really interesting situation occurs when $\mathbf{n}\to\infty$. In that case the Bethe roots for the transfer matrices in the Matsubara direction become dense on the half axis $\zeta^2>0$. If we do not introduce additional rescaling as described below, the function $\omega_{\mathrm{rat},\mathbf{n}}(\zeta,\xi)$ goes to the simple limit:

$$\omega_{\mathrm{rat},\mathbf{n}}(\zeta,\xi)\underset{\mathbf{n}\to\infty}{\longrightarrow}4\omega_0(\zeta/\xi,\alpha)+\nabla\omega(\zeta/\xi,\alpha)\,,\tag{7}$$

where

$$4\omega_0(\zeta,\alpha)=-\int_{-i\infty}^{i\infty}\zeta^u\frac{\sin\frac{\pi}{2}((1-\nu)u-\alpha)}{\sin\frac{\pi}{2}(u-\alpha)\cos\frac{\pi\nu}{2}u}du\,,\tag{8}$$

$$\nabla\omega(\zeta,\alpha)=-\psi(\zeta q,\alpha)+\psi(\zeta q^{-1},\alpha)+2i\zeta^{\alpha}\tan\Big(\frac{\pi\nu\alpha}{2}\Big)\,,$$

$$\psi(\zeta,\alpha)=\zeta^{\alpha}\frac{\zeta^2+1}{2(\zeta^2-1)}\,.$$

The reason for extracting the elementary $\nabla\omega$ is due to the fact that the function ω_0 satisfies the relation typical for CFT

$$\omega_0(\zeta,\alpha)=\omega_0(\zeta^{-1},2-\alpha)\,.\tag{9}$$

Notice that $(\xi/\zeta)^{\alpha}\omega_0(\zeta/\xi,\alpha)$ is not a single-valued function of ζ^2 and ξ^2. So, the property of rationality is lost in the limit. We define Ω_0 and $\nabla\Omega$ in the same way as $\Omega_{\mathbf{n}}$ replacing $\omega_{\mathrm{rat},\mathbf{n}}(\zeta,\xi)$ by respectively $4\omega_0(\zeta/\xi,\alpha)$ and $\nabla\omega(\zeta/\xi,\alpha)$.

Following IV we denote the original creation operators introduced in II by $\mathbf{b}^*_{\mathrm{rat}}(\zeta)$ and $\mathbf{c}^*_{\mathrm{rat}}(\zeta)$. They satisfy the property:

$$\mathrm{Tr}_S\left(\mathbf{b}^*_{\mathrm{rat}}(\zeta)(X)\right)=0,\quad \mathrm{Tr}_S\left(\mathbf{c}^*_{\mathrm{rat}}(\zeta)(X)\right)=0\,,$$

for all quasi-local operators X. In the present paper it is useful to replace these operators by the following Bogolubov transformed ones:

$$\mathbf{b}^*_0(\zeta)=e^{-\nabla\Omega}\mathbf{b}^*_{\mathrm{rat}}(\zeta)e^{\nabla\Omega},\quad \mathbf{c}^*_0(\zeta)=e^{-\nabla\Omega}\mathbf{c}^*_{\mathrm{rat}}(\zeta)e^{\nabla\Omega}\,.\tag{10}$$

Obviously, the functional $Z_{\mathbf{n}}^{\kappa,s}$ calculated on the descendants generated by these operators is expressed as determinant constructed from function $\omega_{\mathrm{rat},\mathbf{n}}(\zeta,\xi) - \nabla\omega(\zeta/\xi,\alpha)$ which in the limit $\mathbf{n}\to\infty$ goes to $4\omega_0(\zeta/\xi,\alpha)$.

As it is explained in IV, the operators $\zeta^{-\alpha}\mathbf{b}_0^*(\zeta)$ and $\zeta^{\alpha}\mathbf{c}_0^*(\zeta)$ are rational functions of ζ^2 as far as they are considered in the functional $Z_{\mathbf{n}}^{\kappa,s}$, and they have the following behaviour at $\zeta^2=0$:

$$\mathbf{b}_0^*(\zeta)=\sum_{j=1}^{\infty}\zeta^{\alpha+2j-2}\mathbf{b}^*_{\mathrm{screen},j}\,,\quad \mathbf{c}_0^*(\zeta)=\sum_{j=1}^{\infty}\zeta^{-\alpha+2j}\mathbf{c}^*_{\mathrm{screen},j}\,.$$

For $\mathbf{b}^*_{\mathrm{screen},j}$, $\mathbf{c}^*_{\mathrm{screen},j}$ we have used the suffix "screen" to stand for "screening" in view of its similarity to the lattice screening operators used in IV.

In IV, another set of operators $\mathbf{b}^*(\zeta)$, $\mathbf{c}^*(\zeta)$ is obtained from $\mathbf{b}^*_{\mathrm{rat}}(\zeta)$, $\mathbf{c}^*_{\mathrm{rat}}(\zeta)$ by a kind of Bogolubov transformation which contains $\mathbf{t}^*(\zeta)$. We shall not write explicitly this Bogolubov transformation, but only the one relating $\mathbf{b}^*(\zeta)$, $\mathbf{c}^*(\zeta)$ to $\mathbf{b}_0^*(\zeta)$, $\mathbf{c}_0^*(\zeta)$, both acting on the quotient space $\mathcal{W}^{(\alpha)}_{\mathrm{quo}}$ because only these operators are used in this paper. The operators $\mathbf{b}^*(\zeta)$, $\mathbf{c}^*(\zeta)$ are important because $Z_{\infty}^{\kappa,s}$ vanishes on their descendants. Notice that we do not allow κ, s to grow together with $\mathbf{n}$. In that case the dependence on κ, s disappears for $\mathbf{n}=\infty$.

Remark. The defining equations for $\omega_{\mathrm{rat},\mathbf{n}}(\zeta,\xi)$ given in IV imply that the limit (7) has a very general nature. Namely, the result is independent not only of κ, s but also of inhomogeneities in the Matsubara chain. The situation is similar to that for the S-matrix which is the same for homogeneous, inhomogeneous XXZ chains or even for the sG model.

If we consider the Bogolubov transformation which connects the operators $\mathbf{b}^*(\zeta)$, $\mathbf{c}^*(\zeta)$ and $\mathbf{b}_0^*(\zeta)$, $\mathbf{c}_0^*(\zeta)$ as acting on the quotient space $\mathcal{W}^{(\alpha)}_{\mathrm{quo}}$, it reduces to

$$\mathbf{b}^*(\zeta)=e^{-\Omega_0}\mathbf{b}_0^*(\zeta)e^{\Omega_0}\,,\quad \mathbf{c}^*(\zeta)=e^{-\Omega_0}\mathbf{c}_0^*(\zeta)e^{\Omega_0}\,. \tag{11}$$

We catch operators acting on $\mathcal{W}^{(\alpha)}$ by developing $\mathbf{b}^*(\zeta)$, $\mathbf{c}^*(\zeta)$ and $\mathbf{b}(\zeta)$, $\mathbf{c}(\zeta)$ around the point $\zeta^2=1$:

$$\mathbf{b}^*(\zeta)\underset{\zeta^2\to1}{\simeq}\sum_{p=1}^{\infty}(\zeta^2-1)^{p-1}\mathbf{b}^*_p\,,\qquad \mathbf{c}^*(\zeta)\underset{\zeta^2\to1}{\simeq}\sum_{p=1}^{\infty}(\zeta^2-1)^{p-1}\mathbf{c}^*_p\,, \tag{12}$$

$$\mathbf{b}(\zeta)=\sum_{p=0}^{\infty}(\zeta^2-1)^{-p}\mathbf{b}_p\,,\qquad \mathbf{c}(\zeta)=\sum_{p=0}^{\infty}(\zeta^2-1)^{-p}\mathbf{c}_p\,. \tag{13}$$

These operators are (quasi-)local in the sense of II, Section 3.3, while $\mathbf{b}^*_{\mathrm{screen},j}$ and $\mathbf{c}^*_{\mathrm{screen},j}$ are highly "non-local".

The operators $\mathbf{b}^*_p$, $\mathbf{c}^*_p$ create quasi-local operators in the sense of II, Section 3.1 by acting on the primary field. Here we slightly change the notation compared to II,III, where the Fourier coefficients are defined after removing an overall power $\zeta^{\pm\alpha}$. The reason was that we wanted that the result of action of $\mathbf{b}^*_p$, $\mathbf{c}^*_p$ is a rational function of q, q^α. This rationality property is irrelevant in this paper, and extracting $\zeta^{\pm\alpha}$ may even cause a confusion.

So far we have been discussing the simple limit $\mathbf{n}\to\infty$. Now we discuss the scaling limit to CFT. In IV we studied the scaling limit of the homogeneous XXZ chain on the cylinder. The key idea is to consider first of all the scaling limit in the Matsubara direction. Namely, denoting the length of the Matsubara chain by $\mathbf{n}$ and introducing the step of the lattice a we consider the limit

$$\mathbf{n}\to\infty\,,\qquad a\to 0\,,\qquad \mathbf{n}a=2\pi R\ \ \text{fixed}\,. \tag{14}$$

The requirement is that if we rescale the spectral parameter as $\zeta=(Ca)^\nu\lambda$, the Bethe roots for the transfer matrix in the Matsubara direction which are close to $\zeta^2=0$ remain finite. The constant C is chosen to have an agreement with CFT as

$$C=\frac{\Gamma\left(\frac{1-\nu}{2\nu}\right)}{2\sqrt{\pi}\,\Gamma\left(\frac{1}{2\nu}\right)}\Gamma(\nu)^{\frac{1}{\nu}}\,. \tag{15}$$

Next we consider the scaling limit in the space direction. We conjecture that under the presence of the background charges effected by the screening operators, the lattice operator $q^{2\kappa S+2\alpha S(0)}$ goes to the limit $\Phi_{1-\kappa'}(-\infty)\Phi_\alpha(0)\Phi_{1+\kappa}(\infty)$ where

$$\kappa'=\kappa+\alpha+2s\frac{1-\nu}{\nu}\,.$$

Furthermore, we have shown that in the weak sense the following limits exist

$$2\boldsymbol{\beta}^*(\lambda)=\lim_{a\to0}\mathbf{b}^*((Ca)^\nu\lambda),\quad 2\boldsymbol{\gamma}^*(\lambda)=\lim_{a\to0}\mathbf{c}^*((Ca)^\nu\lambda)\,. \tag{16}$$

The operators $\boldsymbol{\beta}^*(\lambda)$, $\boldsymbol{\gamma}^*(\lambda)$ have the asymptotics at $\lambda^2\to\infty$:

$$\boldsymbol{\beta}^*(\lambda)\simeq\sum_{j=1}^{\infty}\lambda^{-\frac{2j-1}{\nu}}\boldsymbol{\beta}^*_{2j-1},\quad \boldsymbol{\gamma}^*(\lambda)\simeq\sum_{j=1}^{\infty}\lambda^{-\frac{2j-1}{\nu}}\boldsymbol{\gamma}^*_{2j-1}\,, \tag{17}$$

where $\boldsymbol{\beta}^*_{2j-1}$, $\boldsymbol{\gamma}^*_{2j-1}$ act between different Verma modules as follows

$$\begin{aligned}\boldsymbol{\beta}^*_{2j-1}\ &:\ \mathcal{V}_{\alpha+2\frac{1-\nu}{\nu}(s-1)}\otimes\overline{\mathcal{V}}_\alpha\ \to\ \mathcal{V}_{\alpha+2\frac{1-\nu}{\nu}s}\otimes\overline{\mathcal{V}}_\alpha\,,\\ \boldsymbol{\gamma}^*_{2j-1}\ &:\ \mathcal{V}_{\alpha+2\frac{1-\nu}{\nu}(s+1)}\otimes\overline{\mathcal{V}}_\alpha\ \to\ \mathcal{V}_{\alpha+2\frac{1-\nu}{\nu}s}\otimes\overline{\mathcal{V}}_\alpha\,.\end{aligned}$$

The action on the second component is trivial. The functional $Z_{\mathbf{n}}^{\kappa,s}$ turns in this limit into the three-point function for $c<1$ CFT as explained in IV. The identification of descendants created by $\boldsymbol{\beta}^*$ and $\boldsymbol{\gamma}^*$ with Virasoro descendants is made by studying the function

$$\omega_R(\lambda,\mu)=\frac{1}{4}\big(\lim_{\text{scaling}}\omega_{\text{rat},\mathbf{n}}((Ca)^{\nu}\lambda,(Ca)^{\nu}\mu)-\nabla\omega(\lambda/\mu,\alpha)\big)\,,$$

where $\lim_{\text{scaling}}$ refers to the scaling limit (14). Contrary to the simple-minded limit $\mathbf{n}\to\infty$ (7), the function $(\mu/\lambda)^{\alpha}\omega_R(\lambda,\mu)$ remains a single-valued function of λ^2 and μ^2, but it develops an essential singularity at $\lambda^2=\infty$, $\mu^2=\infty$.

In the present paper we shall consider sG model which requires putting together the two chiralities. To this end we shall need to consider not only the asymptotical region $\lambda^2\to\infty$, but also $\lambda^2\to 0$. Analysing the function $\omega_R(\lambda,\mu)$ one concludes that the following limits exist

$$\tfrac{1}{2}\lim_{a\to 0}\mathbf{b}_0^*((Ca)^{\nu}\lambda)\underset{\lambda^2\to 0}{\simeq}\boldsymbol{\beta}^*_{\text{screen}}(\lambda)=\sum_{j=1}^{\infty}\lambda^{\alpha+2j-2}\boldsymbol{\beta}^*_{\text{screen},j}\,,\tag{18}$$
$$\tfrac{1}{2}\lim_{a\to 0}\mathbf{c}_0^*((Ca)^{\nu}\lambda)\underset{\lambda^2\to 0}{\simeq}\boldsymbol{\gamma}^*_{\text{screen}}(\lambda)=\sum_{j=1}^{\infty}\lambda^{-\alpha+2j}\boldsymbol{\gamma}^*_{\text{screen},j}\,.$$

For the moment we do not know how to use these operators, but one thing is clear: they create highly non-local fields.

The L-operator depends on ζ and ζ^{-1} in a symmetric way. That is why another scaling limit is possible:

$$2\bar{\boldsymbol{\beta}}^*(\lambda)=\lim_{a\to 0}\mathbf{b}^*((Ca)^{-\nu}\lambda),\quad 2\bar{\boldsymbol{\gamma}}^*(\lambda)=\lim_{a\to 0}\mathbf{c}^*((Ca)^{-\nu}\lambda)\,,\tag{19}$$

which allow the power series at $\lambda\to 0$:

$$\bar{\boldsymbol{\beta}}^*(\lambda)\simeq\sum_{j=1}^{\infty}\lambda^{\frac{2j-1}{\nu}}\bar{\boldsymbol{\beta}}^*_{2j-1},\quad \bar{\boldsymbol{\gamma}}^*(\lambda)\simeq\sum_{j=1}^{\infty}\lambda^{\frac{2j-1}{\nu}}\bar{\boldsymbol{\gamma}}^*_{2j-1}\,.\tag{20}$$

The resulting operators act as follows

$$\begin{aligned}\bar{\boldsymbol{\beta}}^*(\lambda)\ &:\ \mathcal{V}_{\alpha}\otimes\overline{\mathcal{V}}_{\alpha+2\frac{1-\nu}{\nu}(s-1)}\ \to\ \mathcal{V}_{\alpha}\otimes\overline{\mathcal{V}}_{\alpha+2\frac{1-\nu}{\nu}s}\,,\\ \bar{\boldsymbol{\gamma}}^*(\lambda)\ &:\ \mathcal{V}_{\alpha}\otimes\overline{\mathcal{V}}_{\alpha+2\frac{1-\nu}{\nu}(s+1)}\ \to\ \mathcal{V}_{\alpha}\otimes\overline{\mathcal{V}}_{\alpha+2\frac{1-\nu}{\nu}s}\,.\end{aligned}$$

These operators are obtainable from the previous ones by the substitution

$$\{\alpha,\ \lambda\}\ \to\{2-\alpha,\ \lambda^{-1}\}\,.$$

The proof goes through considering the scaling limit (14) and studying the function

$$\bar{\omega}_R(\lambda,\mu)=\frac{1}{4}\big(\lim_{\text{scaling}}\omega_{\text{rat},\mathbf{n}}((Ca)^{-\nu}\lambda,(Ca)^{-\nu}\mu)-\nabla\omega(\lambda/\mu,\alpha)\big)\,.$$

The analysis is parallel to the one performed for the first chirality in IV. The function $(\mu/\lambda)^{\alpha}\bar{\omega}_R(\lambda,\mu)$ is a single-valued function of λ^2, μ^2 with essential singularities at $\lambda^2=0$, $\mu^2=0$.

The operators $\bar{\boldsymbol{\beta}}^*_{\mathrm{screen}}(\lambda)$ and $\bar{\boldsymbol{\gamma}}^*_{\mathrm{screen}}(\lambda)$ are introduced similarly to the first chirality.

$$\tfrac{1}{2}\lim_{a\to 0}\mathbf{b}_0^*((Ca)^{-\nu}\lambda) \underset{\lambda^2\to\infty}{\simeq} \bar{\boldsymbol{\beta}}^*_{\mathrm{screen}}(\lambda)=\sum_{j=1}^{\infty}\lambda^{\alpha-2j}\bar{\boldsymbol{\beta}}^*_{\mathrm{screen},j}\,, \tag{21}$$

$$\tfrac{1}{2}\lim_{a\to 0}\mathbf{c}_0^*((Ca)^{-\nu}\lambda) \underset{\lambda^2\to\infty}{\simeq} \bar{\boldsymbol{\gamma}}^*_{\mathrm{screen}}(\lambda)=\sum_{j=1}^{\infty}\lambda^{2-\alpha-2j}\bar{\boldsymbol{\gamma}}^*_{\mathrm{screen},j}\,.$$

For both chiralities we have

$$\omega_R(\lambda,\mu)\underset{R\to\infty}{\longrightarrow}\omega_0(\lambda/\mu,\alpha),\quad \bar{\omega}_R(\lambda,\mu)\underset{R\to\infty}{\longrightarrow}\omega_0(\lambda/\mu,\alpha)\,.$$

So, the naïve $\mathbf{n}\to\infty$ limit is reproduced.

3. Inhomogeneous six vertex model and sine-Gordon model

We want to put two chiral models together and to make them interacting. According to the previous discussion the lattice analogue of chiral CFT is the XXZ model. So, the two non-interacting chiral models correspond to the lattice containing two non-interacting six vertex sublattices. As an example we shall consider the "even" or "odd" sublattices, consisting of lattice points with coordinates $(j,\mathbf{m})$ such that $j,\mathbf{m}$ are both even or both odd. It is well-known[35] how to force these two to interact: one has to consider a six vertex model on the entire lattice with alternating inhomogeneity parameters. We denote $\mathbf{S}=\mathrm{S}\cup\overline{\mathrm{S}}$, $\mathbf{M}=\mathrm{M}\cup\overline{\mathrm{M}}$ and introduce

$$T_{\mathbf{S},\mathbf{M}}=\overset{\curvearrowright}{\prod_{j=-\infty}^{\infty}}T_{j,\mathbf{M}}(\zeta_0^{(-1)^j})\,,\qquad T_{j,\mathbf{M}}(\zeta)=\overset{\curvearrowleft}{\prod_{\mathbf{m}=\mathbf{1}}^{\mathbf{2n}}}L_{j,\mathbf{m}}(\zeta\zeta_0^{-(-1)^{\mathbf{m}}})\,.$$

Here

$$L_{j,\mathbf{m}}(\zeta)=q^{-\frac{1}{2}\sigma_j^3\sigma_{\mathbf{m}}^3}-\zeta^2q^{\frac{1}{2}\sigma_j^3\sigma_{\mathbf{m}}^3}-\zeta(q-q^{-1})(\sigma_j^+\sigma_{\mathbf{m}}^-+\sigma_j^-\sigma_{\mathbf{m}}^+)\,.$$

Note that for $\zeta_0^{\pm1}=\infty$ the inhomogeneous lattice reduces to two non-interacting homogeneous lattices.

The functional $Z^{\mathrm{full}}_{\mathbf{n}}$ is defined by

$$Z^{\mathrm{full}}_{\mathbf{n}}\Big\{q^{2\alpha S(0)}\mathcal{O}\Big\}=\frac{\mathrm{Tr}_{\mathbf{S}}\mathrm{Tr}_{\mathbf{M}}\Big(T_{\mathbf{S},\mathbf{M}}q^{2\alpha S(0)}\mathcal{O}\Big)}{\mathrm{Tr}_{\mathbf{S}}\mathrm{Tr}_{\mathbf{M}}\Big(T_{\mathbf{S},\mathbf{M}}q^{2\alpha S(0)}\Big)}\,. \tag{22}$$

We set $\kappa = 0$ here. The methods of II,III allow one to compute this functional with inhomogeneities in both directions. In particular, from II one concludes that the annihilation operators split into two parts:

$$\mathbf{b}(\zeta) = \mathbf{b}^{+}(\zeta) + \mathbf{b}^{-}(\zeta), \qquad \mathbf{c}(\zeta) = \mathbf{c}^{+}(\zeta) + \mathbf{c}^{-}(\zeta),$$
$$\mathbf{b}^{\pm}(\zeta) = \sum_{p=0}^{\infty} \left(\zeta^2\zeta_0^{\mp 2} - 1\right)^{-p} \mathbf{b}_p^{\pm}, \qquad \mathbf{c}^{\pm}(\zeta) = \sum_{p=0}^{\infty} \left(\zeta^2\zeta_0^{\mp 2} - 1\right)^{-p} \mathbf{c}_p^{\pm}.$$

Let us take $\mathbf{n} = \infty$. Then using the remark from the previous section and II,III one obtains:

$$Z_{\infty}^{\mathrm{full}}\left\{q^{2\alpha S(0)}\mathcal{O}\right\} = \frac{\mathrm{Tr_S}\left(e^{\Omega^{\mathrm{full}}}\left(q^{2\alpha S(0)}\mathcal{O}\right)\right)}{\mathrm{Tr_S}\left(q^{2\alpha S(0)}\right)}. \tag{23}$$

We have

$$\Omega^{\mathrm{full}} = \Omega_0 + \nabla\Omega,$$

where

$$\Omega_0 = \Omega_0^{++} + \Omega_0^{+-} + \Omega_0^{-+} + \Omega_0^{--},$$
$$\Omega_0^{\epsilon\epsilon'} = \frac{4}{(2\pi i)^2}\int_{\Gamma_\epsilon}\int_{\Gamma_{\epsilon'}} \omega_0(\zeta/\xi,\alpha)\mathbf{c}^{\epsilon'}(\xi)\mathbf{b}^{\epsilon}(\zeta)\frac{d\zeta^2}{\zeta^2}\frac{d\xi^2}{\xi^2},$$

and similarly for $\nabla\Omega$. The contour $\Gamma_{\pm}$ goes anticlockwise around $\zeta_0^{\pm 2}$.

As it has been said in the introduction, ideally we would like to start from a non-critical (XYZ or SOS) lattice model and to obtain the relativistic massive model by the usual scaling limit near the critical point. Since we do not have the necessary formulae to do that, we have recourse to the scaling limit of an inhomogeneous model by the procedure of Ref. 35. Here we have an ideal situation from the point of view of QFT. Namely, we have the ultraviolet cutoff (lattice), the infrared cutoff (a finite number $\mathbf{n}$ of sites in the Matsubara direction), and the physical quantities (the values of the the functional $Z_{\mathbf{n}}^{\mathrm{full}}$ on quasi-local fields) are exactly computed with finite cutoffs.

Like in the homogeneous case, let us introduce the step of the lattice a, and consider the scaling limit (14): $\mathbf{n} \to \infty$, $a \to 0$, $\mathbf{n}a = 2\pi R$ fixed. We require further that $\zeta_0^{-1} \to 0$, so that

$$M = 4a^{-1}\zeta_0^{-1/\nu} \quad \text{fixed}. \tag{24}$$

The parameter M is a mass scale which has the meaning of the sG soliton mass.[35] The famous formula relating the soliton mass to the dimensional coupling constant $\boldsymbol{\mu}^{36}$ reads in our notation as

$$\boldsymbol{\mu} = \left[\frac{M}{4C}\right]^{\nu} = (Ca)^{-\nu}\zeta_0^{-1}\,. \tag{25}$$

In this paper we consider a further limit $R \to \infty$. In that case the sG partition function is obtained from Z_{∞}^{full}.

Let us give some explanation at this point. The subject of study in Ref. 35 is the partition function of the sG model on the cylinder. There are two possible approaches to this partition function which correspond to two Hamiltonian pictures. In the first one, the space direction is considered as space and the Matsubara direction as time (space channel). One has scattering of particles and describes the partition function by the Thermodynamic Bethe Ansatz (TBA).[38] This approach has an advantage of dealing with known particle spectrum and S-matrices. It also has a disadvantage, because as usual in the thermodynamics one has to deal with the density matrix, which is a complicated object even in integrable cases.

The paper Ref. 35 uses an alternative picture: the Matsubara direction is space, and the space direction is time. In this approach, the partition function is described by the maximal eigenvalue of the Hamiltonian of the periodic problem for the Matsubara direction (Matsubara channel). The advantage of this approach is clear: one deals with the pure ground state instead of the density matrix. The disadvantage is that describing eigenvalues in the finite volume is a difficult problem. This problem is addressed in Ref. 35.

More precisely, it is proposed in Ref. 35 to obtain the sG partition function as the scaling limit of $\mathrm{Tr}_{\mathbf{S}}\mathrm{Tr}_{\mathbf{M}}(T_{\mathbf{S},\mathbf{M}})$. As in the present paper, it is important to be able to control the computations starting from the lattice and from finite $\mathbf{n}$. In our opinion, the main achievement of Ref. 35 is not in rewriting the Bethe Ansatz equation for the Matsubara transfer matrix in the form of a non-linear integral equation, but in extracting a main linear part and inverting it. The resulting Destri-DeVega equation (DDV) has several nice features. First, it allows the scaling limit and the mass scale (24) appears. Second, after the scaling limit the DDV equation clearly allows the large R expansion. Third, the scattering phase of the sG solitons appears in the DDV equation in the Matsubara channel. The last property allows the identification with the space channel. In particular, M happens to be equal to the mass of soliton.

Now we come to the most important point of this paper. We wish to define the creation operators appropriate for taking the scaling limit to the sG theory. The issue is similar to the one in conformal perturbation theory, where one needs to prescribe a way how to extend the descendants in CFT to the perturbed case. It is claimed in Refs. 18,23 that after subtracting the divergencies these operators are defined uniquely: possible finite counterterms can be dismissed for dimensional reasons, at least in the absence of resonances.[18,23] The latter condition is satisfied in our case if ν and α are generic.

Let us define the operators $\mathbf{b}_0^*(\zeta)$ and $\mathbf{c}_0^*(\zeta)$ by the same formula as in the homogeneous case (10). Starting from these operators we define the creation operators

$$\mathbf{b}_0^*(\zeta) \underset{\zeta^2\to\zeta_0^{\pm 2}}{\simeq} \mathbf{b}_0^{\pm *}(\zeta) = \sum_{p=1}^{\infty}(\zeta^2\zeta_0^{\mp 2}-1)^{p-1}\mathbf{b}_{0,p}^{\pm *}$$

and likewise for $\mathbf{c}_0^*(\zeta)$.

The operators $\mathbf{b}_{0,p}^{\pm *}$ $\mathbf{c}_{0,p}^{\pm *}$ create quasi-local fields. Notice that Z_∞^{full} is defined on the quotient space $\mathcal{W}_{\mathrm{quo}}^{(\alpha)}$ because the $\mathbf{t}^*$-descendants do not contribute to it.

There are two sorts of chiral operators, one living on the even sublattice and the other on the odd sublattice. We have to make some combinations which will give finite answers for the interacting model. At the same time, we want this combination to correspond to our intuitive idea that we have to subtract perturbative series. Let us explain that the correct combinations are given by the following.

$$\begin{aligned}\mathbf{b}^{+*}(\zeta) &= e^{-\Omega_0^{++}}\mathbf{b}_0^{+*}(\zeta)e^{\Omega_0^{++}}, \quad \mathbf{c}^{+*}(\zeta) = e^{-\Omega_0^{++}}\mathbf{c}_0^{+*}(\zeta)e^{\Omega 0^{++}}, \\ \mathbf{b}^{-*}(\zeta) &= e^{-\Omega_0^{--}}\mathbf{b}_0^{-*}(\zeta)e^{\Omega_0^{--}}, \quad \mathbf{c}^{-*}(\zeta) = e^{-\Omega_0^{--}}\mathbf{c}_0^{-*}(\zeta)e^{\Omega_0^{--}}.\end{aligned} \tag{26}$$

Corresponding creation operators which create the quasi-local operators are defined by

$$\mathbf{b}^{\pm *}(\zeta) \underset{\zeta^2\to\zeta_0^{\pm 2}}{\simeq} \sum_{p=1}^{\infty}(\zeta^2\zeta_0^{\mp 2}-1)^{p-1}\mathbf{b}_p^{\pm *}$$

In what follows we shall be interested only in the case of an equal number of $\mathbf{b}^{+*}$ and $\mathbf{c}^{+*}$. Let us examine how the descendants of this form depend on ζ_0, taking the simplest case

$$\mathbf{b}_k^{+*}\ \mathbf{c}_l^{+*}\ \mathbf{b}_r^{-*}\ \mathbf{c}_s^{-*}(q^{2\alpha S(0)}) = e^{-\Omega_0^{++}-\Omega_0^{--}}\ \mathbf{b}_{0,k}^{+*}\ \mathbf{c}_{0,l}^{+*}\ \mathbf{b}_{0,r}^{-*}\ \mathbf{c}_{0,s}^{-*}(q^{2\alpha S(0)})\,.$$

It is easy to see from the definition in II that the second factor in the right hand side is a rational function of ζ_0^2 regular at ∞,

$$\mathbf{b}_{0,k}^{+*}\ \mathbf{c}_{0,l}^{+*}\ \mathbf{b}_{0,r}^{-*}\ \mathbf{c}_{0,s}^{-*}(q^{2\alpha S(0)}) = \left(\mathcal{O}_0+\zeta_0^{-2}\mathcal{O}_1+\zeta_0^{-4}\mathcal{O}_2+\cdots\right)q^{2\alpha S(0)}\,. \tag{27}$$

The rationality holds true after application of $e^{-\Omega_0^{++}-\Omega_0^{--}}$. The expansion (27) looks as a perturbative series for the action (1), because in the scaling limit ζ_0^{-2} comes accompanied by $a^{-2\nu}$, and $\zeta_0^{-2}a^{-2\nu}$ has the dimension of $(\text{mass})^{2\nu}$ i.e. that of $\boldsymbol{\mu}^2$ (see (24)). Notice that the property (27) would be spoiled if we apply $e^{-\Omega_0^{+-}-\Omega_0^{-+}}$ as well, because it will pick up $\omega_0(\zeta,\alpha)$ near $\zeta^2=\zeta_0^{\pm 2}$, and the function $\omega_0(\zeta,\alpha)$ has asymptotics at $\zeta\to\infty$ containing both $\zeta^{\alpha-2m}$ and $\zeta^{-\frac{2n-1}{\nu}}$.

On the descendants created by (26), the value of Z_∞^{full} remains finite in the scaling limit. So we conclude that they create renormalised local operators. We cannot say anything about the finite renormalisation, but it can be taken care of by a dimensional consideration (see below).

We conjecture that the following limits exist

$$\begin{aligned}
\tfrac{1}{2}\lim_{\text{scaling}} \mathbf{b}^{+*}(\zeta) &\underset{\zeta^2\to\infty}{\simeq} \boldsymbol{\beta}^*(\boldsymbol{\mu}\zeta)+\bar{\boldsymbol{\beta}}^*_{\text{screen}}(\zeta/\boldsymbol{\mu}), \qquad (28)\\
\tfrac{1}{2}\lim_{\text{scaling}} \mathbf{c}^{+*}(\zeta) &\underset{\zeta^2\to\infty}{\simeq} \boldsymbol{\gamma}^*(\boldsymbol{\mu}\zeta)+\bar{\boldsymbol{\gamma}}^*_{\text{screen}}(\zeta/\boldsymbol{\mu}),\\
\tfrac{1}{2}\lim_{\text{scaling}} \mathbf{b}^{-*}(\zeta) &\underset{\zeta^2\to 0}{\simeq} \bar{\boldsymbol{\beta}}^*(\zeta/\boldsymbol{\mu})+\boldsymbol{\beta}^*_{\text{screen}}(\boldsymbol{\mu}\zeta),\\
\tfrac{1}{2}\lim_{\text{scaling}} \mathbf{c}^{-*}(\zeta) &\underset{\zeta^2\to 0}{\simeq} \bar{\boldsymbol{\gamma}}^*(\zeta/\boldsymbol{\mu})+\boldsymbol{\gamma}^*_{\text{screen}}(\boldsymbol{\mu}\zeta),
\end{aligned}$$

where by $\lim_{\text{scaling}}$ the scaling (14), (24) and $R\to\infty$ are implied.

In these formulae we denote by the same letters the operators in the sG model as they were denoted in the CFT. We shall not consider the screening operators,while we use the coefficients $\boldsymbol{\beta}^*_{2j-1}$ and so forth, defined as in (17) and (20), to consider the descendants

$$\bar{\boldsymbol{\beta}}^*_{I^+}\bar{\boldsymbol{\gamma}}^*_{\bar{I}^-}\boldsymbol{\beta}^*_{I^+}\boldsymbol{\gamma}^*_{I^-}\Phi_\alpha(0)\,. \qquad (29)$$

This is the field in the interacting model which goes to corresponding descendant in the conformal limit $\boldsymbol{\mu}\to 0$, and which does not develop finite counterterms. The latter are forbidden by dimensional consideration. So, this is exactly the definition which we were supposed to use from the very beginning. Notice that the appearance of $\boldsymbol{\mu}$ in the formulae (28) is a consequence of consistency with the conformal limit due to (25).

Now it is easy to compute the normalised vacuum expectation value of the descendant (29) for the sG model. It is obtained by taking the scaling limit of Z_∞^{full} and computing the asymptotics of $\omega_0(\zeta,\alpha)$ for $\zeta\to\infty$ and $\zeta\to 0$ which is done simply by summing up the residues in appropriate half-planes:

$$\omega_0(\zeta,\alpha)\simeq\frac{i}{\nu}\sum_{n\geq 1}\zeta^{-\frac{2n-1}{\nu}}\cot\frac{\pi}{2\nu}\,(2n-1+\nu\alpha)+i\sum_{n\geq 1}\zeta^{\alpha-2n}\tan\frac{\pi\nu}{2}\,(\alpha-2n)\,.$$

The first part corresponds to the expectation values of operators created by the coefficients of the expansions (17), (20). These are the expectation values in which we are interested in this paper. The second part of the asymptotics corresponds to the operators (18), (21) At present we do not know the meaning of their expectation values, but we hope to return to them in the future.

Introducing the multi-indices as described in the introduction we obtain:

$$\frac{\langle \bar{\boldsymbol{\beta}}^*_{\bar{I}^+}\bar{\boldsymbol{\gamma}}^*_{\bar{I}^-}\boldsymbol{\beta}^*_{I^+}\boldsymbol{\gamma}^*_{I^-}\Phi_\alpha(0)\rangle_{\mathrm{sG}}}{\langle\Phi_\alpha(0)\rangle_{\mathrm{sG}}} = \delta_{\bar{I}^-,I^+}\delta_{\bar{I}^+,I^-}(-1)^{\#(I^+)}\left(\frac{i}{\nu}\right)^{\#(I^+)+\#(I^-)}$$

$$\times\, \boldsymbol{\mu}^{\frac{2}{\nu}(|I^+|+|I^-|)}\prod_{2n-1\in I^+}\cot\tfrac{\pi}{2}\left(\tfrac{2n-1}{\nu}+\alpha\right)\prod_{2n-1\in I^-}\cot\tfrac{\pi}{2}\left(\tfrac{2n-1}{\nu}-\alpha\right). \tag{30}$$

Now we have to recall the definition in IV:

$$\boldsymbol{\beta}^*_{2n-1} = D_{2n-1}(\alpha)\boldsymbol{\beta}^{\mathrm{CFT}*}_{2n-1}, \quad \boldsymbol{\gamma}^*_{2n-1} = D_{2n-1}(2-\alpha)\boldsymbol{\gamma}^{\mathrm{CFT}*}_{2n-1}, \tag{31}$$

where

$$D_{2n-1}(\alpha) = \frac{1}{\sqrt{i\nu}}\, G^{2n-1}\,\frac{\Gamma\left(\frac{\alpha}{2}+\frac{1}{2\nu}(2n-1)\right)}{(n-1)!\,\Gamma\left(\frac{\alpha}{2}+\frac{(1-\nu)}{2\nu}(2n-1)\right)},$$

with

$$G = \Gamma(\nu)^{-1/\nu}\sqrt{1-\nu}\,.$$

Similarly we get for the second chirality

$$\bar{\boldsymbol{\beta}}^*_{2n-1} = D_{2n-1}(2-\alpha)\bar{\boldsymbol{\beta}}^{\mathrm{CFT}*}_{2n-1}, \quad \bar{\boldsymbol{\gamma}}^*_{2n-1} = D_{2n-1}(\alpha)\bar{\boldsymbol{\gamma}}^{\mathrm{CFT}*}_{2n-1}. \tag{32}$$

The main formula (5) follows immediately.

Before concluding this paper, let us say that in principle our approach can be applied to the computation of one-point functions of descendants for finite radius in the Matsubara direction (finite temperature, in other words). However, this would require a detailed study of the DDV equation and the equations for the function ω_R for the sG model in finite volume.

Acknowledgements. Research of MJ is supported by the Grant-in-Aid for Scientific Research B-20340027. Research of TM is supported by the Grant-in-Aid for Scientific Research B-22340017. Research of FS is supported by RFBR-CNRS grant 09 02 93106.

Taking this occasion, MJ and FS wish to offer TM congratulations on his sixtieth birthday.

References

1. H. Boos, M. Jimbo, T. Miwa, F. Smirnov, Hidden Grassmann Structure in the XXZ Model IV: CFT limit, arXiv:0911.3731, *to appear in Commun. Math. Phys.*
2. L.D. Faddeev, V.E. Korepin, Quantum Theory of Solitons, *Phys. Rept.* **42** (1978) 1–87
3. R.F. Dashen, B. Hasslacher, A. Neveu, The particle spectrum in model field theories from semiclassical functional integral techniques, *Phys. Rev.* **D11** (1975) 3424
4. A. Zamolodchikov, Exact two particle S matrix of quantum sine-Gordon solitons, *Pisma Zh. Eksp. Teor. Fiz.* **25** (1977) 499–502
5. A. Zamolodchikov, Al. Zamolodchikov, Factorized S matrices in two-Dimensions as the exact solutions of certain relativistic quantum field models, *Annals Phys.* **120** (1979) 253–291
6. F.A. Smirnov, The general formula for solitons form factors in sine-Gordon model, *J. Phys A* **19** (1986) L575
7. F.A. Smirnov, Form factors in completely integrable models of quantum field theory, *Advanced Series in Mathematical Physics* **14**, World Scientific, Singapore (1992) 208 pp.
8. S. Lukyanov, Form-factors of exponential fields in the sine-Gordon model, *Mod. Phys. Lett.* **A12** (1997) 2543–2550
9. M. Jimbo, T. Miwa. Algebraic analysis of solvable lattice models, *Reg. Conf. Ser. in Math.* **85** (1995)
10. M. Jimbo, T. Miwa, QKZ Equation with $|q| = 1$ and correlation functions of the XXZ model in the gapless regime, *J. Phys. A* **29** (1996) 2923–2958
11. E. K. Sklyanin, Quantum version of the method of inverse scattering problem, *Dokl. Akad. Nauk SSSR* **244** (1979) 1337
12. E. Sklyanin, L. Takhtajan, L. Faddeev, Quantum inverse scattering method, *Theor. Math. Phys.***40** (1979) 194
13. R.J. Baxter, Exactly solved models in statistical mechanics, *Academic Press* 1982, 486pp.
14. F. Smirnov, unpublished.
15. M. Sato, T. Miwa, M. Jimbo, Holonomic quantum fields. IV *Publ. RIMS, Kyoto Univ.* **15** (1978) 871–972.
16. T.T. Wu, B.M. McCoy, C.A. Tracy and E. Barouch, Spin-spin correlation functions for the two-dimensional Ising model: Exact theory in the scaling region, *Phys. Rev. B* **13** (1976) 316–374
17. A.S. Fokas, A.R. Its, A.A. Kapaev, V.Yu Novokshenov, Painlevé transcendents: the Riemann-Hilbert approach, *AMS Publ.* **128** (2006) 553pp.
18. Al.B. Zamolodchikov, Two point correlation function in scaling Lee-Yang model, *Nucl. Phys. B* **348** (1991) 619–641
19. V. Dotsenko and V. Fateev, Conformal algebra and multipoint correlator functions in 2D statistical models, *Nucl. Phys. B* **240** (1984) 312
20. S. Lukyanov, A. Zamolodchikov, Exact expectation values of local fields in

quantum sine-Gordon model *Nucl. Phys. B* **493** (1997) 571–587
21. V. Fateev, S. Lukyanov, A. Zamolodchikov, Al. Zamolodchikov, Expectation values of local fields in Bullough-Dodd model and integrable perturbed conformal field theories, *Nucl. Phys. B* **516** (1998) 652–674
22. A.B. Zamolodchikov, Al.B. Zamolodchikov, Structure constants and conformal bootstrap in Liouville field theory, *Nucl. Phys. B* **477** (1996) 577–605
23. V. Fateev, D. Fradkin, S. Lukyanov, A. Zamolodchikov, Al. Zamolodchikov, Expectation values of descendent fields in the sine-Gordon model, *Nucl. Phys. B* **540** (1999) 587–609
24. A.A Belavin, A.M. Polyakov, A.B. Zamolodchikov, Infinite conformal symmetry in two-dimensional quantum field theory, *Nucl. Phys. B* **241** (1984) 333–380
25. F. Smirnov. Reductions of the sine-Gordon model as a perturbation of minimal models of conformal field theory, *Nucl. Phys. B* **337** (1990) 156
26. F. Smirnov, N. Reshetikhin, Hidden quantum group symmetry and integrable perturbations of conformal field theories, *Comm. Math. Phys.* (1990) **131** 157
27. H. Boos, M. Jimbo, T. Miwa, F. Smirnov, and Y. Takeyama, Hidden Grassmann structure in the XXZ model, *Commun. Math. Phys.* **272** (2007) 263–281
28. H. Boos, M. Jimbo, T. Miwa, F. Smirnov, and Y. Takeyama, Hidden Grassmann structure in the XXZ model II: Creation operators, *Commun. Math. Phys.* **286** (2009) 875–932
29. M. Jimbo, T. Miwa, and F. Smirnov, Hidden Grassmann structure in the XXZ model III: Introducing Matsubara direction, *J. Phys. A* **42** (2009) 304018 (31pp)
30. V. Bazhanov, S. Lukyanov, and A. Zamolodchikov, Integrable structure of conformal field theory, quantum KdV theory and thermodynamic Bethe ansatz, *Commun. Math. Phys.* **177** (1996) 381–398
31. V. Bazhanov, S. Lukyanov, and A. Zamolodchikov, Integrable structure of conformal field theory II. Q-operator and DDV equation, *Commun. Math. Phys.* **190** (1997) 247–278
32. A.B. Zamolodchikov, Higher order integrals of motion in two-dimensional models of the field theory with a broken conformal symmetry, *Pisma Zh. Eksp. Teor. Fiz.* **46** (1987) 129–132
33. H. Boos, M. Jimbo, T. Miwa, F. Smirnov, and Y. Takeyama, Traces on the Sklyanin algebra and correlation functions of the eight-vertex model, *J. Phys.* **A38** (2005) 7629–7660
34. H. Boos, M. Jimbo, T. Miwa, F. Smirnov, and Y. Takeyama, Algebraic representation of correlation functions in integrable spin chains. *Annales Henri Poincare* **7** (2006) 1395–1428
35. C. Destri and H.J. de Vega, Unified approach to thermodynamic Bethe Ansatz and finite size corrections for lattice models and field theories, *Nucl. Phys.B* **438** (1995) 413–454
36. Al. Zamolodchikov, Mass scale in Sine-Gordon model and its reductions. *Int.*

J. Mod. Phys. A **10**(1995)1125-1150
37. H. Boos and F. Göhmann, On the physical part of the factorized correlation functions of the XXZ chain, *J. Phys. A* **42** (2009) 1–27
38. Al.B. Zamolodchikov, Thermodynamic Bethe Ansatz in relativistic models. Scaling three state Potts and Lee-Yang models, *Nucl. Phys. B* **342** (1990) 695-720

PERFECT CRYSTALS FOR THE QUANTUM AFFINE ALGEBRA $U_q(C_n^{(1)})$

SEOK-JIN KANG * †

Department of Mathematical Sciences and Research Institute of Mathematics
Seoul National University
599 Gwanak-ro, Gwanak-gu
Seoul 151-747, Korea
E-mail: sjkang@math.snu.ac.kr

MYUNGHO KIM * and INHA LEE * ‡

Department of Mathematical Sciences
Seoul National University
599 Gwanak-ro, Gwanak-gu
Seoul 151-747, Korea
E-mails: mkim@math.snu.ac.kr, inha30633@snu.ac.kr

KAILASH C. MISRA §

Department of Mathematics
North Carolina State University
Raleigh, NC 27695-8205, USA
E-mail: misra@math.ncsu.edu

Dedicated to Professor Tetsuji Miwa on the occasion of his 60th birthday

In this paper, we give an explicit isomorphism between the adjoint crystal $\mathbb{B}_\ell$ given in Ref. 13 and the Kirillov-Reshetikhin crystals $\mathbf{B}^{1,2\ell}$ presented in Ref. 2 for $C_n^{(1)}$. Consequently we have a complete proof of perfectness of the crystals $\mathbb{B}_\ell$. In the sequel we also present an explicit description of the involution σ given in Ref. 2 for this special case.

*This research was supported by KRF Grant # 2007-341-C00001.

†This research was supported by National Institute for Mathematical Sciences (2010 Thematic Program, TP1004).

‡This research was supported by BK21 Mathematical Sciences Division.

§This research was supported by NSA Grant # H98230-08-1-0080

Introduction

The notion of crystal base (L, B) for an integrable module of a quantized enveloping algebra $U_q(\mathfrak{g})$ was introduced by Kashiwara in Ref. 16. The crystal B has interesting combinatorial properties. It can be thought of as the $q = 0$ limit of the canonical base[20] or global crystal base.[17] Explicit realizations of the crystals for irreducible highest weight modules for the finite dimensional simple Lie algebras of types A, B, C, D (resp. G_2) in terms of certain Young tableaux are given in Ref. 18 (resp. Ref. 11). In Ref. 21, the crystals for the level one integrable highest weight modules for the quantum affine algebra $U_q(\widehat{sl}(n))$ were realized in terms of certain extended Young diagrams. This construction was generalized to arbitrary integrable highest weight modules for $U_q(\widehat{sl}(n))$ in Ref. 7.

The attempt to extend the construction in Ref. 7 to other quantum affine algebras,[10] led to the theory of affine and perfect crystals.[14] In particular, it was shown in Ref. 14 that the affine crystal $B(\lambda)$ for the level $\ell \in \mathbb{Z}_{>0}$ integrable highest weight module $V(\lambda)$ of the quantum affine algebra $U_q(\mathfrak{g})$ can be realized as the semi-infinite tensor product $\cdots \otimes B \otimes B \otimes B$, where B is a perfect crystal of level ℓ. This is known as the path realization. Thus for path realization of affine crystals it is necessary to have explicit constructions of perfect crystals. In Ref. 15, at least one perfect crystal at level ℓ was constructed for each quantum affine algebra of classical type. Subsequently it was noticed in Ref. 13 that one needs a coherent family of perfect crystals $\{B_\ell\}$ in order to give a path realization of the crystal for the Verma module $M(\lambda)$ (or $U_q^-(\mathfrak{g})$). To show that a family of perfect crystals is a coherent family one needs to have explicit actions of all Kashiwara operators including the operators corresponding to the 0-node. For example, the family of perfect crystals for $U_q(C_n^{(1)})$ given in Ref. 15 did not have explicit 0-action. In Ref. 13, a family $\{\mathbb{B}_\ell\}_{\ell \in \mathbb{Z}_{>0}}$ of combinatorial crystals (now known as adjoint crystals) for $U_q(C_n^{(1)})$ with explicit 0-action was introduced and it was stated without a proof that they are perfect crystals. Of course, now using Refs. 15 and 22, one can argue that the adjoint crystals $\mathbb{B}_\ell$ presented in Ref. 13 are perfect.

More recently, it has been realized that a perfect crystal is indeed a crystal for certain finite dimensional module called Kirillov-Reshetikhin module (KR-module)[19] of a quantum affine algebra. The KR-modules are parameterized by two integers (r, s), where r corresponds to a vertex of the associated finite Dynkin diagram and s is a positive integer. It was conjectured by Hatayama *et al.*[4,5] that any KR-module $W^{r,s}$ admit a crystal base $\mathbf{B}^{r,s}$ in the sense of Kashiwara.[17] This conjecture has been proved recently by

Okado and Schilling[22] for quantum affine algebras of classical types. In Ref. 2 an explicit combinatorial description is given for each such crystal $\mathbf{B}^{r,s}$. It was also conjectured by Hatayama *et al.*[4,5] that if s is a multiple of $t_r = \max\{1, 2/(\alpha_r, \alpha_r)\}$, then the crystal $\mathbf{B}^{r,s}$ of the corresponding KR-module is perfect of level s/t_r. This conjecture has been proved in Ref. 3 for the KR-crystals $\mathbf{B}^{r,s}$ constructed in Ref. 2. For example, in Ref. 2, the explicit realization of the level ℓ perfect crystal $\mathbf{B}^{1,2\ell}$ for the quantum affine algebra $U_q(C_n^{(1)})$ is given as a subset of a suitable KR-crystal for the quantum affine algebra $U_q(A_{2n+1}^{(2)})$. The 0-action on $\mathbf{B}^{1,2\ell}$ is given by using an involution σ which is quite complicated in the general case. We give an explicit description of the involution σ in this special case. The main result of this paper is the explicit isomorphism between the adjoint crystal $\mathbb{B}_\ell$ given in Ref. 13 and the KR-crystals $\mathbf{B}^{1,2\ell}$ given in Ref. 2. As a byproduct, we have a complete proof of perfectness of the crystal $\mathbb{B}_\ell$.

This paper is organized as follows. In Section 1, we introduce some of the basic definitions and notations related to quantum affine algebras and perfect crystals. In Section 2, we recall the definition of adjoint crystals $\mathbb{B}_\ell$ of type $C_n^{(1)}$ from Ref. 13. In Section 3, the Kirillov-Reshetikhin crystals $\mathbf{B}^{1,2\ell}$ in Ref. 2 of type $C_n^{(1)}$ are reviewed. They are realized as subsets of certain Kirillov-Reshetikhin crystals of type $A_{2n+1}^{(2)}$. In Section 4, we prove that the crystals $\mathbb{B}_\ell$ and $\mathbf{B}^{1,2\ell}$ are isomorphic.

1. Quantum affine algebras and perfect crystals

Let $I = \{0, 1, ..., n\}$ be the index set and let $A = (a_{ij})_{i,j \in I}$ be a generalized Cartan matrix of affine type. The *dual weight lattice* $P^\vee$ is defined to be the free abelian group $P^\vee = \mathbb{Z}h_0 \oplus \mathbb{Z}h_1 \oplus \cdots \oplus \mathbb{Z}h_n \oplus \mathbb{Z}d$ of rank $n+2$, whose complexification $\mathfrak{h} = \mathbb{C} \otimes P^\vee$ is called the *Cartan subalgebra*. We define the linear functionals α_i and Λ_i $(i \in I)$ on $\mathfrak{h}$ by

$$\alpha_i(h_j) = a_{ji}, \quad \alpha_i(d) = \delta_{i0}, \quad \Lambda_i(h_j) = \delta_{ij}, \quad \Lambda_i(d) = 0 \quad (i, j \in I).$$

The α_i are called the *simple roots* and the Λ_i are called the *fundamental weights*. We denote by $\Pi = \{\alpha_i \mid i \in I\}$ the set of simple roots. We also define the *affine weight lattice* to be $P = \{\lambda \in \mathfrak{h}^* \mid \lambda(P^\vee) \subset \mathbb{Z}\}$. The quadruple $(A, P^\vee, \Pi, P)$ is called an *affine Cartan datum*. We denote by $\mathfrak{g}$ the affine Kac-Moody algebra corresponding to the affine Cartan datum $(A, P^\vee, \Pi, P)$ (see Ch. 1 of Ref. 9). Let δ denote the *null root* and c denote the canonical central element for $\mathfrak{g}$ (see Ch. 4 of Ref. 9). Now the affine weight lattice can be written as $P = \mathbb{Z}\Lambda_0 \oplus \mathbb{Z}\Lambda_1 \oplus \cdots \oplus \mathbb{Z}\Lambda_n \oplus \mathbb{Z}\delta$. Let $P^+ = \{\lambda \in P \mid \lambda(h_i) \geq 0$ for all $i \in I\}$. The elements of P are called the *affine weights* and the elements of P^+ are called the *affine dominant integral weights.*

Let $\bar{P}^\vee = \mathbb{Z}h_0 \oplus \cdots \oplus \mathbb{Z}h_n$, $\bar{\mathfrak{h}} = \mathbb{C} \otimes_{\mathbb{Z}} \bar{P}^\vee$, $\bar{P} = \mathbb{Z}\Lambda_0 \oplus \mathbb{Z}\Lambda_1 \oplus \cdots \oplus \mathbb{Z}\Lambda_n$ and $\bar{P}^+ = \{\lambda \in \bar{P} \mid \lambda(h_i) \geq 0 \text{ for all } i \in I\}$. The elements of $\bar{P}$ are called the *classical weights* and the elements of $\bar{P}^+$ are called the *classical dominant integral weights.* The *level* of a (classical) dominant integral weight λ is defined to be $\ell = \lambda(c)$. We call the quadruple $(A, \bar{P}^\vee, \Pi, \bar{P})$ the *classical Cartan datum.*

For the convenience of notations, we define $[k]_x = \dfrac{x^k - x^{-k}}{x - x^{-1}}$, where k is an integer and x is a symbol. We also define $\begin{bmatrix} m \\ k \end{bmatrix}_x = \dfrac{[m]_x!}{[k]_x! \, [m-k]_x!}$, where m and k are nonnegative integers, $m \geq k \geq 0$, $[k]_x! = [k]_x \, [k-1]_x \cdots [1]_x$ and $[0]_x! = 1$.

The *quantum affine algebra* $U_q(\mathfrak{g})$ is the quantum group associated with the affine Cartan datum $(A, P^\vee, \Pi, P)$. That is, it is the associative algebra over $\mathbb{C}(q)$ with unity generated by $e_i, f_i (i \in I)$ and $q^h (h \in P^\vee)$ satisfying the following defining relations:

(i) $q^0 = 1, q^h q^{h'} = q^{h+h'}$ for all $h, h' \in P^\vee$,

(ii) $q^h e_i q^{-h} = q^{\alpha_i(h)} e_i$, $q^h f_i q^{-h} = q^{-\alpha_i(h)} f_i$ for $h \in P^\vee$,

(iii) $e_i f_j - f_j e_i = \delta_{ij} \dfrac{K_i - K_i^{-1}}{q_i - q_i^{-1}}$ for $i, j \in I$, where $q_i = q^{s_i}$ and $K_i = q^{s_i h_i}$,

(iv) $\displaystyle\sum_{k=0}^{1-a_{ij}} (-1)^k \begin{bmatrix} 1-a_{ij} \\ k \end{bmatrix}_{q_i} e_i^{1-a_{ij}-k} e_j e_i^k = 0$ for $i \neq j$,

(v) $\displaystyle\sum_{k=0}^{1-a_{ij}} (-1)^k \begin{bmatrix} 1-a_{ij} \\ k \end{bmatrix}_{q_i} f_i^{1-a_{ij}-k} f_j f_i^k = 0$ for $i \neq j$.

Here, $D = \mathrm{diag}(s_0, s_1, \ldots, s_n)$ is a diagonal matrix with all $s_i \in \mathbb{Z}_{>0}$ such that DA is symmetric. We denote by $U'_q(\mathfrak{g})$ the subalgebra of $U_q(\mathfrak{g})$ generated by $e_i, f_i, q^{h_i} (i \in I)$. The algebra $U'_q(\mathfrak{g})$ can be regarded as the quantum group associated with the classical Cartan datum $(A, \bar{P}^\vee, \Pi, \bar{P})$.

In this paper, we focus on the quantum affine algebras of type $C_n^{(1)} (n \geq 2)$ and $A_{2n+1}^{(2)} (n \geq 2)$. We will use $I = \{0, 1, ..., n\}$ and $I = \{0, 1, ..., n, n+1\}$ as index sets for $C_n^{(1)} (n \geq 2)$ and $A_{2n+1}^{(2)} (n \geq 2)$, respectively. Thus the *null root* and the *canonical central element* are given by

$$\delta = \begin{cases} \alpha_0 + 2\alpha_1 + 2\alpha_2 + \cdots + 2\alpha_{n-1} + \alpha_n & \text{for } C_n^{(1)} (n \geq 2), \\ \alpha_0 + \alpha_1 + 2\alpha_1 + \cdots + 2\alpha_n + \alpha_{n+1} & \text{for } A_{2n+1}^{(2)} (n \geq 2), \end{cases}$$

$$c = \begin{cases} c = h_0 + h_1 + \cdots + h_n & \text{for } C_n^{(1)} (n \geq 2), \\ c = h_0 + h_1 + 2h_2 + \cdots + 2h_n + 2h_{n+1} & \text{for } A_{2n+1}^{(2)} (n \geq 2). \end{cases}$$

(See Ch. 4 of Ref. 9.)

Definition 1.1. An *affine crystal* (respectively, a *classical crystal*) is a set $\mathcal{B}$ together with the maps wt : $\mathcal{B} \to P$ (respectively, wt : $\mathcal{B} \to \bar{P}$), $\tilde{e}_i, \tilde{f}_i : \mathcal{B} \to \mathcal{B} \cup \{0\}$ and $\varepsilon_i, \varphi_i : \mathcal{B} \to \mathbb{Z} \cup \{-\infty\}$ $(i \in I)$ satisfying the following conditions:

(i) $\varphi_i(b) = \varepsilon_i(b) + \langle h_i, \mathrm{wt}(b) \rangle$ for all $i \in I$,
(ii) $\mathrm{wt}(\tilde{e}_i b) = \mathrm{wt}(b) + \alpha_i$ if $\tilde{e}_i b \in \mathcal{B}$,
(iii) $\mathrm{wt}(\tilde{f}_i b) = \mathrm{wt}(b) - \alpha_i$ if $\tilde{f}_i b \in \mathcal{B}$,
(iv) $\varepsilon_i(\tilde{e}_i b) = \varepsilon_i(b) - 1$, $\varphi_i(\tilde{e}_i b) = \varphi_i(b) + 1$ if $\tilde{e}_i b \in \mathcal{B}$,
(v) $\varepsilon_i(\tilde{f}_i b) = \varepsilon_i(b) + 1$, $\varphi_i(\tilde{f}_i b) = \varphi_i(b) - 1$ if $\tilde{f}_i b \in \mathcal{B}$,
(vi) $\tilde{f}_i b = b'$ if and only if $b = \tilde{e}_i b'$ for $b, b' \in \mathcal{B}, i \in I$,
(vii) If $\varphi_i(b) = -\infty$ for $b \in \mathcal{B}$, then $\tilde{e}_i b = \tilde{f}_i b = 0$.

Example 1.1. Let $\lambda \in P^+$. Then the crystal graph $\mathcal{B}(\lambda)$ of the irreducible highest weight module $V(\lambda)$ is an affine crystal. It can be regarded as a classical crystal by forgetting the action of its weights on d.

Definition 1.2. Let $\mathcal{B}_1$ and $\mathcal{B}_2$ be affine or classical crystals. A *crystal morphism* (or *morphism of crystals*) $\Psi : \mathcal{B}_1 \to \mathcal{B}_2$ is a map $\Psi : \mathcal{B}_1 \cup \{0\} \to \mathcal{B}_2 \cup \{0\}$ such that

(i) $\Psi(0) = 0$,
(ii) if $b \in \mathcal{B}_1$ and $\Psi(b) \in \mathcal{B}_2$, then $\mathrm{wt}(\Psi(b)) = \mathrm{wt}(b)$, $\varepsilon_i(\Psi(b)) = \varepsilon_i(b)$, and $\varphi_i(\Psi(b)) = \varphi_i(b)$ for all $i \in I$,
(iii) if $b, b' \in \mathcal{B}_1$, $\Psi(b), \Psi(b') \in \mathcal{B}_2$ and $\tilde{f}_i b = b'$, then $\tilde{f}_i \Psi(b) = \Psi(b')$ and $\Psi(b) = \tilde{e}_i \Psi(b')$ for all $i \in I$.

A crystal morphism $\Psi : \mathcal{B}_1 \to \mathcal{B}_2$ is called an *isomorphism* if it is a bijection from $\mathcal{B}_1 \cup \{0\}$ to $\mathcal{B}_2 \cup \{0\}$.

For crystals $\mathcal{B}_1$ and $\mathcal{B}_2$, we define the *tensor product* $\mathcal{B}_1 \otimes \mathcal{B}_2$ to be the set $\mathcal{B}_1 \times \mathcal{B}_2$ whose crystal structure is given as follows:

$$\tilde{e}_i(b_1 \otimes b_2) = \begin{cases} \tilde{e}_i b_1 \otimes b_2 & \text{if } \varphi_i(b_1) \geq \varepsilon_i(b_2), \\ b_1 \otimes \tilde{e}_i b_2 & \text{if } \varphi_i(b_1) < \varepsilon_i(b_2), \end{cases}$$

$$\tilde{f}_i(b_1 \otimes b_2) = \begin{cases} \tilde{f}_i b_1 \otimes b_2 & \text{if } \varphi_i(b_1) > \varepsilon_i(b_2), \\ b_1 \otimes \tilde{f}_i b_2 & \text{if } \varphi_i(b_1) \leq \varepsilon_i(b_2), \end{cases}$$

$$\begin{aligned} \mathrm{wt}(b_1 \otimes b_2) &= \mathrm{wt}(b_1) + \mathrm{wt}(b_2), \\ \varepsilon_i(b_1 \otimes b_2) &= \max(\varepsilon_i(b_1), \varepsilon_i(b_2) - \langle h_i, \mathrm{wt}(b_1) \rangle), \\ \varphi_i(b_1 \otimes b_2) &= \max(\varphi_i(b_2), \varphi_i(b_1) + \langle h_i, \mathrm{wt}(b_2) \rangle). \end{aligned}$$

Let $\mathcal{B}$ be a classical crystal. For an element $b \in \mathcal{B}$, we define

$$\varepsilon(b) = \sum_{i \in I} \varepsilon_i(b) \Lambda_i, \ \varphi(b) = \sum_{i \in I} \varphi_i(b) \Lambda_i.$$

Definition 1.3. Let ℓ be a positive integer. A classical crystal $\mathcal{B}$ is called a *perfect crystal of level* ℓ if

(P1) there exists a finite dimensional $U_q'(\mathfrak{g})$-module with a crystal basis whose crystal graph is isomorphic to $\mathcal{B}$,
(P2) $\mathcal{B} \otimes \mathcal{B}$ is connected,
(P3) there exists a classical weight $\lambda_0 \in \bar{P}$ such that $\mathrm{wt}(\mathcal{B}) \subset \lambda_0 + \sum_{i \neq 0} \mathbb{Z}_{\leq 0}\, \alpha_i$, $\#(\mathcal{B}_{\lambda_0}) = 1$, where $\mathcal{B}_{\lambda_0} = \{b \in \mathcal{B} \mid \mathrm{wt}(b) = \lambda_0\}$,
(P4) for any $b \in \mathcal{B}$, $\langle c, \varepsilon(b) \rangle \geq \ell$,
(P5) for any $\lambda \in \bar{P}^+$ with $\lambda(c) = \ell$, there exist unique $b^\lambda, b_\lambda \in \mathcal{B}$ such that $\varepsilon(b^\lambda) = \lambda = \varphi(b_\lambda)$.

The following crystal isomorphism theorem plays a fundamental role in the theory of perfect crystals.

Theorem 1.1. *(Ref. 14) Let $\mathcal{B}$ be a perfect crystal of level ℓ ($\ell \in \mathbb{Z}_{\geq 0}$). For any $\lambda \in \bar{P}^+$ with $\lambda(c) = \ell$, there exists a unique classical crystal isomorphism*

$$\Psi : \mathcal{B}(\lambda) \xrightarrow{\sim} \mathcal{B}(\varepsilon(b_\lambda)) \otimes \mathcal{B} \quad \text{given by} \quad u_\lambda \longmapsto u_{\varepsilon(b_\lambda)} \otimes b_\lambda,$$

where u_λ is the highest weight vector in $\mathcal{B}(\lambda)$ and b_λ is the unique vector in $\mathcal{B}$ such that $\varphi(b_\lambda) = \lambda$.

Set $\lambda_0 = \lambda,\ \lambda_{k+1} = \varepsilon(b_{\lambda_k}),\ b_0 = b_{\lambda_0},\ b_{k+1} = b_{\lambda_{k+1}}$. Appying the above crystal isomorphism repeatedly, we get a sequence of crystal isomorphisms

$$\begin{array}{ccccccc} \mathcal{B}(\lambda) & \xrightarrow{\sim} & \mathcal{B}(\lambda_1)\otimes\mathcal{B} & \xrightarrow{\sim} & \mathcal{B}(\lambda_2)\otimes\mathcal{B}\otimes\mathcal{B} & \xrightarrow{\sim} & \cdots \\ u_\lambda & \longmapsto & u_{\lambda_1}\otimes b_0 & \longmapsto & u_{\lambda_2}\otimes b_1\otimes b_0 & \longmapsto & \cdots. \end{array}$$

In this process, we get an infinite sequence $\mathbf{p}_\lambda = (b_k)_{k=0}^\infty \in \mathcal{B}^{\otimes\infty}$, which is called the *ground-state path of weight* λ. Let $\mathcal{P}(\lambda) := \{\mathbf{p} = (\mathbf{p}(k))_{k=0}^\infty \in \mathcal{B}^{\otimes\infty} \mid \mathbf{p}(k)\in\mathcal{B}, \mathbf{p}(k) = b_k \text{ for all } k \gg 0\}$. The elements of $\mathcal{P}(\lambda)$ are called the λ-*paths*. Let $\mathbf{p} = (\mathbf{p}(k))_{k=0}^\infty$ be a λ-path and let $N > 0$ be the smallest positive integer such that $\mathbf{p}(k) = b_k$ for all $k \geq N$. For each $i\in I$, we define

$$\begin{aligned} \operatorname{wt}\mathbf{p} &= \lambda_N + \sum_{k=0}^{N-1} \operatorname{wt}\mathbf{p}(k), \\ \tilde{e}_i\mathbf{p} &= \cdots\otimes\mathbf{p}(N+1)\otimes\tilde{e}_i(\mathbf{p}(N)\otimes\cdots\otimes\mathbf{p}(0)), \\ \tilde{f}_i\mathbf{p} &= \cdots\otimes\mathbf{p}(N+1)\otimes\tilde{f}_i(\mathbf{p}(N)\otimes\cdots\otimes\mathbf{p}(0)), \\ \varepsilon_i(\mathbf{p}) &= \max(\varepsilon_i(\mathbf{p}') - \varphi(b_N), 0), \\ \varphi_i(\mathbf{p}) &= \varphi_i(\mathbf{p}') + \max(\varphi_i(b_N) - \varepsilon(\mathbf{p}'), 0), \end{aligned}$$

where $\mathbf{p}' = \mathbf{p}(N)\otimes\cdots\otimes\mathbf{p}(1)\otimes\mathbf{p}(0)$. The above maps define a classical crystal structure on $\mathcal{P}(\lambda)$. The classical crystal $\mathcal{P}(\lambda)$ gives the *path realization* of $\mathcal{B}(\lambda)$.

Proposition 1.1. *(Ref. 14) There exists an isomorphism of classical crystals*

$$\Psi_\lambda : \mathcal{B}(\lambda) \xrightarrow{\sim} \mathcal{P}(\lambda) \quad \textit{given by} \quad u_\lambda \longmapsto \mathbf{p}_\lambda,$$

where u_λ *is the highest weight vector in* $\mathcal{B}(\lambda)$.

2. The adjoint crystals of type $C_n^{(1)}$

Let $\{\mathbb{B}_\ell\}_{\ell\in\mathbb{N}}$ be the coherent family of classical crystals type $C_n^{(1)}$ introduced in Ref. 13. (See also Ref. 1 for level 1 case.) In Ref. 8, they were referred to as *adjoint crystals* of type $C_n^{(1)}$.

Fix a positive number ℓ and set

$$\begin{aligned} \mathbb{B}_\ell = \{(x_1, x_2, \ldots, x_n, \bar{x}_n, \ldots, \bar{x}_2, \bar{x}_1) \in \mathbb{Z}^{2n} \mid x_i, \bar{x}_i \geq 0, \\ \sum_{i=1}^n (x_i + \bar{x}_i) \leq 2\ell,\ \sum_{i=1}^n (x_i + \bar{x}_i) \equiv 0 \pmod 2\}. \end{aligned}$$

For $b = (x_1, x_2, \ldots, x_n, \bar{x}_n, \ldots, \bar{x}_2, \bar{x}_1) \in \mathbb{B}_\ell$, the actions of $\tilde{e}_i$, $\tilde{f}_i$ are defined as follows:

$$\tilde{e}_0 b = \begin{cases} (x_1 - 2, x_2, \ldots, \bar{x}_2, \bar{x}_1) & \text{if } x_1 \geq \bar{x}_1 + 2, \\ (x_1 - 1, x_2, \ldots, \bar{x}_2, \bar{x}_1 + 1) & \text{if } x_1 = \bar{x}_1 + 1, \\ (x_1, x_2, \ldots, \bar{x}_2, \bar{x}_1 + 2) & \text{if } x_1 \leq \bar{x}_1, \end{cases} \tag{1}$$

$$\tilde{f}_0 b = \begin{cases} (x_1 + 2, x_2, \ldots, \bar{x}_2, \bar{x}_1) & \text{if } x_1 \geq \bar{x}_1, \\ (x_1 + 1, x_2, \ldots, \bar{x}_2, \bar{x}_1 - 1) & \text{if } x_1 = \bar{x}_1 - 1, \\ (x_1, x_2, \ldots, \bar{x}_2, \bar{x}_1 - 2) & \text{if } x_1 \leq \bar{x}_1 - 2, \end{cases} \tag{2}$$

$$\tilde{e}_i b = \begin{cases} (x_1, \ldots, x_i + 1, x_{i+1} - 1, \ldots, \bar{x}_1) & \text{if } x_{i+1} > \bar{x}_{i+1} \\ (x_1, \ldots, \bar{x}_{i+1} + 1, \bar{x}_i - 1, \ldots, \bar{x}_1) & \text{if } x_{i+1} \leq \bar{x}_{i+1} \end{cases} \tag{3}$$

$$\text{for } i = 1, 2, \ldots, n-1,$$

$$\tilde{f}_i b = \begin{cases} (x_1, \ldots, x_i - 1, x_{i+1} + 1, \ldots, \bar{x}_1) & \text{if } x_{i+1} \geq \bar{x}_{i+1} \\ (x_1, \ldots, \bar{x}_{i+1} - 1, \bar{x}_i + 1, \ldots, \bar{x}_1) & \text{if } x_{i+1} < \bar{x}_{i+1} \end{cases} \tag{4}$$

$$\text{for } i = 1, 2, \ldots, n-1,$$

$$\tilde{e}_n b = (x_1, \ldots, x_n + 1, \bar{x}_n - 1, \ldots, \bar{x}_1), \tag{5}$$

$$\tilde{f}_n b = (x_1, \ldots, x_n - 1, \bar{x}_n + 1, \ldots, \bar{x}_1), \tag{6}$$

where the right-hand side is regarded as 0 if it does not satisfy the conditions for $\mathbb{B}_\ell$. The classical weights $\varepsilon(b)$ and $\varphi(b)$ are given as follows:

$$\varepsilon(b) = \Big(\frac{2\ell - s(b)}{2} + (x_1 - \bar{x}_1)_+\Big)\Lambda_0 + \sum_{i=1}^{n-1} (\bar{x}_i + (x_{i+1} - \bar{x}_{i+1})_+)\Lambda_i + \bar{x}_n \Lambda_n,$$

$$\varphi(b) = \Big(\frac{2\ell - s(b)}{2} + (\bar{x}_1 - x_1)_+\Big)\Lambda_0 + \sum_{i=1}^{n-1} (x_i + (\bar{x}_{i+1} - x_{i+1})_+)\Lambda_i + x_n \Lambda_n,$$

where $s(b) = \sum_{i=1}^{n} (x_i + \bar{x}_i)$ and $(x)_+ = \max\{x, 0\}$. Define $\mathrm{wt} : \mathbb{B}_\ell \to \bar{P}$ by $\mathrm{wt}(b) = \sum_{i=1}^{n} (x_i - \bar{x}_i)(\Lambda_i - \Lambda_{i-1})$. Then $(\mathbb{B}_\ell, \mathrm{wt}, \varphi, \varepsilon, \tilde{f}_i, \tilde{e}_i)$ is a classical crystal.[13]

Example 2.1. The adjoint crystal $\mathbb{B}_1$ for $C_2^{(1)}$:

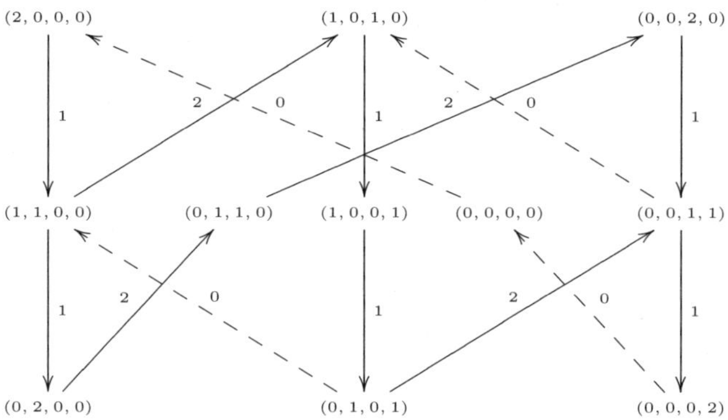

Remark 2.1. Note that $\mathbb{B}_\ell$ is isomorphic to $\mathcal{B}(2\ell\Lambda_1) \oplus \mathcal{B}((2\ell-2)\Lambda_1) \oplus \cdots \oplus \mathcal{B}(0)$ as a $U_q(C_n)$-crystal. Indeed, each element $b = (x_1, x_2, \dots, x_n, \bar{x}_n, \dots, \bar{x}_2, \bar{x}_1) \in \mathbb{B}_\ell$ can be identified with the Kashiwara-Nakashima tableau

$$\overbrace{\boxed{1}\cdots\boxed{1}}^{x_1}\overbrace{\boxed{2}\cdots\boxed{2}}^{x_2}\cdots\overbrace{\boxed{n}\cdots\boxed{n}}^{x_n}\overbrace{\boxed{\bar{n}}\cdots\boxed{\bar{n}}}^{\bar{x}_n}\cdots\overbrace{\boxed{\bar{1}}\cdots\boxed{\bar{1}}}^{\bar{x}_1}$$

in $\mathcal{B}(s(b)\Lambda_1)$. (For the definition of Kashiwara-Nakashima tableaux of type C_n, see Refs. 2,6,18.) It is easy to show that this correspondence is a $U_q(C_n)$-crystal isomorphism.

Proposition 2.1. *$\mathbb{B}_\ell$ satisfies conditions (P2)-(P5) in Definition 1.3.*

Proof.

(P2) By Remark 2.1

$$\mathbb{B}_\ell \otimes \mathbb{B}_\ell = \bigsqcup_{0 \le s,t \le \ell} \mathcal{B}(2s\Lambda_1) \otimes \mathcal{B}(2t\Lambda_1)$$

as a $U_q(C_n)$-crystal. Since every vector in $\mathbb{B}_\ell \otimes \mathbb{B}_\ell$ is connected to some maximal vector for $U_q(C_n)$ (i.e., a vector that is annihilated by $\tilde{e}_i$ for all $i = 1, 2, \dots, n$), it is enough to show that all maximal vectors are connected to each other by various i-arrows for $i = 0, 1, 2, \dots, n$.
Let $b_1 \otimes b_2$ be a maximal vector in $\mathcal{B}(2s\Lambda_1) \otimes \mathcal{B}(2t\Lambda_1)$. Then, by the tensor product rule, b_1 must be $(2s, 0, \dots, 0)$, the highest weight vector in $\mathcal{B}(2s\Lambda_1)$. After applying $\tilde{e}_0$ on $(2s, 0, \dots, 0) \otimes b_2$ repeatedly, by the tensor product rule, we get $(0, 0, \dots, 0) \otimes b_2'$ for some

$b'_2 \in \mathcal{B}(2t'\Lambda_1)(0 \le t' \le \ell)$. Since $(0,0,\ldots,0) \otimes \mathcal{B}(2t'\Lambda_1)$ is isomorphic to $\mathcal{B}(2t'\Lambda_1)$ as $U_q(C_n)$-crystals, by applying appropriate $\tilde{e}_i$'s with $i = 1, 2, \ldots, n$, we will get $(0,0,\ldots,0) \otimes (2t', 0, \ldots, 0)$. Finally, applying $\tilde{f}_0^{\ell - t'}$ to this vector yields the vector $(0,0,\ldots,0) \otimes (2\ell, 0, \ldots, 0)$, which proves our assertion.

(P3) Set $\lambda_0 = 2\ell(\Lambda_1 - \Lambda_0) \in \bar{P}$. Then for $b = (x_1, \ldots, x_n, \bar{x}_n, \ldots, \bar{x}_1) \in \mathbb{B}_\ell$, we have

$$\begin{aligned}
\mathrm{wt}(b) &= \sum_{i=1}^{n} (x_i - \bar{x}_i)(\Lambda_i - \Lambda_{i-1}) \\
&= \Big(s(b) - \sum_{i=2}^{n}(x_i + \bar{x}_i) - 2\bar{x}_1\Big)(\Lambda_1 - \Lambda_0) + \sum_{i=2}^{n}(x_i - \bar{x}_i)(\Lambda_i - \Lambda_{i-1}) \\
&= s(b)(\Lambda_1 - \Lambda_0) - \sum_{i=2}^{n} x_i(-\Lambda_0 + \Lambda_1 + \Lambda_{i-1} - \Lambda_i) \\
&\quad - \sum_{i=2}^{n} \bar{x}_i(-\Lambda_0 + \Lambda_1 - \Lambda_{i-1} + \Lambda_i) - 2\bar{x}_1(\Lambda_1 - \Lambda_0) \\
&= s(b)(\Lambda_1 - \Lambda_0) - \sum_{i=2}^{n} x_i(\alpha_1 + \cdots + \alpha_{i-1}) \\
&\quad - \sum_{i=1}^{n} \bar{x}_i(\alpha_1 + \cdots + \alpha_n + \alpha_{n-1} + \cdots + \alpha_i) \\
&= \lambda_0 - \Big(\frac{2\ell - s(b)}{2}\Big)(2\alpha_1 + \cdots + 2\alpha_{n-1} + \alpha_n) \\
&\quad - \sum_{i=2}^{n} x_i(\alpha_1 + \cdots + \alpha_{i-1}) - \sum_{i=1}^{n} \bar{x}_i(\alpha_1 + \cdots + \alpha_n + \alpha_{n-1} + \cdots + \alpha_i).
\end{aligned}$$

Clearly, we have $(\mathbb{B}_\ell)_{\lambda_0} = \{(2\ell, 0, \ldots, 0)\}$.

(P4) Since $c = h_0 + h_1 + \cdots + h_n$, we have

$$\begin{aligned}
\langle c, \varepsilon(b) \rangle &= \frac{2\ell - s(b)}{2} + (x_1 - \bar{x}_1)_+ + \sum_{i=1}^{n-1}\Big(x_i + (x_{i+1} - \bar{x}_{i+1})_+\Big) + \bar{x}_n \\
&= \ell + \sum_{i=1}^{n-1}\Big(-\frac{1}{2}x_i - \frac{1}{2}\bar{x}_i + \bar{x}_i\Big) - \frac{1}{2}(x_n + \bar{x}_n) + \bar{x}_n + \sum_{i=1}^{n}(x_i - \bar{x}_i)_+ \\
&= \ell + \sum_{i=1}^{n}\Big(-\frac{1}{2}(x_i - \bar{x}_i) + (x_i - \bar{x}_i)_+\Big) \qquad (7) \\
&\ge \ell.
\end{aligned}$$

(P5) Let $(\mathbb{B}_\ell)^{\min} = \{b \in \mathbb{B}_\ell \mid \langle c, \varepsilon(b)\rangle = \ell\}$. By (7), we have $(\mathbb{B}_\ell)^{\min} = \{(x_1, x_2, \dots, x_n, x_n, \dots, x_2, x_1) \in \mathbb{B}_\ell\}$. Now it is easy to show that $\varepsilon : (\mathbb{B}_\ell)^{\min} \to \bar{P}_\ell^+ = \{\lambda \in \bar{P}^+ \mid \langle c, \lambda\rangle = \ell\}$ is a bijection. Indeed, for a $\lambda \in \bar{P}_\ell^+$, $\varepsilon^{-1}(\lambda)$ is given by

$$(\lambda(h_1), \lambda(h_2), \dots, \lambda(h_n), \lambda(h_n), \dots, \lambda(h_2), \lambda(h_1)).$$

Since $\varphi : (\mathbb{B}_\ell)^{\min} \to \bar{P}_\ell^+$ coincides with ε, we have the desired result. □

Remark 2.2. To prove that $\mathbb{B}_\ell$ is perfect, it remains to show that there is a $U_q'(C_n^{(1)})$-module with crystal base $\mathbb{B}_\ell$. In Section 4, instead of constructing such a module directly, we will show that $\mathbb{B}_\ell$ is isomorphic to a crystal which admits a $U_q'(C_n^{(1)})$-module structure.

Example 2.2. The crystal $\mathcal{B}(\Lambda_0)$ for $C_2^{(1)}$:

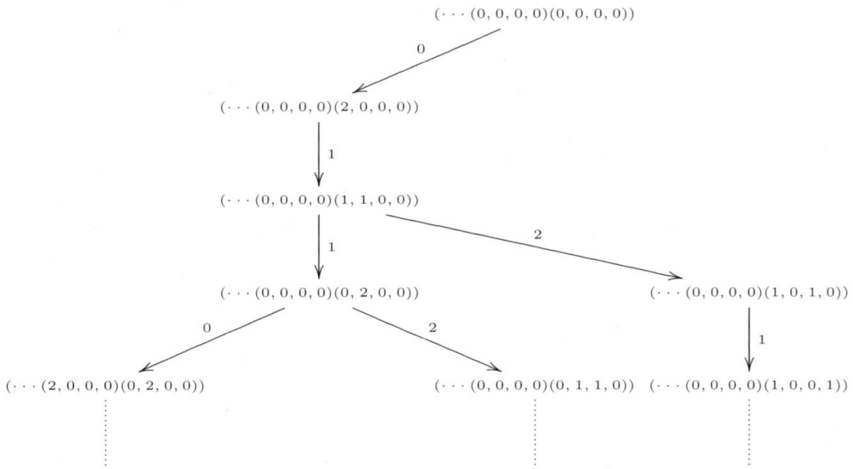

3. Kirillov-Reshetikhin crystals $\mathbf{B}^{1,2\ell}_{A^{(2)}_{2n+1}}$ and $\mathbf{B}^{1,2\ell}_{C^{(1)}_n}$

In Ref. 19, the notion of *Kirillov-Reshetikhin modules* was introduced. They are certain finite dimensional $U_q'(\mathfrak{g})$-modules $W^{r,s}$ labeled by a positive integer s and $r \in \{1, 2, \dots, n\}$. See Ref. 4 for the precise definition. In Ref. 22, it is proved that every Kirillov-Reshetikhin module of classical affine type has a crystal basis. The crystal of $U_q'(\mathfrak{g})$-module $W^{r,s}$ will be called the *Kirillov-Reshetikhin crystal of type* $\mathfrak{g}$ and will be denoted by $\mathbf{B}^{r,s}_{\mathfrak{g}}$.

In this section, we will give a description of the Kirillov-Reshetikhin crystal $\mathbf{B}^{1,2\ell}_{C^{(1)}_n}$. In Section 4, we will show that $\mathbf{B}^{1,2\ell}_{C^{(1)}_n}$ is isomorphic to $\mathbb{B}_\ell$

as a $U_q'(C_n^{(1)})$-crystal, thereby we conclude that there exists a $U_q'(C_n^{(1)})$-module with crystal basis $\mathbb{B}_\ell$. Following Ref. 2, the crystal $\mathbf{B}^{1,2\ell}_{C_n^{(1)}}$ will be realized as a subset of the Kirillov-Reshetikhin crystal $\mathbf{B}^{1,2\ell}_{A^{(2)}_{2n+1}}$. Since $\mathbf{B}^{1,2\ell}_{A^{(2)}_{2n+1}}$ is isomorphic to $\mathcal{B}(2\ell\Lambda_1)$ as a $U_q(C_{n+1})$-crystal,[2,4] it is the set of Kashiwara-Nakashima tableaux of shape $\overbrace{\boxed{}\boxed{}\boxed{\cdots}\boxed{}\boxed{}}^{2\ell}$ with entries from $\{1, 2, \ldots, n, n+1, \overline{n+1}, \overline{n}, \ldots, \overline{1}\}$. As in Remark 2.1, we will identify the tableau

$$\overbrace{\boxed{1}\cdots\boxed{1}}^{x_1}\overbrace{\boxed{2}\cdots\boxed{2}}^{x_2}\cdots\overbrace{\boxed{n+1}\cdots\boxed{n+1}}^{x_{n+1}}\overbrace{\boxed{\overline{n+1}}\cdots\boxed{\overline{n+1}}}^{\bar{x}_{n+1}}\cdots\overbrace{\boxed{\bar{1}}\cdots\boxed{\bar{1}}}^{\bar{x}_1}$$

with $(x_1, \ldots, x_{n+1}, \bar{x}_{n+1}, \ldots, \bar{x}_1)$. Using this parametrization, the Kashiwara operators $\hat{e}_i, \hat{f}_i$ for $i = 1, 2, \cdots, n+1$ on $\mathcal{B}(2\ell\Lambda_1)$ are given as follows: for $b = (x_1, \ldots, x_{n+1}, \bar{x}_{n+1}, \ldots, \bar{x}_1)$, we have

$$\hat{e}_i b = \begin{cases} (x_1, \ldots, x_i + 1, x_{i+1} - 1, \ldots, \bar{x}_1) & \text{if} \quad x_{i+1} > \bar{x}_{i+1}, \\ (x_1, \ldots, \bar{x}_{i+1} + 1, \bar{x}_i - 1, \ldots, \bar{x}_1) & \text{if} \quad x_{i+1} \le \bar{x}_{i+1}, \end{cases} \tag{8}$$

$$\hat{f}_i b = \begin{cases} (x_1, \ldots, x_i - 1, x_{i+1} + 1, \ldots, \bar{x}_1) & \text{if} \quad x_{i+1} \ge \bar{x}_{i+1}, \\ (x_1, \ldots, \bar{x}_{i+1} - 1, \bar{x}_i + 1, \ldots, \bar{x}_1) & \text{if} \quad x_{i+1} < \bar{x}_{i+1}, \end{cases} \tag{9}$$

for $i = 1, 2, \ldots, n$,

$$\hat{e}_{n+1} b = (x_1, \ldots, x_{n+1} + 1, \bar{x}_{n+1} - 1, \ldots, \bar{x}_1), \tag{10}$$

$$\hat{f}_{n+1} b = (x_1, \ldots, x_{n+1} - 1, \bar{x}_{n+1} + 1, \ldots, \bar{x}_1). \tag{11}$$

Since $\mathcal{B}(2\ell\Lambda_1)$ is the crystal basis of the irreducible $U_q(C_{n+1})$-module $V(2\ell\Lambda_1)$, which is completely reducible over the subalgebra of $U_q(C_{n+1})$ corresponding to the Dynkin diagram obtained by removing 1-node from the original one, $\mathcal{B}(2\ell\Lambda_1)$ is a direct sum of highest weight crystals as a $\{2, 3, \ldots, n+1\}$-crystal. Let us describe the $\{2, 3, \ldots, n+1\}$-highest weight vectors in $\mathcal{B}(2\ell\Lambda_1)$.

Lemma 3.1. *The set of* $\{2, 3, \ldots, n+1\}$*-highest weight vectors in* $\mathcal{B}(2\ell\Lambda_1)$ *is given by* $\{(x_1, x_2, 0, \ldots, 0, \bar{x}_1) \in \mathcal{B}(2\ell\Lambda_1) \mid x_1 + x_2 + \bar{x}_1 = 2\ell\}$.

Proof. Clearly, $(x_1, x_2, 0, \ldots, 0, \bar{x}_1)$ is a $\{2, 3, \ldots, n+1\}$-highest weight vector. Suppose that a vector $b = (x_1, \ldots, x_{n+1}, \bar{x}_{n+1}, \ldots, \bar{x}_1) \in \mathcal{B}(2\ell\Lambda_1)$ is a $\{2, 3, \ldots, n+1\}$-highest weight vector. Since $\hat{e}_i b = 0$ for $i = 2, 3, \ldots, n$, from

(8), we get $x_{i+1} \leq \bar{x}_{i+1}$ and $\bar{x}_i = 0$. Thus we see that $x_3 = \cdots = x_n = 0$. From the fact that $\hat{e}_{n+1}b = 0$, we have $\bar{x}_{n+1} = 0$ by (10) and thus $x_{n+1} = 0$. □

Now, following Ref. 2, we define an involution σ on $\mathcal{B}(2\ell\Lambda_1)$ as follows :

(i) $\sigma(x_1, x_2, 0, \ldots, 0, \bar{x}_1) := (\bar{x}_1, x_2, 0, \ldots, 0, x_1)$,

(ii) For $b \in \mathcal{B}(2\ell\Lambda_1)$, let $\hat{e}_{\vec{\mathbf{a}}}(b) := \hat{e}_{a_1}\hat{e}_{a_2}\cdots\hat{e}_{a_k}(b)$ $(a_i \in \{2, 3, \ldots, n+1\})$ be such that $\hat{e}_{\vec{\mathbf{a}}}(b)$ is a $\{2, 3, \ldots, n+1\}$-highest weight vector. Set $\hat{f}_{\overleftarrow{\mathbf{a}}} := \hat{f}_{a_k}\hat{f}_{a_{k-1}}\cdots\hat{f}_{a_1}$, and define $\sigma(b) := \hat{f}_{\overleftarrow{\mathbf{a}}} \circ \sigma \circ \hat{e}_{\vec{\mathbf{a}}}(b)$.

Since the connected component of $(x_1, x_2, 0, \ldots, 0, \bar{x}_1)$ is isomorphic to the connected component of $(\bar{x}_1, x_2, 0, \ldots, 0, x_1)$ as $\{2, 3, \ldots, n+1\}$-crystals, one can see that σ is a well-defined involution on $\mathcal{B}(2\ell\Lambda_1)$. By definition, σ commutes with $\hat{f}_i$ and $\hat{e}_i$ for $i = 2, 3, \ldots, n+1$.

Remark 3.1. Note that σ in the above definition coincides with the one in Definition 4.1 of Ref. 2. Indeed, they are defined in the same way on the vectors which are not $\{2, 3, \ldots, n+1\}$-highest weight vectors. In Ref. 2, σ on the set of $\{2, 3, \ldots, n+1\}$-highest weight vectors is given by $\Phi \circ \mathfrak{S} \circ \Phi^{-1}$, where Φ is a bijective map from the set of $\pm$-diagrams of outer shape $2\ell\Lambda_1$ to the set of $\{2, 3, \ldots, n+1\}$-highest weight vectors and $\mathfrak{S}$ is a permutation on the set of $\pm$-diagrams of outer shape $2\ell\Lambda_1$ (for the precise definitions of $\pm$-diagrams, Φ and $\mathfrak{S}$, see sections 3.2 and 4.2 in Ref. 2). It is easy to show the following properties:

(1) The numbers of $+$'s and $-$'s in a $\pm$-diagram of outer shape $2\ell\Lambda_1$ determine the $\pm$-diagram uniquely.

(2) Φ sends the $\pm$-diagram of outer shape $2\ell\Lambda_1$ with m-many $+$'s and n-many $-$'s to the tableau $(m, 2\ell - m - n, 0, \ldots, 0, n)$.

(3) $\mathfrak{S}$ sends the $\pm$-diagram of outer shape $2\ell\Lambda_1$ with m-many $+$'s and n-many $-$'s to the $\pm$-diagram with n-many $+$'s and m-many $-$'s.

Thus we have

$$\Phi \circ \mathfrak{S} \circ \Phi^{-1}(x_1, x_2, 0, \ldots, 0, \bar{x}_1) = (\bar{x}_1, x_2, 0, \ldots, 0, x_1).$$

Lemma 3.2. *The involution σ exchanges the number of 1's and the number of $\bar{1}$'s in a Kashiwara-Nakashima tableau in $\mathcal{B}(2\ell\Lambda_1)$. That is,*

$$\sigma(x_1, x_2, \ldots, x_{n+1}, \bar{x}_{n+1}, \ldots, \bar{x}_2, \bar{x}_1) = (\bar{x}_1, x_2, \ldots, x_{n+1}, \bar{x}_{n+1}, \ldots, \bar{x}_2, x_1).$$

Proof. Let $b \in \mathcal{B}(2\ell\Lambda_1)$ and let $\hat{e}_{\vec{\mathbf{a}}}(b) := \hat{e}_{a_1}\hat{e}_{a_2}\cdots\hat{e}_{a_k}(b)$ $(a_i \in \{2, 3, \ldots, n+1\})$ be such that $\hat{e}_{\vec{\mathbf{a}}}(b)$ is a $\{2, 3, \ldots, n+1\}$-highest weight vector. Then $\sigma(b) = \sigma(\hat{f}_{a_k}\hat{e}_{a_k}(b)) = \hat{f}_{a_k}\sigma(\hat{e}_{a_k}(b))$. If $b = (x_1, x_2, \ldots, x_{n+1}, \bar{x}_{n+1}, \ldots, \bar{x}_1)$, we know that

$$\hat{e}_{a_k}(b) = (x_1, \ldots, x_{a_{k-1}}, x'_{a_k}, x'_{a_{k+1}}, x_{a_{k+2}}, \ldots, \bar{x}'_{a_{k+1}}, \bar{x}'_{a_k}, \ldots, \bar{x}_2, \bar{x}_1)$$

for some x'_{a_k}, $x'_{a_{k+1}}$, $\bar{x}'_{a_{k+1}}$ and $\bar{x}'_{a_k}$. That is, the tableau $\hat{e}_{a_k}b$ differs from b by the number of a_k's, a_{k+1}'s, $\overline{a_{k+1}}$'s and $\overline{a_k}$'s only. By using induction on the length of $\vec{\mathbf{a}}$, we may assume that

$$\sigma(\hat{e}_{a_k}(b)) = (\bar{x}_1, \ldots, x_{a_{k-1}}, x'_{a_k}, x'_{a_{k+1}}, x_{a_{k+2}}, \ldots, \bar{x}'_{a_{k+1}}, \bar{x}'_{a_k}, \ldots, \bar{x}_2, x_1).$$

From (8)-(11), we conclude

$$\sigma(b) = \hat{f}_{a_k}\sigma(\hat{e}_{a_k}(b)) = (\bar{x}_1, x_2, \ldots, x_{a_k}, x_{a_{k+1}}, \ldots, \bar{x}_{a_{k+1}}, \bar{x}_{a_k}, \ldots, \bar{x}_2, x_1),$$

as desired. □

Using σ, one can define affine crystal operators as follows:

$$\hat{f}_0 := \sigma \circ \hat{f}_1 \circ \sigma, \quad \hat{e}_0 := \sigma \circ \hat{e}_1 \circ \sigma.$$

Equipped with these affine operators, $\mathcal{B}(2\ell\Lambda_1)$ becomes the Kirillov-Reshetikhin crystal $\mathbf{B}^{1,2\ell}_{A^{(2)}_{2n+1}}$ (Theorem 5.1 of Ref. 2).

Example 3.1. The Kirillov-Reshetikhin crystal $\mathbf{B}^{1,2}_{A^{(2)}_5}$:

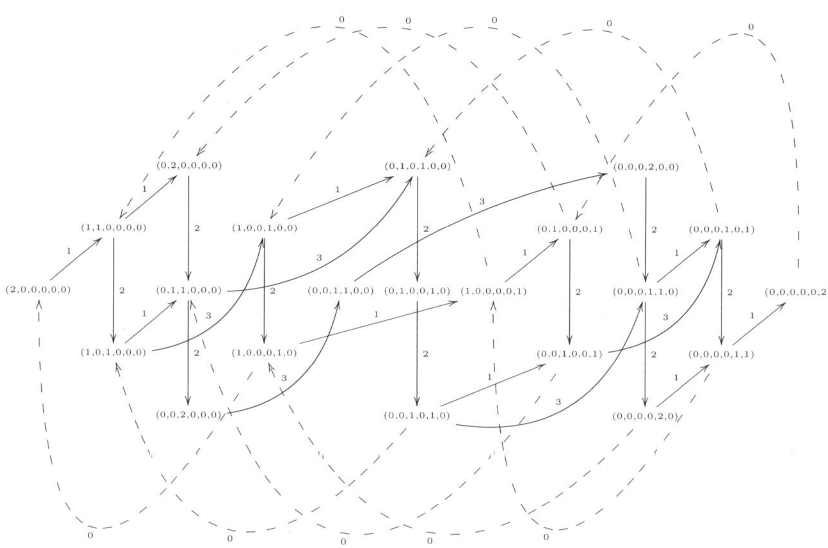

Let $\mathbf{B} = \{b \in \mathbf{B}^{1,2\ell}_{A^{(2)}_{2n+1}} \mid \sigma(b) = b\}$. By Lemma 3.2, we see that $\mathbf{B} = \{(x_1, \ldots, \bar{x}_1) \in \mathbf{B}^{1,2\ell}_{A^{(2)}_{2n+1}} \mid x_1 = \bar{x}_1\}$. We define the crystal operators on $\mathbf{B}$ as follows:

$$\tilde{e}_i := \begin{cases} \hat{e}_0\hat{e}_1 & \text{if} \quad i = 0, \\ \hat{e}_{i+1} & \text{if} \quad i = 1, 2, \ldots, n \end{cases} \qquad \tilde{f}_i := \begin{cases} \hat{f}_0\hat{f}_1 & \text{if} \quad i = 0, \\ \hat{f}_{i+1} & \text{if} \quad i = 1, 2, \ldots, n. \end{cases}$$

With these operators, $\mathbf{B}$ is isomorphic to the Kirillov-Reshetikhin crystal $\mathbf{B}^{1,2\ell}_{C^{(1)}_n}$ (Theorem 5.7 of Ref. 2).

Example 3.2. The Kirillov-Reshetikhin crystal $\mathbf{B}^{1,2}_{C^{(1)}_2}$:

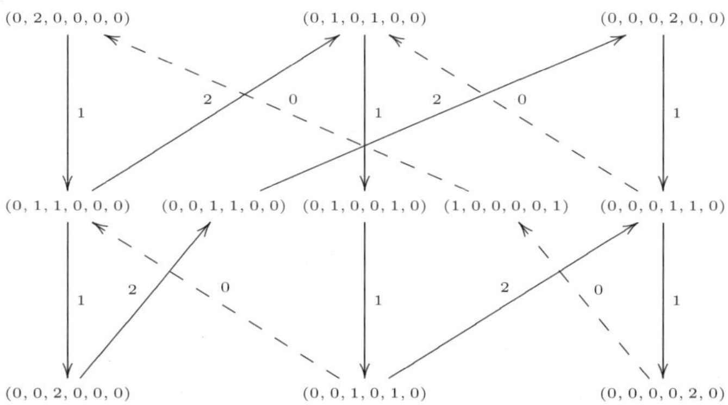

4. Main Theorem

As can be seen in Examples 2.1 and 3.2, the level 1 adjoint crystal $\mathbb{B}_1$ of type $C_2^{(1)}$ and the level 1 KR crystal $\mathbf{B}^{1,2}_{C^{(1)}_2}$ have the same crystal structures. This isomorphism can be extended to the crystals with arbitrary level and arbitrary rank.

Consider a map Ψ from $\mathbf{B} \cup \{0\}$ to $\mathbb{B}_\ell \cup \{0\}$ given by $\Psi(0) := 0$ and

$$\Psi(x_1, x_2, \ldots, x_{n+1}, \bar{x}_{n+1}, \ldots, \bar{x}_2, \bar{x}_1) := (x_2, \ldots, x_{n+1}, \bar{x}_{n+1}, \ldots, \bar{x}_2).$$

Theorem 4.1. *The map Ψ is a $U_q'(C_n^{(1)})$-crystal isomorphism between $\mathbf{B}^{1,2\ell}_{C^{(1)}_n}$ and $\mathbb{B}_\ell$.*

Proof.

(i) The inverse of Ψ is given by

$$(y_1, \ldots, y_n, \bar{y}_n, \ldots, \bar{y}_1) \mapsto$$
$$\Big(\frac{2\ell - \sum_{i=1}^n (y_i + \bar{y}_i)}{2}, y_1, \ldots, y_n, \bar{y}_n, \ldots, \bar{y}_1, \frac{2\ell - \sum_{i=1}^n (y_i + \bar{y}_i)}{2}\Big).$$

Thus Ψ is a bijection.

(ii) It is straightforward to verify that

$$\Psi \circ \tilde{f}_i = \tilde{f}_i \circ \Psi, \quad \Psi \circ \tilde{e}_i = \tilde{e}_i \circ \Psi \quad (1 \leq i \leq n).$$

For example, let $b = (x_1, \ldots, x_{n+1}, \bar{x}_{n+1}, \ldots, \bar{x}_1) \in \mathbf{B}$ and let i be an index with $1 \leq i \leq n-1$ such that $x_{i+2} \geq \bar{x}_{i+2}$. Then we have

$$\begin{aligned}
\Psi \circ \tilde{f}_i(b) &= \Psi \circ \hat{f}_{i+1}(b) \\
&= \Psi(x_1, \ldots, x_{i+1} - 1, x_{i+2} + 1, \ldots, \bar{x}_1) \text{ by (9)} \\
&= (x_2, \ldots, x_{i+1} - 1, x_{i+2} + 1, \ldots, \bar{x}_2) \\
&= \tilde{f}_i(x_2, \ldots, x_{i+1}, x_{i+2}, \ldots, \bar{x}_2) \qquad \text{by (4)} \\
&= \tilde{f}_i \circ \Psi(b).
\end{aligned}$$

The other cases can be checked in a similar manner.

(iii) It remains to show that $\Psi \circ \tilde{f}_0 = \tilde{f}_0 \circ \Psi$ and $\Psi \circ \tilde{e}_0 = \tilde{e}_0 \circ \Psi$. Let $b = (x_1, \ldots, x_{n+1}, \bar{x}_{n+1}, \ldots, \bar{x}_1) \in \mathbf{B}$. Then we have

$$\begin{aligned}
\tilde{f}_0(b) = \hat{f}_0 \hat{f}_1(b) &= \begin{cases} \hat{f}_0(x_1 - 1, x_2 + 1, \ldots, \bar{x}_2, \bar{x}_1) & \text{if } \ x_2 \geq \bar{x}_2, \\ \hat{f}_0(x_1, x_2, \ldots, \bar{x}_2 - 1, \bar{x}_1 + 1) & \text{if } \ x_2 < \bar{x}_2 \end{cases} \\
&= \begin{cases} \sigma \circ \hat{f}_1(\bar{x}_1, x_2 + 1, \ldots, \bar{x}_2, x_1 - 1) & \text{if } \ x_2 \geq \bar{x}_2, \\ \sigma \circ \hat{f}_1(\bar{x}_1 + 1, x_2, \ldots, \bar{x}_2 - 1, x_1) & \text{if } \ x_2 < \bar{x}_2 \end{cases} \\
&= \begin{cases} \sigma(\bar{x}_1 - 1, x_2 + 2, \ldots, \bar{x}_2, x_1 - 1) & \text{if } \ x_2 \geq \bar{x}_2, \\ \sigma(\bar{x}_1, x_2 + 1, \ldots, \bar{x}_2 - 1, x_1) & \text{if } \ x_2 = \bar{x}_2 - 1, \\ \sigma(\bar{x}_1 + 1, x_2, \ldots, \bar{x}_2 - 2, x_1 + 1) & \text{if } \ x_2 \leq \bar{x}_2 - 2 \end{cases} \\
&= \begin{cases} (x_1 - 1, x_2 + 2, \ldots, \bar{x}_2, \bar{x}_1 - 1) & \text{if } \ x_2 \geq \bar{x}_2, \\ (x_1, x_2 + 1, \ldots, \bar{x}_2 - 1, \bar{x}_1) & \text{if } \ x_2 = \bar{x}_2 - 1, \\ (x_1 + 1, x_2, \ldots, \bar{x}_2 - 2, \bar{x}_1 + 1) & \text{if } \ x_2 \leq \bar{x}_2 - 2. \end{cases}
\end{aligned}$$

Thus we have

$$\Psi \circ \tilde{f}_0(b) = \begin{cases} (x_2+2,\ldots,\bar{x}_2) & \text{if} \quad x_2 \geq \bar{x}_2, \\ (x_2+1,\ldots,\bar{x}_2-1) & \text{if} \quad x_2 = \bar{x}_2 - 1, \\ (x_2,\ldots,\bar{x}_2-2) & \text{if} \quad x_2 \leq \bar{x}_2 - 2 \end{cases}$$
$$= \tilde{f}_0 \circ \Psi(b) \qquad \text{by (2).}$$

By the same argument, one can show that $\Psi \circ \tilde{e_0} = \tilde{e_0} \circ \Psi$. □

References

1. G. Benkart, I. Frenkel, S.-J. Kang, H. Lee, *Level 1 perfect crystals and path realizations of basic representations at* $q = 0$, Int. Math. Res. Not. 2006, Art. ID 10312, 1–28.
2. G. Fourier, M. Okado, A. Schilling, *Kirillov-Reshetikhin crystals for nonexceptional types*, Adv. Math. **222** (3) (2009), 1080–1116.
3. ______, *Perfectness of Kirillov-Reshetikhin crystals for nonexceptional types*, arXiv:0811.1604v1[math.RT].
4. G. Hatayama, A. Kuniba, M. Okado, T. Tagaki, Z. Tsuboi, *Paths, crystals and fermionic formulae*, MathPhys Odyssey 2001, 205–272. Prog. Math. Phys. **23**, Birkhäuser Boston, Boston, MA, 2002.
5. G. Hatayama, A. Kuniba, M. Okado, T. Takagi, Y. Yamada, *Remarks on fermionic formula*, Recent developments in quantum affine algebras and related topics (Raleigh, NC, 1998), Contemp. Math., **248** (1999), 243–291,
6. J. Hong, S.-J. Kang, *Introduction to Quantum Groups and Crystal Bases*, Graduate Studies in Mathematics **42**, American Mathematical Society, 2002.
7. M. Jimbo, K. Misra, T. Miwa, M. Okado, *Combinatorics of representations of* $U_q(\widehat{sl}(n))$ *at* $q = 0$, Commun. Math. Phys. **136** (1991), 543–566.
8. J. H. Jung, S.-J. Kang, M. Kim, S. Kim, J.-Y. Yu, *Adjoint crystals and Young walls for* $U_q(\widehat{sl_2})$, Euro. J. Combin., in press, doi:10.1016/j.ejc.2009.10.004.
9. V. Kac, *Infinite Dimensional Lie algebras*, 3rd ed., Cambridge University Press, Cambridge, 1990.
10. S-J. Kang, K.C. Misra, T. Miwa, *Fock space representations of the quantized universal enveloping algebras* $U_q(C_l^{(1)})$, $U_q(A_{2l}^{(2)})$ *and* $U_q(D_{l+1}^{(2)})$, J. Algebra, **155** (1993), 238–251.
11. S-J. Kang, K.C. Misra, *Crystal bases and tensor product decompostions of* $U_q(G_2)$*-modules*, J. Algebra **163** (1994), 675–691.
12. S-J. Kang, K.C. Misra, *The quantum affine Lie algebra* $U_q(C_n^{(1)})$ *and crystal base*, (unpublished).
13. S.-J. Kang, M. Kashiwara, K. C. Misra, *Crystal bases of Verma modules for quantum affine Lie algebras*, Compositio Math. **92** (1994), 299–325.
14. S.-J. Kang, M. Kashiwara, K. C. Misra, T. Miwa, T. Nakashima, A. Nakayashiki, *Affine crystals and vertex models*, Int. J. Mod. Phys. A. **Suppl. 1A** (1992), 449–484.

15. ______, *Perfect crystals of quantum affine Lie algebras*, Duke Math. J. **68** (1992), 499–607.
16. M. Kashiwara, *Crystalizing the q-analogue of universal enveloping algebras*, Commun. Math. Phys. **133** (1990), 249–260.
17. ______, *On crystal bases of the q-analogue of universal enveloping algebras*, Duke Math. J. **63** (1991), 465–516.
18. M. Kashiwara, T. Nakashima, *Crystal graphs for representations of q-analogue of classical Lie algebras*, J. Alg. **165** (1994), 295–345.
19. A. N. Kirillov, N. Yu. Reshetikhin, *Representations of Yangians and multiplicity of occurrence of the irreducible components of the tensor product of representations of simple Lie algebras*, J. Sov. Math. **52** (1990), 3156–3164.
20. G. Lusztig, *Canonical bases arising from quantized enveloping algebras*, J. Amer. Math. Soc. **3** (1990), 447–498.
21. K. C. Misra, T. Miwa, *Crystal bases for the basic representation of $U_q(\widehat{sl}(n))$*, Commun. Math. Phys. **10** (1990), 79–88.
22. M. Okado, A. Schilling, *Existence of Kirillov-Reshetikhin crystals for nonexceptional types*, Represent. Theory **12** (2008), 186–207.

BOUNDARY QUANTUM KNIZHNIK-ZAMOLODCHIKOV EQUATION

MASAHIRO KASATANI

Graduate School of Mathematical Sciences, the University of Tokyo
Tokyo 153-8914, Japan
E-mail: kasatani@ms.u-tokyo.ac.jp

We introduce partial reflection K-matrices, and boundary qKZ equation with 6 parameters. We give a method to construct its solution from a solution of certain eigenvalue problem. The eigenvalue probrem is described in terms of Noumi representation of the affine Hecke algebra of type C. We also show concrete solutions in terms of non-symmetric Koornwinder polynomials. This is an anouncement of a joint work with K. Shigechi.[3]

1. Introduction

The quantum Knizhnik-Zamolodchikov (qKZ) equation, introduced by Frenkel and Reshetikhin,[1] is the system of difference equations satisfied by matrix elements of the vertex operators in the representation theory of the quantum affine algebra.

In the paper[4] by the author and Takeyama, they gave polynomial solutions of the qKZ equation on the tensor product $V^{\otimes n}$ of the vector representation V of the quantum algebra $U_q(sl_N)$. They formulated a method to construct solutions of the qKZ equation from those of certain eigenvalue problem on Dunkl-Cherednik operators and Demazure-Lusztig operators. They found that non-symmetric Macdonald polynomials with certain condition are solutions for the eigenvalue problem.

In the present paper, we introduce a qKZ equation with boundaries. This is a system of difference equations for $V^{\otimes n}$-valued functions defined by a product of R-matrices and K-matrices. (Explicit definition is given in Section 2.) The R-matrix is a linear operator on $V \otimes V$. It stands for an interaction of two spaces. The K-matrix we introduce is a linear operator on V. It stands for a *partial* reflection on a boundary. Totally, these matrices and the equation depend on 6 parameters.

We connect the boundary qKZ equation with a polynomial representation of the affine Hecke algebras (AHA) of type C_n. The representation is so-called Noumi representation.[7] It is a restriction of a polynomial representation of the double affine Hecke algebra (DAHA) of type $C^\vee C_n$ with 6 parameters. We formulate a method to construct solutions of the boundary qKZ equation from those of certain eigenvalue problem on certain operators.

To find solutions of the eigenvalue problem, we use non-symmetric Koornwinder polynomials with 6 parameters. Our construction of the solutions works not only for generic parameters but also for specialized parameters.

In the case where the parameters are specialized, we give a factorized solution. This is a generalization of the level 1 solution of the qKZ equation given in the papers.[6,4]

Recently, Stokman[8] generalized the result[4] to arbitrary root systems. His result is similar to that of the present paper, but there are some different points. He did not treat the case of type $C^\vee C_n$ where 6 parameters appear. In his formulation, corresponding K-matrix only stands for *total* reflection on a boundary. He did not give explicit examples such as Section 4.

Let us give a sketch of our construction of solutions. We use the standard basis $\{v_{-M}, \ldots, v_M\}$ of V. Expand an unknown $V^{\otimes n}$-valued function into a linear combination of the tensor products $v_{\epsilon_1} \otimes \cdots \otimes v_{\epsilon_n}$. We consider the functions which appear in the expansion as coefficients. The boundary qKZ equation can be regarded as constraints for the functions. In the present paper, we introduce a stronger condition than the boudary qKZ equation itself, and call a set of functions satisfying the condition a *qKZ family* (see §3.2).

The defining condition of qKZ family is described in terms of the action of the AHA $\mathcal{H}_n$ on the space of functions. $\mathcal{H}_n$ is generated by elements $T_i \,(0 \leq i \leq n)$ with some defining relations (see §3.1). The action of T_i is nothing but Noumi representation. We can obtain any member of the qKZ family form another member by acting some T_i's.

For each qKZ family, we pick up a special member. We show that the member is a joint eigenfunction of operators given by products of T_i. Conversely, an eigenfunction of these products of operators generates a qKZ family. Thus construction of a qKZ family is reduced to that of an eigenfunction of the products of T_i.

In the definition of the eigenvalue problem (see Def. 3.3), there appears an product of the form:

$$Y_i := T_i \dots T_{n-1}(T_n \dots T_0)T_1^{-1} \dots T_{i-1}^{-1}.$$

It is known that non-symmetric Koornwinder polynomials are joint eigenfunctions of $Y_1, \dots, Y_n$ (see, e.g. Ref. 5). Therefore we construct special solutions of the boundary qKZ equation from non-symmetric Koornwinder polynomials.

The plan of the present paper is as follows. In Section 2, we introduce R- and K-matrices, and define the boundary qKZ equation. In Section 3 we recall AHA and Noumi representation (§3.1), and define the qKZ family (§3.2). We construct solutions of the equation from the qKZ family in Theorem 3.2. The eigenvalue problem explained above is given in §3.3. We show its equivalence to the problem to find qKZ families in Theorem 3.7. In Section 4, we give explicit solutions of the eigenvalue problem in the case where the parameters are generic or special.

2. Boundary quantum Knizhnik-Zamolodchikov equation

In this section, we define linear operators on a vector space, so-called R-matrix and K-matrices. We also define boundary quantum Knizhnik-Zamolodchikov equation using a product of the R-matrices and K-matrices.

Let V be a finite dimensional vector space, and fix its basis by

$$V = \bigoplus_{-M \le \epsilon \le M, \epsilon \neq 0} \mathbb{C} v_\epsilon \quad (\text{if } \dim V = 2M),$$

$$V = \bigoplus_{-M \le \epsilon \le M} \mathbb{C} v_\epsilon \quad (\text{if } \dim V = 2M+1).$$

Define the linear operator $R(z)$ with a parameter q acting on $V \otimes V$ by

$$R(z)\left(v_{\epsilon_1} \otimes v_{\epsilon_2}\right) = \sum_{\epsilon_1', \epsilon_2'} R(z)_{\epsilon_1' \epsilon_2'}^{\epsilon_1 \epsilon_2} v_{\epsilon_1'} \otimes v_{\epsilon_2'},$$

where

$$R(z)_{ii}^{ii} = 1, \quad R(z)_{ij}^{ij} = \frac{(1-z)q}{1-q^2 z}, \quad R(z)_{ij}^{ji} = \frac{1-q^2}{1-q^2 z} z^{\theta(i>j)} \quad (i \neq j)$$

and $R(z)_{i'j'}^{ij} = 0$ otherwise. Here

$$\theta(P) = \begin{cases} 1 \text{ if } P \text{ is true,} \\ 0 \text{ if } P \text{ is false.} \end{cases}$$

Then $R(z)$ satisfies the Yang-Baxter equation on $V^{\otimes 3}$:

$$R_{1,2}\left(\frac{z_1}{z_2}\right) R_{1,3}\left(\frac{z_1}{z_3}\right) R_{2,3}\left(\frac{z_2}{z_3}\right) = R_{2,3}\left(\frac{z_2}{z_3}\right) R_{1,3}\left(\frac{z_1}{z_3}\right) R_{1,2}\left(\frac{z_1}{z_2}\right).$$

Let P be an operator given by $P(u \otimes v) = v \otimes u$, and put $\check{R}(z) = PR(z)$.

Let α and β be non-negative integers such that $0 \le \alpha, \beta \le M$. Define linear operators $K(z)$ and $\overline{K}(z)$ with parameters $q_n^{1/2}, u_n^{1/2}, q_0^{1/2}, u_0^{1/2}, s^{1/2}$ acting on V by:

$$K(z)v_i = \sum_{i'} K^i_{i'}(z)v_{i'}, \qquad \overline{K}(z)v_i = \sum_{i'} \overline{K}^i_{i'}(z)v_{i'},$$

$$\begin{aligned}
K^i_i(z) &= 1 \qquad (|i| \le \alpha),\\
K^i_i(z) &= q_n \frac{1-z^2}{(1-az)(1-bz)} \qquad (|i| > \alpha),\\
K^i_{-i}(z) &= -q_n \frac{(q_n - q_n^{-1})z^{2\theta(i<0)} + (u_n^{1/2} - u_n^{-1/2})z}{(1-az)(1-bz)} \qquad (|i| > \alpha)\\
&\qquad (\text{where } a = q_n^{1/2}u_n^{1/2},\ b = -q_n^{1/2}u_n^{-1/2}).
\end{aligned}$$

$$\begin{aligned}
\overline{K}^i_i(z) &= 1 \qquad (|i| \le \beta),\\
\overline{K}^i_i(z) &= q_0 \frac{1-sz^{-2}}{(1-cz^{-1})(1-dz^{-1})} \qquad (|i| > \beta),\\
\overline{K}^i_{-i}(z) &= -c_i q_0 \frac{(q_0 - q_0^{-1})s^{\theta(i>0)}z^{-2\theta(i>0)} + (u_0^{1/2} - u_0^{-1/2})s^{1/2}z^{-1}}{(1-cz^{-1})(1-dz^{-1})}\\
&\qquad\qquad (|i| > \beta)\\
&\qquad (\text{where } c = s^{1/2}q_0^{1/2}u_0^{1/2},\ d = -s^{1/2}q_0^{1/2}u_0^{-1/2}).
\end{aligned}$$

and $K^i_j(z) = \overline{K}^i_j(z) = 0$ otherwise. Then $K(z)$ and $\overline{K}(z)$ satisfy the reflection equation on $V^{\otimes 2}$:

$$K_2(w)R_{2,1}(zw)K_1(z)R_{1,2}\left(\frac{z}{w}\right) = R_{1,2}\left(\frac{z}{w}\right)K_1(z)R_{2,1}(zw)K_2(w)$$

$$\overline{K}_1(z)R_{2,1}\left(\frac{s}{zw}\right)\overline{K}_2(w)R_{1,2}\left(\frac{z}{w}\right) = R_{1,2}\left(\frac{z}{w}\right)\overline{K}_2(w)R_{2,1}\left(\frac{s}{zw}\right)\overline{K}_1(z).$$

Let P_n and P_0 be operators given by

$$P_n(v_i) = \begin{cases} v_i & (|i| \le \alpha) \\ v_{-i} & (|i| > \alpha), \end{cases} \qquad P_0(v_i) = \begin{cases} v_i & (|i| \le \beta) \\ c_i v_{-i} & (|i| > \beta), \end{cases}$$

and put $\check{K}(z) = P_n K(z)$, $\check{\overline{K}}(z) = P_0 \overline{K}(z)$.

We define signed $\check{R}$- and $\check{K}$-matrices $\check{R}^{\pm}$ and $\check{K}^{\pm}$ by

$$\begin{aligned} \check{R}_i^+(z) &= \check{R}_{i,i+1}(z), & \check{R}_i^-(z) &= f(z)\check{R}_{i,i+1}(z), \\ \check{K}^+(z) &= \check{K}(z), & \check{K}^-(z) &= f^n(z)\check{K}(z), \\ \overline{\check{K}}^+(z) &= \overline{\check{K}}(z), & \overline{\check{K}}^-(z) &= f^0(z)\overline{\check{K}}(z). \end{aligned}$$

Here, f, f^n, f^0 are rational functions given by:

$$f(z) = \frac{q^2 z - 1}{q^2 - z}, \quad f^n(z) = \frac{1 - q_n^2 z^2 - (u_n^{1/2} - u_n^{-1/2})z}{-q_n^2 + z^2 - (u_n^{1/2} - u_n^{-1/2})z},$$

$$f^0(z) = \frac{1 - s q_0^2 z^{-2} - s^{1/2} q_0 (u_0^{1/2} - u_0^{-1/2}) z^{-1}}{-q_0^2 + s z^{-2} - s^{1/2} q_0 (u_0^{1/2} - u_0^{-1/2}) z^{-1}}.$$

Note that f, f^n, f^0 satisfy $f(z)f(1/z) = 1$, $f^n(z)f^n(1/z) = 1$, $f^0(z)f^0(s/z) = 1$.

Take three signs $\sigma, \sigma_n, \sigma_0$. For simplicity, we denote by $Q_i^{\sigma}(z)$, the operator $\check{R}_{i,i+1}^{\sigma}(z)$ acting on i-th and $(i+1)$-th components of $V^{\otimes n}$. Similarly, denote by $Q_n^{\sigma_n}(z)$, the operator $\check{K}^{\sigma_n}(z)$ acting on the last component of $V^{\otimes n}$, and denote by $Q_0^{\sigma_0}(z)$, the operator $\overline{\check{K}}^{\sigma_0}(z)$ acting on the first component of $V^{\otimes n}$.

Definition 2.1. For a $V^{\otimes n}$-valued function $F(z_1, \ldots, z_n)$, the boundary quantum Knizhnik-Zamolodchikov (qKZ) equaion is a system of s-difference equations given as follows: for $1 \le i \le n$

$$\begin{aligned} F(z_1, \ldots, s z_i, \ldots, z_n) &= Q_{i-1}^{\sigma}(s z_i/z_{i-1}) \cdots Q_1^{\sigma}(s z_i/z_1) Q_0^{\sigma_0}(z_i) \\ &\times Q_1^{\sigma}(z_1 z_i) \cdots Q_i^{\sigma}(z_i z_{i+1}) \cdots Q_{n-1}^{\sigma}(z_n z_i) Q_n^{\sigma_n}(z_i) \\ &\times Q_{n-1}^{\sigma}(z_i/z_n) \cdots Q_i^{\sigma}(z_i/z_{i+1}) F(z_1, \ldots, z_n). \end{aligned}$$

3. Eigenvalue problem

In this section, we introduce Noumi representation of the affine Hecke algebra $\mathcal{H}_n$ of type C_n, and introduce a family of Laurent polynomials satisfying some properties described by $\mathcal{H}_n$-action. Combining members of the family with basis vectors in $V^{\otimes n}$, we obtain $V^{\otimes n}$-valued Laurent polynomials. We will see that it gives a solution of the boundary qKZ equation. We also show that finding a qKZ family is equivalent to solving an eigenvalue problem.

3.1. *Affine Hecke algebra and Noumi representation*

The affine Hecke algebra $\mathcal{H}_n = \mathcal{H}_n(t^{1/2}, t_n^{1/2}, t_0^{1/2})$ of type C_n is a unital associative $\mathbb{C}$-algebra generated by $T_0, \ldots, T_n$ with defining relations as follows:

$$(T_0 - t_0^{1/2})(T_0 + t_0^{-1/2}) = 0,$$
$$(T_i - t^{1/2})(T_i + t^{-1/2}) = 0 \qquad 1 \leq i \leq n-1,$$
$$(T_n - t_n^{1/2})(T_n + t_n^{-1/2}) = 0,$$

$$T_0 T_1 T_0 T_1 = T_1 T_0 T_1 T_0,$$
$$T_i T_{i+1} T_i = T_{i+1} T_i T_{i+1} \qquad 1 \leq i \leq n-2,$$
$$T_{n-1} T_n T_{n-1} T_n = T_n T_{n-1} T_n T_{n-1},$$
$$T_i T_j = T_j T_i \qquad |i-j| \geq 2.$$

Note that the elements Y_i $(1 \leq i \leq n)$ defined by

$$Y_i := T_j \dots T_{n-1}(T_n \dots T_0) T_1^{-1} \dots T_{j-1}^{-1}$$

are mutually commutative.

Let $W = \langle s_0, \dots, s_n \rangle$ be the affine Weyl group of type C_n. Define the action of W with a parameter s on n-variable functions by

$$s_i f(\dots, z_i, z_{i+1}, \dots) = f(\dots, z_{i+1}, z_i, \dots)$$
$$s_n f(\dots, z_n) = f(\dots, 1/z_n)$$
$$s_0 f(z_1, \dots) = f(s/z_1, \dots).$$

For additional parameters u_n and u_0, put

$$a := t_n^{1/2} u_n^{1/2}, b := -t_n^{1/2} u_n^{-1/2}, c := s^{1/2} t_0^{1/2} u_0^{1/2}, d := -s^{1/2} t_0^{1/2} u_0^{-1/2}$$

and define linear operators on $\mathbb{C}[z_1^{\pm 1}, \dots, z_n^{\pm 1}]$ as follows:

$$\widehat{T}_0^{\pm 1} = t_0^{\pm 1/2} + t_0^{-1/2} \frac{(1 - c z_1^{-1})(1 - d z_1^{-1})}{1 - s z_1^{-2}} (s_0 - 1)$$
$$\widehat{T}_i^{\pm 1} = t_i^{\pm 1/2} + t_i^{-1/2} \frac{1 - t_i z_i z_{i+1}^{-1}}{1 - z_i z_{i+1}^{-1}} (s_i - 1)$$
$$\widehat{T}_n^{\pm 1} = t_n^{\pm 1/2} + t_n^{-1/2} \frac{(1 - a z_n)(1 - b z_n)}{1 - z_n^2} (s_n - 1).$$

Then the map $T_i \mapsto \hat{T}_i$ $(0 \leq i \leq n)$ gives a representation of $\mathcal{H}_n$. This is so-called Noumi representation.

3.2. *qKZ family*

We split the vector space $V^{\otimes n}$ into orbits of the actions of the R- and K-matrices. Put $\gamma := \min(\alpha, \beta)$ and take positive integers $d_{-M}, d_{-M+1}, \ldots, d_\gamma$ such that $\sum_{i=-M}^{\gamma} d_i = n$. Let

$$\delta := ((-M)^{d_{-M}}, \ldots, \ldots, (-\gamma-1)^{d_{-\gamma-1}}, (-\gamma)^{d_{-\gamma}}, \ldots, \gamma^{d_\gamma}) \tag{1}$$

$$\begin{aligned} I_d := \{(m_1, \ldots, m_n) \in \mathbb{Z}^n; & \\ & \sharp\{j; m_j = i\} = d_i \quad (-\gamma \leq i \leq \gamma) \\ & \sharp\{j; m_j = i\} + \sharp\{j; m_j = -i\} = d_i \quad (-M \leq i \leq -\gamma - 1)\}. \end{aligned} \tag{2}$$

I_d is an index set of each orbit and δ is a representative of I_d.

Define the action $\cdot$ of W on $\mathbb{Z}^n$ as follows:

$$\begin{aligned} s_0 \cdot (m_1, m_2, \ldots) &= (-m_1, m_2, \ldots) \\ s_i \cdot (\ldots, m_{i-1}, m_i, m_{i+1}, m_{i+2}, \ldots) &= (\ldots, m_{i-1}, m_{i+1}, m_i, m_{i+2}, \ldots) \\ s_n \cdot (\ldots, m_{n-1}, m_n) &= (\ldots, m_{n-1}, -m_n). \end{aligned}$$

Definition 3.1. A family of Laurent polynomials $\{f_\epsilon; \epsilon \in I_d\}$ is called a qKZ family with signs $(\sigma, \sigma_n, \sigma_0)$ and exponents $c_1, \ldots, c_M$ if

for $1 \leq i \leq n-1$

$$\hat{T}_i f_\epsilon = q f_\epsilon \quad \text{if } \epsilon_i = \epsilon_{i+1} \tag{3}$$

$$\hat{T}_i f_\epsilon = f_{s_i \cdot \epsilon} \quad \text{if } \epsilon_i > \epsilon_{i+1} \tag{4}$$

for $i = n$

$$\hat{T}_n f_\epsilon = q_n f_\epsilon \quad \text{if } |\epsilon_n| \leq \alpha \tag{5}$$

$$\hat{T}_n f_\epsilon = f_{s_n \cdot \epsilon} \quad \text{if } \epsilon_n > \alpha \tag{6}$$

for $i = 0$

$$\hat{T}_0 f_\epsilon = q_0 f_\epsilon \quad \text{if } |\epsilon_1| \leq \beta \tag{7}$$

$$\hat{T}_0 f_\epsilon = c_{-\epsilon_1} f_{s_0 \cdot \epsilon} \quad \text{if } \epsilon_1 < -\beta \tag{8}$$

where $(q, q_n, q_0) = (\sigma t^{\sigma/2}, \sigma_n t_n^{\sigma_n/2}, \sigma_0 t_0^{\sigma_0/2})$, $c_0 := 1$, and $c_i := c_{-i}^{-1}$ if $i < 0$.

The condition for the polynomials given above is a sufficient condition for the boundary qKZ equation. Thus we obtain one of the main theorems as follows:

Theorem 3.2. *Under the above notation, let $\{f_\epsilon; \epsilon \in I_d\}$ be a qKZ family with signs $(\sigma, \sigma_n, \sigma_0)$. Then,*

$$F(z_1, \ldots, z_n) = \sum_{\epsilon \in I_d} f_\epsilon v_\epsilon$$

is a solution of the boundary qKZ equation with the same signs $(\sigma, \sigma_n, \sigma_0)$. *The correspondence of the parameters is given by* $(q, q_n, q_0) = (\sigma t^{\sigma/2}, \sigma_n t_n^{\sigma_n/2}, \sigma_0 t_0^{\sigma_0/2})$.

3.3. *Eigenvalue problem*

From now on, for simplicity, we often identify the elements of $\mathcal{H}_n$ and their images given by the representation, and omit the hat symbol $\hat{}$.

Recall the mutually commutative elements Y_i $(1 \le i \le n)$:

$$Y_i = T_i \dots T_{n-1}(T_n \dots T_0)T_1^{-1} \dots T_{i-1}^{-1}.$$

Definition 3.3. Let $(\sigma, \sigma_n, \sigma_0)$ be signs and $d_{-M}, d_{-M+1}, \dots, d_\gamma$ be positive integers such that $\sum_{i=-M}^{\gamma} d_i = n$. Fix δ and I_d by (1) and (2). We define the following eigenvalue problem for unknown polynomial E:

$$\begin{aligned}
&Y_i E = \chi_i E \quad \text{if } \delta_i < -\max(\alpha, \beta) \\
&T_i E = \sigma t^{\sigma/2} E \quad \text{if } \delta_i = \delta_{i+1} \\
&T_{i-1} \cdots T_1 T_0 T_1^{-1} \cdots T_{i-1}^{-1} E = \sigma_0 t_0^{\sigma_0/2} E \quad \text{if } |\delta_i| \le \beta \\
&T_i \cdots T_{n-1} T_n T_{n-1}^{-1} \cdots T_i^{-1} E = \sigma_n t_n^{\sigma_n/2} E \quad \text{if } |\delta_i| \le \alpha \\
&(T_{n-1} \cdots T_1 T_0 T_1^{-1} \cdots T_{i-1}^{-1})^{-1} T_n (T_{n-1} \cdots T_1 T_0 T_1^{-1} \cdots T_{i-1}^{-1}) E \\
&\quad = \sigma_n t_n^{\sigma_n/2} E \quad \text{if } -\alpha \le \delta_i < -\beta \\
&(T_1^{-1} \cdots T_{n-1}^{-1} T_n^{-1} T_{n-1}^{-1} \cdots T_i^{-1})^{-1} T_0 (T_1^{-1} \cdots T_{n-1}^{-1} T_n^{-1} T_{n-1}^{-1} \cdots T_i^{-1}) E \\
&\quad = \sigma_0 t_0^{\sigma_0/2} E \quad \text{if } -\beta \le \delta_i < -\alpha.
\end{aligned}$$

For a qKZ family $\{f_\epsilon; \epsilon \in I_d\}$ with signs $(\sigma, \sigma_n, \sigma_0)$ and exponents $c_1, \dots, c_M$, the member f_δ is a solution of the eigenvalue problem above with the eigenvalues

$$\begin{aligned}
\chi_i = c_{-\delta_i} &(\sigma t^{\sigma/2})^{n(\delta, >i) - n(\delta, <i)} \\
&\text{where} \quad n(\delta, < i) := \sharp\{j; j < i, \delta_j = \delta_i\} \\
&\text{and} \quad n(\delta, > i) := \sharp\{j; j > i, \delta_j = \delta_i\}.
\end{aligned}$$

Conversely, by acting T_i's on a solution E of the eigenvalue problem, we can obtain a qKZ family.

Before giving an explicit statement (Theorem 3.7), we introduce some notions.

Lemma 3.4. *Fix* $\epsilon \in I_d$. *Take an element* $w \in W$ *and let* $w = s_{i_r} \cdots s_{i_1}$ *be a reduced expression. For* $1 \le m \le r$, *put* $\epsilon^{(m)} := s_{i_m} \cdots s_{i_1} \cdot \epsilon$. *Define*

$T^{\epsilon}_{\emptyset} = 1$ *and* $T^{\epsilon}_{i_m,\ldots,i_1}$ *inductively as follows:*

for $1 \le i_m \le n-1$

$$T^{\epsilon}_{i_m,\ldots,i_1} := T_{i_m} T^{\epsilon}_{i_{m-1},\ldots,i_1} \qquad \text{if } \epsilon^{(m-1)}_{i_m} > \epsilon^{(m-1)}_{i_m+1}$$
$$T^{\epsilon}_{i_m,\ldots,i_1} := \sigma t^{-\sigma/2} T_{i_m} T^{\epsilon}_{i_{m-1},\ldots,i_1} \qquad \text{if } \epsilon^{(m-1)}_{i_m} = \epsilon^{(m-1)}_{i_m+1}$$
$$T^{\epsilon}_{i_m,\ldots,i_1} := T^{-1}_{i_m} T^{\epsilon}_{i_{m-1},\ldots,i_1} \qquad \text{if } \epsilon^{(m-1)}_{i_m} < \epsilon^{(m-1)}_{i_m+1}$$

for $i_m = n$

$$T^{\epsilon}_{i_m,\ldots,i_1} := T_{i_m} T^{\epsilon}_{i_{m-1},\ldots,i_1} \qquad \text{if } \epsilon^{(m-1)}_{n} > \alpha$$
$$T^{\epsilon}_{i_m,\ldots,i_1} := \sigma_n t_n^{-\sigma_n/2} T_{i_m} T^{\epsilon}_{i_{m-1},\ldots,i_1} \qquad \text{if } |\epsilon^{(m-1)}_{n}| \le \alpha$$
$$T^{\epsilon}_{i_m,\ldots,i_1} := T^{-1}_{i_m} T^{\epsilon}_{i_{m-1},\ldots,i_1} \qquad \text{if } \epsilon^{(m-1)}_{n} < -\alpha$$

for $i_m = 0$

$$T^{\epsilon}_{i_m,\ldots,i_1} := c^{-1}_{-\epsilon^{(m-1)}_1} T_{i_m} T^{\epsilon}_{i_{m-1},\ldots,i_1} \qquad \text{if } \epsilon^{(m-1)}_{1} < -\beta$$
$$T^{\epsilon}_{i_m,\ldots,i_1} := \sigma_0 t_0^{-\sigma_0/2} T_{i_m} T^{\epsilon}_{i_{m-1},\ldots,i_1} \qquad \text{if } |\epsilon^{(m-1)}_{1}| \le \beta$$
$$T^{\epsilon}_{i_m,\ldots,i_1} := c_{\epsilon^{(m-1)}_1} T^{-1}_{i_m} T^{\epsilon}_{i_{m-1},\ldots,i_1} \qquad \text{if } \epsilon^{(m-1)}_{1} > \beta.$$

Then $T^{\epsilon}_{i_r,\ldots,i_1}$ *does not depend on a choice of reduced expression of* w*. We denote it by* T^{ϵ}_w*.*

For an element $\epsilon \in I_d$, we call a sequence $(i_r, \ldots i_1) \in \{0, \ldots, n\}^r$ *is* ϵ*-good* if the elements $\epsilon^{(m)} := s_{i_m} \cdots s_{i_1} \cdot \epsilon$ $(1 \le m \le r)$ satisfy

$$\epsilon^{(m-1)}_{i_m} \ne \epsilon^{(m-1)}_{i_m+1} \qquad (1 \le i_m \le n-1) \tag{9}$$
$$|\epsilon^{(m-1)}_{n}| > \alpha \qquad (i_m = n) \tag{10}$$
$$|\epsilon^{(m-1)}_{1}| > \beta \qquad (i_m = 0). \tag{11}$$

Lemma 3.5. *For* $w \in W$*, let* $w = s_{i_r} \cdots s_{i_1} = s_{j_r} \cdots s_{j_1}$ *be reduced expressions. If* $(i_r, \ldots i_1)$ *is* ϵ*-good, then* $(j_r, \ldots j_1)$ *is also* ϵ*-good. For such situation, we call* w *is an* ϵ*-good element.*

For any $1 \le i \le n$ such that $\delta_i < -\max(\alpha, \beta)$, let

$$(j_{2n}, \ldots j_1) := (i, \ldots, n-1, n, n-1, \ldots, 1, 0, 1, \ldots, i-1).$$

We see that $\delta^{(m)} := s_{j_m} \cdots s_{j_1} \cdot \delta$ $(1 \le m \le 2n)$ satisfies (10) and (11). If $\delta^{(m)}$ does not satisfy (9) for some m, then eliminate m-th component j_m

from the sequence. So that we obtain a subsequence $(k_\ell, \dots k_1)$ which is δ-good. In fact, $(k_\ell, \dots k_1)$ is given by

$$(k_\ell, \dots k_1) = (i + n(\delta, > i), \dots, n-1, n, \\ n-1, \dots, 1, 0, 1, \dots, i-1-n(\delta, < i)).$$

Using $(k_\ell, \dots k_1)$, we define $\tau_i \in W$ and $\Gamma_d \subset W$ as follows:

$$\tau_i := s_{k_\ell} \cdots s_{k_1}, \\ \Gamma_d := \langle \tau_i^{\pm 1} \ ; \ \delta_i < -\max(\alpha, \beta) \rangle.$$

Then we have

Lemma 3.6. *(i) For any δ-good element $w \in W$ satisfying $\delta = w \cdot \delta$, we have $w \in \Gamma_d$. Conversely, for any $w \in \Gamma_d$, we have $\delta = w \cdot \delta$.*

(ii) For any $\epsilon \in I_d$, there exists $w \in W$ such that w is δ-good and $\epsilon = w \cdot \delta$.

(iii) Fix any $\epsilon \in I_d$. Take any w_1 and $w_2 \in W$ such that w_i is δ-good and $\epsilon = w_i \cdot \delta$ $(i = 1, 2)$. Then we have $w_1^{-1} w_2 \in \Gamma_d$.

Now we construct a qKZ family.

Theorem 3.7. *Let E be a solution of the eigenvalue problem in Definition 3.3. For any $\epsilon \in I_d$, take a δ-good element $w_\epsilon \in W$ such that $w_\epsilon \cdot \delta = \epsilon$. Then $T^\delta_{w_\epsilon} E$ does not depend on a choice of w_ϵ. Putting $f_\epsilon := T^\delta_{w_\epsilon} E$, the family $\{f_\epsilon ; \epsilon \in I_d\}$ forms a qKZ family with exponents*

$$c_{-\delta_i} := \chi_i (\sigma t^{\sigma/2})^{-n(\delta, > i) + n(\delta, < i)}.$$

4. Special solutions

In this section, we review non-symmetric Koornwinder polynomials, which are joint Y-eigenfunctions with 6 parameters. We will see that specific non-symmetric Koornwinder polynomials can be also T_i-eigenfunctions. We show that they are special solutions of the eigenvalue problem in Definition 3.3, so that they give polynomial solutions of the boundary qKZ equation.

4.1. *Non-symmetric Koornwinder polynomials*

Let $\lambda \in \mathbb{Z}^n$ be any element. Denote by λ^+ the unique dominant element in $W_0 \cdot \lambda$ (where $W_0 = \langle s_1, \dots, s_n \rangle$ is finite Weyl group of type C_n). That is, $\lambda_1^+ \geq \lambda_2^+ \geq \cdots \geq \lambda_n^+ \geq 0$. Define partial orderings $\lambda \geq \mu$ and $\lambda \succeq \mu$ in $\mathbb{Z}^n$

as follows:

$$\begin{aligned} \lambda \geq \mu \quad & \text{if } \textstyle\sum_{j=1}^{i} \lambda_j \geq \sum_{j=1}^{i} \mu_j \text{ for any } 1 \leq i \leq n, \\ \lambda \succeq \mu \quad & \text{if } \lambda^+ > \mu^+, \text{ or } \lambda^+ = \mu^+ \text{ and } \lambda \geq \mu. \end{aligned}$$

Take the shortest element $w \in W_0$ such that $w \cdot \lambda^+ = \lambda$ and denote it by w_λ^+. Put $\rho = (n-1, n-2, \ldots, 1, 0)$, $\rho(\lambda) = w_\lambda^+ \cdot \rho$, $\sigma(\lambda) = (\mathrm{sgn}(\lambda_1), \ldots, \mathrm{sgn}(\lambda_n))$ where $\mathrm{sgn}(0) = +1$.

Definition 4.1. For $\lambda \in \mathbb{Z}^n$, the non-symmetric Koornwinder polynomial E_λ is defined by the following properties:

$$\begin{aligned} Y_i E_\lambda &= y(\lambda)_i E_\lambda \qquad (12) \\ \text{where} \quad y(\lambda)_i &:= s^{\lambda_i} t^{\rho(\lambda)_i} (t_n t_0)^{\sigma(\lambda)_i/2} \\ E_\lambda &= x^\lambda + \sum_{\mu \prec \lambda} c_{\lambda\mu} x^\mu \quad (c_{\lambda\mu} \in \mathbb{C}). \end{aligned}$$

We call the parameters s, t, t_n, t_0 *generic* if they do not satisfy either of

$$\begin{aligned} & s^{r-1} t^{k+1} = 1 \quad (n-1 \geq k+1 \geq 0, r-1 \geq 1) \\ & s^{r-1} t^{k+1} t_n t_0 = 1 \quad (2n-2 \geq k+1 \geq 0, r-1 \geq 1). \end{aligned}$$

If the parameters are generic, then for any $\lambda \in \mathbb{Z}^n$, the set of eigenvalues $y(\lambda)_i$ is different. Thus all non-symmetric Koornwinder polynomial E_λ is well-defined.

We have the action of T_i $(1 \leq i \leq n)$ on E_λ. If $1 \leq i \leq n-1$ and $\lambda_i < \lambda_{i+1}$, then

$$T_i E_\lambda = -\frac{t^{1/2} - t^{-1/2}}{y(\lambda)_{i+1}/y(\lambda)_i - 1} E_\lambda + t^{1/2} E_{s_i \cdot \lambda}.$$

If $1 \leq i \leq n-1$ and $\lambda_i = \lambda_{i+1}$, then

$$T_i E_\lambda = t^{1/2} E_\lambda. \qquad (13)$$

If $1 \leq i \leq n-1$ and $\lambda_i > \lambda_{i+1}$, then

$$T_i E_\lambda = -\frac{t^{1/2} - t^{-1/2}}{y(\lambda)_{i+1}/y(\lambda)_i - 1} E_\lambda + t^{-1/2} \frac{N_i^+ N_i^-}{D_i^+ D_i^-} E_{s_i \cdot \lambda} \qquad (14)$$

where

$$\begin{aligned} D_i^\pm &:= (y(\lambda)_{i+1}/y(\lambda)_i)^{\pm 1} - 1 \qquad (1 \leq i \leq n-1) \\ N_i^\pm &:= t^{1/2}((y(\lambda)_{i+1}/y(\lambda)_i)^{\pm 1} - t^{-1}) \qquad (1 \leq i \leq n-1). \end{aligned}$$

If $\lambda_n < 0$, then

$$T_n E_\lambda = -\frac{(t_n^{1/2} - t_n^{-1/2}) + (t_0^{1/2} - t_0^{-1/2}) y(\lambda)_n^{-1}}{y(\lambda)_n^{-2} - 1} E_\lambda + t_n^{1/2} E_{s_n \cdot \lambda}.$$

If $\lambda_n = 0$, then

$$T_n E_\lambda = t_n^{1/2} E_\lambda. \tag{15}$$

If $\lambda_n > 0$, then

$$T_n E_\lambda = -\frac{(t_n^{1/2} - t_n^{-1/2}) + (t_0^{1/2} - t_0^{-1/2}) y(\lambda)_n^{-1}}{y(\lambda)_n^{-2} - 1} E_\lambda + t_n^{-1/2} \frac{N_n^+ N_n^-}{D_n^+ D_n^-} E_{s_n \cdot \lambda} \tag{16}$$

where

$$D_n^\pm := y(\lambda)_n^{\mp 2} - 1$$
$$N_n^\pm := t_n^{1/2}(y(\lambda)_n^{\mp 1} - t_n^{-1/2} t_0^{-1/2})(y(\lambda)_n^{\mp 1} + t_n^{-1/2} t_0^{1/2}).$$

4.2. *Generic case*

Suppose the parameters are generic. From (12), (13) and (15), we see that E_λ is a solution of the eigenvalue problem with sign $\sigma = +$ or $\sigma_n = +$.

Proposition 4.1. *Take $d_{-M}, \ldots, d_\gamma$ such that $d_i = 0$ if $|i| \leq \beta$, and fix δ by (1). Take λ such that $\lambda_i = \lambda_{i+1}$ if $\delta_i = \delta_{i+1}$ and $\lambda_i = 0$ if $-\beta > \delta_i \geq -\alpha$. Then E_λ is a solution of the eigenvalue problem with sign $(\sigma, \sigma_n, \sigma_0) = (+, +, \pm)$. (Note that there are no conditions including σ_0 in the eigenvalue problem.)*

4.3. *Specialized case 1*

Fix $2 \leq k+1 \leq n$ and $1 \leq r-1$. we assume that the parameters only satisfy

$$s^{r-1} t^{k+1} = 1. \tag{17}$$

Definition 4.2. For any $\lambda \in \mathbb{Z}^n$ with no negative components, we call λ *admissible* if it satisfies the following two properties:

$$\lambda_i^+ - \lambda_{i+k}^+ \leq r - 1, \text{ and}$$
$$\lambda_i^+ - \lambda_{i+k}^+ = r - 1 \Rightarrow w_\lambda^+(i) < w_\lambda^+(i+k).$$

For any $\lambda \in \mathbb{Z}^n$ with possibly negative components, let $i_1 < \ldots < i_p$ be indexes of non-negative components and $j_1 < \ldots < j_{n-p}$ be that of negative components. Set

$$\lambda^0 := (\lambda_{i_1}, \ldots, \lambda_{i_p}, -\lambda_{j_{n-p}}, \ldots, -\lambda_{j_1}).$$

We call λ *admissible* if λ^0 is admissible.

Lemma 4.3. *(Ref. 2) We specialize the parameters at (17). Fix any admissible $\lambda \in \mathbb{Z}^n$. Then*
(i) E_λ is well-defined,
(ii) $T_i E_\lambda = -t^{-1/2} E_\lambda$ if and only if $s_i\lambda$ is not admissible.

From this lemma, we obtain a solution of the eigenvalue problem.

Theorem 4.4. *Take integers ℓ and m such that $n = k\ell + m$ and $0 \le m \le k-1$. Take $d_{-M}, \ldots, d_\gamma$ such that $d_i = 0$ if $|i| \le \max(\alpha, \beta)$, and non-zero components of $d_{-M}, \ldots, d_\gamma$ be a permutation of $((m+1)^\ell, m^{k-\ell})$. Fix δ by (1). Take $a = (a_1, \ldots, a_k)$ satisfying (i), (ii), and (iii) below.*
(i) If $d_i > d_j$, then $|a_i| > |a_j|$, or
$|a_i| = |a_j|$ and $(\mathrm{sgn}(a_i), \mathrm{sgn}(a_j)) = (+,+)$ and $i < j$, or
$|a_i| = |a_j|$ and $(\mathrm{sgn}(a_i), \mathrm{sgn}(a_j)) = (+,-)$, or
$|a_i| = |a_j|$ and $(\mathrm{sgn}(a_i), \mathrm{sgn}(a_j)) = (+,+)$ and $i < j$.
(ii) $\max_{i,j}(\big||a_i| - |a_j|\big|) \le r-1$.
(iii) $a_i - (r-1)(d_i-1) \ge 0$ or $a_i + (d_i-1)(r-1) < 0$.
For any $1 \le i \le k$, put $\mathbf{a}^{(i)} \in \mathbb{Z}^{d_i}$ as follows. If $a_i - (d_i-1)(r-1) \ge 0$ then

$$\mathbf{a}^{(i)} := (a_i, a_i - (r-1), \ldots, a_i - (d_i-1)(r-1))$$

and if $a_i + (d_i-1)(r-1) < 0$ then

$$\mathbf{a}^{(i)} := (a_i + (d_i-1)(r-1), \ldots, a_i + (r-1), a_i).$$

Take

$$\lambda := (\mathbf{a}^{(1)}, \ldots, \mathbf{a}^{(k)}).$$

Then E_λ is a solution of the eigenvalue problem with the sign $(\sigma, \sigma_n, \sigma_0) = (-,\pm,\pm)$. (Note that there are no conditions including σ_n or σ_0 in the eigenvalue problem.)

Example of Theorem 4.4. (Ref. 2) For simplicity, suppose $n = Mm$. Specialize the parameters at $s = t^{-M-1}$. Then

$$\mu = (m-1, m-2, \ldots, 1, 0, m-1, m-2, \ldots, 1, 0, \ldots, m-1, m-2, \ldots, 1, 0)$$

is admissible. Hence E_μ is well-defined at $s = t^{-M-1}$ and $T_i E_\mu = -t^{-1/2} E_\mu$ if $i \in \{1, \dots, n\} \setminus \{m, 2m, \dots, Mm\}$. Therefore E_μ is a solution of the eigenvalue problem for the case $\alpha = \beta = 0$, $\sigma = -$, $(d_{-M}, \dots, d_{-1}, d_0) = (m, \dots, m, 0)$. Moreover,

$$E_\mu(z_1, \dots, z_n; s = t^{-M-1}) = \prod_{\ell=1}^{k} \prod_{m(\ell-1)<i<j\leq m\ell} (z_i - t^{-1} z_j)(1 - \frac{t^{\ell-M-1}}{z_i z_j}).$$

4.4. *Specialized case 2*

Assume that the parameters only satisfy

$$t_n = -s^{-\ell}.$$

Since such a specialization is generic, E_λ is well-defined for any $\lambda \in \mathbb{Z}^n$.

Let $E = E_{(\ell,\dots,\ell)}(z_1, \dots, z_n; t_n = -s^{-\ell})$. Then from (13) and (16), we see that

$$\begin{aligned} T_i E &= t^{1/2} E \qquad (1 \leq i \leq n-1), \\ T_n E &= -t_n^{-1/2} E, \\ T_0 E &= t_0^{1/2} E. \end{aligned}$$

Hence it gives a solution for the case $\sigma_n = -$.

Proposition 4.2. *$E_{(\ell,\dots,\ell)}(z_1, \dots, z_n; t_n = -s^{-\ell})$ is a solution of the eigenvalue problem with sign $(\sigma, \sigma_n, \sigma_0) = (+, -, +)$.*

Acknowledgements.

The research is partially supported by Grant-in-Aid for Young Scientists (B) No. 21740005. The author is deeply grateful to Tetsuji Miwa for valuable comments. The author also thanks to Yoshihiro Takeyama for many comments.

References

1. I. B. Frenkel and N. Yu. Reshetikhin, Quantum affine algebras and holonomic difference equations, *Comm. Math. Phys.* **146** (1992), no. 1, 1–60.
2. M. Kasatani, The polynomial representation of the double affine Hecke algebra of type $(C_n^\vee, C_n)$ for specialized parameters, arXiv:math/0608773.
3. M. Kasatani, K. Shigechi, in preparation.

4. M. Kasatani, Y. Takeyama, The quantum Knizhnik-Zamolodchikov equation and non-symmetric Macdonald polynomials, *Funkcialaj Ekvacioj* **50** (2007), 491–509.
5. I. G. Macdonald, *Affine Hecke algebras and orthogonal polynomials* (Cambridge University Press, 2003).
6. A. Nakayashiki, Trace construction of a basis for the solution space of sl_N qKZ equation, *Comm. Math. Phys.* **212** (2000), no. 1, 29–61.
7. M. Noumi, Macdonald-Koornwinder polynomials and affine Hecke rings, *Sūrikaisekikenkyūsho Kōkyūroku* **919** (1995), 44–55 (in Japanese).
8. J. V. Stokman, Quantum affine Knizhnik-Zamolodchikov equations and quantum spherical functions, I, arXiv:1001.2645.

A PENTAGON OF IDENTITIES, GRADED TENSOR PRODUCTS, AND THE KIRILLOV-RESHETIKHIN CONJECTURE

RINAT KEDEM

Department of Mathematics, University of Illinois
Urbana, IL 61821, USA
E-mail: rinat@illinois.edu

Dedicated to T. Miwa on the occasion of his 60th birthday

This paper provides a brief review of the relations between the Feigin-Loktev conjecture on the dimension of graded tensor products of $\mathfrak{g}[t]$-modules, the Kirillov-Reshetikhin conjecture, the combinatorial "$M = N$" conjecture, their proofs for all simple Lie algebras, and a pentagon of identities which results from the proof.

Keywords: Fusion products; KR modules.

1. Introduction

This paper reviews work[1–3] which followed the author's fruitful collaboration with T. Miwa and colleagues.[4,5] This work was inspired by the work of Feigin and Loktev on fusion products.[6] The series of results described here finally culminated in a proof[2,3] of the Feigin-Loktev conjecture concerning the graded character of the (non-level restricted) fusion product, in the case of special modules known as Kirillov-Reshetikhin modules. The purpose of this article is to make clear the sequence of connections and relations between the various results which lead to the proof.

The fusion product character first appeared in the '80's, in the work on the completeness conjecture for Bethe Ansatz states[7] for the generalized Heisenberg models. The completeness conjecture is one version of what later became known as the Kirillov-Reshetikhin conjecture, and involves the first appearance of an object called the generalized Kostka polynomial. The (generalized) Kostka number gives the decomposition coefficients of tensor products of KR-modules into irreducible components. Although much work was later published on the subject, the conjecture in its original, combinatorial form – counting solutions of the Bethe equations – was only proved to be true in special cases. In other cases, a similar but not manifestly positive formula[8–10] was shown to hold.

The key to proving both the Kirillov-Reshetikhin conjecture and the Feigin-Loktev conjecture is a combinatorial identity, the equality of two polynomials in q, one written as an alternating sum, and the other as a sum of positive terms. The deeper meaning of this identity remains mysterious, but its proof[3] using purely combinatorial means finally implies several equalities, proving the conjectures above for any simple Lie algebras.

1.1. *The objects of interest*

We describe several objects and relations between their dimensions. (Section 2 contains a fuller discussion of several of these as necessary).

1.1.1. *Kirillov-Reshetikhin modules*

These are finite-dimensional modules of the quantum affine algebra $U_q(\widehat{\mathfrak{g}})$ or the Yangian $Y(\mathfrak{g})$. Let $\mathfrak{g}$ be a simple Lie algebra of rank r with Cartan matrix C. Consider the irreducible, finite-dimensional $\mathfrak{g}$-module V with a highest weight which is a non-negative multiple of a fundamental weight. The Yangian $Y(\mathfrak{g})$ contains $\mathfrak{g}$ as a subalgebra. The irreducible $Y(\mathfrak{g})$-module induced from V is called a Kirillov-Reshetikhin module.[7] It is finite-dimensional, and its $\mathfrak{g}$-highest weight is that of V. In the case where $\mathfrak{g} = A_n$, it is equal to V as a $\mathfrak{g}$-module. In other cases, the restriction to a $\mathfrak{g}$-module may or may not be irreducible, but in that case, V is always a component in the decomposition, with multiplicity 1, and with the highest weight.

Equivalently, one may consider Kirillov-Reshetikhin modules for the quantum affine algebra $U_q(\widehat{\mathfrak{g}})$ and their similar decomposition into $U_q(\mathfrak{g})$-modules.[11] These are also referred to as KR-modules.

We denote the KR-modules by $\mathrm{KR}_{\alpha,m}(\zeta)$, where $1 \le \alpha \le r$ and m is a positive integer. These have a $\mathfrak{g}$-highest weight of the form $m\omega_\alpha$, where ω_α is a fundamental weight of $\mathfrak{g}$. The parameter ζ is a complex number which is called the spectral parameter.

1.1.2. *Chari's graded $\mathfrak{g}[t]$-modules*

These are modules of the current algebra $\mathfrak{g}[t]$, defined as a quotient of $U(\mathfrak{n}_-[t])$ by an ideal generated by relations[11] (see Equations (12),(11)).

The relations are the $q \to 1$ limits of the similar relations which hold in the quantum case for Kirillov-Reshetikhin modules. These modules also have a $\mathfrak{g}$-highest weight equal to a multiple of one of the fundamental weights of $\mathfrak{g}$, as in the quantum algebra case. We denote this module by $C_{\alpha,m}(\zeta)$, where the highest weight is again $m\omega_\alpha$.

1.1.3. *Decomposition of tensor products*

We observe that by general deformation arguments, the dimension of KR-modules is bounded from above by that of Chari's modules.

More precisely, the decomposition coefficients of the KR-modules, and therefore their tensor products at generic values of the spectral parameters, into irreducible $\mathfrak{g}$-modules are bounded from above by those of Chari's modules. That is, choose a sequence of non-negative integers $\mathbf{n} = \{n_m^{(\alpha)} : 1 \le \alpha \le r, m > 0\}$ and consider the multiplicities $M^Y_{\lambda,\mathbf{n}}$ and $M^{\mathfrak{g}[t]}_{\lambda,\mathbf{n}}$, defined by

$$M^Y_{\lambda,\mathbf{n}} = \dim \mathrm{Hom}_{\mathfrak{g}} \left(\otimes_{\alpha,m} \mathrm{KR}_{\alpha,m}^{\otimes n_m^{(\alpha)}}, V(\lambda) \right),$$

where $V(\lambda)$ is the irreducible highest weight $\mathfrak{g}$-module with highest weight λ, and

$$M^{\mathfrak{g}[t]}_{\lambda,\mathbf{n}} = \dim \mathrm{Hom}_{\mathfrak{g}} \left(\otimes_{\alpha,m} C_{\alpha,m}^{\otimes n_m^{(\alpha)}}, V(\lambda) \right).$$

(Here, we omitted the dependence of the modules on the spectral parameter: We assume that all spectral parameters are taken at generic values with respect to each other). Then we have the first inequality:

$$M^Y_{\lambda,\mathbf{n}} \le M^{\mathfrak{g}[t]}_{\lambda,\mathbf{n}}, \tag{1}$$

which simply follows by general deformation arguments: Both are defined as quotients by some ideal, and the ideal in the limit $q \to 1$ may be smaller than that for generic values of q.

1.1.4. *The combinatorial KR-conjecture: The M-sum formula*

This is a conjecture that $M^Y_{\lambda,\mathbf{n}}$ is equal to the number of Bethe vectors. The generalized, inhomogeneous Heisenberg spin chain has a Hilbert space which is equal, by definition, to the tensor product of Yangian modules,

$$\mathcal{H}_{\mathbf{n}} = \prod_{\alpha,m} \mathrm{KR}_{\alpha,m}^{\otimes n_m^{(\alpha)}}.$$

Again, the modules are taken at generic values of the spectral parameters, that is, pairwise not separated by an integer. (Note that the model is also well-defined for any other finite-dimensional $Y(\mathfrak{g})$-modules, but no Bethe Ansatz solution is known generically.)

This model has a Bethe Ansatz solution. The completeness conjecture of Kirillov and Reshetikhin[7] states that there is a Bethe vector for each $\mathfrak{g}$-highest weight vector in $\mathcal{H}_{\mathbf{n}}$. In particular, there is an explicit formula for the number of Bethe vectors, and in fact, the authors wrote down a graded formula (which we now know has a direct interpretation as a grading by the linearized energy function of the model), although at the time, the meaning of the refinement was unknown. For the $\mathfrak{g}$-highest weight λ, the multiplicity is the number $M_{\lambda,\mathbf{n}}$ obtained as the $q \to 1$ limit of the following, grading-endowed formula:[7]

$$M_{\lambda,\mathbf{n}}(q) = \sum_{\mathbf{m}} q^{Q(\mathbf{m},\mathbf{n})} \prod_{\alpha,i} \begin{bmatrix} m_i^{(\alpha)} + P_i^{(\alpha)} \\ m_i^{(\alpha)} \end{bmatrix}_q \tag{2}$$

where the sum is taken over the non-negative integers $\mathbf{m} = \{m_i^{(\alpha)} : 1 \leq \alpha \leq r, i \geq 1\}$ such that $\sum_i m_i^{(\alpha)} = m^{(\alpha)}$, where $m^{(\alpha)}$ are integers fixed by the "zero weight condition"

$$\sum_{\beta} C_{\alpha,\beta} m^{(\alpha)} = \sum_i n_i^{(\alpha)} - \ell^{(\alpha)}, \tag{3}$$

$\ell^{(\alpha)}$ being the coefficient of ω_α in the weight λ. Note that this sum has only a finite number of non-vanishing terms. Let us define

$$B_{i,j}^{(\alpha,\beta)} = \text{sign}(C_{\alpha,\beta}) \min(|C_{\alpha,\beta}|j, |C_{\beta,\alpha}|i).$$

Then the vacancy numbers $P_i^{(\alpha)}$ are defined as

$$P_i^{(\alpha)} = \sum_i \min(i,j) n_j^{(\alpha)} - (B\mathbf{m})_i^{(\alpha)}, \tag{4}$$

and the quadratic form $Q(\mathbf{m},\mathbf{n})$ is

$$Q(\mathbf{m},\mathbf{n}) = \frac{1}{2}(\mathbf{m} \cdot (\mathbf{P} + A\mathbf{n})). \tag{5}$$

where A is the matrix with entries $A_{i,j}^{(\alpha,\beta)} = \delta_{\alpha,\beta} \min(i,j)$.

The q-binomial coefficient is defined as usual, and in the limit $q \to 1$ becomes the usual binomial coefficient. In particular, *the sum is taken over the restricted set of integers* $\mathbf{m}$ *such that* $P_j^{(\alpha)} \geq 0$.

This provides a formula for the tensor multiplicities $M_{\lambda,\mathbf{n}}^Y$. It was proved in several special cases using combinatorial means.[7,12–16] In general, a similar but not equivalent formula was known to be true, as explined below.

1.1.5. *The HKOTY N-sum formula*

For general Lie algebras, and for generic KR-modules, the following formula was conjectured:[8]

$$M^Y_{\lambda,\mathbf{n}} = \lim_{q\to 1} N_{\lambda,\mathbf{n}}(q), \tag{6}$$

where $N_{\lambda,\mathbf{n}}(q)$ is a modified form of the formula (2), obtained by simply removing the restriction $P_j^{(\alpha)} \geq 0$. Both the usual and the q-binomial coefficients are defined when $P_j^{(\alpha)} < 0$, but they carry a sign in that case. The authors conjectured, after extensive testing, that all terms coming from sets $\mathbf{m}$ such that $P_j^{(\alpha)} < 0$ for some j, α cancel, so that

$$M_{\lambda,\mathbf{n}}(q) = N_{\lambda,\mathbf{n}}(q). \tag{7}$$

The conjecture (6) holds provided that the so-called Q-system[7] is satisfied by the characters of KR-modules. It was shown by Nakajima (for simply-laced algebras)[9] and Hernandez for all other Lie algebras[17] that the q-characters of KR-modules satisfy the more general T-system,[18] from which the Q-system follows. Hence, Equation (6) had achieved the status of a Theorem.

1.1.6. *Feigin-Loktev fusion products*

The Feigin-Loktev fusion product is a graded $\mathfrak{g}[t]$-module,[6] which is a refinement of the usual tensor product of $\mathfrak{g}$-module (cyclic, finite-dimensional). One chooses a finite-dimensional cyclic $\mathfrak{g}$-module V, from which one induces an action of the current algebra $\mathfrak{g}[t]$ localized at some complex number ζ. More specifically, one defines a graded tensor product by choosing N $\mathfrak{g}$-modules V_i, each with a cyclic vector v_i, localized at N distinct points in $\mathbb{C}P$. One then defines the fusion product as the associated graded space of the filtered space, generated by the action of $U(\mathfrak{g}[t])$ on the tensor product of cyclic vectors, with the grading defined by degree in t. This is a graded space, and the graded components are $\mathfrak{g}$-modules.

Feigin and Loktev conjectured that the fusion product as a graded space is independent of the localization parameters for sufficiently well-behaved $\mathfrak{g}$-modules. Moreover, they conjectured a relation between the graded coefficients of the $\mathfrak{g}$-module $V(\lambda)$ in the fusion product, and the generalized Kostka polynomials.[19] This conjecture was proved for $\mathfrak{sl}_2$ in Ref. 5, and in greater generality in several other works.

In particular, in Ref. 2, we proved the following inequality, using techniques generalized from Ref. 5. Let $\mathcal{F}^*_{\mathbf{n}}$ be the fusion product of the modules $C_{\alpha,m}$ with multiplicity $n_m^{(\alpha)}$. This is a graded space. We define the q-dimension to be the Hilbert polynomial of the graded space. Then

$$q\text{-}\dim \mathrm{Hom}_{\mathfrak{g}}\left(\mathcal{F}_{\mathbf{n}}^*, V(\lambda)\right) \leq M_{\lambda,\mathbf{n}}(q), \tag{8}$$

where $M_{\lambda,\mathbf{n}}$ is the fermionic formula of Kirillov and Reshetikhin for the number of Bethe vectors in the generalized, inhomogeneous Heisenberg spin chain corresponding to KR-modules $KR_{\alpha,m}$ with the same multiplicities.

Remark 1.1. The inequality in (8) arises from the following sequence of maps: One may completely characterize the dual space of functions of the fusion product in terms of symmetric functions with certain zeros and poles (we do this in Section 3). Actual calculation of the Hilbert polynomial of this space requires another injective mapping into another filtered space, whose Hilbert polynomial is the polynomial $M_{\lambda,\mathbf{n}}(q)$. We do not prove the surjectivity of the map, resulting in the inequality in Equation (8).

Moreover, the space $\mathcal{F}_{\mathbf{n}}^*$ is the associated graded space of the tensor product of Chari modules, which are defined as a quotient of $U(\mathfrak{g}[t])$ by a certain ideal. Again, by a general deformation argument, we have that

$$\dim \mathrm{Hom}_{\mathfrak{g}}(\otimes C_{\alpha,i}^{\otimes n_i^{(\alpha)}}, V(\lambda)) \leq \dim \mathrm{Hom}_{\mathfrak{g}}(\mathcal{F}_{\mathbf{n}}^*, V(\lambda)).$$

Note that the sum on the right hand side of Equation (8) is manifestly positive, and therefore if, in the $q \to 1$ limit, it is equal to the tensor product multiplicity, then we have the equality of graded spaces also, since the left-hand side has a dimension which is greater than or equal to the tensor product multiplicity by the deformation argument.

1.2. *A pentagon of identities*

We have a sequence of identities and inequalities:

$$\begin{array}{ccccc} & & \left|\mathrm{Hom}_{\mathfrak{g}}\left(\mathcal{F}_{\mathbf{n}}^*\,,\ V(\lambda)\right)\right| & & \\ & \rotatebox{45}{$\leq$} & & \rotatebox{-45}{$\leq$} & \\ \left|\mathrm{Hom}_{\mathfrak{g}}\left(\bigotimes\limits_{\alpha,m} C_{\alpha,m}^{\otimes n_m^{(\alpha)}}\,,\ V(\lambda)\right)\right| & & & & M_{\lambda,\mathbf{n}} \\ \rotatebox{90}{$\leq$} & & & & \| \leftarrow \text{final step} \\ \left|\mathrm{Hom}_{U_q(\mathfrak{g})}\left(\bigotimes\limits_{\alpha,m} KR_{\alpha,m}^{\otimes n_m^{(\alpha)}}\,,\ V(\lambda)\right)\right| & & = & & N_{\lambda,\mathbf{n}} \end{array}$$

The "final step" remaining in this pentagon was to prove the conjectured identity (7). The proof turns all the inequalities in the pentagon to equalities. This conjecture was proven by combinatorial means[3] for all simple Lie algebras. Therefore, this provides a proof of the completeness problem in the Bethe Ansatz known as the Kirillov-Reshetikhin conjecture, as well as the Feigin-Loktev conjecture for the cases of Kirillov-Reshetikhin conjecture.

1.3. *Plan of the paper*

In the following sections, we will summarize the proof[2] of the inequality (8) and the proof of the $M = N$ conjecture.[3] In Section 2, we give a definition of the Feigin-Loktev fusion product of Chari's modules. In Section 3, we summarize the proof of the inequality (8) which is obtained via a functional realization of the multiplicity space, following the ideas of B. Feigin. In Section 4, we explain the combinatorial proof of the $M = N$ conjecture.[3]

2. Definitions

Here, we add some details to the definitions of the representation-theoretical objects which are important in the theorems below.

Let $\mathfrak{g}$ be a simple Lie algebra of rank r and Cartan matrix C. Let $\mathfrak{g}[t] = \mathfrak{g} \otimes \mathbb{C}[t]$ be the corresponding Lie algebra of polynomials in t with coefficients in $\mathfrak{g}$.

2.1. *Finite-dimensional $\mathfrak{g}[t]$-modules and the fusion action*

Given a complex number ζ, any $\mathfrak{g}$-module V can be extended to a $\mathfrak{g}[t]$-module evaluation module $V(\zeta)$, with t evaluated at ζ. The generators $x[n] := x \otimes t^n$ $(x \in \mathfrak{g})$ act on $v \in V$ as $\pi(x[n])v = \zeta^n xv$.

The dimension of the evaluation module is the same as that of V. If V is irreducible as a $\mathfrak{g}$-module, so is $V(\zeta)$.

More generally, given a $\mathfrak{g}[t]$-module V, the $\mathfrak{g}[t]$-module localized at ζ, $V(\zeta)$, is the module on which $\mathfrak{g}[t]$ acts by expansion in the local parameter $t_\zeta := t - \zeta$. If $v \in V(\zeta)$, then

$$x[n]v = x \otimes (t_\zeta + \zeta)^n v = \sum_j \binom{n}{j} \zeta^j x[n-j]_\zeta v,$$

where $x[n]_\zeta := x \otimes t_\zeta$ and $x[n]_\zeta$ acts on $v \in V(\zeta)$ in the same way that $x[n]$ acts on $v \in V$.

Another way to write this is in terms of generating functions, for any

$x \in \mathfrak{g}$,

$$x(z) = \sum_{n\in\mathbb{Z}} x[n]z^{-n-1}.$$

Then if $\zeta \in \mathbb{C}$,

$$x[n]_\zeta = \frac{1}{2\pi i}\oint_{z=\zeta}(z-\zeta)^n x(z)dz. \tag{9}$$

We will also need to be able to localize modules at infinity. In that case,

$$x[n]_\infty = \frac{1}{2\pi i}\oint_{z=\infty} z^{-n}x(z)dz = \frac{-1}{2\pi i}\oint_{z=0} z^{n-2}x(z^{-1})dz. \tag{10}$$

An evaluation module $V(\zeta)$ is a special case of a localized module, on which the positive modes $x[n]_\zeta$ with $n > 0$ and $x \in \mathfrak{g}$ act trivially.

Let V be a cyclic $\mathfrak{g}[t]$-module with cyclic vector v. Then V is endowed with a $\mathfrak{g}$-equivariant grading inherited from the grading of $U := U(\mathfrak{g}[t])$. The filtred components of V are $\mathcal{F}(n) = U^{\leq n}v$, where $U^{\leq n}$ is the subspace of homogeneous degree in t bounded by n. The grading on V is the associated graded space of this filtration, $\underset{n\geq 0}{\oplus}\mathcal{F}(n)/\mathcal{F}(n-1)$.

As the filtration is $\mathfrak{g}$-equivariant, the graded components are $\mathfrak{g}$-modules.

2.2. *Chari's KR-modules of* $\mathfrak{g}[t]$

A special case of the construction described in the previous subsection is given as follows. Consider $\mathfrak{g}[t]$-modules with a highest weight $m\omega_\alpha$ ($m \geq 0$ and ω_α a fundamental $\mathfrak{g}$-weight) defined as the cyclic module generated by a highest weight vector v, with relations

$$\begin{aligned} x[n]_\zeta v &= 0 \quad \text{if } x \in \mathfrak{n}_+ \text{ and } n \geq 0; \\ h_\beta[n]_\zeta v &= \delta_{n,0}\delta_{\alpha,\beta}mv; \\ f_\beta[n]_\zeta v &= 0 \quad \text{if } n \geq \delta_{\alpha,\beta}; \qquad (11) \\ f_\alpha[0]_\zeta^{m+1}v &= 0. \qquad (12) \end{aligned}$$

We refer to these modules as $C_{\alpha,m}(\zeta)$.[11] Their graded version has been previously considered by Chari and Moura.[20] The graded components of the associated graded space corresponding to the filtration by homogeneous degree are, of course, $\mathfrak{g}$-modules.

Except in the case of $\mathfrak{g} = A_r$, these modules are not necessarily irreducible under restriction to the action of $\mathfrak{g}$. However, $C_{\alpha,m}(\zeta)$ does have a highest weight component isomorphic to the $\mathfrak{g}$-module $V(m\omega_\alpha)$, which appears with multiplicity 1, all other components having a smaller highest weights in the total ordering.

It has not been directly proven (except in special cases) that these modules have the same $\mathfrak{g}$-decomposition as the Yangian KR-modules, but this theorem will follow from the proof of the Feigin-Loktev conjecture below.

2.3. *Fusion products and the Fegin-Loktev conjecture*

Consider a set of cyclic $\mathfrak{g}[t]$-modules $\{V_1(\zeta_1), ..., V_N(\zeta_N)\}$ localized at pairwise distinct points in $\mathbb{C}$, $\{\zeta_1, ..., \zeta_N\}$. Denote the chosen cyclic vector of $V_i(\zeta_i)$ by v_i. If $V_i(\zeta_i)$ are finite-dimensional, so is the space $U(\mathfrak{g}[t])v_1 \otimes \cdots \otimes v_N$. Moreover, it has a finite filtration by homogeneous degree in t. The Feigin-Loktev fusion product[6] is the associated graded space of this filtration. We denote the fusion product by $\mathcal{F}^*_{\mathbf{V}}$. As the grading of $\mathcal{F}^*_{\mathbf{V}}$ is $\mathfrak{g}$-equivariant, the graded components are $\mathfrak{g}$-modules. The graded multiplicity of the irreducible $\mathfrak{g}$-module $V(\lambda)$ in the fusion product is a certain polynomial generating function for the multiplicities in the graded components.

The Feigin-Loktev conjecture is that this polynomial is independent of the localization parameters ζ_i for sufficiently well-behaved $\mathfrak{g}$-modules, and that in the case that V_i are KR-modules, the graded multiplicity of $V(\lambda)$ is equal to the M-sum formula (2). The equality was proven for $\mathfrak{sl}_2$-modules in Ref. 5 and for symmetric power representations of $\mathfrak{sl}_n$ in Ref. 21.

In this paper, we consider only fusion products of KR-modules. They are generated by the highest weight vector v localized at ζ and the relations are those in (12). Let $\mathbf{V} = (V_1, ..., V_N)$ be a collection of KR-modules of $\mathfrak{g}[t]$ localized at distinct complex numbers $\boldsymbol{\zeta} = (\zeta_1, ..., \zeta_N)$.

We parametrize the collection $\mathbf{V}$ by their highest weights $\mathbf{n} = (n_j^{(\alpha)} : 1 \leq \alpha \leq r, j \geq 0)$, meaning that in $\mathbf{V}$ there are exactly $n_j^{(\alpha)}$ KR-modules with highest weight $j\omega_\alpha$. We call this fusion product $\mathcal{F}^*_{\mathbf{n}}$.

3. Functional realization of fusion spaces

We make use of the fact that $\mathfrak{g}[t] \subset \widehat{\mathfrak{g}}$, therefore given a $\mathfrak{g}[t]$-module V, we can consider the integrable modules induced at some fixed integer level k. We choose k to be sufficiently large, so that the tensor products we consider below have a decomposition determined by the Littlewood-Richardson rule rather than the Verlinde rule. We choose integrable $\widehat{\mathfrak{g}}$-modules since they have the property that they are completely reducible. Note that smaller values of k are also of interest, and computing the graded fusion product at finite k is still an open problem for the most part.

Consider the action of products of (generating functions of) elements of the affine algebra $\widehat{\mathfrak{g}}$ on the tensor product of highest weight vectors v_i of KR-modules localized at distinct points ζ_i. We use the generating fuctions $f_\alpha(t) := \sum_{n\in\mathbb{Z}} f_\alpha[n]t^{-n-1}$, where f_α is the element of $\mathfrak{n}_-$ corresponding to the simple root α. We define $\mathcal{F}^*_{\lambda,\mathbf{n}} = \mathrm{Hom}(\mathcal{F}^*_{\mathbf{n}}, V(\lambda))$, where $\mathbf{n}$ parametrizes the set of KR-modules in the fusion product.

The dual space of $\mathcal{F}^*_{\lambda,\mathbf{n}}$ is the associated graded space of $\mathcal{C}_{\lambda,\mathbf{n}}$, consisting of all correlation functions the form

$$\langle u_\lambda | f_{\alpha_1}(t_1)\cdots f_{\alpha_M}(t_M)|v_1\otimes\cdots\otimes v_N\rangle \tag{13}$$

Here, $M\geq 0$ and $\boldsymbol{\alpha}=(\alpha_1,...,\alpha_M)\in I_r^M$ where $I_r=[1,...,r]$. The action of the currents is the fusion action of the previous section. The vector u_λ is the lowest weight vector of the module V localized at $\zeta=\infty$, dual to the highest weight module localized at 0 with $\mathfrak{g}$-highest weight λ.

This space has a filtration by the homogeneous degree in t_j, and its associated graded space is the graded multiplicity space of the $\mathfrak{g}$-module $V(\lambda)$ in the fusion product $\mathcal{F}^*_{\mathbf{n}}$.

3.1. *Characterization of functions in* $\mathcal{C}_{\lambda,\mathbf{n}}$

We now fix λ and $\mathbf{n}$, and characterize functions in the space $\mathcal{C}_{\lambda,\mathbf{n}}$ according to their symmetry, pole and zero structure:[2]

(1) **Zero weight condition:** The correlation function (13) is $\mathfrak{g}$-invariant. Therefore it must have total $\mathfrak{g}$-weight equal to 0, which means that

$$0=\ell^{(\alpha)}+\sum_{\beta,j}C_{\alpha,\beta}m_j^{(\beta)}-\sum_j jn_j^{(\alpha)},\quad 1\leq\alpha\leq r, \tag{14}$$

where $\lambda=\sum_\alpha \ell^{(\alpha)}\omega_\alpha$. This fixes $\{m^{(1)},...,m^{(r)}\}$.
For convenience, we rename the variables to keep track of the root α of the generating function in which they appear. Thus, we have functions in the variables $\{t_1,...,t_M\}=\{t_i^{(\alpha)}:\ \alpha\in I_r, 1\leq i\leq m^{(\alpha)}\}$, where $m^{(\alpha)}$ is the number of generators with root α. Note that the space $\mathcal{C}_{\lambda,\mathbf{n}}$ is the direct sum of spaces of the with fixed $\mathbf{m}=(m^{(1)},...,m^{(r)})$.

(2) **Pole structure:** Functions in $\mathcal{C}_{\lambda,\mathbf{n}}$ have at most a simple pole when $t_i^{(\alpha)}=t_j^{(\beta)}$ if $C_{\alpha,\beta}<0$. This is due to the relations in the algebra, which, in the language of generating functions, means that $f_\alpha(t)f_\beta(u)\sim\frac{f_{\alpha+\beta}(t)}{t-u}$ + non-singular terms. We are therefore led to define the less singular function $g(t)$ for each $f(t)\in\mathcal{C}_{\lambda,\mathbf{n}}$:

$$g(\mathbf{t}):=\prod_{\alpha<\beta,C_{\alpha,\beta}<0}\prod_{i,j}(t_i^{(\alpha)}-t_j^{(\beta)})f(\mathbf{t}),\quad f(\mathbf{t})\in\mathcal{C}_{\lambda,\mathbf{n}}. \tag{15}$$

(3) **Symmetry:** The function $g(\mathbf{t})$ is symmetric under the exchange $t_i^{(\alpha)} \leftrightarrow t_j^{(\alpha)}$. This is due to the fact that $[f_\alpha(t_1), f_\alpha(t_2)] = 0$.

(4) **Serre condition:** Let α, β be simple roots such that $C_{\alpha,\beta} < 0$ and define $m_{\alpha,\beta} = 1 - C_{\alpha,\beta}$. Then there is a Serre relation in $\mathfrak{g}$ (hence a corresponding relation in $\widehat{\mathfrak{g}}$) of the form $\mathrm{ad}(f_\alpha)^{m_{\alpha,\beta}} f_\beta = 0$. In generating function language,

$$f_\alpha(t_1^{(\alpha)}) \cdots f_\alpha(t_{m_{\alpha,\beta}}^{(\alpha)}) t_\beta(t_1^{(\beta)})$$

has no singularity when all the variables are set equal to each other. This implies that the function $g(\mathbf{t})$ of (15) has the following vanishing property:

$$g(\mathbf{t})\Big|_{t_1^{(\alpha)} = \cdots = t_{m_{\alpha,\beta}}^{(\alpha)} = t_1^{(\beta)}} = 0.$$

This cancels out the pole which would otherwise appear in the function $f(\mathbf{t})$.

(5) **Degree restriction:** As u_λ is a lowest weight vector of the module localized at infinity, positive currents $f_\alpha[n]_\infty$ with $n \geq 0$ act on it trivially. The action is given by taking the contour integral at infinity (see Equation (10)), or equivalently, a residue taken at 0. That is, there should be no residue when integrating $t^{n-2} f(t^{-1})$, with $n \geq 0$, at $t = 0$. This gives a degree restriction on the function $f(\mathbf{t}) \in \mathcal{C}_{\lambda,\mathbf{n}}$ for each of the variables:

$$\deg_{t_i^{(\alpha)}} f(\mathbf{t}) \leq -2.$$

(6) **Poles at ζ_i:** The relation (11) implies that $f(\mathbf{t}) \in \mathcal{C}_{\lambda,\mathbf{n}}$ may have a simple pole at $t_i^{(\alpha)} = \zeta_j$ only if the highest weight of the module localized at ζ_j is a multiple of ω_α, in accordance with the Equation (9). Otherwise, $f[n]_{\zeta_i} v(\zeta_i) = 0$ if $n \geq 0$. Moreover, we have

(7) **Integrability condition:** We assume each module V_k has highest weight $\ell_k \omega_{\alpha_k}$. The relation (12) requires that $g_2(\mathbf{t}) = \left(\prod_{\alpha,j,k} (t_j^{(\alpha)} - \zeta_k)^{\delta_{\alpha,\alpha_k}}\right) g(\mathbf{t})$ has the following vanishing property:

$$g_2(\mathbf{t})\Big|_{t_1^{(\alpha)} = \cdots = t_{\ell_k+1}^{(\alpha)} = \zeta_k} = 0.$$

These conditions characterize the space $\mathcal{C}_{\lambda,\mathbf{n}}$ completely. The only difficulty is to compute its Hilbert polynomial. This is done by introducing another filtration on the space of functions. The idea for such a filtration was first introduced by Feigin and Stoyanovsky.[22]

3.2. *Filtration of the space of functions*

Let $\boldsymbol{\lambda} = (\lambda^{(1)}, ..., \lambda^{(r)})$ where $\lambda^{(\alpha)}$ is a partition of $m^{(\alpha)}$. Let $m_a^{(\alpha)}$ denote the number of parts of $\lambda^{(\alpha)}$ equal to a. Thus, $\sum_a a m_a^{(\alpha)} = m^{(\alpha)}$. Fix a standard tableau for each partition (the result is independent of the choice of tableaux, and when we discuss a partition below we always refer to the fixed tableau) and define the evaluation map $\mathrm{ev}_\lambda : \mathcal{C}_{\lambda,\mathbf{n}}[m^{(1)}, ..., m^{(r)}] \to \mathcal{H}[\boldsymbol{\lambda}]$, where $\mathcal{H}[\boldsymbol{\lambda}]$ is the space of functions in several variables: one variable for each row of each partition in $\boldsymbol{\lambda}$.

The evaluation map is defined as follows. If the letter i appears in the jth row of length a in $\lambda^{(\alpha)}$, then $\mathrm{ev}_{\boldsymbol{\lambda}}(t_i^{(\alpha)}) = u_{a,j}^{(\alpha)}$. This is extended by linearity to $\mathcal{C}_{\lambda,\mathbf{n}}$.

We order multipartitions lexicographically, and define

$$\Gamma_{\boldsymbol{\lambda}} = \bigcap_{\boldsymbol{\mu}>\boldsymbol{\lambda}} \ker \mathrm{ev}_{\boldsymbol{\mu}}.$$

This gives a finite filtration of $\mathcal{C}_{\lambda,\mathbf{n}}$, with $\Gamma_{\boldsymbol{\mu}} \subset \Gamma_{\boldsymbol{\lambda}}$ if $\boldsymbol{\mu} < \boldsymbol{\lambda}$. We consider the image of the graded components $\Gamma_{\boldsymbol{\lambda}}/(\Gamma_{\boldsymbol{\lambda}} \cap \ker \mathrm{ev}_{\boldsymbol{\lambda}})$ under the evaluation map $\mathrm{ev}_{\boldsymbol{\lambda}}$.

Again, this is a space of functions, isomorphic to a subspace of $\mathcal{H}[\boldsymbol{\lambda}]$. Let us denote its image by $\widetilde{\mathcal{H}}[\boldsymbol{\lambda}]$. Its characterization is as follows.

(1) **Symmetry:** Functions in $\Gamma_{\boldsymbol{\lambda}}$ are symmetric in the variables $\{t_1^{(\alpha)}, ..., t_{m^{(\alpha)}}^{(\alpha)}\}$ for each α. The full symmetry is lost under the evaluation map, but the functions are still symmetric with respect to the variables labeled by rows of the same length in $\lambda^{(\alpha)}$. That is, they are symmetric with respect to the exchange of variables $\{u_{a,1}^{(\alpha)}, ..., u_{a,m_a^{(\alpha)}}^{(\alpha)}\}$ for each a, α.

(2) Functions in $\Gamma_{\boldsymbol{\lambda}}$ are in the kernel of any evaluation $ev_{\boldsymbol{\mu}}$ with $\boldsymbol{\mu} > \boldsymbol{\lambda}$, which means that functions in the image vanish whenever we set the variables corresponding to different rows of the same partition equal to each other. In fact, one can prove that

Lemma 3.1. *Functions in* $\widetilde{\mathcal{H}}[\boldsymbol{\lambda}]$ *have a factor* $(u_{a,j}^{(\alpha)} - u_{a,k}^{(\alpha)})^{2\min(a,b)}$ *for all* $j < k$.

(3) **Pole structure and Serre condition:** The pole at $t_i^{(\alpha)} = t_j^{(\beta)}$ when $C_{\alpha,\beta} < 0$, together with the vanishing condition of $g(\mathbf{t})$ which follows from from the Serre relation, implies that functions in $\widetilde{\mathcal{H}}[\boldsymbol{\lambda}]$ have a pole of order at most $\min(|C_{\alpha,\beta}|b, |C_{\alpha,\beta}|a)$ whenever $u_{a,i}^{(\alpha)} = u_{b,j}^{(\alpha)}$ (inherited from conditions (1) and (3) of the previous subsection).

(4) **Poles at ζ_j:** The pole at $t_i^{(\alpha)} = \zeta_j$ in case V_j has highest weight proportional to ω_α, together with the integrability of that module, translate to the following statement for $f \in \widetilde{\mathcal{H}}[\boldsymbol{\lambda}]$: There is a pole of order at most $\min(\ell, a)$ at $u_{a,i}^{(\alpha)} = \zeta_j$ if V_j has a highest weight equal to $\ell_j \omega_\alpha$. We define $\delta(j,\alpha) = 1$ if the highest weight of V_j is a multiple of ω_α, and $\delta(j,\alpha) = 0$ otherwise.

(5) **Degree restriction:** Functions in $\widetilde{\mathcal{H}}[\boldsymbol{\lambda}]$ have a degree in $u_{a,j}^{(\alpha)}$ which is bounded from above by $-2a$.

We do not know that these are all the conditions on functions in $\widetilde{\mathcal{H}}[\boldsymbol{\lambda}]$: The map $ev_{\boldsymbol{\lambda}}$ is injective by definition but not necessarily surjective. However, we can compute the Hilbert polynomial of the space $\mathcal{F}$ defined by the conditions above, which gives an upper bound on the Hilbert polynomial of $\widetilde{\mathcal{H}}[\boldsymbol{\lambda}]$.

To summarize, we know that $f(\mathbf{u}) \in \mathcal{F}$ has the form

$$\frac{\prod\limits_{(a,i)\neq(b,j)} (u_{a,i}^{(\alpha)} - u_{b,j}^{(\alpha)})^{\min(a,b)} \times f^0(\mathbf{u})}{\prod\limits_{a,i,j} (u_{a,i}^{(\alpha)} - \zeta_j)^{\delta(j,\alpha)\min(\ell_j,a)} \prod\limits_{a,i,b,j} \prod\limits_{\alpha<\beta:C_{\alpha,\beta}<0} (u_{a,i}^{(\alpha)} - u_{b,j}^{(\alpha)})^{\min(|C_{\alpha,\beta}|b,|C_{\beta,\alpha}|a)}},$$

where $f^0(\mathbf{u})$ is a polynomial in $\mathbf{u}$, symmetric under the exchange $u_{a,i}^{(\alpha)} \leftrightarrow u_{a,j}^{(\alpha)}$, of degree such that

$$\deg_{u_{a,j}^{(\alpha)}} f(\mathbf{u}) \leq -2a.$$

To compute the Hilbert polynomial we set all $\zeta_j = 0$ so that the function above is homogeneous in $\mathbf{u}$. That is, we compute the Hilbert polynomial of the associated graded space. It is very important to note that the values of ζ_j do not affect the value of the Hilbert polynomial, that is, there is no change in the q-dimensions of the space when we take the associated graded space.

The degree in $u_{a,j}^{(\alpha)}$ of the prefactor of $f^0(\mathbf{u})$ is $-2a - P_a^{(\alpha)}$, where $P_a^{(\alpha)}$ is defined in Equation (4). Moreover, the overall homogeneous degree of the prefactor is $Q(\mathbf{m}, \mathbf{n})$ as defined in equation (5). The Hilbert polynomial of the space of symmetric functions in m variables of degree less than or equal to p is the q-binomial coefficient,

$$\begin{bmatrix} m+p \\ m \end{bmatrix}_q = \frac{\prod_{i=1}^{m+p}(1-q^i)}{\prod_{i=1}^{m}(1-q^i)\prod_{i=1}^{p}(1-q^i)}.$$

Therefore, the Hilbert polynomial of $\mathcal{F}$ is

$$q^{Q(\mathbf{m},\mathbf{n})}\prod_{\alpha,j}\begin{bmatrix} m_j^{(\alpha)}+p_j^{(\alpha)} \\ m_j^{(\alpha)} \end{bmatrix}_q,$$

which is the upper bound (at each degree in q) of the Hilbert polynomial of $\widetilde{\mathcal{H}}[\boldsymbol{\lambda}]$, since it is a polynomial with positive coefficients.

Summing over the graded components, there follows the main Theorem:

Theorem 3.1. *(Ref. 2) The Hilbert polynomial of $\mathcal{C}_{\lambda,\mathbf{n}}$, which is the Feigin-Loktev fusion product of KR-modules, is bounded from above by $M_{\lambda,\mathbf{n}}(q)$ defined in Equation* (2).

4. Proof of the $M = N$ conjecture

In this section, we explain the proof[3] of the identity (7). For ease of readability, we explain the technique explicitly for the Lie algebra $\mathfrak{sl}_2$, and then state the key ingredients necessary in the generalization to arbitrary Lie algebras. The only difficulty in this generalization is the rapid proliferation of indices.

4.1. *The case of* $\mathfrak{sl}_2$

As explained in the introduction, one need only prove the $M = N$ identity only for the case $q = 1$ for the pentagon of identities to hold, due to the positivity of the M-sum. In the case of $\mathfrak{sl}_2$, we drop the root superscript (α) in the vacancy numbers $P_i^{(\alpha)}$ and so forth.

Fix $\mathbf{n} = (n_1, ..., n_k) \in \mathbb{Z}_+^k$ and an $\mathfrak{sl}_2$-highest weight $\ell\omega_1$ with $\ell \in \mathbb{Z}_+$. Consider the following generating function:

$$Z_{\ell,\mathbf{n}}^{(k)}(x_0, x_1) = \sum_{\mathbf{m}\in\mathbb{N}^k} x_1^{-q_0} x_0^{q_1} \prod_{i=1}^{k} \binom{m_i + q_i}{m_i} \tag{16}$$

Here, we have defined

$$q_i = \ell + \sum_{j=i+1}^{k}(j-i)(2m_j - n_j), \quad i \geq 0.$$

In particular, notice that when $q_0 = 0$, $q_i = P_i$ for all $i > 0$.

The binomial coefficient is defined as usual

$$\binom{m+p}{m} = \frac{(m+p)(m+p-1)\cdots(p+1)}{m!}.$$

This is well-defined for both negative and positive values of p.

This N-sum can be obtained from this generating function as follows. First, here and below, we note that in the N and M-sums, $m_j = 0$ if $j > k$ in (2). However all the identities we prove are valid under this restriction; since only a finite number of the m_j make a non-trivial contribution to the summation (2), one can take $k \to \infty$ at the end of the day with no loss of generality.

Second, in both the N and M sums, there is a "weight restriction" on the **m**-summation. This is equivalent to setting $q = 0$, or alternatively, considering only the constant term in x_1 in the generating function. We do not restrict the sum to $P_i \geq 0$ yet, but in the M and N sums, the variable x_0 must be set to 1.

Lemma 4.1. *There is a recursion relation,*

$$Z^{(k)}_{\ell,(n_1,..,n_k)}(x_0, x_1) = \frac{x_1^{n_1+2}}{x_0 x_2} Z^{(k-1)}_{\ell;(n_2,...,n_k)}(x_1, x_2),$$

where x_i are solutions of the A_1 Q-system or cluster algebra mutation[23] *with arbitrary boundary conditions:*

$$x_{i+1}x_{i-1} = x_i^2 - 1, \quad i \in \mathbb{Z}.$$

Proof. The variable m_1 is not part of the expression for q_1 so we can perform the summation over m_1, using the identity

$$\sum_{m_1 \geq 0} x_1^{-2m_1} \binom{m_1 + q_1}{m_1} = \left(\frac{x_1^2}{x_1^2 - 1}\right)^{q_1+1} = \left(\frac{x_1^2}{x_0 x_2}\right)^{q_1+1},$$

where we have used the Q-system in the second equality.

We separate out the dependence on m_1 in the summand, and note that

$$2q_i - q_{i-1} + 2m_i - n_i = q_{i+1}.$$

Moreover, if we denote by $q_i^{(j)}$ the function q_i with arguments being of the last $j - i$ variables in the list $(m_1, ..., m_k)$ (so that $q_i^{(k)} = q_i$), then $q_i^{(k-1)} = q_{i+1}^{(k)}$, or $q_{i-1}^{(k-1)} = q_i^{(k)}$.

We have

$$\begin{aligned}
Z^{(k)}_{\ell,(n_1,\ldots,n_k)}(x_0,x_1) &= \sum_{m_1,\ldots,m_k} x_1^{-q_0}x_0^{q_1}\prod_{j=1}^{k}\binom{m_j+q_j}{m_j} \\
&= \sum_{m_2,\ldots,m_k} x_0^{q_1}\prod_{j\geq 2}\binom{m_j+q_j}{m_j} x_1^{n_1+q_2-2q_1}\sum_{m_1} x_1^{-2m_1}\binom{m_1+q_1}{m_1} \\
&= \sum_{m_2,\ldots,m_k} x_0^{-1}x_1^{n_1+q_2+2}x_2^{-1-q_1}\prod_{j=2}^{k}\binom{m_j+q_{j-1}^{(k-1)}}{m_j} \\
&= \frac{x_1^{n_1+2}}{x_0x_2}\sum_{m_2,\ldots,m_k} x_2^{-q_0^{(k-1)}}x_1^{q_1^{(k-1)}}\prod_{j=1}^{k-1}\binom{m_{j+1}+q_j^{(k-1)}}{m_{j+1}} \\
&= \frac{x_1^{n_1+2}}{x_0x_2}Z^{(k-1)}_{\ell,(n_2,\ldots,n_k)}(x_1,x_2).
\end{aligned}$$

(Here, the superscript $(k-1)$ on q_0, q_1 means we take these variables as defined for the $k-1$ variables with indices $2, \ldots, k$.) □

Using the Lemma, by induction, we see that the generating function factorizes:

$$Z^{(k)}_{\ell,\mathbf{n}}(x_0,x_1) = \frac{x_1x_k^{\ell+1}}{x_0x_{k+1}^{\ell+1}}\prod_{i=1}^{k}x_i^{n_i}. \tag{17}$$

In particular,

$$Z^{(k)}_{\ell,\mathbf{n}}(x_0,x_1) = Z^{(p-1)}_{0,(n_1,\ldots,n_p)}(x_0,x_1)Z^{(k-p+1)}_{\ell,(n_{p+1},\ldots,n_k)}(x_{p-1},x_p). \tag{18}$$

We are interested in the constant term in x_1 in $Z^{(k)}_{\ell,\mathbf{n}}(x_0,x_1)$. We use the factorization Lemma for the first factor, and the definition via summation for the second factor:

$$Z^{(k)}_{\ell,\mathbf{n}}(x_0,x_1) = \frac{x_1x_{p-1}}{x_0x_p}\prod_{j=1}^{p-1}x_j^{n_j}\sum_{m_p,\ldots,m_k} x_p^{-q_{p-1}}x_{p-1}^{q_p}\prod_{j=p}^{k}\binom{m_j+q_j}{m_j}. \tag{19}$$

Suppose we restrict the summation in the second factor to $q_p \geq 0$ only. Moreover, we are interested in the generating function when $x_0 = 1$. In this case, all x_i are polynomials in x_1 (Chebyshev polynomials of the second type). Terms in the summation in which $q_{p-1} < 0$ are therefore products of polynomials in x_1 since there are no factors of x_i in the denominator in this case. Moreover, there is an overall factor of x_1, so that there is no constant term in x_1 in this case. Thus,

Lemma 4.2. *If the summation over* $(m_p, ..., m_k)$ *in* (19), *is restricted to* $q_p \geq 0$, *then only terms with* $q_{p-1} \geq 0$ *contribute to the constant term in* x_1 *when* $x_0 = 1$.

We use an induction argument, where the base step is clear ($q_k = \ell$), to conclude that the only terms which contribute to the constant term in x_1 are terms from the restricted summation, $q_i \geq 0$ ($i > 0$). When $q_0 = 0$, this is the $N = M$ identity, since $q_i = P_i$ in that case.

4.2. *The simply-laced case*

This case is a straightforward generalization of the $\mathfrak{sl}_2$ case[a].

We now define the generating function

$$Z^{(k)}_{\lambda,\mathbf{n}}(\mathbf{x}_0, \mathbf{x}_1) = \sum_{\mathbf{m}} \mathbf{x}_0^{\mathbf{q}_1} \mathbf{x}_1^{-\mathbf{q}_0} \prod_{\alpha,j} \binom{m_j^{(\alpha)} + q_j^{(\alpha)}}{m_j^{(\alpha)}},$$

(as is the norm, when $\mathbf{x}$ and $\mathbf{q}$ represent vectors indexed by the same set, we write $\mathbf{x}^{\mathbf{q}}$ for the product over the components.) Here, $\lambda = \sum_{\alpha=1}^{r} \ell^{(\alpha)} \omega_\alpha$, $\mathbf{n} = (n_j^{(\alpha)})_{\alpha \in I_r, j \in I_k}$, the summation is over $\mathbf{m} = \{m_j^{(\alpha)},\ \alpha \in I_r, j \in I_k\}$ non-negative integers, and we define

$$q_i^{(\alpha)} = \ell^{(\alpha)} + \sum_{j=i+1}^{k} \sum_{\beta \in I_r} (j - i)(C_{\alpha,\beta} m_j^{(\alpha)} - \delta_{\alpha,\beta} n_j^{(\alpha)}).$$

When $q_{\alpha,0} = 0$ for all α, this corresponds to the "weight restriction" (3) in the M and N-sums, and in that case, $q_i^{(\alpha)} = P_i^{(\alpha)}$ if $i > 0$. We have now $2r$ variables $\mathbf{x}_0 = (x_{1,0}, ..., x_{r,0})$ and $\mathbf{x}_1 = (x_{1,1}, ..., x_{r,1})$. The generating function is related to the M or N-sums when we evaluate the sum at $x_{\alpha,0} = 1$ and consider the constant term in $\mathbf{x}_1$.

Following the steps outlined for $\mathfrak{sl}_2$ we derive a recursion relation for the generating function:

$$Z^{(k)}_{\lambda,\mathbf{n}}(\mathbf{x}_0, \mathbf{x}_1) = \frac{\mathbf{x}_1^{2+\mathbf{n}_1}}{\mathbf{x}_0 \mathbf{x}_2} Z^{(k-1)}_{\lambda,\mathbf{n}^{(k-1)}}(\mathbf{x}_1, \mathbf{x}_2)$$

where $\mathbf{n}^{(k-1)}$ is $\mathbf{n}$ with $\mathbf{n}_1 = \mathbf{0}$. Here, we have defined $x_{\alpha,i}$ to be the solutions of the following system:

$$x_{\alpha,i+1} x_{\alpha,i-1} = x_{\alpha,i}^2 - \prod_{C_{\alpha,\beta}=-1} x_{\beta,i}.$$

[a]Below, we have two sets of indices for $\mathbf{x}$, $\mathbf{n}$ etc. When we write $\mathbf{x}_0$ we mean the collection of r elements $(x_{1,0}, ..., x_{r,0})$, and so forth.

This is called the Q-system for the simply-laced Lie algebra $\mathfrak{g}$, provided we set the initial conditions $x_{\alpha,0} = 1$. Otherwise it is a cluster algebra mutation,[23] and therefore, under these special initial conditions, all its solutions are polynomials in the variables $x_{\beta,1}$.[24]

We again repeat the arguments of the previous section to factorize the generating function:

$$Z_{\lambda,\mathbf{n}}^{(k)}(\mathbf{x}_0, \mathbf{x}_1) = \prod_{\alpha} \frac{x_{\alpha,1} x_{\alpha,k}^{\ell_\alpha+1}}{x_{\alpha,0} x_{\alpha,k+1}^{\ell_\alpha+1}} \prod_{j=1}^{k} x_{\alpha,j}^{n_j^{(\alpha)}},$$

from which we deduce that

$$Z_{\lambda,\mathbf{n}}^{(k)}(\mathbf{x}_0, \mathbf{x}_1) = \frac{\mathbf{x}_1 \mathbf{x}_{p-1}}{\mathbf{x}_0 \mathbf{x}_p} \prod_{j=1}^{p-1} \mathbf{x}_j^{\mathbf{n}_j} \sum_{\mathbf{m}^{(p)}} \mathbf{x}_p^{-\mathbf{q}_{p-1}} \mathbf{x}_{p-1}^{\mathbf{q}_p} \prod_{j=p}^{k} \binom{\mathbf{m}_j + \mathbf{q}_j}{\mathbf{q}_j}.$$

Here, $\mathbf{m}^{(p)}$ are the last $k-p+1$ components of the list $(\mathbf{m}_1, ..., \mathbf{m}_k)$. A binomial coefficient with vector entries is notation for the product of binomial coefficients over the components.

Suppose we restrict the summation to $\mathbf{m}^{(p)}$ such that $q_p^{(\alpha)} \geq 0$ for some α, and such that $q_{p-1}^{(\alpha)} < 0$ for the same α. We look for a contribution to the constant term in $x_{\alpha,1}$. All $\mathbf{x}_i$ are polynomials in $\mathbf{x}_1$ after evaluation at $x_{\alpha,0} = 1$ for all α. Terms with $q_p^{(\alpha)} < 0$ do not have a factor $x_{\alpha,p}$ in the denominator, and are therefore polynomials in $x_{\alpha,i}$ for several i and fixed α. One can show that $\prod_{\beta\neq\alpha} x_{\beta,p}^{-1}$ has no negative powers of $x_{\alpha,1}$ (see Ref. 3, Lemma 4.8). Therefore we have a polynomial in $x_{\alpha,1}$, with an overall power of $x_{\alpha,1}$, hence there is no constant term in $x_{\alpha,1}$. We repeat this argument for each α and inductively for each p starting from $p = k$, until we get

Lemma 4.3. *There is no contribution to the constant term in* $\mathbf{x}_1$ *in the summation from terms with* $q_j^{(\alpha)} < 0$ *for any* p, j, *hence from terms with* $P_j^{(\alpha)} < 0$ *when we consider the terms with* $q_0^{(\alpha)} = 0$.

This implies that $M = N$ for the simply-laced Lie algebras.

4.3. *The non-simply laced case*

This case is less elegantly derived, as it requires the introduction of even more variables in the generating function, and each case must be treated separately. Nevertheless, the argument goes through in the same (more involved) manner. In the process we must define the set of variables which satisfy the following system of equations:

$$x_{\alpha,i+1}x_{\alpha,i-1} = x_{\alpha,i} - \prod_{\beta:C_{\alpha,\beta}<0} \prod_{j=0}^{-C_{\alpha,\beta}-1} x_{\beta,\lfloor (C_{\beta,\alpha}i+j)/|C_{\alpha,\beta}|\rfloor}.$$

If $x_{\alpha,0} = 1$ for all α, then the equation for $i > 0$ is known as the Q-system (for the simple Lie algebra with Cartan matrix C), and it is known to be satisfied by the characters $x_{\alpha,i}$ of the KR-modules $\mathrm{KR}_{\alpha,i}$ if $x_{\alpha,1}$ is the character of the fundamental module.

We find in the $k \to \infty$ limit that

Theorem 4.1. *(Ref. 3) For any simple Lie algebra and λ a dominant weight, $\mathbf{n}$ a vector in $\mathbb{Z}_+^{r\times k}$, $M_{\lambda,\mathbf{n}} = N_{\lambda,\mathbf{n}}$.*

5. Summary

Prior to the work described in the previous section, it was known that for any simple Lie algebra, the multiplicity of the $U_q(\mathfrak{g})$-module with $\mathfrak{g}$ highest weight λ in the tensor product of Kirillov-Reshetikhin modules is the N-sum formula. This followed theorems of Hatayama et al[8] and Nakajima's theorem about the q-characters for T-systems corresponding to simply-laced Lie algebras,[9] as well as the extension by Hernandez for other algebras.[10]

We now have all the equalities in the pentagon of identities. That is, since we have proven that $M = N$, we have proven also the following:

Corollary 5.1. *The multiplicity of the $\mathfrak{g}$-module $V(\lambda)$ in the tensor product of Chari's KR modules of $\mathfrak{g}[t]$ is equal to the multiplicity of the $U_q(\mathfrak{g})$ module with $\mathfrak{g}$-highest weight λ in the corresponding tensor product of $U_q(\widehat{\mathfrak{g}})$ of Kirillov-Reshetikhin type.*

Corollary 5.2. *The Hilbert polynomial of the graded multiplicity space of $V(\lambda)$ in the Feigin-Loktev fusion product is the fermionic M-sum (generalized Kostka polynomials in the case of A_n). This is the Feigin-Loktev conjecture.*

Corollary 5.3. *The Bethe integer sets (parametrizing Bethe vectors) in the generalized Heisenberg model as solved by Kirillov and Reshetikhin are in bijection with vectors in the Hilbert space of the model, and therefore the completeness conjecture holds.*

We should remark that although it is well known that not all Bethe states come from the so-called "string hypothesis" in these models, nevertheless this gives a good counting of the states.

The proof described in the previous section shows that the vanishing of the "non-positive" components of the N-sum formula is due to the fact that the solutions of the Q-system with the KR-boundary condition are polynomials in the initial data $x_{\alpha,1}$. This fact is clear, once one refers to the theorem that the solutions $x_{\alpha,n}$ with $n > 0$ are characters of KR-modules, which are in the Grothendieck group generated by $\{x_{1,1}, ..., x_{r,1}\}$ (the characters of the r fundamental representations). However these facts are not immediately obvious without resorting to the proven theorems on the subject. The cluster algebra formulation of the Q-system gives an entirely combinatorial interpretation for this fact.[23,24]

The polynomiality property is quite general for a much larger class of cluster algebras, under even more general boundary conditions, which give a certain vanishing of the numerators.[24] For example, the same property holds for the T-systems.

6. Acknowlegments

I am deeply indebted to T. Miwa for so generously sharing his knowledge and experience over the span of many years. I also thank my collaborators in completing the series of proofs, Eddy Ardonne and Philippe Di Francesco. This work is supported by NSF grant DMS-0802511.

References

1. E. Ardonne, R. Kedem and M. Stone, *J. Phys. A* **38**, 9183 (2005).
2. E. Ardonne and R. Kedem, *J. Algebra* **308**, 270 (2007).
3. P. D. Francesco and R. Kedem, *Int. Math. Res. Not. IMRN* , Art. ID rnn006, 57 (2008).
4. B. Feigin, R. Kedem, S. Loktev, T. Miwa and E. Mukhin, *Compositio Math.* **134**, 193 (2002).
5. B. Feigin, M. Jimbo, R. Kedem, S. Loktev and T. Miwa, *Duke Math. J.* **125**, 549 (2004).
6. B. Feigin and S. Loktev, On generalized Kostka polynomials and the quantum Verlinde rule, in *Differential topology, infinite-dimensional Lie algebras, and applications*, , Amer. Math. Soc. Transl. Ser. 2 Vol. 194 (Amer. Math. Soc., Providence, RI, 1999) pp. 61–79.
7. A. N. Kirillov and N. Y. Reshetikhin, *Zap. Nauchn. Sem. Leningrad. Otdel. Mat. Inst. Steklov. (LOMI)* **160**, 211 (1987).
8. G. Hatayama, A. Kuniba, M. Okado, T. Takagi and Y. Yamada, Remarks on fermionic formula, in *Recent developments in quantum affine algebras and*

related topics (Raleigh, NC, 1998), , Contemp. Math. Vol. 248 (Amer. Math. Soc., Providence, RI, 1999) pp. 243–291.
9. H. Nakajima, *Represent. Theory* **7**, 259 (2003).
10. D. Hernandez, *J. Reine Angew. Math.* **596**, 63 (2006).
11. V. Chari, *Internat. Math. Res. Notices* , 629 (2001).
12. S. V. Kerov, A. N. Kirillov and N. Y. Reshetikhin, *Zap. Nauchn. Sem. Leningrad. Otdel. Mat. Inst. Steklov. (LOMI)* **155**, 50 (1986).
13. A. N. Kirillov, A. Schilling and M. Shimozono, *Selecta Math. (N.S.)* **8**, 67 (2002).
14. A. Schilling, *J. Algebra* **285**, 292 (2005).
15. M. Okado, A. Schilling and M. Shimozono, A crystal to rigged configuration bijection for nonexceptional affine algebras, in *Algebraic combinatorics and quantum groups*, (World Sci. Publ., River Edge, NJ, 2003) pp. 85–124.
16. A. Schilling and M. Shimozono, *J. Algebra* **295**, 562 (2006).
17. D. Hernandez, *J. Reine Angew. Math.* **596**, 63 (2006).
18. A. Kuniba, T. Nakanishi and J. Suzuki, *Internat. J. Modern Phys. A* **9**, 5215 (1994).
19. A. N. Kirillov and M. Shimozono, *J. Algebraic Combin.* **15**, 27 (2002).
20. V. Chari and A. Moura, *Comm. Math. Phys.* **266**, 431 (2006).
21. R. Kedem, *Int. Math. Res. Not.* , 1273 (2004).
22. A. V. Stoyanovskiĭ and B. L. Feĭgin, *Funktsional. Anal. i Prilozhen.* **28**, 68 (1994).
23. R. Kedem, *J. Phys. A* **41**, 194011, 14 (2008).
24. P. Di Francesco and R. Kedem, *Lett. Math. Phys.* **89**, 183 (2009).

TBA FOR THE TODA CHAIN

KAROL KAJETAN KOZLOWSKI and JÖRG TESCHNER
DESY Theory, Notkestr. 85, 22603 Hamburg, Germany

We give a direct derivation of a proposal of Nekrasov-Shatashvili concerning the quantization conditions of the Toda chain. The quantization conditions are formulated in terms of solutions to a nonlinear integral equation similar to the ones coming from the thermodynamic Bethe ansatz. This is equivalent to extremizing a certain function called Yang's potential. It is shown that the Nekrasov-Shatashvili formulation of the quantization conditions follows from the solution theory of the Baxter equation, suggesting that this way of formulating the quantization conditions should indeed be applicable to large classes of quantized algebraically integrable models.

1. Introduction

The N-body quantum mechanical Hamiltonian

$$\mathbf{H}=\sum_{\ell=1}^{N}\frac{\boldsymbol{p}_\ell^2}{2}+\left(\kappa g^2\right)^{\hbar}\mathrm{e}^{\boldsymbol{x}_N-\boldsymbol{x}_1}+\sum_{k=2}^{N}g^{2\hbar}\mathrm{e}^{\boldsymbol{x}_{k-1}-\boldsymbol{x}_k}\,, \tag{1}$$

is known as the quantum Toda chain. Above, $\boldsymbol{p}_\ell$ and $\boldsymbol{x}_k$ are quantum observables satisfying the canonical commutation relations $[\boldsymbol{p}_\ell,\boldsymbol{x}_k]=i\hbar\delta_{\ell,k}$, g and κ are coupling constant. When $\kappa=1$ one deals with the so-called closed Toda chain and, when $\kappa=0$, with the open Toda chain.

This model appears to be a prototype for an interesting class of integrable models called algebraically integrable models. It was introduced and solved, on the classical level, by Toda[12] in 1967. Then, in 1977, Olshanetsky and Perelomov[9] constructed the set of N commuting and independent integrals of motion for the closed chain, thus proving the so-called quantum integrability of the model. In 1980-81, Gutzwiller[6] was able to build explicitly the eigenfunctions and write down the quantization conditions for small numbers of particles ($N=2,3,4$). In particular he expressed the eigenfunctions of the closed chain with N-sites as a linear combination of the eigenfunctions of the open chain with $N-1$ particles.

A particularly succesful approach to the solution of the quantum Toda chain was initiated by Sklyanin. In 1985, Sklyanin applied the quantum inverse scattering method (QISM) to the study of the Toda chain. This led to the development of the so-called quantum separation of variables method. In this novel framework, he was able to obtain the Baxter equations for the model, from which Gutzwiller's quantization conditions can be obtained. The Baxter equation was later re-derived by Pasquier and Gaudin[5] with the help of an explicit construction of the so-called Q-operator, similar to the method developed by Baxter for the solution of the eight vertex model.[2]

In '99 Kharchev and Lebedev[7] constructed the multiple integral representations for the eigenfunctions of the closed N-particle Toda chain. Their construction can be seen as a generalization of Gutzwiller's solution allowing one to express the eigenfunctions of the closed N-particle Toda chain in terms of those of the open chain with $N-1$ particles and this for all values of N. In '09, An[1] completed the picture by proving rigorously that Gutzwiller's quantization conditions are necessary and sufficient for obtaining a state in the spectrum.

However, the form of the quantization conditions obtained in the abovementioned works appears to be rather involved. Recently Nekrasov and Shatashvili proposed in Ref. 8 that the quantization conditions for the Toda chain can be reformulated in terms of the solutions to a nonlinear integral equation (NLIE) similar to the equations originating in the thermodynamic Bethe ansatz method. With the help of the solutions to the relevant nonlinear integral equation, Nekrasov and Shatashvili defined a function $\mathcal{W}$ whose critical points are in a one-to-one correspondence with the simultaneous eigenstates of the conserved quantitites. This formulation not only seems to be in some respects more efficient than the previous one, it also indicates an amazing universality of the form the quantization conditions may take in integrable models.

The proposal of Ref. 8 was based on rather indirect arguments coming from the study of supersymmetric gauge theories. It seems desirable to derive the proposal more directly from the integrable structure of the model. Our main aim in this note is to give such a derivation. It is obtained from the solution theory of the Baxter equation. In other integrable models there are known connections between the Baxter equation and nonlinear integral equations that look similar to the one that appears here, see e.g. Refs. 3,11,14. However, the precise form of the NLIE depends heavily on the analytic properties that the relevant solutions of the Baxter equation must have in the different models. In the present case of a particle system we encounter an interesting new feature: the quantization conditions are not formulated as equations on the zeros of the solutions of the Baxter equation, but instead, they are equations on the poles of the so-called quantum Wronskian formed from two linearly independent solutions of the Baxter equation. The positions $\delta = (\delta_1, \ldots, \delta_N)$ of these poles are the variables that the Yang's potential $\mathcal{W} = \mathcal{W}(\delta)$ depends on.

It is worth stressing that the Baxter equation or generalizations thereof have a good chance to figure as a universal tool for the study of the spectrum of quantum integrable models. Our method of derivation strongly indicates that similar formulations of the quantization conditions should exist for large classes of quantized *algebraically* integrable models.

This article is organized as follows. We first recall how the Separation of Variables method reduces the problem of finding the eigenstates of the Toda chain to the problem to find a certain set of solutions to the Baxter functional equation specified by strong conditions on the analyticity and the asymptotics of its elements. Then, in Section 3 we explain how Gutzwiller's quantization conditions can be reformulated in terms of the solutions to a certain NLIE and in terms of the Yang's potential $\mathcal{W}(\delta)$. The derivation of this reformulation is sketched. The proofs of our claims are presented in the Appendices. In Appendix A, we establish the relevant properties of Gutzwiller's basis of fundamental solutions to the T-Q equation. Then in Appendix B we prove the existence and uniqueness of solutions to the NLIE introduced in Section 3. In Appendix C we give rigorous proofs of the main results presented in Section 3.

Acknowledgements. The authors gratefully acknowledge support from the EC by the Marie Curie Excellence Grant MEXT-CT-2006-042695.

We are happy to dedicate this paper to T. Miwa on the occasion of his 60th birthday.

2. Separation of variables approach to the Toda chain

2.1. *Integrability of the Toda chain*

The integrability of the Toda chain follows from the existence of Lax matrices

$$L_n(\lambda) = \begin{pmatrix} \lambda - \boldsymbol{p}_n & g^{\hbar}\,\mathrm{e}^{-\boldsymbol{x}_n} \\ -g^{\hbar}\,\mathrm{e}^{\boldsymbol{x}_n} & 0 \end{pmatrix} \qquad [\,\boldsymbol{x}_n\,,\,\boldsymbol{p}_n\,] = i\,, \tag{2}$$

satisfying a Yang-Baxter equation with a rational, six-vertex type, R-matrix. Thus, the set of κ-twisted monodromy matrices

$$M(\lambda) = \begin{pmatrix} 1 & 0 \\ 0 & \kappa^{\hbar} \end{pmatrix} L_N(\lambda) \dots L_1(\lambda) , \tag{3}$$

allows one to build the transfer matrix $\mathbf{T}(\lambda) = \mathrm{tr}[M(\lambda)]$ which is the generating function of the set of N commuting Hamiltonians associated with the Toda chain

$$\mathbf{T}(\lambda) = \lambda^N + \sum_{k=0}^{N-1} (-1)^k \lambda^{N-k} \mathbf{H}_k . \tag{4}$$

The first two Hamiltonians read $\mathbf{H}_1 = \sum_{k=1}^{N} \boldsymbol{p}_k = \mathbf{P}$ and

$$\mathbf{H}_2 = \frac{\mathbf{P}^2}{2} - \left\{ \sum_{\ell=1}^{N} \frac{\boldsymbol{p}_\ell^2}{2} + g^{2\hbar} \kappa^{\hbar} \mathrm{e}^{\boldsymbol{x}_N - \boldsymbol{x}_1} + \sum_{k=2}^{N} g^{2\hbar} \mathrm{e}^{\boldsymbol{x}_{k-1} - \boldsymbol{x}_k} \right\} . \tag{5}$$

The eigenvalues $\boldsymbol{t}(\lambda)$ of the transfer matrix $\mathbf{T}(\lambda)$ are thus polynomials $\boldsymbol{t}(\lambda) = \prod_{k=1}^{N} (\lambda - \tau_k)$. The N commuting Hamiltonians are self-adjoint, hence the set $\{\tau\}$ is necessarily self conjugated: $\{\tau_k\} = \{\overline{\tau}_k\}$.

2.2. *Separation of variables*

The Separation of Variables (SOV) method was developed for the Toda chain in Refs. 1,5,7,10. The main results of these works may be summarized as follows:

The wave-functions $\Psi_{\boldsymbol{t}}(x)$, $x = (x_1, \dots, x_N)$ of any eigenstate to the transfer matrix $\mathbf{T}(\lambda)$ with eigenvalue $\boldsymbol{t}(\lambda)$ can be represented by means of an integral transformation of the form

$$\Psi_{\boldsymbol{t}}(x) = \int_{\mathbb{R}^{N-1}} \mathrm{d}\mu(\gamma) \, \Phi_{\boldsymbol{t}}(\gamma) \, \Xi_P(\gamma|x) , \tag{6}$$

where integration is over vectors $\gamma = (\gamma_1, \dots, \gamma_{N-1}) \in \mathbb{R}^{N-1}$ with respect to a measure $\mathrm{d}\mu(\gamma)$ first found in Ref. 10, $\Xi_P(\gamma|x)$ is an integral kernel for which the explicit expression can be found in Ref. 7, P is the eigenvalue of the center of mass momentum $\mathbf{P}$ in the state $\Psi_{\boldsymbol{t}}$, and $\Phi_{\boldsymbol{t}}(\gamma)$ is the wave-function in the so-called SOV-representation. The key feature of the SOV representation is that $\Phi_{\boldsymbol{t}}(\gamma)$ takes a factorized form

$$\Phi_{\boldsymbol{t}}(\gamma) = \prod_{k=1}^{N-1} q_{\boldsymbol{t}}(\gamma_k) . \tag{7}$$

The integral transformation (6) is constructed in such a way that the eigenvalue equation for the family of operators $\mathbf{T}(\lambda)$ is equivalent to the fact that the function $q_{\boldsymbol{t}}(y)$ which represents the state $\Psi_{\boldsymbol{t}}$ *via* (6) and (7) satisfies the so-called Baxter equation,

$$\boldsymbol{t}(\lambda)q_{\boldsymbol{t}}(\lambda) = i^N g^{N\hbar} q_{\boldsymbol{t}}(\lambda + i\hbar) + \kappa^{\hbar}(-i)^N g^{N\hbar} q_{\boldsymbol{t}}(\lambda - i\hbar)\,. \tag{8}$$

The integral transformation (6) can be inverted to express $\Phi_{\boldsymbol{t}}(\gamma)$ in terms of $\Psi_{\boldsymbol{t}}(x)$. In this way it becomes possible to find the necessary and sufficient conditions that $q_t(y)$ has to satisfy in order to represent an eigenstate of $\mathbf{T}(\lambda)$ *via* (6) and (7). The conditions are[1,7]

(i) $\boldsymbol{t}(\lambda)$ is a polynomial of the form $\boldsymbol{t}(\lambda) = \prod_{k=1}^{N}(\lambda - \tau_k)$, with $\{\tau_k\} = \{\overline{\tau}_k\}$.

(ii) $q(\lambda)$ is entire and has asymptotic behavior

$$q\,(\lambda) = \mathrm{O}\big(\mathrm{e}^{-\frac{N\pi}{2\hbar}|\Re(\lambda)|}\,|\lambda|^{\frac{N}{2\hbar}(2|\Im(\lambda)|-\hbar)}\,\big).$$

Above, the O symbol is uniform in the strip $\{z : |\Im(z)| \leq \hbar/2\}$. This reduces the problem to construct all the eigenfunctions and finding the complete spectrum of the Toda chain to finding the set $\mathbb{S}$ of all solutions $(\boldsymbol{t}\,(\lambda)\,, q_{\boldsymbol{t}}\,(\lambda))$ to the Baxter equation (8) that satisfy the conditions (i) and (ii) above.

3. Quantization conditions

It turns out that the Baxter equation (8) admits solutions within the class $\mathbb{S}$ described above only for a discrete set of choices for the polyonomial $\boldsymbol{t}(\lambda)$. This is what expresses the quantization of the spectrum of $\mathbf{T}(\lambda)$ within the SOV-framework. Our first aim in this section will be to outline how to reformulate the resulting conditions on $\boldsymbol{t}(\lambda)$ more concretely, following the approaches initiated by Gutzwiller and Pasquier-Gaudin.

It gives useful insight to divide the problem to construct and classify the solutions to the Baxter equation (8) which satisfy (i) and (ii) into two steps. In the first step, one weakens the analytic requirements (ii) on $q(\lambda)$ by allowing $q(\lambda)$ to have a certain number of poles. In this case, it will be possible to find two linearly independent solutions $q_{\boldsymbol{t}}^{\pm}(\lambda)$ to (8) for arbitrary $\boldsymbol{t}(\lambda)$ satisfying (i). In the second step, one constructs the solution $q(\lambda)$ satisfying (i) and (ii) in the form

$$q(\lambda) = P_+ q_{\boldsymbol{t}}^{+}(\lambda) + P_- q_{\boldsymbol{t}}^{-}(\lambda)\,, \tag{9}$$

where $P_\pm$ are constants. The requirement that $q(\lambda)$ is entire means that the poles of $q_{\boldsymbol{t}}^{\pm}(\lambda)$ must cancel each other in (9) which is only possible if

$\boldsymbol{t}(\lambda)$ is fine-tuned in a suitable way. This is the origin of the quantization of the spectrum of $\mathbf{T}(\lambda)$.

3.1. *Gutzwiller's formulation of the quantization conditions*

It turns out that there is a canonical minmal choice for the set of poles of $q_{\boldsymbol{t}}^{\pm}(\lambda)$ that one needs to allow. One needs to allow N poles $\delta_1, \ldots, \delta_N$ whose positions are determined by the choice of $\boldsymbol{t}(\lambda)$. More precisely, out of $\boldsymbol{t}(\lambda)$ one constructs the so-called Hill determinant

$$\mathcal{H}(\lambda) = \det \begin{bmatrix} \cdots & \ddots & \ddots & \ddots & \cdots & \cdots \\ \cdots & \dfrac{\rho^{\hbar}}{\boldsymbol{t}(\lambda - i\hbar)} & 1 & \dfrac{1}{\boldsymbol{t}(\lambda - i\hbar)} & 0 & \cdots \\ \cdots & 0 & \dfrac{\rho^{\hbar}}{\boldsymbol{t}(\lambda)} & 1 & \dfrac{1}{\boldsymbol{t}(\lambda)} & 0 \ldots \\ \ddots & \ddots & \ddots & \ddots & \ddots & \ddots \end{bmatrix}, \tag{10}$$

where $\rho := \kappa g^{2N}$. It can be shown that $\mathcal{H}(\lambda)$ admits the representation

$$\mathcal{H}(\lambda) = \prod_{a=1}^{N} \frac{\sinh \frac{\pi}{\hbar}(\lambda - \delta_a)}{\sinh \frac{\pi}{\hbar}(\lambda - \tau_a)}, \tag{11}$$

where one chooses $|\Im(\delta_k)| < \hbar/2$. This defines $\delta_1, \ldots, \delta_N$ in terms of $\boldsymbol{t}(\lambda)$.

Let us then, instead of $\mathbb{S}$ consider the class $\mathbb{S}'$ of solutions to (8) which satisfy the conditions (i) and (ii)$'$,

(ii)$'$ $q(\lambda)$ is meromorphic with set of poles contained in $\{\delta_1, \ldots, \delta_N\}$ and it has an asymptotic behavior $|q(\lambda)| = \mathrm{O}\big(\mathrm{e}^{-\frac{N\pi}{2\hbar}|\Re(\lambda)|}\,|\lambda|^{\frac{N}{2\hbar}(2|\Im(\lambda)|-\hbar)}\big)$.

The Baxter equation (8) has two linearly independent solutions $q_{\boldsymbol{t}}^{\pm}(\lambda)$ within $\mathbb{S}'$ for arbitrary $\boldsymbol{t}(\lambda)$. One possible construction of the solutions $q_{\boldsymbol{t}}^{\pm}(\lambda)$ goes back to Gutzwiller's work on the Toda chain. They may be defined as follows:

$$q_{\boldsymbol{t}}^{\pm}(\lambda) = \frac{Q_{\boldsymbol{t}}^{\pm}(\lambda)}{\prod_{a=1}^{N} \{\mathrm{e}^{-\frac{\pi\lambda}{\hbar}} \sinh \frac{\pi}{\hbar}(\lambda - \delta_a)\}}, \tag{12}$$

where

$$
\begin{aligned}
Q_{\boldsymbol{t}}^{+}(\lambda) &= \frac{\left(\kappa g^{N}\right)^{-i\lambda} K_{+}(\lambda)\, \mathrm{e}^{-N\frac{\pi}{\hbar}\lambda}}{\prod_{k=1}^{N} \hbar^{-i\frac{\lambda}{\hbar}} \Gamma(1-i(\lambda-\tau_k)/\hbar)}\,, \\
Q_{\boldsymbol{t}}^{-}(\lambda) &= \frac{g^{iN\lambda} K_{-}(\lambda)\, \mathrm{e}^{-N\frac{\pi}{\hbar}\lambda}}{\prod_{k=1}^{N} \hbar^{i\frac{\lambda}{\hbar}} \Gamma(1+i(\lambda-\tau_k)/\hbar)}\,,
\end{aligned}
\tag{13}
$$

with $K_{\pm}(\lambda)$ being half-infinite determinants:

$$
K_{+}(\lambda) = \det \begin{bmatrix} 1 & \boldsymbol{t}^{-1}(\lambda+i\hbar) & 0 & \cdots & \\ \frac{\rho^{\hbar}}{\boldsymbol{t}(\lambda+2i\hbar)} & 1 & \boldsymbol{t}^{-1}(\lambda+2i\hbar) & 0 & \cdots \\ 0 & \ddots & \ddots & \ddots & \ddots \;\cdots \end{bmatrix}
\tag{14}
$$

and $K_{-}(\lambda) = \overline{K_{+}(\overline{\lambda})}$ (recall that $\{\tau_k\} = \{\overline{\tau}_k\}$). For the reader's convenience we have included a self-contained proof that $q_{\boldsymbol{t}}^{\pm} \in \mathbb{S}'$ in Appendix A. It's worth noting that $Q_{\boldsymbol{t}}^{\pm}$ are linearly independent entire functions whose Wronskian can be evaluated explicitly, *cf.* Lemma 1:

$$
\begin{aligned}
&Q_{\boldsymbol{t}}^{+}(\lambda) Q_{\boldsymbol{t}}^{-}(\lambda+i\hbar) - Q_{\boldsymbol{t}}^{+}(\lambda+i\hbar)\, Q_{\boldsymbol{t}}^{-}(\lambda) = \\
&\quad = \kappa^{-i\lambda} g^{-N\hbar} \mathrm{e}^{-2N\frac{\pi}{\hbar}\lambda} \prod_{a=1}^{N} \left\{ \frac{\hbar}{i\pi} \sinh \frac{\pi}{\hbar}(\lambda-\tau_k)\, \mathcal{H}(\lambda) \right\}.
\end{aligned}
\tag{15}
$$

It can be shown that the most general solution $q(\lambda) \in \mathbb{S}'$ to the Baxter equation (8) may be represented in the form (9). The additional requirement that $q(\lambda)$ should be entire implies

$$
Q_{\boldsymbol{t}}^{+}(\delta_a) - \zeta\, Q_{\boldsymbol{t}}^{-}(\delta_a) = 0\,, \quad \text{for} \quad a = 1, \ldots, N \text{ and } \zeta \in \mathbb{C},\ |\zeta| = 1\,, \tag{16}
$$

to be supplemented by the condition that $\sum_{k=1}^{N} \delta_k = P$.[5,6] This formulation of the quantization conditions looks fairly involved. It may be considered as a highly transcendental system of equations on the parameters τ_k determining $\boldsymbol{t}(\lambda)$, in which both $Q_{\boldsymbol{t}}^{\pm}(\lambda)$ and $\delta_1, \ldots, \delta_N$ have to be constructed from $\boldsymbol{t}(\lambda)$ by means of (13) and (10), (11) respectively.

3.2. *Reformulation in terms of solutions to a nonlinear integral equation*

One may note that the set of parameters in $\delta = (\delta_1, \ldots, \delta_N)$ is just as big as the set of parameters in $\tau = (\tau_1, \ldots, \tau_N)$ characterizing the polynomial $\boldsymbol{t}(\lambda)$ appearing on the left hand side of the Baxter equation. The form of the quantization conditions (16) suggests that it may be useful to formulate these conditions directly in terms of the parameters $\delta = (\delta_1, \ldots, \delta_N)$ with $\boldsymbol{t}(\lambda) = \boldsymbol{t}(\lambda|\tau(\delta))$ being determined in terms of δ by inverting the relation $\delta = \delta(\tau)$. A more convenient representation of the quantization conditions (16) would then be obtained if one was able to construct the solutions $Q_{\boldsymbol{t}}^{\pm}(\lambda)$ more directly as functions of the parameters $\delta = (\delta_1, \ldots, \delta_N)$. In the following, for a given polynomial $\vartheta(\lambda) = \prod_{k=1}^{N}(\lambda - \delta_k)$ with complex conjugated roots, we will construct functions $Q_{\delta}^{\pm}(\lambda)$. These will be shown to yield solutions the Baxter equation (8) *via* (12), with $t_{\delta}(\lambda)$ being a polynomial whose coefficients depend on the parameters δ.

The functions $Q_{\delta}^{\pm}(\lambda)$ will be build out of the solutions $Y_{\delta}(\lambda)$ to the following NLIE,

$$\log Y_{\delta}(\lambda) = \int_{\mathbb{R}} d\mu \, K(\lambda - \mu) \log\left(1 + \frac{\rho^{\hbar} \, Y_{\delta}(\mu)}{\vartheta(\mu - i\hbar/2)\vartheta(\mu + i\hbar/2)}\right), \tag{17}$$

where

$$K(\lambda) = \frac{\hbar}{\pi(\lambda^2 + \hbar^2)} \,. \tag{18}$$

It will be shown in Appendix B that the solutions $Y_{\delta}(\lambda)$ to (17) are unique, and that they exist for all tuples $\delta = (\delta_1, \ldots, \delta_N)$ of zeros of Hill determinants $\mathcal{H}(\lambda)$ constructed from polynomials $\boldsymbol{t}(\lambda)$ whose zeroes τ_k satisfy $|\Im(\tau_k)| < \hbar/2$. The function $Y_{\delta}(\lambda)$ is meromorphic, with its poles accumulating in the direction $|\arg(\lambda)| = \pi/2$ and such that $Y_{\delta} \to 1$ if $\lambda \to \infty$ for λ uniformly away from its set of poles. The properties of Y_{δ} allow one to define two auxiliary functions:

$$\begin{aligned}
\log v_{\uparrow}(\lambda) &= -\int_{\mathbb{R}} \frac{\mathrm{d}\mu}{2i\pi} \frac{1}{\lambda - \mu + i\hbar/2} \log\left(1 + \frac{\rho^{\hbar} Y_{\delta}(\mu)}{\vartheta(\mu - i\hbar/2)\,\vartheta(\mu + i\hbar/2)}\right), \\
\log v_{\downarrow}(\lambda - i\hbar) &= \int_{\mathbb{R}} \frac{\mathrm{d}\mu}{2i\pi} \frac{1}{\lambda - \mu - i\hbar/2} \log\left(1 + \frac{\rho^{\hbar} Y_{\delta}(\mu)}{\vartheta(\mu - i\hbar/2)\,\vartheta(\mu + i\hbar/2)}\right).
\end{aligned} \tag{19}$$

Out of $v_{\uparrow}(\lambda)$ and $v_{\downarrow}(\lambda - i\hbar)$, we may then construct

$$\begin{aligned}
Q_{\delta}^{+}(\lambda) &= \frac{\left(\kappa g^{N}\right)^{-i\lambda} \hbar^{i\frac{N\lambda}{\hbar}} \mathrm{e}^{-\frac{N\pi}{\hbar}\lambda} v_{\uparrow}(\lambda)}{\prod_{k=1}^{N} \Gamma\left(1 - i(\lambda - \delta_k)/\hbar\right)}, \\
Q_{\delta}^{-}(\lambda) &= \frac{g^{iN\lambda} \hbar^{-i\frac{N\lambda}{\hbar}} \mathrm{e}^{-\frac{N\pi}{\hbar}\lambda} v_{\downarrow}(\lambda - i\hbar)}{\prod_{k=1}^{N} \Gamma\left(1 + i(\lambda - \delta_k)/\hbar\right)}.
\end{aligned} \tag{20}$$

It is shown in Appendix C that the functions $Q_\delta^\pm$ are entire (*cf.* Lemma 4). It is also shown there that the functions $q_\delta^\pm(\lambda)$ defined from $Q_\delta^\pm(\lambda)$ by relations like (12) are solutions to the Baxter equation (8) which have the correct asymptotic behavior so as to be contained in $\mathbb{S}'$.

One may therefore take the construction of $Q_\delta^\pm(\lambda)$ from the solutions of the nonlinear integral equation (17) as a replacement for the construction based on Gutzwillers solutions. The quantization conditions (16) may now be rewritten in terms of $Y_\delta(\lambda)$ in the form

$$2\pi n_k = \frac{N\delta_k}{\hbar}\log\hbar - \delta_k\log\rho + i\log\zeta - i\sum_{p=1}^{N}\log\frac{\Gamma(1+i\,(\delta_k-\delta_p)\,/\hbar)}{\Gamma(1-i\,(\delta_k-\delta_p)\,/\hbar)}$$
$$+\int_{\mathbb{R}}\frac{\mathrm{d}\tau}{2\pi}\left\{\frac{1}{\delta_k-\tau+i\hbar/2}+\frac{1}{\delta_k-\tau-i\hbar/2}\right\} \tag{21}$$
$$\times\log\left(1+\frac{\rho^\hbar Y_\delta\,(\tau)}{\vartheta\,(\tau-i\hbar/2)\,\vartheta\,(\tau+i\hbar/2)}\right),$$

as is fully demonstrated in Appendix C. This form of the quantization condition may be more convenient for many applications than the ones previously obtained, equations (16).

3.3. *Solutions to the Baxter equation from the solutions to a NLIE*

In order to understand how the connection between nonlinear integral equations and the Baxter equation comes about, the key observation is that the two functions $q_\delta^\pm$ defined above constitute a system of two linearly independent solutions of the Baxter equation (8). This fact can be deduced from the so-called quantum Wronskian equation satisfied by $Q_\delta^\pm$:

$$Q_\delta^+\,(\lambda)\,Q_\delta^-\,(\lambda+i\hbar) - Q_\delta^-\,(\lambda)\,Q_\delta^+\,(\lambda+i\hbar) \tag{22}$$
$$= \kappa^{-i\lambda}\left(\frac{\hbar e^{-\frac{2\pi\lambda}{\hbar}}}{i\pi g^\hbar}\right)^N\prod_{k=1}^{N}\sinh\frac{\pi}{\hbar}\,(\lambda-\delta_k)\;.$$

The quantum Wronskian equation allows one to show that $Q_\delta^\pm(\lambda)$ satisfy a Baxter-type equation

$$t_\delta(\lambda)Q_\delta^\pm(\lambda) = i^{-N}g^{N\hbar}Q_\delta^\pm(\lambda+i\hbar) + \kappa^\hbar(i)^N g^{N\hbar}Q_\delta^\pm(\lambda-i\hbar)\,, \tag{23}$$

where the polynomial $t_\delta\,(\lambda)$ is defined by

$$t_\delta\,(\lambda) = \left(i\kappa g^N\right)^\hbar\frac{Q_\delta^+\,(\lambda-i\hbar)\,Q_\delta^-\,(\lambda+i\hbar) - Q_\delta^+\,(\lambda+i\hbar)\,Q_\delta^-\,(\lambda-i\hbar)}{Q_\delta^+\,(\lambda)\,Q_\delta^-\,(\lambda+i\hbar) - Q_\delta^+\,(\lambda+i\hbar)\,Q_\delta^-\,(\lambda)}\,. \tag{24}$$

On the one hand, the Wronskian relation (22) allows one to show that $t_\delta(\lambda)$ and $Q_\delta^\pm(\lambda)$ are related by the Baxter equation (23). On the other hand, it also ensures that the residues of the possible poles of t_δ (24) vanish. Then, the polynomiality of t_δ is a consequence of the asymptotic behavior of the functions $Q_\delta^\pm$. The details of the arguments are found in Appendix C.

In order to see how the quantum Wronskian relation is connected to the NLIE (17), let us, starting from a solution $Y_\delta(\lambda)$ to (17), introduce two functions $v_\uparrow(\lambda)$ and $v_\downarrow(\lambda)$ *via* (19). On the one hand, noting that the kernel $K(\lambda)$ defined in (18) can be written as

$$K(\lambda) = \frac{1}{2\pi i}\left(\frac{1}{\lambda - i\hbar} - \frac{1}{\lambda + i\hbar}\right)$$

it is easy to see that (17) implies

$$\log Y_\delta(\lambda) = \log\left(v_\uparrow(\lambda + i\hbar/2)\right) + \log\left(v_\downarrow(\lambda - 3i\hbar/2)\right). \tag{25}$$

On the other hand, note that

$$\begin{aligned}
&\log\left[v_\uparrow\left(\lambda - i\frac{\hbar}{2} + i0\right)\right] + \log\left[v_\downarrow\left(\lambda - i\frac{\hbar}{2} - i0\right)\right] \qquad (26)\\
&= \left(\int_{\mathbb{R}+i0} - \int_{\mathbb{R}-i0}\right)\frac{d\mu}{2i\pi}\frac{1}{\lambda-\mu}\log\left(1 + \frac{\rho^\hbar Y_\delta(\mu)}{\vartheta(\mu - i\hbar/2)\,\vartheta(\mu + i\hbar/2)}\right)\\
&= \log\left(1 + \frac{\rho^\hbar Y_\delta(\lambda)}{\vartheta(\lambda - i\hbar/2)\,\vartheta(\lambda + i\hbar/2)}\right).
\end{aligned}$$

Thus, using that $v_{\uparrow/\downarrow}$ are meromorphic on $\mathbb{C}$, we are able to continue the obtained relation everywhere on $\mathbb{C}$, leading to the functional relation

$$v_\uparrow(\lambda)\,v_\downarrow(\lambda) = 1 + \frac{\rho^\hbar}{\vartheta(\lambda)\,\vartheta(\lambda + i\hbar)}v_\uparrow(\lambda + i\hbar)\,v_\downarrow(\lambda - i\hbar)\;. \tag{27}$$

Rewriting this in terms of $Q_\delta^\pm$ by means of (20) yields the quantum Wronskian equation (22).

At the moment we don't have a direct proof that a solution to (17) exists for all choices of $\vartheta(\lambda)$. We are able, however, to prove that all functions $Y_\delta(\lambda)$ that can be constructed from Gutzwiller's solutions are in fact solutions to (17). This implies that all functions $Y_\delta(\lambda)$ needed for the formulation of the quantization conditions (21) can be obtained in this way.

Let us finally note that there is a more direct way (*cf.* Proposition 5) to reconstruct the Newton polynomials in the zeroes $\{\tau_k\}$ of t_δ from the solution Y_δ to (8):

$$\sum_{p=1}^{N} \tau_p^k = \sum_{p=1}^{N} \delta_p^k - k \int_{\mathbb{R}} \frac{d\tau}{2i\pi} \left\{ (\tau + i\hbar/2)^{k-1} - (\tau - i\hbar/2)^{k-1} \right\} \\ \times \log\left(1 + \frac{\rho^\hbar Y_\delta(\tau)}{\vartheta(\tau - i\hbar/2)\,\vartheta(\tau + i\hbar/2)}\right) . \tag{28}$$

As one may reconstruct the eigenvalues h_k of the conserved quantities $\mathbf{H}_k$ from the $\sum_{p=1}^{N} \tau_p^k$, this essentially amounts to a reconstruction of the h_k.

3.4. *Definition of Yang's potential*

It is interesting to notice that the quantization conditions (21) characterize the extrema of a certain function $\mathcal{W}(\delta)$ called Yang's potential in Ref. 8. This Yang's potential is defined as $\mathcal{W}(\delta) = \mathcal{W}^{\text{inst}}(\delta) + \mathcal{W}^{\text{pert}}(\delta)$, where

$$\mathcal{W}^{\text{pert}}(\delta) = i \sum_{k=1}^{N} \frac{\delta_k^2}{2} \log\left(\frac{\hbar^{N/\hbar}}{\rho}\right) - \log\zeta \sum_{k=1}^{N} \delta_k + \sum_{j,k=1}^{N} \varpi(\delta_k - \delta_j) - 2i\pi \sum_{k=1}^{N} n_k , \tag{29}$$

$\varpi'(\lambda) = \log\Gamma(1 + i\lambda/\hbar)$, n_k are some integers paremeterizing the eigenstate, and

$$\mathcal{W}^{\text{inst}}(\delta) = \\ -\int_{\mathbb{R}} \left\{ \frac{\log Y_\delta(\mu)}{2} \log\left(1 + \frac{\rho^\hbar Y_\delta(\mu)}{|\vartheta(\mu - i\hbar/2)|^2}\right) + \mathrm{Li}_2\left(\frac{-\rho^\hbar Y_\delta(\mu)}{|\vartheta(\mu - i\hbar/2)|^2}\right) \right\} \frac{d\mu}{2i\pi}, \tag{30}$$

with

$$\mathrm{Li}_2(z) = \int_z^0 \frac{\log(1-t)}{t} dt . \tag{31}$$

It seems worth emphasizing that, in our treatment, the reformulation of the quantization conditions in terms of Yang's potential was obtained from the solution theory of the Baxter equation (8). This may be seen as a hint towards a more direct understanding of the claim in Ref. 8 that the quantization conditions for large classes of quantized algebraically integrable models can be formulated in this way. The claim should follow quite generally from the solution theory of the Baxter equation.

Appendix A. Properties of Gutzwiller's solutions

Appendix A.1. *Analytic properties of Gutzwiller's solution*

The explicit construction of a set of two linearly independent entire solutions $Q_{\boldsymbol{t}}^{\pm}$ of (8) with arbitrary monic polynomial $\boldsymbol{t}(\lambda)$ goes back to Gutzwiller.[6] Prior to writing down these two solutions, we recall the definition of the Wronskian of two solutions q_1 and q_2

$$W[q_1, q_2](\lambda) = q_1(\lambda)\, q_2(\lambda + i\hbar) - q_2(\lambda)\, q_1(\lambda + i\hbar)\ . \tag{A.1}$$

It is straightforward to see, using (8), that $W[q_1, q_2]$ is $i\hbar$ quasi-periodic:

$$W[q_1, q_2](\lambda + i\hbar) = (-1)^N \kappa^{\hbar} W[q_1, q_2](\lambda)\ . \tag{A.2}$$

Proposition 1. *Let* $\boldsymbol{t}(\lambda) = \prod_{k=1}^{N}(\lambda - \tau_k)$ *be a monic polynomial of degree* N *with roots appearing in complex-conjugate pairs* $\{\tau_k\} = \{\overline{\tau}_k\}$*. Then, the two functions below are entire solutions to the Baxter equation (23),*

$$\begin{aligned} Q_{\boldsymbol{t}}^{+}(\lambda) &= \frac{\left(\kappa g^N\right)^{-i\lambda} K_{+}(\lambda)\, \mathrm{e}^{-N\frac{\pi}{\hbar}\lambda}}{\prod_{k=1}^{N} \hbar^{-i\frac{\lambda}{\hbar}} \Gamma(1 - i(\lambda - \tau_k)/\hbar)}\,, \\ Q_{\boldsymbol{t}}^{-}(\lambda) &= \frac{g^{iN\lambda} K_{-}(\lambda)\, \mathrm{e}^{-N\frac{\pi}{\hbar}\lambda}}{\prod_{k=1}^{N} \hbar^{i\frac{\lambda}{\hbar}} \Gamma(1 + i(\lambda - \tau_k)/\hbar)}\,. \end{aligned} \tag{A.3}$$

Here $K_{\pm}(\lambda)$ *correspond to the unique meromorphic solutions to difference equations*

$$K_{+}(\lambda - i\hbar) = K_{+}(\lambda) - \frac{\rho^{\hbar} K_{+}(\lambda + i\hbar)}{\boldsymbol{t}(\lambda)\, \boldsymbol{t}(\lambda + i\hbar)}\,, \tag{A.4}$$

$$K_{-}(\lambda + i\hbar) = K_{-}(\lambda) - \frac{\rho^{\hbar} K_{-}(\lambda - i\hbar)}{\boldsymbol{t}(\lambda)\, \boldsymbol{t}(\lambda - i\hbar)}\,, \tag{A.5}$$

that go to 1*, when* $\lambda \to \infty$ *uniformly away from their set of poles. The solutions to these recurrence relations are given explicitly by the determinant formula* (14) *and its complex conjugate. These two solutions are entire and posses the asymptotic behavior*

$$\left|Q_{\boldsymbol{t}}^{\pm}(\lambda)\right| = \mathrm{e}^{-\frac{N\pi}{\hbar}\Re(\lambda)} \mathrm{O}\left(\mathrm{e}^{+\frac{N\pi}{2\hbar}|\Re(\lambda)|}\, |\lambda|^{\frac{N}{2\hbar}(\mp 2\Im(\lambda) - \hbar)}\right) \ \text{for}\ \Re(\lambda) \to \pm\infty\,. \tag{A.6}$$

There the O *is uniform in any strip of* $\mathbb{C}$ *of bounded width.*

Proof. The only non-trivial part concerns the asymptotic behavior of $K_{\pm}$. It follows as a corollary of Lemma 3. □

Lemma 1. *The functions $Q_t^{\pm}$ are linearly independent and their Wronskian is expressed in terms of the Hill determinant (10) by* (15). *The latter is closely related to K_+ and K_-*

$$\mathcal{H}(\lambda) = K_+(\lambda) K_-(\lambda + i\hbar) - \rho^{\hbar} \frac{K_+(\lambda + i\hbar) K_-(\lambda)}{\boldsymbol{t}(\lambda) \boldsymbol{t}(\lambda + i\hbar)} . \tag{A.7}$$

Its zeroes form complex conjugated pairs $\{\delta_k\} = \{\overline{\delta}_k\}$, belong to the fundamental strip $\{z \, : \, |\Im(z)| < \hbar/2\}$ and fulfill $\sum_{p=1}^{N} \tau_p = \sum_{p=1}^{N} \delta_p$.

Proof.

The fact that $Q_t^{\pm}$ are linearly independent is a consequence of the fact that their Wronskian does not vanish identically. The explicit expression for this Wronskian follows after some algebra.

As we have assumed that the set $\{\tau_k\}$ is self-conjugated, it follows from the determinant representation for $\mathcal{H}$ that $\overline{\mathcal{H}(\overline{\lambda})} = \mathcal{H}(\lambda)$, *ie*, the set $\{\delta_k\}$ is self-conjugated. In its turn, this implies a particular relation between the set of δ's and τ's. Namely, computing the $\Re(\lambda) \to +\infty$ asymptotics of $\mathcal{H}$ yields

$$\sum_{p=1}^{N} \tau_p = \sum_{p=1}^{N} \delta_p + in\hbar , \quad \text{for some} \quad n \in \mathbb{N} . \tag{A.8}$$

However, as $\sum \tau_k \in \mathbb{R}$ and $\sum \delta_k \in \mathbb{R}$, the only possibility is $n = 0$. □

We are now in position to prove the

Lemma 2. *Let q be any meromorphic solution to* (8). *Then, there exists two meromorphic $i\hbar$-periodic functions $P_{\pm}(\lambda)$ such that*

$$q(\lambda) = P_+(\lambda) Q_t^+(\lambda) + P_-(\lambda) Q_t^-(\lambda) . \tag{A.9}$$

Proof.

Let q be any meromorphic solution to Baxter's T-Q equation (8). Then consider

$$\widetilde{q}(\lambda) = q(\lambda) - \frac{W\left[q, Q_t^-\right](\lambda)}{W\left[Q_t^+, Q_t^-\right](\lambda)} \cdot Q_t^+(\lambda) + \frac{W\left[q, Q_t^+\right](\lambda)}{W\left[Q_t^+, Q_t^-\right](\lambda)} \cdot Q_t^-(\lambda) . \tag{A.10}$$

The ratio of two Wronskian being $i\hbar$ periodic, one gets that, by construction

$$W\left[\widetilde{q}, Q_t^+\right](\lambda) = W\left[\widetilde{q}, Q_t^-\right](\lambda) = 0 . \tag{A.11}$$

This leads to the system of equations for $\widetilde{q}(\lambda)$:

$$\begin{pmatrix} Q_t^+(\lambda) & Q_t^+(\lambda + i\hbar) \\ Q_t^-(\lambda) & Q_t^-(\lambda + i\hbar) \end{pmatrix} \begin{pmatrix} -\widetilde{q}(\lambda + i\hbar) \\ \widetilde{q}(\lambda) \end{pmatrix} = 0. \tag{A.12}$$

Given any fixed λ, there exist non-trivial solutions to (A.12) if only if the determinant of the matrix defining the system vanishes, *ie* $W\left[Q_{\boldsymbol{t}}^{+},Q_{\boldsymbol{t}}^{-}\right](\lambda)=0$. However, it follows from (15) that $W\left[Q_{\boldsymbol{t}}^{+},Q_{\boldsymbol{t}}^{-}\right](\lambda)$ is an entire function that is non-identically zero. Therefore, it can only vanish at isolated points. Hence, we get that $\widetilde{q}(\lambda)\neq 0$ only at an at most countable set. As $\widetilde{q}(\lambda)$ is meromorphic on $\mathbb{C}$, $\widetilde{q}=0$. □

We now provide a rough characterization of the set of zeroes of $Q_{\boldsymbol{t}}^{\pm}$. As the Γ function has no zeroes on $\mathbb{C}$, the only zeroes of $Q_{\boldsymbol{t}}^{\pm}$ are those of $K_{\pm}(\lambda)$.

Proposition 2. *Assume that $|\Im(\tau_k)|<\hbar/2$ and that the set $\{\tau_k\}$ is invariant under complex conjugation. Then, the set of zeroes of $K_{+}(\lambda-i\hbar/2)$ belongs to the half-plane $\{z\in\mathbb{C}\,:\,\Im(z)<-\hbar/2\}$ and*

$$\frac{|K_{+}(\lambda-i\hbar/2)|^{2}}{\mathcal{H}(\lambda-i\hbar/2)}>1\,,\qquad \text{for}\quad \lambda\in\mathbb{R}\,. \tag{A.13}$$

A similar statement holds for $K_{-}(\lambda+i\hbar/2)$, namely the set of zeroes of $K_{-}(\lambda+i\hbar/2)$ lies in the half-plane $\{z\in\mathbb{C}\,:\,\Im(z)>+\hbar/2\}$ and $K_{-}(\lambda+i\hbar/2)$ does not vanish on $\mathbb{R}$.

Proof.

It follows from the determinant representations that $K_{+}(\lambda-i\hbar/2)$ has poles at $\tau_k-i(2n+1)\hbar/2$, $n\in\mathbb{N}$, in particular they all belong to the half-plane $\{z\in\mathbb{C}\,:\,\Im(z)<-\hbar/2\}$. Also, since the zeroes and poles of the Hill determinant are self-conjugated,

$$\mathcal{H}(\lambda-i\hbar/2)=\prod_{k=1}^{N}\frac{\cosh\frac{\pi}{\hbar}(\lambda-\delta_k)}{\cosh\frac{\pi}{\hbar}(\lambda-\tau_k)}>0\quad,\quad\forall\lambda\in\mathbb{R}\,. \tag{A.14}$$

As the set $\{\tau_k\}$ is self-conjugate, it is easy to see that for $\lambda\in\mathbb{R}$

$$\overline{K_{+}(\lambda)}=K_{-}(\lambda)\qquad\text{and}\qquad \boldsymbol{t}(\lambda-i\hbar/2)\,\boldsymbol{t}(\lambda+i\hbar/2)=|\boldsymbol{t}(\lambda-i\hbar/2)|^{2}\,, \tag{A.15}$$

This allows us to rewrite (A.7) in the form

$$\frac{|K_{+}(\lambda-i\hbar/2)|^{2}}{\mathcal{H}(\lambda-i\hbar/2)}=1+\frac{\rho^{\hbar}}{\mathcal{H}(\lambda-i\hbar/2)}\left|\frac{K_{+}(\lambda+i\hbar/2)}{\boldsymbol{t}(\lambda-i\hbar/2)}\right|^{2}\,. \tag{A.16}$$

It follows from (A.14) and (A.16) that there exists a $c>0$ such that $|K_{+}(\lambda-i\hbar/2)|>c$ for $\lambda\in\mathbb{R}$. Hence, $K_{+}(\lambda-i\hbar/2)$ has no zeroes on $\mathbb{R}$. As $K_{+}(\lambda-i\hbar/2)$ has manifestly no poles in the half-plane

$\{z : \Im(z) > -\hbar/2\}$, one has that

$$f(\rho) = \int_{\mathbb{R}-i\hbar/2} \frac{\mathrm{d}\tau}{2i\pi} \frac{K'_+(\tau)}{K_+(\tau)} = \#\left\{ z \in \mathbb{C} : \Im(z) > -\frac{\hbar}{2} \quad \text{and} \quad K_+(z) = 0 \right\} . \tag{A.17}$$

Here, we remind that ρ is the deformation parameter appearing in (14). We also specify that the function $f(\rho)$ is well defined as $|K_+|_{|\mathbb{R}-i\hbar/2} > c$ and the ratio K'_+/K_+ decays at least as $\lambda^{-(2N+1)}$ at infinity, uniformly in ρ, *cf* lemma 3. Thus, applying the dominated convergence theorem we obtain that $f(\rho)$ is continuous in ρ. As it is integer valued, it is constant. The value of this constant is fixed from $f(0) = 0$ (as then $K_+ = 1$). This shows that $K_+(\lambda)$ has all of its zeros lying below the line $\mathbb{R} - i\hbar/2$. □

Appendix A.2. *Bounds for K_+*

Lemma 3. *Let $\{\tau_k\} = \{\overline{\tau}_k\}$ and $|\Im(\tau_k)| < \hbar/2$, $K_\pm(\lambda)$ is bounded uniformly away from its set of poles, $K_\pm \to 1$ for $\lambda \to \infty$ with $|\arg(\lambda) \mp \pi/2| > \epsilon$ for any $\epsilon > 0$ and*

$$\frac{K'_\pm}{K_\pm}(\lambda) = \mathrm{O}\left(|\Re(\lambda)|^{-(2N+1)} \right) , \tag{A.18}$$

where the O is uniform in ρ as long as ρ belongs to some fixed compact subset of $\mathbb{C}$.

Proof.

As the T-Q equations can be solved explicitly when $N = 1$ in terms of Bessel functions, it is enough to consider the case $N \geq 2$. We focus on K_+ as the behavior of K_- follows by complex conjugation.

K_+ admits the discrete Fredholm series representation:

$$K_+(\lambda) = 1 + \sum_{n \geq 1} \frac{1}{n!} \sum_{\substack{h_1,\dots,h_n \\ \in \mathbb{N}}} \det_n \left[M_{h_a h_b}(\lambda) \right] , \tag{A.19}$$

where we have

$$M_{ab}(\lambda) = \frac{\delta_{a,b+1}}{\boldsymbol{t}(\lambda - ia\hbar)} + \frac{\delta_{a,b-1}\rho^\hbar}{\boldsymbol{t}(\lambda - ia\hbar)} . \tag{A.20}$$

Then, by Hadamard's inequality

$$\left| \sum_{\substack{h_1,\ldots,h_n \\ \in \mathbb{N}}} \det_n \left[M_{h_a h_b} (\lambda) \right] \right| \tag{A.21}$$

$$\leq \sum_{\substack{h_1,\ldots,h_n \\ \in \mathbb{N}}} \prod_{a=1}^{n} \left\{ \sum_{b=1}^{n} \left| M_{h_a h_b} \right|^2 \right\}^{\frac{1}{2}} \leq \sum_{\substack{h_1,\ldots,h_n \\ \in \mathbb{N}}} \prod_{a=1}^{n} \left\{ \sum_{b=1}^{+\infty} \left| M_{h_a b} \right|^2 \right\}^{\frac{1}{2}}$$

$$\leq \sum_{\substack{h_1,\ldots,h_n \\ \in \mathbb{N}}} \prod_{a=1}^{n} \left\{ \sum_{b=1}^{+\infty} \left| M_{h_a b} \right| \right\} \leq \left\{ \sum_{a,b=1}^{n} \left| M_{ab} \right| \right\}^{n} \leq u^n (\lambda) \left(1 + |\rho|^{\hbar} \right)^n .$$

Hence, we get that

$$|K_+ (\lambda) - 1| \leq \mathrm{e}^{u(\lambda)\left(1+|\rho|^{\hbar}\right)} - 1 \,, \quad \text{with} \quad u(\lambda) = \sum_{a=1}^{+\infty} \left| \boldsymbol{t} (\lambda - ia\hbar) \right|^{-1} . \tag{A.22}$$

We will now show that $u(\lambda)$ is bounded uniformly away from the set of the poles of $K_+(\lambda)$ and that $u(\lambda) \to 0$ for $\lambda \to \infty$ in any sector $|\arg(\lambda) - \pi/2| > \epsilon$ with ϵ. This fact ensures that $K_+ \to 1$ for $\lambda \to \infty$ in such sectors and that K_+ is bounded uniformly away from the set of its poles.

Let b be such that $\lambda - ib\hbar \in \{|\Im(z)| \leq \hbar/2\}$. Let $\eta > 0$ and $1 > \eta' > 0$ be such that, for all k,

$$|\lambda - \tau_k - ib\hbar| > \eta\hbar > 0 \quad \text{and} \quad \eta'\hbar > |\Im(\lambda - \tau_k - ib\hbar)| \ . \tag{A.23}$$

Such η and η' exist by definition of b and $\{\tau_k\}$. Setting $\hbar\eta_k = \lambda - \tau_k - ib\hbar$, we get

$$|\lambda - \tau_k - ia\hbar|^2 \geq \hbar^2 \left[\Re^2 (\eta_k) + \left(|a-b| - |\Im\eta_k|\right)^2 \right]$$
$$\geq \hbar^2 \left[\delta_{a,b}\eta^2 + (1-\delta_{a,b}) \left(|a-b| - \eta'\right)^2 \right] .$$

Thus we obtain the boudedness of u for λ uniformly away from the set of poles of K_+:

$$u(\lambda) \leq \sum_{a=1}^{+\infty} \left(\frac{\hbar^{-2}}{\delta_{a,b}\eta^2 + (1-\delta_{a,b})\left(|a-b| - \eta'\right)^2} \right)^{\frac{N}{2}}$$
$$\leq \left(\frac{1}{(\hbar\eta)^N} + 2 \sum_{a=1}^{+\infty} \frac{1}{\left[\hbar(a - \eta')\right]^N} \right) .$$

There, in order to get the last estimate, we have extended the summation to $a \in \mathbb{Z}$ and separated the contributions from $a = b$ and $a \neq b$.

Also, using that for $\Re(\lambda)$ large $|\lambda - \tau_k - ia\hbar| \geq |\Re(\lambda) - \Re(\tau_k)| \geq |\Re(\lambda)|/2$, we obtain

$$\begin{aligned}
&\frac{1}{|\boldsymbol{t}(\lambda - ia\hbar)|} \qquad\qquad (A.24)\\
&= \frac{1}{|\lambda - \tau_1 - ia\hbar|} \cdot \frac{1}{|\lambda - \tau_2 - ia\hbar|^{\frac{1}{2}}} \left\{ \frac{1}{|\lambda - \tau_2 - ia\hbar|^{\frac{1}{2}}} \prod_{k=3}^{N} \frac{1}{|\lambda - \tau_k - ia\hbar|} \right\}\\
&\leq \left\{ \frac{2}{|\Re(\lambda)|} \right\}^{N-\frac{3}{2}} \hbar^{-\frac{3}{2}} \left[\delta_{a,b}\eta^2 + (1 - \delta_{a,b}) \left(|a-b| - \eta'\right)^2 \right]^{-3/2} .
\end{aligned}$$

Thus,

$$u(\lambda) \leq \left(\frac{2}{|\Re(\lambda)|} \right)^{N-\frac{3}{2}} \left(\frac{1}{(\hbar\delta)^{\frac{3}{2}}} + 2 \sum_{a=1}^{+\infty} \frac{1}{[\hbar(a - \delta')]^{\frac{3}{2}}} \right) . \qquad (A.25)$$

It remains to prove the estimates for $K'_+(\lambda)$ at $|\Re(\lambda)| \to +\infty$. Termwise differentiation in (A.19) leads to

$$\begin{aligned}
\left|K'_+(\lambda)\right| &\leq \sum_{n \geq 1} \frac{n}{n!} \left(u(\lambda) \left(1 + |\rho|^{\hbar}\right)^{n-1} \right) \widetilde{u}(\lambda) \left(1 + |\rho|^{\hbar}\right)\\
&\leq \widetilde{u}(\lambda) \left(1 + |\rho|^{\hbar}\right) \mathrm{e}^{u(\lambda)(1+|\rho|^{\hbar})} ,
\end{aligned}$$

where $\widetilde{u}(\lambda) = \sum_{n \geq 1} \left| \boldsymbol{t}'/\boldsymbol{t}^2(\lambda - in\hbar) \right| = \mathrm{O}\left(|\Re(\lambda)|^{-1} \right)$. One then takes the derivative of the Hill's determinant relation (A.7) at $\lambda - i\hbar/2$. This leads to

$$K'_+(\lambda) = \mathcal{H}'(\lambda) + \partial_\lambda \left\{ \rho^{\hbar} \frac{K_+(\lambda + i\hbar) K_-(\lambda)}{\boldsymbol{t}(\lambda) \boldsymbol{t}(\lambda + i\hbar) K_-(\lambda + i\hbar)} \right\} . \qquad (A.26)$$

The uniform estimates in ρ that we have established combined with the fact that $\mathcal{H}'(\lambda) = \mathrm{O}\left(|\Re(\lambda)|^{-\infty} \right)$ uniformly in ρ lead to the desired form of the estimates, with a O that is uniform as long as ρ belongs to some compact subset of $\mathbb{C}$ and λ to any fixed angular sector $|\arg(\lambda) - \pi/2| \, |\arg(\lambda) + \pi/2| > \epsilon$, with $\epsilon > 0$. □

Appendix B. Existence and uniqueness of solutions to the non-linear integral equation

In this Appendix we prove the existence and uniqueness of solutions to the TBA non-linear integral equation (23). Let $\vartheta(\lambda) = \prod_{k=1}^{N} (\lambda - \delta_k)$ have its zeroes given by the N zeroes of the Hill determinant built out of $\boldsymbol{t}(\lambda)$. Then set

$$Y_{\boldsymbol{t}}(\lambda)=\frac{K_{+}(\lambda+i\hbar/2)\,K_{-}(\lambda-i\hbar/2)}{\mathcal{H}(\lambda-i\hbar/2)}\cdot\frac{\vartheta(\lambda-i\hbar/2)\,\vartheta(\lambda+i\hbar/2)}{\boldsymbol{t}(\lambda-i\hbar/2)\,\boldsymbol{t}(\lambda+i\hbar/2)}\,. \tag{B.1}$$

It is a straightforward consequence of Lemma 3 that $Y_{\boldsymbol{t}}$ is a meromorphic function whose poles accumulate in the direction $|\arg(\lambda)|=\pi/2$. $Y_{\boldsymbol{t}}$ is also bounded for $\lambda\to\infty$ uniformly away from the set of its poles and $Y_{\boldsymbol{t}}\to 1$ for $\lambda\to\infty$ in any sector $|\arg(\lambda)-\pi/2|\,|\arg(\lambda)+\pi/2|>\epsilon$ for some fixed $\epsilon>0$.

Proposition 3. *The function* $\log Y_{\boldsymbol{t}}$ *defined in (B.1) is continuous, positive and bounded on* $\mathbb{R}$*. It is the unique solution in this class to the non-linear integral equation (17).*

Proof. We first prove the uniqueness of solutions. Let $\|\cdot\|_{\infty}$ stand for the sup norm on bounded and continuous functions on $\mathbb{R}$. We set $\mathcal{F}=\big\{f\in\mathscr{C}^{0}(\mathbb{R})\ :\ f\geq 0\quad\text{and}\quad\|f\|_{\infty}<+\infty\big\}$. Then we define the operator L on $\mathcal{F}$ by

$$L[f](\lambda)=\int_{\mathbb{R}}\mathrm{d}\mu\,K(\lambda-\mu)\log\left(1+\frac{\rho^{\hbar}\mathrm{e}^{f(\mu)}}{|\vartheta(\lambda-i\hbar/2)|^{2}}\right)\,. \tag{B.2}$$

The mapping L stabilizes $\mathcal{F}$. Indeed,

$$\begin{aligned}|L[f](\lambda)|&\leq\int_{\mathbb{R}}\mathrm{d}\mu\,K(\lambda-\mu)\log\left(1+\frac{\rho^{\hbar}\mathrm{e}^{f(\mu)}}{|\vartheta(\lambda-i\hbar/2)|^{2}}\right)\\&\leq\log\left(1+\rho^{\hbar}\mathrm{e}^{\|f\|_{\infty}}J^{-1}\right)\,,\end{aligned} \tag{B.3}$$

where $J=\inf_{\lambda\in\mathbb{R}}|\vartheta(\lambda-i\hbar/2)|^{2}>0$, due to $|\Im(\delta_k)|<\hbar/2$.

Any solution to the NLIE appears as a fixed point of L in $\mathcal{F}$. We shall now prove that L can have at most one fixed point. This settles the question of uniqueness of solutions to (17). This part goes as in Ref. 4. Let $f,g\in\mathcal{F}$, then

$$\begin{aligned}&|L[f]-L[g]|(\lambda) \qquad\qquad (\mathrm{B.4})\\&=\left|\int_{0}^{1}\mathrm{d}t\int_{\mathbb{R}}\mathrm{d}\tau K(\lambda-\tau)\frac{\rho^{\hbar}\mathrm{e}^{g(\tau)+t(f-g)(\tau)}}{|\vartheta(\tau-i\hbar/2)|^{2}+\rho^{\hbar}\mathrm{e}^{g(\tau)+t(f-g)(\tau)}}(f-g)(\tau)\right|\\&\leq c\,\|f-g\|_{\infty}\quad\text{with}\quad c=\frac{\rho^{\hbar}\mathrm{e}^{\max\left(\|g\|_{\infty},\|f\|_{\infty}\right)}}{J+\rho^{\hbar}\mathrm{e}^{\max\left(\|g\|_{\infty},\|f\|_{\infty}\right)}}<1\,.\end{aligned}$$

Hence, L admits a unique fixed point.

A direct proof of existence of the solutions to (17) is possible if $\rho^{\hbar}/J < 1$. In this case it is easily seen that $L[f](\lambda)$ is a bounded mapping in the sense that it stabilizes all balls in $\mathcal{F}$ of radius $R \geq -\log\left(1-\rho^{\hbar}/J\right)$. In such a case, (B.4) implies that $L[f]$ is a contractive map on a Banach space. It thus admits a unique fixed point.

However, it is always possible to construct a solution to (17) in terms of the the half-infinite determinants $K_{\pm}$. Recall that $\{\tau_k\}$ and hence $\{\delta_k\}$ are invariant under complex conjugation. Let

$$v_{\uparrow}(\lambda) = K_{+}(\lambda)\prod_{k=1}^{N}\frac{\Gamma(1-i(\lambda-\delta_k)/\hbar)}{\Gamma(1-i(\lambda-\tau_k)/\hbar)} \tag{B.5}$$

$$v_{\downarrow}(\lambda) = K_{-}(\lambda+i\hbar)\prod_{k=1}^{N}\frac{\Gamma(i(\lambda-\delta_k)/\hbar)}{\Gamma(i(\lambda-\tau_k)/\hbar)}\,. \tag{B.6}$$

It follows from Proposition 2 that $v_{\uparrow/\downarrow}(\lambda-i\hbar/2)$ are holomorphic and non-vanishing in $\overline{\mathbb{H}}_{+/-}$. Moreover, as $\sum\delta_k = \sum\tau_k$, we get that $v_{\uparrow/\downarrow}(\lambda-i\hbar/2) = 1+\mathrm{O}\left(\lambda^{-1}\right)$ in their respective domains of holomorphy. Also, due to the Hill determinant identity (A.7)

$$\frac{|K_{+}(\lambda-i\hbar/2)|^2}{\mathcal{H}(\lambda)} = v_{\uparrow}(\lambda-i\hbar/2)\,v_{\downarrow}(\lambda-i\hbar/2) = 1+\frac{\rho^{\hbar}Y_{\boldsymbol{t}}(\lambda)}{|\vartheta(\lambda-i\hbar/2)|^2}. \tag{B.7}$$

We agree upon choosing a determination of $v_{\uparrow/\downarrow}(\lambda-i\hbar/2)$ such that

$$\log v_{\uparrow/\downarrow}(\lambda-i\hbar/2) \underset{\lambda\to+\infty}{\longrightarrow} 0$$
$$\Rightarrow \log[v_{\uparrow}v_{\downarrow}](\lambda-i\hbar/2) = \log v_{\uparrow}(\lambda-i\hbar/2)+\log v_{\downarrow}(\lambda-i\hbar/2)\,.$$

Thus, for $\lambda\in\mathbb{R}$, by computing the residues in the upper or lower half-plane and using the decay properties of the integrand at infinity, one sees that $v_{\uparrow}$ and $v_{\downarrow}$ are recovered from $Y_{\boldsymbol{t}}(\lambda)$ *via* the integral representations (19). Hence, with the same choice of branches of logarithm as before (the one that goes to 0 when $\Re(\lambda)$ goes to $+\infty$) we get, on the one hand, that

$$\log[v_{\uparrow}(\lambda+i\hbar/2)\,v_{\downarrow}(\lambda-3i\hbar/2)] = \int_{\mathbb{R}}\mathrm{d}\tau K(\lambda-\tau)\log\left(1+\frac{\rho^{\hbar}Y_{\boldsymbol{t}}(\tau)}{|\vartheta(\tau-i\hbar/2)|^2}\right)\,.$$

On the other hand, it is straightforward to check that

$$v_{\uparrow}(\lambda+i\hbar/2)\,v_{\downarrow}(\lambda-3i\hbar/2) = Y(\lambda)\,.$$

This proves the existence of the relevant set of solutions to (17). □

Appendix C. Baxter Equation and quantization conditions from TBA

We first prove basic properties of the functions $Q_\delta^\pm$. Then we derive the T-Q equation generated by $Q_\delta^\pm$ and finally obtain the quantization conditons.

Appendix C.1. *Analytic properties of $Q_\delta^\pm$*

Lemma 4. *The functions $Q_\delta^\pm$ defined in (20) are entire and have the asymptotic behavior*

$$\left|Q_\delta^\pm\right| = \mathrm{e}^{-\frac{N\pi\lambda}{\hbar}} \cdot \mathrm{O}\left(\mathrm{e}^{\frac{N\pi}{2\hbar}|\Re(\lambda)|}\,|\lambda|^{\frac{N}{2\hbar}(\pm 2\Im(\lambda)-\hbar)}\right) \qquad \Re(\lambda) \to \pm\infty \qquad \text{(C.1)}$$

where the O *symbol is uniform in* $\{z\ :\ |\Im(z)| \leq \hbar/2\}$.

Proof.

It is readily seen from the asymptotic behavior in the strip $\{z\ :\ |\Im(z)| < \hbar\}$ of the solution Y_δ to (17), that $v_\uparrow(\lambda) \to 1$ and $v_\uparrow(\lambda - i\hbar) \to 1$ when $\Re(\lambda) \to \pm\infty$ in the strip $\{z\ :\ |\Im(z)| \leq \hbar/2\}$. Then a straightforward computation leads to (C.1). We assume that all the δ's are distinct and, if necessary, take the limit of coinciding δ's at the end of the calculation.

It remains to prove that $Q_\delta^\pm$ are entire. For this, we show that the products of Γ-functions cancel the poles of $v_{\uparrow/\downarrow}$. We need to construct a meromorphic continuation to $\mathbb{C}$ of $v_{\uparrow/\downarrow}$ starting from the strip $\mathcal{B}^{(1)}$, with $\mathcal{B}^{(n)} = \{z\ :\ |\Im(z)| < n\hbar\}$. It follows from the very form of the NLIE (17) that the solution Y_δ is holomorphic in $\mathcal{B}^{(1)}$. Let us introduce the notation

$$V_\delta(\mu) = 1 + \frac{\rho^{\hbar}\, Y_\delta(\mu)}{|\vartheta(\mu - i\hbar/2)|^2}\,. \qquad \text{(C.2)}$$

The only singularities of $V_\delta'(\mu)/V_\delta(\mu)$ in $\mathcal{B}^{(1)}$ correspond to the zeroes of V_δ and to its poles. The latter are located at $\mu = \delta_k \pm i\hbar/2$, $k = 1, \ldots, N$.

Hence, as $V_\delta'(\mu)/V_\delta(\mu)$ is decaying sufficiently fast at infinity, one gets

$$\frac{Y_\delta'}{Y_\delta}(\tau) = \sum_{\substack{z\in\mathcal{B}_\uparrow^{(n)} \\ V_\delta(z)=0}} \frac{n_z}{\lambda - z - i\hbar} + \sum_{\substack{z\in\mathcal{B}_\downarrow^{(n)} \\ V_\delta(z)=0}} \frac{n_z}{\lambda - z + i\hbar} \tag{C.3}$$
$$-\sum_{p=1}^{n}\sum_{k=1}^{N} \frac{1}{\lambda - \delta_k - i(2p+1)\hbar/2} - \sum_{p=1}^{n}\sum_{k=1}^{N} \frac{1}{\lambda - \delta_k + i(2p+1)\hbar/2}$$
$$+ \int\limits_{\mathbb{R}+in\hbar-i0^+} \frac{\mathrm{d}\mu}{2i\pi} \frac{V_\delta'(\mu)/V_\delta(\mu)}{\lambda - \mu - i\hbar} - \int\limits_{\mathbb{R}-in\hbar+i0^+} \frac{\mathrm{d}\mu}{2i\pi} \frac{V_\delta'(\mu)/V_\delta(\mu)}{\lambda - \mu + i\hbar}.$$

Above, we have denoted by n_z the multiplicity of a zero z of V_δ and $\mathcal{B}_\uparrow^{(n)}$, resp. $\mathcal{B}_\downarrow^{(n)}$, stands for $\mathcal{B}^{(n)} \cap \mathbb{H}_+$, resp. $\mathcal{B}^{(n)} \cap \mathbb{H}_-$. It thus follows that V_δ has simple poles at $\delta_k + i\frac{\hbar}{2}(2n+1)$, $n \in \mathbb{Z}$. Exactly the same reasoning as before shows that $v_\uparrow(\lambda - i\hbar)$ has its only simple poles at $\delta_k - in\hbar$, $k = 1, \ldots, N$ and $n \in \mathbb{N}^*$. Similarly, $v_\downarrow(\lambda)$ has its only simple poles at $\delta_k + in\hbar$, $k = 1, \ldots, N$ and $n \in \mathbb{N}$. Therefore, these poles are canceled out by the zeroes of the Γ-functions and $Q_\delta^\pm$ are both entire. □

Appendix C.2. *Baxter equation*

Proposition 4. *The polynomial $t_\delta(\lambda)$ given in (24) is a monic real valued polynomial of degree N. It is such that the functions $Q_\delta^\pm$ solve the Baxter equation*

$$t_\delta(\lambda)\, Q_\delta^\pm(\lambda) = i^N g^{N\hbar}\, Q_\delta^\pm(\lambda + i\hbar) + \kappa^\hbar (-i)^N g^{N\hbar}\, Q_\delta^\pm(\lambda - i\hbar)\,, \tag{C.4}$$

Finally, given any monic real valued polynomial of degree N with roots in the strip $\{z : |\Im(z)| < \hbar/2\}$, it is always possible to find a complex-conjugation invariant set of parameters $\{\delta_k\}$ such that t_δ equals to this polynomial.

Proof. The proof that Q_δ^+ satisfies (C.4) with t_δ being given by (24) can be done by straightforward calculation, using the definition of t_δ and the Wronskian equation (22).

Now we show that t_δ, as defined in (24), is indeed a monic real valued polynomial of degree N. We first assume that the δ_k's are pairwise distinct. The case when several δ_k's coincide follows by taking the limit in the final formulae. As $Q_\delta^\pm$ are both entire, we get that the only potential poles of $t_\delta(\lambda)$ are located at $\lambda = \delta_k + in\hbar$, $n \in \mathbb{Z}$. The set of zeroes of $v_{\uparrow/\downarrow}$ differs necessarily from its set of poles (as follows readily from (C.3) and similar representations for $v_{\uparrow/\downarrow}$). Hence, $Q_\delta^{\pm}(\delta_k + in\hbar) \neq 0$, for $k = 1, \ldots, N$ and $n \in \mathbb{Z}$. Therefore, it follows from the Wronskian relation (22) , that

$$\frac{Q_\delta^{+}(\delta_k)}{Q_\delta^{-}(\delta_k)} = \frac{Q_\delta^{+}(\delta_k + in\hbar)}{Q_\delta^{-}(\delta_k + in\hbar)} \quad \text{for} \quad k \in [\![1\,;\,N]\!] \text{ and } \quad n \in \mathbb{N}\,. \tag{C.5}$$

This implies that the possible poles of the expression in (24) get canceled. It follows that t_δ is entire. It remains to control its asymptotic behavior. We may express t_δ in terms of $v_{\uparrow/\downarrow}$,

$$t_\delta(\lambda) \tag{C.6}$$
$$= v_\uparrow(\lambda - i\hbar)\, v_\downarrow(\lambda) \prod_{a=1}^{N}(\lambda - \delta_a) - \frac{\left(\kappa g^{2N}\right)^{2\hbar} v_\uparrow(\lambda + i\hbar)\, v_\downarrow(\lambda - 2i\hbar)}{\prod\limits_{a=1}^{N}(\lambda - \delta_a)\left(\hbar^2 + (\lambda - \delta_a)^2\right)}\,.$$

Due to the asymptotic behavior of Y_δ at ∞, one can deform the integration contour in the definition $v_{\uparrow/\downarrow}$ so as to obtain its asymptotic behavior in the whole plane $\lambda \to \infty$, for λ uniformly away from the set of poles of $v_{\uparrow/\downarrow}$.

As we have that $v_{\uparrow/\downarrow} \to 1$ when $\lambda \to \infty$, we get that $t_\delta \simeq \lambda^N$ when $\lambda \to \infty$. Hence, t_δ is a monic polynomial of degree N. It is real valued for $\lambda \in \mathbb{R}$ as for such λ's, $V_\delta(\lambda)$ defined in (C.2) belongs to $\mathbb{R}$, what implies that $v_\uparrow(\lambda - i\hbar) = v_\downarrow(\lambda)$, *ie* $\overline{t_\delta(\lambda)} = t_\delta(\overline{\lambda})$.

The fact that, in this way, one is able to generate any monic polynomial with roots in the strip $\{z \,:\, |\Im(z)| < \hbar/2\}$ follows form the uniqueness of solutions to the TBA-NLIE and the construction of the function $v_{\uparrow/\downarrow}$ in terms of determinants, as given in (B.5)-(B.6). □

Appendix C.3. *The quantization conditions*

We now prove that the quantization conditions for the model (conditions on the zeroes of the polynomial $t_\delta(\lambda)$ for (8) to have entire solutions with a prescribed decay as given in point (ii)) can be written down in a TBA-like form. Moreover, as opposed to the Gutzwiller form of the quantization conditions, the ones that will follow only involve one set of parameters. Namely, the zeroes $\{\delta_k\}$ of the Hill determinant associated with $t_\delta(\lambda)$ given in (24). We show that under certain reasonable assumptions, it is possible to reconstruct the Newton polynomials in the zeroes of t_δ and hence the spectrum of the model. This proves the Nekrasov-Shatashvili conjecture.[8] We first reconstruct the zeroes of t_δ.

Proposition 5. *Let* $t_\delta(\lambda) = \prod_{p=1}^{N}(\lambda - \tau_a)$ *be a polynomial whose zeroes* τ_k *lie in the strip* $\{z \in \mathbb{C} : |\Im(z)| < \hbar/2\}$. *If* $\{\delta_k\}$ *is the associated set of zeroes of the Hill determinant and* Y_δ *the unique solution to the NLIE (17), then the Newton polynomials* $\mathcal{E}_k = \sum_{p=1}^{N} \tau_p^k$ *in the zeroes of* t_δ *are reconstructed by means of formula (28) above. The convergence of these integrals is part of the conclusion.*

Proof.

Due to the uniqueness of solutions to the NLIE (17), one has that the solution Y_δ can be expressed, as in (B.1) in terms of $K_\pm$, $\mathcal{H}$. The latter determinants are parameterized by the zeroes $\{\tau_k\}$ of t_δ, and the parameters $\{\delta_k\}$ appearing in the NLIE (17) coincide with the set of zeroes of the Hill determinant. By invoking the continuity of the logarithm on $\mathbb{R}$ and its decay at infinity, we get

$$
\begin{aligned}
& k\int_{\mathbb{R}} \frac{\mathrm{d}\mu}{2i\pi}\left\{(\mu+i\hbar/2)^{k-1}-(\mu-i\hbar/2)^{k-1}\right\}\log\left(1+\frac{\rho^{\hbar}Y_\delta(\mu)}{|\vartheta(\mu-i\hbar/2)|^2}\right) \\
&= -\int_{\mathbb{R}-i\hbar/2} \frac{\mathrm{d}\mu}{2i\pi}\left\{(\mu+i\hbar)^k-\mu^k\right\}\left[\frac{K'_+}{K_+}(\mu)+\frac{K'_-}{K_-}(\mu+i\hbar)-\frac{\mathcal{H}'}{\mathcal{H}}(\mu)\right] \\
&= \int_{\substack{\mathbb{R}+i\hbar/2\rightarrow \\ \mathbb{R}-i\hbar/2\leftarrow}} \frac{\mathrm{d}\mu}{2i\pi}\mu^k\frac{\mathcal{H}'}{\mathcal{H}}(\mu) = \sum_{p=1}^{N}\left(\tau_p^k-\delta_p^k\right) .
\end{aligned}
\tag{C.7}
$$

In the intermediate steps, we have used the quick decay at infinity of the integrand

$$
\frac{K'_\pm}{K_\pm}(\lambda) = \mathrm{O}\left(\lambda^{-2N-1}\right) \qquad \text{and} \qquad \frac{\mathcal{H}'}{\mathcal{H}}(\lambda) = \mathrm{O}\left(\lambda^{-\infty}\right) . \tag{C.8}
$$

This allows us to split the integral in three and compute the parts involving K_+, resp. K_-, by the residues in the upper/lower half plane (thus giving 0). Hence, the only part that gives a non-trivial contribution is the contour integral involving $\mathcal{H}'/\mathcal{H}$. The only poles that contribute to the result are located at the zeroes δ_k of the Hill determinant (they have residue +1) and at the poles τ_k of the Hill determinant (they have residue -1) that are located in the strip $|\Im(z)| < \hbar/2$. □

This result offers a direct way to recover the spectum of the model from a solution to the TBA equation (17). It remains to derive the set of quantization conditions on the parameters δ_k.

Theorem 1. *There exists a unique entire solution $q(\lambda)$ to the T-Q equation* (8) *whose asymptotic behavior is as stated in (ii) if and only if the parameters $\{\delta_k\}$ appearing in the TBA NLIE* (17) *satisfy to the quantization conditions given in* (21).

Remark 1. The solvability of the quantization conditions, the occurrence of complex solutions ($\Im(\delta_k) \neq 0$, $\delta_k \in \{z : |\Im(z)| < \hbar/2\}$, the uniqueness of solutions for a given choice of integers $n_k \in \mathbb{Z}$ are all open questions.

Proof.

According to lemma 2, any meromorphic solution q to the T-Q equation takes the form

$$q(\lambda) = \frac{W\left[q, Q_{\delta}^{-}\right](\lambda)}{W\left[Q_{\delta}^{+}, Q_{\delta}^{-}\right](\lambda)} \cdot Q_{\delta}^{+}(\lambda) - \frac{W\left[q, Q_{\delta}^{+}\right](\lambda)}{W\left[Q_{\delta}^{+}, Q_{\delta}^{-}\right](\lambda)} \cdot Q_{\delta}^{-}(\lambda) . \qquad \text{(C.9)}$$

Recall that the Wronskian $W\left[Q_{\delta}^{+}, Q_{\delta}^{-}\right](\lambda)$ is given by (22). It is possible to compute the Wronskians $W\left[q, Q_{\delta}^{\pm}\right](\lambda)$ by using the asymptotic behavior of $q(\lambda)$ and $Q_{\delta}^{\pm}$. Due to their $i\hbar$ quasi-periodicity, these Wronskians take the form $W\left[q, Q_{\delta}^{\pm}\right](\lambda) = \mathrm{e}^{-N\frac{\pi}{\hbar}\lambda}\kappa^{-i\lambda}w_{\pm}(\lambda)$, where $w_{\pm}(\lambda)$ are entire $i\hbar$-periodic functions. However, using the asymptotic behavior of $q(\lambda)$ and $Q_{\delta}^{\pm}$ we get that $w_{\pm}(\lambda)$ are bounded at infinity in the strip $|\Im(\lambda)| \leq \hbar/2$, and hence on $\mathbb{C}$. They are thus constant. This proves the uniqueness of solutions for a given choice of τ_k's and hence δ_k's. Indeed, up to a normalization constant, any solution $q(\lambda)$ satisfying to the requirements stated in point (ii), is of the form

$$q(\lambda) = \mathrm{e}^{\frac{N\pi}{\hbar}\lambda} \frac{Q_{\delta}^{+}(\lambda) - \zeta Q_{\delta}^{-}(\lambda)}{\prod_{k=1}^{N} \sinh\frac{\pi}{\hbar}(\lambda - \delta_k)} . \qquad \text{(C.10)}$$

As the solution $q(\lambda)$ is entire, it has a vanishing residue at $\lambda = \delta_k$, $k = 1\ldots, N$. Therefore, the quantization conditions for the Toda chain appear as the set of $N-1$ conditions that $q(\lambda)$ has a vanishing residue at δ_k, $k = 1, \ldots, N$ supplemented with the Nth quantization condition for the overall momentum : $\sum_{p=1}^{N} \tau_p = \sum_{p=1}^{N} \delta_p = P$. Note that it follows from $W\left[Q_{\delta}^{+}, Q_{\delta}^{-}\right](\delta_k + in\hbar) = 0$ that if $q(\lambda)$ has a vanishing residue at δ_k then it also has a vanishing residue at $\delta_k + in\hbar$, $n \in \mathbb{Z}$. Therefore, there is indeed only a finite number N of constraints of the parameters δ_a. The explicit form of these quantization conditions is then indeed as given in (21).

Conversely, if the quantization conditions are satisfied, then by taking $q(\lambda)$ as in (C.10), one obtains an entire solution with the desired asymptotics. □

References

1. D. An, *"Complete set of Eigenfunctions of the quantum Toda chain."*, Lett. Math. Phys. **87** (2009), 209–223.
2. R. J. Baxter, *Exactly solved models in statistical mechanics*, Academic Press, 1984.
3. V. Bazhanov, S. Lukyanov, A.B. Zamolodchikov, *"Integrable Structure of Conformal Field Theory II. Q-operator and DDV equation."* Comm. Math. Phys. **190**, (1997), 247-278.
4. A. Fring, C. Korff, and B.J. Schulz, *"The ultraviolet Behaviour of Integrable Quantum Field Theories, Affine Toda Field Theory."*, Nucl.Phys. B **549** (1999), 579–612.
5. M. Gaudin and V. Pasquier, *"The periodic Toda chain and a matrix generalization of the Bessel function recursion relations."*, J. Phys. A: Math. Gen **25** (1992), 5243–5252.
6. M. C. Gutzwiller, *"The quantum mechanical Toda lattice."*, Ann. Phys. **124** (1980), 347–381; *"The quantum mechanical Toda lattice II."*, Ann. Phys. **133** (1981), 304–331.
7. S. Kharchev and D. Lebedev, *"Integral representation for the eigenfunctions of a quantum periodic Toda chain."*, Lett. Math. Phys. **50** (1999), 53–77; *"Integral representations for the eigenfunctions of quantum open and periodic Toda chains from QISM formalism"*. J. Phys. **A34** (2001) 2247-2258
8. S. L. Shatashvili N. A. Nekrasov, *"Quantization of Integrable Systems and Four Dimensional Gauge Theories."*, Preprint [arXiv:0908.4052] (hep-th)
9. M.A. Olshanetsky and A.M. Perelomov, *"Quantum completely integrable systems connected with semi-simple Lie algebras."*, Lett. Math. Phys. **2** (1977), 7–13.
10. E.K. Sklyanin, *"The quantum Toda chain."*, Lect. Notes in Phys. **226** (1985), 196–233.
11. J. Teschner, *"On the spectrum of the Sinh-Gordon model in finite volume."* Nucl. Phys. B**799**, (2008), 403-429.
12. W. Toda, *"Wave propagation in anharmonic lattices."*, J. Phys. Soc. Jap. **23** (1967), 501–506.
13. E.T. Whittaker, G.N. Watson, *A Course of Modern Analysis*, Cambridge University Press, 1927.
14. Al.B. Zamolodchikov *"On the Thermodynamic Bethe Ansatz Equation in Sinh-Gordon Model"*. J. Phys. **A39**, (2006), 12863-12887.

GENERALIZED ENERGIES AND INTEGRABLE $D_n^{(1)}$ CELLULAR AUTOMATON

ATSUO KUNIBA

Institute of Physics, University of Tokyo, Komaba
Tokyo 153-8902, Japan

REIHO SAKAMOTO

Department of Physics, Tokyo University of Science
Tokyo 162-8601, Japan

YASUHIKO YAMADA

Department of Mathematics, Faculty of Science, Kobe University
Hyogo 657-8501, Japan

Dedicated to Professor Tetsuji Miwa on his 60th birthday

We introduce generalized energies for a class of $U_q(D_n^{(1)})$ crystals by using the piecewise linear functions that are building blocks of the combinatorial R. They include the conventional energy in the theory of affine crystals as a special case. It is shown that the generalized energies count the particles and anti-particles in a quadrant of the two dimensional lattice generated by time evolutions of an integrable $D_n^{(1)}$ cellular automaton. Explicit formulas are conjectured for some of them in the form of ultradiscrete tau functions.

Keywords: Crystal base; integrable cellular automaton; generalized energy; combinatorial Bethe ansatz; inverse scattering method; ultradiscrete tau function.

1. Introduction

Let B_l be the crystal of the l-fold symmetric tensor representation of the quantum affine algebra $U_q(D_n^{(1)})$.[9,11] The combinatorial $R : x \otimes y \mapsto y' \otimes x'$ is the isomorphism of crystals $B_l \otimes B_m \xrightarrow{\sim} B_m \otimes B_l$ corresponding to the quantum R at $q = 0$.[10] In Ref. 15, an explicit formula of the combinatorial R was obtained in terms of several piecewise linear functions $g_i(x \otimes y) \in \mathbb{Z}$ on $B_l \otimes B_m$. See Theorem 2.1. Among them is the *local energy*, which plays an essential role in the theory of affine crystals.[10] The family of piecewise linear functions $\{g_i\}$, which we call *generalized local energies* in this paper, are ultradiscretization of the subtraction-free rational functions that have emerged as building blocks of the tropical R[15,16] of the geometric crystal.[3] They may be viewed as local energies in a principal picture rather than in the conventional homogeneous picture.

From the local energy, one can form the integer-valued function called *energy* on the tensor product $\mathcal{P} = B_{l_1} \otimes \cdots \otimes B_{l_L}$. Its generating function is the one dimensional configuration sum that originates in the corner transfer matrix method.[1,2]

In this paper we introduce *generalized energies* $\mathcal{E}_{g_i} : \mathcal{P} \to \mathbb{Z}_{\geq 0}$ corresponding to g_i's, and study them from the viewpoint of the integrable cellular automaton of type $D^{(1)}_n$.[6,7] The latter is an integrable $U_q(D^{(1)}_n)$ vertex model at $q = 0$. It is a dynamical system on $\mathcal{P}$ equipped with commuting time evolutions $\{T_l\}_{l\geq 1}$. Elements of $\mathcal{P}$ are naturally regarded as arrays of particles and anti-particles, and T_l induces their factorized scattering involving pair creation and annihilation. See Examples 3.1 and 3.2.

Our main result is Theorem 4.1, which states that $\mathcal{E}_{g_i}(p) = \rho_{g_i}(p)$ for any $p = p_1 \otimes \cdots \otimes p_L \in \mathcal{P}$. Here $\rho_{g_i}(p)$ is a *counting function* giving the number of certain particles and anti-particles specified by g_i in the region (40) under the time evolutions $p, T_\infty(p), T^2_\infty(p), \ldots$. As such, the counting functions are non-local variables attached to a quadrant of the 2 dimensional lattice. However, it will also be shown in Theorem 3.1 that the combined data $\{\rho_{g_i}(p_1 \otimes \cdots \otimes p_j) \mid j = k-1, k\}$ in turn reproduces the local variable $p_k \in B_{l_k}$ completely in agreement with the spirit of the corner transfer matrix method. Therefore the joint spectrum $\{\mathcal{E}_{g_i}(p_1 \otimes \cdots \otimes p_k)\}$ of the generalized energies with $1 \leq k \leq L$ is equivalent to $p = p_1 \otimes \cdots \otimes p_L \in \mathcal{P}$ itself. This extends a similar result on type $A^{(1)}_n$ (Proposition 4.6 in Ref. 17) which is related to the katabolism.[22] A supplementary result (Proposition 4.1) is parallel with Theorem 4.1 and treats generalized local energies with opposite chirality (cf. Remark 2.2).

The layout of the paper is as follows. In Section 2, generalized (local) energies are extracted from the piecewise linear formula of the combinatorial R.[15] In Section 3, the integrable $D^{(1)}_n$ cellular automaton[6,7] is recalled and the counting functions are defined. In Section 4, the main Theorem 4.1 of the paper is stated and proved. In Section 5, aspects related to combinatorial Bethe ansatz are discussed. In Section 5.1 we give the inverse scattering formalism of the $D^{(1)}_n$ cellular automaton like Ref. 14. In Section 5.2, we conjecture piecewise linear formulas for some generalized energies in terms of *ultradiscrete tau functions*. This is also motivated by the $A^{(1)}_n$ case,[17] where analogous results have led to a piecewise linear formula for the Kerov-Kirillov-Reshetikhin map.[13] Although the conjecture is yet to cover the full family of generalized energies, the last one (57) is already rather intriguing. We expect that the extention and the solution of Conjecture 5.1 will uncover an interplay among combinatorial Bethe ansatz, ultradiscretization of the DKP hierarchy[8] and the bilinearlization of the tropical R.[16]

2. Generalized energies for $D_n^{(1)}$ crystal

2.1. *Crystals and combinatorial R*

Let us recall the basic facts on crystal and combinatorial R briefly. For a more information, see Refs. 9–11 and 19. For a positive integer l, let

$$B_l = \{\zeta = (\zeta_1, \ldots, \zeta_n, \overline{\zeta}_n, \ldots, \overline{\zeta}_1) \in \mathbb{Z}_{\geq 0}^{2n} \mid \sum_{i=1}^{n} (\zeta_i + \overline{\zeta}_i) = l, \;\; \zeta_n \overline{\zeta}_n = 0\} \quad (1)$$

be the crystal of the l-fold symmetric tensor representation of $U_q(D_n^{(1)})$.[9] We assume $n \geq 3$. As for the functions ε_i, φ_i, the tensor product rule and the action of Kashiwara operators $\tilde{e}_i$ and $\tilde{f}_i$ $(0 \leq i \leq n)$, see Ref. 19.

The *affinization* of the crystal B_l is defined by $\mathrm{Aff}(B_l) = \{b[d] \mid d \in \mathbb{Z}, b \in B_l\}$ with the crystal structure $\tilde{e}_i(b[d]) = (\tilde{e}_i b)[d + \delta_{i0}]$ and $\tilde{f}_i(b[d]) = (\tilde{f}_i b)[d - \delta_{i0}]$. We call b and d the classical and the affine part of $b[d]$, respectively. There exists the unique bijection (crystal isomorphism) $B_l \otimes B_m \overset{\sim}{\to} B_m \otimes B_l$ that commutes with all Kashiwara operators. It is lifted up to a map $\mathrm{Aff}(B_l) \otimes \mathrm{Aff}(B_m) \overset{\sim}{\to} \mathrm{Aff}(B_m) \otimes \mathrm{Aff}(B_l)$ called the *combinatorial R*, which has the following form:

$$\begin{aligned} R \;:\; \mathrm{Aff}(B_l) \otimes \mathrm{Aff}(B_m) &\longrightarrow \mathrm{Aff}(B_m) \otimes \mathrm{Aff}(B_l) \\ b[d] \otimes b'[d'] &\longmapsto \tilde{b}'[d' + H(b \otimes b')] \otimes \tilde{b}[d - H(b \otimes b')], \end{aligned}$$

where $b \otimes b' \mapsto \tilde{b}' \otimes \tilde{b}$ under the isomorphism $B_l \otimes B_m \overset{\sim}{\to} B_m \otimes B_l$.[a] The quantity $H(b \otimes b')$ is called the *local energy* and determined up to a global additive constant by

$$H(\tilde{e}_i(b \otimes b')) = \begin{cases} H(b \otimes b') + 1 & \text{if } i = 0, \; \varphi_0(b) \geq \varepsilon_0(b'), \varphi_0(\tilde{b}') \geq \varepsilon_0(\tilde{b}), \\ H(b \otimes b') - 1 & \text{if } i = 0, \; \varphi_0(b) < \varepsilon_0(b'), \varphi_0(\tilde{b}') < \varepsilon_0(\tilde{b}), \\ H(b \otimes b') & \text{otherwise.} \end{cases}$$

The Yang-Baxter equation

$$(R \otimes 1)(1 \otimes R)(R \otimes 1) = (1 \otimes R)(R \otimes 1)(1 \otimes R) \quad (2)$$

[a]This classical part of the combinatorial R will also be referred as combinatorial R and denoted by $R(b \otimes b') = \tilde{b'} \otimes \tilde{b}$.

is satisfied on $\mathrm{Aff}(B_l) \otimes \mathrm{Aff}(B_m) \otimes \mathrm{Aff}(B_k)$.

2.2. *Generalized local energies*

Let us give an explicit piecewise linear formula of the combinatorial R that originates in the tropical R for geometric crystals of type $D_n^{(1)}$.[15] First we make a slight variable change. The set B_l (1) is in one to one correspondence with another set

$$B'_l = \{x = (x_1, \ldots, x_n, \overline{x}_{n-1}, \ldots, \overline{x}_1) \in \mathbb{Z}^{2n-1} \mid x_i, \overline{x}_i \geq 0 \text{ for } 1 \leq i \leq n-1,$$
$$x_n \geq -\min(x_{n-1}, \overline{x}_{n-1}), \sum_{i=1}^{n-1}(x_i + \overline{x}_i) + x_n = l\} \tag{3}$$

by the relations

$$x_i = \zeta_i, \quad \overline{x}_i = \overline{\zeta}_i \quad (1 \leq i \leq n-2), \tag{4}$$
$$x_{n-1} = \zeta_{n-1} + \overline{\zeta}_n, \; x_n = \zeta_n - \overline{\zeta}_n, \; \overline{x}_{n-1} = \overline{\zeta}_{n-1} + \overline{\zeta}_n, \tag{5}$$
$$\zeta_n = \max(0, x_n), \; \overline{\zeta}_n = \max(0, -x_n), \tag{6}$$
$$\zeta_{n-1} = x_{n-1} + \min(0, x_n), \; \overline{\zeta}_{n-1} = \overline{x}_{n-1} + \min(0, x_n). \tag{7}$$

Note that x_n can be negative. We naturally use the notations like $x[d] \in \mathrm{Aff}(B'_l)$ and $R(x[d] \otimes x'[d']) = \tilde{x}'[d' + H(x \otimes x')] \otimes \tilde{x}[d - H(x \otimes x')]$, etc. Set

$$\ell(\zeta) = \sum_{i=1}^{n}(\zeta_i + \overline{\zeta}_i) \quad (\zeta \in B_l), \quad \ell(x) = \sum_{i=1}^{n-1}(x_i + \overline{x}_i) + x_n \quad (x \in B'_l), \tag{8}$$

so that $\ell(\zeta) = \ell(x) = l$ for $\zeta \in B_l$ and $x \in B'_l$.

Let $x = (x_1, \ldots, \overline{x}_1) \in B'_l$ and $y = (y_1, \ldots, \overline{y}_1) \in B'_m$. On the pair (x, y) we introduce mutually commuting involutions σ_1, σ_n and $*$ by

$$(x, y)^{\sigma_1} = (x^{\sigma_1}, y^{\sigma_1}), \quad (x, y)^{\sigma_n} = (x^{\sigma_n}, y^{\sigma_n}), \quad (x, y)^* = (y^*, x^*), \tag{9}$$
$$\begin{aligned} \sigma_1 &: x_1 \longleftrightarrow \overline{x}_1, \\ \sigma_n &: x_{n-1} \to x_{n-1} + x_n, \quad \overline{x}_{n-1} \to \overline{x}_{n-1} + x_n, \quad x_n \to -x_n, \\ * &: x_i \longleftrightarrow \overline{x}_i \quad (1 \leq i \leq n-1). \end{aligned} \tag{10}$$

The coordinates not included in the above rules are left unchanged. These involutions are naturally defined on $(\xi, \zeta) \in B_l \times B_m$ as well by the correspondence (4)-(7). For instance, one has $(\xi, \zeta)^* = (\zeta^*, \xi^*)$ with $\zeta^* = (\overline{\zeta}_1, \ldots, \overline{\zeta}_{n-1}, \zeta_n, \overline{\zeta}_n, \zeta_{n-1}, \ldots, \zeta_1)$ for $\zeta = (\zeta_1, \ldots, \zeta_n, \overline{\zeta}_n, \ldots, \overline{\zeta}_1)$.

For any function $g = g(x, y)$, we write $g^{\sigma_1} = g^{\sigma_1}(x, y) = g(x^{\sigma_1}, y^{\sigma_1})$, etc. Introduce the piecewise linear functions $V_i = V_i(x, y)$ and $W_i = W_i(x, y)$

for $0 \le i \le n-1$ as follows.

$$V_i = \max\left(\{\theta_{i,j}, \theta'_{i,j} | 1 \le j \le n-2\} \cup \{\eta_{i,j}, \eta'_{i,j} | 1 \le j \le n\}\right), \tag{11}$$
$$W_0 = 2V_0, \quad W_1 = V_0 + V_0^{\sigma_1}, \quad W_{n-1} = V_{n-1} + V_{n-1}^*, \tag{12}$$
$$W_i = \max\left(V_i + V_{i-1}^* - y_i, V_{i-1} + V_i^* - \overline{x}_i\right) + \min(x_i, \overline{y}_i), \tag{13}$$

where $2 \le i \le n-2$ in the last line. The functions $\theta_{i,j} = \theta_{i,j}(x,y), \theta'_{i,j} = \theta'_{i,j}(x,y), \eta_{i,j} = \eta_{i,j}(x,y), \eta'_{i,j} = \eta'_{i,j}(x,y)$ are defined by

$$\theta_{i,j}(x,y) = \begin{cases} \ell(x) + \displaystyle\sum_{k=j+1}^{i} (\overline{y}_k - \overline{x}_k) & \text{for } 1 \le j \le i, \\ \ell(y) + \displaystyle\sum_{k=i+1}^{j} (\overline{x}_k - \overline{y}_k) & \text{for } i+1 \le j \le n-2, \end{cases}$$

$$\theta'_{i,j}(x,y) = \ell(x) + \sum_{k=1}^{i} (\overline{y}_k - \overline{x}_k) + \sum_{k=1}^{j} (y_k - x_k) \text{ for } 1 \le j \le n-2,$$

$$\eta_{i,j}(x,y) = \begin{cases} \ell(x) + \displaystyle\sum_{k=j+1}^{i} (\overline{y}_k - \overline{x}_k) + \overline{y}_j - x_j & \text{for } 1 \le j \le i, \\ \ell(y) + \displaystyle\sum_{k=i+1}^{j} (\overline{x}_k - \overline{y}_k) + \overline{y}_j - x_j & \text{for } i+1 \le j \le n-1, \\ \ell(y) + \displaystyle\sum_{k=i+1}^{n-1} (\overline{x}_k - \overline{y}_k) + x_n & \text{for } j = n, \end{cases}$$

$$\eta'_{i,j}(x,y) = \begin{cases} \ell(x) + \displaystyle\sum_{k=1}^{i} (\overline{y}_k - \overline{x}_k) + \sum_{k=1}^{j} (y_k - x_k) + x_j - \overline{y}_j \quad \text{for } 1 \le j \le n-1, \\ \ell(x) + \delta_{i,n-1}(\ell(x) - \ell(y)) + \displaystyle\sum_{k=1}^{i} (\overline{y}_k - \overline{x}_k) + \sum_{k=1}^{n-1} (y_k - x_k) - x_n \\ \hfill \text{for } j = n. \end{cases}$$

Theorem 2.1 (Ref. 15, Theorem 4.28 and Remark 4.29). *The image $y' \otimes x' = R(x \otimes y)$ of the combinatorial R is given by*

$$\begin{aligned} &x'_i = x_i + V_{i-1}^* - V_i^*, \quad \overline{x}'_i = \overline{x}_i + V_{i-1}^* + W_i - V_i^* - W_{i-1} \quad (1 \le i \le n-1), \\ &x'_n = x_n + V_{n-1}^* - V_{n-1}, \quad y'_n = y_n + V_{n-1} - V_{n-1}^*, \\ &y'_i = y_i + V_{i-1} + W_i - V_i - W_{i-1}, \quad \overline{y}'_i = \overline{y}_i + V_{i-1} - V_i \quad (1 \le i \le n-1). \end{aligned} \tag{14}$$

Moreover, the local energy is given by

$$H(x \otimes y) = V_0(x, y) \tag{15}$$

up to a constant shift.

The functions $V_1, \ldots, V_{n-1}, W_1, \ldots, W_{n-1}$ and σ_1, σ_n and $*$ of them are relatives of the local energy. In addition to the involutions σ_1, σ_n and $*$, the combinatorial R naturally acts on them by $(RV_0)(x, y) = V_0(y', x')$ with $y' \otimes x' = R(x \otimes y)$, etc. Their transformation properties under $\sigma_1, \sigma_n, *$ and R are summarized in Table 1.[15] These involutions are commutative, thus for instance $R(V_0^{\sigma_1}) = (R(V_0))^{\sigma_1} = V_0^{\sigma_1}$.

Table 1. Transformation by $\sigma_1, \sigma_n, *$ and R.

	V_0	$V_i\ (1 \le i \le n-2)$	V_{n-1}	$W_i\ (1 \le i \le n-1)$
σ_1	$V_0^{\sigma_1}$	V_i	V_{n-1}	W_i
σ_n	V_0	V_i	V_{n-1}^*	W_i
$*$	V_0	V_i^*	V_{n-1}^*	W_i
R	V_0	$W_i - V_i^*$	V_{n-1}	W_i

Due to these properties, there are a few simplifications in (14) as

$$\begin{aligned} x_1' &= x_1 + V_0 - V_1^*, \quad \overline{x}_1' = \overline{x}_1 + V_0^{\sigma_1} - V_1^*, \\ y_1' &= y_1 + V_0^{\sigma_1} - V_1, \quad \overline{y}_1' = \overline{y}_1 + V_0 - V_1. \end{aligned} \tag{16}$$

We write

$$u_l = (l, 0, \ldots, 0) \in B_l. \tag{17}$$

By using Theorem 2.1, one can show for any $\zeta \in B_m$ that

$$B_l \otimes B_m \ni u_l \otimes \zeta \overset{\sim}{\mapsto} u_m \otimes \xi' \in B_m \otimes B_l \ \text{ if } l \ge m \tag{18}$$

for some ξ' under the combinatorial R. In particular

$$u_l \otimes u_m \simeq u_m \otimes u_l$$

holds. The functions in Table 1 attain their maximum $V_0 = V_0^{\sigma_1} = V_i = V_i^* = l + m$ and $W_i = 2(l + m)$ for $1 \le i \le n - 1$ at $(x, y) = (u_l, u_m)$.

For $\xi \otimes \zeta \in B_l \otimes B_m$, let $x \in B_l'$ and $y \in B_m'$ be the elements corresponding to ξ and ζ, respectively. We set

$$v_i(\xi \otimes \zeta) = \ell(\xi) + \ell(\zeta) - V_i(x, y) \ \ (0 \le i \le n-1) \tag{19}$$

$$v_0^{\sigma_1}(\xi \otimes \zeta) = \ell(\xi) + \ell(\zeta) - V_0^{\sigma_1}(x, y), \tag{20}$$

$$v_i^*(\xi \otimes \zeta) = \ell(\xi) + \ell(\zeta) - V_i^*(x, y) \ \ (1 \le i \le n-1), \tag{21}$$

$$w_i(\xi \otimes \zeta) = 2\ell(\xi) + 2\ell(\zeta) - W_i(x, y) \ \ (1 \le i \le n-1), \tag{22}$$

and call them *generalized local energies*. Note that $w_{n-1} - v_{n-1} = v_{n-1}^*$. They are building blocks of the piecewise linear formula of the combinatorial R (14). From the above remark, generalized local energies are all nonnegative and normalized so that

$$g(u_l \otimes u_m) = 0 \text{ for any } g = v_i, v_0^{\sigma_1}, v_i^* \text{ and } w_i. \tag{23}$$

For $\zeta = (\zeta_1, \dots, \zeta_n, \overline{\zeta}_n, \dots, \overline{\zeta}_1) \in B_l$, we introduce

$$\mathfrak{a}(\zeta) = \zeta_2 + \cdots \zeta_n + \overline{\zeta}_n + \cdots + \overline{\zeta}_2 + 2\overline{\zeta}_1 = \ell(\zeta) + \overline{\zeta}_1 - \zeta_1, \tag{24}$$

$$\gamma_{v_a}(\zeta) = \zeta_2 + \cdots \zeta_n + \overline{\zeta}_n + \cdots + \overline{\zeta}_{a+1} + \overline{\zeta}_1 \quad (0 \le a \le n-2), \tag{25}$$

$$\gamma_{v_{n-1}}(\zeta) = \zeta_2 + \cdots + \zeta_{n-1} + \zeta_n + \overline{\zeta}_1, \tag{26}$$

$$\gamma_{v_{n-1}^*}(\zeta) = \zeta_2 + \cdots + \zeta_{n-1} + \overline{\zeta}_n + \overline{\zeta}_1, \tag{27}$$

$$\gamma_{w_a - v_a}(\zeta) = \zeta_2 + \cdots + \zeta_a + \overline{\zeta}_1 \quad (1 \le a \le n-2), \tag{28}$$

$$\gamma_{v_0^{\sigma_1}}(\zeta) = 0. \tag{29}$$

Note that $\gamma_{v_0}(\zeta) = \mathfrak{a}(\zeta)$.

Lemma 2.1. *Let $\xi = (\xi_1, \dots, \overline{\xi}_1) \in B_l$ and $\zeta = (\zeta_1, \dots, \overline{\zeta}_1) \in B_m$. Set $\zeta' \otimes \xi' = R(\xi \otimes \zeta) \in B_m \otimes B_l$. For ξ_1 (hence l as well) sufficiently large, the relation*

$$g(\xi \otimes \zeta) = \gamma_g(\zeta) + \mathfrak{a}(\zeta') - \gamma_g(\zeta')$$

is valid for g appearing in (25)–(29), where the left hand side with $g = w_a - v_a$ is to be understood as $w_a(\xi \otimes \zeta) - v_a(\xi \otimes \zeta)$.

Proof. One can check that $\xi_1 \ge m - \zeta_1 + \overline{\zeta}_1$ is sufficient to guarantee that V_0 (11) is equal to $\eta'_{0,1} = l + \zeta_1 - \overline{\zeta}_1$. This implies that $v_0(\xi \otimes \zeta) = m - \zeta_1 + \overline{\zeta}_1 = \gamma_{v_0}(\zeta)$ showing the $g = v_0$ case. All the other cases are deduced from this, (14) and (4)–(7) without using the concrete forms of $\theta_{i,j}, \theta'_{i,j}, \eta_{i,j}$ and $\eta'_{i,j}$. □

For $g = w_a$ $(1 \le a \le n-2)$, an analogue of Lemma 2.1 holds with

$$\begin{aligned} w_a(\xi \otimes \zeta) &= \gamma_{w_a}(\zeta) + 2\mathfrak{a}(\zeta') - \gamma_{w_a}(\zeta'), \\ \gamma_{w_a}(\zeta) &= \gamma_{w_a - v_a}(\zeta) + \gamma_{v_a}(\zeta) \\ &= \mathfrak{a}(\zeta) + \zeta_2 + \cdots + \zeta_a - (\overline{\zeta}_a + \cdots + \overline{\zeta}_2). \end{aligned} \tag{30}$$

As ξ_1 gets large, ζ' stabilizes since it is a piecewise linear function of ξ_1 staying in a finite set B_m. ($\xi_1 \ge m$ seems sufficient for the convergence.) Therefore Lemma 2.1 ensures that all the generalized local energies $g(\xi \otimes \zeta)$ are well defined in the limit $\xi_1 \to \infty$.

2.3. *Generalized energies*

For $p = p_1 \otimes \cdots \otimes p_L \in B_{l_1} \otimes \cdots \otimes B_{l_L}$, define $p_j^{(i)} \in B_{l_j}$ $(i < j)$ by

$$\begin{aligned} (B_{l_i} \otimes \cdots \otimes B_{l_{j-1}}) \otimes B_{l_j} &\overset{\sim}{\to} B_{l_j} \otimes (B_{l_i} \otimes \cdots \otimes B_{l_{j-1}}) \\ p_i \otimes \cdots \otimes p_{j-1} \otimes p_j &\mapsto p_j^{(i)} \otimes p_i' \otimes \cdots \otimes p_{j-1}', \end{aligned} \tag{31}$$

sending p_j to the left by successive applications of the combinatorial R. We set $p_j^{(j)} = p_j$. For any generalized local energy in (19)–(22), we define the *generalized energy* of $p = p_1 \otimes \cdots \otimes p_L \in B_{l_1} \otimes \cdots \otimes B_{l_L}$ by

$$\mathcal{E}_g(p) = \sum_{0 \le i < j \le L} g(p_i \otimes p_j^{(i+1)}) \tag{32}$$

by taking $p_0 = u_l$ with sufficiently large l. This is well defined (finite) due to Lemma 2.1 and the comment following it. In the rest of the paper, we will simply write $p_0 = u_\infty \in B_\infty$.

When $g = v_0$, (32) is the energy introduced in Refs. 18 and 5 up to a sign and a constant shift. If furthermore $l_1, \ldots, l_L$ are all equal, then $p_j^{(i+1)} = p_{i+1}$ holds and (32) reduces to

$$\mathcal{E}_{v_0}(p) = \sum_{0 \le i < L} (L - i)\, v_0(p_i \otimes p_{i+1}).$$

Its generating function $\sum_p q^{\mathcal{E}_{v_0}(p)}$ is a version of the one dimensional configuration sum going back to Refs. 2 and 1, which is the essential ingredient in the corner transfer matrix method.

Any quantity $G(p_\alpha \otimes p_{\alpha+1} \otimes \cdots \otimes p_\beta)$ will be said *R-invariant* if $G(\cdots \otimes p_i \otimes p_{i+1} \otimes \cdots) = G(\cdots \otimes R(p_i \otimes p_{i+1}) \otimes \cdots)$ for any $\alpha \le i < \beta$.

Remark 2.1. Due to the transformation property under R in Table 1, the generalized energy $\mathcal{E}_g(p)$ is R-invariant for $g = v_0, v_0^{\sigma_1}, v_{n-1}, v_{n-1}^*$ and $w_1, \ldots, w_{n-1}$. On the other hand, $\mathcal{E}_g$ with $g = v_1, \ldots, v_{n-2}$ and $v_1^*, \ldots, v_{n-2}^*$ are *not* R-invariant.

Let us depict the relation $R(b \otimes c) = \tilde{c} \otimes \tilde{b}$ as

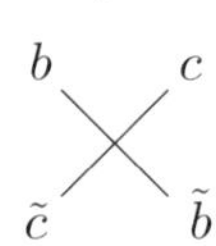

Then the Yang-Baxter equation (2) takes the well known form:

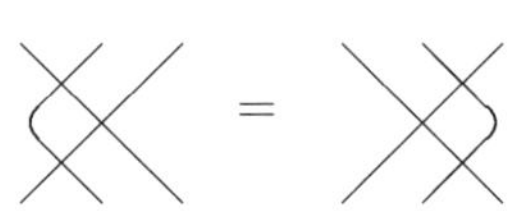

The defining relation (31) of $p_j^{(i)}$ looks as

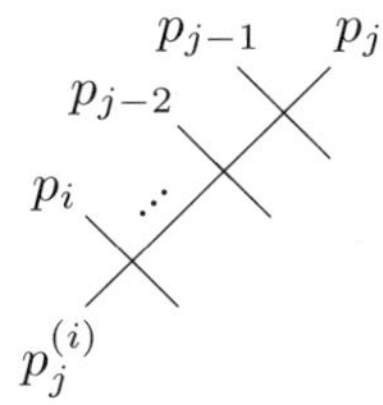

(33)

Remember that each vertex is associated with various generalized local energies $g(b \otimes c)$. Let

$$I_g = I_g(p_i \otimes \cdots \otimes p_{j-1} \otimes p_j) = \sum_{i \le k < j} g(p_k \otimes p_j^{(k+1)}) \tag{34}$$

be the sum of generalized local energy g over all the vertices in (33). Then the generalized energy (32) is expressed as

$$\begin{aligned} \mathcal{E}_g(p_1 \otimes \cdots \otimes p_L) &= \mathcal{E}_g(p_1 \otimes \cdots \otimes p_{L-1}) + I_g(u_\infty \otimes p_1 \otimes \cdots \otimes p_L) \\ &= \sum_{1 \le j \le L} I_g(u_\infty \otimes p_1 \otimes \cdots \otimes p_{j-1} \otimes p_j). \end{aligned} \tag{35}$$

From (18) and (23), it follows that

$$I_g(u_\infty \otimes u_\infty \otimes p_1 \otimes \cdots \otimes p_j) = I_g(u_\infty \otimes p_1 \otimes \cdots \otimes p_j), \tag{36}$$

$$\mathcal{E}_g(u_\infty \otimes p_1 \otimes \cdots \otimes p_L) = \mathcal{E}_g(p_1 \otimes \cdots \otimes p_L). \tag{37}$$

Lemma 2.2. *In (33), the following quantities are R-invariant as the functions of $p_i \otimes \cdots \otimes p_{j-1}$. (i) The element $p_j^{(i)}$. (ii) I_g (34) for any $g = v_a$ $(0 \le a \le n-1)$, $w_a - v_a$ $(1 \le a \le n-2)$, v_{n-1}^* and $v_0^{\sigma_1}$.*

Proof. (i) This is due to the classical part of the Yang-Baxter equation (2). (ii) The R-invariance of I_{v_0} follows from (19), (15) and the affine part of the Yang-Baxter equation. Let $y, y' \in B'_{l_j}$ be the elements corresponding to $p_j, p_j^{(i)} \in B_{l_j}$, respectively. From (16), we have $\overline{y}'_1 - \overline{y}_1 = I_{v_1} - I_{v_0}$. Since the left hand side is R-invariant by (i), this relation implies the R-invariance of I_{v_1}. By similarly using the R-invariance of $\overline{y}'_i - \overline{y}_i$ and $y'_i - y_i$ in (16) and (14), one can verify the R-invariance of the other I_g. □

We note that Lemma 2.2 is applicable to the situation $i = 0$, i.e., $p_i \otimes \cdots \otimes p_{j-1} = u_\infty \otimes p_1 \otimes \cdots \otimes p_{j-1}$.

Remark 2.2. Lemma 2.2 (ii) does not concern $I_{v_1^*}, \ldots, I_{v_{n-1}^*}$. In fact the proof does not persist since $V_1^*, \ldots, V_{n-2}^*$ are not contained in any difference of the components of y' and y in (14). Similarly, $V_1, \ldots, V_{n-2}$ do not appear in the differences of x and x'. This *chirality* of the combinatorial R is a characteristic feature of the $D_n^{(1)}$ case. In contrast, I_g's with all the generalized local energies g for $A_n^{(1)}$ (so called ith (un)winding number[17]) are R-invariant. We shall come back to this point again in Section 4.2.

The following proposition and its proof are parallel with Lemma 4.4 in Ref. 17 for type $A_n^{(1)}$.

Proposition 2.1. *For* $g = v_i\,(0 \le i \le n-1), w_i\,(1 \le i \le n-1)$, $v_0^{\sigma_1}$ *and* v_{n-1}^*, *the generalized energy* $\mathcal{E}_g(p_1 \otimes \cdots \otimes p_L)$ *(32) is equal to the sum of the generalized local energy* g *attached to all the vertices in the following diagram* ($L = 3$ *example*):

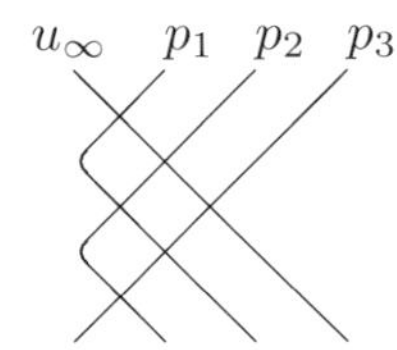

Proof. We invoke the induction on L. For $L = 1$, one has $\mathcal{E}_g(p_1) = g(u_\infty \otimes p_1)$, and the assertion is obviously true. We illustrate the induction step from $L = 2$ to $L = 3$. Consider the following identity obtained by successive applications of the Yang-Baxter equation:

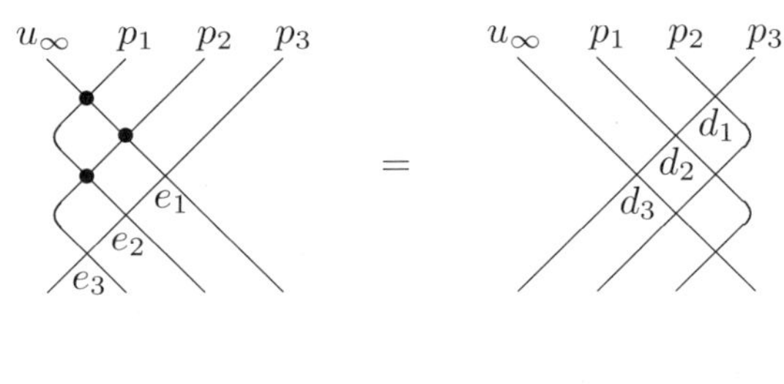

Here $\bullet, e_i, d_i$ stand for the values of g at the attached vertices. By the induction assumption, the sum of the three $\bullet$ in the left hand side is equal to $\mathcal{E}_g(p_1 \otimes p_2)$. In view of the recursion relation (35), we are to verify $e_1 + e_2 + e_3 = I_g(u_\infty \otimes p_1 \otimes p_2 \otimes p_3)$. By the definition, $I_g(u_\infty \otimes p_1 \otimes p_2 \otimes p_3) = d_1 + d_2 + d_3$ in the right diagram, where $d_1 = g(p_2 \otimes p_3^{(3)}), d_2 = g(p_1 \otimes p_3^{(2)}), d_3 = g(u_\infty \otimes p_3^{(1)})$. Thanks to the R-invariance of I_g in Lemma 2.2 (ii), this is equal to $e_1 + e_2 + e_3$. $\square$

3. Integrable $D_n^{(1)}$ cellular automaton

3.1. *States and time evolution*

Let us recall the integrable $D_n^{(1)}$ cellular automaton associated with B_l.[6,7] Consider the crystal $B_{l_1} \otimes \cdots \otimes B_{l_L}$. Its elements are called states. We regard each component $(\zeta_1, \ldots, \overline{\zeta}_1) \in B_l$ as a capacity l box containing ζ_a particles a and $\overline{\zeta}_a$ anti-particles $\overline{a}$ for $2 \le a \le n$, and furthermore $\overline{\zeta}_1$ extra $\overline{1}$'s which we call bound pairs. The remaining 1's represent empty space in the box. Thus u_l stands for an empty box. The indices $a, \overline{a}$ will be referred as color of particles and anti-particles, respectively. A state $p_1 \otimes \cdots \otimes p_L \in B_{l_1} \otimes \cdots \otimes B_{l_L}$ represents a configuration of particles, anti-particles and bound pairs in an array of boxes with capacity $l_1, \ldots, l_L$. Because of the constraint $\zeta_n \overline{\zeta}_n = 0$ in (1), particles n and anti-particles $\overline{n}$ do not coexist within a box. We shall denote the element $(3,0,1,0,2,0,1,0) \in B_7$ of $D_4^{(1)}$, for example, by $1113\overline{442}$, etc.

For a positive integer l, we define the time evolution $T_l(p) = p'_1 \otimes \cdots \otimes p'_L$ of a state $p = p_1 \otimes \cdots \otimes p_L$ by

$$u_l \otimes p_1 \otimes \cdots \otimes p_L \simeq p'_1 \otimes \cdots \otimes p'_L \otimes \xi$$

under the isomorphism $B_l \otimes (B_{l_1} \otimes \cdots \otimes B_{l_L}) \simeq (B_{l_1} \otimes \cdots \otimes B_{l_L}) \otimes B_l$. Here $\xi \in B_l$ as well as $T_l(p)$ are uniquely determined from p by the combinatorial R. It can be shown that

$$\xi = u_l \ \text{ if } p_j = u_{l_j} \text{ for } L' \le j \le L \text{ with sufficiently large } L - L'. \tag{38}$$

The time evolutions $\{T_l\}$ form a commuting family, and T_l stabilizes as l gets large, which will be denoted by T_∞.

When $l_1 = \cdots = l_L = 1$, T_∞ is factorized as

$$T_\infty = K_2 K_3 \cdots K_n K_{\overline{n}} \cdots K_{\overline{3}} K_{\overline{2}}$$

with K_a given by the following algorithm (we understand $\overline{\overline{a}} = a$).[7]

(i) Replace each $\overline{1}$ by a pair $a, \overline{a}$ within a box.
(ii) Pick the leftmost a (if any) and move it to the nearest right box which is empty or containing just $\overline{a}$. (Boxes involving the pair $a, \overline{a}$ are prohibited as the destination.)
(iii) Repeat (ii) for those a's that are not yet moved until all of a's are moved once.
(iv) Replace the pair $a, \overline{a}$ within a box (if any) by $\overline{1}$.

In the above, taking some $b\,(\neq 1) \in B_1$ away from a box means the change of the local state $b \to 1$. Similarly, putting $b\,(\neq 1) \in B_1$ into an empty box means the change $1 \to b$. The steps (i) and (iv) can be viewed as pair creation and annihilation, respectively.

Example 3.1. We consider $D_4^{(1)}$ and states from $B_1^{\otimes L}$ with $L = 48$. Successive time evolutions of the initial state on the first line under T_∞ is presented downward, where $b_1 \otimes \cdots \otimes b_L$ is simply denoted by a horizontal array $b_1 b_2 \ldots b_L$. It shows a collision of solitons with amplitudes 6 and 3.

$$
\begin{array}{l}
111\bar{3}\bar{3}\bar{4}322 11111\bar{2}\bar{2}\bar{4}1111111111111111111111111111111 \\
111111111\bar{3}\bar{3}\bar{4}32211\bar{2}\bar{2}\bar{4}1111111111111111111111111111 \\
111111111111111\bar{3}\bar{3}\bar{4}32\bar{1}\bar{2}\bar{4}1111111111111111111111111 \\
111111111111111111111\bar{3}4\bar{1}\bar{1}\bar{4}\bar{4}\bar{4}11111111111111111111 \\
1111111111111111111111111422\bar{2}\bar{2}\bar{3}\bar{4}\bar{4}\bar{4}11111111111111 \\
1111111111111111111111111111422111\bar{2}\bar{2}\bar{3}\bar{4}\bar{4}\bar{4}11111111 \\
1111111111111111111111111111111422111111\bar{2}\bar{2}\bar{3}\bar{4}\bar{4}\bar{4}11
\end{array}
$$

For $l_1, \ldots, l_L$ general, T_∞ still admits a similar, although slightly more involved, algorithm. We omit it here and give an example instead.

Example 3.2. We consider $D_4^{(1)}$ and states from $B_6 \otimes B_3 \otimes B_4 \otimes B_4 \otimes B_2^{\otimes 8}$.

$$
\begin{array}{l}
124\overline{321} \cdot 234 \cdot 2\overline{321} \cdot 13\overline{44} \cdot 11 \cdot 11 \cdot 11 \cdot 11 \cdot 11 \cdot 11 \cdot 11 \cdot 11 \\
111111 \cdot 123 \cdot 12\overline{32} \cdot 344\bar{1} \cdot \overline{32} \cdot \overline{43} \cdot 3\bar{4} \cdot 12 \cdot 11 \cdot 11 \cdot 11 \cdot 11 \\
111111 \cdot 111 \cdot 1123 \cdot 1113 \cdot \overline{33} \cdot 44 \cdot 23 \cdot 1\bar{2} \cdot \overline{32} \cdot \overline{43} \cdot 3\bar{4} \cdot 12 \\
111111 \cdot 111 \cdot 1111 \cdot 1123 \cdot 13 \cdot 11 \cdot 11 \cdot \overline{33} \cdot 44 \cdot 23 \cdot 11 \cdot 1\bar{2} \\
111111 \cdot 111 \cdot 1111 \cdot 1111 \cdot 23 \cdot 13 \cdot 11 \cdot 11 \cdot 11 \cdot 11 \cdot \overline{33} \cdot 44 \\
111111 \cdot 111 \cdot 1111 \cdot 1111 \cdot 11 \cdot 23 \cdot 13 \cdot 11 \cdot 11 \cdot 11 \cdot 11 \cdot 11 \\
111111 \cdot 111 \cdot 1111 \cdot 1111 \cdot 11 \cdot 11 \cdot 23 \cdot 13 \cdot 11 \cdot 11 \cdot 11 \cdot 11 \\
111111 \cdot 111 \cdot 1111 \cdot 1111 \cdot 11 \cdot 11 \cdot 11 \cdot 23 \cdot 13 \cdot 11 \cdot 11 \cdot 11 \\
111111 \cdot 111 \cdot 1111 \cdot 1111 \cdot 11 \cdot 11 \cdot 11 \cdot 11 \cdot 23 \cdot 13 \cdot 11 \cdot 11 \\
111111 \cdot 111 \cdot 1111 \cdot 1111 \cdot 11 \cdot 11 \cdot 11 \cdot 11 \cdot 11 \cdot 23 \cdot 13 \cdot 11 \\
111111 \cdot 111 \cdot 1111 \cdot 1111 \cdot 11 \cdot 11 \cdot 11 \cdot 11 \cdot 11 \cdot 11 \cdot 23 \cdot 13
\end{array}
$$

Here, $\cdot$ represents $\otimes$.

Remark 3.1. Suppose $p_j = u_{l_j}$ for $1 \le j \le k$ in a state $p = p_1 \otimes \cdots \otimes p_L$. Then in the state $T_\infty(p) = p'_1 \otimes \cdots \otimes p'_L$, $p'_j = u_{l_j}$ is valid for $1 \le j \le k+1$.

We postpone the inverse scattering formalism of the dynamics to Theorem 5.2.

3.2. *Counting particles and anti-particles*

Recall that $\mathfrak{a}(\zeta)$ is defined in (24). In our present context, it is the number of all the particles and anti-particles within a box specified by $\zeta \in B_l$, where the term $2\overline{\zeta}_1$ means that a bound pair is regarded as a pair of a particle and an anti-particle (whose color is unspecified). The symbol $\mathfrak{a}$ means $\mathfrak{a}$ll kinds of (anti-)particles.

Let $p = p_1 \otimes \cdots \otimes p_L$ be a state and write its time evolution as

$$T_\infty^t(p_1 \otimes \cdots \otimes p_L) = p_1^t \otimes \cdots \otimes p_L^t,$$

where $p_j^t \in B_{l_j}$. We write $p_j = p_j^0 = (\zeta_{j,1}, \ldots, \zeta_{j,n}, \overline{\zeta}_{j,n}, \ldots, \overline{\zeta}_{j,1}) \in B_{l_j}$. For any elements $a_1, \ldots, a_r$ of $\{2, 3, \ldots, n, \overline{n}, \ldots, \overline{2}, \overline{1}\}$, we define the *counting function*

$$\rho_{a_1,\ldots,a_r}(p) = \sum_{j=1}^{L}(\zeta_{j,a_1} + \cdots + \zeta_{j,a_r}) + \sum_{t\geq 1}\sum_{j=1}^{L}\mathfrak{a}(p_j^t), \tag{39}$$

where $\zeta_{j,\overline{3}} = \overline{\zeta}_{j,3}$, etc. The dependence on $a_1, \ldots, a_r$ enters the first term only. The indices in $\rho_{a_1,\ldots,a_r}$ will always be arranged in the order $2, 3, \ldots, n, \overline{n}, \ldots, \overline{3}, \overline{2}, \overline{1}$. The second term is finite due to Remark 3.1. In fact the double sum may well be restricted to $\sum_{t=1}^{L-1}\sum_{j=t+1}^{L}$ where the nonzero contributions are contained. This region is depicted as the SW quadrant of the time evolution patterns like Examples 3.1 and 3.2.

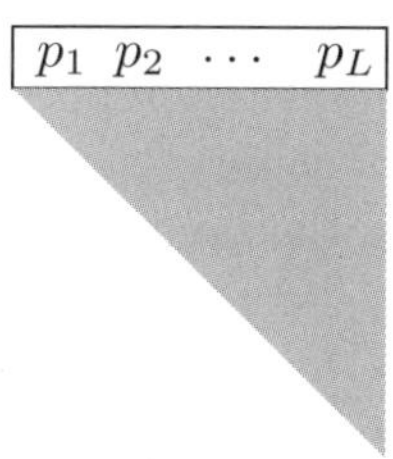

(40)

The first term in (39) is the number of (anti-)particles with colors $a_1, \ldots, a_r$ contained in the top row which is the state p itself. The second term counts all kinds of particles and anti-particles in the hatched domain in (40).[b] By the definition it follows that

$$\rho_{\emptyset}(p) = \rho_{2,\ldots,n,\overline{n},\ldots,\overline{2},\overline{1},\overline{1}}(T_\infty(p)). \tag{41}$$

Given a state $p = p_1 \otimes \cdots \otimes p_L$, we write

$$p_{[k]} = p_1 \otimes \cdots \otimes p_k \quad (1 \leq k \leq L). \tag{42}$$

[b]More precisely, it should be hatched in a staircase shape.

Example 3.3. Let p be the state in the first line in Example 3.2. Then, the counting function $\rho_{a_1,\dots,a_r}(p_{[k]})$ for $1 \le k \le 9$ takes the following values. (The middle column shows g such that $\rho_{a_1,\dots,a_r} = \rho_g$ in (44)–(47).)

$a_1,\dots,a_r$	g	$k=1$	2	3	4	5	6	7	8	9
$234\overline{4}\overline{3}\overline{2}\overline{1}\overline{1}$	v_0	6	11	21	32	39	46	53	60	67
$234\overline{4}\overline{3}\overline{2}\overline{1}$	v_1	5	10	19	30	37	44	51	58	65
$234\overline{4}\overline{3}\overline{1}$	v_2	4	9	17	28	35	42	49	56	63
$234\overline{1}$	v_3	3	8	15	24	31	38	45	52	59
$23\overline{4}\overline{1}$	v_3^*	2	6	13	24	31	38	45	52	59
$2\overline{1}$	$w_2 - v_2$	2	5	12	20	27	34	41	48	55
$\overline{1}$	$w_1 - v_1$	1	3	9	17	24	31	38	45	52
$\emptyset$	$v_0^{\sigma_1}$	0	2	7	15	22	29	36	43	50

We set $\rho_{a_1,\dots,a_r}(p_{[0]}) = 0$ for any $a_1,\dots,a_r$. Consider the difference $\rho_{2\overline{1}}(p_{[k]}) - \rho_{\overline{1}}(p_{[k]})$ for example. By the definition (39), it is the number of color 2 particles contained in $p_{[k]}$. Thus we have

$$\begin{aligned}&\sharp(\text{color 2 particles}) \text{ in } p_k \; (= \zeta_{k,2})\\ &= \rho_{2\overline{1}}(p_{[k]}) - \rho_{\overline{1}}(p_{[k]}) - \rho_{2\overline{1}}(p_{[k-1]}) + \rho_{\overline{1}}(p_{[k-1]}).\end{aligned}$$

This is an example of the relations that reproduces a local variable from non-local counting functions. Given l_k, the set of counting functions that are necessary and sufficient to completely reproduce the local state $p_k \in B_{l_k}$ is not unique. However there is a choice that is linked with the generalized energies in Section 2.3. By using the function γ_g in (25)–(29), we set

$$\rho_g(p) = \sum_{j=1}^{L} \gamma_g(p_j) + \sum_{t\ge 1}\sum_{j=1}^{L} \mathfrak{a}(p_j^t) \tag{43}$$

for $g = v_a$ $(0 \le a \le n-1)$, $w_a - v_a$ $(1 \le a \le n-2)$, v_{n-1}^* and $v_0^{\sigma_1}$. Although the notations ρ_g here and $\rho_{a_1,\dots,a_r}$ in (39) are somewhat confusing, we dare to use the both in the sequel supposing the resemblance is not too serious. Then (43) is explicitly given as follows:

$$\rho_{v_a}(p) = \rho_{2,\dots,n,\overline{n},\dots,\overline{a+1},\overline{1}}(p) \quad (0 \le a \le n-2), \tag{44}$$

$$\rho_{v_{n-1}}(p) = \rho_{2,3,\dots,n-1,n,\overline{1}}(p), \quad \rho_{v_{n-1}^*}(p) = \rho_{2,3,\dots,n-1,\overline{n},\overline{1}}(p), \tag{45}$$

$$\rho_{w_a - v_a}(p) = \rho_{2,3,\dots,a,\overline{1}}(p) \quad (1 \le a \le n-2), \tag{46}$$

$$\rho_{v_0^{\sigma_1}}(p) = \rho_{\emptyset}(p). \tag{47}$$

The last one is subsidiary in that $\rho_{v_0^{\sigma_1}}(p) = \rho_{w_1-v_1}(p) - \rho_{v_0}(p) + \rho_{v_1}(p)$ holds reflecting (12). One may also additionally introduce

$$\rho_{w_a}(p) = \rho_{w_a-v_a}(p) + \rho_{v_a}(p) = \sum_{j=1}^{L} \gamma_{w_a}(p_j) + 2\sum_{t\geq 1}\sum_{j=1}^{L} \mathfrak{a}(p_j^t)$$

for $1 \leq a \leq n-2$. See (30). For $D_4^{(1)}$, the counting functions (44)–(47) are precisely those listed in Example 3.3.

Theorem 3.1. *For $p_{[k]}$ in (42), set $\delta\rho_g = \rho_g(p_{[k]}) - \rho_g(p_{[k-1]})$. The counting functions (44)–(46) reproduce the local state $p_k = (\zeta_1, \dots, \zeta_n, \overline{\zeta}_n, \dots, \overline{\zeta}_1) \in B_{l_k}$ by*

$$\begin{aligned}
\zeta_1 &= l_k - \delta\rho_{v_0} + \delta\rho_{w_1-v_1},\\
\zeta_a &= \delta\rho_{w_a-v_a} - \delta\rho_{w_{a-1}-v_{a-1}} \quad (2 \leq a \leq n-2),\\
\zeta_{n-1} &= \min(\delta\rho_{v_{n-1}}, \delta\rho_{v^*_{n-1}}) - \delta\rho_{w_{n-2}-v_{n-2}},\\
\zeta_n &= \max(\delta\rho_{v_{n-1}} - \delta\rho_{v^*_{n-1}}, 0),\\
\overline{\zeta}_n &= \max(\delta\rho_{v^*_{n-1}} - \delta\rho_{v_{n-1}}, 0),\\
\overline{\zeta}_{n-1} &= -\max(\delta\rho_{v_{n-1}}, \delta\rho_{v^*_{n-1}}) + \delta\rho_{v_{n-2}},\\
\overline{\zeta}_a &= \delta\rho_{v_{a-1}} - \delta\rho_{v_a} \quad (1 \leq a \leq n-2).
\end{aligned}$$

Proof. Straightforward by using (39), (44)–(46) and $\min(\zeta_n, \overline{\zeta}_n) = 0$. □

4. Main result

4.1. *Counting functions and generalized energies*

Theorem 4.1. *For any state $p \in B_{l_1} \otimes \cdots \otimes B_{l_L}$, the counting functions and the generalized energies (32) coincide, namely,*

$$\mathcal{E}_g(p) = \rho_g(p) \tag{48}$$

for $g = v_a$ $(0 \leq a \leq n-1)$, $w_a - v_a$ $(1 \leq a \leq n-2)$, v^*_{n-1} and $v_0^{\sigma_1}$.

Here, $\mathcal{E}_{w_a-v_a}(p)$ should be understood as $\mathcal{E}_{w_a}(p) - \mathcal{E}_{v_a}(p)$, and the same convention is assumed for $I_{w_a-v_a}(p)$ in the sequel. Of course $\mathcal{E}_{w_a}(p) = \rho_{w_a}(p)$ follows as a corollary. The g's in Theorem 4.1 are the same as those considered in Lemma 2.2 (ii) and (43). By substituting (44)–(47) into (48), the theorem may be rephrased as

$$\begin{aligned}
\mathcal{E}_{v_a}(p) &= \rho_{2,\ldots,n,\overline{n},\ldots,\overline{a+1},\overline{1}}(p) \quad (0 \le a \le n-2),\\
\mathcal{E}_{v_{n-1}}(p) &= \rho_{2,3,\ldots,n-1,n,\overline{1}}(p), \quad \mathcal{E}_{v^*_{n-1}}(p) = \rho_{2,3,\ldots,n-1,\overline{n},\overline{1}}(p),\\
\mathcal{E}_{w_a - v_a}(p) &= \rho_{2,3,\ldots,a,\overline{1}}(p) \quad (1 \le a \le n-2),\\
\mathcal{E}_{v_0^{\sigma_1}}(p) &= \rho_{\emptyset}(p).
\end{aligned}$$

For the proof we need one more Lemma.

Lemma 4.1. *Let $p = p_1 \otimes \cdots \otimes p_L \in B_{l_1} \otimes \cdots \otimes B_{l_L}$. For those g's in Theorem 4.1, the following equality is valid:*

$$\mathcal{E}_g(p) - \mathcal{E}_g(T_\infty(p)) = \sum_{i=1}^{L} g(\xi^{(i)} \otimes p_i), \tag{49}$$

where $\xi^{(i)} \in B_\infty$ is defined by $u_\infty \otimes p_1 \otimes \cdots \otimes p_{i-1} \simeq p'_1 \otimes \cdots \otimes p'_{i-1} \otimes \xi^{(i)}$.

Proof. The following proof simplifies the one for Proposition 4.6 in Ref. 17 in that the assumption $l_1 \ge \cdots \ge l_L$ is not needed. We illustrate it for $L = 3$. Set $p = p_1 \otimes p_2 \otimes p_3$ and $T_\infty(p) = p'_1 \otimes p'_2 \otimes p'_3$. Then, Proposition 2.1 tells that $\mathcal{E}_g(T_\infty(p))$ is the sum of g at all $\bullet$ in the following diagram.

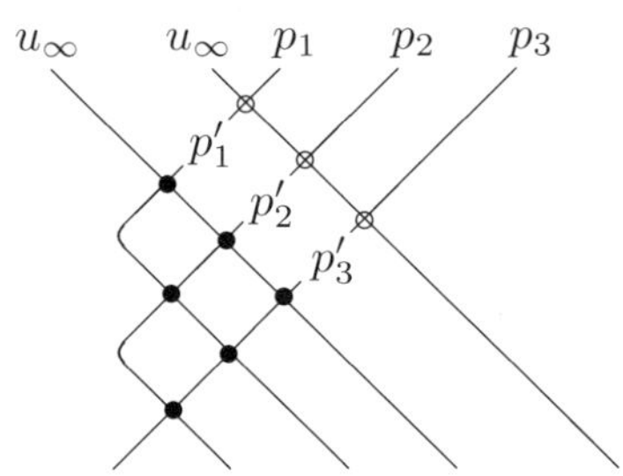

On the other hand, the right hand side of (49) is equal to the sum of g at all $\circ$. Thus from Lemma 2.2 (ii), we find

$$\begin{aligned}
\mathcal{E}_g(T_\infty(p)) + \sum_{i=1}^{L} g(\xi^{(i)} \otimes p_i) &= \sum_{1 \le j \le 3} I_g(u_\infty \otimes u_\infty \otimes p_1 \otimes \cdots \otimes p_j)\\
&\overset{(36)}{=} \sum_{1 \le j \le 3} I_g(u_\infty \otimes p_1 \otimes \cdots \otimes p_j) \overset{(35)}{=} \mathcal{E}_g(p),
\end{aligned}$$

completing the proof. □

Proof of Theorem 4.1. From Remark 3.1, $T^t_\infty(p) = u_{l_1} \otimes \cdots \otimes u_{l_L}$ for $t \ge L$. For such a state $\mathcal{E}_g = 0$ and $\rho_g = 0$ hold due to (23) and (44)–(47), respectively. Thus it suffices to show

$$\mathcal{E}_g(p) - \mathcal{E}_g(T_\infty(p)) = \rho_g(p) - \rho_g(T_\infty(p)).$$

By applying (49) and (43) to the left and the right hand sides, respectively, this becomes

$$\sum_{i=1}^{L} g(\xi^{(i)} \otimes p_i) = \sum_{i=1}^{L} (\gamma_g(p_i) + \mathfrak{a}(p'_i) - \gamma_g(p'_i)),$$

where we have set $T_\infty(p) = p'_1 \otimes \cdots \otimes p'_L$. (This was denoted by $p^1_1 \otimes \cdots \otimes p^1_L$ in (43).) From the definition of $\xi^{(i)}$ in (49), we have $\xi^{(i)} \otimes p_i \simeq p'_i \otimes \xi^{(i+1)}$. Therefore Lemma 2.1 tells that $g(\xi^{(i)} \otimes p_i) = \gamma_g(p_i) + \mathfrak{a}(p'_i) - \gamma_g(p'_i)$ holds for each i, finishing the proof. □

4.2. *$*$-transformed correspondence*

Let us give an analogous result on $g = v^*_a$ $(1 \le a \le n-2)$ which is not included in Theorem 4.1. Our presentation in this subsection is brief since the essential features are the same as the previous case. To state the result, let us introduce a $*$-transformed generalized energy and a $*$-transformed $D_n^{(1)}$ cellular automaton. (See (9) and (10) for the original definition of $*$.)

Let $u^*_l = (0, \ldots, 0, l) \in B_l$. See (17). The $*$-transformed generalized energy $\mathcal{E}^*_{v^*_a}(p_L \otimes \cdots \otimes p_1)$ of an element $p_L \otimes \cdots \otimes p_1 \in B_{l_L} \otimes \cdots \otimes B_{l_1}$ is the sum of v^*_a for all the vertices in the following diagram ($L = 3$ example):

$$\mathcal{E}^*_{v^*_a}(p_3 \otimes p_2 \otimes p_1) \quad =$$

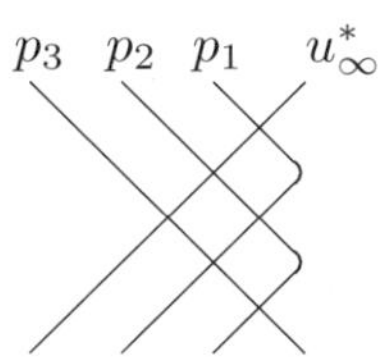

Compare this with Proposition 2.1. One can show that $\mathcal{E}^*_{v^*_a}$ is well defined and R-invariant.

The $*$-transformed $D_n^{(1)}$ cellular automaton is the dynamical system on $B_{l_L} \otimes \cdots \otimes B_{l_1}$ endowed with the commuting time evolutions $T^*_l (l \ge 1)$ defined by $p_L \otimes \cdots \otimes p_1 \otimes u^*_l \simeq \xi \otimes T^*_l(p_L \otimes \cdots \otimes p_1)$. ($\xi \in B_l$ is determined by this relation.) T^*_∞ is well defined. Moreover, under the time evolution $p'_L \otimes \cdots \otimes p'_1 = T^*_\infty(p_L \otimes \cdots \otimes p_1)$, the equality $p'_j = u^*_{l_j}$ is valid for $1 \le j \le k+1$ if $p_j = u^*_{l_j}$ for $1 \le j \le k$, which is parallel with Remark 3.1. For $\zeta \in B_l$, introduce the charge conjugation of (24)–(25) by

$$\mathfrak{a}^*(\zeta) := \mathfrak{a}(\zeta^*) = 2\zeta_1 + \zeta_2 + \cdots + \zeta_n + \overline{\zeta}_n + \cdots + \overline{\zeta}_2,$$
$$\gamma^*_{v^*_a}(\zeta) := \gamma_{v_a}(\zeta^*) = \zeta_1 + \zeta_{a+1} + \cdots + \zeta_n + \overline{\zeta}_n + \cdots + \overline{\zeta}_2 \quad (1 \le a \le n-2).$$

Writing the time evolutions of a state $p = p_L \otimes \cdots \otimes p_1 \in B_{l_L} \otimes \cdots \otimes B_{l_1}$ as $(T^*_\infty)^t(p) = p^t_L \otimes \cdots \otimes p^t_1$, we define the counting function $(1 \le a \le n-2)$:

$$\rho^*_{v^*_a}(p) = \sum_{j=1}^{L} \gamma^*_{v^*_a}(p_j) + \sum_{t\ge 1}\sum_{j=1}^{L} \mathfrak{a}^*(p^t_j) \quad =$$

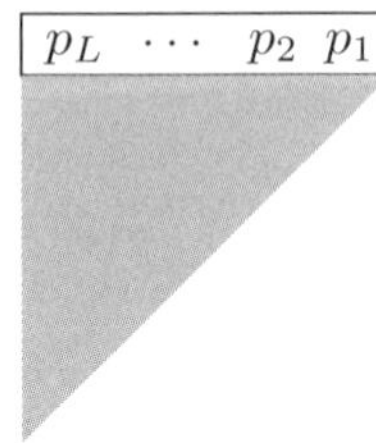

The counting is done by $\gamma^*_{v^*_a}$ for the top row and by $\mathfrak{a}^*$ for the SE quadrant generated by T^*_∞ beneath it. Thanks to the commutativity of $*$ and the combinatorial R (Prop.4.4 in Ref. 15), Theorem 4.1 implies the following:

Proposition 4.1. *For any state $p \in B_{l_L} \otimes \cdots \otimes B_{l_1}$, the following equality is valid:*

$$\mathcal{E}^*_{v^*_a}(p) = \rho^*_{v^*_a}(p) \quad (1 \le a \le n-2).$$

This completes our interpretation of the ($*$-transformed) generalized energies associated with all the generalized local energies in terms of the ($*$-transformed) $D^{(1)}_n$ cellular automaton.

5. Connection with combinatorial Bethe ansatz

Combinatorial Bethe ansatz was initiated by Kerov-Kirillov-Reshetikhin (KKR)[12,13] to establish a fermionic formula of the Kostka-Foulkes polynomials with the invention of rigged configurations and the KKR bijection. Their fermionic formula generalized Bethe's formula[4] for some simplest Kostka numbers that originates in the completeness issue.

Rigged configurations are combinatorial analogue of solutions to the Bethe equations. The KKR bijection maps them to the combinatorial analogue of Bethe vectors which may be viewed as elements of (a subset of) $B_{l_1} \otimes \cdots \otimes B_{l_L}$. The combinatorial Bethe ansatz has flourished in the fermionic formulas for general affine Lie algebras,[5,19–21] the solution of the initial value problem of integrable $A^{(1)}_n$ cellular automata by the inverse scattering method[14] and a connection with the classical soliton theory[8] via ultradiscrete tau functions[17] and so forth. Our aim in this section is to present an inverse scattering formalism of the $D^{(1)}_n$ cellular automaton and to conjecture explicit formulas for some generalized energies in the form of ultradiscrete tau functions associated with the $D^{(1)}_n$ rigged configurations.

5.1. *Inverse scattering formalism*

Set $\mathcal{P}_+ = \{p \in B_{l_1} \otimes \cdots \otimes B_{l_L} \mid \tilde{e}_i p = 0 \text{ for } i = 1,2,\ldots,n\}$. A state belonging to $\mathcal{P}_+$ is called *highest*. It is known that there is a bijection between $\mathcal{P}_+$ and the set of *rigged configurations*.[20,21] Consider a set

$$S = \{(a_i, j_i, r_i) \in \{1,2,\ldots,n\} \times \mathbb{Z}_{\geq 1} \times \mathbb{Z}_{\geq 0} \mid i = 1,2,\ldots,N\}, \tag{50}$$

where $N \geq 0$ is arbitrary and each triplet $s = (a,j,r)$ called *string* possesses color, length and rigging which will be denoted by $\mathrm{cl}(s) = a, \mathrm{lg}(s) = j$ and $\mathrm{rg}(s) = r$, respectively.[c] S is a rigged configuration if $\mathrm{rg}(s) \leq p^{(\mathrm{cl}(s))}_{\mathrm{lg}(s)}$ is satisfied for all $s \in S$. Here $p^{(a)}_j = \delta_{a,1} \sum_{k=1}^{L} \min(j, l_k) - \sum_{t \in S} C_{a,\mathrm{cl}(t)} \min(j, \mathrm{lg}(t))$, where $(C_{a,b})_{1 \leq a,b \leq n}$ is the Cartan matrix of D_n. Note that $p^{(\mathrm{cl}(s))}_{\mathrm{lg}(s)} \geq 0$ has to be satisfied for all $s \in S$, which imposes a stringent condition on the set $\{(a_i, j_i) \mid i = 1,\ldots,N\}$. Set $\mathrm{RC} = \{S :$ rigged configuration$\}$.

Theorem 5.1 (Refs. 20 and 21). *There is a bijection $\Phi : \mathcal{P}_+ \to \mathrm{RC}$.*

An explicit algorithm to determine the image of $\Phi^{\pm 1}$ is known. It is a $D_n^{(1)}$ analogue of the Kerov-Kirillov-Reshetikhin bijection[13] for $A_n^{(1)}$, which plays a central role in the combinatorial Bethe ansatz. Our convention here is the one adopted in Ref. 14.

For any highest state $p \in \mathcal{P}_+$, its time evolution $T_l(p)$ is again highest. Thus T_l induces a time evolution on RC via Φ. Let L be large enough and assume the situation in (38). Then we have

Theorem 5.2. *$\Phi(T_l(p)) = \tilde{T}_l \Phi(p)$ holds, where $\tilde{T}_l : \{(a_i, j_i, r_i)\} \mapsto \{(a_i, j_i, r_i + \delta_{a_i,1} \min(j_i, l))\}$ is a linear flow on rigged configurations.*

Proof. The proof uses Theorem 8.6 of Ref. 21 and is similar to that of Proposition 2.6 in Ref. 14 for type $A_n^{(1)}$. □

Thus the composition $\Phi^{-1} \circ \tilde{T}_l \circ \Phi$ linearizes the original time evolution T_l and solves the initial value problem in the $D_n^{(1)}$ cellular automaton by the inverse scattering method. See Ref. 14 for an analogous result for $A_n^{(1)}$.

5.2. *Conjecture on ultradiscrete tau functions*

For a rigged configuration S (50), let $T \subseteq S$ be a (possibly empty) subset of S. In general, T is no longer a rigged configuration. We introduce the

---

[c]Colors $1,2,\ldots,n$ of strings in rigged configurations should not be confused with colors $1,2,\ldots,n,\overline{n},\ldots,\overline{2},\overline{1}$ of (anti)-particles.

piecewise linear functions ($0 \le k \le L$ and $0 \le d \le n$)

$$c(T) = \frac{1}{2} \sum_{s,t \in T} C_{\mathrm{cl}(s),\mathrm{cl}(t)} \min(\mathrm{lg}(s), \mathrm{lg}(t)) + \sum_{s \in T} \mathrm{rg}(s),$$

$$c_k^{(d)}(T) = c(T) - \sum_{i=1}^{k} \sum_{s \in T, \mathrm{cl}(s)=1} \min(l_i, \mathrm{lg}(s)) + \sum_{s \in T, \mathrm{cl}(s)=d} \mathrm{lg}(s)$$

By the definition, the last term in $c_k^{(d)}(T)$ is 0 when $d = 0$, and the relation

$$c_k^{(d)}(T) = c_k^{(0)}(T)|_{\mathrm{rg}(s) \to \mathrm{rg}(s) + \mathrm{lg}(s)\delta_{\mathrm{cl}(s),d}} \tag{51}$$

holds. Obviously we have $c(\emptyset) = c_k^{(d)}(\emptyset) = 0$. On the other hand, $c(S)$ is known as the (co)charge of the rigged configuration S.[5,13,20,21]

We define a $\mathbb{Z}_{\ge 0}$-valued piecewise linear function on S as follows:

$$\tau_k^{(d)}(S) = -\min_{T \subseteq S} \left(c_k^{(d)}(T) \right) \quad (0 \le k \le L,\ 0 \le d \le n). \tag{52}$$

For S in (50), the minimum extends over 2^N candidates and reminds us of the structure of tau functions in the theory of solitons.[8] In fact, for type $A_n^{(1)}$, analogous functions have been identified[17] as ultradiscretization of the tau functions in KP hierarchy. Although such an origin is yet to be clarified, we call (52) *ultradiscrete tau function.* Guided by the results in $A_n^{(1)}$ and supported by computer experiments, we propose

Conjecture 5.1. *For any highest state $p \in \mathcal{P}_+$, let $S = \Phi(p)$ be the corresponding rigged configuration. Then, the following equalities hold for $p_{[k]}$ (42) with $0 \le k \le L$.*

$$\tau_k^{(0)}(S) = \mathcal{E}_{v_0}(p_{[k]}), \tag{53}$$

$$\tau_k^{(1)}(S) = \mathcal{E}_{v_0^{\sigma 1}}(p_{[k]}), \tag{54}$$

$$\tau_k^{(n-1)}(S) = \mathcal{E}_{v_{n-1}^*}(p_{[k]}), \tag{55}$$

$$\tau_k^{(n)}(S) = \mathcal{E}_{v_{n-1}}(p_{[k]}), \tag{56}$$

$$\tau_k^{(2)}(S) = \mathcal{E}_{w_2}(p_{[k]}) - \mathcal{E}_{v_0}(p_{[k]}) + \varphi_0(p_{[k]}). \tag{57}$$

In (57), $\varphi_0(p_{[k]})$ is the standard notation in crystal theory meaning $\max\{j \ge 0 \mid \tilde{f}_0^j p_{[k]} \ne 0\}$. By using (41), (44) with $a = 0$, (47), (51) and Theorem 5.2, one can show that (53) and (54) are equivalent.

The algorithm[20,21] for $\Phi^{\pm 1}$ seems valid not only for highest but *arbitrary* states if one allows negative rigging. With such a generalization, we expect that Theorem 5.2 and Conjecture 5.1 hold for any state, which was indeed the case for type $A_n^{(1)}$.[17]

Example 5.1. For the initial state p in Example 3.2, its rigged configuration is $S = \Phi(p) = \{(1,8,-2),(1,6,0),(1,2,-1),(1,1,-1),(2,8,0),(2,6,-1),(2,2,-1),(3,8,-3),(4,8,-1)\}$. The ultradiscrete tau function $\tau_k^{(d)}(S)$ and $\varphi_0(p_{[k]})$ take the following values.

	$k=1$	2	3	4	5	6	7	8	9
$\tau_k^{(0)}(S)$	6	11	21	32	39	46	53	60	67
$\tau_k^{(1)}(S)$	0	2	7	15	22	29	36	43	50
$\tau_k^{(2)}(S)$	1	3	9	16	23	30	37	44	51
$\tau_k^{(3)}(S)$	2	6	13	24	31	38	45	52	59
$\tau_k^{(4)}(S)$	3	8	15	24	31	38	45	52	59
$\varphi_0(p_{[k]})$	1	0	1	0	0	0	0	0	0

Comparing this with Example 3.3, one can check Conjecture 5.1.

Still many generalized energies in previous sections await formulas as in Conjecture 5.1 to be discovered. They are ultradiscrete analogue of the so called $X = M$ conjecture[5,19] in the sense that the generalized energies from crystal theory acquire explicit formulas of a fermionic nature originating in the combinatorial Bethe ansatz. Such results combined with Theorems 3.1 and 4.1 will lead to a piecewise linear formula for the bijection Φ^{-1} as was done for $A_n^{(1)}$.[17]

Let us end by raising a closely related question as another future problem. Let $\mathcal{P}_+(\lambda)$ denote the subset of $\mathcal{P}_+$ having the prescribed weight λ. Then, does the generating function

$$X_g(\lambda) = \sum_{p \in \mathcal{P}_+(\lambda)} q^{\mathcal{E}_g(p)}$$

of the generalized energy admit a fermionic formula like Refs. 13 and 5?

Acknowledgments

The authors thank Masato Okado for discussion. This work is supported by Grants-in-Aid for Scientific Research No. 19540393, No. 21740114 and No. 17340047 from JSPS.

References

1. G. E. Andrews, R. J. Baxter and P. J. Forrester, *J. Stat. Phys.* **35** 193 (1984).

2. R. J. Baxter, *Exactly solved models in statistical mechanics*, (Dover, 2007).
3. A. Berenstein and D. Kazhdan, *Geometric and unipotent crystals*, GAFA 2000 (Tel Aviv, 1999), Geom. Funct. Anal. 2000, Special Volume, Part I, pp188–236.
4. H. A. Bethe, *Z. Physik* **71** 205 (1931).
5. G. Hatayama, A. Kuniba, M. Okado, T. Takagi and Y. Yamada, *Remarks on fermionic formula*, in *Recent Developments in Quantum Affine Algebras and Related Topics*, eds. N. Jing and K.C. Misra, Contemp. Math **248** (AMS 1999) pp243-291.
6. G. Hatayama, A. Kuniba and T. Takagi, *Nucl. Phys. B* **577** [PM] 619 (2000).
7. G. Hatayama, A. Kuniba and T. Takagi, *J. Phys. A* **34** 10697 (2001).
8. M. Jimbo and T. Miwa, *Publ. RIMS.* **19** (Kyoto Univ. 1983) pp.943–1001.
9. S-J. Kang, M. Kashiwara and K. C. Misra, *Compositio Math.* **92** 299 (1994).
10. S-J. Kang, M. Kashiwara, K. C. Misra, T. Miwa, T. Nakashima and A. Nakayashiki, *Int. J. Mod. Phys. A* **7** (suppl. 1A) 449 (1992).
11. S-J. Kang, M. Kashiwara, K. C. Misra, T. Miwa, T. Nakashima and A. Nakayashiki, *Duke Math. J.* **68** 499 (1992).
12. S. V. Kerov, A. N. Kirillov and N. Yu. Reshetikhin, *J. Soviet Math.* **41** 916 (1988).
13. A. N. Kirillov and N. Yu. Reshetikhin, *J. Soviet Math.* **41** 925 (1988).
14. A. Kuniba, M. Okado, R. Sakamoto, T. Takagi and Y. Yamada: *Nucl. Phys. B* **740** [PM] 299 (2006).
15. A. Kuniba, M. Okado, T. Takagi and Y. Yamada, *Int. Math. Res. Notices.* **48** 2565 (2003).
16. A. Kuniba, M. Okado, T. Takagi and Y. Yamada, *Commun. Math. Phys.* **245** 491 (2004).
17. A. Kuniba, R. Sakamoto and Y. Yamada, *Nucl. Phys. B* **786** [PM] 207 (2007).
18. A. Nakayashiki and Y. Yamada, *Selecta Math.*, New Ser. **3** 547 (1997).
19. M. Okado, *$X = M$ conjecture*, in *Combinatorial Aspect of Integrable Systems*, eds. A. Kuniba and M. Okado, MSJ Memoir Vol. 17 (MSJ 2007) pp43–73.
20. M. Okado, A. Schilling and M. Shimozono, *A crystal to rigged configuration bijection for nonexceptional affine algebras*, in *Algebraic Combinatorics and Quantum Groups*, eds. N. Jing, (World Scientific 2003), pp85–124.
21. A. Schilling and M. Shimozono, *J. Alg.* **295** 562 (2006).
22. M. Shimozono, private communication.

DYSON'S CONSTANT FOR THE HYPERGEOMETRIC KERNEL

OLEG LISOVYY*

Laboratoire de Mathématiques et Physique Théorique CNRS/UMR 6083
Université de Tours, Parc de Grandmont, 37200 Tours, France
E-mail: lisovyi@lmpt.univ-tours.fr

We study a Fredholm determinant of the hypergeometric kernel arising in the representation theory of the infinite-dimensional unitary group. It is shown that this determinant coincides with the Palmer–Beatty–Tracy tau function of a Dirac operator on the hyperbolic disk. Solution of the connection problem for Painlevé VI equation allows to determine its asymptotic behavior up to a constant factor, for which a conjectural expression is given in terms of Barnes functions. We also present analogous asymptotic results for the Whittaker and Macdonald kernel.

Keywords: Painlevé equations; tau function; Fredholm determinant.

1. Introduction

Connections between Painlevé equations and Fredholm determinants have long been a subject of great interest, mainly because of their applications in random matrix theory and integrable systems.[17,27,29] One of the most famous examples is concerned with the Fredholm determinant $F(t) = \det(1 - K_{\text{sine}})$, where K_{sine} is the integral operator with the sine kernel $\frac{\sin(x-y)}{\pi(x-y)}$ on the interval $[0,t]$. It is well-known that $F(t)$ is equal to the gap probability for the Gaussian Unitary Ensemble (GUE) in the bulk scaling limit. As shown in Ref. 17, the function $\sigma(t) = t\frac{d}{dt}\ln F(t)$ satisfies the σ-form of a Painlevé V equation,

$$\left(t\sigma''\right)^2 + 4\left(t\sigma' - \sigma\right)\left(t\sigma' - \sigma + (\sigma')^2\right) = 0. \quad (1)$$

Equation (1) and the obvious leading behavior $F(t \to 0) = 1 - t + O\left(t^2\right)$ provide an efficient method of numerical computation of $F(t)$ for all t.

*On leave from Bogolyubov Institute for Theoretical Physics, 03680 Kyiv, Ukraine

Further, as $t \to \infty$, one has

$$F(2t) = f_0\, t^{-\frac{1}{4}} e^{-t^2/2} \Big(1 + \sum_{k=1}^{N} f_k t^{-k} + O\big(t^{-N-1}\big)\Big).$$

The coefficients $f_1, f_2, \dots$ in this expansion can in principle be determined from (1). It was conjectured by Dyson[12] that the value of the remaining unknown constant is $f_0 = 2^{\frac{1}{12}} e^{3\zeta'(-1)}$, where $\zeta(z)$ is the Riemann ζ-function.

Dyson's conjecture was rigorously proved only recently.[9,13,19] Similar results[3,8] were also obtained for the Airy-kernel determinant describing the largest eigenvalue distribution for GUE in the edge scaling limit.[26]

The present paper is devoted to the asymptotic analysis of the Fredholm determinant of the hypergeometric kernel on $L^2(0,t)$ with $t \in (0,1)$. This determinant, to be denoted by $D(t)$, arises in the representation theory of the infinite-dimensional unitary group[5] and provides a 4-parameter class of solutions to Painlevé VI (PVI) equation.[6] Rather surprisingly, it turns out to coincide with the Palmer–Beatty–Tracy (PBT) τ-function of a Dirac operator on the hyperbolic disk[20,23] under suitable identification of parameters. Relation to PVI allows to give a complete description of the behavior of $D(t)$ as $t \to 1$ up to a constant factor analogous to Dyson's constant f_0 in the sine-kernel asymptotics. Relation to the PBT τ-function, on the other hand, suggests a conjectural expression for this constant in terms of Barnes functions.

The paper is planned as follows. In Sec. 2, we recall basic facts on Painlevé VI and the associated linear system. The ${}_2F_1$ kernel determinant $D(t)$ and the PBT τ-function are introduced in Secs. 3 and 4. Section 5 gives a simple proof of a result of Ref. 6, relating $D(t)$ to Painlevé VI. In Sec. 6, we discuss Jimbo's asymptotic formula for PVI and determine the monodromy corresponding to the ${}_2F_1$ kernel solution. Section 7 contains the main results of the paper: the asymptotics of $D(t)$ as $t \to 1$, obtained from the solution of PVI connection problem (Proposition 7.1) and a conjecture for the unknown constant (Conjecture 7.1). Numerical and analytic tests of the conjecture are discussed in Secs. 8 and 9. Similar asymptotic results for the Whittaker and Macdonald kernel are presented in Sec. 10. Appendix contains a brief summary of formulas for the Barnes function.

2. Painlevé VI and JMU τ-function

Consider the linear system

$$\frac{d\Phi}{d\lambda} = \left(\frac{A_0}{\lambda} + \frac{A_1}{\lambda - 1} + \frac{A_t}{\lambda - t}\right)\Phi, \tag{2}$$

where $A_\nu \in \mathfrak{sl}_2(\mathbb{C})$ $(\nu = 0, 1, t)$ are independent of λ with eigenvalues $\pm\theta_\nu/2$ and

$$A_0 + A_1 + A_t = \begin{pmatrix} -\theta_\infty/2 & 0 \\ 0 & \theta_\infty/2 \end{pmatrix}, \qquad \theta_\infty \neq 0.$$

The fundamental matrix solution $\Phi(\lambda)$ is a multivalued function on $\mathbb{P}^1 \backslash \{0, 1, t, \infty\}$. Fix the basis of loops as shown in Fig. 1 and denote by $M_0, M_t, M_1, M_\infty \in SL(2, \mathbb{C})$ the corresponding monodromy matrices. Clearly, one has $M_\infty M_1 M_t M_0 = \mathbf{1}$.

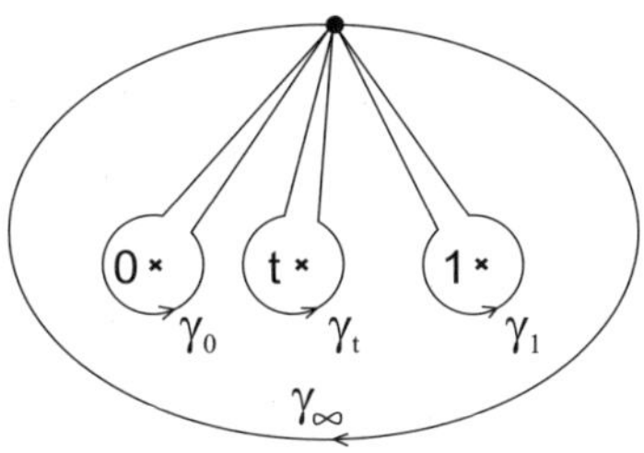

Fig. 1. Generators of $\pi_1\left(\mathbb{P}^1 \backslash \{0, 1, t, \infty\}\right)$.

Since the monodromy is defined up to overall conjugation, it is convenient to introduce, following Ref. 16, a 7-tuple of invariant quantities

$$p_\nu = \operatorname{Tr} M_\nu = 2\cos \pi\theta_\nu, \qquad \nu = 0, 1, t, \infty, \tag{3}$$

$$p_{\mu\nu} = \operatorname{Tr}(M_\mu M_\nu) = 2\cos \pi\sigma_{\mu\nu}, \qquad \mu, \nu = 0, 1, t. \tag{4}$$

These data uniquely fix the conjugacy class of the triple (M_0, M_1, M_t) unless the monodromy is reducible. The traces (3)–(4) satisfy Jimbo–Fricke relation

$$\begin{aligned} p_{0t}p_{1t}p_{01} + p_{0t}^2 + p_{1t}^2 + p_{01}^2 - (p_0 p_t + p_1 p_\infty) p_{0t} \\ -(p_1 p_t + p_0 p_\infty) p_{1t} - (p_0 p_1 + p_t p_\infty) p_{01} = 4. \end{aligned}$$

As a consequence, for fixed $\{p_\nu\}$, p_{0t}, p_{1t} there are at most two possible values for p_{01}.

It is well-known that the monodromy preserving deformations of the system (2) are described by the so-called Schlesinger equations

$$\frac{dA_0}{dt} = \frac{[A_t, A_0]}{t}, \qquad \frac{dA_1}{dt} = \frac{[A_t, A_1]}{t-1}, \tag{5}$$

which are equivalent to the sixth Painlevé equation:

$$\frac{d^2q}{dt^2} = \frac{1}{2}\left(\frac{1}{q}+\frac{1}{q-1}+\frac{1}{q-t}\right)\left(\frac{dq}{dt}\right)^2 - \left(\frac{1}{t}+\frac{1}{t-1}+\frac{1}{q-t}\right)\frac{dq}{dt} \tag{6}$$
$$+\frac{q(q-1)(q-t)}{2t^2(t-1)^2}\left((\theta_\infty-1)^2-\frac{\theta_0^2 t}{q^2}+\frac{\theta_1^2(t-1)}{(q-1)^2}+\frac{(1-\theta_t^2)t(t-1)}{(q-t)^2}\right).$$

Relation between $A_{0,1,t}(t)$ and $q(t)$ is given by

$$\left(\frac{A_0}{\lambda}+\frac{A_1}{\lambda-1}+\frac{A_t}{\lambda-t}\right)_{12} = \frac{k(t)(\lambda-q(t))}{\lambda(\lambda-1)(\lambda-t)}. \tag{7}$$

Jimbo–Miwa–Ueno (JMU) τ-function[18] of Painlevé VI is defined as

$$\frac{d}{dt}\ln\tau_{JMU}(t;\boldsymbol{\theta}) = \frac{\operatorname{tr}(A_0A_t)}{t}+\frac{\operatorname{tr}(A_1A_t)}{t-1}, \tag{8}$$

where $\boldsymbol{\theta}=(\theta_0,\theta_1,\theta_t,\theta_\infty)$. Introducing a logarithmic derivative

$$\sigma(t) = t(t-1)\frac{d}{dt}\ln\tau_{JMU}(t;\boldsymbol{\theta})+\frac{t\left(\theta_t^2-\theta_\infty^2\right)}{4}-\frac{\theta_t^2+\theta_0^2-\theta_1^2-\theta_\infty^2}{8}, \tag{9}$$

it can be deduced from the Schlesinger system (5) that $\sigma(t)$ satisfies the following 2nd order ODE (σ-form of Painlevé VI):

$$\sigma'\left(t(t-1)\sigma''\right)^2+\left[2\sigma'(t\sigma'-\sigma)-(\sigma')^2-\frac{(\theta_t^2-\theta_\infty^2)(\theta_0^2-\theta_1^2)}{16}\right]^2 \tag{10}$$
$$=\left(\sigma'+\frac{(\theta_t+\theta_\infty)^2}{4}\right)\left(\sigma'+\frac{(\theta_t-\theta_\infty)^2}{4}\right)\left(\sigma'+\frac{(\theta_0+\theta_1)^2}{4}\right)\left(\sigma'+\frac{(\theta_0-\theta_1)^2}{4}\right).$$

In terms of $q(t)$, the definition of $\sigma(t)$ reads

$$\sigma(t) = \frac{t^2(t-1)^2}{4q(q-1)(q-t)}\left(q'-\frac{q(q-1)}{t(t-1)}\right)^2-\frac{\theta_0^2 t}{4q}+\frac{\theta_1^2(t-1)}{4(q-1)}-\frac{\theta_t^2 t(t-1)}{4(q-t)}$$
$$-\frac{\theta_\infty^2(q-1)}{4}-\frac{\theta_t^2 t}{4}+\frac{\theta_t^2+\theta_0^2-\theta_1^2-\theta_\infty^2}{8}. \tag{11}$$

3. Hypergeometric kernel determinant

It was shown in Ref. 5 that the spectral measure associated to the decomposition of a remarkable 4-parameter family of characters of the infinite-dimensional unitary group $U(\infty)$ gives rise to a determinantal point process with correlation kernel

$$K(x,y) = \lambda\,\frac{A(x)B(y)-B(x)A(y)}{y-x}, \qquad x,y\in(0,1),$$

where

$$\lambda = \frac{\sin \pi z \sin \pi z'}{\pi^2}\,\Gamma\left[\begin{matrix} 1+z+w, 1+z+w', 1+z'+w, 1+z'+w' \\ 1+z+z'+w+w', 2+z+z'+w+w' \end{matrix}\right], \tag{12}$$

$$A(x) = x^{\frac{z+z'+w+w'}{2}} (1-x)^{-\frac{z+z'+2w'}{2}} {}_2F_1\left[\begin{matrix} z+w', z'+w' \\ z+z'+w+w' \end{matrix}\middle|\, \frac{x}{x-1}\right], \tag{13}$$

$$B(x) = x^{\frac{z+z'+w+w'+2}{2}} (1-x)^{-\frac{z+z'+2w'+2}{2}} {}_2F_1\left[\begin{matrix} z+w'+1, z'+w'+1 \\ z+z'+w+w'+2 \end{matrix}\middle|\, \frac{x}{x-1}\right]. \tag{14}$$

Note that our notation slightly differs from the standard one;[5,6] to shorten some formulas from Painlevé theory, the interval $\left(\frac{1}{2}, \infty\right)$ of Refs. 5,6 is mapped to $(0,1)$ by $x \mapsto 1/\left(\frac{1}{2}+x\right)$.

The kernel $K(x,y)$ has a number of symmetries:

(S1) It is invariant under transformations $z \leftrightarrow z'$ and $w \leftrightarrow w'$; the latter symmetry follows from ${}_2F_1\left[\begin{matrix} a,b \\ c \end{matrix}\middle|\, z\right] = (1-z)^{c-a-b}{}_2F_1\left[\begin{matrix} c-a, c-b \\ c \end{matrix}\middle|\, z\right]$.

(S2) It is also straightforward to check that $K(x,y)$ is invariant under transformation $z \mapsto -z$, $z' \mapsto -z'$, $w \mapsto w'+z+z'$, $w' \mapsto w+z+z'$.

(S3) We can simultaneously shift $z \mapsto z \pm 1$, $z' \mapsto z' \pm 1$, $w \mapsto w \mp 1$, $w' \mapsto w' \mp 1$; together with (S2), this allows to assume without loss of generality that $0 \leq \operatorname{Re}(z+z') \leq 1$.

We are interested in the Fredholm determinant

$$D(t) = \det\left(1 - K\big|_{(0,t)}\right), \qquad t \in (0,1). \tag{15}$$

Assume that the parameters $z, z', w, w' \in \mathbb{C}$ satisfy the conditions:

(C1) $z' = \bar{z} \in \mathbb{C}\backslash\mathbb{Z}$ or $k < z, z' < k+1$ for some $k \in \mathbb{Z}$,
(C2) $w' = \bar{w} \in \mathbb{C}\backslash\mathbb{Z}$ or $l < w, w' < l+1$ for some $l \in \mathbb{Z}$,
(C3) $z+z'+w+w' > 0$, $|z+z'| < 1$, $|w+w'| < 1$.

Then, as was shown by Borodin and Deift,[6] the determinant (15) is well-defined and $D(t) = \tau_{JMU}(t; \boldsymbol{\theta})$ for the following choice of PVI parameters:

$$\boldsymbol{\theta} = (z+z'+w+w', z-z', 0, w-w'). \tag{16}$$

The original proof[6] that $D(t)$ satisfies Painlevé VI is rather involved. In Sec. 5, we give an alternative simple derivation of this result in the spirit of Ref. 27.

Lemma 3.1. *Assume (C1)–(C3). Then the asymptotic expansion of* $D(t)$

as $t \to 0$ has the form

$$D(t) = 1 - \kappa \cdot t^{1+z+z'+w+w'} + O\left(t^{2+z+z'+w+w}\right), \tag{17}$$

where

$$\kappa = \frac{\sin \pi z \sin \pi z'}{\pi^2} \Gamma \left[\begin{matrix} 1+z+w, 1+z+w', 1+z'+w, 1+z'+w' \\ 2+z+z'+w+w', 2+z+z'+w+w' \end{matrix} \right]. \tag{18}$$

Proof. As $t \to 0$, one has $D(t) \sim 1 - \int_0^t K(x,x)\,dx$. The result then follows from

$$A(x) \sim x^{\frac{z+z'+w+w'}{2}}, \qquad B(x) \sim x^{\frac{z+z'+w+w'}{2}+1} \qquad \text{as } x \to 0.$$

Note that in the expression for κ given in Remark 7.2 of Ref. 6 the gamma product is missing, which seems to be a typesetting error. □

The asymptotics (17) and σPVI equation (10) uniquely fix $D(t)$ by a result of Ref. 7. Gamma product in (18) is a function of θ_0, θ_1, θ_∞ only, but $\frac{\sin \pi z \sin \pi z'}{\pi^2}$ depends on an additional parameter (e.g. $z + z'$); hence we are dealing with a 1-parameter family of initial conditions.

The results of Ref. 6 can be extended to a larger set of parameters. This follows already from the observation that the subset of $\mathbb{C}^4$ defined by (C1)–(C3) is not stable under the transformations (S1)–(S3). However, instead of trying to identify all admissible values of z, z', w, w', in the remainder of this paper we simply replace (C1)–(C3) by a much weaker (invariant) condition

(C4) $z + w, z + w', z' + w, z' + w' \notin \mathbb{Z}_{<0}$ and $\mathrm{Re}\,(z + z' + w + w') > 0$,

and *define* $D(t)$ as the JMU τ-function of Painlevé VI with parameters (16), whose leading behavior as $t \to 0$ is specified by (17)–(18). Our aim in the next sections is to determine the asymptotics of $D(t)$ as $t \to 1$.

4. PBT τ-function

Palmer–Beatty–Tracy τ-function[20,23] is a regularized determinant of the quantum hamiltonian of a massive Dirac particle moving on the hyperbolic disk in the superposition of a uniform magnetic field B and the field of two non-integer Aharonov–Bohm fluxes $2\pi\nu_{1,2}$ ($-1 < \nu_{1,2} < 0$) located at the points $a_{1,2}$.

Denote by m and E the particle mass and energy, by $-4/R^2$ the disk curvature and write $b = \frac{BR^2}{4}$, $\mu = \frac{\sqrt{(m^2-E^2)R^2+4b^2}}{2}$, $s = \tanh^2 \frac{d(a_1,a_2)}{R}$, where $d(a_1, a_2)$ denotes the geodesic distance between a_1 and a_2. Then $\tau_{PBT}(s)$ can be expressed[23] in terms of a solution $u(s)$ of the sixth Painlevé equation (6):

$$\frac{d}{ds}\ln\tau_{PBT}(s) = \frac{s(1-s)}{4u(1-u)(u-s)}\left(\frac{du}{ds} - \frac{1-u}{1-s}\right)^2 - \frac{1-u}{1-s}\left(\frac{(\theta_\infty - 1)^2}{4s} - \frac{(\theta_0+1)^2}{4u} + \frac{\theta_t^2}{4(u-s)}\right), \tag{19}$$

where the corresponding PVI parameters are given by $\boldsymbol{\theta} = (1+\nu_1+\nu_2-2b, 0, 2\mu, 1+\nu_1-\nu_2)$. The initial conditions are specified by the asymptotics of $\tau_{PBT}(s)$ as $s \to 1$, computed in Ref. 20:

$$\tau_{PBT}(s) = 1 - \kappa_{PBT}(1-s)^{1+2\mu} + O\left((1-s)^{2+2\mu}\right), \tag{20}$$

$$\kappa_{PBT} = \frac{\sin\pi\nu_1 \sin\pi\nu_2}{\pi^2}\,\Gamma\left[\begin{matrix} 2+\mu+\nu_1-b, \mu-\nu_1+b, 2+\mu+\nu_2-b, \mu-\nu_2+b \\ 2+2\mu, 2+2\mu \end{matrix}\right].$$

Some resemblance between (11) and (19) suggests that $\tau_{PBT}(s)$ is a special case of the JMU τ-function. Indeed, consider the following transformation:

$$s \mapsto 1-t, \qquad u \mapsto \frac{1-t}{1-q}.$$

In the notation of Table 1 of Ref. 21, this corresponds to Bäcklund transformation $r_x P_{xy}$ for Painlevé VI. If $u(s)$ is a solution with parameters $\boldsymbol{\theta} = (\theta_0, \theta_1, \theta_t, \theta_\infty)$, then $q(t)$ solves PVI with parameters $\boldsymbol{\theta}' = (\theta_t, \theta_\infty - 1, \theta_1, \theta_0 + 1)$. Straightforward calculation then shows that $\tau_{PBT}(1-t) = \tau_{JMU}(t; \boldsymbol{\theta}')$ provided $\theta_1 = 0$.

Lemma 4.1. *Under the following identification of parameters*

$$z + z' + w + w' = 2\mu,\ z - z' = \nu_1 - \nu_2,\ w - w' = 2 + \nu_1 + \nu_2 - 2b, \tag{21}$$

$$\cos\pi(z+z') = \cos\pi(\nu_1+\nu_2), \tag{22}$$

we have $D(t) = \tau_{PBT}(1-t)$.

Proof. It was shown above that if (21) holds, then both $D(t)$ and $\tau_{PBT}(1-t)$ are JMU τ-functions with the same $\boldsymbol{\theta}$. To show the equality, it suffices to verify that (22) implies $\kappa = \kappa_{PBT}$. □

Symmetries of $D(t)$ imply that $\tau_{PBT}(s)$ is invariant under transformations

(S1) $\mu \mapsto \mu$, $\nu_{1,2} \mapsto \nu_{1,2}$, $b \mapsto 2 + \nu_1 + \nu_2 - b$;
(S2) $\mu \mapsto \mu$, $\nu_{1,2} \mapsto -2 - \nu_{1,2}$, $b \mapsto -b$.

These symmetries of $\tau_{PBT}(s)$ are by no means manifest, although they can also be deduced from the Fredholm determinant representation in Ref. 20, Theorem 1.1.

5. Painlevé VI from Tracy-Widom equations

5.1. *Basic notation*

Tracy and Widom[27] have developed a systematic approach for deriving differential equations satisfied by Fredholm determinants of the form

$$D_I = \det (1 - K_I), \tag{23}$$

where K_I is an integral operator with the kernel

$$K_I(x,y) = \frac{\varphi(x)\psi(y) - \psi(x)\varphi(y)}{x-y}, \tag{24}$$

on $L^2(J)$, with $J = \bigcup_{j=1}^{M} (a_{2j-1}, a_{2j})$. The kernels of the form (24) are called integrable and possess rather special properties: e.g. the kernel of the resolvent $(1-K_I)^{-1} K_I$ is also integrable.[15]

The method of Ref. 27 requires that φ, ψ in (24) obey a system of linear ODEs of the form

$$m(x)\frac{d}{dx}\begin{pmatrix}\varphi\\ \psi\end{pmatrix} = \left(\sum_{k=0}^{N} \mathcal{A}_k x^k\right)\begin{pmatrix}\varphi\\ \psi\end{pmatrix}, \tag{25}$$

where $m(x)$ is a polynomial and $\mathcal{A}_k \in \mathfrak{sl}_2(\mathbb{C})$ $(k = 0, \ldots, N)$. Note that a linear transformation $\begin{pmatrix}\varphi\\ \psi\end{pmatrix} \mapsto G\begin{pmatrix}\varphi\\ \psi\end{pmatrix}$ leaves $K_I(x,y)$ invariant provided $\det G = 1$, and therefore $\{\mathcal{A}_k\}$ can be conjugated by an arbitrary $SL(2,\mathbb{C})$-matrix.

Our aim is to show that in the special case

$$m(x) = x(1-x), \qquad N = 1, \qquad J = (0,t)$$

the determinant (23) (i) coincides with the ${}_2F_1$ kernel determinant $D(t)$ and (ii) considered as a function of t, is a Painlevé VI τ-function.

Let us temporarily switch to the notation of Ref. 27 and introduce the quantities

$$q = \left[(1-K_I)^{-1}\varphi\right](t), \qquad p(t) = \left[(1-K_I)^{-1}\psi\right](t),$$

$$u = \langle\varphi|(1-K_I)^{-1}|\varphi\rangle,\ v = \langle\varphi|(1-K_I)^{-1}|\psi\rangle,\ w = \langle\psi|(1-K_I)^{-1}|\psi\rangle,$$

where the inner products $\langle\,|\,\rangle$ are taken over J. Then

$$D_I^{-1}D_I{}' = qp' - pq',$$

with primes denoting derivatives with respect to t. Tracy-Widom approach gives a system of nonlinear first order ODEs for q, p, u, v, w, which we are about to examine.

5.2. *Derivation*

Let $\mathcal{A}_1$ be diagonalizable, so that one can set

$$\mathcal{A}_0 = \begin{pmatrix} \alpha_0 & \beta_0 \\ -\gamma_0 & -\alpha_0 \end{pmatrix}, \qquad \mathcal{A}_1 = \begin{pmatrix} \alpha_1 & 0 \\ 0 & -\alpha_1 \end{pmatrix}.$$

The Tracy-Widom equations then read

$$t(1-t)\begin{pmatrix} q' \\ p' \end{pmatrix} = \begin{pmatrix} \alpha & \beta \\ -\gamma & -\alpha \end{pmatrix}\begin{pmatrix} q \\ p \end{pmatrix}, \tag{26}$$

$$u' = q^2, \qquad v' = pq, \qquad w' = p^2, \tag{27}$$

where

$$\alpha = \alpha_0 + \alpha_1 t + v, \qquad \beta = \beta_0 + (2\alpha_1 - 1)u, \qquad \gamma = \gamma_0 - (2\alpha_1 + 1)w.$$

The system (26)–(27) has two first integrals

$$I_1 = 2\alpha pq + \beta p^2 + \gamma q^2 - 2\alpha_1 v, \tag{28}$$

$$I_2 = (v+\alpha_0)^2 - \beta\gamma - 2\alpha_1 t(1-t)pq + 2\alpha_1(1-t)v - I_1 t. \tag{29}$$

Consider the logarithmic derivative $\zeta(t) = t(t-1)D_I^{-1}D_I{}'$. It can be easily checked that

$$\zeta = 2\alpha pq + \beta p^2 + \gamma q^2 = 2\alpha_1 v + I_1, \tag{30}$$

$$\zeta' = 2\alpha_1 pq, \tag{31}$$

$$t(1-t)\zeta'' = 2\alpha_1(\beta p^2 - \gamma q^2). \tag{32}$$

Note that v, α are expressible in terms of ζ and pq in terms of ζ'. Using (29) and (30) one may also write $\beta\gamma$ and $\beta p^2 + \gamma q^2$ in terms of ζ and ζ'. Now squaring (32) we find a second order equation for ζ:

$$\begin{aligned}\left(t(1-t)\zeta''\right)^2 + 4\left(\zeta' - \alpha_1^2\right)\left(t\zeta' - \zeta\right)^2 - 4\zeta'\left(t\zeta' - \zeta\right)\left(\zeta' + 2\alpha_0\alpha_1 - I_1\right) \\ = 4(I_1 + I_2)\left(\zeta'\right)^2.\end{aligned} \tag{33}$$

If we parameterize the integrals I_1, I_2 as

$$I_1 = -k_1k_2 + \alpha_1(2\alpha_0 + \alpha_1), \qquad I_2 = \frac{(k_1+k_2)^2}{4} - \alpha_1(2\alpha_0+\alpha_1),$$

and define

$$\sigma(t) = \zeta(t) - \alpha_1^2 t + \frac{\alpha_1^2 + k_1k_2}{2}, \tag{34}$$

then (33) transforms into σPVI equation (10) with parameters $\boldsymbol{\theta} = (k_1 - k_2, k_1 + k_2, 0, 2\alpha_1)$. Moreover, (34) and the definition of $\zeta(t)$ imply that $D_I(t)$ coincides with the corresponding JMU τ-function.

The system (25) has two linearly independent solutions, only one of which can be chosen to be regular as $x \to 0$. This is the only solution of interest here, as if φ, ψ have an irregular part, the operator K_I fails to be trace-class. The regularity further implies that q, p, u, v, w vanish as $t \to 0$, and therefore the integrals I_1, I_2 are given by $I_1 = 0$, $I_2 = \alpha_0^2 - \beta_0\gamma_0$.

Choosing $\mathcal{A}_0$, $\mathcal{A}_1$ as above, one can still conjugate them by a diagonal matrix. Use this freedom to parameterize α_0, β_0, γ_0, α_1 as follows:

$$\alpha_0 = -\frac{c}{2} - \frac{ab}{c-a-b}, \qquad \beta_0 = -\frac{(c-a)(c-b)}{c-a-b},$$
$$\gamma_0 = -\frac{ab}{c-a-b}, \qquad \alpha_1 = \frac{c-a-b}{2},$$

so that $I_2 = \dfrac{c^2}{4}$ and therefore $(k_1+k_2)^2 = (a-b)^2$, $(k_1-k_2)^2 = c^2$. Now if $\operatorname{Re} c > 0$, the regular solution of (25) is given by

$$\begin{pmatrix} \varphi \\ \psi \end{pmatrix}(x) = \begin{pmatrix} 1 & \frac{(c-a)(c-b)}{c(1+c)} \\ -1 & -\frac{ab}{c(1+c)} \end{pmatrix} \begin{pmatrix} \frac{x^{\frac{c}{2}}}{(1-x)^{\frac{a+b}{2}}}\, {}_2F_1\left[\begin{matrix} a, b \\ c \end{matrix} \middle| \frac{x}{x-1}\right] \\ \frac{x^{1+\frac{c}{2}}}{(1-x)^{1+\frac{a+b}{2}}}\, {}_2F_1\left[\begin{matrix} 1+a, 1+b \\ 2+c \end{matrix} \middle| \frac{x}{x-1}\right] \end{pmatrix}. \tag{35}$$

Setting $a = z + w'$, $b = z' + w'$, $c = z + z' + w + w'$ and comparing (35) with (13)–(14) we see that $K_I(t)$ coincides, up to an adjustable constant factor, with the ${}_2F_1$ kernel of Section 3.

Remark 5.1. A system similar to (26)–(27) has already appeared in the Tracy-Widom analysis of the Jacobi kernel, see Section V.C of Ref. 27. As the integral I_2 was not noticed there, the final result of Ref. 27 was a *third* order ODE (as one may well guess, it represents the first derivative of (33) in a disguised form). Later Haine and Semengue[14] have derived another third order equation for the Jacobi kernel determinant using the Virasoro approach of Ref. 2, and obtained Painlevé VI as the compatibility condition of the two equations. Our calculation gives, among other things, an elementary proof of this result.

Remark 5.2. For non-diagonalizable $\mathcal{A}_1$ it can be assumed that $\mathcal{A}_1 = \begin{pmatrix} 0 & 1 \\ 0 & 0 \end{pmatrix}$. Equations (27) remain unchanged, whereas instead of (26) we get

$$t(1-t)\begin{pmatrix} q' \\ p' \end{pmatrix} = \begin{pmatrix} \tilde{\alpha} & \tilde{\beta} \\ -\tilde{\gamma} & -\tilde{\alpha} \end{pmatrix}\begin{pmatrix} q \\ p \end{pmatrix},$$

where

$$\tilde{\alpha} = \alpha_0 + v - w, \qquad \tilde{\beta} = \beta_0 + s - u + 2v, \qquad \tilde{\gamma} = \gamma_0 - w.$$

As before, we have two first integrals,

$$\begin{aligned} I_1 &= 2\tilde{\alpha}pq + \tilde{\beta}p^2 + \tilde{\gamma}(q^2+1), \\ I_2 &= \tilde{\alpha}^2 - \tilde{\beta}\tilde{\gamma} - t(1-t)p^2 + (2t-1)\tilde{\gamma} - I_1 t. \end{aligned}$$

The rest of the computation is completely analogous to the diagonalizable case. As a final result, one finds that the determinant $D(t)$ with $w = w'$ is a τ-function of Painlevé VI with parameters $\boldsymbol{\theta} = (z + z' + 2w, z - z', 0, 0)$.

6. Jimbo's asymptotic formula

A remarkable result of Jimbo[16] relates the asymptotic behavior of the JMU τ-function (8) near the singular points $t = 0, 1, \infty$ to the monodromy of the associated linear system (2).

Theorem 6.1 (Theorem 1.1 in Ref. 16). *Assume that*

$$\theta_0, \theta_1, \theta_t, \theta_\infty \notin \mathbb{Z}, \tag{J1}$$

$$0 \le \operatorname{Re}\sigma_{0t} < 1, \tag{J2}$$

$$\theta_0 \pm \theta_t \pm \sigma_{0t}, \theta_\infty \pm \theta_1 \pm \sigma_{0t} \notin 2\mathbb{Z}. \tag{J3}$$

Then $\tau_{JMU}(t)$ has the following asymptotic expansion as $t \to 0$:

$$\begin{aligned} \tau_{JMU}(t) = \text{const} \cdot t^{\frac{\sigma_{0t}^2 - \theta_0^2 - \theta_t^2}{4}} \Bigg[1 &- \frac{(\theta_0^2 - (\theta_t - \sigma_{0t})^2)(\theta_\infty^2 - (\theta_1 - \sigma_{0t})^2)}{16\sigma_{0t}^2(1+\sigma_{0t})^2}\, \hat{s}\, t^{1+\sigma_{0t}} \\ &- \frac{(\theta_0^2 - (\theta_t + \sigma_{0t})^2)(\theta_\infty^2 - (\theta_1 + \sigma_{0t})^2)}{16\sigma_{0t}^2(1-\sigma_{0t})^2}\, \hat{s}^{-1} t^{1-\sigma_{0t}} \\ &+ \frac{(\theta_0^2 - \theta_t^2 - \sigma_{0t}^2)(\theta_\infty^2 - \theta_1^2 - \sigma_{0t}^2)}{8\sigma_{0t}^2}\, t + O\left(|t|^{2(1-\operatorname{Re}\sigma_{0t})}\right)\Bigg], \end{aligned} \tag{36}$$

where $\sigma_{0t} \neq 0$ *and*

$$\hat{s} = \Gamma \begin{bmatrix} 1-\sigma_{0t}, 1-\sigma_{0t}, 1+\frac{\theta_0+\theta_t+\sigma_{0t}}{2}, 1-\frac{\theta_0-\theta_t-\sigma_{0t}}{2}, 1+\frac{\theta_\infty+\theta_1+\sigma_{0t}}{2}, 1-\frac{\theta_\infty-\theta_1-\sigma_{0t}}{2} \\ 1+\sigma_{0t}, 1+\sigma_{0t}, 1+\frac{\theta_0+\theta_t-\sigma_{0t}}{2}, 1-\frac{\theta_0-\theta_t+\sigma_{0t}}{2}, 1+\frac{\theta_\infty+\theta_1-\sigma_{0t}}{2}, 1-\frac{\theta_\infty-\theta_1+\sigma_{0t}}{2} \end{bmatrix} s,$$

$$\begin{aligned} s^{\pm 1}\big(\cos\pi(\theta_t \mp \sigma_{0t}) - \cos\pi\theta_0\big)\big(\cos\pi(\theta_1 \mp \sigma_{0t}) - \cos\pi\theta_\infty\big) = \\ = (\pm\, i \sin\pi\sigma_{0t}\cos\pi\sigma_{1t} - \cos\pi\theta_t\cos\pi\theta_\infty - \cos\pi\theta_0\cos\pi\theta_1)\, e^{\pm i\pi\sigma_{0t}} \\ \pm\, i\sin\pi\sigma_{0t}\cos\pi\sigma_{01} + \cos\pi\theta_t\cos\pi\theta_1 + \cos\pi\theta_\infty\cos\pi\theta_0. \end{aligned}$$

If $\sigma_{0t} = 0$, *then*

$$\begin{aligned} \tau_{JMU}(t) = \text{const}\cdot t^{-\frac{\theta_0^2+\theta_t^2}{4}} \Big[1 - \frac{\theta_1\theta_t}{2}\, t - \frac{(\theta_\infty^2-\theta_1^2)(\theta_0^2-\theta_t^2)}{16}\, t(\Omega^2+2\Omega+3) \\ + \frac{\theta_t(\theta_\infty^2-\theta_1^2)+\theta_1(\theta_0^2-\theta_t^2)}{4}\, t(\Omega+1) + o(|t|)\Big], \end{aligned}$$

where $\Omega = 1 - \hat{s}' - \ln t$ *and*

$$\begin{aligned} \hat{s}' = s' + \psi\left(1+\frac{\theta_0+\theta_t}{2}\right) + \psi\left(1+\frac{\theta_t-\theta_0}{2}\right) \\ + \psi\left(1+\frac{\theta_\infty+\theta_1}{2}\right) + \psi\left(1+\frac{\theta_1-\theta_\infty}{2}\right) - 4\psi(1). \end{aligned}$$

Here $\psi(x)$ *denotes the digamma function and*

$$s' = i\pi\, \frac{\cos\pi\sigma_{1t} + \cos\pi\sigma_{01} - \cos\pi\theta_0\, e^{i\pi\theta_1} - \cos\pi\theta_\infty\, e^{i\pi\theta_t} + i\sin\pi(\theta_1+\theta_t)}{\big(\cos\pi\theta_t - \cos\pi\theta_0\big)\big(\cos\pi\theta_1 - \cos\pi\theta_\infty\big)}.$$

When one tries to determine from Theorem 6.1 the monodromy associated to the ${}_2F_1$ kernel solution $D(t)$ of σPVI, it turns out that all three assumptions (J1)–(J3) are not satisfied:

- Firstly, (J1) does not hold since in our case $\theta_t = 0$. This requirement can nevertheless be relaxed as the appropriate non-resonancy condition for (2) is $\theta_0, \theta_1, \theta_t, \theta_\infty \notin \mathbb{Z}\backslash\{0\}$. The proof of asymptotic formulas when some θ's are equal to zero differs from that in Ref. 16 only in technical details; see e.g. Ref. 11.
- If we blindly accept (36) then from $D(t \to 0) \sim 1$ follows $\sigma_{0t} = \theta_0 = z + z' + w + w'$. Thus (J2) is violated unless $\operatorname{Re}\theta_0 < 1$ and (J3) does not hold in any case. Note, however, that (36) admits a well-defined limit as $\theta_t = 0$, $\sigma_{0t} \to \theta_0$. In this limit, the coefficients of t and $t^{1-\sigma_{0t}}$ vanish; we also have

$$\cos\pi\sigma_{01} \to \cos\pi\theta_\infty + (\cos\pi\theta_1 - \cos\pi\sigma_{1t})\, e^{i\pi\theta_0},$$

$$s(\theta_0 - \sigma_{0t}) \to \frac{1}{\pi}\cdot\frac{\sin\pi\theta_0\,(\cos\pi\theta_1 - \cos\pi\sigma_{1t})}{\sin\frac{\pi}{2}(\theta_\infty - \theta_0 + \theta_1)\sin\frac{\pi}{2}(\theta_\infty + \theta_0 - \theta_1)},$$

and hence the coefficient of $t^{1+\sigma_{0t}}$ becomes

$$\frac{\cos\pi\theta_1 - \cos\pi\sigma_{1t}}{2\pi^2}\,\Gamma\left[\begin{matrix} 1+\frac{\theta_0+\theta_1+\theta_\infty}{2}, 1+\frac{\theta_0+\theta_1-\theta_\infty}{2}, 1+\frac{\theta_0-\theta_1+\theta_\infty}{2}, 1+\frac{\theta_0-\theta_1-\theta_\infty}{2} \\ 2+\theta_0, 2+\theta_0 \end{matrix}\right]. \tag{37}$$

- Suppose that in our case the error estimate in (36) can be improved to $O\left(t^{2+\theta_0}\right)$ (or at least to $o\left(t^{1+\theta_0}\right)$). Then, assuming that $0 \leq \operatorname{Re}(z+z') \leq 1$ and comparing (37) with (18), (16) one would conclude that $\sigma_{1t} = z+z'$.

The above steps can indeed be justified — after some tedious analysis going into the depths of Jimbo's proof. Alternatively, the monodromy can be extracted from Sections 3, 4 of Ref. 6, where σPVI equation for $D(t)$ has itself been derived from a Riemann-Hilbert problem.

7. Asymptotics of $D(t)$ as $t \to 1$

Once the monodromy is known, the asymptotics of $\tau_{JMU}(t)$ as $t \to 1$ can be determined from Jimbo's formula after substitutions $t \leftrightarrow 1-t$, $\theta_0 \leftrightarrow \theta_1$, $\sigma_{0t} \leftrightarrow \sigma_{1t}$, $\sigma_{01} \to \tilde{\sigma}_{01}$, where

$$2\cos\pi\tilde{\sigma}_{01} = \operatorname{Tr}\left(M_0 M_t^{-1} M_1 M_t\right) = p_0 p_1 + p_t p_\infty - p_{0t} p_{1t} - p_{01}. \tag{38}$$

Remark 7.1. The transformation $\sigma_{01} \to \tilde{\sigma}_{01}$ is missing in Ref. 16 due to an incorrectly stated symmetry: the relation $\tau_{JMU}\left(1-t; M_0, M_t, M_1\right) = \text{const}\cdot\tau_{JMU}\left(t; M_1, M_t, M_0\right)$ on p. 1144 of Ref. 16 should be replaced by

$$\begin{aligned}&\tau_{JMU}\left(1-t; M_0, M_t, M_1\right) = \\ &= \text{const}\cdot\tau_{JMU}\left(t; (M_t M_0)^{-1} M_1 M_t M_0, (M_0)^{-1} M_t M_0, M_0\right),\end{aligned}$$

which can be understood from Fig. 2.

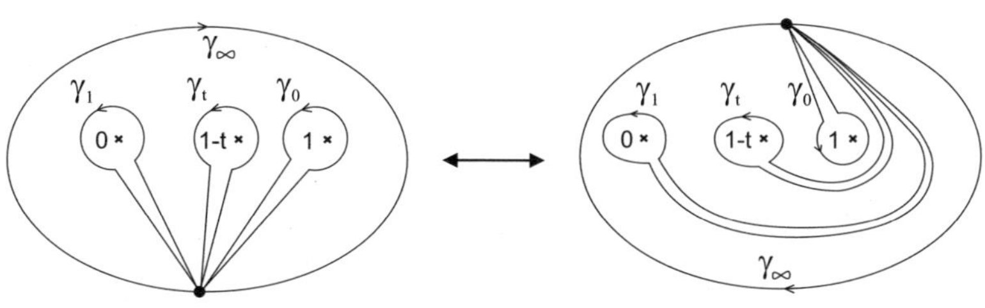

Fig. 2. Homotopy basis after transformation $\lambda \mapsto 1-\lambda$, $t \mapsto 1-t$.

Proposition 7.1. *Assume that* $0 \le \operatorname{Re}(z+z') < 1$ *and*

$$z, z', w, w', z+z'+w, z+z'+w' \notin \mathbb{Z}.$$

(1) If $z+z' \neq 0$*, then the following asymptotics is valid as* $t \to 1$*:*

$$D(t) = C\,(1-t)^{zz'}\left[1 + \frac{zz'\left((z+z'+w)(z+z'+w')+ww'\right)}{(z+z')^2}(1-t) - a^+(1-t)^{1+z+z'} - a^-(1-t)^{1-z-z'} + O\left((1-t)^{2-2\operatorname{Re}(z+z')}\right)\right], \tag{39}$$

where C *is a constant and*

$$a^{\pm} = \Gamma\left[\begin{matrix} \mp z \mp z', \mp z \mp z', 1 \pm z, 1 \pm z', 1+w+\frac{z+z'}{2} \pm \frac{z+z'}{2}, 1+w'+\frac{z+z'}{2} \pm \frac{z+z'}{2} \\ 2 \pm z \pm z', 2 \pm z \pm z', \mp z, \mp z', w+\frac{z+z'}{2} \mp \frac{z+z'}{2}, w'+\frac{z+z'}{2} \mp \frac{z+z'}{2} \end{matrix}\right].$$

(2) Similarly, if $z+z'=0$*, then*

$$D(t) = C(1-t)^{-z^2}\left[1 + z^2 ww'\,(1-t)(\tilde{\Omega}^2 + 2\tilde{\Omega} + 3) + z^2(w+w')(1-t)(\tilde{\Omega}+1) + o(1-t)\right], \tag{40}$$

where $\tilde{\Omega} = 1 - a' - \ln(1-t)$ *and*

$$a' = \psi(1+z) + \psi(1-z) + \psi(1+w) + \psi(1+w') - 4\psi(1).$$

Proof. Take into account that in our case $\theta_t = 0$, $\sigma_{0t} = \theta_0$ and replace $\theta_0 \leftrightarrow \theta_1$, $\sigma_{0t} \leftrightarrow \sigma_{1t}$, $\sigma_{01} \to \tilde{\sigma}_{01}$. Different quantities in Theorem 6.1 then transform into

$$s \to s_{1t} = 1\,, \qquad s' \to s'_{1t} = 0,$$

$$\hat{s} \to \hat{s}_{1t} = \Gamma\left[\begin{matrix} 1-\sigma_{1t}, 1-\sigma_{1t}, 1+\frac{\theta_1+\sigma_{1t}}{2}, 1-\frac{\theta_1-\sigma_{1t}}{2}, 1+\frac{\theta_0+\theta_\infty+\sigma_{1t}}{2}, 1+\frac{\theta_0-\theta_\infty+\sigma_{1t}}{2} \\ 1+\sigma_{1t}, 1+\sigma_{1t}, 1+\frac{\theta_1-\sigma_{1t}}{2}, 1-\frac{\theta_1+\sigma_{1t}}{2}, 1+\frac{\theta_0+\theta_\infty-\sigma_{1t}}{2}, 1+\frac{\theta_0-\theta_\infty-\sigma_{1t}}{2} \end{matrix}\right],$$

$$\hat{s}' \to \hat{s}'_{1t} = \psi\left(1+\frac{\theta_1}{2}\right) + \psi\left(1-\frac{\theta_1}{2}\right) + \psi\left(1+\frac{\theta_0+\theta_\infty}{2}\right) + \psi\left(1+\frac{\theta_0-\theta_\infty}{2}\right) - 4\psi(1).$$

The statement now follows from $\sigma_{1t} = z+z'$ and (16). □

The constant C in (39)–(40) remains as yet undetermined. We will find an expression for it using Lemma 4.1 and earlier results of Doyon,[10] who conjectured that for vanishing magnetic field $\tau_{PBT}(s)$ coincides with a correlation function of twist fields in the theory of free massive Dirac fermions on the hyperbolic disk. The asymptotics of $\tau_{PBT}(s)$ as $s \to 0$ and $s \to 1$ is then fixed, respectively, by conformal behavior of the correlator and its form factor expansion. The basic statement of Ref. 10 (supported by numerics) is that there indeed exists a solution of the appropriate σPVI equation which interpolates between the two asymptotics.

Although the proof that the correlator of twist fields satisfies σPVI has not yet been found, there are further confirmations of Doyon's hypothesis: long-distance asymptotics (20) with $b = 0$ and the exponent zz' in the short-distance power law (39) reproduce the conjectures of Ref. 10.

The QFT analogy also implies that for real $z, z' \in (0, 1)$ such that $0 < z + z' < 1$ and $w' = w - z - z'$ (this corresponds to $b = 0$) the constant C in (39) can be expressed in terms of vacuum expectation values of twist fields, which have been computed in Ref. 10 (see also Ref. 22). The resulting conjectural evaluation is:

$$C\Big|_{w'=w-z-z'} = G\left[\begin{array}{c} 1-z,1+z,1-z',1+z',1+w,1+w,1+z+z'+w,1-z-z'+w \\ 1-z-z',1+z+z',1+z+w,1-z+w,1+z'+w,1-z'+w \end{array}\right], \qquad (41)$$

where $G\left[\begin{array}{c} a_1,\ldots,a_m \\ b_1,\ldots,b_n \end{array}\right] = \dfrac{\prod_{k=1}^{m} G(a_k)}{\prod_{k=1}^{n} G(b_k)}$ and $G(x)$ denotes the Barnes function:

$$G(x+1) = (2\pi)^{x/2} \exp\left\{\frac{\psi(1)x^2 - x(x+1)}{2}\right\} \prod_{n=1}^{\infty} \left[\left(1 + \frac{x}{n}\right)^n \exp\left\{-x + \frac{x^2}{2n}\right\}\right].$$

In spite of what one might expect, extension of the above approach to the case $b \neq 0$ turns out to be rather complicated. However, the simple structure of (41) and the symmetries of the ${}_2F_1$ kernel suggest the following:

Conjecture 7.1. *Under assumptions of Proposition 7.1, the constant C in the asymptotic expansions (39), (40) is given by*

$$C = G\left[\begin{array}{c} 1-z,1+z,1-z',1+z',1+w,1+w',1+z+z'+w,1+z+z'+w' \\ 1-z-z',1+z+z',1+z+w,1+z+w',1+z'+w,1+z'+w' \end{array}\right]. \qquad (42)$$

The formula (42) is clearly compatible with (41) and (S1)–(S2). It has been checked both numerically and analytically as described below.

8. Numerics

To verify Conjecture 7.1, one can proceed in the following way:

(1) The solution of PVI associated to the ${}_2F_1$ kernel solution $D(t)$ of σPVI (uniquely determined by (17), (18)) has the following asymptotic behavior as $t \to 0$:

$$q(t) = t - \lambda_0\, t^{1+z+z'+w+w'} + O(t^{2+z+z'+w+w'}), \qquad (43)$$

$$\lambda_0 = \frac{(1+z+z'+w+w')^2}{(z+w)(z'+w)}\,\kappa.$$

(2) In fact one can show that in this case

$$q(t) = t - \lambda_0\, t^{1+z+z'+w+w'}(1-t)^{1+z-z'}\,{}_2F_1\left[\begin{array}{c} z+w,1+z+w' \\ 1+z+z'+w+w' \end{array}\middle|\, t\right]^2 + O(t^{2+2(z+z'+w+w')}). \tag{44}$$

(3) Use this asymptotics as initial condition and integrate the corresponding PVI equation numerically for some admissible choice of $\boldsymbol{\theta}$. It is then instructive to check Proposition 7.1 by verifying that for $0 < \mathrm{Re}(z+z') < 1$ the asymptotic expansion of $q(t)$ as $t \to 1$ is given by

$$q(t) = 1 - \lambda_1\,(1-t)^{1-z-z'} + o\left((1-t)^{1-\mathrm{Re}(z+z')}\right),$$

where

$$\lambda_1 = \Gamma\left[\begin{array}{c} z+z',z+z',1-z,1-z',w,1+w' \\ 1-z-z',1-z-z',z,z',z+z'+w,1+z+z'+w' \end{array}\right] = \frac{(1-z-z')^2}{w\,(z+z'+w')}\,a^- .$$

Similarly, for $z+z'=0$ one has a logarithmic behavior,

$$q(t) = 1 + (1-t)\left[z^2\left(\tilde{\Omega} + w^{-1} - 1\right)^2 - 1\right] + O\left((1-t)^2 \ln^4(1-t)\right).$$

(4) Finally, use $q(t)$ and the initial condition $D(t) \sim 1$ as $t \to 0$ to compute $D(t)$ from (9), (11). Looking at the asymptotics of $D(t)$ as $t \to 1$, one can numerically check the formula (42) for C.

9. Special solutions check

For special choices of parameters and initial conditions Painlevé VI equation can be solved explicitly. All explicit solutions found so far are either algebraic or of Picard or Riccati type. Algebraic solutions have been classified in Ref. 21; up to parameter equivalence, their list consists of 3 continuous families and 45 exceptional solutions.

It turns out that the parameters of exceptional algebraic solutions cannot be transformed to satisfy ${}_2F_1$ kernel constraints $p_0 = p_{0t}$, $p_t = 2$. Continuous families, however, do contain representatives verifying these conditions. Explicit computation of the corresponding τ-functions provides a number of analytic tests of Conjecture 7.1, some of which are presented below. Our notation for PVI Bäcklund transformations follows Table 1 in Ref. 21.

Example 9.1. Painlevé VI equation with parameters $\boldsymbol{\theta} = (1, \theta_1, 0, \theta_1)$ is satisfied by

$$q(t) = 1 - \frac{(2\theta_1 - 1) - (2\theta_1 + 1)\sqrt{1-t}}{(2\theta_1 - 3) - (2\theta_1 - 1)\sqrt{1-t}}\sqrt{1-t}.$$

This two-branch solution is obtained by applying Bäcklund transformation $s_\delta s_x s_y s_z s_\delta s_z s_\delta P_{xy}$ to Solution II in Ref. 21 (set $\theta_a = 1, \theta_b = \theta_1$). An explicit formula for the corresponding JMU τ-function can be found from (9), (11):

$$\tau_{JMU}(t) = \left[\frac{2\,(1-t)^{1/4}}{1+\sqrt{1-t}}\right]^{\frac{1-4\theta_1^2}{4}}.$$

Note that $\tau_{JMU}(t \to 0) = 1 - \frac{1-4\theta_1^2}{128}\,t^2 + O\left(t^3\right)$, and therefore $\tau_{JMU}(t)$ coincides with the hypergeometric kernel determinant $D(t)$ if we set $z = w = \frac{1+2\theta_1}{4}$, $z' = w' = \frac{1-2\theta_1}{4}$.

The asymptotics of $\tau_{JMU}(t)$ as $t \to 1$ has the form

$$\tau_{JMU}(t) = 2^{\frac{1-4\theta_1^2}{4}}\,(1-t)^{\frac{1-4\theta_1^2}{16}}\left(1 + O\left(\sqrt{1-t}\right)\right),$$

which implies that $C = 2^{\frac{1-4\theta_1^2}{4}}$. To verify that this coincides with the expression

$$C = G\left[\begin{matrix} \frac{3+2\theta_1}{4}, \frac{3-2\theta_1}{4}, \frac{5+2\theta_1}{4}, \frac{5+2\theta_1}{4}, \frac{5-2\theta_1}{4}, \frac{5-2\theta_1}{4}, \frac{7+2\theta_1}{4}, \frac{7-2\theta_1}{4} \\ \frac{1}{2}, \frac{3}{2}, \frac{3}{2}, \frac{3}{2}, \frac{3+2\theta_1}{2}, \frac{3-2\theta_1}{2} \end{matrix}\right],$$

given by Conjecture 7.1, one can use the recursion relation $G(z+1) = \Gamma(z)G(z)$, the duplication formulas (A.1), (A.3) for Barnes and gamma functions, and the value of $G\left(\frac{1}{2}\right)$ from Appendix A.

Example 9.2. Consider the rational curve

$$q = \frac{(s+1)(s-2)(5s^2+4)}{s(s-1)(5s^2-4)}, \qquad t = \frac{(s+1)^2(s-2)}{(s-1)^2(s+2)}.$$

It defines a three-branch solution of PVI with parameters $\boldsymbol{\theta} = (2, 0, 0, 2/3)$, which can be obtained from Solution III in Ref. 21 (with $\theta = 0$) by the transformation $t_x = s_x s_\delta \left(s_y s_z s_\infty s_\delta\right)^2$.

The associated τ-function is given by

$$\tau_{JMU}(t(s)) = \frac{3^{\frac{15}{8}}}{2^{\frac{25}{9}}} \cdot \frac{s\,(s+2)^{\frac{8}{9}}}{(s+1)^{\frac{15}{8}}\,(s-1)^{\frac{7}{72}}},$$

where the normalization constant is introduced for convenience. The map $t(s)$ bijectively maps the interval $(2,\infty)$ onto $(0,1)$. Choosing the corresponding solution branch one finds that

$$\tau_{JMU}(t\to 0) = 1 - \frac{16}{19683}\,t^3 + O\left(t^4\right),$$
$$\tau_{JMU}(t\to 1) \sim 3^{\frac{15}{8}}\cdot 2^{-\frac{17}{6}}\cdot(1-t)^{\frac{1}{36}}.$$

First asymptotics implies that $\tau_{JMU}(t)$ coincides with $D(t)$ provided $z = z' = \frac{1}{6}$, $w = \frac{7}{6}$, $w' = \frac{1}{2}$. From the second asymptotics we obtain $C = 3^{\frac{15}{8}}\cdot 2^{-\frac{17}{6}}$, whereas Conjecture 7.1 gives

$$C = G\left[\begin{matrix}\frac{3}{2},\frac{5}{2},\frac{5}{6},\frac{5}{6},\frac{7}{6},\frac{7}{6},\frac{11}{6},\frac{13}{6}\\ \frac{2}{3},\frac{4}{3},\frac{5}{3},\frac{5}{3},\frac{7}{3},\frac{7}{3}\end{matrix}\right].$$

Equality of both expressions can be shown using the known evaluations of $G\left(\frac{k}{6}\right)$, $k = 1\ldots 5$, see Ref. 1 or Appendix A.

Example 9.3. Applying the transformation $\left(s_\delta s_x s_y\right)^3 s_z s_\infty s_\delta r_x$ to Solution IV in Ref. 21 and setting $\theta = 0$, one obtains a four-branch solution of PVI with $\boldsymbol{\theta} = (1, 1/2, 0, 1)$ parameterized by

$$q = \frac{s(2-s)(5s^2-15s+12)}{(3-s)(3-2s)},\qquad t = \frac{s(2-s)^3}{3-2s}.$$

The corresponding τ-function has the form

$$\tau_{JMU}(t(s)) = \frac{2^{\frac{5}{12}}}{3^{\frac{15}{16}}}\cdot\frac{(3-s)^{\frac{15}{16}}}{(2-s)^{\frac{5}{12}}(1-s)^{\frac{5}{48}}}.$$

Choose the solution branch with $s\in(0,1)$. From the asymptotics $\tau_{JMU}(t\to 0) = 1 + \frac{15}{2048}\,t^2 + O\left(t^3\right)$ follows that $\tau_{JMU}(t)$ coincides with $D(t)$ provided $z = \frac{5}{12}$, $z' = -\frac{1}{12}$, $w = \frac{5}{6}$, $w' = -\frac{1}{6}$. Leading term in the asymptotic behavior of $\tau_{JMU}(t)$ as $t\to 1$ is

$$\tau_{JMU}(t\to 1) \sim 2^{\frac{25}{18}}\cdot 3^{-\frac{15}{16}}\cdot(1-t)^{-\frac{5}{144}},$$

so that we have $C = 2^{\frac{25}{18}}\cdot 3^{-\frac{15}{16}}$. On the other hand, Conjecture 7.1 implies that

$$C = G\left[\begin{matrix}\frac{5}{6},\frac{7}{6},\frac{11}{6},\frac{13}{6},\frac{7}{12},\frac{11}{12},\frac{13}{12},\frac{17}{12}\\ \frac{2}{3},\frac{4}{3},\frac{3}{4},\frac{5}{4},\frac{7}{4},\frac{9}{4}\end{matrix}\right].$$

To prove that these expressions are equivalent, (i) use the multiplication formula (A.1) with $n = 2$ and $z = \frac{1}{12},\frac{5}{12}$ to compute $G\left(\frac{1}{12}\right)G\left(\frac{5}{12}\right)G\left(\frac{7}{12}\right)G\left(\frac{11}{12}\right)$ and (ii) combine the resulting expression with the evaluations of $G\left(\frac{k}{4}\right)$, $G\left(\frac{k}{6}\right)$.

10. Limiting kernels

10.1. *Flat space limit: PVI → PV*

The interpretation of $D(t)$ as a determinant of a Dirac operator (Sec. 4) suggests to consider the flat space limit $R \to \infty$. This corresponds to the following scaling limit of the ${}_2F_1$ kernel:

$$w' \to +\infty, \qquad 1-t \sim \frac{s}{w'}, \qquad s \in (0,\infty).$$

In this limit, $D(t)$ transforms into the Fredholm determinant $D_L(s) = \det\left(1 - K_L\big|_{(s,\infty)}\right)$ with the kernel

$$\begin{aligned} K_L(x,y) &= \lim_{w'\to+\infty} \frac{1}{w'} K\left(1-\frac{x}{w'}, 1-\frac{y}{w'}\right) \\ &= \lambda_L \frac{A_L(x)B_L(y) - B_L(x)A_L(y)}{x-y}, \\ \lambda_L &= \frac{\sin \pi z \sin \pi z'}{\pi^2} \Gamma\left[1+z+w, 1+z'+w\right] \end{aligned}$$

$$A_L(x) = x^{-\frac12} W_{\frac12 - \frac{z+z'+2w}{2}, \frac{z-z'}{2}}(x), \quad B_L(x) = x^{-\frac12} W_{-\frac12 - \frac{z+z'+2w}{2}, \frac{z-z'}{2}}(x),$$

where $W_{\alpha,\beta}(x)$ denotes the Whittaker's function of the 2nd kind. $K_L(x,y)$ is the so-called Whittaker kernel,[4] which plays the same role in the harmonic analysis on the infinite symmetric group as the ${}_2F_1$ kernel does for $U(\infty)$.

The function $\sigma_L(s) = s\,\frac{d}{ds} \ln D_L(s)$ satisfies a Painlevé V equation written in σ-form:

$$\left(s\,\sigma_L''\right)^2 = \left(2\left(\sigma_L'\right)^2 - (z+z'+2w+s)\sigma_L' + \sigma_L\right)^2 - 4\left(\sigma_L'\right)^2 \left(\sigma_L' - z - w\right)\left(\sigma_L' - z' - w\right). \tag{45}$$

This can be shown by considering the appropriate limit of the σPVI equation for $D(t)$. An initial condition for (45) is provided by the asymptotics

$$D_L(s\to\infty) = 1 - \lambda_L\, e^{-s} s^{-z-z'-2w-2}\left(1 + O\left(s^{-1}\right)\right).$$

To link our notation with the one used in the PV part of Jimbo's paper,[16] we should set $(\theta_0, \theta_t, \theta_\infty)^{(V)}_{\text{Jimbo}} = (z'+w, -z-w, z-z')$, which gives $D_L(s) = e^{\frac{(z+w)s}{2}} \tau^{(V)}_{\text{Jimbo}}(s)$. This in turn allows to obtain from Theorem 3.1 in Ref. 16 the asymptotics of $D_L(s)$ as $s \to 0$:

Proposition 10.1. *Assume that* $0 \le \operatorname{Re}(z+z') < 1$ *and* $z, z', w, z+z'+w \notin \mathbb{Z}$.

(1) If $z+z'\neq 0$, *then*

$$D_L(s)=C_L s^{zz'}\left[1+\frac{zz'(z+z'+2w)}{(z+z')^2}s-a_L^+s^{1+z+z'}-a_L^-s^{1-z-z'}\right.$$
$$\left.+O\left(s^{2-2\,\mathrm{Re}(z+z')}\right)\right]$$

with $a_L^\pm=\Gamma\left[\begin{array}{c}\mp z\mp z',\mp z\mp z',1\pm z,1\pm z',1+w+\frac{z+z'}{2}\pm\frac{z+z'}{2}\\ 2\pm z\pm z',2\pm z\pm z',\mp z,\mp z',w+\frac{z+z'}{2}\mp\frac{z+z'}{2}\end{array}\right]$.

(2) If $z+z'=0$, *then*

$$D_L(s)=C_L s^{-z^2}\left[1+z^2ws\left(\tilde\Omega_L^2+2\tilde\Omega_L+3\right)+z^2s\left(\tilde\Omega_L+1\right)+o(s)\right],$$

where $\tilde\Omega_L=1-a_L'-\ln s$ *and* $a_L'=\psi(1+z)+\psi(1-z)+\psi(1+w)-4\psi(1)$.

Note that the same result is obtained by considering the formal limit of the leading terms in the asymptotics of $D(t)$. This further suggests an expression for constant C_L:

Conjecture 10.1. *Under assumptions of Proposition 10.1, we have*

$$C_L=\lim_{w'\to\infty}(w')^{-zz'}C=G\left[\begin{array}{c}1-z,1+z,1-z',1+z',1+w,1+z+z'+w\\ 1-z-z',1+z+z',1+z+w,1+z'+w\end{array}\right].$$

10.2. *Zero field limit: PV → PIII*

Next we consider the limit of vanishing magnetic field, $B\to 0$. In terms of the parameters of the Whittaker kernel, this translates into

$$w\to+\infty,\qquad s\sim\frac{\xi}{w},\qquad \xi\in(0,\infty).$$

The scaled kernel is given by

$$K_M(x,y)=\lim_{w\to+\infty}\frac{1}{w}K_L\left(\frac{x}{w},\frac{y}{w}\right)$$
$$=\frac{\sin\pi z\sin\pi z'}{\pi^2}\cdot\frac{A_M(x)B_M(y)-B_M(x)A_M(y)}{x-y},$$

$$A_M(x)=2\sqrt{x}\,K_{z'-z+1}\left(2\sqrt{x}\right),\qquad B_M(x)=2\,K_{z'-z}\left(2\sqrt{x}\right),$$

where $K_\alpha(x)$ is the Macdonald function.

Denote $D_M(\xi)=\det\left(1-K_M\big|_{(s,\infty)}\right)$ and introduce $\sigma_M(\xi)=\xi\frac{d}{d\xi}\ln D_M(\xi)$. Then $\sigma_M(\xi)$ solves the σ-version of a particular Painlevé III equation:

$$\left(\xi\sigma_M''\right)^2=4\sigma_M'(\sigma_M'-1)(\sigma_M-\xi\sigma_M')+(z-z')^2\left(\sigma_M'\right)^2. \qquad (46)$$

To match the notation in Ref. 16, we have to set $(\theta_0, \theta_\infty)^{(III)}_{\text{Jimbo}} = (z' - z, z - z')$, which gives $D_L(s) = e^{\xi} \tau^{(III)}_{\text{Jimbo}}(\xi)$. The appropriate initial condition for this σPIII is given by

$$D_M(\xi \to \infty) = 1 - \frac{\sin \pi z \sin \pi z'}{4\pi} \cdot \frac{e^{-4\sqrt{\xi}}}{\sqrt{\xi}} \left(1 + \frac{4(z-z')^2 - 3}{8\sqrt{\xi}} + O\left(\xi^{-1}\right)\right). \tag{47}$$

The asymptotics of $D_M(\xi)$ as $\xi \to 0$ can now be obtained from Theorem 3.2 in Ref. 16:

Proposition 10.2. *Assume that* $0 \leq \operatorname{Re}(z + z') < 1$ *and* $z, z' \notin \mathbb{Z}$.

(1) If $z + z' \neq 0$, *then*

$$D_M(\xi \to 0) = C_M\, \xi^{zz'} \Big[1 + \frac{2zz'}{(z+z')^2}\xi - a_M^+ \xi^{1+z+z'} - a_M^- \xi^{1-z-z'} + O\left(\xi^{2-2\operatorname{Re}(z+z')}\right) \Big],$$

with $a_M^{\pm} = \Gamma\left[\begin{matrix} \mp z \mp z', \mp z \mp z', 1 \pm z, 1 \pm z' \\ 2 \pm z \pm z', 2 \pm z \pm z', \mp z, \mp z' \end{matrix}\right]$.

(2) If $z + z' = 0$, *then*

$$D_M(\xi \to 0) = C_M\, \xi^{-z^2} \Big[1 + z^2 \xi (\tilde{\Omega}_M^2 + 2\tilde{\Omega}_M + 3) + o(\xi) \Big],$$

where $\tilde{\Omega}_M = 1 - a'_M - \ln \xi$ *and* $a'_M = \psi(1+z) + \psi(1-z) - 4\psi(1)$.

Analogously to the above, we suggest a conjectural expression for C_M:

Conjecture 10.2. *Under assumptions of Proposition 10.2, we have*

$$C_M = \lim_{w \to \infty} w^{-zz'} C_L = G\left[\begin{matrix} 1-z, 1+z, 1-z', 1+z' \\ 1-z-z', 1+z+z' \end{matrix}\right].$$

Partial proof. This formula can in fact be proved for real $z = z' \in \left[0, \frac{1}{2}\right)$, though in an indirect way. Consider the solution $\psi(r)$ of the radial sinh-Gordon equation

$$\frac{d^2\psi}{dr^2} + \frac{1}{r}\frac{d\psi}{dr} = \frac{1}{2}\sinh 2\psi,$$

satisfying the boundary condition $\psi(r, \nu) \sim 2\nu K_0(r)$ as $r \to +\infty$. Define the function

$$\tau(r, \nu) = \exp\left\{ \frac{1}{2} \int_r^\infty u \left[\sinh^2 \psi(u, \nu) - \left(\frac{d\psi}{du}\right)^2 \right] du \right\}.$$

and consider the logarithmic derivative $\tilde{\sigma}(\xi) = \xi\frac{d}{d\xi}\ln\tau(2\sqrt{\xi},\nu)$. It is straightforward to show that $\tilde{\sigma}(\xi)$ satisfies σPIII equation (46) with $z = z'$. Further, a little calculation shows that, as $r \to +\infty$,

$$\tau(r,\nu) = 1 - \pi\nu^2\,\frac{e^{-2r}}{2r}\left(1 - \frac{3}{4r} + O\left(r^{-2}\right)\right).$$

Comparing this asymptotics with (47), we conclude that $D_M(\xi)\Big|_{z=z'} = \tau\Big(2\sqrt{\xi}, \pm\frac{\sin\pi z}{\pi}\Big)$.

On the other hand, $\tau(r,\nu) = \tau_B^{-1}(r,\nu)$, where $\tau_B(r,\nu)$ is a special case of the bosonic 2-point tau function of Sato, Miwa and Jimbo, which can be represented as an infinite series of integrals (formulas (4.5.30)–(4.5.31) in Ref. 24 with $l_1 = l_2$). By direct asymptotic analysis of this series, Tracy[25] has proved that for $\nu \in \left[0, \frac{1}{\pi}\right)$ it has the following behavior as $r \to 0$:

$$\tau_B(r,\nu) = e^{\beta(\nu)}r^{-\alpha(\nu)}\left(1 + o(1)\right),$$

with

$$\alpha(\nu) = \frac{\sigma^2(\nu)}{2}, \qquad \sigma(\nu) = \frac{2}{\pi}\arcsin\pi\nu,$$

$$\beta(\nu) = 3\alpha(\nu)\ln 2 + \frac{1}{2}\ln\frac{1-\pi^2\nu^2}{\cos^4\frac{\pi\sigma(\nu)}{2}} - 2\ln\left(G\left[\begin{array}{c}\frac{1}{2},\ \frac{1}{2}\\ \frac{1+\sigma(\nu)}{2},\ \frac{1-\sigma(\nu)}{2}\end{array}\right]\right).$$

From $\nu = \pm\frac{\sin\pi z}{\pi}$ one readily obtains $\sigma^2 = 2\alpha = 4z^2$. Thus, in order to show that $\beta(\nu)$ reproduces the conjectured expression for C_M with $z = z'$, it is sufficient to prove the identity

$$G\left[\begin{array}{c}1+z,1+z,1-z,1-z\\ 1+2z,1-2z\end{array}\right] = 2^{-4z^2}\cos\pi z\; G\left[\begin{array}{c}\frac{1}{2},\frac{1}{2},\frac{1}{2},\frac{1}{2}\\ \frac{1}{2}+z,\frac{1}{2}+z,\frac{1}{2}-z,\frac{1}{2}-z\end{array}\right].$$

This, however, is a simple consequence of the duplication formula for Barnes function and the known evaluation of $G\left(\frac{1}{2}\right)$. □

Appendix A.

Multiplication formula for Barnes function:[28]

$$\ln G(nx) = \left(\frac{n^2x^2}{2} - nx\right)\ln 2 - \frac{(n-1)(nx-1)}{2}\ln 2\pi - \frac{n^2-1}{12} + \frac{5}{12}\ln n$$
$$+ \left(n^2-1\right)\ln A + \sum_{j=0}^{n-1}\sum_{k=0}^{n-1}\ln G\left(x + \frac{j+k}{n}\right), \tag{A.1}$$

where $A = \exp\left(\frac{1}{12} - \zeta'(-1)\right)$ denotes Glaisher's constant.

Asymptotic expansion as $|z| \to \infty$, $\arg z \neq \pi$:

$$\ln G(1+z) = \left(\frac{z^2}{2} - \frac{1}{12}\right)\ln z - \frac{3z^2}{4} + \frac{z}{2}\ln 2\pi - \ln A + \frac{1}{12} + O\left(\frac{1}{z^2}\right). \quad \text{(A.2)}$$

Special values (see, e.g. Ref. 1):

$$\ln G\left(\frac{1}{2}\right) = \frac{\ln 2}{24} - \frac{\ln \pi}{4} - \frac{3}{2}\ln A + \frac{1}{8},$$

$$\ln G\left(\frac{1}{3}\right) = \frac{\ln 3}{72} + \frac{\pi}{18\sqrt{3}} - \frac{2}{3}\ln\Gamma\left(\frac{1}{3}\right) - \frac{4}{3}\ln A - \frac{1}{12\pi\sqrt{3}}\psi'\left(\frac{1}{3}\right) + \frac{1}{9},$$

$$\ln G\left(\frac{2}{3}\right) = \frac{\ln 3}{72} + \frac{\pi}{18\sqrt{3}} - \frac{1}{3}\ln\Gamma\left(\frac{2}{3}\right) - \frac{4}{3}\ln A - \frac{1}{12\pi\sqrt{3}}\psi'\left(\frac{2}{3}\right) + \frac{1}{9},$$

$$\ln G\left(\frac{1}{6}\right) = -\frac{\ln 12}{144} + \frac{\pi}{20\sqrt{3}} - \frac{5}{6}\ln\Gamma\left(\frac{1}{6}\right) - \frac{5}{6}\ln A - \frac{1}{40\pi\sqrt{3}}\psi'\left(\frac{1}{6}\right) + \frac{5}{72},$$

$$\ln G\left(\frac{5}{6}\right) = -\frac{\ln 12}{144} + \frac{\pi}{20\sqrt{3}} - \frac{1}{6}\ln\Gamma\left(\frac{5}{6}\right) - \frac{5}{6}\ln A - \frac{1}{40\pi\sqrt{3}}\psi'\left(\frac{5}{6}\right) + \frac{5}{72},$$

$$\ln G\left(\frac{1}{4}\right) = -\frac{3}{4}\ln\Gamma\left(\frac{1}{4}\right) - \frac{9}{8}\ln A + \frac{3}{32} - \frac{K}{4\pi},$$

$$\ln G\left(\frac{3}{4}\right) = -\frac{1}{4}\ln\Gamma\left(\frac{3}{4}\right) - \frac{9}{8}\ln A + \frac{3}{32} + \frac{K}{4\pi},$$

where K is Catalan's constant.

When checking Conjecture 7.1 with explicit examples, one also needs the relations

$$\Gamma(nx) = (2\pi)^{-\frac{n-1}{2}} n^{nx-\frac{1}{2}} \prod_{k=0}^{n-1} \Gamma\left(x + \frac{k}{n}\right), \qquad \psi'(x) + \psi'(1-x) = \frac{\pi^2}{\sin^2 \pi x}. \quad \text{(A.3)}$$

References

1. V. S. Adamchik, *On the Barnes function*, Proc. 2001 Int. Symp. Symbolic and Algebraic Computation, Academic Press, (2001), 15–20.
2. M. Adler, T. Shiota, P. van Moerbeke, *Random matrices, vertex operators and the Virasoro algebra*, Phys. Lett. **A208**, (1995), 67–78.
3. J. Baik, R. Buckingham, J. DiFranco, *Asymptotics of Tracy-Widom distributions and the total integral of a Painlevé II function*, Comm. Math. Phys. **280**, (2008), 463–497; preprint `arXiv:0704.3636 [math.FA]`.
4. A. Borodin, *Harmonic analysis on the infinite symmetric group and the Whittaker kernel*, St. Petersburg Math. J. **12**, (2001), 733–759.
5. A. Borodin, G. Olshanski, *Harmonic analysis on the infinite-dimensional unitary group and determinantal point processes*, Ann. Math. **161**, (2005), 1319–1422; preprint `math/0109194 [math.RT]`.

6. A. Borodin, P. Deift, *Fredholm determinants, Jimbo-Miwa-Ueno tau-functions, and representation theory*, Comm. Pure Appl. Math. **55**, (2002), 1160–1230; preprint math-ph/0111007.
7. O. Costin, R. D. Costin, *Asymptotic properties of a family of solutions of the Painlevé equation P_{VI}*, Int. Math. Res. Notices **22**, (2002), 1167-1182; preprint math/0202235 [math.CA].
8. P. Deift, A. Its, I. Krasovsky, *Asymptotics of the Airy-kernel determinant*, Comm. Math. Phys. **278**, (2008), 643–678; preprint math/0609451 [math.FA].
9. P. Deift, A. Its, I. Krasovsky, X. Zhou, *The Widom-Dyson constant for the gap probability in random matrix theory*, J. Comput. Appl. Math. **202**, (2007), 26–47; preprint math/0601535 [math.FA].
10. B. Doyon, *Two-point correlation functions of scaling fields in the Dirac theory on the Poincaré disk*, Nucl. Phys. **B675**, (2003), 607–630; preprint hep-th/0304190.
11. B. Dubrovin, M. Mazzocco, *Monodromy of certain Painlevé VI transcendents and reflection groups*, Inv. Math. **141**, (2000), 55–147; preprint math.AG/9806056.
12. F. J. Dyson, *Fredholm determinants and inverse scattering problems*, Comm. Math. Phys. **47**, (1976), 171–183.
13. T. Ehrhardt, *Dyson's constant in the asymptotics of the Fredholm determinant of the sine kernel*, Comm. Math. Phys. **262**, (2006), 317–341; preprint math/0401205 [math.FA].
14. L. Haine, J.-P. Semengue, *The Jacobi polynomial ensemble and the Painlevé VI equation*, J. Math. Phys. **40**, (1999), 2117–2134.
15. A. R. Its, A. G. Izergin, V. E. Korepin, N. A. Slavnov, *Differential equations for quantum correlation functions*, Int. J. Mod. Phys. **B4**, (1990), 1003–1037.
16. M. Jimbo, *Monodromy problem and the boundary condition for some Painlevé equations*, Publ. RIMS, Kyoto Univ. **18**, (1982), 1137–1161.
17. M. Jimbo, T. Miwa, Y. Môri, M. Sato, *Density matrix of an impenetrable Bose gas and the fifth Painlevé transcendent*, Physica **1D**, (1980), 80–158.
18. M. Jimbo, T. Miwa, K. Ueno, *Monodromy preserving deformations of linear ordinary differential equations with rational coefficients I*, Physica **2D**, (1981), 306–352.
19. I. V. Krasovsky, *Gap probability in the spectrum of random matrices and asymptotics of polynomials orthogonal on an arc of the unit circle*, Int. Math. Res. Not. **2004**, (2004), 1249–1272; preprint math/0401258 [math.FA].
20. O. Lisovyy, *On Painlevé VI transcendents related to the Dirac operator on the hyperbolic disk*, J. Math. Phys. **49**, (2008), 093507; preprint arXiv:0710.5744 [math-ph].
21. O. Lisovyy, Yu. Tykhyy, *Algebraic solutions of the sixth Painlevé equation*, preprint arXiv:0809.4873 [math.CA].
22. S. Lukyanov, A. B. Zamolodchikov, *Exact expectation values of local fields in quantum sine-Gordon model*, Nucl. Phys. **B493**, (1997), 571–587; preprint hep-th/9611238.

23. J. Palmer, M. Beatty, C. A. Tracy, *Tau functions for the Dirac operator on the Poincaré disk*, Comm. Math. Phys. **165**, (1994), 97–173; preprint `hep-th/9309017`.
24. M. Sato, T. Miwa, M. Jimbo, *Holonomic quantum fields IV*, Publ. RIMS, Kyoto Univ. **15**, (1979), 871–972.
25. C. A. Tracy, *Asymptotics of a τ-function arising in the two-dimensional Ising model*, Comm. Math. Phys. **142**, (1991), 297–311.
26. C. A. Tracy, H. Widom, *Level-spacing distributions and the Airy kernel*, Comm. Math. Phys. **159**, (1994), 151–174; preprint `hep-th/9211141`.
27. C. A. Tracy, H. Widom, *Fredholm determinants, differential equations and matrix models*, Comm. Math. Phys. **163**, (1994), 33–72 ; preprint `hep-th/9306042`.
28. I. Vardi, *Determinants of Laplacians and multiple gamma functions*, SIAM J. Math. Anal. **19**, (1988), 493-507.
29. T. T. Wu, B. M. McCoy, C. A. Tracy, E. Barouch, *Spin-spin correlation functions for the two-dimensional Ising model: exact theory in the scaling region*, Phys. Rev. **B13**, (1976), 316–374.

REPRESENTATION THEORY OF A SMALL RAMIFIED PARTITION ALGEBRA

PAUL P. MARTIN

Department of Pure Mathematics, University of Leeds
Leeds, LS2 9JT, UK
E-mail: p.p.martin@leeds.ac.uk

Here $n \in \mathbb{N}$, k is a field, S_n is the symmetric group, $\delta \in k$, and $P_n(\delta)$ is the partition algebra over k. Our aim in this note is to study the representation theory of a subalgebra $P_n^{\ltimes}$ of $kS_n \otimes_k P_n(\delta)$ with certain interesting combinatorial and representation theoretic properties.

In Section 1 we discuss the motivating combinatorial background. In Section 2 we define $P_n^{\ltimes}$ (see Proposition 1). In Section 3 we determine its complex representation theory.

Keywords: Partition algebra; symmetric group wreath product.

Foreword. Tetsuji Miwa, together with Michio Jimbo, kindly invited me to RIMS Kyoto in the spring of 1989. That was an exciting time in Mathematical Physics. In the inspiring environment which Miwa and Jimbo created at RIMS, I had some fun with Potts transfer matrix algebras. (Miwa–Jimbo themselves, of course, were working on more important things,[6] but their encouragement was vital to me.) Twenty years on, there is still fun to be had with partition algebras.

1. Introduction

The Young graph[13] has vertex set the set Λ of all finite Young diagrams (equivalently of all integer partitions), and encodes the induction and restriction rules for ordinary irreducible modules of the sequence $... \subset S_n \subset S_{n+1} \subset ...$ of symmetric groups.[11] That is, the Young graph is the Bratteli diagram for this sequence [13, §1.1]. It can be considered to lie at the heart of the analysis of these groups, and much of combinatorics.[13,15] The multiplicity free graph (see Figure 1) and simple associated combinatorics allows a gentle build up of what, eventually, becomes a deep and powerful representation theory.[10,11,14,26]

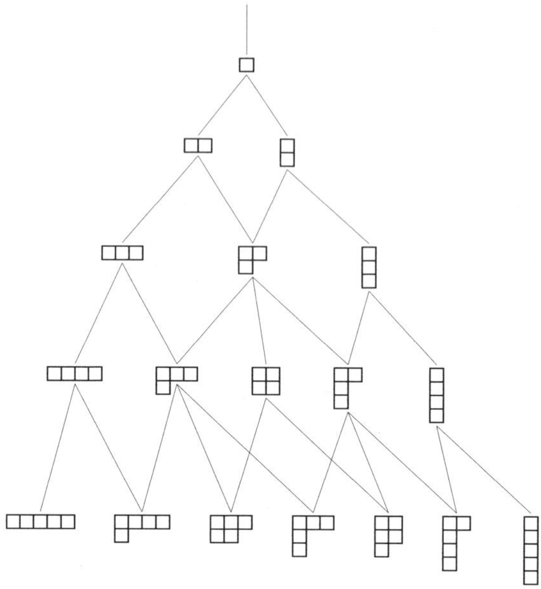

Fig. 1. The Young graph up to rank 5

In various areas of Physics,[2,19] algebra[4,11,27] and analysis[1,25] one is led also to study the *wreath products* of symmetric groups:

$$\begin{array}{ccc} \vdots & & \vdots \\ \cup & & \cup \\ S_2 \wr S_1 & \subset & S_2 \wr S_2 \subset \ldots \\ \cup & & \cup \\ S_1 \wr S_1 & \subset & S_1 \wr S_2 \subset \ldots \end{array}$$

Here however, no such multiplicity free Young graph can exist in general (at least, without further refinement), and one confronts a much more rapid onset of combinatorial complexity. By working at the level of suitable Morita equivalents, we might aim to bypass this obstruction and assemble an analogue of Young's theory of comparable reach. The challenge is to find a sequence of algebras with suitable properties.

For example, as we shall see here, in the Bratteli diagram for the sequence ... $\subset \overline{P}_n^{\ltimes} \subset \overline{P}_{n+1}^{\ltimes} \subset$ ..., the vertex set Λ is replaced by Λ^{Λ^*}, the set of functions from Λ^* to Λ (where Λ^* is the set of finite Young diagrams excluding the empty diagram); while the combinatorial representation theory can also be closely tied to that of wreaths. (We do not claim here that this sequence, encountered by chance while working on a different problem, is the ideal tool for this purpose, but at least that it is worth studying.) The original idea for this approach (inflating, in the sense of Ref. 16, from one combinatorial structure to another through Morita equivalences) comes serendipitously from observations on the representation theory of the ramified partition algebra.[23] We suspend all technical details for now, until we have reviewed the relevant components of representation theory — see Section 3.5.

Partition algebras play potentially important roles in Statistical Mechanics, in combinatorics, and in invariant theory. This is partly captured by the Schur-Weyl duality diagram here:

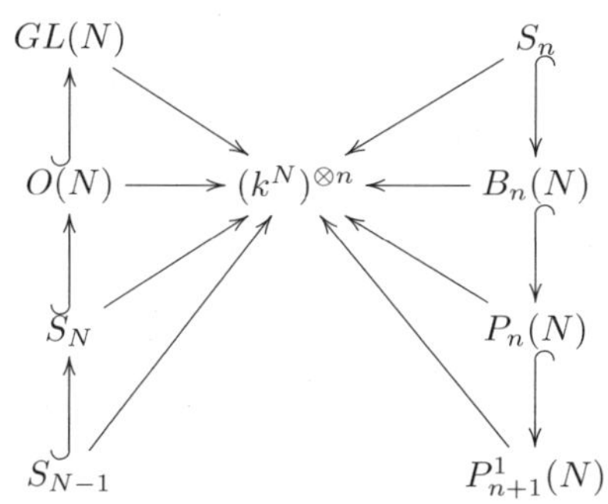

where the groups/algebras in each layer give a dual pair of (left) actions on tensor space. In each successive layer the action shown on the left-hand side is included in the one above (and the action on the right correspondingly includes the one above). Thus $O(N)$ and $B_n(N)$ are the orthogonal group and the Brauer algebra respectively;[8] S_N acts by permuting the standard ordered basis of k^N and $P_n(N)$ acts by the Potts representation (§8.2.1 of Ref. 20, Ref. 12). The S_{N-1} layer corresponds to breaking the global S_N symmetry of the N-state Potts model by applying a magnetic field.[22] (From an invariant theory perspective this dual pair sequence has been extended below $(S_N, P_n(N))$ in a number of ways. See for example Refs. 4,24.)

The complex reductive representation theory (i.e. Cartan decomposition matrices and so forth, in case $k = \mathbb{C}$) of all the algebras appearing in this diagram is reasonably well understood. It has been noted that *ramified* partition algebras (RPAs) have applications in similar areas,[23] but these are much less well understood. Particularly intriguing is the relationship between RPAs and wreaths (which, independently, also have roles in Physics[2,19] and combinatorics[18]). The ramified partition algebras $P_n^{\frac{2}{n}}(\delta', \delta)$ are physically motivated subalgebras of $P_n(\delta') \otimes_k P_n(\delta)$ (see Ref. 23 for a definition; δ, δ' are independently chosen parameters). As we shall see in Section 2, we have algebra inclusions

$$P_n(\delta') \otimes_k P_n(\delta) \begin{array}{l} \supset kS_n \otimes_k P_n(\delta) \supset \\ \supset P_n^{\frac{2}{n}}(\delta', \delta) \qquad \supset \end{array} P_n^{\ltimes}$$

and the representation theory of $P_n^{\ltimes}$ provides, from one perspective, a kind of approximation to that of $P_n^{\frac{2}{n}}$ (and hence also to that of the assembly of wreaths). Here, focussing on the representation theory of $P_n^{\ltimes}$, we are able to get pleasingly complete results on this representation theory (see the Theorems in the main section, §3.4).

The Bratteli diagram sought for the connection to wreath combinatorics is then discussed in the final section (we determine the simple restriction rules in low rank, and give a conjecture for the general form).

2. Definitions

Set $\underline{n} = \{1, 2, ..., n\}$ and $\underline{n}' = \{1', 2', ..., n'\}$ and so on. Write

$$\text{add}' : \underline{n} \cup \underline{n}' \to \underline{n}' \cup \underline{n}''$$

for the map that adds a prime; and $\text{cor}^{-\prime} : \underline{n} \cup \underline{n}'' \to \underline{n} \cup \underline{n}'$ for the map that removes a prime when necessary (i.e. when there are two).

For S a set, $\mathcal{P}_S$ is the set of partitions of S, and $\mathfrak{P}(S)$ the power set. Thus $|\mathcal{P}_{\underline{n}}| = B_n$, the Bell number.[17] We write $(\mathcal{P}_S, >)$ for the usual refinement order on $\mathcal{P}_S$, that is $p > q$ if each part of p is a union of parts of q. This order is a lattice.

Define $\mathcal{P}_n = \mathcal{P}_{\underline{n} \cup \underline{n}'}$. We write $\mathcal{P}'_n$ for the subset of partitions in $\mathcal{P}_n$ in which every part contains both primed and unprimed elements.

The algebra $P_n(\delta)$ has a basis $\mathcal{P}_n$. We now briefly recall the algebra product. (We refer the reader to Ref. 21 or Ref. 23 for a gentler exposition.) For $a \subset \mathfrak{P}(S)$ (some S) write $\overline{a} \in \mathcal{P}_S$ for the most refined (lowest) partition such that each part of $\overline{a}$ is a union of elements of a. Thus for example $a = \{\{1, 2\}, \{2, 3\}, \{4\}\}$ gives $\overline{a} = \{\{1, 2, 3\}, \{4\}\}$. Note that if $p, q \in \mathcal{P}_{\underline{n} \cup \underline{n}'}$ then $p \cup \text{add}'(q) \subset \mathfrak{P}(\underline{n} \cup \underline{n}' \cup \underline{n}'')$ and we can define

$$p \nabla q := \overline{p \cup \text{add}'(q)} \in \mathcal{P}_{\underline{n} \cup \underline{n}' \cup \underline{n}''}.$$

For $r \in \mathcal{P}_{\underline{n} \cup \underline{n}' \cup \underline{n}''}$ we write $\text{res}(r)$ for the restriction of this partition to $\mathcal{P}_{\underline{n} \cup \underline{n}''}$ (so that $\text{cor}^{-\prime}(\text{res}(r)) \in \mathcal{P}_{\underline{n} \cup \underline{n}'}$); and $c(r)$ for the number of parts containing only elements of $\underline{n}'$. Then the multiplication in $P_n(\delta)$ is defined on pairs p, q from the basis $\mathcal{P}_n$ by

$$p.q \;=\; \delta^{c(p\nabla q)}\ \mathrm{cor}^{-\prime}(\mathrm{res}(p\nabla q)).$$

Note from this construction that the set $\mathcal{P}'_n$ forms a submonoid in $P_n(\delta)$, and that this submonoid contains an isomorphic image of S_n, defined by identifying the transposition $\sigma_i = (i, i+1) \in S_n$ with the partition

$$\sigma_i = \{\{1,1'\},\{2,2',\},...,\{i,(i+1)'\},\{(i+1),i'\},...,\{n,n'\}\}$$

Write diag-$\mathcal{P}_n$ for the subset of $\mathcal{P}'_n$ of partitions such that i, i' are in the same part for all i. Such partitions are in natural bijection with the partitions of $\underline{n}$, so $|\text{diag-}\mathcal{P}_n| = |\mathcal{P}_{\underline{n}}|$. For example

$$A^{i,j} \;:= \{\{1,1'\},\{2,2',\},...,\{i,i',j,j'\},...,\{n,n'\}\}$$

is in diag-$\mathcal{P}_n$.

We write M_n^b for the monoid generated by $\{A^{ij}\}_{ij}$ and M_n^d for that generated by $S_n \cup \{A^{ij}\}_{ij}$. Define subalgebras of $P_n(\delta)$ generated by subsets: $P_n^d = k\langle S_n, A^{i,j}\rangle_{i,j}$ and $P_n^b = k\langle A^{i,j}\rangle_{i,j}$. (Note that neither subalgebra depends on δ.) These are simply the monoid algebras of the monoids above.

From the form of the partition algebra product we have

Lemma 1. *P_n^b is a commutative algebra with basis diag-$\mathcal{P}_n$ of idempotents. Indeed P_n^b is isomorphic (via the natural bijection) to the monoid algebra of the monoid $(\mathcal{P}_{\underline{n}}, \wedge)$, where $\wedge$ is the meet operation on $(\mathcal{P}_{\underline{n}}, >)$.* □

Note that this remark completely determines the reductive representation theory of P_n^b (as for any finite commutative monoid of idempotents).

The tensor product algebra $kS_n \otimes_k P_n(\delta)$ has basis $S_n \times \mathcal{P}_n$. Just as for S_n[11] and P_n^b, the complex representation theory of $P_n(\delta)$ is well understood,[22] and hence so are the tensor products.[5] We get a more challenging new algebra, however, if we proceed as follows. Define an injective map

$$\ltimes : S_n \times \text{diag-}\mathcal{P}_n \to S_n \times \mathcal{P}_n$$

$$(a, b) \mapsto (a, ba)$$

Write $P_n^{\ltimes}$ for the free k-submodule of $kS_n \otimes_k P_n^d$ with basis $\ltimes(S_n \times \text{diag-}\mathcal{P}_n)$.

Proposition 1. *The k-submodule $P_n^{\ltimes}$ is a subalgebra of $kS_n \otimes_k P_n^d$.*

Proof: Multiplication is given by $(a, ba)(c, dc) = (ac, badc)$, but $badc = bada^{-1}ac$, and $bada^{-1} \in P_n^b$. □

Proposition 2. *The algebra $P_n^{\ltimes}$ is generated by $(1, A^{ij})$ and (σ_i, σ_i) (all i, j), and hence by $(1, A^{12})$ and (σ_i, σ_i).* □

We will write $[a, b] = \ltimes(a, b)$. Thus

$$[a, b][c, d] = [ac, bada^{-1}] \tag{1}$$

and in particular

$$[a, 1][1, d] = [a, ada^{-1}]$$

Note that $A^{ij} \mapsto (1, A^{ij})$ defines a natural injection of P_n^b into $P_n^{\ltimes}$; and $\sigma_i \mapsto (\sigma_i, \sigma_i)$ a natural injection of kS_n into $P_n^{\ltimes}$.

Define the set of ($\underline{2}$-)ramified partitions

$$\mathcal{P}_n^{\underline{2}} = \{(a, b) \mid a, b \in \mathcal{P}_n;\ a < b\}$$

From Ref. 23 this is a basis for the RPA $P_n^{\underline{2}}(\delta, \delta') \subset P_n(\delta) \otimes_k P_n(\delta')$. Note also from the definition of $P_n^{\underline{2}}(\delta, \delta')$ in Ref. 23 that $kS_n \otimes_k P_n(\delta')$ is not a subalgebra of $P_n^{\underline{2}}(\delta, \delta')$ (for example, any non-identical pair of permutations lies outside $\mathcal{P}_n^{\underline{2}}$). However

Proposition 3. *We have an algebra inclusion $P_n^{\ltimes} \hookrightarrow P_n^{\underline{2}}$.*

Proof: It is easy to see that elements of form $[1, b]$ and $[s, 1]$ are ramified, and these generate $P_n^{\ltimes}$. □

Remark: We shall not make explicit use of it here, but for those comfortable with the ramified diagram calculus (see in particular Fig. 2 of Ref. 23) it might well be helpful to note that the diagrams for these generators may be exemplified as follows (in case $n = 5$):

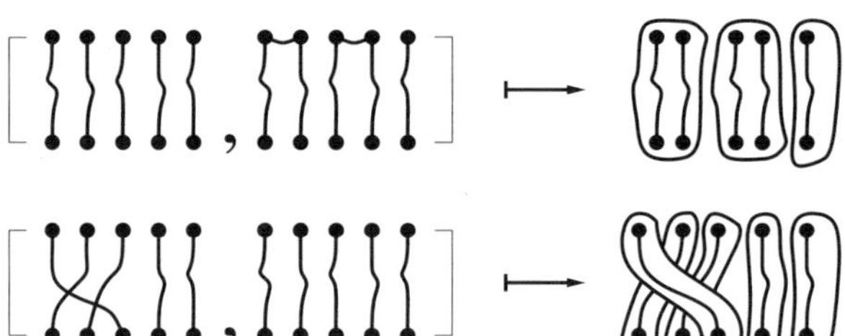

3. Representation theory of $P_n^{\ltimes}$

3.1. *Shapes and combinatorics*

The *shape* of a set partition is the list of sizes of parts in non-increasing order. Thus the shape of a partition of $\underline{n}$ is an integer partition of n. We will write $b \Vdash \mu$ if set partition b has shape μ.

By convention we shall express shapes in power notation:

$$\mu = (\underbrace{\lambda_1, \lambda_1, ..., \lambda_1}_{p_1}, \underbrace{\lambda_2, \lambda_2, ..., \lambda_2}_{p_2}, ...) \rightsquigarrow \qquad \lambda^p = (\lambda_1^{p_1}, \lambda_2^{p_2}, ...)$$

In particular

$$\lambda^p{}_i = \lambda_i^{p_i}$$

Via this notation a shape can be considered as a pair of a strictly descending integer partition $(\lambda_1, \lambda_2, ...)$ and a composition $(p_1, p_2, ...)$ of the same length.

There is a natural action of S_n on $\mathcal{P}_{\underline{n}}$. For each $b \in \mathcal{P}_{\underline{n}}$ define $S(b)$ as the subgroup that fixes b. We mention two subgroups in $S(b)$: $S^0(b)$ is the group the permutes within parts: $S^0(b) \cong (S_{\lambda_1})^{\times p_1} \times (S_{\lambda_2})^{\times p_2} \times ... \subset S_n$ (in case $b \Vdash \lambda^p$); and $S^1(b)$ permutes parts of equal order: $S^1(b) \cong S_{p_1} \times S_{p_2} \times ... \subset S_n$. We have

$$S(b) \cong \times_i (S_{\lambda_i} \wr S_{p_i}) \tag{2}$$

Considering $S(b)$ or otherwise (see e.g. Ref. 21), the number of partitions of given shape is

$$\mathcal{D}_{\lambda^p} = \frac{n!}{\prod_i((\lambda_i!)^{p_i} p_i!)} = \frac{n!}{|S(b \Vdash \lambda^p)|} \tag{3}$$

Write T_b^L (resp. T_b^R) for a traversal of the left (resp. right) cosets of $S(b)$ in S_n. I.e. $\cup_{w \in T_b^L} wS(b)$ is a partition of S_n.

3.2. *On representations of wreaths*

We shall establish later a construction of irreducible representations of our algebra $P_n^{\ltimes}$ directly in terms of representations of $S(b)$. Accordingly we mention these now. (However the reader may safely skip all the standard material in this section.)

Write Λ for the set of all integer partitions including the empty partition, and Λ_n for the subset of partitions of degree n. For G a group, write $\Lambda_{\mathbb{C}}(G)$ for an index set for ordinary irreducible representations (together, in principle, with a map to explicit representations), so that $\Lambda_{\mathbb{C}}(S_n) = \Lambda_n$. (We will use the analogous notation, $\Lambda_{\mathbb{C}}(A)$, for any algebra A over the complex field.) Set $s_G = |\Lambda_{\mathbb{C}}(G)|$ and assume there is a natural counting. For S, T any sets, write $\mathrm{Hom}(S, T)$ for the set of maps $f : S \to T$. Thus an element V of $\mathrm{Hom}(\Lambda_{\mathbb{C}}(G), \Lambda)$ may be expressed as an s_G-tuple $(V_1, V_2, ..., V_{s_G})$ of integer partitions (a *multipartition*). For any $\mathrm{Hom}(S, \Lambda)$, write $\mathrm{Hom}(S, \Lambda)_n$ for the subset of multipartitions of total degree n.

The ordinary irreducible representation theory of $S(b)$ is, in effect, fairly well understood. Since $\mathbb{C}$ is a splitting field it is enough to study the wreath factors. Now see Ref. 11. In particular we have

Theorem 1. *(Cf. [11, COR.4.4.4] or [18, §1.Appendix B])*

$$\Lambda_{\mathbb{C}}(G \wr S_n) = Hom(\Lambda_{\mathbb{C}}(G), \Lambda)_n$$

The construction of irreducible L_V, $V \in \Lambda_{\mathbb{C}}(S_l \wr S_n)$ is then as follows. The datum V consists (see Ref. 11 or, say, Ref. 2) of a map $V : \Lambda_{\mathbb{C}}(G = S_l) \to \Lambda$ such that $\sum_i |V_i| = n$. We set $v = (v_1, v_2, ...) = (|V_1|, |V_2|, ...)$ and form a traversal T_v of the left cosets of S_v in S_n. Let B^i be a basis for the irreducible representation $\mathcal{S}_i$ in our numbering scheme for irreducible representations of S_l (*lex* order of Λ_l, say); and B^{V_i} a basis for the irreducible representation $\mathcal{S}(V_i)$ of S_{v_i} (note that $V_i \vdash v_i$, so this is via the usual labelling scheme). Thus

$$B^v = \times_i((B^i)^{\times v_i})$$

is a basis for the irreducible representation of $S_l^{\times n}$ obtained from the representations $(\underbrace{\mathcal{S}_1, \mathcal{S}_1, .., \mathcal{S}_1}_{v_1\ copies}, \underbrace{\mathcal{S}_2, \mathcal{S}_2, .., \mathcal{S}_2}_{v_2\ copies}, ..., \mathcal{S}_s)$ of S_l. Set

$$B_V^v = \times_i((B^i)^{\times v_i} \times B^{V_i})$$

or rather in the order

$$B_V^v = (\times_i(B^i)^{\times v_i}) \times (\times_i B^{V_i})$$

Then $B_V^v \times T_v$ can be equipped with the property of basis for an (irreducible) representation of $S_l \wr S_n$.

Let $b_1 \otimes ... \otimes b_n \otimes (b_{n+1}..) \otimes [t]$ be an element of this basis. If $\sigma \in S_v$, $t' \in T_v$, then the action of $(g_1, g_2, .., g_n; t'\sigma)$ is given by

$$(g_1, g_2, .., g_n; t'\sigma)\ \ b_1 \otimes ... \otimes b_{v_1} \otimes b_{v_1+1} \otimes ... \otimes b_n \otimes (b_{n+1}..) \otimes [t]$$

$$= g_1 b_{\sigma^{-1}(1)} \otimes ... \otimes g_n b_{\sigma^{-1}(n)} \otimes \sigma(b_{n+1}..) \otimes [t'\sigma t]$$

where $[t'\sigma t]$ is understood as the coset representative of the coset containing this element. (See Ref. 11 for a much more detailed exposition, but) Note that the dimension of L_V is clear:

$$\dim L_V = \frac{n!}{\prod_i v_i!} \prod_i d_{V_i} (d_i)^{v_i} \tag{4}$$

where we write d_λ for the dimension of the S_{v_i} Specht module $\mathcal{S}_\lambda$, and d_i for Specht dimensions for S_l labeled using our numbering scheme.

Recall

$$n! = \sum_{\lambda \vdash n} d_\lambda^2 \tag{5}$$

3.3. *Useful decompositions of* Λ^{Λ^*}

The following will be useful later.

Another way to express an integer partition in an (ascending) power notation is simply as an element α of $\mathrm{Hom}(\mathbb{N}, \mathbb{N}_0)$ of finite support. The construct $(1^{\alpha(1)}, 2^{\alpha(2)}, ...)$ determines an integer partition in ordinary power notation on omitting all terms i such that $\alpha(i) = 0$ and then reversing the order of the remaining terms.

For example $\alpha : (1, 2, 3, 4, ...) = (2, 4, 0, 0, ...)$ becomes $(2^4, 1^2)$.

More generally, to specify a function $\mu \in \mathrm{Hom}(S, T)$, given $\underline{x}$ an ordered list of the elements of S, we may write $\mu : \underline{x} = \underline{y}$, meaning $\mu(x_i) = y_i$ (as in the example immediately above). But if almost all $\mu(x_i) = t_0$, with t_0 some given element of T, then it is convenient to write $\mu = (x_{i_1}, \mu(x_{i_1}))(x_{i_2}, \mu(x_{i_2}))...$ where $\{i_1, i_2, ...\}$ is the set of i such that $\mu(x_i) \neq t_0$. Depending on circumstances, the alternative layout

$$\mu = \frac{x_{i_1}}{\mu(x_{i_1})} \frac{x_{i_2}}{\mu(x_{i_2})} ... \tag{6}$$

may also be useful.

In this notation our example above becomes $\alpha = \frac{2}{4}\frac{1}{2}$ (with $t_0 = 0$).

Let us write $\mathrm{Hom}^f(\Lambda^*, \Lambda)$ for the set of functions

$$\mu : \Lambda^* \to \Lambda$$

with only finitely many $\lambda \in \Lambda^*$ such that $\mu(\lambda) \neq \emptyset$. We also emphasise that Λ is the set of integer partitions of finite integers. Thus the *degree* of $\mu \in \mathrm{Hom}^f(\Lambda^*, \Lambda)$

$$|\mu| = \sum_{\lambda} |\lambda||\mu(\lambda)|$$

is well defined. Write $\mathrm{Hom}_N(\Lambda^*, \Lambda)$ for the subset of $\mathrm{Hom}^f(\Lambda^*, \Lambda)$ of functions of degree N.

For example, $\mathrm{Hom}_3(\Lambda^*, \Lambda) = \{\frac{(3)}{(1)}, \frac{(21)}{(1)}, \frac{(1^3)}{(1)}, \frac{(2)}{(1)}\frac{(1)}{(1)}, \frac{(1^2)}{(1)}\frac{(1)}{(1)}, \frac{(1)}{(3)}, \frac{(1)}{(21)}, \frac{(1)}{(1^3)}\}$

The *shape* of $\mu \in \mathrm{Hom}^f(\Lambda^*, \Lambda)$ is an integer partition $\kappa(\mu)$ defined as follows. We specify via ascending power notation, in terms of which the partition is given by

$$\alpha(i) = \sum_{\lambda \vdash i} |\mu(\lambda)|$$

and then recast in ordinary power notation as described above.
Example: $\mu : ((1),(2),(1^2),(3),...) = (\emptyset,(1),(1^3),\emptyset,...)$ has $\kappa(\mu) = (2^4)$.
We write $\mathrm{Hom}_{\lambda^p}(\Lambda^*,\Lambda)$ for the subset of functions of shape λ^p. We have

$$\mathrm{Hom}_N(\Lambda^*,\Lambda) = \bigcup_{\lambda^p \vdash N} \mathrm{Hom}_{\lambda^p}(\Lambda^*,\Lambda)$$

In the simple case in which κ has just a single 'factor' i^m then $\mathrm{Hom}_{(i^m)}(\Lambda^*,\Lambda)$ is just the set of maps from Λ_i to Λ of total degree m. By Theorem 1 then,

$$\Lambda_{\mathbb{C}}(S_n \wr S_m) = \mathrm{Hom}_{(n^m)}(\Lambda^*,\Lambda)$$

Thus with $b \Vdash \lambda^p$

$$\Lambda_{\mathbb{C}}(S(b)) = \mathrm{Hom}_{\lambda^p}(\Lambda^*,\Lambda)$$

We now have the notation to assert (as we shall show in Theorem 4)

$$\Lambda_{\mathbb{C}}(P_n^{\ltimes}) = \mathrm{Hom}_n(\Lambda^*,\Lambda)$$

3.4. *Decomposing the regular $P_n^{\ltimes}$-module*

We will say that the shape of $[a,b] = (a,ba)$ is the shape of b. It follows from (1) that the shape of $[a,b] = (a,ba)$ is unchanged by left or right multiplication by $[\sigma_i,1] = (\sigma_i,\sigma_i)$.

As shapes of set partitions, integer partitions inherit a partial order from the order on set partitions themselves. E.g.

$$\begin{aligned}(1^4) < (2,1^2) < (3,1) &< (4)\\ &< (2^2) <\end{aligned}$$

Thus left or right multiplication by $[1,A^{ij}]$ either acts like 1 or takes the shape up in this order. Altogether, then, the left regular $P_n^{\ltimes}$-module is filtered by a poset of submodules (indeed ideals) labelled by shape. Set

$$e_{\lambda^p} := \sum_{b \Vdash \lambda^p} [1,b]$$

and note that these are central elements in $P_n^{\ltimes}$. For example $e_{1^n} = [1,1]$. We have

$$P_n^{\ltimes} e_{\lambda^p} \subset P_n^{\ltimes} e_{\lambda^{p'}} \iff \lambda^p > \lambda^{p'}$$

The sections $\mathcal{M}_{\lambda^p}$ of this poset each have basis the set of elements of $\ltimes(S_n \times \text{diag-}\mathcal{P}_n)$ of fixed shape. The number of basis elements of shape λ^p is $n!\mathcal{D}_{\lambda^p}$.

We want to decompose the sections as far as possible.

As a vector space we have

$$\mathcal{M}_{\lambda^p} = \bigoplus_{b \Vdash \lambda^p} k[S_n, b] = \bigoplus_{b \Vdash \lambda^p} \bigoplus_{w \in T_b^R} k[S(b)w, b] \tag{7}$$

Note that the $S(b)$-module $k[S_n, b]$ is isomorphic to kS_n as an $S(b)$-module, and hence is simply $\frac{n!}{|S(b)|}$ copies of the regular module.

Consider the quotient algebra of $P_n^{\ltimes}$ by all the ideals $P_n^{\ltimes} e_{\lambda^{p'}}$ below λ^p. The central element e_{λ^p} is idempotent in this quotient. Thus we can regard $\mathcal{M}_{\lambda^p}$ as an idempotent subalgebra of the quotient, with identity element e_{λ^p}. The category of left $\mathcal{M}_{\lambda^p}$-modules thus fully embeds in the category of left $P_n^{\ltimes}$-modules (§6.2 of Ref. 9), with the simple modules not hit by this embedding coming from the other $\mathcal{M}_{\lambda^{p'}}$.

Now consider the idempotent $[1, b_0]$, $b_0 \Vdash \lambda^p$, and note that in the algebra $\mathcal{M}_{\lambda^p}$ we have $[1, b_0][1, b] = \delta_{b_0, b}[1, b_0]$. We have

$$[1, b_0]\mathcal{M}_{\lambda^p} = [1, b_0] \bigoplus_{w \in S_n; b \Vdash \lambda^p} k[w, b] = \bigoplus_{w \in S_n; b \Vdash \lambda^p} k[1, b_0][w, b] = \bigoplus_{w \in S_n} k[w, b_0]$$

Thus

$$[1, b_0]\mathcal{M}_{\lambda^p}[1, b_0] = \bigoplus_{w \in S_n} k[w, b_0][1, b_0] = \bigoplus_{w \in S_n} k[w, b_0 w b_0 w^{-1}]$$

$$= \bigoplus_{w \in S(b_0)} k[w, b_0] \cong kS(b_0)$$

and

$$\mathcal{M}_{\lambda^p}[1, b_0]\mathcal{M}_{\lambda^p} = \mathcal{M}_{\lambda^p} \bigoplus_{w \in S_n} k[w, b_0] = \left(\bigoplus_{x \in S_n;\ b \Vdash \lambda^p} k[x, b] \right) \bigoplus_{w \in S_n} k[w, b_0]$$

$$= \bigoplus_{x \in S_n;\ b \Vdash \lambda^p} \bigoplus_{w \in S_n} k[x, b][w, b_0] = \bigoplus_{x \in S_n;\ b \Vdash \lambda^p;\ w \in S_n} k[xw, bxb_0x^{-1}] = \mathcal{M}_{\lambda^p}$$

Thus

Theorem 2. *The algebras $\mathcal{M}_{\lambda^p}$ and $kS(b_0)$ (with $b_0 \Vdash \lambda^p$) are Morita equivalent.*

Recall that $P_n^{\ltimes}$ has a subalgebra isomorphic to P_n^b. By restricting to this we see that no two sections contain any isomorphic factors. Thus each simple factor will appear in its section with multiplicity given by the dimension of its projective cover (with this dimension bounded from below, ab initio, by the dimension of the simple itself).

It also follows that

$$\Lambda_{\mathbb{C}}(P_n^{\ltimes}) = \bigcup_{\lambda^p \vdash n} \Lambda_{\mathbb{C}}(\mathcal{M}_{\lambda^p})$$

so we have determined $\Lambda_{\mathbb{C}}(P_n^{\ltimes})$ (by Theorem 2 and the results in §3.2 – equation(2) and Theorem 1). We will unpack the details shortly.

Next we compute the dimensions of these simple modules, and the overall algebra structure.

Consider the left submodule generated by an arbitrary non-zero element $\sum_{ij} c_{ij}[x_i, y_j]$ of the λ^p-th section, $\mathcal{M}_{\lambda^p}$. Choosing l so that some scalar $c_{il} \neq 0$, then in the section,

$$[1, y_l] \sum_{ij} c_{ij}[x_i, y_j] = \sum_{ij} c_{ij}[1, y_l][x_i, y_j] = \sum_{ij} c_{ij}[x_i, y_l y_j] = \sum_i c_{il}[x_i, y_l]$$

Thus this submodule itself contains a submodule generated by $\sum_i c_{il}[x_i, y_l]$. Further, by (1) this submodule contains, for every partition of shape λ^p, an element of this form whose partition part is that partition. (These elements are of course all linearly independent.) Thus

Lemma 2. *Any submodule of $\mathcal{M}_{\lambda^p}$ contains a non-vanishing element of form $\sum_i c_i[x_i, b]$, with $b \Vdash \lambda^p$.*

How does $P_n^{\ltimes} = \langle [1, A^{12}], [S_n, 1] \rangle$ act on this element? As noted, $[1, A^{12}]$ acts as 1 or 0. We consider the action of $[S_n, 1]$ in two parts: $[S(b), 1]$; and a traversal. The first part is simply a copy of $S(b) \hookrightarrow P_n^{\ltimes}$, so the element in Lemma 2 generates at least a simple $S(b)$-module. But since $S(b)$ fixes b, the $S(b)$-module generated will be spanned by elements of this form, so there will be an element of this form which generates precisely a simple $S(b)$-module. Meanwhile the action of an element w of a traversal is

$$w \sum_i c_i[x_i, b] = [w, 1] \sum_i c_i[x_i, b] = \sum_i c_i[wx_i, b^w]$$

Note that the right hand side generates an $S(b^w)$-module that is isomorphic (via the natural group isomorphism) to the original $S(b)$-module. This tells us that every $P_n^{\ltimes}$-submodule of $\mathcal{M}_{\lambda^p}$ decomposes as a vector space in to summands, indexed by $b \Vdash \lambda^p$, the b-th of which is an $S(b)$-module isomorphic (via the various group isomorphisms) to all the other summands. Clearly then, in particular every *simple* $P_n^{\ltimes}$-submodule is at least a sum (as a vector space) of $\mathcal{D}_{\lambda^p}$ spaces each of which is an (isomorphic) simple module for $S(b)$ for the appropriate b.

In particular

Proposition 4. *For each inequivalent simple $S(b)$-module L_μ (i.e. with $\mu \in Hom_{\lambda^p}(\Lambda^*, \Lambda)$ and $\lambda^p \dashv\vdash b$) of dimension m_μ and basis $\{g_i^\mu x_\mu | i = 1, .., m_\mu\}$, say (see §3.2), there is a simple $P_n^\ltimes$-module $L_\mu^\ltimes$ of dimension*

$$\dim L_\mu^\ltimes = m_\mu \mathcal{D}_{\lambda^p} \tag{8}$$

and basis $\{[wg_i^\mu x_\mu, b^w] \mid i = 1, .., m_\mu,\ w \in T_b^L\}$. The modules $\{L_\mu^\ltimes\}$ are pairwise inequivalent. □

Similarly,

Theorem 3. *The decomposition of the b-th summand (any b) of $\mathcal{M}_{\lambda^p}$ itself, $S(b)[1, b]$, into a series of simple $S(b)$-modules passes to a complete decomposition of $\mathcal{M}_{\lambda^p}$ into a series of simple $P_n^\ltimes$-modules of this construction.*

That is, every simple $P_n^\ltimes$-module arises this way (for some λ^p).

Working over k such that $kS(b)$ is split semisimple for every shape (e.g. over the complex numbers), the multiplicity of L_μ in the b-th summand is $m_\mu \mathcal{D}_{\lambda^p}$, since the summand is $\mathcal{D}_{\lambda^p}$ copies of the regular $S(b)$-module. Thus (or by Theorem 2) each $\mathcal{M}_{\lambda^p}$ is semisimple (cf. §1.7 of Ref. 3 for example), and hence

Theorem 4. *Let $n \in \mathbb{N}$. Over k as above, $P_n^\ltimes$ is split semisimple. The simple modules may be indexed by the set $Hom_n(\Lambda^*, \Lambda)$. The dimensions of the simple modules are given by (8), using (2) and (4).* □

We give some concrete illustrative examples in the next Section. For each λ^p we have

$$n!\mathcal{D}_{\lambda^p} = \sum_{\mu \in \mathrm{Hom}_{\lambda^p}(\Lambda^*, \Lambda)} (m_\mu^{\lambda^p} \mathcal{D}_{\lambda^p})^2$$

$$n! = \sum_\mu (m_\mu^{\lambda^p})^2 \mathcal{D}_{\lambda^p} \tag{9}$$

$$\prod_i ((\lambda_i!)^{p_i} p_i!) = \sum_\mu (m_\mu^{\lambda^p})^2$$

This is not a trivial identity, but it is simply the $S(b)$ version of the hook dimension formula (cf. Ref. 15). Note that the solution to this when $\mathcal{D}_\lambda = 1$ is given *by* the hook dimension formula.

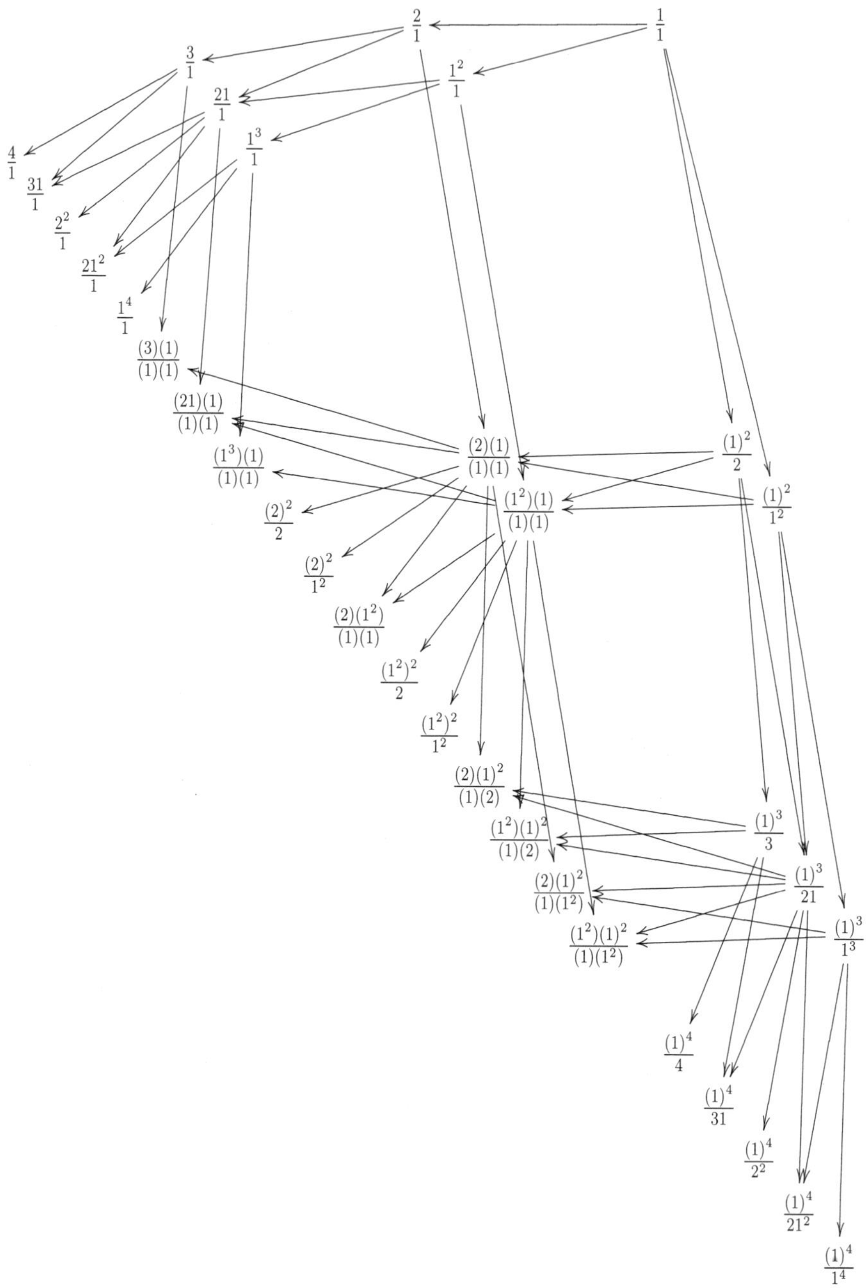

Fig. 2. Bratelli diagram for $P_n^{\ltimes}$ (drawn rotated to fit on the page).

3.5. *Examples and combinatorial restriction*

First we unpack Theorem 4 a little. The explicit simple module index sets for the first few algebras $P_n^{\ltimes}$ $(n = 1, 2, 3, 4, ...)$ are given in figure 2. In the figure they appear as the vertices in the n-th layer of a certain directed graph. In a vertex, i.e. in an element $\mu \in \mathrm{Hom}_n(\Lambda^*, \Lambda)$, we use here the notation $\frac{\lambda^{|\mu(\lambda)|}}{\mu(\lambda)}$ for each 'factor' (i.e. each remaining $(\lambda, \mu(\lambda))$-pair after removing pairs of form $\frac{\lambda}{\emptyset}$). (We further omit the brackets, if there is only a single factor in μ.) The slight redundancy here (the exponent on the 'numerator' – a redundant addition to our notation in (6)) facilitates some useful consistency checking in practical calculations.

To explicitly illustrate the application of the structure Theorem we compute the dimensions of certain modules $\dim L_\mu^{\ltimes}$ explicitly. For $\mu = \frac{(2)^2(1)^2}{(2)(1^2)}$ we need $\mathcal{D}_{2^21^2} = \frac{6!}{(2!)^2 2!(1!)^2 2!} = 45$, $\dim L_{\frac{(2)}{(2)}} = \frac{2!}{2!} d_{(2)} d_{(2)}^2 = 1$, and $\dim L_{\frac{(1)}{(1^2)}} = \frac{2!}{2!} d_{(1^2)} d_{(1)}^2 = 1$ (from the formula (4)), giving $\dim L_\mu^{\ltimes} = 45$. For $\frac{(2)}{(1)}\frac{(1^3)}{(21)}$, $\frac{(2)}{(1)}\frac{(1^3)}{(1^3)}$ and $\frac{(2)^2}{(2)}\frac{(1)}{(1)}$ we have $\mathcal{D}_{21^3} = \frac{5!}{2!3!} = 10$, and $\mathcal{D}_{2^21} = \frac{5!}{(2!)^2 2!} = 15$. We have $\dim L_{\frac{(1^3)}{(21)}} = \frac{3!}{3!} d_{(21)} d_{(1)}^3 = 2$, and all the other $\dim L_V$s are 1. Altogether then $\dim L^{\ltimes}_{\frac{(2)}{(1)}\frac{(1^3)}{(21)}} = 20$, $\dim L^{\ltimes}_{\frac{(2)}{(1)}\frac{(1^3)}{(1^3)}} = 10$, and $\dim L^{\ltimes}_{\frac{(2)^2}{(2)}\frac{(1)}{(1)}} = 15$.

Next we turn attention to restriction rules. The graph in the figure shows the Bratteli diagram of the sequence $P_{n-1}^{\ltimes} \subset P_n^{\ltimes}$ for $n \leq 4$ (computed by direct calculation of characters). A *heuristic* explanation for the general process of restriction is as follows. (We assume that the reader is familiar with induction and restriction rules for S_n.) A representation of the 'right-hand end' of a diagram in a typical simple $P_n^{\ltimes}$-module is as in Figure 3. Here we shall call a collection of 'strings' with symmetrisation λ a λ-string. Thus the last string in line in any diagram (i.e. string n) is involved in a λ-string for some λ. The action of $P_{n-1}^{\ltimes}$ on this $P_n^{\ltimes}$-module excludes the last string, which has the following effect. Firstly it 'destroys' a λ-string into $\mu(\lambda)$, so we need to restrict $\mu(\lambda) \rightsquigarrow \sum_j \mu(\lambda) - e_j$ (cf. the first term in (10) below). At the same time this creates a new $\lambda - e_k$-string for each suitable k. And for each such k this extra string gives rise to an induction on $\mu(\lambda - e_k)$, hence $\mu(\lambda - e_k) \rightsquigarrow \sum_l \mu(\lambda - e_k) + e_l$ for each suitable l. (Note that the overall degree of every term produced in this way is $n - 1$, as required.)

In light of the above heuristic, we define another directed graph $\mathcal{G}$ with vertex set $\mathrm{Hom}(\Lambda^*, \Lambda)$ as follows. Consider an element μ. Each non-trivial factor is of the form $\frac{\lambda}{\mu(\lambda)}$ (or $\frac{\lambda^{|\mu(\lambda)|}}{\mu(\lambda)}$ in the redundant notation) as noted.

We first define a linear map M from $\mathbb{Z}\mathrm{Hom}(\Lambda^*, \Lambda)$ to itself by

$$M\mu = {\sum_{\lambda}}' \sum_{j} \frac{\lambda}{\mu(\lambda) - e_j} \sum_{k} \sum_{l} \frac{\lambda - e_k}{\mu(\lambda - e_k) + e_l} \mu|_{\lambda, \lambda - e_k} \tag{10}$$

where the sum $\sum'_{\lambda}$ is over partitions not mapped to $\emptyset$ by μ; and all the sums involving rows of partitions are restricted to the appropriate addable or subtractable rows as usual (if $\lambda = (1)$ then the $\sum_k$ nominally consists in a single summand contributing a factor with 'numerator' $(1) - e_1 = \emptyset$ and 'denominator' undefined — this overall-undefined factor is omitted, but the term is kept); and $\mu|_{\lambda,\lambda-e_k}$ means μ with the images of $\lambda, \lambda - e_k$ both omitted (NB, they are replaced by the explicitly given factors). We draw an edge between μ and μ' in $\mathcal{G}$ if μ' appears in $M\mu$ above.

The edges of $\mathcal{G}$ up to level 4 are again as in figure 2. For example $M\frac{(2)}{(1)} = \frac{(2)}{\emptyset}\frac{(1)}{\emptyset+e_1} = \frac{(1)}{(1)}$, and $M\frac{(1)^2}{(2)} = \frac{(1)}{(1)}\frac{\emptyset}{-} = \frac{(1)}{(1)}$ omitting the undefined factor. A more challenging example is $\frac{(2)^2(1)^2}{(2)(1^2)}$. Here we have

$$M\mu = \frac{(2)(1)^3}{(1)(21)} + \frac{(2)(1)^3}{(1)(1^3)} + \frac{(2)^2(1)}{(2)(1)} \tag{11}$$

Note that the dimensions of the corresponding simple modules were computed in our examples above, and indeed obey $45 = 20 + 10 + 15$.

We conjecture that the graphs coincide. (If the conjecture is true then the Bratteli diagram is multiplicity-free and, noting the results at the end of Section 3.3, our 'unification' of wreaths would have the property sought in our first motivating example.)

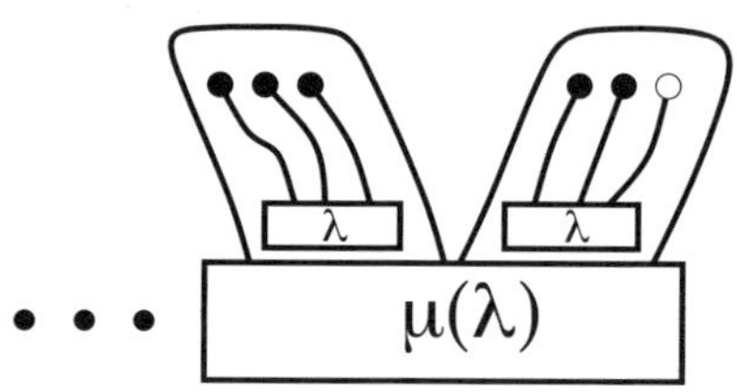

Fig. 3. Right-hand end of a representative diagram in a simple module over $\mathbb{C}$, with the last string marked.

3.6 *Discussion*

We remark that there is an established setting in which the Faa di Bruno coefficients $\mathcal{D}_{\lambda^p}$ appear *ensemble* in a way intriguingly analogous to their role in $P_n^{\ltimes}$. This is in the combinatorics of *Bell matrices* (that is, of Taylor series of composite functions).[1] Let $g(x) = \sum_{i=1} g_i \frac{x^i}{i!}$ be a formal power series with $g(0) = 0$. The *Bell matrix* is the matrix whose j^{th} column contains the coefficients of the corresponding power series for $\frac{g^j(x)}{j!}$ (see e.g. (13.66) of Ref. 1). This begins

$$B[g] = \begin{pmatrix} g_1 & 0 & & & & \\ g_2 & g_1^2 & 0 & & & \\ g_3 & 3g_1g_2 & g_1^3 & 0 & & \\ g_4 & 4g_1g_3 + 3g_2^2 & 6g_1^2g_2 & g_1^4 & 0 & \\ g_5 & 5g_1g_4 + 10g_2g_3 & 10g_1^2g_3 + 15g_1g_2^2 & 10g_1^3g_2 & g_1^5 & 0 \\ g_6 & \begin{matrix} 6g_1g_5 + 15g_2g_4 \\ +10g_3^2 \end{matrix} & \begin{matrix} 15g_1^2g_4 + 60g_1g_2g_3 \\ +15g_2^3 \end{matrix} & \begin{matrix} 20g_1^3g_3 \\ +45g_1^2g_2^2 \end{matrix} & 15g_1^4g_2 \; g_1^6 & 0 \\ \cdots & & & & & \end{pmatrix}$$

The coefficients (*within* the entries) in $B[g]$ are the Faa di Bruno coefficients. The intriguing point is that these coincide with the multiplicities $\mathcal{D}_{\lambda^p}$ from (3), specifically with $\mathcal{D}_{\lambda^p}$ the coefficient of $\prod_i g_{\lambda_i}^{p_i}$. It would be interesting to extend this connection to the dimensions of the simple modules of $P_n^{\ltimes}$ (cf. Ref. 7 for example).

Acknowledgements.
The earlier version of this paper that has been hosted on my webpage for the last year or so (http://www.maths.leeds.ac.uk/~ppmartin/pdf/baby08.pdf) is not fun to read, so warm thanks to Robert Marsh for reading it, and for many valuable comments; and to Mark Wildon for reading the newer draft.

References

1. R Aldrovandi, *Special matrices of mathematical physics*, World Scientific, 2001.
2. P Bantay and G Zala, *Ultrametric matrices and representation theory*, cond-mat/9705041v1 (1997).
3. D J Benson, *Representations and cohomology I*, Cambridge, 1995.
4. M Bloss, *g-colored partition algebras as centralizer algebras of wreath products*, Journal of Algebra **265** (2003), 690–710.
5. C W Curtis and I Reiner, *Representation theory of finite groups and associative algebras*, Wiley Interscience, New York, 1962.
6. B Davies, O Foda, M Jimbo, T Miwa, and A Nakayashiki, *Diagonalization of the XXZ Hamiltonian by vertex operators*, Commun. Math. Phys. **151** (1993), 89–153.
7. S Frisch, *Polynomial functions on finite commutative rings*, Advances in commutative ring theory (D E Dobbs et al, ed.), 1999.

8. R Goodman and N R Wallach, *Representations and invariants of the classical groups*, Cambridge, 1998.
9. J A Green, *Polynomial representations of* GL_n, Springer-Verlag, Berlin, 1980.
10. M Hamermesh, *Group theory*, Pergamon, Oxford, 1962.
11. G D James and A Kerber, *The representation theory of the symmetric group*, Addison-Wesley, London, 1981.
12. V F R Jones, *The Potts model and the symmetric group*, in Subfactors: Proceedings of the Taniguchi Symposium on Operator Algebras, World Scientific, Singapore, 1994.
13. S V Kerov, *Asymptotic representation theory of the symmetric group and its applications in analysis*, Translations, vol. 219, AMS, 2003.
14. A Kleshchev, *Branching rules for modular representations of symmetric groups, IV*, J. Algebra **201** (1998), 547–572.
15. D E Knuth, *Sorting and searching*, 2 ed., The Art of Computer Programming, vol. 3, Addison Wesley, 1998.
16. S Koenig and C C Xi, *Cellular algebras: inflations and Morita equivalences*, Journal of the LMS **60** (1999), 700–722.
17. C L Liu, *Introduction to combinatorial mathematics*, McGraw–Hill, New York, 1968.
18. I Macdonald, *Symmetric functions and Hall polynomials*, 2nd ed., Oxford, 1995.
19. P P Martin, *Block spin transformations in the operator formulation of two–dimensional Potts models*, J Phys A **22** (1989), 3991–4005.
20. ———, *Potts models and related problems in statistical mechanics*, World Scientific, Singapore, 1991.
21. ———, *Temperley–Lieb algebras for non–planar statistical mechanics — the partition algebra construction*, Journal of Knot Theory and its Ramifications **3** (1994), no. 1, 51–82.
22. ———, *The partition algebra and the Potts model transfer matrix spectrum in high dimensions*, J Phys A **32** (2000), 3669–3695.
23. P P Martin and A Elgamal, *Ramified partition algebras*, Math. Z. **246** (2004), 473–500, (math.RT/0206148).
24. M Parvathi, *Signed partition algebras*, Commun. Algebra **32** (2004), 1865–1880.
25. P Sniady, *Gaussian fluctuations of representations of wreath products*, QUANTUM PROBABILITY AND RELATED TOPICS **9** (2006), 529.
26. A M Vershik and A Y Okunkov, *An inductive method of expounding the representation theory of symmetric groups*, Russian Mathematical Surveys **51** (1996), 1237–1239.
27. Weiqiang Wang, *Vertex algebras and the class algebras of wreath products*, Proceedings of the London Mathematical Society **88** (2004), 381.

THE SAGA OF THE ISING SUSCEPTIBILITY

BARRY M. MCCOY[1], MICHAEL ASSIS[1], SALAH BOUKRAA[2], SAOUD HASSANI[3], JEAN-MARIE MAILLARD[4], WILLIAM P. ORRICK[5], and NADJAH ZENINE[3]

1. CN Yang Institute for Theoretical Physics, State University of New York Stony Brook, NY 11794, USA
2. LPTHIRM and Département d'Aeronautique, Université de Blida. Algeria
3. Centre de Recherche Nucléaire d'Alger, 2 Bd. Frantz Fanon BP 399, 16000 Alger, Algeria
4. LPTMC, Université de Paris 6, Tour 24, 4éme étage, case 121, 4 Place Jussieu 75252 Paris Cedex 05, France
5. Dept. of Math. Indiana University, Bloomington, Indiana, 47405 USA

We review developments made since 1959 in the search for a closed form for the susceptibility of the Ising model. The expressions for the form factors in terms of the nome q and the modulus k are compared and contrasted. The λ generalized correlations $C(M,N;\lambda)$ are defined and explicitly computed in terms of theta functions for $M = N = 0, 1$.

1. Introduction

There are three important thermodynamic properties of any magnetic system in zero magnetic field: the partition function from which free energy and the specific heat are obtained; the magnetization; and the magnetic susceptibility. For the two dimensional Ising model in zero field defined by

$$\mathcal{E}_0 = -\sum_{j,k}\{E^v\sigma_{j,k}\sigma_{j+1,k} + E^h\sigma_{j,k}\sigma_{j,k+1}\} \tag{1}$$

with $\sigma_{j,k} = \pm 1$ the free energy was first computed by Onsager[1] in 1944 and the spontaneous magnetization was announced by Onsager in 1948[2] and proven by Yang[3] in 1952. To this day a closed form for the magnetic susceptibility has never been found. We will here trace the saga of the quest for this susceptibility.

If we could solve the Ising model in the presence of a magnetic field H

which interacts with the total spin of the system as

$$\mathcal{E} = \mathcal{E}_0 - H \sum_{j,k} \sigma_{j,k} \tag{2}$$

then the magnetic susceptibility would be computed as

$$\chi(H) = \frac{\partial M(H)}{\partial H} \tag{3}$$

where the magnetization is

$$M(H) = \frac{1}{Z(H)} \sum_{\sigma_{j,k}=\pm 1} \sigma_{0,0}\, e^{-\mathcal{E}/k_B T} \tag{4}$$

with the partition function defined by

$$Z(H) = \sum_{\sigma_{j,k}=\pm 1} e^{-\mathcal{E}/k_B T} \tag{5}$$

However, because the Ising model has only been solved for $H = 0$ we are forced to restrict our attention to $\chi(0)$ which from (2)-(5) is given in terms of the two point correlation functions as

$$k_B T \chi(0) = \sum_{M,N} \{\langle \sigma_{0,0}\, \sigma_{M,N} \rangle - M(0)^2\} \tag{6}$$

where $M(0)$ is the spontaneous magnetization of the system which is zero for $T > T_c$ and for $T < T_c$

$$M(0) = (1 - k^2)^{1/8} \tag{7}$$

where

$$k = (\sinh 2K^v \sinh 2K^h)^{-1} \tag{8}$$

with $K^{v,h} = E^{v,h}/k_B T$ and T_c is defined by

$$k = 1. \tag{9}$$

The first exact result for the susceptibility was given in 1959 by Fisher[4] who used results of Kaufmann and Onsager[5] to argue that as $T \to T_c$ the susceptibility diverges as $|T - T_c|^{-7/4}$. The saga may be said to begin with the concluding remark of this paper:

In conclusion we note that the relatively simple results (1) and (3) suggest strongly that there is a closed expression for the susceptibility in terms of elliptic integrals. It is to be hoped that such a formula will be discovered,$\cdots$.

One year later Syozi and Naya,[6] on the basis of short series expansions, proposed such a formula for $T > T_c$ which does not involve elliptic integrals

$$k_B T\chi(0) = \frac{(1 - \sinh^2 2K^v \sinh^2 2K^h)^{1/4}}{\cosh 2K^v \cosh 2K^h - \sinh 2K^v - \sinh 2K^h} \tag{10}$$

However, when higher order terms were computed this conjecture was shown not to be exact.

To this day the "closed expression" for the susceptibility hoped for in Ref. 4 has not been found.

2. Form factor expansion and the λ extension

To proceed further a systematic understanding of the two point correlation function is required. For short distances the correlations are well represented by determinants[5,7] whose size grows with the separation of the spins. However, in order to execute the sum over all separations required by (6) an alternative form of the correlations which is efficient for large distances is needed. The study of this alternative form was initiated in 1966 by Wu[8] who discovered that for the row correlation $\langle\sigma_{0,0}\sigma_{0,N}\rangle$ that when $N|T - T_c| \gg 1$ for $T < T_c$

$$\langle\sigma_{0,0}\sigma_{0,N}\rangle \;=\; (1-t)^{1/4} \cdot \{1 + f^{(2)}_{0,N} + \cdots\} \tag{11}$$

with $t = k^2$ with k given by (8). For $T > T_c$ we define $k = \sinh 2K^v \sinh 2K^h$ and find

$$\langle\sigma_{0,0}\sigma_{0,N}\rangle \;=\; (1-t)^{1/4} \cdot \{f^{(1)}_{0,N} + \cdots\} \tag{12}$$

with $t = k^2$. In (11) and (12) $f^{(n)}_{0,N}$ is an n fold integral which exponentially decays for large N. The results (11) and (12) are the leading terms in what has become known as the form factor representation of the correlations which in general for $T < T_c$ is

$$\langle\sigma_{0,0}\sigma_{M,N}\rangle \;=\; (1-t)^{1/4} \cdot \{1 + \sum_{n=1}^{\infty} f^{(2n)}_{M,N}\} \tag{13}$$

and for $T > T_c$

$$\langle\sigma_{0,0}\sigma_{M,N}\rangle \;=\; (1-t)^{1/4} \cdot \sum_{n=0}^{\infty} f^{(2n+1)}_{M,N} \tag{14}$$

where $f^{(n)}_{M,N}$ is an n dimensional integral. For general M, N these $f^{(n)}_{M,N}$ were computed in 1976 by Wu, McCoy, Tracy and Barouch[9] and related forms are given in Refs. 10,11. However, for the diagonal correlations an alternative and simpler form is available which was announced in Ref. 12 and proven in Ref. 13. For the diagonal form factor for $T < T_c$

$$f_{N,N}^{(2n)}(t) = \frac{t^{n(N+n)}}{(n!)^2\pi^{2n}} \int_0^1 \prod_{k=1}^{2n} dx_k x_k^N \prod_{j=1}^{n} \left(\frac{(1-tx_{2j})(x_{2j}^{-1}-1)}{(1-tx_{2j-1})(x_{2j-1}^{-1}-1)} \right)^{1/2}$$
$$\prod_{1\le j\le n} \prod_{1\le k\le n} \left(\frac{1}{1-tx_{2k-1}x_{2j}} \right)^2 \prod_{1\le j<k\le n} (x_{2j-1}-x_{2k-1})^2 (x_{2j}-x_{2k})^2 \tag{15}$$

for $T > T_c$

$$f_{N,N}^{(2n+1)}(t) =$$
$$\frac{t^{(n+1/2)N+n(n+1)}}{n!(n+1)!\pi^{2n+1}} \int_0^1 \prod_{k=1}^{2n+1} dx_k x_k^N \prod_{j=1}^{n+1} x_{2j-1}^{-1}[(1-tx_{2j-1})(x_{2j-1}^{-1}-1)]^{-1/2}$$
$$\prod_{j=1}^{n} x_{2j}[(1-tx_{2j})(x_{2j}^{-1}-1)]^{1/2} \prod_{1\le j\le n+1} \prod_{1\le k\le n} \left(\frac{1}{1-tx_{2j-1}x_{2k}} \right)^2$$
$$\prod_{1\le j<k\le n+1} (x_{2j-1}-x_{2k-1})^2 \prod_{1\le j<k\le n} (x_{2j}-x_{2k})^2 \tag{16}$$

In particular

$$f_{N,N}^{(1)}(t) = t^{N/2} \cdot \frac{\Gamma(N+1/2)}{\pi^{1/2} N!} \cdot F\left(\frac{1}{2}, N+\frac{1}{2}; N+1; t\right) \tag{17}$$

where $F(a,b;c;t)$ is the hypergeometric function.

It is often useful and instructive to extend the form factor expansions (13) and (14) by weighting $f_{M,N}^{(n)}$ by λ^n and thus we define "λ generalized correlations"

$$C_-(M,N;\lambda) = (1-t)^{1/4} \cdot \{1 + \sum_{n=1}^{\infty} \lambda^{2n} f_{M,N}^{(2n)}\} \tag{18}$$

and for $T > T_c$

$$C_+(M,N;\lambda) = (1-t)^{1/4} \cdot \sum_{n=0}^{\infty} \lambda^{2n+1} f_{M,N}^{(2n+1)} \tag{19}$$

This λ extension was first introduced in 1977 by McCoy, Tracy and Wu[14] in the context of the scaling limit.

3. Leading divergence as $T \to T_c$

These form factor expansions may now be used in (6) where the sums over M, N are easily executed under the integral signs to produce a corresponding expansion of the susceptibility[9] which we write for $T < T_c$ as

$$k_B T \chi(0) = (1-t)^{1/4} \cdot \sum_{n=1}^{\infty} \hat{\chi}^{(2n)} \tag{20}$$

and for $T > T_c$

$$k_B T \chi(0) = (1-t)^{1/4} \cdot \sum_{n=0}^{\infty} \hat{\chi}^{(2n+1)} \tag{21}$$

where

$$\tilde{\chi}^{(n)} = \sum_{M=-\infty}^{\infty} \sum_{N=-\infty}^{\infty} f_{M,N}^{(n)}. \tag{22}$$

For $n = 1, 2$ the $\hat{\chi}^{(n)}$ are explicitly evaluated.[9] For the isotropic lattice we have

$$\hat{\chi}^{(1)} = \frac{1}{(1-k^{1/2})^2} \tag{23}$$

$$\hat{\chi}^{(2)} = \frac{(1+k^2)E - (1-k^2)K}{3\pi(1-k)(1-k^2)} \tag{24}$$

where K and E are the complete elliptic integrals of the first and second kind. The form (10) of Syozi and Naya[6] with $K^v = K^h$ is seen to be the first term in (21). It is quite clear that unlike the 1959 argument of Ref. 4 the behavior of the susceptibility as $T \to T_c$ will be different depending on whether T approaches T_c from above or below. This dramatic difference was first seen in 1973 in Ref. 15 where it is shown that for $T \to T_c\pm$

$$k_B \chi(0) \sim C_{0\pm} \cdot |T - T_c|^{-7/4} \tag{25}$$

where

$$C_{0\pm} = 2^{-1/2} \cdot \coth 2K_c^v \coth 2K_c^h \cdot [K_c^v \coth 2K_c^v + K_c^h \coth 2K_c^h]^{-7/4} \cdot I_\pm$$

and $I_\pm$ have been numerically evaluated to 52 digits in Ref. 16:

$$I_+ = 1.000815260440212647119476363047210236937534925597789 \cdots$$

$$I_- = \frac{1}{12\pi} \cdot 1.000960328725262189480934955172097320572505951770117 \cdots$$

4. The singularities of Nickel

The next advance in the understanding of the analytic structure of the susceptibility came in 1996 when Guttmann and Enting,[17] using resummed high temperature series expansions of the anisotropic Ising model, argued that the susceptibility cannot satisfy a finite order differential equation and raised the question of the occurrence of a natural boundary. This natural boundary argument was made very concrete for the isotropic case in 1999[18] and 2000[19] by Nickel who analyzed the singularities of the n fold integrals $\tilde{\chi}^{(n)}$. These integrals, of course, have singularities at $T = T_c$ where the individual correlation functions $\langle \sigma_{0,0}\sigma_{M,N}\rangle$ have singularities. However, Nickel made the remarkable discovery that the integrals, for $\hat{\chi}^{(n)}$, contain many more singularities. In particular he found that, for the isotropic lattice, $\hat{\chi}^{(n)}$ has singularities in the complex temperature variable $s = \sinh 2E/k_B T$ at

$$s = \; s_{j,k} = \; e^{i\theta_{j,k}} \tag{26}$$

where

$$2\cos(\theta_{j,k}) \; = \; \cos(2\pi k/n) + \cos(2\pi j/n) \tag{27}$$

For n odd $(T > T_c)$ the behavior of $\hat{\chi}^{(n)}$ near the singularity is

$$\hat{\chi}^{(2n+1)} \; \sim \; \epsilon^{2n(n+1)-1} \cdot \ln \epsilon \tag{28}$$

with

$$\epsilon = 1 \; - \; s/s_{j,k} \tag{29}$$

and for even n $(T < T_c)$

$$\hat{\chi}^{(2n)} \; \sim \; \epsilon^{2n^2 - 3/2} \tag{30}$$

The discovery of these singularities demonstrates that the magnetic susceptibility is a far more complicated object than either the free energy or the spontaneous magnetization and that the hope expressed in Ref. 4 of a closed form in terms of a few elliptic integrals is far too simple.

5. The theta function expressions of Orrick, Nickel, Guttmann and Perk

In the following year a major advance was made by Orrick, Nickel, Guttmann and Perk[16] who studied both the form factors and the susceptibility by means of generating on the computer series of over 300 terms. From these series they then made several remarkable conjectures for the form factors.

To present these conjectures we define theta functions as

$$\theta_1(u,q) \;=\; 2\sum_{n=0}^{\infty}(-1)^n\, q^{(n+1/2)^2}\ \sin[(2n+1)u] \tag{31}$$

$$\theta_2(u,q) \;=\; 2\sum_{n=0}^{\infty} q^{(n+1/2)^2}\ \cos[(2n+1)u] \tag{32}$$

$$\theta_3(u,q) \;=\; 1+2\sum_{n=1}^{\infty} q^{n^2}\ \cos 2nu \tag{33}$$

$$\theta_4(u,q) \;=\; 1+2\sum_{n=1}^{\infty}(-1)^n\, q^{n^2}\ \cos 2nu \tag{34}$$

and for $u=0$ we use the short hand

$$\theta_2 = \theta_2(0,q), \qquad \theta_3 = \theta_3(0,q), \qquad \theta_4 = \theta_4(0,q) \tag{35}$$

The quantity q is the nome of the elliptic functions and is related to the modulus k by the relation

$$k \;=\; 4\,q^{1/2}\cdot\prod_{n=1}^{\infty}\left[\frac{1+q^{2n}}{1+q^{2n-1}}\right]^4 \tag{36}$$

In terms of these theta functions, conjectures for form factors are given in sec. 5.2 of Ref. 16 by defining an operator Φ_0 which converts a power series in z to a power series in q as

$$\Phi_0\Big(\sum_{n=0}^{\infty} c_n z^n\Big) \;=\; \sum_{n=0}^{\infty} c_n q^{n^2/4} \tag{37}$$

Conjectures are then given for $f_{0,0}^{(n)}$, $f_{1,1}^{(n)}, f_{1,0}^{(n)}$, $f_{2,0}^{(n)}$ and $f_{2,1}^{(n)}$. In particular we note

$$2^{-n}\cdot(1-k^2)^{1/4}\cdot f_{0,0}^{(n)} \;=\; \frac{(1,k^{-1/2})}{\theta_3}\cdot\Phi_0\Big(\frac{z^n(1-z^2)}{(1+z^2)^{n+1}}\Big) \tag{38}$$

and

$$2^{-n}\cdot(1-k^2)^{1/4} f_{1,1}^{(n)} = \frac{2\ (n+1)(1,k^{-1/2})}{\theta_2\theta_3^2}\cdot\Phi_0\Big(\frac{z^{n+1}(1-z^2)}{(1+z^2)^{n+2}}\Big) \tag{39}$$

where

$$\begin{aligned}(1,k^{-1/2}) =\ & 1 \qquad \text{for} \qquad T<T_c \ \ (n \text{ even})\\ & k^{-1/2} \qquad \text{for} \qquad T>T_c \ \ (n \text{ odd}).\end{aligned} \tag{40}$$

6. Linear differential equations

A second approach to the form factors and susceptibility was initiated in 2004 in Ref. 20 and subsequently greatly developed in Refs. 21–26. These studies are similar to Ref. 16 in that they expand the form factors and susceptibility in long series. However, instead of the nome q the expansion is in the (modular) variable t. The goal of these studies is to characterize the n particle contributions $\hat{\chi}^{(n)}(t)$ to the susceptibility in terms of finding a Fuchsian linear ordinary differential equation satisfied by $\hat{\chi}^{(n)}(t)$. Such a linear differential equation always exists for an n-fold integral with an algebraic integrand in some well-suited choice of integration variable and in the parameter t. However the order and the degree of the equation rapidly become large for increasing n and it may take series of many thousands of terms to find the differential equation. Such a study can only be done by computer.

There are several features of these differential equations to be noted. In particular the operator which annihilates $\hat{\chi}^{(n)}$ factorizes and furthermore the operator has a direct sum decomposition such that $\hat{\chi}^{(n-2j)}$ for $j = 1, \cdots [n/2]$ are "contained" in $\hat{\chi}^{(n)}$.

7. Diagonal form factors

With the observation of factorization, direct sum decomposition and Nickel singularities of the n particle contributions to the bulk susceptibility $\hat{\chi}^{(n)}$, it has become clear that the susceptibility is far more complicated than what was envisaged by Fisher[4] in 1959. Because of this complexity the question was asked if there could be a simpler object to study which would yet be able to give insight into the structures which had been observed. Several such "simplified" objects have been studied[27] which consist of more or less forcibly modifying parts of the integrals for the $\hat{\chi}^{(n)}$. However, there is one "simplified" model which commands interest in its own right. This is the "diagonal susceptibility" which is defined[28] by restricting the sum in (6) to the correlation of spins on the diagonal

$$k_B T \cdot \chi_d = \sum_{N=-\infty}^{\infty} \{\langle \sigma_{0,0} \sigma_{N,N} \rangle - M^2(0)\}. \tag{41}$$

In statistical language this diagonal susceptibility is the susceptibility for a magnetic field interacting only with the spins on one diagonal. In magnetic language this is the $p = 0$ value of the groundstate structure function

$$S^x(p) = \sum_{j=-\infty}^{\infty} e^{ipj}\{\langle\sigma_0^x\sigma_j^x\rangle - M_x^2\} \tag{42}$$

of the transverse Ising model

$$H_{TI} = \sum_{j=-\infty}^{\infty} \{\sigma_j^x\sigma_{j+1}^x + H^z\sigma_j^z\} \tag{43}$$

These interpretations give the diagonal susceptibility a physical interpretation which the other "simplified" models do not have.[27] Furthermore much more analytic information is available for the diagonal Ising correlations than for correlations off the diagonal. Firstly it is known from the work of Jimbo and Miwa[29] that the diagonal correlations are characterized by the solutions of a particular sigma form of Painlevé VI equation and secondly the integral representation of the diagonal form factors (15) and (16) is more tractable than the representation of the general off diagonal correlations.

The diagonal form factors have been extensively studied in Ref. 12 by means of processing the differential equations obtained from long series expansions by use of Maple. Diagonal form factors $f_{N,N}^{(n)}$ for n as large as 9 and N as large as 4 have been studied and many examples are given in Ref. 12 where they have all been reduced to expressions in the elliptic integrals E and K. A few such examples are as follows:

For $n = 1$ (when the hypergeometric function of (17) is reduced to the basis of E and K by use of the contiguous relations)

$$f_{0,0}^{(1)} = (2/\pi)\cdot K \tag{44}$$

$$t^{1/2}\, f_{1,1}^{(1)} = (2/\pi)\cdot\{K - E\} \tag{45}$$

$$3\,t\, f_{2,2}^{(1)} = (2/\pi)\cdot\{(t+2)K - 2(t+1)\,E\} \tag{46}$$

$$15\,t^{3/2}\, f_{3,3}^{(1)} = (2/\pi)\cdot\{(4t^2+3t+8)\,K - (8t^2+7t+8)\,E\} \tag{47}$$

$$105\,t^2\, f_{4,4}^{(1)} = (2/\pi)\cdot\{(24t^3+17t^2+16t+48)K - (48t^3+40t^2+40t+48)\,E\}; \tag{48}$$

for $n = 2$

$$2\, f_{0,0}^{(2)} \;=\; (2/\pi)^2 \cdot K\,(K - E) \tag{49}$$

$$2\, f_{1,1}^{(2)} \;=\; 1 \;-\, (2/\pi)^2 \cdot K \cdot \{(t-2)\, K \;+ 3E\} \tag{50}$$

$$6\, t\, f_{2,2}^{(2)} \;=\; 6t$$

$$-(2/\pi)^2 \cdot \{6t^2 - 11t + 2)\, K^2 \;+ (15t - 4)\, KE \;+ 2(t+1)\, E^2\} \tag{51}$$

$$90\, t^2\, f_{3,3}^{(2)} \;=\; 135t^2 - (2/\pi)^2 \cdot \{(137t^3 - 242t^2 + 52t + 8)\, K^2 \tag{52}$$

$$-(8t^3 - 319t^2 + 112t + 16)\, KE \;+ 4(t+1)(2t^2 + 13t + 2)E^2\} \tag{53}$$

$$3150\, t^3\, f_{4,4}^{(2)} \;=\; 6300\, t^3$$

$$-(2/\pi)^2 \cdot \{(32t^5 + 6440t^4 - 11191t^3 + 2552t^2 + 464t + 128)\, K^2$$

$$-(128t^5 + 576t^4 - 14519t^3 + 5648t^2 + 1056t + 256)\, KE$$

$$+8(1+t)(16t^4 + 58t^3 + 333t^2 + 58t + 16)\, E^2\}; \tag{54}$$

for $n = 3$

$$6\, f_{0,0}^{(3)} \;=\; (2/\pi) \cdot\, K \;-\, (2/\pi)^3 \cdot\, K^2\{(t-2)\, K \;+ 3E\} \tag{55}$$

$$6\, t^{1/2}\, f_{1,1}^{(3)} \;=\; 4\,(2/\pi) \cdot (K - E) - (2/\pi)^3 \cdot K\{(2t-3)\, K^2 \;+ 6KE \;- 3E^2\} \tag{56}$$

$$18\, t\, f_{2,2}^{(3)} \;=\; 7\,(2/\pi) \cdot \{(t+2)\, K \;- 2\,(t+1)\, E\} \tag{57}$$

$$-(2/\pi)^3 \cdot \{3(t^2 - 2)\, K^3 \;- 3\,(2t^2 - 11t + 2)\, K^2 E$$

$$-36\,(t-1)\, KE^2 \;- 24E^3\} \tag{58}$$

$$270\, t^{5/2}\, f_{3,3}^{(3)} \;=\; 30(2/\pi)\{(4t^2 + 3t + 8)\, K \;- (8t^2 + 7t + 8)\, t\, E\}$$

$$-(2/\pi)^3 \cdot \{(72t^4 - 158t^3 + 189t^2 - 156t + 8)\, K^3$$

$$6(24t^4 - 108t^3 + 29t^2 - 6t + 4)\, K^2 E$$

$$-3\,(232t^3 \;- 111t^2 - 180\, t \;- 8)\, KE^2$$

$$-4\,(t+1)(2t^2 + 103t + 2)\, t\, E^3\}; \tag{59}$$

for $n = 4$

$$24\, f_{0,0}^{(4)} \;=\; 4\,(2/\pi)^2 \cdot K\,(K - E)$$
$$-(2/\pi)^4 \cdot K^2\{(2t-3)\,K^2 \, + 6KE \, - 3E^2\} \tag{60}$$
$$24\, f_{1,1}^{(4)} \;=\; 9 \; - (2/\pi)^2 \cdot 10\,K\{(t-2)\,K \, + 3E\}$$
$$+(2/\pi)^4 \cdot K^2\{(t^2 - 6t + 6)\,K^2 \, + 10\,(t-2)\,KE \, + 15E^2\} \tag{61}$$
$$72\,t\, f_{2,2}^{(4)} \;=\; 72t$$
$$-(2/\pi)^2 \cdot 16 \cdot \{(6t^2 - 11t + 2)\,K^2 \, + (15t-4)\,KE \, + 2(t+1)\,E^2\}$$
$$+(2/\pi)^4 \cdot \{24t^3 - 98t^2 + 113t - 36)\,K^4 \, + 2\,(74t^2 - 157t + 66)\,K^3E$$
$$+3\,(71t - 60)\,K^2E^2 \, + 12\,(t+9)\,KE^3 \, - 24E^4\}. \tag{62}$$

These examples are sufficient to illustrate the following phenomena which hold for all examples considered in Ref. 12 and which are certainly true in general:

$$f_{N,N}^{(2n)} \;=\; \sum_{j=0}^{n} c_{j;n}^{-} g_{N,N}^{(2j)}(t) \tag{63}$$
$$f_{N,N}^{(2n+1)} \;=\; \sum_{j=0}^{n} c_{j;n}^{+} g_{N,N}^{(2j+1)}(t) \tag{64}$$

where $c_{j;n}^{\pm}$ are constants independent of t and $g_{N,N}^{(j)}(t)$ for even j are of the form

$$g_{0,0}^{(2n)}(t) \;=\; \sum_{j=0}^{n} P_{j,n;0}^{-}(t)\, K^{2n-j} E^j \tag{65}$$
$$g_{1,1}^{(2n)}(t) \;=\; \sum_{j=0}^{n} P_{j,n;1}^{-}(t)\, K^{2n-j} E^j \tag{66}$$
$$g_{N,N}^{(2n)}(t) \;=\; t^{-N+1} \sum_{j=0}^{2n} P_{j,n;N}^{-}(t)\, K^{2n-j} E^j \qquad \text{for} \quad N \geq 2 \tag{67}$$

and for odd j

$$g_{0,0}^{(2n+1)}(t) = \sum_{j=0}^{n} P_{j,n;0}^{+}(t)\, K^{2n+1-j} E^j \qquad (68)$$

$$g_{1,1}^{(2n+1)}(t) = t^{-1/2} \sum_{j=0}^{n+1} P_{j,n;1}^{+}(t)\, K^{2n+1-j} E^j \qquad (69)$$

$$g_{N,N}^{(2n+1)}(t) = t^{-N/2} \sum_{j=0}^{2n+1} P_{j,n;N}^{+}(t)\, K^{2n+1-j} E^j \quad \text{for} \quad N \geq 2 \qquad (70)$$

where $P_{j,n;m}^{\pm}(t)$ are polynomials.

The decompositions (63) and (64) represent a direct sum decomposition of the form factors.[28] The functions $g_{N,N}^{(j)}$ individually are annihilated by Fuchsian operators which are equivalent to the $j+1$ symmetric power of the second order operator associated with the complete elliptic integral E (or equivalently K).

We also observe the relation between $f_{1,1}^{(n)}(t)$ and $f_{0,0}^{(n+1)}(t)$

$$(2/\pi) \cdot K \cdot f_{1,1}^{(2n)}(t) = (2n+1) \cdot f_{0,0}^{(2n+1)}(t) \qquad (71)$$

$$(2/\pi) \cdot t^{1/2} \cdot K \cdot f_{1,1}^{(2n+1)}(t) = 2\,(n+1) \cdot f_{0,0}^{(2n+2)}(t) \qquad (72)$$

8. Nome q-representation versus modulus k-representation

We will need the following identities which relate functions of the nome $q = e^{i\pi\tau}$ where $\tau = iK(k')/K(k)$ with functions of the modulus k

$$k = \frac{\theta_2^2}{\theta_3^2}, \qquad k' = (1-k^2)^{1/2} = \frac{\theta_4^2}{\theta_3^2}, \qquad (73)$$

$$\frac{2}{\pi} K = \theta_3^2, \qquad \text{and} \qquad \frac{dq}{dk} = \frac{\pi^2}{2} \frac{q}{kk'^2K^2} \qquad (74)$$

which we will use as

$$q\,\frac{d}{dq} = \frac{2}{\pi^2}\, k\, k'^2 \cdot K^2 \cdot \frac{d}{dk}. \qquad (75)$$

We will also use

$$\frac{dK}{dk} = \frac{E - k'^2 K}{kk'^2}, \qquad \frac{dE}{dk} = \frac{E-K}{k} \qquad (76)$$

8.1. $f_{0,0}^{(2n)}$

We first write (38) for $j = 2n$ using (73) as

$$f_{0,0}^{(2n)} = \frac{1}{\theta_4} \cdot \Phi_0 \Big(\frac{2^{2n} z^{2n} (1-z^2)}{(1+z^2)^{2n+1}} \Big) \qquad (77)$$

Thus, by use of the elementary expansion

$$\frac{2^{2n} \cdot z^{2n} \cdot (1-z^2)}{(1+z^2)^{2n+1}} = 2 \, \frac{(-1)^n}{(2n)!} \sum_{j=0}^{\infty} (-1)^j \, z^{2j} \prod_{m=0}^{n-1} 4 \, [j^2 - m^2] \qquad (78)$$

and the definition (37) of the operator Φ_0 we find that in terms of the nome q

$$f_{0,0}^{(2n)} = 2 \, \frac{(-1)^n \, 4^n}{\theta_4 (2n)!} \cdot \sum_{j=0}^{\infty} (-1)^j \, q^{j^2} \prod_{m=0}^{n-1} [j^2 - m^2]$$

$$= 2 \frac{(-1)^n \, 4^n}{\theta_4 (2n)!} \cdot \sum_{j=0}^{\infty} (-1)^j \prod_{m=0}^{n-1} [q \, \frac{d}{dq} - m^2] q^{j^2}$$

$$= \frac{(-1)^n \, 4^n}{\theta_4 (2n)!} \cdot \prod_{m=1}^{n-1} [q \, \frac{d}{dq} - m^2] \cdot q \frac{d}{dq} \theta_4. \qquad (79)$$

To convert this to an expression in terms of the modulus k we first use (73) to write

$$\theta_4^2 = \frac{2}{\pi} \cdot k' \cdot K \qquad (80)$$

and thus using (74)

$$q \frac{d}{dq} \theta_4^2 = \frac{2}{\pi^2} \cdot k \, k'^2 \cdot K^2 \cdot \frac{d}{dk} \left(\frac{2}{\pi} k' K \right) \qquad (81)$$

which using (76) reduces to

$$q \, \frac{d}{dq} \theta_4^2 = \frac{2}{\pi^2} \cdot k' \cdot K^2 \cdot \frac{2}{\pi} \{E - K\}. \qquad (82)$$

Using (80) on the right hand side we find

$$2 \, \theta_4 \cdot q \frac{d}{dq} \theta_4 = \frac{2}{\pi^2} \cdot \theta_4^2 \cdot K \cdot \{E - K\} \qquad (83)$$

and thus

$$\frac{1}{\theta_4} \cdot q \, \frac{d}{dq} \theta_4 = \frac{1}{\pi^2} \cdot K \cdot \{E - K\}. \qquad (84)$$

To evaluate $f_{0,0}^{(2)}$ we use (84) in (79) with $n=1$ to obtain

$$f_{0,0}^{(2)} = \frac{2}{\pi^2}\cdot K\cdot\{K-E\} \tag{85}$$

which is in agreement with (49). For arbitrary n the form factor $f_{0,0}^{(2n)}$ is obtained from (79) by repeated use of (84) and (76).

8.2. $f_{0,0}^{(2n+1)}$

To study $f_{0,0}^{(2n+1)}$ we first use (73) to write (38) as

$$f_{0,0}^{(2n+1)} = \frac{\theta_3}{\theta_2\theta_4}\cdot\Phi_0\Big(\frac{2^{2n+1}z^{2n+1}(1-z^2)}{(1+z^2)^{2n+2}}\Big) \tag{86}$$

and then, using the elementary expansion

$$\begin{aligned}&\frac{2^{2n+1}\cdot z^{2n+1}\cdot(1-z^2)}{(1+z^2)^{2n+2}} \\ &= 2\,\frac{(-1)^n}{(2n+1)!}\cdot\sum_{j=0}^{\infty}(2j+1)\cdot(-1)^j\cdot z^{2j+1}\cdot\prod_{m=0}^{n-1}[(2j+1)^2-(2m+1)^2]\end{aligned} \tag{87}$$

and the definition (37) of the operator Φ_0 we find

$$\begin{aligned}&f_{0,0}^{(2n+1)} \\ &= \frac{\theta_3}{\theta_2\theta_4}\frac{2(-1)^n}{(2n+1)!}\cdot\sum_{j=0}^{\infty}(2j+1)\cdot(-1)^j\cdot q^{(2j+1)^2/4}\cdot\prod_{m=0}^{n-1}[(2j+1)^2-(2m+1)^2] \\ &= \frac{\theta_3}{\theta_2\theta_4}\frac{2\,(-1)^n}{(2n+1)!}\cdot\prod_{m=0}^{n-1}[4\,q\frac{d}{dq}-(2m+1)^2]\sum_{j=0}^{\infty}(2j+1)(-1)^j q^{(2j+1)^2/4}.\end{aligned} \tag{88}$$

Thus, if we write

$$2\sum_{j=0}^{\infty}(2j+1)\cdot(-1)^j\cdot q^{(2j+1)^2/4} = \frac{\partial}{\partial u}\theta_1(u,q)|_{u=0} = \theta_2\theta_3\theta_4 \tag{89}$$

where in the last line we have used a well known identity, we find the result

$$f_{0,0}^{(2n+1)} = \frac{\theta_3}{\theta_2\theta_4}\frac{(-1)^n}{(2n+1)!}\cdot\prod_{m=0}^{n-1}[4\,q\frac{d}{dq}-(2m+1)^2]\,\theta_2\theta_3\theta_4. \tag{90}$$

We may now use (73)–(76) to reduce (90) from a function of q to a function of k.

For $n = 0$ we use (74) to find

$$f_{0,0}^{(1)} = \theta_3^2 = \frac{2}{\pi} \cdot K \tag{91}$$

which agrees with (44).

For $n \geq 3$ we need an expression analogous to (84) for the product $\theta_2\theta_3\theta_4$. From (73), (74)

$$\theta_2^2\theta_3^2\theta_4^2 = k\,k' \cdot (2\,K/\pi)^3 \tag{92}$$

and thus

$$q\frac{d}{dq}\theta_2^2\theta_3^3\theta_4^2 = 2\,\theta_2\theta_3\theta_4 \cdot q\frac{d}{dq}\theta_2\theta_3\theta_4 = \frac{2}{\pi^2} \cdot k\,k'^2\,K^2 \cdot \frac{d}{dk}\{k\,k'(2\,K/\pi)^3\}$$
$$= \frac{2}{\pi^2} \cdot k\,k' \cdot (2K/\pi)^3 \cdot K \cdot \{(k^2-2)\,K + 3E\}$$
$$= \frac{2}{\pi^2} \cdot \theta_2^2\,\theta_3^2\,\theta_4^2 \cdot K \cdot \{(k^2-2)\,K + 3E\}. \tag{93}$$

Therefore we obtain

$$q\frac{d}{dq}\theta_2\theta_3\theta_4 = \frac{1}{\pi^2}\theta_2\theta_3\theta_4 \cdot K\,\{(k^2-2)\,K + 3E\} \tag{94}$$

which when used in (90) with $n = 1$ gives

$$f_{0,0}^{(3)} = \frac{1}{3!} \cdot \{(2/\pi) \cdot K - (2/\pi)^3\,K \cdot [(k^2-2)\,K + 3E]\} \tag{95}$$

which is in agreement with (55).

8.3. $f_{1,1}^{(n)}$

The equalities (71) and (72) which express $f_{1,1}^{(n)}$ in terms of $f_{0,0}^{(m)}$ follow immediately from (38) and (39) by use of (73) and (74).

9. The λ generalized correlations

For $N = 0, 1$ the diagonal λ generalized correlations defined by (18) and (19) may be obtained by using the expressions for $(1-t)^{1/4} f_{N,N}^{(n)}$ in terms of the operator Φ_0 as given by (38) and (39). In this form the sums over n are easily done as geometric series and the operator Φ_0 is then used to convert the series in z to series in the nome q which can then be expressed in terms of θ functions as was done in the previous section. Then, setting

$$\lambda = \cos u \tag{96}$$

we obtain the following results

$$C_-(0,0;\lambda) = \frac{\theta_3(u;q)}{\theta_3(0;q)} \tag{97}$$

$$C_+(0,0;\lambda) = \frac{\theta_2(u;q)}{\theta_2(0;q)} \tag{98}$$

$$C_-(1,1;\lambda) = \frac{-\theta_2'(u;q)}{\sin(u)\theta_2(0;q)\theta_3(0;q)^2} \tag{99}$$

$$C_+(1,1;\lambda) = \frac{-\theta_3'(u;q)}{\sin(u)\theta_2^2(0;q)\theta_3(0;q)} \tag{100}$$

where prime indicates the derivative with respect to u. The result (97) was first reported in Ref. 12. The results (98)–(100) have recently been given in Ref. 30. For u a rational multiple of π these diagonal generalized correlations reduce to algebraic functions of the modulus k. Several examples for $C_-(0,0;\lambda)$, $C_-(1,1;\lambda)$ and $C_-(2,2;\lambda)$ are given in Ref. 12.

10. Diagonal susceptibility

We may now explicitly obtain[28] the diagonal susceptibility by using the form factor expansion (13)–(16) in the definition (41) and evaluate the sum on N as a geometric series. We obtain for $T < T_c$

$$kT \cdot \chi_{d-}(t) = (1-t)^{1/4} \cdot \sum_{n=1}^{\infty} \tilde{\chi}_d^{(2n)}(t) \tag{101}$$

with

$$\begin{aligned}\tilde{\chi}_d^{(2n)}(t) = & \frac{t^{n^2}}{(n!)^2} \frac{1}{\pi^{2n}} \cdot \int_0^1 \cdots \int_0^1 \prod_{k=1}^{2n} dx_k \cdot \frac{1+t^n x_1 \cdots x_{2n}}{1-t^n x_1 \cdots x_{2n}} \\ & \times \prod_{j=1}^{n} \left(\frac{x_{2j-1}(1-x_{2j})(1-tx_{2j})}{x_{2j}(1-x_{2j-1})(1-t\,x_{2j-1})} \right)^{1/2} \\ & \times \prod_{1\le j\le n} \prod_{1\le k\le n} (1-t\,x_{2j-1}\,x_{2k})^{-2} \\ & \times \prod_{1\le j<k\le n} (x_{2j-1}-x_{2k-1})^2\,(x_{2j}-x_{2k})^2 \end{aligned} \tag{102}$$

and for $T > T_c$

$$kT \cdot \chi_{d+}(t) = (1-t)^{1/4} \cdot \sum_{n=0}^{\infty} \tilde{\chi}_d^{(2n+1)}(t) \tag{103}$$

with

$$\begin{aligned}\tilde{\chi}_d^{(2n+1)}(t) &= \frac{t^{n(n+1))}}{\pi^{2n+1} n!\,(n+1)!} \cdot \int_0^1 \cdots \int_0^1 \prod_{k=1}^{2n+1} dx_k \\ &\times \frac{1 + t^{n+1/2}\, x_1 \cdots x_{2n+1}}{1 - t^{n+1/2}\, x_1 \cdots x_{2n+1}} \cdot \prod_{j=1}^{n} \Big((1 - x_{2j})(1 - t\, x_{2j}) \cdot x_{2j} \Big)^{1/2} \\ &\times \prod_{j=1}^{n+1} \Big((1 - x_{2j-1})(1 - t\, x_{2j-1}) \cdot x_{2j-1} \Big)^{-1/2} \\ &\times \prod_{1 \le j \le n+1} \prod_{1 \le k \le n} (1 - t\, x_{2j-1}\, x_{2k})^{-2} \\ &\times \prod_{1 \le j < k \le n+1} (x_{2j-1} - x_{2k-1})^2 \prod_{1 \le j < k \le n} (x_{2j} - x_{2k})^2. \end{aligned} \tag{104}$$

This diagonal susceptibility has been extensively studied in Ref. 28. The integrals for $\tilde{\chi}_d^{(1)}(t)$ and $\tilde{\chi}_d^{(2)}(t)$ are explicitly evaluated as

$$\tilde{\chi}_d^{(1)}(t) = \frac{1}{1 - t^{1/2}} \tag{105}$$

and

$$\tilde{\chi}_d^{(2)}(t) = \frac{1}{8\pi i} \oint dz_1 \frac{t}{(1 - t^{1/2} z_1)(z_1 - t^{1/2})} = \frac{t}{4\,(1-t)}. \tag{106}$$

Fuchsian equations have been obtained for $\tilde{\chi}_d^{(3)}(t)$, $\tilde{\chi}_d^{(4)}(t)$ and $\tilde{\chi}_d^{(5)}(t)$.

From these equations we find that $\tilde{\chi}_d^{(3)}(t)$ has a direct sum decomposition into the sum of three terms. One term is just $\tilde{\chi}_d^{(1)}(t)$ as given by (105); the second is

$$\frac{1}{k-1} \cdot (2K/\pi) + \frac{1}{(k-1)^2} \cdot (2E/\pi) \tag{107}$$

and the three solutions to the differential equation for the third term are two Meijer G functions and

$$\begin{aligned}\frac{(1+2k)(k+2)}{(1-k)(1+k+k^2)} &\cdot \{F(1/6, 1/3; 1; Q)^2 \\ &+ \frac{2Q}{9} \cdot F(1/6, 1/3; 1; Q) F(7/6, 4/3; 2; Q)\}\end{aligned} \tag{108}$$

where

$$Q = \frac{27}{4} \frac{(1+k)^2 k^2}{(k^2+k+1)^2} \tag{109}$$

Furthermore the $\tilde{\chi}_d^{(n)}(t)$ have singularities on $|t| = 1$ which are the analog of the Nickel singularities for the bulk susceptibility. For $T < T_c$ the singularities in $\tilde{\chi}^{(2n)}(t)$ are at $t^n = 1$ and are of the form $\epsilon^{2n^2-1}\ln\epsilon$ and for $T > T_c$ the singularities in $\tilde{\chi}^{(2n+1)}$ are at $t^{n+1/2} = 1$ and are of the form $\epsilon^{(n+1)^2-1/2}$.

11. Natural boundary

The most intriguing feature in both the bulk and the diagonal susceptibility are the singularities on the unit circle of the modular variable $k = 1$. As n increases the number of these singularities increases and becomes dense as $n \to \infty$. Therefore, unless a massive cancellation occurs the susceptibility will have a natural boundary on the circle $|k| = 1$. Recently further arguments in favor of such a natural boundary were given in Ref. 23. In terms of the nome q the circle $|k| = 1$ corresponds to the curve in Fig. 1.

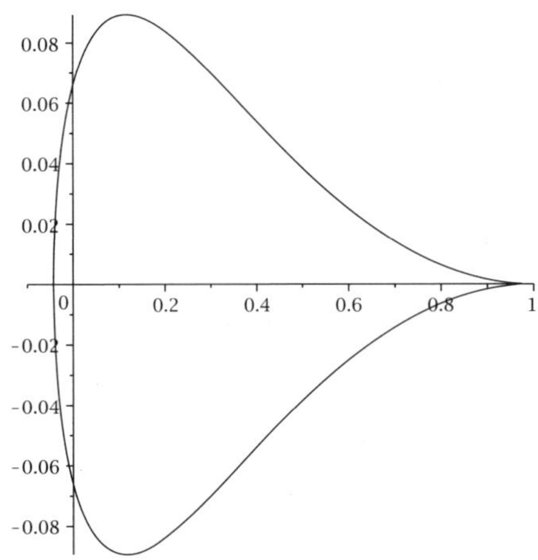

Fig. 1. The curve in the plane of the nome q of the unit circle $|k| = 1$

12. Conclusion

We have seen that since 1959 a great deal of progress has been made in understanding the susceptibility of the Ising model and that the analytic structure is vastly more complicated than was envisaged 50 years ago in Ref. 4. In particular the existence of a natural boundary is a completely new phenomenon which has never before appeared in the study of critical behavior. The connections with elliptic modular functions are profound and extensive and much of the structure still remains to be discovered. It is quite remarkable that in 50 years the problem has not been solved.

Acknowledgments

One of us, BMM, wishes to thank the organizers for the opportunity to participate in "New trends in quantum integrable systems" in honor of the 60th birthday of Tetsuji Miwa. JMM thanks the Simons Center for Geometry and Physics for support which helped in the finalizing of the manuscript. BMM also wishes to thank Prof. Miwa for 35 years worth in inspiration.

References

1. L. Onsager, Crystal statistics,I. A two-dimensional model with an order disorder transition, Phys. Rev. 65 (1944) 117–149.
2. L. Onsager, discussion, Nuovo Cimento 6 suppl. (1949) 261.
3. C.N. Yang, The spontaneous magnetization of the two dimensional Ising model, Phys. Rev. 85 (1952) 808–816.
4. M.E. Fisher, The susceptibility of the plane Ising model, Physics 25 (1959) 521-524.
5. B. Kaufmann and L. Onsager, Crystal statistics III. short range order in a binary Ising lattice, Phys. Rev. 76 (1949) 1244–1252.
6. I. Syozi and S. Naya, Symmetrical properties of two-dimensional Ising lattices, Prog. Theo. Phys. 24 (1960) 829–839.
7. E.W. Montroll, R.B. Potts and J.C. Ward, Correlations and spontaneous magnetization of the two dimensional Ising model, J. Math. Phys. 4 (1963) 308-322.
8. T.T. Wu, Theory of Toeplitz determinants and spin correlations of the two dimensional Ising model, Phys. Rev. 149 (1966) 380–401.
9. T.T. Wu. B.M. McCoy, C.A. Tracy and E. Barouch, Spin-spin correlation functions for the two dimensional Ising model: exact theory in the scaling region, Phys. Rev. B13 (1976) 315–374.
10. J. Palmer and C. Tracy, Two-dimensional Ising correlations: convergence of the scaling limit, Adv. Appl. Math. 2 (1981) 329–388.
11. K. Yamada, On the spin-spin correlation functions in the Ising square lattice and the zero field susceptibility, Prog. Theo. Phys. 71 (1984) 1416-1421.
12. S. Boukraa, S. Hassani, J-M. Maillard, B.M. McCoy, W.P. Orrick and N. Zenine, Holonomy of the Ising model form factors, J. Phys. A40 (2007) 75-112.
13. I. Lyberg and B.M. McCoy, Form factor expansion of the row and diagonal correlation functions of the two dimensional Ising model, J. Phys. A 40 (2007) 3329–3346.

14. B.M. McCoy, C.A. Tracy and T.T. Wu, Painlevé equations of the third kind, J. Math. Phys. 18 (1977) 1058–1092.
15. E. Barouch, B.M. McCoy and T.T. Wu, Zero field susceptibility of the two dimensional Ising model near T_c, Phys. Rev. Letts. 31 (1973) 1409-1411.
16. W.P. Orrick, B.G. Nickel, A.J. Guttmann and J.H.H. Perk, The susceptibility of the square lattice Ising model: new developements, J. Stat. Phys. 102 (2001) 795–841.
17. A.J. Guttmann and I.G. Enting, Solvability of some statistical mechanical systems, Phys. Rev. Letts. 76 (1996) 344–347.
18. B.G. Nickel, On the singularity structure of the 2D Ising model, J. Phys. A32 (1999) 3889–3906.
19. B.G. Nickel, Addendum to "On the singularity structure of the 2D Ising model", J. Phys. A 33 (2000) 1693–1711.
20. N. Zenine, S. Boukraa, S. Hassani and J-M. Maillard, The Fuchsian differential equation of the square Ising $\chi^{(3)}$ susceptibility, J. Phys. A 37 (2004) 9651–9668.
21. N. Zenine, S. Boukraa, S. Hassani and J-M. Maillard, Square lattice Ising model susceptibility: series expansion method and differential equation for $\chi^{(3)}$, J. Phys. A 38 (2005) 1975–1899.
22. N. Zenine, S. Boukraa, S. Hassani and J-M. Maillard, Ising model susceptibility: The Fuchsian equation for $\chi^{(4)}$ and its factorization properties, J. Phys. A 38 (2005) 4149–4173.
23. S. Boukraa, A.J. Guttmann, S. Hassani, I. Jensen, J-M. Maillard, B. Nickel and N. Zenine, Experimental mathematics on the magnetic susceptibility of the square lattice Ising model, J. Phys. A 41 (2008) 455202.
24. A. Bostan, S. Boukraa, A.J. Guttmann, S. Hassani, I. Jensen, J-M. Maillard and N. Zenine, High order Fuchsian equations for the square lattice Ising model: $\tilde{\chi}^{(5)}$, J. Phys. A 42 (2009) 275209.
25. S. Boukraa, S. Hassani, I. Jensen, J-M. Maillard, N.J. Zenine, High order Fuchsian equations for the square lattice Ising model: $\chi^{(6)}$, in press J. Phys.A: Math. Theor. (2010), arXiv:0912.4968v1 [math-ph].
26. A. Bostan, S. Boukraa, S. Hassani, J-M. Maillard, J-A. Weil and N. Zenine, Global nilpotent differential operators and the square Ising model, J. Phys. A 40 (2007) 2583–2614.
27. S. Boukraa, S. Hassani, J-M. Maillard, and N. Zenine, Landau singularities and singularities of holonomic integrals of the Ising class, J. Phys. A 40 (2007) 2583-2614
28. S. Boukraa, S. Hassani, J-M. Maillard, B.M. McCoy, J-A. Weil and N. Zenine, The diagonal Ising susceptibility, J. Phys. A 40 (2007) 8219–8236.
29. M. Jimbo and T. Miwa, Studies on holonomic quantum fields XVII, Proc. Jpn. Acad. 56A (1980) 405: 57A (1981) 347.
30. V.V. Mangazeev and A.J. Guttmann, Form factor expansions in the 2D Ising model and Painlevé VI, ArXiv.1002.2490

BETHE ALGEBRA OF THE $\mathfrak{gl}_{N+1}$ GAUDIN MODEL AND ALGEBRA OF FUNCTIONS ON THE CRITICAL SET OF THE MASTER FUNCTION

EVGENY MUKHIN*, VITALY TARASOV*,⋆ and ALEXANDER VARCHENKO**

*Department of Mathematical Sciences, Indiana University Purdue University Indianapolis, 402 North Blackford St, Indianapolis, IN 46202-3216, USA

⋆St. Petersburg Branch of Steklov Mathematical Institute Fontanka 27, St. Petersburg, 191023, Russia

**Department of Mathematics, University of North Carolina at Chapel Hill Chapel Hill, NC 27599-3250, USA

Consider a tensor product of finite-dimensional irreducible $\mathfrak{gl}_{N+1}$-modules and its decomposition into irreducible modules. The $\mathfrak{gl}_{N+1}$ Gaudin model assigns to each multiplicity space of that decomposition a commutative (Bethe) algebra of linear operators acting on the multiplicity space. The Bethe ansatz method is a method to find eigenvectors and eigenvalues of the Bethe algebra. One starts with a critical point of a suitable (master) function and constructs an eigenvector of the Bethe algebra.

In this paper we consider the algebra of functions on the critical set of the associated master function and show that the action of this algebra on itself is isomorphic to the action of the Bethe algebra on a suitable subspace of the multiplicity space.

As a byproduct we prove that the Bethe vectors corresponding to different critical points of the master function are linearly independent and, in particular, nonzero.

Keywords: Bethe ansatz; Gaudin model.

1. Introduction

Let $L_{\boldsymbol{\lambda}}$ be the irreducible finite-dimensional $\mathfrak{gl}_{N+1}$-module of highest weight $\boldsymbol{\lambda}$. Let $L_{\boldsymbol{\Lambda}} = \otimes_{s=1}^{n} L_{\boldsymbol{\lambda}^{(s)}}$ be a tensor product of such modules, and $L_{\boldsymbol{\Lambda}} = \oplus_{\boldsymbol{\lambda}^{(\infty)}} L_{\boldsymbol{\lambda}^{(\infty)}} \otimes W_{\boldsymbol{\lambda}^{(\infty)}}$ the decomposition into irreducible representations. The multiplicity space $W_{\boldsymbol{\lambda}^{(\infty)}}$ of $L_{\boldsymbol{\lambda}^{(\infty)}}$ can be identified with $\operatorname{Sing} L_{\boldsymbol{\Lambda}}[\boldsymbol{\lambda}^{(\infty)}] \subset L_{\boldsymbol{\Lambda}}$, the subspace of singular vectors of weight $\boldsymbol{\lambda}^{(\infty)}$. To each multiplicity space $\operatorname{Sing} L_{\boldsymbol{\Lambda}}[\boldsymbol{\lambda}^{(\infty)}]$ and distinct complex numbers $z_1, \dots, z_n$, the $\mathfrak{gl}_{N+1}$ Gaudin model assigns a commutative subalgebra of

End (Sing $L_{\mathbf{\Lambda}}[\boldsymbol{\lambda}^{(\infty)}]$) called the Bethe algebra and denoted by $\mathcal{B}_{\mathbf{\Lambda},\boldsymbol{\lambda}^{(\infty)},\boldsymbol{z}}$.

The $\mathcal{B}_{\mathbf{\Lambda},\boldsymbol{\lambda}^{(\infty)},\boldsymbol{z}}$-module Sing $L_{\mathbf{\Lambda}}[\boldsymbol{\lambda}^{(\infty)}]$ has an interesting geometric realization. In Ref. 6 we constructed an isomorphism of the $\mathcal{B}_{\mathbf{\Lambda},\boldsymbol{\lambda}^{(\infty)},\boldsymbol{z}}$-module Sing $L_{\mathbf{\Lambda}}[\boldsymbol{\lambda}^{(\infty)}]$ and the regular representation of the algebra of functions on the scheme-theoretical intersection of suitable Schubert cycles. This isomorphism can be viewed as the geometric Langlands correspondence in the $\mathfrak{gl}_{N+1}$ Gaudin model. In Ref. 7 we argued that this geometric Langlands correspondence extends to the third, equally important, player — the algebra of functions on the critical set of the corresponding master function. In this paper we prove another result supporting that principle.

The master and weight functions are useful objects associated with each multiplicity space Sing $L_{\mathbf{\Lambda}}[\boldsymbol{\lambda}^{(\infty)}]$, see Ref. 13. They are functions of some auxiliary variables $\boldsymbol{t} = (t_j^{(i)})$. The master function $\Phi(\boldsymbol{t})$ is a scalar function and the weight function $\omega(\boldsymbol{t})$ is an $L_{\mathbf{\Lambda}}$-valued function. They are used in the Bethe ansatz method to construct eigenvectors of the Bethe algebra $\mathcal{B}_{\mathbf{\Lambda},\boldsymbol{\lambda}^{(\infty)},\boldsymbol{z}}$. Namely, if $\boldsymbol{p}$ is a critical point of the master function then the vector $\omega(\boldsymbol{p})$ lies in Sing $L_{\mathbf{\Lambda}}[\boldsymbol{\lambda}^{(\infty)}]$ and is an eigenvector of the Bethe algebra, see Ref. 5.

In this paper we consider the algebra A_Φ of functions on the critical set of the master function. With the help of the weight function, we construct a linear embedding $\alpha : A_\Phi \to$ Sing $L_{\mathbf{\Lambda}}[\boldsymbol{\lambda}^{(\infty)}]$ and show that $\alpha(A_\Phi)$ is a $\mathcal{B}_{\mathbf{\Lambda},\boldsymbol{\lambda}^{(\infty)},\boldsymbol{z}}$-submodule of Sing $L_{\mathbf{\Lambda}}[\boldsymbol{\lambda}^{(\infty)}]$. We denote the image of $\mathcal{B}$ in End$(\alpha(A_\Phi))$ by A_B. We construct an algebra isomorphism $\beta : A_\Phi \to A_B$ and show that the A_B-module $\alpha(A_\Phi)$ is isomorphic to the regular representation of A_Φ. That statement is our main result, see Theorems 5.5 and 7.1. As a byproduct we show that for any critical point $\boldsymbol{p}$ of the master function, the vector $\omega(\boldsymbol{p})$ is nonzero. For a nondegenerate critical point that fact was proved in Ref. 10 and Ref. 15.

The paper is organized as follows. In Section 2 we define the master function Φ and the algebra A_Φ of functions on the critical set C_Φ of the master function. The algebra A_Φ is the direct sum of local algebras $A_{\boldsymbol{p},\Phi}$ corresponding to points $\boldsymbol{p} \in C_\Phi$. In Theorem 2.1 we describe useful generators of the algebra $A_{\boldsymbol{p},\Phi}$. We prove Theorem 2.1 in Section 3. In Section 4 the Bethe algebra is introduced. We define the weight function in Section 5 and formulate our first main result Theorem 5.5. We prove Theorem 5.5 in Section 6. Our second main result, Theorem 7.1 is formulated and proved in Section 7.

The authors thank P. Belkale for help in proving Theorem 2.1. The first author is supported in part by NSF grant DMS-0900984. The second author is supported in part by NSF grant DMS-0901616. The third author is supported in part by NSF grant DMS-0555327.

2. Algebra A_Φ

2.1. *Lie algebra* $\mathfrak{gl}_{N+1}$

Let e_{ij}, $i,j = 1,\dots,N+1$, be the standard generators of the Lie algebra $\mathfrak{gl}_{N+1}$ satisfying the relations $[e_{ij}, e_{sk}] = \delta_{js}e_{ik} - \delta_{ik}e_{sj}$. Let $\mathfrak{h} \subset \mathfrak{gl}_{N+1}$ be the Cartan subalgebra generated by e_{ii}, $i = 1,\dots,N+1$. Let $\mathfrak{h}^*$ be the dual space. Let ϵ_i, $i = 1,\dots,N+1$, be the basis of $\mathfrak{h}^*$ dual to the basis e_{ii}, $i = 1,\dots,N+1$, of $\mathfrak{h}$. Let $\alpha_1,\dots,\alpha_N \in \mathfrak{h}^*$ be simple roots, $\alpha_i = \epsilon_i - \epsilon_{i+1}$. Let $(\,,\,)$ be the standard scalar product on $\mathfrak{h}^*$ such that the basis ϵ_i, $i = 1,\dots,N+1$, is orthonormal.

A sequence of integers $\boldsymbol{\lambda} = (\lambda_1,\dots,\lambda_{N+1})$ such that $\lambda_1 \geqslant \lambda_2 \geqslant \dots \geqslant \lambda_{N+1} \geqslant 0$ is called a *partition with at most* $N+1$ *parts*. Denote $|\boldsymbol{\lambda}| = \sum_{i=1}^{N+1}\lambda_i$. We identify partitions $\boldsymbol{\lambda}$ with vectors $\lambda_1\epsilon_1 + \dots + \lambda_{N+1}\epsilon_{N+1}$ of $\mathfrak{h}^*$.

2.2. *Master function*

Let $\boldsymbol{\Lambda} = (\boldsymbol{\lambda}^{(1)},\dots,\boldsymbol{\lambda}^{(n)})$ be a collection of partitions, where $\boldsymbol{\lambda}^{(i)} = (\lambda_1^{(i)},\dots,\lambda_{N+1}^{(i)})$ and $\lambda_{N+1}^{(i)} = 0$. Let $\boldsymbol{l} = (l_1,\dots,l_N)$ be nonnegative integers such that

$$\boldsymbol{\lambda}^{(\infty)} = \sum_{i=1}^{n}\boldsymbol{\lambda}^{(i)} - \sum_{j=1}^{N} l_j\alpha_j$$

is a partition. Denote $l = l_1 + \dots + l_N$,

$$\boldsymbol{t} = (t_1^{(1)},\dots,t_{l_1}^{(1)},t_1^{(2)},\dots,t_{l_2}^{(2)},\dots,t_1^{(N)},\dots,t_{l_N}^{(N)})\,.$$

Fix a collection of distinct complex numbers $\boldsymbol{z} = (z_1,\dots,z_n)$. Let $\Phi(\boldsymbol{t})$ be the master function associated with this data,

$$\Phi(\boldsymbol{t}) = \prod_{i=1}^{N}\prod_{1\leqslant j<j'\leqslant l_i}(t_j^{(i)} - t_{j'}^{(i)})^2$$
$$\times\prod_{i=1}^{N-1}\prod_{j=1}^{l_i}\prod_{j'=1}^{l_{i+1}}(t_j^{(i)} - t_{j'}^{(i+1)})^{-1}\prod_{i=1}^{N}\prod_{j=1}^{l_i}\prod_{s=1}^{n}(t_j^{(i)} - z_s)^{-(\boldsymbol{\lambda}^{(s)},\alpha_i)}.$$

Denote

$$U = \{\boldsymbol{p} \in \mathbb{C}^l \mid \Phi \text{ is well-defined at } \boldsymbol{p} \text{ and } \Phi(\boldsymbol{p}) \neq 0\}. \tag{1}$$

The set U is the complement in $\mathbb{C}^l$ to a union of hyperplanes. The master function is a rational function regular on U.

Denote by $\mathbb{C}(\boldsymbol{t})_U$ the algebra of rational functions on $\mathbb{C}^l$ regular on U. The partial derivatives

$$\Psi_{ij} = \partial(\log\Phi)/\partial t_j^{(i)}, \qquad i = 1,\dots,N, \quad j = 1,\dots,l_i\,,$$

are elements of $\mathbb{C}(\boldsymbol{t})_U$. Denote by $I_\Phi \subset \mathbb{C}(\boldsymbol{t})_U$ the ideal generated by Ψ_{ij}, $i = 1,\dots,N$, $j = 1,\dots,l_i$, and set

$$A_\Phi = \mathbb{C}(\boldsymbol{t})_U/I_\Phi. \tag{2}$$

Denote by C_Φ the zero set of the ideal. The zero set is finite, Ref. 9. The algebra A_Φ is finite-dimensional and is the direct sum of local algebras,

$$A_\Phi = \oplus_{\boldsymbol{p}\in C_\Phi} A_{\boldsymbol{p},\Phi}$$

corresponding to points $\boldsymbol{p} \in C_\Phi$. For $\boldsymbol{p} \in C_\Phi$, the local algebra $A_{\boldsymbol{p},\Phi}$ may be defined as the quotient of the algebra of germs at $\boldsymbol{p}$ of holomorphic functions modulo the ideal $I_{\boldsymbol{p},\Phi}$ generated by all the functions Ψ_{ij}. The algebra $A_{\boldsymbol{p},\Phi}$ contains the maximal ideal $\mathfrak{m}_{\boldsymbol{p}}$ generated by the germs of functions equal to zero at $\boldsymbol{p}$.

2.3. *Generators of the local algebra of a critical point*

Let u be a variable. Define an N-tuple of polynomials $T_1,\dots,T_N \in \mathbb{C}[u]$,

$$T_i(u) = \prod_{s=1}^{n} (u - z_s)^{(\boldsymbol{\lambda}^{(s)},\alpha_i)}\,,$$

an N-tuple of polynomials $y_1,\dots,y_N \in \mathbb{C}[u,\boldsymbol{t}]$,

$$y_i(u,\boldsymbol{t}) = \prod_{j=1}^{l_i} (u - t_j^{(i)})\,,$$

and the differential operator

$$\begin{aligned}\mathcal{D}_\Phi &= (\partial_u - \log'(\frac{T_1\dots T_N}{y_1}))\\ &\times (\partial_u - \log'(\frac{y_1T_2\dots T_N}{y_2}))\dots(\partial_u - \log'(\frac{y_{N-1}T_N}{y_N}))(\partial_u - \log'(y_N)),\end{aligned}$$

where $\partial_u = d/du$ and $\log' f$ denotes $(df/du)/f$. We have

$$\mathcal{D}_\Phi = \partial_u^{N+1} + \sum_{i=1}^{N+1} G_i\,\partial_u^{N+1-i}, \qquad G_i = \sum_{j=i}^{\infty} G_{ij}u^{-j}, \tag{3}$$

where $G_{ij} \in \mathbb{C}[\boldsymbol{t}]$.

For $\boldsymbol{p} \in C_\Phi$, $f \in \mathbb{C}(\boldsymbol{t})_U$ denote by $\bar{f}$ the image of f in $A_{\boldsymbol{p},\Phi}$. Denote

$$\bar{\mathcal{D}}_\Phi = \partial_u^{N+1} + \sum_{i=1}^{N+1} \bar{G}_i \, \partial_u^{N+1-i},$$

where $\bar{G}_i = \sum_{j=i}^{\infty} \bar{G}_{ij} u^{-j}$.

Theorem 2.1. *For any* $\boldsymbol{p} \in C_\Phi$, *the elements* $\bar{G}_{ij}$, $i = 1, \ldots, N$, $j \geqslant i$, *generate* $A_{\boldsymbol{p},\Phi}$.

Theorem 2.1 is proved in Section 3.4.

2.4. *Polynomials* h_i

Let A be a commutative algebra. For $g_1, \ldots, g_i \in A[u]$, denote by $\mathrm{Wr}(g_1(u), \ldots, g_i(u))$ the Wronskian,

$$\mathrm{Wr}(g_1(u), \ldots, g_i(u)) = \det \begin{pmatrix} g_1(u) & g_1'(u) & \ldots & g_1^{(i-1)}(u) \\ g_2(u) & g_2'(u) & \ldots & g_2^{(i-1)}(u) \\ \ldots & \ldots & \ldots & \ldots \\ g_i(u) & g_i'(u) & \ldots & g_i^{(i-1)}(u) \end{pmatrix},$$

where $g^{(j)}(u)$ denotes the j-th derivative of $g(u)$ with respect to u.

Introduce a set

$$P = \{d_1, d_2, \ldots, d_{N+1}\}, \qquad d_i = \lambda_i^{(\infty)} + N + 1 - i. \tag{4}$$

Theorem 2.2. *There exist unique polynomials* $h_1, \ldots, h_{N+1} \in A_{\boldsymbol{p},\Phi}[u]$ *of the form*

$$h_i = u^{d_i} + \sum_{j=1,\ d_i - j \notin P}^{d_i} h_{ij} u^{d_i - j} \tag{5}$$

such that $h_{N+1} = y_N$ *and*

$$\mathrm{Wr}(h_{N+1}, h_N \ldots, h_{N+1-j}) = y_{N-j} T_N^j T_{N-1}^{j-1} \ldots T_{N-j+1}^1 \prod_{N+1-j \leqslant i < i' \leqslant N+1} (d_i - d_{i'}) \tag{6}$$

for $j = 1, \ldots, N$, *where* $y_0 = 1$. *Moreover, each of the polynomials* $h_1, \ldots, h_{N+1}$ *is a solution of the differential equation* $\bar{\mathcal{D}}_\Phi h(u) = 0$. $\square$

Proof. The existence of unique polynomials h_i satisfying (6) is proved in Ref. 3 generalizing the corresponding result in Section 5 of Ref. 9. The fact that the polynomials h_i satisfy the differential equation $\bar{\mathcal{D}}_\Phi h(u) = 0$ is proved like in Section 5 of Ref. 9. □

Lemma 2.3. *The subalgebra of $A_{\boldsymbol{p},\Phi}$ generated by elements $\bar{G}_{ij}$, $i = 1, \dots, N$, $j \geqslant i$, contains all the coefficients h_{ij}, $i = 1, \dots, N+1$, $j = 1, \dots, d_i$, $d_i - j \notin P$.*

The proof of the lemma is the same as the proof of Lemma 3.4 in Ref. 6.

3. Algebra A_{Gr}

3.1. *Algebra* $\mathcal{O}_{\boldsymbol{\lambda}^{(\infty)}}$

Let $\boldsymbol{\Lambda} = (\boldsymbol{\lambda}^{(1)}, \dots, \boldsymbol{\lambda}^{(n)})$, $\boldsymbol{\lambda}^{(\infty)}$, $\boldsymbol{z} = (z_1, \dots, z_n)$ be partitions and numbers as in Section 2.2. Let d be a natural number such that $d - N - 1 \geqslant \lambda_1^{(\infty)}$ and $d - N - 1 \geqslant \lambda_1^{(i)}$ for $i = 1, \dots, n$.

Let $\mathbb{C}_d[u]$ be the space of polynomials in u of degree less than d. Let $\mathrm{Gr}(N+1, d)$ be the Grassmannian of all $N+1$-dimensional subspaces of $\mathbb{C}_d[u]$.

For a complete flag $\mathcal{F} = \{0 \subset F_1 \subset F_2 \subset \dots \subset F_d = \mathbb{C}_d[u]\}$ and a partition $\boldsymbol{\lambda} = (\lambda_1, \dots, \lambda_{N+1})$ with $\lambda_1 \leqslant d - N - 1$, define a Schubert cell $\Omega_{\boldsymbol{\lambda}}(\mathcal{F}) \subset \mathrm{Gr}(N+1, d)$,

$$\begin{aligned}\Omega_{\boldsymbol{\lambda}}(\mathcal{F}) = \{\boldsymbol{q} \in \mathrm{Gr}(N+1, d) \mid \dim(\boldsymbol{q} \cap F_{d-j-\lambda_j}) &= N + 1 - j\,, \\ \dim(\boldsymbol{q} \cap F_{d-j-\lambda_j-1}) &= N - j\}\,.\end{aligned}$$

We have $\operatorname{codim} \Omega_{\boldsymbol{\lambda}}(\mathcal{F}) = |\boldsymbol{\lambda}|$.

Let $P = \{d_1, d_2, \dots, d_{N+1}\}$ be defined in (4). Introduce a new partition

$$\boldsymbol{\lambda}^{(\vee)} = (d - N - 1 - \lambda_N^{(\infty)}, d - N - 1 - \lambda_{N-1}^{(\infty)}, \dots, d - N - 1 - \lambda_1^{(\infty)})\,. \tag{7}$$

Denote

$$\mathcal{F}(\infty) = \{0 \subset \mathbb{C}_1[u] \subset \mathbb{C}_2[u] \subset \dots \subset \mathbb{C}_d[u]\}\,.$$

Consider the Schubert cell $\Omega_{\boldsymbol{\lambda}^{(\vee)}}(\mathcal{F}(\infty))$. We have $\dim \Omega_{\boldsymbol{\lambda}^{(\vee)}}(\mathcal{F}(\infty)) = |\boldsymbol{\lambda}^{(\infty)}|$.

The Schubert cell $\Omega_{\boldsymbol{\lambda}^{(\vee)}}(\mathcal{F}(\infty))$ consists of $N+1$-dimensional subspaces $\boldsymbol{q} \subset \mathbb{C}_d[u]$ with a basis $\{f_1, \dots, f_{N+1}\}$ of the form

$$f_i = u^{d_i} + \sum_{j=1,\ d_i - j \notin P}^{d_i} f_{ij} u^{d_i - j}\,. \tag{8}$$

Such a basis is unique.

Denote by $\mathcal{O}_{\boldsymbol{\lambda}^{(\infty)}}$ the algebra of regular functions on $\Omega_{\boldsymbol{\lambda}^{(\vee)}}(\mathcal{F}(\infty))$. The cell $\Omega_{\boldsymbol{\lambda}^{(\vee)}}(\mathcal{F}(\infty))$ is an affine space with coordinate functions f_{ij}. The algebra $\mathcal{O}_{\boldsymbol{\lambda}^{(\infty)}}$ is the polynomial algebra in variables f_{ij},

$$\mathcal{O}_{\boldsymbol{\lambda}^{(\infty)}} = \mathbb{C}[f_{ij},\ i=1,\dots,N+1,\ j=1,\dots,d_i,\ d_i-j\notin P]. \qquad (9)$$

3.2. *Intersection of Schubert cells*

For $z\in\mathbb{C}$, consider the complete flag

$$\mathcal{F}(z) = \left\{0\subset(u-z)^{d-1}\mathbb{C}_1[u]\subset(u-z)^{d-2}\mathbb{C}_2[u]\subset\dots\subset\mathbb{C}_d[u]\right\}.$$

Denote by $\Omega_{\boldsymbol{\Lambda},\boldsymbol{\lambda}^{(\infty)},\boldsymbol{z}}$ the set-theoretic intersection and by A_{Gr} the scheme-theoretic intersection of the $n+1$ Schubert cells $\Omega_{\boldsymbol{\lambda}^{(\vee)}}(\mathcal{F}(\infty))$, $\Omega_{\boldsymbol{\lambda}^{(s)}}(\mathcal{F}(z_s))$, $s=1,\dots,n$, see Section 4 of Ref. 6. The set-theoretic intersection $\Omega_{\boldsymbol{\Lambda},\boldsymbol{\lambda}^{(\infty)},\boldsymbol{z}}$ is a finite set and the scheme-theoretic intersection A_{Gr} is a finite-dimensional algebra ("of functions on $\Omega_{\boldsymbol{\Lambda},\boldsymbol{\lambda}^{(\infty)},\boldsymbol{z}}$"). The algebra of functions on $\Omega_{\boldsymbol{\Lambda},\boldsymbol{\lambda}^{(\infty)},\boldsymbol{z}}$ is the direct sum of local algebras,

$$A_{\mathrm{Gr}} = \oplus_{\boldsymbol{q}\in\Omega_{\boldsymbol{\Lambda},\boldsymbol{\lambda}^{(\infty)},\boldsymbol{z}}} A_{\boldsymbol{q},\mathrm{Gr}}\ ,$$

corresponding to points $\boldsymbol{q}\in\Omega_{\boldsymbol{\Lambda},\boldsymbol{\lambda}^{(\infty)},\boldsymbol{z}}$. The algebra A_{Gr} is the quotient of the algebra $\mathcal{O}_{\boldsymbol{\lambda}^{(\infty)}}$ of functions on $\Omega_{\boldsymbol{\lambda}^{(\vee)}}(\infty)$ by a suitable ideal. For $\boldsymbol{q}\in\Omega_{\boldsymbol{\Lambda},\boldsymbol{\lambda}^{(\infty)},\boldsymbol{z}}$ and $f\in\mathcal{O}_{\boldsymbol{\lambda}^{(\infty)}}$ denote by $\bar f$ the image of f in $A_{\boldsymbol{q},\mathrm{Gr}}$.

Lemma 3.1. *For any* $\boldsymbol{q}\in\Omega_{\boldsymbol{\Lambda},\boldsymbol{\lambda}^{(\infty)},\boldsymbol{z}}$, *the elements* $\bar f_{ij}$, *where* $i=1,\dots,N+1$, $j=1,\dots,d_i$, $d_i-j\notin P$, *generate* $A_{\boldsymbol{q},\mathrm{Gr}}$. □

3.3. *Isomorphism of algebras*

Theorem 3.2 (Ref. 9). *Let* $\boldsymbol{p}\in C_\Phi$. *Let* $h_1,\dots,h_{N+1}\in A_{\boldsymbol{p},\Phi}[u]$ *be polynomials defined in Theorem 2.2. Denote by* $\tilde h_1,\dots,\tilde h_{N+1}$ *the projection of the polynomials to* $A_{\boldsymbol{p},\Phi}/\mathfrak{m}_{\boldsymbol{p}}[u]=\mathbb{C}[u]$. *Then* $\langle\tilde h_1,\dots,\tilde h_{N+1}\rangle\in\Omega_{\boldsymbol{\Lambda},\boldsymbol{\lambda}^{(\infty)},\boldsymbol{z}}$. □

Denote $\boldsymbol{q}=\langle\tilde h_1,\dots,\tilde h_{N+1}\rangle$. Let $\bar f_{ij}\in A_{\boldsymbol{q},\mathrm{Gr}}$ be elements of Lemma 3.1. Let h_{ij} be coefficients of the polynomials $h_1,\dots,h_{N+1}$ in Theorem 3.2.

Theorem 3.3 (Ref. 3). *The map* $\bar f_{ij}\mapsto h_{ij}$, *where* $i=1,\dots,N+1$, $j=1,\dots,d_i$, $d_i-j\notin P$, *extends uniquely to an algebra isomorphism* $A_{\boldsymbol{q},\mathrm{Gr}}\to A_{\boldsymbol{p},\Phi}$. □

Corollary 3.4. *The elements* h_{ij}, *where* $i=1,\dots,N+1$, $j=1,\dots,d_i$, $d_i-j\notin P$, *generate* $A_{\boldsymbol{p},\Phi}$. □

3.4. *Proof of Theorem 2.1*

By Lemma 2.3 the subalgebra of $A_{\boldsymbol{p},\Phi}$ generated by all the elements $\bar{G}_{ij}$ contains all the coefficients h_{ij}. By Corollary 3.4 the coefficients h_{ij} generate $A_{\boldsymbol{p},\Phi}$. Theorem 2.1 is proved.

4. Bethe algebra

4.1. *Lie algebra* $\mathfrak{gl}_{N+1}[t]$

Let $\mathfrak{gl}_{N+1}[t] = \mathfrak{gl}_{N+1} \otimes \mathbb{C}[t]$ be the Lie algebra of $\mathfrak{gl}_{N+1}$-valued polynomials with the pointwise commutator. For $g \in \mathfrak{gl}_{N+1}$, we set $g(u) = \sum_{s=0}^{\infty}(g \otimes t^s)u^{-s-1}$.

We identify $\mathfrak{gl}_{N+1}$ with the subalgebra $\mathfrak{gl}_{N+1} \otimes 1$ of constant polynomials in $\mathfrak{gl}_{N+1}[t]$. Hence, any $\mathfrak{gl}_{N+1}[t]$-module has a canonical structure of a $\mathfrak{gl}_{N+1}$-module.

For each $a \in \mathbb{C}$, there exists an automorphism ρ_a of $\mathfrak{gl}_{N+1}[t]$, $\rho_a : g(u) \mapsto g(u-a)$. Given a $\mathfrak{gl}_{N+1}[t]$-module M, we denote by $M(a)$ the pull-back of M through the automorphism ρ_a. As $\mathfrak{gl}_{N+1}$-modules, M and $M(a)$ are isomorphic by the identity map.

We have the evaluation homomorphism, $\mathfrak{gl}_{N+1}[t] \to \mathfrak{gl}_{N+1}$, $g(u) \mapsto gu^{-1}$. Its restriction to the subalgebra $\mathfrak{gl}_{N+1} \subset \mathfrak{gl}_{N+1}[t]$ is the identity map. For any $\mathfrak{gl}_{N+1}$-module M, we denote by the same letter the $\mathfrak{gl}_{N+1}[t]$-module, obtained by pulling M back through the evaluation homomorphism.

4.2. *Definition of row determinant*

Given an algebra A and an $(N+1) \times (N+1)$-matrix $C = (c_{ij})$ with entries in A, we define its *row determinant* to be

$$\operatorname{rdet} C = \sum_{\sigma \in \Sigma_{N+1}} (-1)^{\sigma} c_{1\sigma(1)} c_{2\sigma(2)} \dots c_{N+1\,\sigma(N+1)}\,.$$

4.3. *Definition of Bethe algebra*

Define the universal differential operator $\mathcal{D}_{\mathcal{B}}$ by the formula

$$\mathcal{D}_{\mathcal{B}} = \operatorname{rdet} \begin{pmatrix} \partial_u - e_{11}(u) & -e_{21}(u) & \dots & -e_{N+1\,1}(u) \\ -e_{12}(u) & \partial_u - e_{22}(u) & \dots & -e_{N+1\,2}(u) \\ \dots & \dots & \dots & \dots \\ -e_{1\,N+1}(u) & -e_{2\,N+1}(u) & \dots & \partial_u - e_{N+1\,N+1}(u) \end{pmatrix}.$$

We have

$$\mathcal{D}_{\mathcal{B}} = \partial_u^{N+1} + \sum_{i=1}^{N+1} B_i \,\partial_u^{N+1-i}, \quad B_i = \sum_{j=i}^{\infty} B_{ij} u^{-j}, \quad B_{ij} \in U(\mathfrak{gl}_{N+1}[t]) \,. \tag{10}$$

The unital subalgebra of $U(\mathfrak{gl}_{N+1}[t])$ generated by B_{ij}, $i = 1, \ldots, N+1$, $j \geqslant i$, is called the *Bethe algebra* and denoted by $\mathcal{B}$.

By Ref. 14, cf. Ref. 5, the algebra $\mathcal{B}$ is commutative, and $\mathcal{B}$ commutes with the subalgebra $U(\mathfrak{gl}_{N+1}) \subset U(\mathfrak{gl}_{N+1}[t])$.

As a subalgebra of $U(\mathfrak{gl}_{N+1}[t])$, the algebra $\mathcal{B}$ acts on any $\mathfrak{gl}_{N+1}[t]$-module M. Since $\mathcal{B}$ commutes with $U(\mathfrak{gl}_{N+1})$, it preserves the $\mathfrak{gl}_{N+1}$ weight subspaces of M and the subspace $\operatorname{Sing} M$ of $\mathfrak{gl}_{N+1}$-singular vectors.

If L is a $\mathcal{B}$-module, then the image of $\mathcal{B}$ in $\operatorname{End}(L)$ is called the *Bethe algebra of* L.

4.4. *Bethe algebra of* Sing $L_{\boldsymbol{\Lambda}}[\boldsymbol{\lambda}^{(\infty)}]$

For a partition $\boldsymbol{\lambda}$ with at most $N+1$ parts denote by $L_{\boldsymbol{\lambda}}$ the irreducible $\mathfrak{gl}_{N+1}$-module with highest weight $\boldsymbol{\lambda}$.

Let $\boldsymbol{\Lambda} = (\boldsymbol{\lambda}^{(1)}, \ldots, \boldsymbol{\lambda}^{(n)})$, $\boldsymbol{\lambda}^{(\infty)}$, $\boldsymbol{z} = (z_1, \ldots, z_n)$ be partitions and numbers as in Section 2.2. Denote $L_{\boldsymbol{\Lambda}} = L_{\boldsymbol{\lambda}^{(1)}} \otimes \cdots \otimes L_{\boldsymbol{\lambda}^{(n)}}$. Let

$$\begin{aligned} L_{\boldsymbol{\Lambda}}[\boldsymbol{\lambda}^{(\infty)}] &= \{v \in L_{\boldsymbol{\Lambda}} \mid e_{ii} v = \lambda_i^{(\infty)} v \text{ for } i = 1, \ldots, N+1\}, \\ \operatorname{Sing} L_{\boldsymbol{\Lambda}}[\boldsymbol{\lambda}^{(\infty)}] &= \{v \in L_{\boldsymbol{\Lambda}}[\boldsymbol{\lambda}^{(\infty)}] \mid e_{ij} v = 0 \text{ for } i < j\} \end{aligned}$$

be the subspace of vectors of $\mathfrak{gl}_{N+1}$-weight $\boldsymbol{\lambda}^{(\infty)}$ and the subspace of $\mathfrak{gl}_{N+1}$-singular vectors of $\mathfrak{gl}_{N+1}$-weight $\boldsymbol{\lambda}^{(\infty)}$, respectively. Consider on $L_{\boldsymbol{\Lambda}}$ the $\mathfrak{gl}_{N+1}[t]$-module structure of the tensor product of evaluation modules, $L_{\boldsymbol{\Lambda}} = \otimes_{s=1}^n L_{\boldsymbol{\lambda}^{(s)}}(z_s)$. Then $\operatorname{Sing} L_{\boldsymbol{\Lambda}}[\boldsymbol{\lambda}^{(\infty)}]$ is a $\mathcal{B}$-submodule. We denote by $\mathcal{B}_{\boldsymbol{\Lambda},\boldsymbol{\lambda}^{(\infty)},\boldsymbol{z}}$ the Bethe algebra of $\operatorname{Sing} L_{\boldsymbol{\Lambda}}[\boldsymbol{\lambda}^{(\infty)}]$.

4.5. *Shapovalov Form*

Let $\tau : \mathfrak{gl}_{N+1} \to \mathfrak{gl}_{N+1}$ be the anti-involution sending e_{ij} to e_{ji} for all (i,j). Let M be a highest weight $\mathfrak{gl}_{N+1}$-module with a highest weight vector w. The Shapovalov form S on M is the unique symmetric bilinear form such that

$$S(w,w) = 1, \qquad S(xu,v) = S(u,\tau(x)v)$$

for all $u, v \in M$ and $x \in \mathfrak{gl}_{N+1}$.

Fix highest weight vectors $v_{\boldsymbol{\lambda}^{(s)}} \in L_{\boldsymbol{\lambda}^{(s)}}$, $s = 1, \dots, n$. Define a symmetric bilinear form on the tensor product $L_{\boldsymbol{\Lambda}} = L_{\boldsymbol{\lambda}^{(1)}} \otimes \cdots \otimes L_{\boldsymbol{\lambda}^{(n)}}$ by the formula

$$S_{\boldsymbol{\Lambda}} = S_1 \otimes \cdots \otimes S_n, \tag{11}$$

where S_s is the Shapovalov form on $L_{\boldsymbol{\lambda}^{(s)}}$. The form $S_{\boldsymbol{\lambda}}$ is called the tensor Shapovalov form.

Theorem 4.1 (Ref. 5). *Consider the space $L_{\boldsymbol{\Lambda}}$ as the $\mathfrak{gl}_{N+1}[t]$-module $\otimes_{s=1}^{n} L_{\boldsymbol{\lambda}^{(s)}}(z_s)$. Then any element $B \in \mathcal{B}$ acts on $L_{\boldsymbol{\Lambda}}$ as a symmetric operator with respect to the tensor Shapovalov form, $S_{\boldsymbol{\Lambda}}(Bu, v) = S_{\boldsymbol{\Lambda}}(u, Bv)$ for any $u, v \in L_{\boldsymbol{\Lambda}}$.* □

5. Weight function

5.1. *Definition of the weight function*

Let $\boldsymbol{\Lambda} = (\boldsymbol{\lambda}^{(1)}, \dots, \boldsymbol{\lambda}^{(n)})$, $\boldsymbol{\lambda}^{(\infty)}$, $\boldsymbol{z} = (z_1, \dots, z_n)$ be partitions and numbers as in Section 2.2. Recall the construction of a rational map

$$\omega \; : \; \mathbb{C}^l \; \to L_{\boldsymbol{\Lambda}}[\boldsymbol{\lambda}^{(\infty)}]$$

called the weight function, see Ref. 13, cf. Ref. 4, Ref. 12.

Denote by $P(\boldsymbol{l}, n)$ the set of sequences

$$C = (c_1^1, \dots, c_{b_1}^1; \; \dots; \; c_1^n, \dots, c_{b_n}^n)$$

of integers from $\{1, \dots, N\}$ such that for every $i = 1, \dots, N$, the integer i appears in C precisely l_i times.

Denote by $\Sigma(C)$ the set of all bijections σ of the set $\{1, \dots, l\}$ onto the set of variables $\{t_1^{(1)}, \dots, t_{l_1}^{(1)}, t_1^{(2)}, \dots, t_{l_2}^{(2)}, \dots, t_1^{(N)}, \dots, t_{l_N}^{(N)}\}$ with the following property. For every $a = 1, \dots, l$ the a-th element of the sequence C equals i, if $\sigma(a) = t_j^{(i)}$.

To every $C \in P(\boldsymbol{l}, n)$ we assign a vector

$$e_C v = e_{c_1^1+1, c_1^1} \cdots e_{c_{b_1}^1+1, c_{b_1}^1} v_{\boldsymbol{\lambda}^{(1)}} \otimes \cdots \otimes e_{c_1^n+1, c_1^n} \cdots e_{c_{b_n}^n+1, c_{b_n}^n} v_{\boldsymbol{\lambda}^{(n)}},$$

$e_C v \in L_{\boldsymbol{\Lambda}}[\boldsymbol{\lambda}^{(\infty)}]$.

To every $C \in P(\boldsymbol{l}, n)$ and $\sigma \in \Sigma(C)$, we assign a rational function

$$\omega_{C,\sigma} = \omega_{\sigma;1,2,\dots,b_1}(z_1) \cdots \omega_{\sigma;b_1+\cdots+b_{n-1}+1, b_1+\cdots+b_{n-1}+2, \dots, b_1+\cdots+b_{n-1}+b_n}(z_n),$$

where

$$\omega_{\sigma;a,a+1,\dots,a+j}(z) = \frac{1}{(\sigma(a)-\sigma(a+1))\dots(\sigma(a+j-1)-\sigma(a+j))(\sigma(a+j)-z)}.$$

We set

$$\omega(\boldsymbol{t}) = \sum_{C\in P(\boldsymbol{l},n)} \sum_{\sigma\in\Sigma(C)} \omega_{C,\sigma} e_C v\,. \tag{12}$$

Examples. If $n=2$ and $(l_1,l_2,\dots,l_N)=(1,1,0,\dots,0)$, then

$$\begin{aligned}
\omega(\boldsymbol{t}) &= \frac{1}{(t_1^{(1)}-t_1^{(2)})(t_1^{(2)}-z_1)}\, e_{21}e_{32}v_{\boldsymbol{\lambda}^{(1)}}\otimes v_{\boldsymbol{\lambda}^{(2)}} \\
&+ \frac{1}{(t_1^{(2)}-t_1^{(1)})(t_1^{(1)}-z_1)}\, e_{32}e_{21}v_{\boldsymbol{\lambda}^{(1)}}\otimes v_{\boldsymbol{\lambda}^{(2)}} \\
&+ \frac{1}{(t_1^{(1)}-z_1)(t_1^{(2)}-z_2)}\, e_{21}v_{\boldsymbol{\lambda}^{(1)}}\otimes e_{32}v_{\boldsymbol{\lambda}^{(2)}} \\
&+ \frac{1}{(t_1^{(2)}-z_1)(t_1^{(1)}-z_2)}\, e_{32}v_{\boldsymbol{\lambda}^{(1)}}\otimes e_{21}v_{\boldsymbol{\lambda}^{(2)}} \\
&+ \frac{1}{(t_1^{(1)}-t_1^{(2)})(t_1^{(2)}-z_2)}\, v_{\boldsymbol{\lambda}^{(1)}}\otimes e_{21}e_{32}v_{\boldsymbol{\lambda}^{(2)}} \\
&+ \frac{1}{(t_1^{(2)}-t_1^{(1)})(t_1^{(1)}-z_2)}\, v_{\boldsymbol{\lambda}^{(1)}}\otimes e_{32}e_{21}v_{\boldsymbol{\lambda}^{(2)}}\,.
\end{aligned}$$

If $n=2$ and $(l_1,l_2,\dots,l_N)=(2,0,\dots,0)$, then

$$\begin{aligned}
\omega(\boldsymbol{t}) &= \Big(\frac{1}{(t_1^{(1)}-t_2^{(1)})(t_2^{(1)}-z_1)} + \frac{1}{(t_2^{(1)}-t_1^{(1)})(t_1^{(1)}-z_1)}\Big)\, e_{21}^2 v_{\boldsymbol{\lambda}^{(1)}}\otimes v_{\boldsymbol{\lambda}^{(2)}} \\
&+ \Big(\frac{1}{(t_1^{(1)}-z_1)(t_2^{(1)}-z_2)} + \frac{1}{(t_2^{(1)}-z_1)(t_1^{(1)}-z_2)}\Big)\, e_{21}v_{\boldsymbol{\lambda}^{(1)}}\otimes e_{21}v_{\boldsymbol{\lambda}^{(2)}} \\
&+ \Big(\frac{1}{(t_1^{(1)}-t_2^{(1)})(t_2^{(1)}-z_2)} + \frac{1}{(t_2^{(1)}-t_1^{(1)})(t_1^{(1)}-z_2)}\Big)\, v_{\boldsymbol{\lambda}^{(1)}}\otimes e_{21}^2 v_{\boldsymbol{\lambda}^{(2)}}\,.
\end{aligned}$$

Lemma 5.1 (Lemma 2.1 in Ref. 10). *The weight function is regular on U.* □

5.2. *Grothendieck residue and Hessian*

Let

$$\operatorname{Hess}\log\Phi = \det\left(\frac{\partial^2}{\partial t_j^{(i)}\partial t_{j'}^{(i')}}\log\Phi\right)$$

be the Hessian of $\log \Phi$. Let $\boldsymbol{p} \in U$ be a critical point of Φ. Denote by $H_{\boldsymbol{p}}$ the image of the Hessian in the local algebra $A_{\boldsymbol{p},\Phi}$. It is known that $H_{\boldsymbol{p}}$ is nonzero and the one-dimensional ideal $\mathbb{C}H_{\boldsymbol{p}} \subset A_{\boldsymbol{p},\Phi}$ is the annihilator of the maximal ideal $\mathfrak{m}_{\boldsymbol{p}} \subset A_{\boldsymbol{p},\Phi}$.

Let $\rho_{\boldsymbol{p}} : A_{\boldsymbol{p},\Phi} \to \mathbb{C}$, be the Grothendieck residue,

$$f \;\mapsto\; \frac{1}{(2\pi i)^l}\,\mathrm{Res}_{\boldsymbol{p}}\,\frac{f}{\prod_{ij}\Psi_{ij}}\;.$$

It is known that $\rho_{\boldsymbol{p}}(H_{\boldsymbol{p}}) = \mu_{\boldsymbol{p}}$, where $\mu_{\boldsymbol{p}} = \dim A_{\boldsymbol{p},\Phi}$ is the Milnor number of the critical point $\boldsymbol{p}$. Let $(\ ,\ )_{\boldsymbol{p}}$ be the Grothendieck residue bilinear form on $A_{\boldsymbol{p},\Phi}$,

$$(f, g)_{\boldsymbol{p}} \;=\; \rho_{\boldsymbol{p}}(fg)\;.$$

It is known that $(\ ,\ )_{\boldsymbol{p}}$ is nondegenerate. These facts see for example in Section 5 of Ref. 1.

5.3. *Projection of the weight function*

Let $\boldsymbol{p} \in C_\Phi$ be a critical point of Φ. Let

$$\omega_{\boldsymbol{p}} \in L_{\boldsymbol{\Lambda}}[\boldsymbol{\lambda}^{(\infty)}] \otimes A_{\boldsymbol{p},\Phi}$$

be the element induced by the weight function. Let $S_{\boldsymbol{\Lambda}}$ be the tensor Shapovalov form on $L_{\boldsymbol{\Lambda}}$.

Theorem 5.2 (Ref. 10, Ref. 15). *We have*

$$S_{\boldsymbol{\Lambda}}(\omega_{\boldsymbol{p}}, \omega_{\boldsymbol{p}}) = H_{\boldsymbol{p}}\;. \tag{13}$$

□

Theorem 5.3 (Ref. 13). *The element $\omega_{\boldsymbol{p}}$ belongs to* $\mathrm{Sing}\, L_{\boldsymbol{\Lambda}}[\boldsymbol{\lambda}^{(\infty)}] \otimes A_{\boldsymbol{p},\Phi}$*, where* $\mathrm{Sing}\, L_{\boldsymbol{\Lambda}}[\boldsymbol{\lambda}^{(\infty)}] \subset L_{\boldsymbol{\Lambda}}[\boldsymbol{\lambda}^{(\infty)}]$ *is the subspace of singular vectors.* □

Theorem 5.3 is a direct corollary of Theorem 6.16.2 in Ref. 13, see also Ref. 11 and Ref. 2.

5.4. *Bethe ansatz*

Let $\boldsymbol{p} \in C_\Phi$ be a critical point of Φ. Consider the differential operator

$$\mathcal{D}_\Phi = \partial_u^{N+1} + \sum_{i=1}^{N+1} G_i\,\partial_u^{N+1-i}, \quad G_i = \sum_{j=i}^{\infty} G_{ij} u^{-j}, \quad G_{ij} \in \mathbb{C}[\boldsymbol{t}],$$

described by (3), and projections $\bar{G}_{ij}$ of its coefficients to $A_{\boldsymbol{p},\Phi}$. Consider the differential operator

$$\mathcal{D}_{\mathcal{B}} = \partial_u^{N+1} + \sum_{i=1}^{N+1} B_i\, \partial_u^{N+1-i}, \quad B_i = \sum_{j=i}^{\infty} B_{ij} u^{-j}, \quad B_{ij} \in U(\mathfrak{gl}_{N+1}[t]) ,$$

described by (10).

Theorem 5.4 (Ref. 5). *For any $i = 1, \ldots, N+1$, $j \geqslant i$, we have*

$$(B_{ij} \otimes 1)\, \omega_{\boldsymbol{p}} = (1 \otimes \bar{G}_{ij})\, \omega_{\boldsymbol{p}} \tag{14}$$

in $\operatorname{Sing} L_{\boldsymbol{\Lambda}}[\boldsymbol{\lambda}^{(\infty)}] \otimes A_{\boldsymbol{p},\Phi}$. □

This statement is the Bethe ansatz method to construct eigenvectors of the Bethe algebra in the $\mathfrak{gl}_{N+1}$ Gaudin model starting with a critical point of the master function.

5.5. *Main result*

Let $g_1, \ldots, g_{\mu_{\boldsymbol{p}}}$ be a basis of $A_{\boldsymbol{p},\Phi}$ considered as a $\mathbb{C}$-vector space. Write $\omega_{\boldsymbol{p}} = \sum_i v_i \otimes g_i$, with $v_i \in \operatorname{Sing} L_{\boldsymbol{\Lambda}}[\boldsymbol{\lambda}^{(\infty)}]$. Denote by $M_{\boldsymbol{p}} \subset \operatorname{Sing} L_{\boldsymbol{\Lambda}}[\boldsymbol{\lambda}^{(\infty)}]$ the vector subspace spanned by $v_1, \ldots, v_{\mu_{\boldsymbol{p}}}$. Define a linear map

$$\alpha : A_{\boldsymbol{p},\Phi} \to M_{\boldsymbol{p}} , \qquad f \mapsto (f, \omega_{\boldsymbol{p}})_{\boldsymbol{p}} = \sum_{i=1}^{\mu_{\boldsymbol{p}}} (f, g_i)_{\boldsymbol{p}}\, v_i \, . \tag{15}$$

Theorem 5.5. *Let $\boldsymbol{p} \in C_\Phi$. Then the following statements hold:*

(i) The subspace $M_{\boldsymbol{p}} \subset \operatorname{Sing} L_{\boldsymbol{\Lambda}}[\boldsymbol{\lambda}^{(\infty)}]$ is a $\mathcal{B}$-submodule. Let $A_{\boldsymbol{p},B} \subset \operatorname{End}(M_{\boldsymbol{p}})$ be the Bethe algebra of $M_{\boldsymbol{p}}$. Denote by $\bar{B}_{ij}$ the image in $A_{\boldsymbol{p},B}$ of generators $B_{ij} \in \mathcal{B}$.
(ii) The map $\alpha : A_{\boldsymbol{p},\Phi} \to M_{\boldsymbol{p}}$ is an isomorphism of vector spaces.
(iii) The map $\bar{G}_{ij} \mapsto \bar{B}_{ij}$ extends uniquely to an algebra isomorphism $\beta : A_{\boldsymbol{p},\Phi} \to A_{\boldsymbol{p},B}$.
(iv) The isomorphisms α and β identify the regular representation of $A_{\boldsymbol{p},\Phi}$ and the $\mathcal{B}$-module $M_{\boldsymbol{p}}$, that is, for any $f, g \in A_{\boldsymbol{p},\Phi}$ we have $\alpha(fg) = \beta(f)\alpha(g)$.

Corollary 5.6. *Let $\boldsymbol{p} \in C_\Phi$. Then the value $\omega(\boldsymbol{p})$ of the weight function at $\boldsymbol{p}$ is a nonzero vector of* $\operatorname{Sing} L_{\boldsymbol{\Lambda}}[\boldsymbol{\lambda}^{(\infty)}]$.

Theorem 5.5 and Corollary 5.6 are proved in Section 6.4.

6. Proof of Theorem 5.5

6.1. *Proof of part (i) of Theorem 5.5*

It is sufficient to show that for any $f \in A_{\boldsymbol{p},\Phi}$ and any (i,j) we have $B_{ij}\alpha(f) \in M_{\boldsymbol{p}}$. Indeed, we have

$$\begin{aligned} B_{ij}\alpha(f) = \sum_{l=1}^{\mu_{\boldsymbol{p}}} (f, g_l)_{\boldsymbol{p}} B_{ij} v_l &= \sum_{l=1}^{\mu_{\boldsymbol{p}}} (f, \bar{G}_{ij} g_l)_{\boldsymbol{p}} \bar{v}_l \\ &= \sum_{l=1}^{\mu_{\boldsymbol{p}}} (\bar{G}_{ij} f, g_l)_{\boldsymbol{p}} \bar{v}_l = \alpha(\bar{G}_{ij} f). \qquad (16) \end{aligned}$$

Here the second equality follows from Theorem 5.4 and the third equality follows from properties of the Grothendieck residue form.

6.2. *Bilinear form* $(\, , \,)_S$

Define a symmetric bilinear form $(\, , \,)_S$ on $A_{\boldsymbol{p},\Phi}$,

$$(f, g)_S = S_{\boldsymbol{\Lambda}}(\alpha(f), \alpha(g)) = \sum_{i,j=1}^{\mu_{\boldsymbol{p}}} S_{\boldsymbol{\Lambda}}(v_i, v_j)\, (f, g_i)_{\boldsymbol{p}}\, (g, g_j)_{\boldsymbol{p}}$$

for all $f, g \in A_{\boldsymbol{p},\Phi}$.

Lemma 6.1. *For all* $f, g, h \in A_{\boldsymbol{p},\Phi}$ *we have* $(fg, h)_S = (f, gh)_S$.

Proof. By Theorem 2.1 the elements $\bar{G}_{ij}$ generate $A_{\boldsymbol{p},\Phi}$. We have

$$\begin{aligned} (\bar{G}_{ij} f, h)_S &= S_{\boldsymbol{\Lambda}}(\alpha(\bar{G}_{ij} f), \alpha(h)) = S_{\boldsymbol{\Lambda}}(\bar{B}_{ij}\alpha(f), \alpha(h)) \\ &= S_{\boldsymbol{\Lambda}}(\alpha(f), \bar{B}_{ij}\alpha(h)) = S_{\boldsymbol{\Lambda}}(\alpha(f), \alpha(\bar{G}_{ij} h)) = (f, \bar{G}_{ij} h)_S. \end{aligned}$$

Here the third equality follows from Theorem 4.1. □

Lemma 6.2. *There exists* $F \in A_{\boldsymbol{p},\Phi}$ *such that* $(f, h)_S = (Ff, h)_{\boldsymbol{p}}$ *for all* $f, h \in A_{\boldsymbol{p},\Phi}$.

Proof. Consider the linear function $A_{\boldsymbol{p},\Phi} \to \mathbb{C}$, $h \mapsto (1, h)_S$. The form $(\, , \,)_{\boldsymbol{p}}$ is nondegenerate. Hence there exits $F \in A_{\boldsymbol{p},\Phi}$ such that $(1, h)_S = (F, h)_{\boldsymbol{p}}$ for all $h \in A_{\boldsymbol{p},\Phi}$. Now the lemma follows from Lemma 6.1. □

6.3. *Auxiliary lemmas*

Lemma 6.3. *For any $f \in A_{\boldsymbol{p},\Phi}$, we have*

$$fH_{\boldsymbol{p}} = \frac{1}{\mu_{\boldsymbol{p}}}(f, H_{\boldsymbol{p}})_{\boldsymbol{p}} H_{\boldsymbol{p}} \,. \tag{17}$$

Proof. The lemma follows from the fact that formula (17) evidently holds for $1 \in A_{\boldsymbol{p},\Phi}$ and for any element of the maximal ideal. □

For $f \in A_{\boldsymbol{p},\Phi}$, denote by L_f the linear operator $A_{\boldsymbol{p},\Phi} \to A_{\boldsymbol{p},\Phi}$, $h \mapsto fh$.

Lemma 6.4. *We have* $\operatorname{tr} L_f = (f, H_{\boldsymbol{p}})_{\boldsymbol{p}}$.

Proof. The linear function $A_{\boldsymbol{p},\Phi} \to \mathbb{C}$, $f \mapsto \operatorname{tr} L_f$, is such that $1 \mapsto \mu_{\boldsymbol{p}}$ and $f \mapsto 0$ for all $f \in \mathfrak{m}_{\boldsymbol{p}}$. Hence this function equals the linear function $f \mapsto (f,\, H_{\boldsymbol{p}})_{\boldsymbol{p}}$. □

Let $g_1^*, \dots, g_{\mu_{\boldsymbol{p}}}^*$ be the basis of $A_{\boldsymbol{p},\Phi}$ dual to the basis $g_1, \dots, g_{\mu_{\boldsymbol{p}}}$ with respect to the form $(\,,\,)_{\boldsymbol{p}}$. Then $H_{\boldsymbol{p}} = \sum_{i=1}^{\mu_{\boldsymbol{p}}} (H_{\boldsymbol{p}}, g_i^*)_{\boldsymbol{p}} g_i$. Indeed for any $f \in A_{\boldsymbol{p},\Phi}$, we have $f = \sum_i (f, g_i^*)_{\boldsymbol{p}} g_i$.

Lemma 6.5. *We have* $\sum_{i=1}^{\mu_{\boldsymbol{p}}} g_i^* g_i = H_{\boldsymbol{p}}$.

Proof. For $f \in A_{\boldsymbol{p},\Phi}$, we have $\operatorname{tr} L_f = \sum_i (g_i^*, f g_i)_{\boldsymbol{p}} = (\sum_i g_i^* g_i, f)_{\boldsymbol{p}}$. By Lemma 6.4, we get $(\sum_i g_i^* g_i, f)_{\boldsymbol{p}} = (H_{\boldsymbol{p}}, f)_{\boldsymbol{p}}$. Hence $\sum_i g_i^* g_i = H_{\boldsymbol{p}}$, since the form $(\,,\,)_{\boldsymbol{p}}$ is nondegenerate. □

Lemma 6.6. *Let $F \in A_{\boldsymbol{p},\Phi}$ be the element defined in Lemma 6.2. Then F is invertible, $FH_{\boldsymbol{p}} = H_{\boldsymbol{p}}$, and the form $(\,,\,)_S$ is nondegenerate.*

Proof. By definitions we have

$$(f, h)_S = \sum_{ij} S_{\boldsymbol{\Lambda}}(v_i, v_j)(g_i, f)_{\boldsymbol{p}} (g_j, h)_{\boldsymbol{p}}$$

and

$$(f, h)_S = (Ff, h)_{\boldsymbol{p}} = \sum_i (g_i, Ff)_{\boldsymbol{p}} (g_i^*, h)_{\boldsymbol{p}} = \sum_i (Fg_i,\, f)_{\boldsymbol{p}} (g_i^*, h)_{\boldsymbol{p}} \,.$$

Hence $\sum_{ij} S_{\boldsymbol{\Lambda}}(v_i, v_j) g_i \otimes g_j = \sum_i F g_i \otimes g_i^*$ and therefore by Lemma 6.5 we get

$$\sum_{ij} S_{\boldsymbol{\Lambda}}(v_i, v_j) g_i g_j = \sum_i F g_i g_i^* = FH_{\boldsymbol{p}}.$$

By Theorem 5.2, $\sum_{ij} S_{\boldsymbol{\Lambda}}(v_i, v_j) g_i g_j = H_{\boldsymbol{p}}$. Hence $FH_{\boldsymbol{p}} = H_{\boldsymbol{p}}$, the element F is invertible, and the form $(\,,\,)_S$ is nondegenerate. □

6.4. *Proof of Theorem 5.5 and Corollary 5.6*

Part (i) of Theorem 5.5 is proved in Section 6.1.

Assume that $\sum_{i=1}^{\mu_{\boldsymbol{p}}} \lambda_i v_i = 0$. Denote $h = \sum_i \lambda_i g_i^*$. Then $\alpha(h) = 0$ and $(f, h)_S = S_{\boldsymbol{\Lambda}}(\alpha(f), \alpha(h)) = 0$ for all $f \in A_{\boldsymbol{p},\Phi}$. Hence $h = 0$ since $(\ ,\ )_S$ is nondegenerate. Therefore, $\lambda_i = 0$ for all i and the vectors $v_1, \dots, v_{\mu_{\boldsymbol{p}}}$ are linearly independent. We have $\alpha(g_i^*) = v_i$ for all i. That proves part (ii) of Theorem 5.5.

Parts (iii-iv) easily follow from part (ii) and formula (16).

We have

$$\mu_{\boldsymbol{p}}\, \omega(\boldsymbol{p}) = (H_{\boldsymbol{p}}, \omega_{\boldsymbol{p}})_{\boldsymbol{p}} \ = \alpha(H_{\boldsymbol{p}}). \tag{18}$$

That implies that $\omega(\boldsymbol{p})$ is a nonzero vector.

7. Concluding remarks

Theorem 7.1. *Let $C_\Phi = \{\boldsymbol{p}_1, \dots, \boldsymbol{p}_k\}$, be the critical set of Φ in U. Let $M_{\boldsymbol{p}_s} \subset \operatorname{Sing} L_{\boldsymbol{\Lambda}}[\boldsymbol{\lambda}^{(\infty)}]$, $s = 1, \dots, k$, be the corresponding subspaces defined in Section 5.5. Then the sum of these subspaces is direct.*

Proof. It follows from Theorem 5.5 that for any s and any (i, j) the operator $\bar{B}_{ij} - G_{ij}(\boldsymbol{p}_s)$ restricted to $M_{\boldsymbol{p}_s}$ is nilpotent. Moreover, the differential operators $\mathcal{D}_\Phi|_{\boldsymbol{t}=\boldsymbol{p}_s}$, $s = 1, \dots, k$, which contain eigenvalues of the operators B_{ij}, are distinct. These observations imply Theorem 7.1. □

Let $\alpha(A_\Phi) = \oplus_{s=1}^k M_{p_s}$. Denote by A_B the image of $\mathcal{B}$ in $\operatorname{End}(\alpha(A_\Phi))$. Consider the isomorphisms

$$\begin{aligned} \alpha = \oplus_{s=1}^k \alpha_s \ &: \ \oplus_{s=1}^k A_{\boldsymbol{p}_s,\Phi} \ \to \ \oplus_{s=1}^k M_{\boldsymbol{p}_s}\,, \\ \beta = \oplus_{s=1}^k \beta_s \ &: \ \oplus_{s=1}^k A_{\boldsymbol{p}_s,\Phi} \ \to \ \oplus_{s=1}^k A_{\boldsymbol{p}_s,B} \end{aligned}$$

of Theorem 5.5.

Corollary 7.2. *We have*

(i) $A_B = \oplus_{s=1}^k A_{p_s,B}$*;*
(ii) The isomorphisms α, β identify the regular representation of the algebra A_Φ and the A_B-module $\alpha(A_\Phi)$.

The corollary follows from Theorems 5.5 and 7.1.

References

1. V.I. Arnol'd, S.M. Gusein-Zade, A.N. Varchenko, *Singularities of differentiable maps, Vol. II. Monodromy and asymptotics of integrals*, Monographs in Mathematics, **83**, Birkhäuser, 1988
2. H. Babujian, *Off-shell Bethe ansatz equations and N-point correlators in the* SU(2) *WZNW theory*, J. Phys. A **26** (1993), no. 23, 6981–6990
3. P. Belkale, E. Mukhin, A. Varchenko, *Multiplicity of critical points of master functions and Schubert calculus*, Differential equations and quantum groups, 59–84, IRMA Lect. Math. Theor. Phys. **9**, Eur. Math. Soc., Zürich, 2007
4. A. Matsuo, *An application of Aomoto-Gelfand hypergeometric functions to the* SU(n) *Knizhnik-Zamolodchikov equation*, Comm. Math. Phys. **134** (1990), no. 1, 65–77
5. E. Mukhin, V. Tarasov, A. Varchenko, *Bethe Eigenvectors of Higher Transfer Matrices*, J. Stat. Mech. Theory Exp. (2006), no. 8, P08002, 44 pp. (electronic)
6. E. Mukhin, V. Tarasov, A. Varchenko, Schubert calculus and representations of the general linear group, J. Amer. Math. Soc. **22** (2009), no. 4, 909–940
7. E. Mukhin, V. Tarasov, A. Varchenko, *Bethe algebra of homogeneous XXX Heisenberg model has simple spectrum*, Comm. Math. Phys. **288** (2009), no. 1, 1–42
8. E. Mukhin, V. Tarasov, A. Varchenko, *Three sides of the geometric Langlands correspondence for* $\mathfrak{gl}_N$ *Gaudin model and Bethe vector averaging maps*, Preprint (2009), 1–26; `arXiv:0907.3266`
9. E. Mukhin, A. Varchenko, *Critical points of master functions and flag varieties*, Commun. Contemp. Math. **6** (2004), no. 1, 111–163
10. E. Mukhin, A. Varchenko, *Norm of a Bethe Vector and the Hessian of the Master Function*, Compos. Math. **141** (2005), no. 4, 1012–1028
11. N. Reshetikhin and A. Varchenko, *Quasiclassical asymptotics of solutions to the KZ equations*, Geometry, Topology and Physics for R. Bott, Intern. Press, 1995, 293–322.
12. R. Rimanyi, L. Stevens, A. Varchenko, *Combinatorics of rational functions and Poincaré-Birkhoff Witt expansions of the canonical*

$U(\mathfrak{n}_-)$-valued differential form, Ann. Comb. **9** (2005), no. 1, 57–74

13. V. Schechtman and A. Varchenko, *Arrangements of hyperplanes and Lie algebra homology*, Invent. Math. **106** (1991), 139–194
14. D. Talalaev, *Quantization of the Gaudin System*, Preprint (2004), 1–19;
`hep-th/0404153`
15. A. Varchenko, *Bethe Ansatz for Arrangements of Hyperplanes and the Gaudin Model*, Mosc. Math. J. **6** (2006), no. 1, 195–210, 223–224

T-SYSTEMS, Y-SYSTEMS, AND CLUSTER ALGEBRAS: TAMELY LACED CASE

TOMOKI NAKANISHI

Graduate School of Mathematics, Nagoya University, Nagoya, 464-8604, Japan
E-mail: nakanisi@math.nagoya-u.ac.jp

Dedicated to Professor Tetsuji Miwa on his 60th birthday

The T-systems and Y-systems are classes of algebraic relations originally associated with quantum affine algebras and Yangians. Recently they were generalized to quantum affinizations of quantum Kac-Moody algebras associated with a wide class of generalized Cartan matrices which we say tamely laced. Furthermore, in the simply laced case, and also in the nonsimply laced case of finite type, they were identified with relations arising from cluster algebras. In this note we generalize such an identification to any tamely laced Cartan matrices, especially to the nonsimply laced ones of nonfinite type.

Keywords: T-systems; Y-systems; quantum groups; cluster algebras.

1. Introduction

The T-systems and Y-systems appear in various aspects for integrable systems. Originally, the T-systems are systems of relations among the Kirillov-Reshetikhin modules in the Grothendieck rings of modules over quantum affine algebras and Yangians. The T and Y-systems are related to each other by certain changes of variables. See, for example, Ref. 1 and references therein for more information and background.

Let $I = \{1, \dots, r\}$ and let $C = (C_{ij})_{i,j \in I}$ be a *(generalized) Cartan matrix* in Ref. 2; namely, it satisfies $C_{ij} \in \mathbb{Z}$, $C_{ii} = 2$, $C_{ij} \leq 0$ for any $i \neq j$, and $C_{ij} = 0$ if and only if $C_{ji} = 0$. We assume that C is *symmetrizable*, i.e., there is a diagonal matrix $D = \mathrm{diag}(d_1, \dots, d_r)$ with $d_i \in \mathbb{N} := \mathbb{Z}_{>0}$ such that $B = DC$ is symmetric. We always assume that there is no common divisor for $d_1, \dots, d_r$ except for 1. Following Ref. 4, we say that a Cartan matrix C is *tamely laced* if it is symmetrizable and satisfies the following condition due to Hernandez:[3]

$$\text{If } C_{ij} < -1, \text{ then } d_i = -C_{ji} = 1. \tag{1}$$

Recently, the T-systems were generalized by Hernandez[3] to the quantum affinizations of the quantum Kac-Moody algebras associated with *tamely laced* Cartan matrices. Subsequently, the corresponding Y-systems were also introduced by Kuniba, Suzuki, and the author.[4]

Remarkably, these T and Y-systems are identified with (a part of) relations among the variables for *cluster algebras*,[5,6] which are a class of commutative algebras closely related to the representation theory of quivers. For the T and Y-systems associated with *simply laced* Cartan matrices of *finite type*, this identification is a topic intensively studied by various authors recently with several reasons (periodicity, categorification, positivity, dilogarithm identities, etc.)[1,6–16] In Ref. 4, such an identification was generalized to the *simply laced* Cartan matrices. In Refs. 15 and 16, it was also extended to the *nonsimply laced* Cartan matrices of *finite type*.

In this note we present a generalization of the above identification to *any tamely laced* Cartan matrices, especially to the *nonsimply laced ones of nonfinite type*, thereby justifying Sec. 6.5 of Ref. 4 which announced that such a generalization is possible. Basically it is a straightforward extension of the simply laced ones[4] and the nonsimply laced ones of finite type,[15,16] but it is necessarily more complicated. At this time we do not have any immediate application of such a generalization. However, we believe that this is a necessary step toward further study of the intriguing interplay of two worlds — the representation theories of quantum groups and quivers — through cluster algebras.

2. T and Y-systems

In this section we recall the definitions of (restricted) T and Y-systems. See Ref. 4 for more detail.

With a tamely laced Cartan matrix C, we associate a *Dynkin diagram* $X(C)$ in the standard way: For any pair $i \neq j \in I$ with $C_{ij} < 0$, the vertices i and j are connected by $\max\{|C_{ij}|, |C_{ji}|\}$ lines, and the lines are equipped with an arrow from j to i if $C_{ij} < -1$. Note that the condition (1) means

(i) the vertices i and j are not connected, if $d_i, d_j > 1$ and $d_i \neq d_j$,
(ii) the vertices i and j are connected by d_i lines with an arrow from i to j or not connected, if $d_i > 1$ and $d_j = 1$,
(iii) the vertices i and j are connected by a single line or not connected, if $d_i = d_j$.

As usual, we say that a Cartan matrix C is *simply laced* if $C_{ij} = 0$ or -1 for any $i \neq j$. If C is simply laced, then it is tamely laced.

For a tamely laced Cartan matrix C, we set integers t and t_a $(a \in I)$ by

$$t = \mathrm{lcm}(d_1, \dots, d_r), \quad t_a = \frac{t}{d_a}. \tag{2}$$

For an integer $\ell \geq 2$, we set

$$\mathcal{I}_\ell := \{(a, m, u) \mid a \in I; m = 1, \dots, t_a\ell - 1; u \in \frac{1}{t}\mathbb{Z}\}. \tag{3}$$

For $a, b \in I$, we write $a \sim b$ if $C_{ab} < 0$, i.e., a and b are adjacent in $X(C)$.

First, we introduce the T-systems and the associated rings.

Definition 2.1. Fix an integer $\ell \geq 2$. For a tamely laced Cartan matrix C, the *level ℓ restricted T-system $\mathbb{T}_\ell(C)$ associated with C (with the unit boundary condition)* is the following system of relations for a family of variables $T_\ell = \{T_m^{(a)}(u) \mid (a, m, u) \in \mathcal{I}_\ell\}$,

$$T_m^{(a)}\left(u - \tfrac{d_a}{t}\right) T_m^{(a)}\left(u + \tfrac{d_a}{t}\right) = T_{m-1}^{(a)}(u) T_{m+1}^{(a)}(u) + \prod_{b:b\sim a} T_{\frac{d_a}{d_b}m}^{(b)}(u) \quad \text{if } d_a > 1, \tag{4}$$

$$T_m^{(a)}\left(u - \tfrac{d_a}{t}\right) T_m^{(a)}\left(u + \tfrac{d_a}{t}\right) = T_{m-1}^{(a)}(u) T_{m+1}^{(a)}(u) + \prod_{b:b\sim a} S_m^{(b)}(u) \quad \text{if } d_a = 1, \tag{5}$$

where $T_0^{(a)}(u) = 1$, and furthermore, $T_{t_a\ell}^{(a)}(u) = 1$ (the *unit boundary condition*) if they occur in the right hand sides in the relations. The symbol $S_m^{(b)}(u)$ is defined as follows. For $m = 0, 1, 2, \dots$ and $0 \leq j < d_b$,

$$\begin{aligned} S_{d_b m+j}^{(b)}(u) = &\left\{ \prod_{k=1}^{j} T_{m+1}^{(b)}\left(u + \frac{1}{t}(j + 1 - 2k)\right) \right\} \\ &\times \left\{ \prod_{k=1}^{d_b - j} T_m^{(b)}\left(u + \frac{1}{t}(d_b - j + 1 - 2k)\right) \right\}. \end{aligned} \tag{6}$$

For the later use, let us formally write (4) and (5) in a unified manner

$$\begin{aligned} T_m^{(a)}\left(u - \tfrac{d_a}{t}\right) T_m^{(a)}\left(u + \tfrac{d_a}{t}\right) = &\, T_{m-1}^{(a)}(u) T_{m+1}^{(a)}(u) \\ &+ \prod_{(b,k,v)\in\mathcal{I}_\ell} T_k^{(b)}(v)^{G(b,k,v;a,m,u)}. \end{aligned} \tag{7}$$

Definition 2.2. Let $\mathcal{T}_\ell(C)$ be the commutative ring over $\mathbb{Z}$ with identity element, with generators $T_m^{(a)}(u)^{\pm 1}$ $((a, m, u) \in \mathcal{I}_\ell)$ and relations $\mathbb{T}_\ell(C)$

together with $T^{(a)}_m(u)T^{(a)}_m(u)^{-1} = 1$. Let $\mathcal{T}^\circ_\ell(C)$ be the subring of $\mathcal{T}_\ell(C)$ generated by $T^{(a)}_m(u)$ $((a,m,u) \in \mathcal{I}_\ell)$.

Similarly, we introduce the Y-systems and the associated groups.

Definition 2.3. Fix an integer $\ell \geq 2$. For a tamely laced Cartan matrix C, the *level ℓ restricted Y-system $\mathbb{Y}_\ell(C)$ associated with C* is the following system of relations for a family of variables $Y = \{Y^{(a)}_m(u) \mid (a,m,u) \in \mathcal{I}_\ell\}$,

$$Y^{(a)}_m\left(u - \tfrac{d_a}{t}\right) Y^{(a)}_m\left(u + \tfrac{d_a}{t}\right) = \frac{\prod_{b:b\sim a} Z^{(b)}_{\frac{d_a}{d_b},m}(u)}{(1+Y^{(a)}_{m-1}(u)^{-1})(1+Y^{(a)}_{m+1}(u)^{-1})} \quad \text{if } d_a > 1, \tag{8}$$

$$Y^{(a)}_m\left(u - \tfrac{d_a}{t}\right) Y^{(a)}_m\left(u + \tfrac{d_a}{t}\right) = \frac{\prod_{b:b\sim a} \left(1 + Y^{(b)}_{\frac{m}{d_b}}(u)\right)}{(1+Y^{(a)}_{m-1}(u)^{-1})(1+Y^{(a)}_{m+1}(u)^{-1})} \quad \text{if } d_a = 1, \tag{9}$$

where $Y^{(a)}_0(u)^{-1} = Y^{(a)}_{t_a\ell}(u)^{-1} = 0$ if they occur in the right hand sides in the relations. Besides, $Y^{(b)}_{m/d_b}(u) = 0$ in (9) if $m/d_b \notin \mathbb{N}$. The symbol $Z^{(b)}_{p,m}(u)$ $(p \in \mathbb{N})$ is defined as follows.

$$Z^{(b)}_{p,m}(u) = \prod_{j=-p+1}^{p-1} \left\{ \prod_{k=1}^{p-|j|} \left(1 + Y^{(b)}_{pm+j}\left(u + \tfrac{1}{t}(p - |j| + 1 - 2k)\right)\right) \right\}. \tag{10}$$

One can write (8) and (9) in a unified manner as

$$Y^{(a)}_m\left(u - \tfrac{d_a}{t}\right) Y^{(a)}_m\left(u + \tfrac{d_a}{t}\right) = \frac{\prod_{(b,k,v)\in\mathcal{I}_\ell} (1 + Y^{(b)}_k(v))^{{}^tG(b,k,v;a,m,u)}}{(1+Y^{(a)}_{m-1}(u)^{-1})(1+Y^{(a)}_{m+1}(u)^{-1})}, \tag{11}$$

where ${}^tG(b,k,v;a,m,u) := G(a,m,u;b,k,v)$.

A *semifield* $(\mathbb{P}, \oplus)$ is an abelian multiplicative group $\mathbb{P}$ endowed with a binary operation of addition $\oplus$ which is commutative, associative, and distributive with respect to the multiplication in $\mathbb{P}$.

Definition 2.4. Let $\mathcal{Y}_\ell(C)$ be the semifield with generators $Y^{(a)}_m(u)$ $((a,m,u) \in \mathcal{I}_\ell)$ and relations $\mathbb{Y}_\ell(C)$. Let $\mathcal{Y}^\circ_\ell(C)$ be the multiplicative subgroup of $\mathcal{Y}_\ell(C)$ generated by $Y^{(a)}_m(u)$, $1+Y^{(a)}_m(u)$ $((a,m,u) \in \mathcal{I}_\ell)$. (Here we use the symbol $+$ instead of $\oplus$ for simplicity.)

3. Cluster algebra with coefficients

In this section we recall the definition of cluster algebras with coefficients following Ref. 6. The description here is minimal to fix convention and notion. See Ref. 6 for more detail and information.

Let I be a finite set, and let $B = (B_{ij})_{i,j\in I}$ be a skew symmetric (integer) matrix. Let $x = (x_i)_{i\in I}$ and $y = (y_i)_{i\in I}$ be I-tuples of formal variables. Let $\mathbb{P} = \mathbb{Q}_{\mathrm{sf}}(y)$ be the *universal semifield* of $y = (y_i)_{i\in I}$, namely, the semifield consisting of the *subtraction-free* rational functions of y with usual multiplication and addition (but no subtraction) in the rational function field $\mathbb{Q}(y)$. Let $\mathbb{QP}$ denote the quotient field of the group ring $\mathbb{ZP}$ of $\mathbb{P}$.

For the above triplet (B, x, y), called the *initial seed*, the *cluster algebra* $\mathcal{A}(B, x, y)$ *with coefficients in* $\mathbb{P}$ is defined as follows.

Let (B', x', y') be a triplet consisting of skew symmetric matrix B', an I-tuple $x' = (x'_i)_{i\in I}$ with $x'_i \in \mathbb{QP}(x)$, and an I-tuple $y' = (y'_i)_{i\in I}$ with $y'_i \in \mathbb{P}$. For each $k \in I$, we define another triplet $(B'', x'', y'') = \mu_k(B', x', y')$, called the *mutation of* (B', x', y') *at* k, as follows.

(i) Mutations of matrix.

$$B''_{ij} = \begin{cases} -B'_{ij} & i = k \text{ or } j = k, \\ B'_{ij} + \frac{1}{2}(|B'_{ik}|B'_{kj} + B'_{ik}|B'_{kj}|) & \text{otherwise.} \end{cases} \tag{12}$$

(ii) Exchange relation of coefficient tuple.

$$y''_i = \begin{cases} {y'_k}^{-1} & i = k, \\ y'_i \left(\dfrac{y'_k}{1 \oplus y'_k}\right)^{B'_{ki}} & i \neq k,\ B'_{ki} \geq 0, \\ y'_i(1 \oplus y'_k)^{-B'_{ki}} & i \neq k,\ B'_{ki} \leq 0. \end{cases} \tag{13}$$

(iii) Exchange relation of cluster.

$$x''_i = \begin{cases} \dfrac{y'_k \prod_{j:B'_{jk}>0} {x'_j}^{B'_{jk}} + \prod_{j:B'_{jk}<0} {x'_j}^{-B'_{jk}}}{(1 \oplus y'_k)x'_k} & i = k, \\ x'_i & i \neq k. \end{cases} \tag{14}$$

It is easy to see that μ_k is an involution, namely, $\mu_k(B'', x'', y'') = (B', x', y')$. Now, starting from the initial seed (B, x, y), iterate mutations and collect all the resulted triplets (B', x', y'). We call (B', x', y') the *seeds*, y' and y'_i a *coefficient tuple* and a *coefficient*, x' and x'_i, a *cluster* and a *cluster variable*, respectively. The *cluster algebra* $\mathcal{A}(B, x, y)$ *with coefficients in* $\mathbb{P}$ is the $\mathbb{ZP}$-subalgebra of the rational function field $\mathbb{QP}(x)$ generated by all the cluster variables. Similarly, the *coefficient group* $\mathcal{G}(B, y)$ *associated with* $\mathcal{A}(B,x,y)$ is the multiplicative subgroup of the semifield $\mathbb{P}$ generated by all the coefficients y'_i together with $1 \oplus y'_i$.

It is standard to identify a skew-symmetric (integer) matrix $B = (B_{ij})_{i,j\in I}$ with a *quiver* Q *without loops or 2-cycles*. The set of the vertices of Q is given by I, and we put B_{ij} arrows from i to j if $B_{ij} > 0$. The mutation $Q'' = \mu_k(Q')$ of a quiver Q' is given by the following rule: For each pair of an incoming arrow $i \to k$ and an outgoing arrow $k \to j$ in Q', add a new arrow $i \to j$. Then, remove a maximal set of pairwise disjoint 2-cycles. Finally, reverse all arrows incident with k.

4. Cluster algebraic formulation: The case $|I| = 2$; t is odd

4.1. *Cartan matrix* M_t

We are going to identify $\mathbb{T}_\ell(C)$ and $\mathbb{Y}_\ell(C)$ as relations for cluster variables and coefficients of the cluster algebra associated with a certain quiver $Q_\ell(C)$.

To begin with, we consider the case $I = \{1,2\}$, which will be used as building blocks of the general case. Without loss of generality we may assume that a Cartan matrix C is *indecomposable*, i.e., $X(C)$ is connected. Thus, we assume that our tamely laced Cartan matrix C has the form $(t = 1, 2, \dots)$

$$C = M_t := \begin{pmatrix} 2 & -1 \\ -t & 2 \end{pmatrix}, \quad D = \begin{pmatrix} t & 0 \\ 0 & 1 \end{pmatrix}. \tag{15}$$

We have the data $d_1 = t$, $d_2 = 1$, $t = \mathrm{lcm}(d_1, d_2)$, $t_1 = 1$, $t_2 = t$, and the corresponding Dynkin diagram looks as follows (t lines in the middle and there is no arrow for $t = 1$):

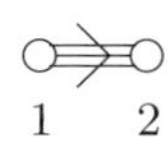

We ask the reader to refer to Refs. 4, 15, and 16, where the cases $t = 1$ (type A_2), 2 (type B_2), and 3 (type G_2), respectively, are treated in detail.

It turns out that we should separate the problem depending on the parity of t. In this section we consider the case when t is odd.

4.2. *Parity decompositions of T and Y-systems*

For a triplet $(a, m, u) \in \mathcal{I}_\ell$, we set the parity conditions $\mathbf{P}_+$ and $\mathbf{P}_-$ by

$$\begin{aligned} &\mathbf{P}_+ : m + tu \text{ is odd for } a = 1;\ m + tu \text{ is even for } a = 2, \\ &\mathbf{P}_- : m + tu \text{ is even for } a = 1;\ m + tu \text{ is odd for } a = 2. \end{aligned} \tag{16}$$

We write, for example, $(a,m,u) : \mathbf{P}_+$ if (a,m,u) satisfies $\mathbf{P}_+$. We have $\mathcal{I}_\ell = \mathcal{I}_{\ell+} \sqcup \mathcal{I}_{\ell-}$, where $\mathcal{I}_{\ell\varepsilon}$ is the set of all $(a,m,u) : \mathbf{P}_\varepsilon$. Define $\mathcal{T}^\circ_\ell(M_t)_\varepsilon$ ($\varepsilon = \pm$) to be the subring of $\mathcal{T}^\circ_\ell(M_t)$ generated by $T^{(a)}_m(u)$ ($(a,m,u) \in \mathcal{I}_{\ell\varepsilon}$). Then, we have $\mathcal{T}^\circ_\ell(M_t)_+ \simeq \mathcal{T}^\circ_\ell(M_t)_-$ by $T^{(a)}_m(u) \mapsto T^{(a)}_m(u + \frac{1}{t})$ and

$$\mathcal{T}^\circ_\ell(M_t) \simeq \mathcal{T}^\circ_\ell(M_t)_+ \otimes_{\mathbb{Z}} \mathcal{T}^\circ_\ell(M_t)_-. \tag{17}$$

For a triplet $(a,m,u) \in \mathcal{I}_\ell$, we introduce another parity conditions $\mathbf{P}'_+$ and $\mathbf{P}'_-$ by

$$\begin{aligned} &\mathbf{P}'_+ : m + tu \text{ is even for } a = 1;\ m + tu \text{ is odd for } a = 2,\\ &\mathbf{P}'_- : m + tu \text{ is odd for } a = 1;\ m + tu \text{ is even for } a = 2. \end{aligned} \tag{18}$$

Since $\mathbf{P}'_\pm = \mathbf{P}_\mp$, it may seem redundant, but we use this notation to make the description unified for both odd and even t. We have

$$(a,m,u) : \mathbf{P}'_+ \iff (a,m,u \pm \tfrac{d_a}{t}) : \mathbf{P}_+. \tag{19}$$

Let $\mathcal{I}'_{\ell\varepsilon}$ be the set of all $(a,m,u) : \mathbf{P}'_\varepsilon$. Define $\mathcal{Y}^\circ_\ell(M_t)_\varepsilon$ ($\varepsilon = \pm$) to be the subgroup of $\mathcal{Y}^\circ_\ell(M_t)$ generated by $Y^{(a)}_m(u)$, $1 + Y^{(a)}_m(u)$ ($(a,m,u) \in \mathcal{I}'_{\ell\varepsilon}$). Then, we have $\mathcal{Y}^\circ_\ell(M_t)_+ \simeq \mathcal{Y}^\circ_\ell(M_t)_-$ by $Y^{(a)}_m(u) \mapsto Y^{(a)}_m(u + \frac{1}{t})$, $1 + Y^{(a)}_m(u) \mapsto 1 + Y^{(a)}_m(u + \frac{1}{t})$, and

$$\mathcal{Y}^\circ_\ell(M_t) \simeq \mathcal{Y}^\circ_\ell(M_t)_+ \times \mathcal{Y}^\circ_\ell(M_t)_-. \tag{20}$$

4.3. *Quiver $Q_\ell(M_t)$*

With the Cartan matrix M_t and $\ell \geq 2$ we associate a quiver $Q_\ell(M_t)$ as below. First, as a rather general example, the case $t = 5$ is given in Fig. 1, where the right columns in the five quivers $Q_1, \dots, Q_5$ are identified. Also we assign the empty or filled circle $\circ/\bullet$ and the sign $+/-$ to each vertex as shown. For a general odd t, the quiver $Q_\ell(M_t)$ is defined by naturally extending the case $t = 5$. Namely, we consider t quivers $Q_1, \dots, Q_t$. In each quiver Q_i there are $\ell - 1$ vertices (with $\circ$) in the left column and $t\ell - 1$ vertices (with $\bullet$) in the right column. The arrows are put as clearly indicated by the example in Fig. 1. The right columns in all the quivers $Q_1, \dots, Q_t$ are identified.

Let us choose the index set $\mathbf{I}$ of the vertices of $Q_\ell(M_t)$ so that $\mathbf{i} = (i, i') \in \mathbf{I}$ represents the vertex at the i'th row (from the bottom) of the left column in Q_i for $i = 1, \dots, t$, and the one of the right column in any quiver for $i = t + 1$. Thus, $i = 1, \dots, t + 1$, and $i' = 1, \dots, \ell - 1$ if $i \neq t + 1$ and $i' = 1, \dots, t\ell - 1$ if $i = t + 1$.

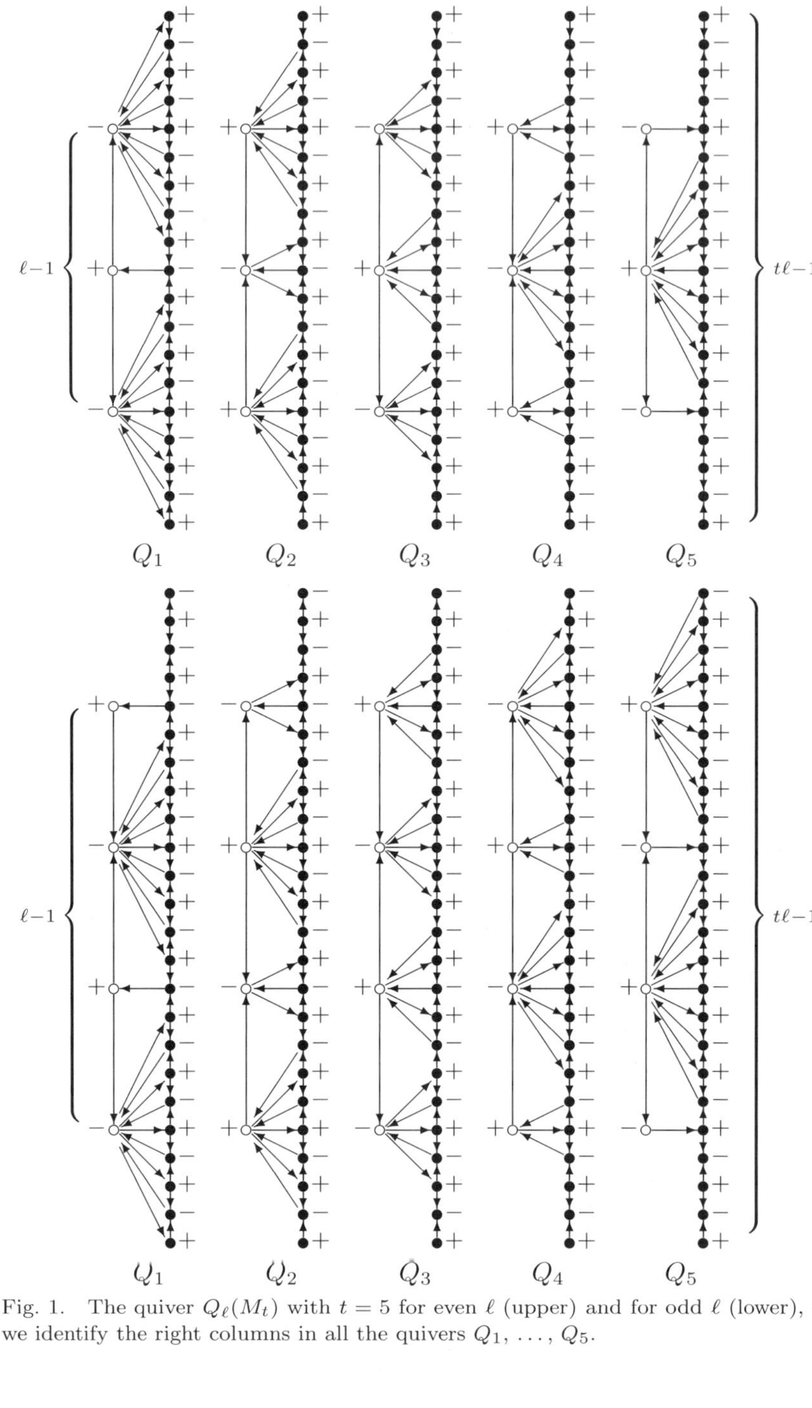

Fig. 1. The quiver $Q_\ell(M_t)$ with $t = 5$ for even ℓ (upper) and for odd ℓ (lower), where we identify the right columns in all the quivers $Q_1, \ldots, Q_5$.

For $k \in \{1, \dots, t\}$, let $\mathbf{I}^{\circ}_{+,k}$ (resp. $\mathbf{I}^{\circ}_{-,k}$) denote the set of the vertices $\mathbf{i}$ in Q_k with property $\circ$ and $+$ (resp. $\circ$ and $-$). Similarly, let $\mathbf{I}^{\bullet}_{+}$ (resp. $\mathbf{I}^{\bullet}_{-}$) denote the set of the vertices $\mathbf{i}$ with property $\bullet$ and $+$ (resp. $\bullet$ and $-$). We define composite mutations,

$$\mu^{\circ}_{+,k} = \prod_{\mathbf{i} \in \mathbf{I}^{\circ}_{+,k}} \mu_{\mathbf{i}}, \quad \mu^{\circ}_{-,k} = \prod_{\mathbf{i} \in \mathbf{I}^{\circ}_{-,k}} \mu_{\mathbf{i}}, \quad \mu^{\bullet}_{+} = \prod_{\mathbf{i} \in \mathbf{I}^{\bullet}_{+}} \mu_{\mathbf{i}}, \quad \mu^{\bullet}_{-} = \prod_{\mathbf{i} \in \mathbf{I}^{\bullet}_{-}} \mu_{\mathbf{i}}. \tag{21}$$

Note that they do not depend on the order of the product.

For a permutation w of $\{1, \dots, t\}$, let $\tilde{w}$ be the permutation of $\mathbf{I}$ such that $\tilde{w}(i, i') = (w(i), i')$ for $i \neq t{+}1$ and $(t{+}1, i')$ for $i = t{+}1$. Let $\tilde{w}(Q_\ell(M_t))$ denote the quiver induced from $Q_\ell(M_t)$ by $\tilde{w}$. Namely, if there is an arrow $\mathbf{i} \to \mathbf{j}$ in $Q_\ell(M_t)$, then, there is an arrow $\tilde{w}(\mathbf{i}) \to \tilde{w}(\mathbf{j})$ in $\tilde{w}(Q_\ell(M_t))$. For a quiver Q, let Q^{op} denote the opposite quiver.

Lemma 4.1. *Let $Q(0) := Q_\ell(M_t)$. We have the following periodic sequence of mutations of quivers:*

$$\begin{array}{lllllllll}
Q(0) & \overset{\mu^{\bullet}_{+}\mu^{\circ}_{+,1}}{\longleftrightarrow} & Q(\frac{1}{t}) & \overset{\mu^{\bullet}_{-}\mu^{\circ}_{+,t-1}}{\longleftrightarrow} & Q(\frac{2}{t}) & \overset{\mu^{\bullet}_{+}\mu^{\circ}_{+,3}}{\longleftrightarrow} & Q(\frac{3}{t}) & \overset{\mu^{\bullet}_{-}\mu^{\circ}_{+,t-3}}{\longleftrightarrow} & Q(\frac{4}{t}) \\
 & \overset{\mu^{\bullet}_{+}\mu^{\circ}_{+,5}}{\longleftrightarrow} & \cdots & & & \overset{\mu^{\bullet}_{-}\mu^{\circ}_{+,2}}{\longleftrightarrow} & Q(\frac{t-1}{t}) & \overset{\mu^{\bullet}_{+}\mu^{\circ}_{+,t}}{\longleftrightarrow} & Q(1) \\
 & \overset{\mu^{\bullet}_{-}\mu^{\circ}_{-,1}}{\longleftrightarrow} & Q(\frac{t+1}{t}) & \overset{\mu^{\bullet}_{+}\mu^{\circ}_{-,t-1}}{\longleftrightarrow} & Q(\frac{t+2}{t}) & \overset{\mu^{\bullet}_{-}\mu^{\circ}_{-,3}}{\longleftrightarrow} & Q(\frac{t+3}{t}) & \overset{\mu^{\bullet}_{+}\mu^{\circ}_{-,t-3}}{\longleftrightarrow} & Q(\frac{t+4}{t}) \\
 & \overset{\mu^{\bullet}_{-}\mu^{\circ}_{-,5}}{\longleftrightarrow} & \cdots & & & \overset{\mu^{\bullet}_{+}\mu^{\circ}_{-,2}}{\longleftrightarrow} & Q(\frac{2t-1}{t}) & \overset{\mu^{\bullet}_{-}\mu^{\circ}_{-,t}}{\longleftrightarrow} & Q(2) = Q(0).
\end{array} \tag{22}$$

Here, the quiver $Q(p/t)$ $(p = 1, \dots, 2t)$ is defined by

$$Q(p/t) := \begin{cases} \tilde{w}_p(Q(0))^{\mathrm{op}} & p\text{: odd} \\ \tilde{w}_p(Q(0)) & p\text{: even,} \end{cases} \tag{23}$$

and w_p is a permutation of $\{1, \dots, t\}$ defined by

$$w_p = \begin{cases} r_+ r_- \cdots r_+ \ (p \text{ terms}) & p\text{: odd} \\ r_+ r_- \cdots r_- \ (p \text{ terms}) & p\text{: even,} \end{cases} \tag{24}$$

$$r_+ = (23)(45) \cdots (r-1, r), \quad r_- = (12)(34) \cdots (r-2, r-1), \tag{25}$$

where (ij) is the transposition of i and j.

Proof. Let Q_1, …, Q_t be the subquivers in the definition of $Q_\ell(M_t)$ as in Fig. 1. By the sequence of mutations (22), one can easily check that Q_1 mutates as

$$Q_1 \leftrightarrow Q_1^{\mathrm{op}} \leftrightarrow Q_2 \leftrightarrow Q_3^{\mathrm{op}} \leftrightarrow \cdots Q_t^{\mathrm{op}} \leftrightarrow Q_t \leftrightarrow \cdots Q_3 \leftrightarrow Q_2^{\mathrm{op}} \leftrightarrow Q_1, \tag{26}$$

Q_2 mutates as

$$Q_2 \leftrightarrow Q_3^{\mathrm{op}} \leftrightarrow \cdots \leftrightarrow Q_t^{\mathrm{op}} \leftrightarrow Q_t \leftrightarrow \cdots \leftrightarrow Q_2^{\mathrm{op}} \leftrightarrow Q_1 \leftrightarrow Q_1^{\mathrm{op}} \leftrightarrow Q_2, \tag{27}$$

Q_3 mutates as

$$Q_3 \leftrightarrow Q_2^{\mathrm{op}} \leftrightarrow Q_1 \leftrightarrow Q_1^{\mathrm{op}} \leftrightarrow \cdots \leftrightarrow Q_t^{\mathrm{op}} \leftrightarrow Q_t \leftrightarrow \cdots \leftrightarrow Q_4^{\mathrm{op}} \leftrightarrow Q_3, \tag{28}$$

and so on. The result is summarized as (23). □

Example 4.2. The mutation sequence (22) for $t = 5$ is explicitly given in Figs. 2 and 3, where only a part of each quiver is presented. (Caution: the mutations of the top and bottom arrows may look erroneous but they are correct because of the effect from the omitted part.) The encircled vertices are the mutation points of (22) in the forward direction.

4.4. *Embedding maps*

Let $B = B_\ell(M_t)$ be the skew-symmetric matrix corresponding to the quiver $Q_\ell(M_t)$. Let $\mathcal{A}(B, x, y)$ be the cluster algebra with coefficients in the universal semifield $\mathbb{Q}_{\mathrm{sf}}(y)$, and let $\mathcal{G}(B, y)$ be the coefficient group associated with $\mathcal{A}(B, x, y)$ as in Section 3.

In view of Lemma 4.1 we set $x(0) = x$, $y(0) = y$ and define clusters $x(u) = (x_{\mathbf{i}}(u))_{\mathbf{i}\in\mathbf{I}}$ $(u \in \frac{1}{t}\mathbb{Z})$ and coefficient tuples $y(u) = (y_{\mathbf{i}}(u))_{\mathbf{i}\in\mathbf{I}}$ $(u \in \frac{1}{t}\mathbb{Z})$ by the sequence of mutations

$$\begin{aligned} \cdots \overset{\mu^\bullet_-\mu^\circ_{-,t}}{\longleftrightarrow} (B(0), x(0), y(0)) \overset{\mu^\bullet_+\mu^\circ_{+,1}}{\longleftrightarrow} (B(\tfrac{1}{t}), x(\tfrac{1}{t}), y(\tfrac{1}{t})) \\ \overset{\mu^\bullet_-\mu^\circ_{+,t-1}}{\longleftrightarrow} \cdots \overset{\mu^\bullet_-\mu^\circ_{-,t}}{\longleftrightarrow} (B(2), x(2), y(2)) \overset{\mu^\bullet_+\mu^\circ_{+,1}}{\longleftrightarrow} \cdots, \end{aligned} \tag{29}$$

where $B(u)$ is the skew-symmetric matrix corresponding to $Q(u)$.

For $(\mathbf{i}, u) \in \mathbf{I} \times \frac{1}{t}\mathbb{Z}$, we set the parity condition $\mathbf{p}_+$ by

$$\mathbf{p}_+ : \begin{cases} \mathbf{i} \in \mathbf{I}^\bullet_+ \sqcup \mathbf{I}^\circ_{+,p+1} & u \equiv \frac{p}{t}, 0 \le p \le t-1, p\text{: even} \\ \mathbf{i} \in \mathbf{I}^\bullet_- \sqcup \mathbf{I}^\circ_{+,t-p} & u \equiv \frac{p}{t}, 0 \le p \le t-1, p\text{: odd} \\ \mathbf{i} \in \mathbf{I}^\bullet_+ \sqcup \mathbf{I}^\circ_{-,2t-p} & u \equiv \frac{p}{t}, t \le p \le 2t-1, p\text{: even} \\ \mathbf{i} \in \mathbf{I}^\bullet_- \sqcup \mathbf{I}^\circ_{-,p+1-t} & u \equiv \frac{p}{t}, t \le p \le 2t-1, p\text{: odd,} \end{cases} \tag{30}$$

where $\equiv$ is modulo $2\mathbb{Z}$. We define the condition $\mathbf{p}_-$ by $(\mathbf{i}, u) : \mathbf{p}_- \iff (\mathbf{i}, u - 1/t) : \mathbf{p}_+$. Plainly speaking, each $(\mathbf{i}, u) : \mathbf{p}_+$ (resp. $\mathbf{p}_-$) is a mutation point of (29) in the forward (resp. backward) direction of u.

There is a correspondence between the parity condition $\mathbf{p}_+$ here and $\mathbf{P}_+$, $\mathbf{P}'_+$ in (16).

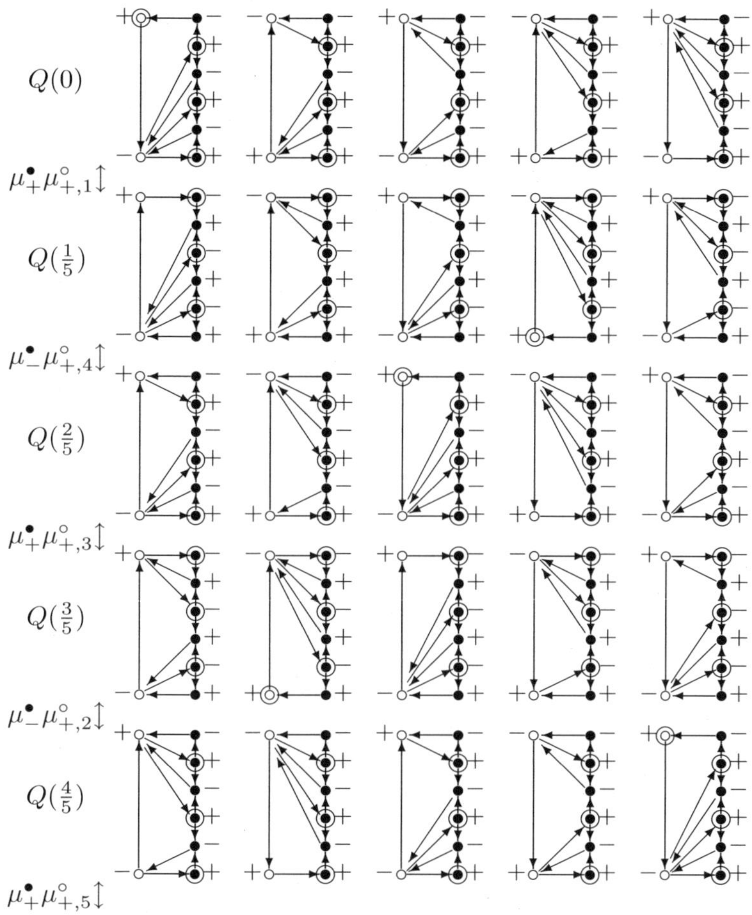

Fig. 2. (Continues to Fig. 3.) The mutation sequence of the quiver $Q_\ell(M_t)$ in (22) for $t = 5$. Only a part of each quiver is presented. The encircle vertices correspond to the mutation points in the forward direction.

Lemma 4.3. *Below* $\equiv$ *means the equivalence modulo* $2\mathbb{Z}$.

(i) The map $g : \mathcal{I}_{\ell+} \to \{(\mathbf{i}, u) : \mathbf{p}_+\}$

$$(a, m, u - \tfrac{d_a}{t}) \mapsto \begin{cases} ((2j+1, m), u) & a = 1; m + u \equiv \frac{2j}{t} \\ & (j = 0, 1, \dots, (t-1)/2) \\ ((2t - 2j, m), u) & a = 1; m + u \equiv \frac{2j}{t} \\ & (j = (t+1)/2, \dots, t-1) \\ ((t+1, m), u) & a = 2 \end{cases} \tag{31}$$

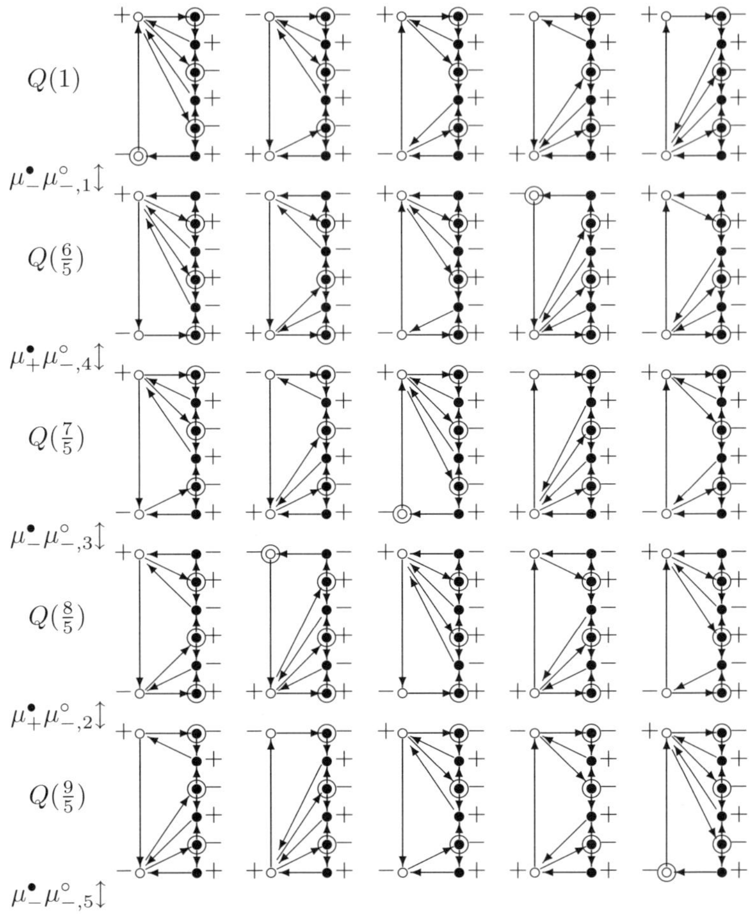

Fig. 3. (Continues from Fig. 2.)

is a bijection.

(ii) The map $g' : \mathcal{I}'_{\ell+} \to \{(\mathbf{i}, u) : \mathbf{p}_+\}$

$$(a, m, u) \mapsto \begin{cases} ((2j+1, m), u) & a = 1; m + u \equiv \frac{2j}{t} \\ & (j = 0, 1, \dots, (t-1)/2) \\ ((2t-2j, m), u) & a = 1; m + u \equiv \frac{2j}{t} \\ & (j = (t+1)/2, \dots, t-1) \\ ((t+1, m), u) & a = 2 \end{cases} \tag{32}$$

is a bijection.

Proof. The fact (i) is equivalent to (ii) due to (19). So, it is enough to prove (ii). Let us examine the meaning of the map (32) in the case $t = 5$ with Fig. 2. Each encircled vertex therein corresponds to $(\mathbf{i}, u) : \mathbf{p}_+$, and some (a, m, u) is attached to it by g'. For example, in $Q(0)$, $(1, m, 0)$ (m: even) are attached to the vertices with $(\circ, +)$ in the first quiver (from the left), and $(2, m, 0)$ (m: odd) are attached to the vertices with $(\bullet, +)$. Similarly, in $Q(1/5)$, $(1, m, 1/5)$ (m: odd) are attached to the vertices with $(\circ, +)$ in the fourth quiver (from the left), and $(2, m, 1/5)$ (m: even) are attached to the vertices with $(\bullet, -)$. Then, one can easily confirm that g' is indeed a bijection. A general case is verified similarly. □

We introduce alternative labels $x_{\mathbf{i}}(u) = x_m^{(a)}(u - d_a/t)$ ($(a, m, u - d_a/t) \in \mathcal{I}_{\ell+}$) for $(\mathbf{i}, u) = g((a, m, u - d_a/t))$ and $y_{\mathbf{i}}(u) = y_m^{(a)}(u)$ ($(a, m, u) \in \mathcal{I}'_{\ell+}$) for $(\mathbf{i}, u) = g'((a, m, u))$, respectively.

Remark 4.4. In the case $t = 1$, i.e., the simply laced case, the map g in Lemma 4.3 reads $(a, m, u) \mapsto ((a, m), u + 1)$, thus, differs from the simpler one $(a, m, u) \mapsto ((a, m), u)$ used in Refs. 1 and 4. Either will serve as a natural parametrization and the transferring from one to the other is easy.

4.5. *T-system and cluster algebra*

We show that the T-system $\mathbb{T}_\ell(M_t)$ naturally appears as a system of relations among the cluster variables $x_{\mathbf{i}}(u)$ in the trivial evaluation of coefficients. (The quiver $Q_\ell(M_t)$ is designed to do so.) Let $\mathcal{A}(B, x)$ be the cluster algebra with trivial coefficients, where (B, x) is the initial seed. Let $\mathbf{1} = \{1\}$ be the *trivial semifield* and $\pi_{\mathbf{1}} : \mathbb{Q}_{\rm sf}(y) \to \mathbf{1}$, $y_{\mathbf{i}} \mapsto 1$ be the projection. Let $[x_{\mathbf{i}}(u)]_{\mathbf{1}}$ denote the image of $x_{\mathbf{i}}(u)$ by the algebra homomorphism $\mathcal{A}(B, x, y) \to \mathcal{A}(B, x)$ induced from $\pi_{\mathbf{1}}$. It is called the *trivial evaluation.*

Lemma 4.5. *Let $G(b, k, v; a, m, u)$ be the one in (7). The family $\{x_m^{(a)}(u) \mid (a, m, u) \in \mathcal{I}_{\ell+}\}$ satisfies a system of relations*

$$x_m^{(a)}\left(u - \tfrac{d_a}{t}\right) x_m^{(a)}\left(u + \tfrac{d_a}{t}\right) = \frac{y_m^{(a)}(u)}{1 + y_m^{(a)}(u)} \prod_{(b,k,v) \in \mathcal{I}_{\ell+}} x_k^{(b)}(v)^{G(b,k,v;\, a,m,u)} + \frac{1}{1 + y_m^{(a)}(u)} x_{m-1}^{(a)}(u) x_{m+1}^{(a)}(u), \tag{33}$$

where $(a, m, u) \in \mathcal{I}'_{\ell+}$. In particular, the family $\{[x_m^{(a)}(u)]_{\mathbf{1}} \mid (a, m, u) \in \mathcal{I}_{\ell+}\}$ satisfies the T-system $\mathbb{T}_\ell(M_t)$ in $\mathcal{A}(B, x)$ by replacing $T_m^{(a)}(u)$ with

$[x_m^{(a)}(u)]_{\mathbf{1}}$.

Proof. This follows from the exchange relation of cluster variables (14) and the property of the sequence (22) which are observed in Figs. 2 and 3.

Let us demonstrate how to obtain these relations in the case $t = 5$ using Figs. 2 and 3. For example, consider the mutation at $((1,2),0)$. Then, the attached variable $x_2^{(1)}(-1)$ is mutated to

$$\frac{1}{x_2^{(1)}(-1)}\left\{\frac{y_2^{(1)}(0)}{1+y_2^{(1)}(0)}x_{10}^{(2)}(0)+\frac{1}{1+y_2^{(1)}(0)}x_1^{(1)}(0)x_3^{(1)}(0)\right\}, \tag{34}$$

which should equal to $x_2^{(1)}(1)$. Also, consider the mutation at, say, $((2,9),0)$. Then, the attached variable $x_9^{(2)}(-1/5)$ is mutated to

$$\frac{1}{x_9^{(2)}(-\frac{1}{5})}\left\{\frac{y_9^{(2)}(0)}{1+y_9^{(2)}(0)}x_1^{(1)}(0)x_2^{(1)}(-\tfrac{3}{5})x_2^{(1)}(-\tfrac{1}{5})x_2^{(1)}(\tfrac{1}{5})x_2^{(1)}(\tfrac{3}{5})\right. \tag{35}$$

$$\left.+\frac{1}{1+y_9^{(2)}(0)}x_8^{(2)}(0)x_{10}^{(2)}(0)\right\}, \tag{36}$$

which should equal to $x_9^{(2)}(1/5)$. They certainly agree with (4) and (5). (The quiver $Q_\ell(M_t)$ is designed to do so.) □

Definition 4.6. The *T-subalgebra* $\mathcal{A}_T(B,x)$ *of* $\mathcal{A}(B,x)$ *associated with the sequence* (29) is the subring of $\mathcal{A}(B,x)$ generated by $[x_{\mathbf{i}}(u)]_{\mathbf{1}}$ $((\mathbf{i},u)\in \mathbf{I}\times \frac{1}{t}\mathbb{Z})$.

By Lemma 4.5, we have the following embedding.

Theorem 4.7. *The ring* $\mathcal{T}_\ell^\circ(M_t)_+$ *is isomorphic to* $\mathcal{A}_T(B,x)$ *by the correspondence* $T_m^{(a)}(u)\mapsto [x_m^{(a)}(u)]_{\mathbf{1}}$.

Proof. The map $\rho : T_m^{(a)}(u) \mapsto [x_m^{(a)}(u)]_{\mathbf{1}}$ is a ring homomorphism due to Lemma 4.5. We can construct the inverse of ρ as follows. For each $\mathbf{i}\in\mathbf{I}$, let $u_{\mathbf{i}} \in \frac{1}{t}\mathbb{Z}$ be the smallest nonnegative $u_{\mathbf{i}}$ such that $(\mathbf{i},u):\mathbf{p}_+$. Then, thanks to Lemma 4.3 (i) there is a unique $(a,m,u_{\mathbf{i}}-d_a/t)\in\mathcal{I}_{\ell+}$ such that $g((a,m,u_{\mathbf{i}}-d_a/t)) = (\mathbf{i},u_{\mathbf{i}})$. We define a ring homomorphism $\tilde{\varphi} : \mathbb{Z}[x_{\mathbf{i}}^{\pm 1}]_{\mathbf{i}\in\mathbf{I}} \to \mathcal{T}_\ell(M_t)$ by $x_{\mathbf{i}}^{\pm 1} \mapsto T_m^{(a)}(u_{\mathbf{i}}-d_a/t)^{\pm 1}$. Thus, we have $\tilde{\varphi} : [x_m^{(a)}(u_{\mathbf{i}}-d_a/t)]_{\mathbf{1}} \mapsto T_m^{(a)}(u_{\mathbf{i}}-d_a/t)$. Furthermore, one can prove that $\tilde{\varphi} : [x_m^{(a)}(u)]_{\mathbf{1}} \mapsto T_m^{(a)}(u)$ for any $(a,m,u)\in\mathcal{I}_{\ell+}$ by induction on the forward and backward mutations, applying the same T-systems for the both sides. By the the restriction of $\tilde{\varphi}$ to $\mathcal{A}_T(B,x)$, we obtain a ring homomorphism $\varphi : \mathcal{A}_T(B,x)\to\mathcal{T}_\ell^\circ(M_t)_+$, which is the inverse of ρ. □

4.6. *Y-system and cluster algebra*

The Y-system $\mathbb{Y}_\ell(M_t)$ also naturally appears as a system of relations among the coefficients $y_{\mathbf{i}}(u)$.

The following lemma follows from the exchange relation of coefficients and the property of the sequence (22).

Lemma 4.8. *The family $\{y_m^{(a)}(u) \mid (a,m,u) \in \mathcal{I}'_{\ell+}\}$ satisfies the Y-system $\mathbb{Y}_\ell(M_t)$ by replacing $Y_m^{(a)}(u)$ with $y_m^{(a)}(u)$.*

Proof. Again, let us demonstrate how to obtain these relations in the case $t = 5$ using Figs. 2 and 3. For example, consider the mutation at $((1,2),0)$. Then, the attached variable $y_2^{(1)}(0)$ is mutated to $y_2^{(1)}(0)^{-1}$. Then, the following factors are multiplied to $y_2^{(1)}(0)^{-1}$ during $u = \frac{1}{5}, \dots, \frac{9}{5}$:

$$\begin{aligned}
&(1 + y_{14}^{(2)}(\tfrac{5}{5})),\\
&(1 + y_{13}^{(2)}(\tfrac{4}{5}))(1 + y_{13}^{(2)}(\tfrac{6}{5})),\\
&(1 + y_{12}^{(2)}(\tfrac{3}{5}))(1 + y_{12}^{(2)}(\tfrac{5}{5}))(1 + y_{12}^{(2)}(\tfrac{7}{5})),\\
&(1 + y_{11}^{(2)}(\tfrac{2}{5}))(1 + y_{11}^{(2)}(\tfrac{4}{5}))(1 + y_{11}^{(2)}(\tfrac{6}{5}))(1 + y_{11}^{(2)}(\tfrac{8}{5})),\\
&(1 + y_{10}^{(2)}(\tfrac{1}{5}))(1 + y_{10}^{(2)}(\tfrac{3}{5}))(1 + y_{10}^{(2)}(\tfrac{5}{5}))(1 + y_{10}^{(2)}(\tfrac{7}{5}))(1 + y_{10}^{(2)}(\tfrac{9}{5})),\\
&(1 + y_{9}^{(2)}(\tfrac{2}{5}))(1 + y_{9}^{(2)}(\tfrac{4}{5}))(1 + y_{9}^{(2)}(\tfrac{6}{5}))(1 + y_{9}^{(2)}(\tfrac{8}{5})),\\
&(1 + y_{8}^{(2)}(\tfrac{3}{5}))(1 + y_{8}^{(2)}(\tfrac{5}{5}))(1 + y_{8}^{(2)}(\tfrac{7}{5})),\\
&(1 + y_{7}^{(2)}(\tfrac{4}{5}))(1 + y_{7}^{(2)}(\tfrac{6}{5})),\\
&(1 + y_{8}^{(2)}(\tfrac{5}{5})),\\
&(1 + y_{1}^{(1)}(1)^{-1})^{-1}(1 + y_{3}^{(1)}(1)^{-1})^{-1}.
\end{aligned}$$

The result should equal to $y_2^{(1)}(2)$. Also, consider the mutation at, say, $((2,9),0)$. Then, the attached variable $y_9^{(2)}(0)$ is mutated to $y_9^{(2)}(0)^{-1}$. Then, the following factors are multiplied to $y_9^{(2)}(0)^{-1}$ at $u = \frac{1}{5}$:

$$(1 + y_8^{(2)}(\tfrac{1}{5})^{-1})^{-1}(1 + y_{10}^{(2)}(\tfrac{1}{5})^{-1})^{-1}.$$

The result should equal to $y_9^{(2)}(\frac{2}{5})$. They certainly agree with (8) and (9). □

Definition 4.9. The *Y-subgroup $\mathcal{G}_Y(B,y)$ of $\mathcal{G}(B,y)$ associated with the sequence* (29) is the subgroup of $\mathcal{G}(B,y)$ generated by $y_{\mathbf{i}}(u)$ $((\mathbf{i},u) \in \mathbf{I} \times \frac{1}{t}\mathbb{Z})$ and $1 + y_{\mathbf{i}}(u)$ $((\mathbf{i},u) : \mathbf{p}_+$ or $\mathbf{p}_-)$.

By Lemma 4.8, we have the following embedding.

Theorem 4.10. *The group* $\mathcal{Y}^{\circ}_{\ell}(M_t)_+$ *is isomorphic to* $\mathcal{G}_Y(B, y)$ *by the correspondence* $Y^{(a)}_m(u) \mapsto y^{(a)}_m(u)$ *and* $1 + Y^{(a)}_m(u) \mapsto 1 + y^{(a)}_m(u)$.

Proof. The map $\rho : Y^{(a)}_m(u) \mapsto y^{(a)}_m(u)$, $1 + Y^{(a)}_m(u) \mapsto 1 + y^{(a)}_m(u)$ is a group homomorphism due to Lemma 4.5. We can construct the inverse of ρ as follows. For each $\mathbf{i} \in \mathbf{I}$, let $u_{\mathbf{i}} \in \frac{1}{t}\mathbb{Z}$ be the largest nonpositive $u_{\mathbf{i}}$ such that $(\mathbf{i}, u_{\mathbf{i}}) : \mathbf{p}_+$. Then, thanks to Lemma 4.3 (ii) there is a unique $(a, m, u_{\mathbf{i}}) \in \mathcal{I}'_{\ell+}$ such that $g'((a, m, u_{\mathbf{i}})) = (\mathbf{i}, u_{\mathbf{i}})$. We define a semifield homomorphism $\tilde{\varphi} : \mathbb{Q}_{\mathrm{sf}}(y_{\mathbf{i}})_{\mathbf{i}\in\mathbf{I}} \to \mathcal{Y}_{\ell}(M_t)$ as follows. If $u_{\mathbf{i}} = 0$, then $y_{\mathbf{i}} \mapsto Y^{(a)}_m(0)$. If $u_{\mathbf{i}} < 0$, we define

$$\tilde{\varphi}(y_{\mathbf{i}}) = Y^{(a)}_m(u_{\mathbf{i}})^{-1} \frac{\prod\limits_{(b,k,v)} (1 + Y^{(b)}_k(v))}{\prod\limits_{(b,k,v)} (1 + Y^{(b)}_k(v)^{-1})}, \tag{37}$$

where the product in the numerator is taken for $(b, k, v) : \mathcal{I}'_{\ell+}$ such that $u_{\mathbf{i}} < v < 0$ and $B_{\mathbf{ji}}(v) < 0$ for $(\mathbf{j}, v) = g'((b, k, v))$, and the product in the denominator is taken for $(b, k, v) : \mathcal{I}'_{\ell+}$ such that $u_{\mathbf{i}} < v < 0$ and $B_{\mathbf{ji}}(v) > 0$ for $(\mathbf{j}, v) = g'((b, k, v))$. Then, we have $\tilde{\varphi} : y^{(a)}_m(u_{\mathbf{i}}) \mapsto Y^{(a)}_m(u_{\mathbf{i}})$. Furthermore, one can prove that $\tilde{\varphi} : y^{(a)}_m(u) \mapsto Y^{(a)}_m(u)$ for any $(a, m, u) \in \mathcal{I}'_{\ell+}$ by induction on the forward and backward mutations, applying the same Y-systems for the both sides. By the restriction of $\tilde{\varphi}$ to $\mathcal{G}_Y(B, x)$, we obtain a group homomorphism $\varphi : \mathcal{G}_Y(B, x) \to \mathcal{Y}^{\circ}_{\ell}(M_t)_+$, which is the inverse of ρ. □

5. Cluster algebraic formulation: The case $|I| = 2$; t is even

In this section we consider the case $|I| = 2$ when t is even. Basically it is parallel to the former case and we omit proofs.

5.1. *Parity decompositions of T and Y-systems*

For a triplet $(a, m, u) \in \mathcal{I}_{\ell}$, we reset the 'parity conditions' $\mathbf{P}_+$ and $\mathbf{P}_-$ by

$$\begin{aligned} \mathbf{P}_+ &: \ tu \text{ is even if } a = 1;\ m + tu \text{ is even if } a = 2,\\ \mathbf{P}_- &: \ tu \text{ is odd if } a = 1;\ m + tu \text{ is odd if } a = 2. \end{aligned} \tag{38}$$

Let $\mathcal{I}_{\ell\varepsilon}$ be the set of all $(a, m, u) : \mathbf{P}_{\varepsilon}$. Define $\mathcal{T}^{\circ}_{\ell}(M_t)_{\varepsilon}$ ($\varepsilon = \pm$) to be the subring of $\mathcal{T}^{\circ}_{\ell}(M_t)$ generated by $T^{(a)}_m(u)$ ($(a, m, u) \in \mathcal{I}_{\ell\varepsilon}$). Then, we have

$\mathcal{T}^\circ_\ell(M_t)_+ \simeq \mathcal{T}^\circ_\ell(M_t)_-$ by $T^{(a)}_m(u) \mapsto T^{(a)}_m(u+\frac{1}{t})$, and the decomposition (17) holds.

For a triplet $(a,m,u) \in \mathcal{I}_\ell$, we set another 'parity conditions' $\mathbf{P}'_+$ and $\mathbf{P}'_-$ by

$$\begin{aligned}\mathbf{P}'_+ &: \ tu \text{ is even if } a=1;\ m+tu \text{ is odd if } a=2,\\ \mathbf{P}'_- &: \ tu \text{ is odd if } a=1;\ m+tu \text{ is even if } a=2.\end{aligned} \tag{39}$$

We have

$$(a,m,u) : \mathbf{P}'_+ \iff (a,m,u \pm \tfrac{d_a}{t}) : \mathbf{P}_+. \tag{40}$$

Let $\mathcal{I}'_{\ell\varepsilon}$ be the set of all (a,m,u) : $\mathbf{P}'_\varepsilon$. Define $\mathcal{Y}^\circ_\ell(M_t)_\varepsilon$ ($\varepsilon = \pm$) to be the subgroup of $\mathcal{Y}^\circ_\ell(M_t)$ generated by $Y^{(a)}_m(u)$, $1+Y^{(a)}_m(u)$ ($(a,m,u) \in \mathcal{I}'_{\ell\varepsilon}$). Then, we have $\mathcal{Y}^\circ_\ell(M_t)_+ \simeq \mathcal{Y}^\circ_\ell(M_t)_-$ by $Y^{(a)}_m(u) \mapsto Y^{(a)}_m(u+\frac{1}{t})$, $1+Y^{(a)}_m(u) \mapsto 1+Y^{(a)}_m(u+\frac{1}{t})$, and the decomposition (20) holds.

5.2. *Quiver $Q_\ell(M_t)$*

With the Cartan matrix M_t and $\ell \geq 2$ we associate the quiver $Q_\ell(M_t)$. Again, as a rather general example, the case $t=4$ is given by Fig. 4, where the right columns in the four quivers $Q_1,\dots,Q_4$ are identified. Also we assign the empty or filled circle $\circ/\bullet$ and the sign $+/-$ to each vertex as shown. For a general even t, the quiver $Q_\ell(M_t)$ is defined by naturally extending the case $t=4$. Even though it looks quite similar to the odd t case in Fig. 1, there is one important difference due to the parity of t; that is, when t is even, any vertex in the right column of Q_i has the sign '$-$' whenever it is connected to a vertex in the left column by a *horizontal* arrow. This is not so when t is odd.

Let us choose the index set $\mathbf{I}$ of the vertices of $Q_\ell(M_t)$ as before so that $\mathbf{i}=(i,i') \in \mathbf{I}$ represents the vertex at the i'th row (from the bottom) of the left column in Q_i for $i=1,\dots,t$, and the one of the right column in any quiver for $i=t+1$. We use the same notations $\mathbf{I}^\circ_{\pm,k}$ $\mathbf{I}^\bullet_\pm$ as before. For $k,k' \in \{1,\dots,t\}$, $k \neq k'$, let $\mathbf{I}^\circ_{\pm,k,k'} = \mathbf{I}^\circ_{\pm,k} \sqcup \mathbf{I}^\circ_{\pm,k'}$. We define composite mutations,

$$\mu^\circ_{+,k,k'} = \prod_{\mathbf{i}\in\mathbf{I}^\circ_{+,k,k'}} \mu_{\mathbf{i}}, \quad \mu^\circ_{-,k,k'} = \prod_{\mathbf{i}\in\mathbf{I}^\circ_{-,k,k'}} \mu_{\mathbf{i}}, \quad \mu^\bullet_+ = \prod_{\mathbf{i}\in\mathbf{I}^\bullet_+} \mu_{\mathbf{i}}, \quad \mu^\bullet_- = \prod_{\mathbf{i}\in\mathbf{I}^\bullet_-} \mu_{\mathbf{i}}. \tag{41}$$

Lemma 5.1. *Let $Q(0) := Q_\ell(M_t)$. We have the following periodic sequence*

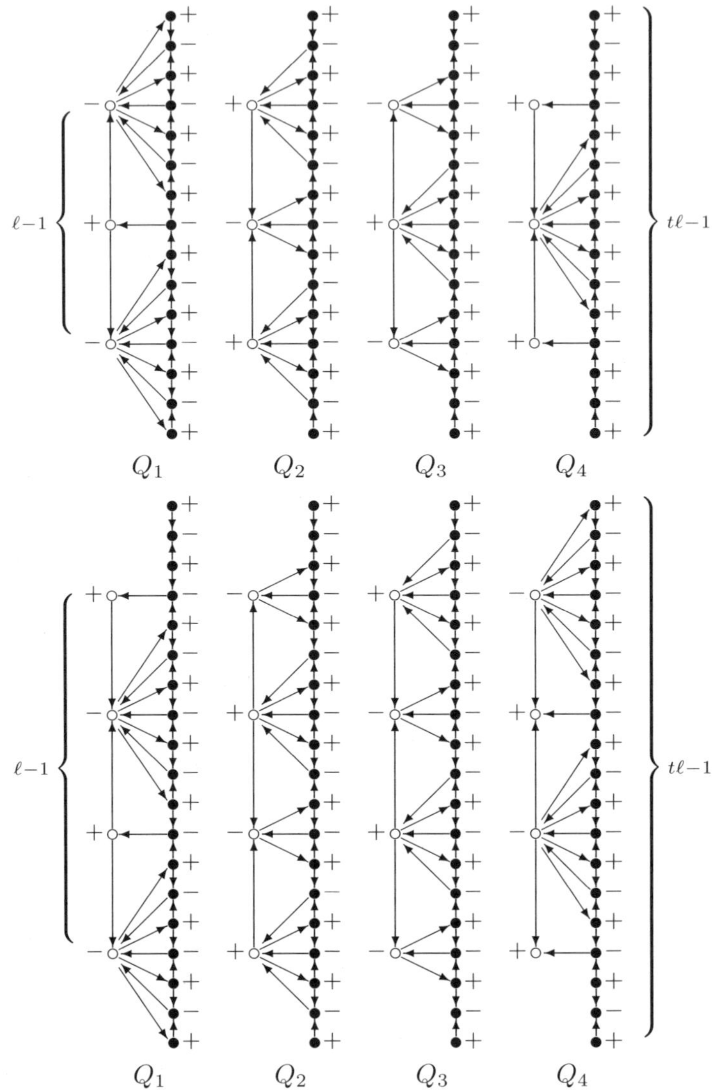

Fig. 4. The quiver $Q_\ell(M_t)$ with $t = 4$ for even ℓ (upper) and for odd ℓ (lower), where we identify the right columns in all the quivers $Q_1, \dots, Q_4$.

of mutations of quivers:

$$
\begin{aligned}
Q(0) &\xleftrightarrow{\mu^\bullet_+\mu^\circ_{+,1,t}} Q(\tfrac{1}{t}) \xleftrightarrow{\mu^\bullet_-} Q(\tfrac{2}{t}) \xleftrightarrow{\mu^\bullet_+\mu^\circ_{+,3,t-2}} Q(\tfrac{3}{t}) \xleftrightarrow{\mu^\bullet_-} Q(\tfrac{4}{t})\\
&\xleftrightarrow{\mu^\bullet_+\mu^\circ_{+,5,t-4}} \cdots \xleftrightarrow{\mu^\bullet_+\mu^\circ_{+,t-1,2}} Q(\tfrac{t-1}{t}) \xleftrightarrow{\mu^\bullet_-} Q(1)\\
&\xleftrightarrow{\mu^\bullet_+\mu^\circ_{-,1,t}} Q(\tfrac{t+1}{t}) \xleftrightarrow{\mu^\bullet_-} Q(\tfrac{t+2}{t}) \xleftrightarrow{\mu^\bullet_+\mu^\circ_{-,3,t-2}} Q(\tfrac{t+3}{t}) \xleftrightarrow{\mu^\bullet_-} Q(\tfrac{t+4}{t})\\
&\xleftrightarrow{\mu^\bullet_+\mu^\circ_{-,5,t-4}} \cdots \xleftrightarrow{\mu^\bullet_+\mu^\circ_{-,t-1,2}} Q(\tfrac{2t-1}{t}) \xleftrightarrow{\mu^\bullet_-} Q(2) = Q(0).
\end{aligned}
\tag{42}
$$

Here, the quiver $Q(p/t)$ $(p = 1, \dots, 2t)$ is defined by

$$Q(p/t) := \begin{cases} \tilde{w}_p(Q)^{\mathrm{op}} & p\text{: odd} \\ \tilde{w}_p(Q) & p\text{: even,} \end{cases} \tag{43}$$

and w_p is a permutation of $\{1, \dots, t\}$ defined by

$$w_p = \begin{cases} r_+ r_- \cdots r_+ \ (p \text{ terms}) & p\text{: odd} \\ r_+ r_- \cdots r_- \ (p \text{ terms}) & p\text{: even,} \end{cases} \tag{44}$$

$$r_+ = (23)(45)\cdots(r-2, r-1), \quad r_- = (12)(34)\cdots(r-1, r), \tag{45}$$

where (ij) is the transposition of i and j.

Example 5.2. The mutation sequence (42) for $t = 4$ is explicitly given in Fig. 5. where only a part of each quiver is presented as before.

5.3. *Embedding maps*

Let $B = B_\ell(M_t)$ be the corresponding skew-symmetric matrix to the quiver $Q_\ell(M_t)$ for even t. Let $\mathcal{A}(B, x, y)$ be the cluster algebra with coefficients in the universal semifield, and let $\mathcal{G}(B, y)$ be the coefficient group associated with $\mathcal{A}(B, x, y)$ as before.

In view of Lemma 5.1 we set $x(0) = x$, $y(0) = y$ and define clusters $x(u) = (x_{\mathbf{i}}(u))_{\mathbf{i} \in \mathbf{I}}$ $(u \in \frac{1}{t}\mathbb{Z})$ and coefficient tuples $y(u) = (y_{\mathbf{i}}(u))_{\mathbf{i} \in \mathbf{I}}$ $(u \in \frac{1}{t}\mathbb{Z})$ by the sequence of mutations

$$\begin{aligned} \cdots \overset{\mu^\bullet_-}{\longleftrightarrow} (B(0), x(0), y(0)) \overset{\mu^\bullet_+ \mu^\circ_{+,1,t}}{\longleftrightarrow} (B(\tfrac{1}{t}), x(\tfrac{1}{t}), y(\tfrac{1}{t})) \\ \overset{\mu^\bullet_-}{\longleftrightarrow} \quad \cdots \quad \overset{\mu^\bullet_-}{\longleftrightarrow} (B(2), x(2), y(2)) \overset{\mu^\bullet_+ \mu^\circ_{+,1,t}}{\longleftrightarrow} \cdots, \end{aligned} \tag{46}$$

where $B(u)$ is the skew-symmetric matrix corresponding to $Q(u)$.

For $(\mathbf{i}, u) \in \mathbf{I} \times \frac{1}{t}\mathbb{Z}$, we set the parity condition $\mathbf{p}_+$ by

$$\mathbf{p}_+ : \begin{cases} \mathbf{i} \in \mathbf{I}^\bullet_+ \sqcup \mathbf{I}^\circ_{+,p+1,t-p} & u \equiv \frac{p}{t}, 0 \le p \le t-1, p\text{: even} \\ \mathbf{i} \in \mathbf{I}^\bullet_+ \sqcup \mathbf{I}^\circ_{-,p+1-t,2t-p} & u \equiv \frac{p}{t}, t \le p \le 2t-1, p\text{: even} \\ \mathbf{i} \in \mathbf{I}^\bullet_- & u \equiv \frac{p}{t}, 0 \le p \le 2t-1, p\text{: odd,} \end{cases} \tag{47}$$

where $\equiv$ is modulo $2\mathbb{Z}$. Again, each $(\mathbf{i}, u) : \mathbf{p}_+$ is a mutation point of (42) in the forward direction of u.

Lemma 5.3. *Below $\equiv$ means the equivalence modulo $2\mathbb{Z}$.*

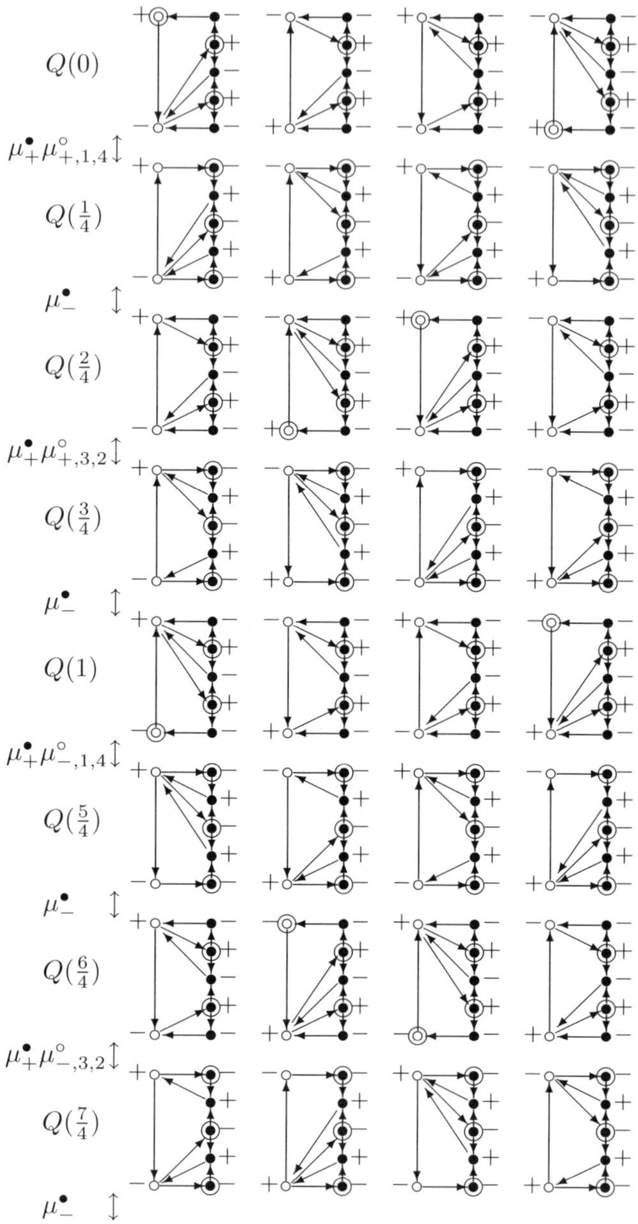

Fig. 5 The mutation sequence of the quiver $Q_\ell(M_t)$ in (42) for $\ell = 4$.

(i) The map $g : \mathcal{I}_{\ell+} \to \{(\mathbf{i}, u) : \mathbf{p}_+\}$

$$(a, m, u - \tfrac{d_a}{t}) \mapsto \begin{cases} ((2j+1, m), u) & a = 1; m + u \equiv \frac{2j}{t} \\ & (j = 0, 1, \dots, t/2 - 1) \\ ((2t - 2j, m), u) & a = 1; m + u \equiv \frac{2j}{t} \\ & (j = t/2, \dots, t - 1) \\ ((t+1, m), u) & a = 2 \end{cases} \tag{48}$$

is a bijection.

(ii) The map $g' : \mathcal{I}'_{\ell+} \to \{(\mathbf{i}, u) : \mathbf{p}_+\}$

$$(a, m, u) \mapsto \begin{cases} ((2j+1, m), u) & a = 1; m + u \equiv \frac{2j}{t} \\ & (j = 0, 1, \dots, t/2 - 1) \\ ((2t - 2j, m), u) & a = 1; m + u \equiv \frac{2j}{t} \\ & (j = t/2, \dots, t - 1) \\ ((t+1, m), u) & a = 2 \end{cases} \tag{49}$$

is a bijection.

5.4. *T-system, Y-system, and cluster algebra*

All the properties depending on the parity of t are now absorbed in the quiver $Q_\ell(M_t)$, the mutation sequence (46), and the embedding maps g and g' in Lemma 4.3. Lemmas 4.5, 4.8, and Theorems 4.7, 4.10 are true for even t.

6. Cluster algebraic formulation: Tree case

In this section we extend Theorems 4.7 and 4.10 to any tamely laced Cartan matrix C whose Dynkin diagram is a tree, by patching parity conditions and quivers introduced in Secs. 4 and 5. This is an intermediate step for treating the most general case in Sec. 7.

6.1. *Parity decompositions of T and Y-systems*

Throughout this section we assume that C is a tamely laced and indecomposable Cartan matrix whose Dynkin diagram $X(C)$ is a tree, i.e., without cycles.

We decompose the index set I of $X(C)$ into two parts $I = I_+ \sqcup I_-$ such that the following two rules are satisfied:

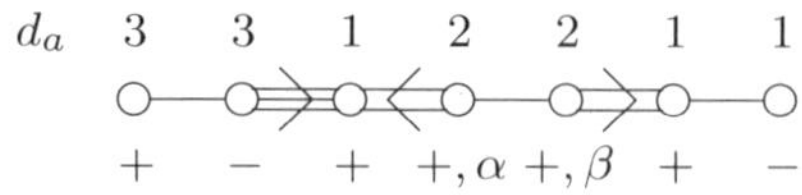

Fig. 6. Example of a decomposition and a coloring of I.

(I) If a and b are adjacent in $X(C)$ and both d_a and d_b are odd, then either $a \in I_+$, $b \in I_-$ or $a \in I_+$, $b \in I_-$ holds.
(II) If a and b are adjacent in $X(C)$ and at least one of d_a and d_b is even, then either $a, b \in I_+$ or $a, b \in I_-$ holds.

To each $a \in I$ with d_a even, we also attach the 'color' $c_a = \alpha$ or β satisfying the following condition:

(III) If a and b are adjacent in $X(C)$, then $c_a \neq c_b$.

See Fig. 6 for an example. (The coloring is not used in this subsection.) For a triplet $(a, m, u) \in \mathcal{I}_\ell$, we set the parity conditions $\mathbf{Q}_+$ as follows.

$$\mathbf{Q}_+ = \begin{cases} \text{(o+) } m + tu \text{ is even} & d_a \text{ is odd, } a \in I_+ \\ \text{(o−) } m + tu \text{ is odd} & d_a \text{ is odd, } a \in I_- \\ \text{(e+) } tu \text{ is odd} & d_a \text{ is even, } a \in I_+ \\ \text{(e−) } tu \text{ is even} & d_a \text{ is even, } a \in I_- \end{cases} \tag{50}$$

Let $\mathbf{Q}_-$ be the negation of $\mathbf{Q}_+$. Suppose that a and b in I with $d_a \geq d_b$ are adjacent in $X(C)$. Due to the condition (1), we have four possibilities: (i) d_a is odd and $d_b = 1$, (ii) $d_a = d_b$, and d_a is odd and not 1, (iii) d_a is even and $d_b = 1$, (iv) $d_a = d_b$, and d_a is even. For (i), the condition $\mathbf{Q}_+$ is compatible with $\mathbf{P}_\pm$ in (16) with $t = d_a$ therein. For (iii), the condition $\mathbf{Q}_+$ is compatible with $\mathbf{P}_\pm$ in (38) with $t = d_a$ therein. For (ii) and (iv), one can directly check that the condition $\mathbf{Q}_+$ is compatible with (4). Therefore, we have the parity decomposition

$$\mathcal{T}^\circ_\ell(C) \simeq \mathcal{T}^\circ_\ell(C)_+ \otimes_\mathbb{Z} \mathcal{T}^\circ_\ell(C)_-, \tag{51}$$

where $\mathcal{T}^\circ_\ell(C)_\varepsilon$ $(\varepsilon = \pm)$ is the subring of $\mathcal{T}^\circ_\ell(C)$ generated by $T^{(a)}_m(u)$ $((a, m, u) : \mathbf{Q}_\varepsilon)$.

Similarly, for a triplet $(a, m, u) \in \mathcal{I}_\ell$, we set the parity conditions $\mathbf{Q}'_+$

as follows.

$$\mathbf{Q}'_+ = \begin{cases} \text{(o+) } m+tu \text{ is odd} & d_a \text{ is odd, } a \in I_+ \\ \text{(o−) } m+tu \text{ is even} & d_a \text{ is odd, } a \in I_- \\ \text{(e+) } tu \text{ is odd} & d_a \text{ is even, } a \in I_+ \\ \text{(e−) } tu \text{ is even} & d_a \text{ is even, } a \in I_- \end{cases} \tag{52}$$

We have

$$(a, m, u) : \mathbf{Q}'_+ \iff (a, m, u \pm \tfrac{d_a}{t}) : \mathbf{Q}_+. \tag{53}$$

Let $\mathbf{Q}'_-$ be the negation of $\mathbf{Q}'_+$. Then, we have the parity decomposition

$$\mathcal{Y}^\circ_\ell(C) \simeq \mathcal{Y}^\circ_\ell(C)_+ \times \mathcal{Y}^\circ_\ell(C)_-, \tag{54}$$

where $\mathcal{Y}^\circ_\ell(C)_\varepsilon$ ($\varepsilon = \pm$) is the subring of $\mathcal{Y}^\circ_\ell(C)$ generated by $Y^{(a)}_m(u)$, $1 + Y^{(a)}_m(u)$ ($(a, m, u) : \mathbf{Q}'_\varepsilon$).

6.2. *Construction of quiver $Q_\ell(C)$*

Let us construct a quiver $Q_\ell(C)$ for C and ℓ. We do it in two steps. In Step 1, to each adjacent pair (a, b) of the Dynkin diagram $X(C)$ we attach a certain quiver $Q(a, b)$. In Step 2, these quivers are 'patched' at each vertex.

Step 1. $Q(a, b)$.

Recall that t is the one in (2). Below suppose that a and b are adjacent in $X(C)$ and $d_a \geq d_b$.

Case (i). d_a **is odd and** $d_b = 1$**.** (a) The case $a \in I_-$. We set the quiver $Q(a, b)$ by the quiver $Q_{\ell'}(M_{t'})$ in Sec. 4.3 with $t' = d_a$ and $\ell' = t\ell/d_a$. We assign $+/-$ as in Sec. 4.3. (We do not need to assign $\bullet/\circ$ here.)

(b) The case $a \in I_+$. We set the quiver $Q(a, b)$ by the quiver $Q(1)$ obtained from $Q(0) = Q_{\ell'}(M_{t'})$ in Sec. 4.3 with $t' = d_a$ and $\ell' = t\ell/d_a$. We assign $+/-$ in the opposite way to Sec. 4.3.

Case (ii). $d_a = d_b$, **and** d_a **is odd and not 1.** We can assume that $a \in I_-$ and $b \in I_+$. We set the quiver $Q(a, b)$ as a disjoint union of quivers $Q_1, \dots, Q_{d_a}$ specified as follows. The quivers $Q_1, Q_3, \dots, Q_{d_a}$ are the quiver $Q_{\ell'}(M_{t'})$ in Sec. 4.3 with $t' = 1$ and $\ell' = t\ell/d_a$. We assign $+/-$ as in Sec. 4.3. The quivers $Q_2, Q_4, \dots, Q_{d_a - 1}$ are the opposite quiver of $Q_{\ell'}(M_{t'})$ in Sec. 4.3 with $t' = 1$ and $\ell' = t\ell/d_a$. We assign $+/-$ in the opposite way to Sec. 4.3.

Case (iii). d_a **is even and** $d_b = 1$**.** (a) The case $a \in I_+$ and $c_a = \alpha$. We set the quiver $Q(a, b)$ by the quiver $Q_{\ell'}(M_{t'})$ in Sec. 5.2 with $t' = d_a$ and $\ell' = t\ell/d_a$. We assign $+/-$ as in Sec. 5.2.

(b) The case $a \in I_+$ and $c_a = \beta$. We set the quiver $Q(a,b)$ by the quiver $Q(1)$ obtained from $Q(0) = Q_{\ell'}(M_{t'})$ in Sec. 5.2 with $t' = d_a$ and $\ell' = t\ell/d_a$. For $\bullet/\circ$ in Sec. 5.2, we assign $+/-$ to vertices with $\bullet$ as in Sec. 5.2, while we assign $+/-$ to vertices with $\circ$ in the opposite way to Sec. 5.2.

(c) The case $a \in I_-$ and $c_a = \alpha$. We set the quiver $Q(a,b)$ by the quiver $Q(-1/t')$ obtained from $Q(0) = Q_{\ell'}(M_{t'})$ in Sec. 5.2 with $t' = d_a$ and $\ell' = t\ell/d_a$. For $\bullet/\circ$ in Sec. 5.2. we assign $+/-$ to vertices with $\circ$ as in Sec. 5.2, while we assign $+/-$ to vertices with $\bullet$ in the opposite way to Sec. 5.2.

(d) The case $a \in I_-$ and $c_a = \beta$. We set the quiver $Q(a,b)$ by the quiver $Q((t'-1)/t')$ obtained from $Q(0) = Q_{\ell'}(M_{t'})$ in Sec. 5.2 with $t' = d_a$ and $\ell' = t\ell/d_a$. We assign $+/-$ in the opposite way to Sec. 5.2.

Case (iv). $d_a = d_b$, and d_a is even. We can assume that $c_a = \alpha$ and $c_b = \beta$. We set the quiver $Q(a,b)$ as a disjoint union of quivers $Q_1,\dots,Q_{d_a}$ specified as follows. The quivers $Q_1, Q_3,\dots, Q_{d_a-1}$ are the quiver $Q_{\ell'}(M_{t'})$ in Sec. 4.3 with $t' = 1$ and $\ell' = t\ell/d_a$. We assign $+/-$ as in Sec. 4.3. The quivers $Q_2, Q_4,\dots, Q_{d_a}$ are the opposite quiver of $Q_{\ell'}(M_{t'})$ in Sec. 4.3 with $t' = 1$ and $\ell' = t\ell/d_a$. We assign $+/-$ in the opposite way to Sec. 4.3.

Throughout Step 1, we regard the left column(s) of $Q(a,b)$ (with length $t\ell/d_a - 1$) as *attached to a* and the right column(s) of $Q(a,b)$ (with length $t\ell/d_b - 1$) as *attached to b*.

Step 2. $Q_\ell(C)$.

The quiver $Q_\ell(C)$ is defined by patching the above quivers $Q(a,b)$ at each vertex. Namely, fix $a \in I$, and take all b's which are adjacent to a. If $d_a = 1$, we identify the columns attached to a in $Q(b,a)$ for all b. If $d_a > 1$, for each $i = 1,\dots,d_a$, we identify the columns attached to a in the ith quivers Q_i of $Q(a,b)$ or $Q(b,a)$ (depending on the sign and color of a) for all b. (For Cases (i) and (ii), Q_i appears in the construction of $Q_\ell(M_t)$.)

Some basic examples are given below.

Example 6.1. The two examples below mostly clarify the situation involving Cases (i) and (ii).

(1) Let C be the Cartan matrix with the following Dynkin diagram.

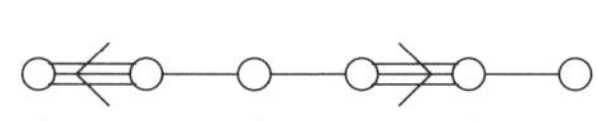

The corresponding quiver $Q_\ell(C)$ for $\ell = 3$ is given as follows.

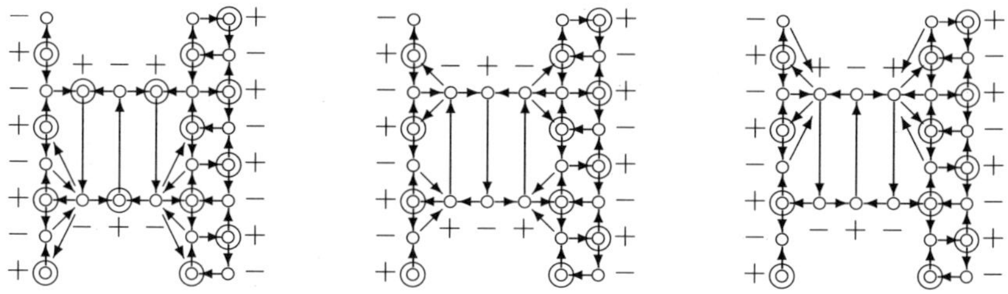

Q_1 Q_2 Q_3

Here, the long columns at the same horizontal positions in three quivers are identified with each other. The encircled vertices are the mutation points at $Q(0) = Q_\ell(C)$, which will be described in the next subsection. The same remark applies below.

(2) Let C be the Cartan matrix with the following Dynkin diagram.

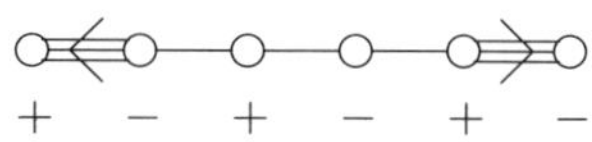

The corresponding quiver $Q_\ell(C)$ for $\ell = 3$ is given as follows.

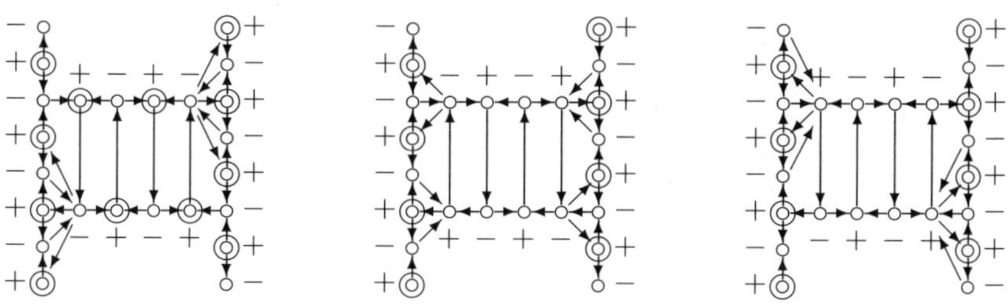

Q_1 Q_2 Q_3

Example 6.2. The four examples below mostly clarify the situation involving Cases (iii) and (iv).

(1) Let C be the Cartan matrix with the following Dynkin diagram.

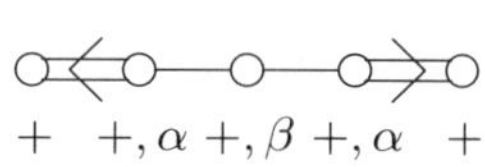

The corresponding quiver $Q_\ell(C)$ for $\ell = 3$ is given as follows.

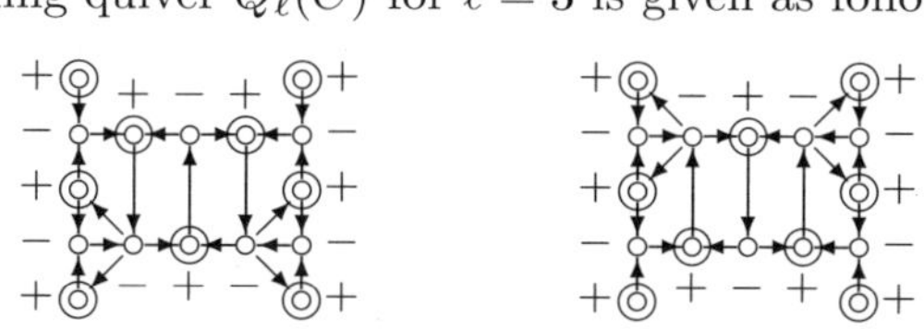

Q_1 Q_2

(2) Let C be the Cartan matrix with the following Dynkin diagram.

The corresponding quiver $Q_\ell(C)$ for $\ell = 3$ is given as follows.

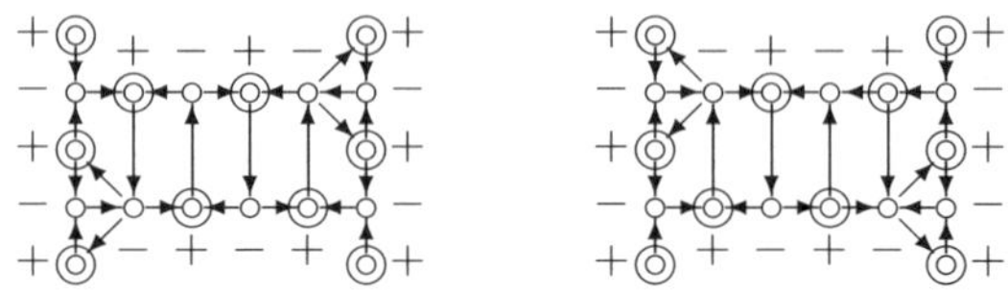

Q_1 Q_2

(3) Let C be the Cartan matrix with the following Dynkin diagram.

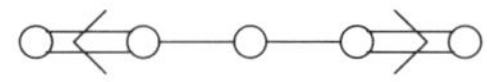

The corresponding quiver $Q_\ell(C)$ for $\ell = 3$ is given as follows.

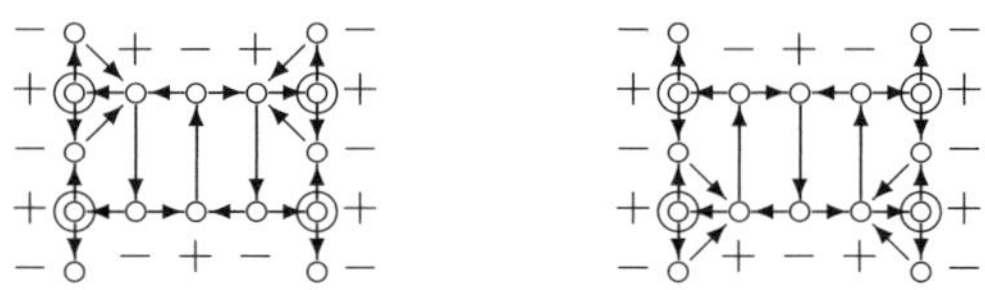

Q_1 Q_2

(4) Let C be the Cartan matrix with the following Dynkin diagram.

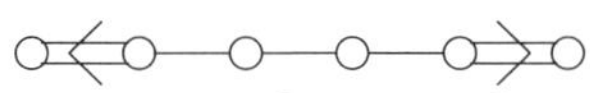

The corresponding quiver $Q_\ell(C)$ for $\ell = 3$ is given as follows.

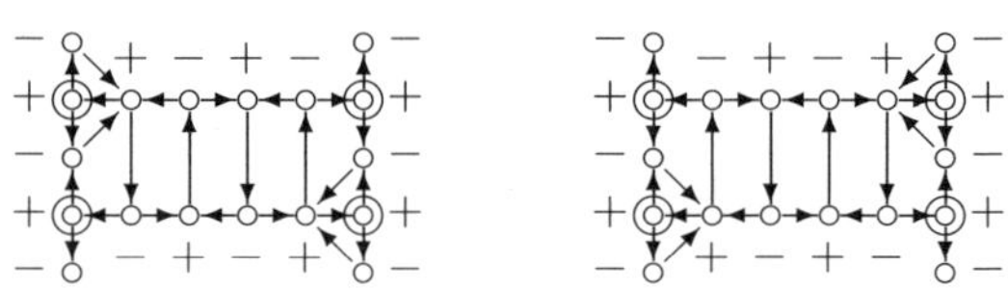

Q_1 Q_2

6.3. *Mutation sequence*

We set $Q(0) = Q_\ell(C)$ and define a periodic sequence of mutations of quivers

$$Q(0) \overset{\mu(0)}{\longleftrightarrow} Q(\tfrac{1}{t}) \overset{\mu(\frac{1}{t})}{\longleftrightarrow} Q(\tfrac{2}{t}) \overset{\mu(\frac{2}{t})}{\longleftrightarrow} \cdots \overset{\mu(\frac{2t-1}{t})}{\longleftrightarrow} Q(2) = Q(0) \qquad (55)$$

by patching the ones in (22) and (42). Let $M_a(k/t)$ ($a \in I$, $k = 0, \dots, 2t-1$) be the set of the mutation points of $\mu(k/t)$ in the columns attached to a.

It is defined as follows. (Below we use the assignment of +/− specified in Sec. 6.2. Also we use the similar notations in the ones in (22) and (42), e.g., $\mathbf{I}^a_{+,i}$ denotes the set of vertices in the column attached to a of ith quiver Q_i with property +.)

(i) d_a: odd. (cf. (22))

$$\begin{aligned}&M_a(0)=\mathbf{I}^a_{+,1},\ M_a(\tfrac{1}{t})=\mathbf{I}^a_{+,d_a-1},\ M_a(\tfrac{2}{t})=\mathbf{I}^a_{+,3},\dots,\\&M_a(\tfrac{2d_a-2}{t})=\mathbf{I}^a_{-,2},\ M_a(\tfrac{2d_a-1}{t})=\mathbf{I}^a_{-,d_a},\ M_a(\tfrac{2d_a}{t})=M_a(0),\dots\end{aligned}\tag{56}$$

In particular, for $d_a=1$,

$$M_a(0)=\mathbf{I}^a_{+,1},\ M_a(\tfrac{1}{t})=\mathbf{I}^a_{-,1},\ M_a(\tfrac{2}{t})=M_0(0),\dots\tag{57}$$

(ii) d_a: even, $a\in I_+$ (cf. (42))

$$\begin{aligned}&M_a(0)=\mathbf{I}^a_{+,1,d_a},\ M_a(\tfrac{1}{t})=\emptyset,\ M_a(\tfrac{2}{t})=\mathbf{I}^a_{+,3,d_a-2},\dots,\\&M_a(\tfrac{2d_a-2}{t})=\mathbf{I}^a_{-,d_a-1,2},\ M_a(\tfrac{2d_a-1}{t})=\emptyset,\ M_a(\tfrac{2d_a}{t})=M_0(0),\dots\end{aligned}\tag{58}$$

(iii) d_a: even, $a\in I_-$ (cf. (42))

$$\begin{aligned}&M_a(0)=\emptyset,\ M_a(\tfrac{1}{t})=\mathbf{I}^a_{+,1,d_a},\ M_a(\tfrac{2}{t})=\emptyset,\ M_a(\tfrac{3}{t})=\mathbf{I}^a_{+,3,d_a-2},\dots,\\&M_a(\tfrac{2d_a-2}{t})=\emptyset,\ M_a(\tfrac{2d_a-1}{t})=\mathbf{I}^a_{-,d_a-1,2},\ M_a(\tfrac{2d_a}{t})=M_0(0),\dots\end{aligned}\tag{59}$$

6.4. *T-system, Y-system, and cluster algebra*

Now it is straightforward to repeat the formulation in Secs. 4 and 5. The compatibility of mutations is the only issue, but it has been already taken care of in the construction of $Q_\ell(C)$ as self-explained in Examples 6.1 and 6.2.

Let $\mathbf{I}$ be the index set of the quiver $Q_\ell(C)$. Let B the skew-symmetric matrix corresponding to $Q_\ell(C)$. Using the sequence (55), we define cluster variables $x_{\mathbf{i}}(u)$ and coefficients $y_{\mathbf{i}}(u)$ ($\mathbf{i}\in\mathbf{I}, u\in\frac{1}{t}\mathbb{Z}$) as before. Define the T-subalgebra $\mathcal{A}_T(B,x)$ and Y-subgroup $\mathcal{A}_T(B,y)$ as parallel to Definitions 4.6 and 4.9.

Repeating the same argument as before, we obtain the conclusion of this section.

Theorem 6.3. *Let C be any tamely laced and indecomposable Cartan matrix whose Dynkin diagram is a tree. Then, the ring $\mathfrak{T}^\circ_\ell(C)_+$ is isomorphic to $\mathcal{A}_T(B,x)$. The group $\mathcal{Y}^\circ_\ell(C)_+$ is isomorphic to $\mathcal{G}_Y(B,y)$.*

7. Cluster algebraic formulation: General case

It is easy to extend Theorem 6.3 to any tamely laced Cartan matrix C with suitable modification. Due to the lack of the space, we concentrate on describing the construction of the quiver $Q_\ell(C)$. Throughout the section we assume that C is a tamely laced and indecomposable Cartan matrix.

Before starting, we introduce some preliminary definitions. We call a subdiagram Y of $X(C)$ an *even block*, if Y is an maximal indecomposable subdiagram of $X(C)$ such that d_a of each vertex a of Y is *even*. Due to the condition (1), d_a is constant for any vertex a of Y. Below we suppose that $X(C)$ has n even blocks $Y_1,\dots, Y_n$. (n may be zero.) Let $X'(C)$ be the diagram obtained from $X(C)$ by shrinking each even block Y_i into a vertex '$\otimes$' while keeping any line from Y_i to its outside. For example, for the following $X(C)$

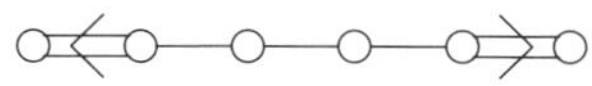
(60)

$X'(C)$ is given by

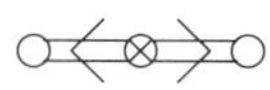
(61)

7.1. *The case $X'(C)$ is bipartite*

Let us assume that $X'(C)$ is *bipartite*, i.e., it contains no odd cycle.

First, consider the case when all the even blocks $Y_1, \dots, Y_n$ are also bipartite. Then, $X(C)$ admits a decomposition and a coloring of I satisfying Conditions (I)–(III) in Sec. 6.1, and one can construct $Q_\ell(C)$ as in Sec. 6.2.

Next, consider the case when some of the even blocks, say, $Y_1, \dots, Y_k$ are nonbipartite. Then, $X(C)$ does not admit a coloring of I satisfying Condition (III) in Sec. 6.1. Following Ref. 4, we define the *bipartite double* $Y^\#$ of any tamely laced Dynkin diagram Y as follows. Let J be the vertex set of Y. The vertex set $J^\#$ of $Y^\#$ is the disjoint union $J^\# = J_+ \sqcup J_-$, where $J_+ = \{j_+ \mid j \in J\}$ and $J_- = \{j_- \mid j \in J\}$; furthermore, we write a line (or multiple line with arrow) in $Y^\#$ from i_+ to j_- and also from i_- to j_+ if and only if there is a line (or multiple line with arrow) from i to j in Y. Let $\tilde{X}(C)$ be the diagram obtained from $X(C)$ by replacing each nonbipartite even block Y_i ($i = 1, \dots, k$) with its bipartite double $Y_i^\#$, while connecting $i_\pm$ in Y_i to any vertex j outside Y_i by a line (or multiple line with arrow) if and only if i and j are connected in $X(C)$ by a line (or multiple line with arrow). The diagram $\tilde{X}(C)$ now admits a decomposition and coloring satisfying Conditions (I)–(III) in Sec. 6.1. See Fig. 7 for an example. Then, we repeat the construction of the quiver $Q_\ell(C)$ in Sec. 6.2 for the diagram

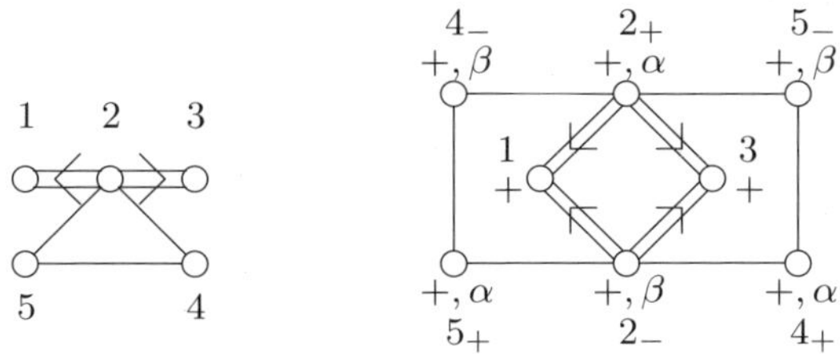

Fig. 7. Example of Dynkin diagram $X(C)$ (left) and $\tilde{X}(C)$ (right).

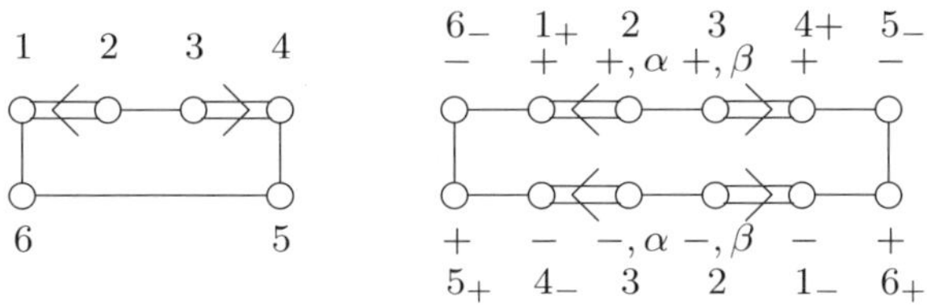

Fig. 8. Example of Dynkin diagram $X(C)$ (left) and $\tilde{X}(C)$ (right).

$\tilde{X}(C)$ with the following modification: *In Step 1 of Sec. 6.2, in Cases (iii) and (iv), we only take the* $d_a/2$ *subquivers* Q_1, $Q_3, \dots,$ Q_{d_a-1} *for those* $Q(a,b)$ *involving the vertices of* $Y_1^{\#}, \dots, Y_k^{\#}$. We write the resulted quiver as $Q_\ell(C)$. Accordingly, we also replace $\mathbf{I}^a_{+,1,d_a}$, $\mathbf{I}^a_{+,3,d_a-2}, \dots$ in (58) and (59) with $\mathbf{I}^a_{+,1}$, $\mathbf{I}^a_{+,3}, \dots$. The rest are defined in the same way as in Sect. 6.4.

Now we have the first main result of the paper.

Theorem 7.1. *Let C be any tamely laced and indecomposable Cartan matrix such that $X'(C)$ is bipartite. Let B the skew-symmetric matrix corresponding to the quiver $Q_\ell(C)$ defined above. Then, the ring $\mathcal{T}^\circ_\ell(C)_+$ is isomorphic to $\mathcal{A}_T(B,x)$. The group $\mathcal{Y}^\circ_\ell(C)_+$ is isomorphic to $\mathcal{G}_Y(B,y)$.*

7.2. *The case $X'(C)$ is nonbipartite*

Let us assume that $X'(C)$ is *nonbipartite*. Then, $X(C)$ does not admit a decomposition of I satisfying Conditions (I) and (II) in Sec. 6.1; consequently, neither $\mathcal{T}^\circ_\ell(C)$ nor $\mathcal{Y}^\circ_\ell(C)$ admits the parity decomposition.

First, consider the case when all the even blocks $Y_1, \dots, Y_n$ of $X(C)$ are bipartite. We take the bipartite double $X'(C)^{\#}$ of $X'(C)$. Then, in $X'(C)^{\#}$, restore each even block of $X(C)$, which appears twice in $X'(C)^{\#}$, in place of $\otimes$. The resulted diagram $\tilde{X}(C)$ now admits a decomposition and a coloring satisfying Conditions (I)—(III) in Sec. 6.1. See Fig. 8 for an example. Now we repeat the construction of the quiver $Q_\ell(C)$ in Sec. 6.2 for the diagram $\tilde{X}(C)$. We write the resulted quiver as $Q_\ell(C)$.

Next, consider the case when some of the even blocks of $X(C)$, say, Y_1, $\dots$, Y_k are nonbipartite. Then, in the above construction of $\tilde{X}(C)$, we further replace each nonbipartite even block Y_i $(i = 1, \dots, k)$ with its bipartite double $Y_i^{\#}$ as in Sec. 7.1. We write the resulted diagram as $\tilde{X}(C)$. Then, repeat the construction of the quiver $Q_\ell(C)$ in Sec. 7.1 for the diagram $\tilde{X}(C)$. We write the resulted quiver as $Q_\ell(C)$.

The rest are defined in the same way as before. Then, as in the simply laced case,[4] we have the counterpart of Theorem 7.1, which is the second main result of the paper.

Theorem 7.2. *Let C be any tamely laced and indecomposable Cartan matrix such that $X'(C)$ is nonbipartite. Let B the skew-symmetric matrix corresponding to the quiver $Q_\ell(C)$ defined above. Then, the ring $\mathcal{T}^\circ_\ell(C)$ is isomorphic to $\mathcal{A}_T(B, x)$. The group $\mathcal{Y}^\circ_\ell(C)$ is isomorphic to $\mathcal{G}_Y(B, y)$.*

Acknowledgments

It is my great pleasure to thank Professor Tetsuji Miwa on the occasion of his sixtieth birthday for his generous support and continuous interest in my works through many years. I thank Rei Inoue, Osamu Iyama, Bernhard Keller, Atsuo Kuniba, and Junji Suzuki for sharing their insights in the preceding joint works.

References

1. R. Inoue, O. Iyama, A. Kuniba, T. Nakanishi and J. Suzuki, Periodicities of T and Y-systems, *Nagoya Math. J.* **197**, 59 (2010).
2. V. Kac, *Infinite dimensional Lie algebras*, 3rd edn. (Cambridge University Press, 1990).
3. D. Hernandez, *Proc. London Math. Soc.* **95**, 567 (2007).
4. A. Kuniba, T. Nakanishi and J. Suzuki, *SIGMA* **5** (2009), 108, 23 pages.
5. S. Fomin and A. Zelevinsky, *J. Amer. Math. Soc.* **15**, 497 (2002).
6. S. Fomin and A. Zelevinsky, *Compositio Mathematica* **143**, 112 (2007).
7. S. Fomin and A. Zelevinsky, *Ann. of Math.* **158**, 977 (2003).
8. P. Di Francesco and R. Kedem, *Lett. Math. Phys.* **89**, 183 (2009).
9. B. Keller, Cluster algebras, quiver representations and triangulated categories, arXiv:0807.1960.
10. D. Hernandez and B. Leclerc, Cluster algebras and quantum affine algebras, arXiv:0903.1452.
11. H. Nakajima, Quiver varieties and cluster algebras, arXiv:0905.0002.
12. P. Di Francesco and R. Kedem, Positivity of the T-system cluster algebra, arXiv:0908.3122.
13. T. Nakanishi, Dilogarithm identities for conformal field theories and cluster algebras: Simply laced case, arXiv:math.0909.5480.

14. B. Keller, The periodicity conjecture for pairs of Dynkin diagrams, arXiv:1001.1531.
15. R. Inoue, O. Iyama, B. Keller, A. Kuniba and T. Nakanishi, Periodicities of T and Y-systems, dilogarithm identities, and cluster algebras I: Type B_r, arXiv:1001.1880.
16. R. Inoue, O. Iyama, B. Keller, A. Kuniba and T. Nakanishi, Periodicities of T and Y-systems, dilogarithm identities, and cluster algebras II: Types C_r, F_4, and G_2, arXiv:1001.1881.

PERIODIC BENJAMIN-ONO EQUATION WITH DISCRETE LAPLACIAN AND 2D-TODA HIERARCHY

JUN'ICHI SHIRAISHI

Graduate School of Mathematical Sciences, The University of Tokyo
3-8-1 Komaba Meguro-ku Tokyo 153-8914, Japan
E-mail: shiraish@ms.u-tokyo.ac.jp

YOHEI TUTIYA

Kanagawa Institute of Technology
1030 Shimo-Ogino Atsugi-city Kanagawa 243-0292, Japan
E-mail: tutiya@gen.kanagawa-it.ac.jp

We study the relation between the periodic Benjamin-Ono equation with discrete Laplacian and the two-dimensional-Toda hierarchy. We introduce the tau-functions $\tau_{\pm}(z)$ for the periodic Benjamin-Ono equation, construct two families of integrals of motion $\{M_1, M_2, \cdots\}$, $\{\overline{M}_1, \overline{M}_2, \cdots\}$, and calculate some examples of the bilinear equations using the Hamiltonian structure. We confirmed that some of the low lying bilinear equations agree with the ones obtained from a certain reduction of the 2D-Toda hierarchy.

Keywords: Benjamin-Ono equation; 2D-Toda hierarchy.

1. Introduction

1.1. *Periodic Benjamin-Ono equation with discrete Laplacian*

Let γ be a complex parameter satisfying $\mathrm{Im}(\gamma) \geq 0$. Let x and t be real independent variables, and $\eta(x,t)$ be an analytic function satisfying the periodicity condition $\eta(x+1,t) = \eta(x,t)$. In Refs. 1 and 2, we considered the integro-differential equation

$$\frac{\partial}{\partial t}\eta(x,t) = \eta(x,t) \cdot \frac{i}{2} \fint_{-1/2}^{1/2} \Big(\Delta_\gamma \cot(\pi(y-x))\Big)\eta(y,t)dy, \tag{1}$$

where the discrete Laplacian Δ_γ is defined by $(\Delta_\gamma f)(x) = f(x-\gamma) - 2f(x) + f(x+\gamma)$, and the integral $\fint$ means the Cauchy principal value. This can be regarded as a periodic version of the Benjamin-Ono[3,4] equation associated with the discrete Laplacian Δ_γ.

For the sake of simplicity, set $z = e^{2\pi i x}$ and $q = e^{2\pi i \gamma}$. By abuse of notation, we use the notation $\eta(z)$ to indicate the dependence on z. From the spatial periodicity, we have the Fourier series expansion $\eta(z) = \sum_{n\in\mathbf{Z}} \eta_n z^{-n}$. Set $\eta_+(z) = \sum_{n>0} \eta_{-n} z^n$ and $\eta_-(z) = \sum_{n>0} \eta_n z^{-n}$. Note that we have $\eta(z) = \eta_+(z) + \eta_0 + \eta_-(z)$. Then the equation (1) can be expressed as

$$\begin{aligned}\partial_t \eta(z) &= \eta(z) \sum_{l\neq 0} \operatorname{sgn}(l)(1-q^{|l|})\eta_{-l} z^l \\ &= \eta(z)(\eta_+(z) - \eta_+(zq) - \eta_-(z) + \eta_-(z/q)). \end{aligned} \tag{2}$$

1.2. *Poisson algebra and Toda field equation*

We show that one can introduce another time $\bar{t}$ and obtain the 2D Toda field equation[5] by using the Poisson Heisenberg algebra for our periodic Benjamin-Ono equation with discrete Laplacian.[1,2] As for the Hamiltonian structure for the usual Benjamin-Ono equation, see Refs. 6 and 7. See Ref. 8 also.

Our Poisson algebra is generated by α_n $(n \in \mathbf{Z}_{\neq 0})$ with the Poisson brackets

$$\{\alpha_n, \alpha_m\} = \operatorname{sgn}(n)(1-q^{|n|})\delta_{n+m,0}, \tag{3}$$

where $\operatorname{sgn}(n) = |n|/n$ for $n \neq 0$ and $\operatorname{sgn}(0) = 0$. Namely, the α_n's are the canonical coordinates being introduced to the study of the Hamiltonian structure associated with (2).

Definition 1.1. Set

$$\tau_+(z) = \exp\left(-\sum_{n>0} \frac{\alpha_{-n}}{1-q^n} z^n\right), \quad \tau_-(z) = \exp\left(-\sum_{n>0} \frac{\alpha_n}{1-q^n} z^{-n}\right). \tag{4}$$

We call $\tau_\pm(z)$ the tau-functions.

Express the dependent variable $\eta(z)$ in terms of the tau-functions and a constant ε as

$$\eta(z) = \sum_{n\in\mathbf{Z}} \eta_n z^{-n} = \varepsilon \exp\left(\sum_{n\neq 0} \alpha_n z^{-n}\right) = \varepsilon \frac{\tau_-(z/q)\tau_+(zq)}{\tau_-(z)\tau_+(z)}. \tag{5}$$

We need to introduce another dependent variable $\xi(z)$ by

$$\xi(z) = \sum_{n\in\mathbf{Z}} \xi_n z^{-n} = \frac{1}{\varepsilon}\exp\left(-\sum_{n\neq 0}\alpha_n q^{-|n|/2} z^{-n}\right)$$
$$= \frac{1}{\varepsilon}\frac{\tau_-(zq^{1/2})\tau_+(zq^{-1/2})}{\tau_-(zq^{-1/2})\tau_+(zq^{1/2})}. \tag{6}$$

The Poisson brackets among our dependent variables can be calculated as follows.

Lemma 1.1. *We have*

$$\{\eta(z),\eta(w)\} = \eta(z)\eta(w)\sum_{l\neq 0}\mathrm{sgn}(l)(1-q^{|l|})\left(\frac{w}{z}\right)^l, \tag{7}$$

$$\{\xi(z),\xi(w)\} = \xi(z)\xi(w)\sum_{l\neq 0}\mathrm{sgn}(l)(q^{-|l|}-1)\left(\frac{w}{z}\right)^l, \tag{8}$$

$$\{\eta(z),\xi(w)\} = \delta(q^{1/2}w/z)\frac{\tau_+(zq)\tau_+(z/q)}{\tau_+(z)\tau_+(z)}$$
$$-\delta(q^{-1/2}w/z)\frac{\tau_-(zq)\tau_-(z/q)}{\tau_-(z)\tau_-(z)}, \tag{9}$$

$$\{\eta(w),\tau_-(z)\} = \eta(w)\tau_-(z)\sum_{n>0}\left(\frac{w}{z}\right)^n, \tag{10}$$

$$\{\eta(w),\tau_+(z)\} = -\eta(w)\tau_+(z)\sum_{n>0}\left(\frac{z}{w}\right)^n, \tag{11}$$

$$\{\xi(w),\tau_-(z)\} = -\xi(w)\tau_-(z)\sum_{n>0}q^{-n/2}\left(\frac{w}{z}\right)^n, \tag{12}$$

$$\{\xi(w),\tau_+(z)\} = \xi(w)\tau_+(z)\sum_{n>0}q^{-n/2}\left(\frac{z}{w}\right)^n, \tag{13}$$

where $\delta(z) = \sum_{n\in\mathbf{Z}} z^n$.

Remark 1.1. In Ref. 9, a deep connection was found between the Macdonald Polynomials[10] $P_\lambda(x;q,t)$ and the level one representation of the quantum algebra of Ding-Iohara $\mathcal{U}(q,t)$. We note that the α_n, $\eta(z)$ and $\xi(z)$ in the present paper are the level one generators of the Ding-Iohara algebra in the classical (namely commutative) limit given by letting the parameter as $t\to 1$. (Note that t here is one of the two parameters q,t for the Macdonald polynomials and should not be confused with the time t.)

As a special case of (7) in Lemma 1.1, we have

$$\{\eta_0, \eta(z)\} = \eta(z) \sum_{l\neq 0} \mathrm{sgn}(l)(1-q^{|l|})\eta_{-l} z^l. \tag{14}$$

By looking at this and (2), one is lead to the following definition.

Definition 1.2. Let η_0 be our Hamiltonian, and define the time evolution equation over the Poisson algebra by setting $\partial_t * = \{\eta_0, *\}$.

Proposition 1.1. *The periodic Benjamin-Ono equation with discrete Laplacian (2) is written in the Hamiltonian form* $\partial_t \eta(z) = \{\eta_0, \eta(z)\}$.

Now we move on to the study of another Hamiltonian derived from $\xi(z)$.

Proposition 1.2. *We have* $\{\eta_0, \xi_0\} = 0$.

Proof. This follows from (9) in Lemma 1.1. □

Remark 1.2. Since our Poisson algebra is the classical limit of the Ding-Iohara algebra, we have two sets of mutually Poisson commutative families, having η_0 and ξ_0 respectively, in the same way as was discussed in Ref. 9. As for the explicit form of them, see (38) and (39) below.

Because of the commutativity $\{\eta_0, \xi_0\} = 0$, we may interpret ξ_0 as another Hamiltonian corresponding to time $\bar{t}$.

Definition 1.3. Define $\partial_{\bar{t}} * = \{*, \xi_0\}$.

Proposition 1.3. *From (9) we have*

$$\partial_{\bar{t}} \eta(z) = \frac{\tau_+(zq)\tau_+(z/q)}{\tau_+(z)\tau_+(z)} - \frac{\tau_-(zq)\tau_-(z/q)}{\tau_-(z)\tau_-(z)}, \tag{15}$$

Next we turn to the equations for the tau-functions $\tau_\pm(z)$.

Proposition 1.4. *From (10), (11), (12) and (13), we have*

$$\partial_t \tau_-(z) = \eta_-(z)\tau_-(z), \qquad \partial_t \tau_+(z) = -\eta_+(z)\tau_+(z), \tag{16}$$
$$\partial_{\bar{t}} \tau_-(zq^{-1/2}) = \xi_-(z)\tau_-(zq^{-1/2}), \qquad \partial_{\bar{t}} \tau_+(zq^{1/2}) = -\xi_+(z)\tau_+(zq^{1/2}),$$

where $\xi_+(z) = \sum_{n>0} \xi_{-n} z^n$ *and* $\xi_-(z) = \sum_{n>0} \xi_n z^{-n}$.

Suitable combinations of these may give us equations written in terms of the Hirota derivatives.

Definition 1.4. Define the Hirota derivative $D_{t_1}, D_{t_2}, \cdots$ by

$$\left(D_{t_1}^{k_1} D_{t_2}^{k_2} \cdots\right) f \cdot g \tag{17}$$
$$= \partial_{a_1}^{k_1} \partial_{a_2}^{k_2} \cdots f(t_1 + a_1, t_2 + a_2, \cdots) g(t_1 - a_1, t_2 - a_2, \cdots)\Big|_{a_1 = a_2 = \cdots = 0}.$$

Proposition 1.5. *We have the Hirota equations*

$$D_t \tau_-(z) \cdot \tau_+(z) = \varepsilon \tau_-(z/q) \tau_+(zq) - \eta_0 \tau_-(z) \tau_+(z), \tag{18}$$
$$D_{\bar{t}} \tau_-(zq^{-1/2}) \cdot \tau_+(zq^{1/2}) \tag{19}$$
$$= \varepsilon^{-1} \tau_-(zq^{1/2}) \tau_+(zq^{-1/2}) - \xi_0 \tau_-(zq^{-1/2}) \tau_+(zq^{1/2}),$$
$$\frac{1}{2} D_t D_{\bar{t}} \tau_\pm(z) \cdot \tau_\pm(z) \tag{20}$$
$$+ \tau_\pm(zq) \cdot \tau_\pm(z/q) - \tau_\pm(z) \cdot \tau_\pm(z) = 0.$$

Proof. From (16), we have $\partial_t \tau_-(z) \cdot \tau_+(z) - \tau_-(z) \cdot \partial_t \tau_+(z) = (\eta(z) - \eta_0) \tau_-(z) \tau_+(z)$. Using (5) we have (18). The equation (19) can be derived in the same way. Eq. (20) is obtained from (15) and (16). □

Remark 1.3. Note that Eq. (20) is nothing but the Toda field equation written in terms of the tau-function.[5]

One finds that the Heisenberg generators correspond to the standard dependent variables of the Toda field theory.

Definition 1.5. Set

$$\phi_+(z) = \sum_{n>0} \alpha_{-n} z^n, \qquad \phi_-(z) = -\sum_{n>0} \alpha_n z^{-n}. \tag{21}$$

Proposition 1.6. *The $\phi_\pm(z)$ satisfy the Toda field equation*

$$\partial_t \partial_{\bar{t}} \phi_\pm(z) = e^{\phi_\pm(z) - \phi_\pm(z/q)} - e^{\phi_\pm(zq) - \phi_\pm(z)}. \tag{22}$$

Proof. This follows from Eqs. (3) and (9). □

Motivated by the appearance of the Toda field equation (22), in this article we will try to understand how the 2D Toda hierarchy appears from the point of view of the Hamiltonian structure.

The present paper is organized as follows. In Section 2, we recall the Hirota-Miwa equation[11,12] (23) for the 2D Toda hierarchy. Based on the n-soliton solutions, we derive some variants of the bilinear equations (Proposition 2.1). In Section 3, two sets of integrals of motion $M_1, M_2, \cdots$ and

$\overline{M}_1, \overline{M}_2, \cdots$ are introduced (Definition 3.2). Since at present we lack enough technologies to handle the evolution equations in general, we need to restrict ourself to some low lying cases. To show some evidences of the agreement, we check up to certain degree that exactly the same equations are obtained both from Proposition 2.1 and the Hamiltonian M_k's. Our observation is summarized in Conjecture 3.2.

2. Hirota-Miwa equation for 2D Toda hierarchy

2.1. *Hirota-Miwa equation*

We briefly recall the Hirota-Miwa equation[11,12] for the 2D Toda hierarchy.[5] Using the n-soliton solution, we derive several variants of bilinear equations which can be connected to Eqs. (18), (19) and (20) obtained in the previous section.

Let s, $t = (t_1, t_2, \cdots)$ and $\bar{t} = (\bar{t}_1, \bar{t}_2, \cdots)$ be independent variables, and $\tau(s, t, \bar{t})$ be the tau-function of the 2D Toda hierarchy. For a parameter λ, we use the standard notation for the infinite vector $[\lambda] = (\lambda, \frac{1}{2}\lambda^2, \frac{1}{3}\lambda^3, \cdots)$.

The Hirota-Miwa equation for the 2D Toda hierarchy is written as follows.

$$\begin{aligned}&(1-\alpha\beta)\tau(s,t,\bar{t})\tau(s,t+[\alpha],\bar{t}+[\beta]) - \tau(s,t+[\alpha],\bar{t})\tau(s,t,\bar{t}+[\beta])\\ &+ \alpha\beta\tau(s+1,t+[\alpha],\bar{t})\tau(s-1,t,\bar{t}+[\beta]) = 0.\end{aligned} \tag{23}$$

It is well known that the n-soliton solution to the Hirota-Miwa equation is given by

$$\begin{aligned}\tau(s,t,\bar{t}) = &\sum_{r=0}^{n} \sum_{\substack{I\subset\{1,2,\cdots,n\}\\ |I|=r}} \prod_{\substack{\{i,j\}\subset I\\ i<j}} \frac{(\lambda_i-\lambda_j)(\mu_i-\mu_j)}{(\lambda_i-\mu_j)(\mu_i-\lambda_j)}\\ &\times \prod_{k\in I} (\lambda_k/\mu_k)^s e^{\sum_{i=1}^{\infty}(t_i\lambda_k^i+\bar{t}_i\lambda_k^{-i}) - \sum_{i=1}^{\infty}(t_i\mu_k^i+\bar{t}_i\mu_k^{-i})}.\end{aligned} \tag{24}$$

Let $a_1, \cdots, a_n$ be parameters. Set $\lambda_k = a_k, \mu_k = qa_k$ for $k = 1, 2, \cdots, n$. Write $z = q^{-s}$. Then we have $(\lambda_k/\mu_k)^s = q^{-s} = z$.

We define $\tau_+(z, t, \bar{t})$ by the n-soliton solution $\tau(s, t, \bar{t})$ of 2D Toda hierarchy under this specialization ($\lambda_k = a_k, \mu_k = qa_k$).

Definition 2.1. Set

$$\begin{aligned}\tau_+(z,t,\bar{t}) = &\sum_{r=0}^{n} z^r \sum_{\substack{I\subset\{1,2,\cdots,n\}\\ |I|=r}} \prod_{\substack{\{i,j\}\subset I\\ i<j}} \frac{(a_i-a_j)^2}{(a_i-qa_j)(a_i-q^{-1}a_j)}\\ &\times \prod_{k\in I} e^{\sum_{i=1}^{\infty}(1-q^i)t_i a_k^i + \sum_{i=1}^{\infty}(1-q^{-i})\bar{t}_i a_k^{-i}}.\end{aligned} \tag{25}$$

Note that $\tau_+(z,t,\bar{t})$ is a polynomial in z whose degree is n.

To introduce $\tau_-(z,t,\bar{t})$, we need a Lemma.

Lemma 2.1. *We have*

$$\begin{aligned}
&\tau_+(z,t,\bar{t}-[\beta]) \\
&= \sum_{r=0}^{n} z^r \sum_{\substack{I\subset\{1,2,\cdots,n\}\\ |I|=r}} \prod_{\substack{\{i,j\}\subset I\\ i<j}} \frac{(a_i-a_j)^2}{(a_i-qa_j)(a_i-q^{-1}a_j)} \\
&\quad \times \prod_{k\in I} \frac{1-\beta/a_k}{1-\beta/qa_k} e^{\sum_{i=1}^{\infty}(1-q^i)t_i a_k^i + \sum_{i=1}^{\infty}(1-q^{-i})\bar{t}_i a_k^{-i}} \\
&= z^n \prod_{1\le i<j\le n} \frac{(a_i-a_j)^2}{(a_i-qa_j)(a_i-q^{-1}a_j)} \\
&\quad \times \prod_{k=1}^{n} \frac{1-\beta/a_k}{1-\beta/qa_k} e^{\sum_{i=1}^{\infty}(1-q^i)t_i a_k^i + \sum_{i=1}^{\infty}(1-q^{-i})\bar{t}_i a_k^{-i}} \\
&\quad \times \sum_{r=0}^{n} z^{-r} \sum_{\substack{I\subset\{1,2,\cdots,n\}\\ |I|=r}} \prod_{\substack{\{i,j\}\subset I\\ i<j}} \frac{(a_i-a_j)^2}{(a_i-qa_j)(a_i-q^{-1}a_j)} \\
&\quad \times \prod_{k\in I} d_k(\beta) e^{-\sum_{i=1}^{\infty}(1-q^i)t_i a_k^i - \sum_{i=1}^{\infty}(1-q^{-i})\bar{t}_i a_k^{-i}},
\end{aligned}$$

where

$$d_k(\beta) = \frac{1-\beta/qa_k}{1-\beta/a_k} \prod_{\substack{j\ne k\\ 1\le j\le n}} \frac{(a_k-qa_j)(a_k-q^{-1}a_j)}{(a_k-a_j)^2}.$$

Now we define $\tau_-(z,t,\bar{t})$ by the following Laurent polynomial.

Definition 2.2. Set

$$\begin{aligned}
&\tau_-(z,t,\bar{t}) \\
&= \tau_+(z,t,\bar{t}-[q^n\varepsilon]) \times z^{-n} \prod_{1\le i<j\le n} \frac{(a_i-qa_j)(a_i-q^{-1}a_j)}{(a_i-a_j)^2} \\
&\quad \times \prod_{k=1}^{n} \frac{1-q^{n-1}\varepsilon/a_k}{1-q^n\varepsilon/a_k} e^{-\sum_{i=1}^{\infty}(1-q^i)t_i a_k^i - \sum_{i=1}^{\infty}(1-q^{-i})\bar{t}_i a_k^{-i}} \\
&= \sum_{r=0}^{n} z^{-r} \sum_{\substack{I\subset\{1,2,\cdots,n\}\\ |I|=r}} \prod_{\substack{\{i,j\}\subset I\\ i<j}} \frac{(a_i-a_j)^2}{(a_i-qa_j)(a_i-q^{-1}a_j)} \qquad (26) \\
&\quad \times \prod_{k\in I} d_k(q^n\varepsilon) e^{-\sum_{i=1}^{\infty}(1-q^i)t_i a_k^i - \sum_{i=1}^{\infty}(1-q^{-i})\bar{t}_i a_k^{-i}}.
\end{aligned}$$

Proposition 2.1. *We have*

$$\begin{aligned}
&\tau_-(z,t+[\alpha],\bar{t})\tau_+(z,t,\bar{t})\\
&=(1-\alpha q^n\varepsilon)\prod_{k=1}^{n}\frac{(1-\alpha a_k)}{(1-\alpha q a_k)}\tau_-(z,t,\bar{t})\tau_+(z,t+[\alpha],\bar{t}) \qquad (27)\\
&\qquad +\alpha\varepsilon\,\tau_-(z/q,t+[\alpha],\bar{t})\tau_+(zq,t,\bar{t}),\\
&\tau_-(z/q,t,\bar{t}+[\beta])\tau_+(z,t,\bar{t})\\
&=(1-\beta/q^n\varepsilon)\prod_{k=1}^{n}\frac{(1-\beta/a_k)}{(1-\beta/q a_k)}\tau_-(z/q,t,\bar{t})\tau_+(z,t,\bar{t}+[\beta]) \qquad (28)\\
&\qquad +(\beta/\varepsilon)\,\tau_-(z,t,\bar{t}+[\beta])\tau_+(z/q,t,\bar{t}),\\
&\tau_\pm(z,t+[\alpha],\bar{t})\tau_\pm(z,t,\bar{t}+[\beta]) \qquad (29)\\
&=(1-\alpha\beta)\tau_\pm(z,t,\bar{t})\tau_\pm(z,t+[\alpha],\bar{t}+[\beta])\\
&\qquad +\alpha\beta\tau_\pm(z/q,t+[\alpha],\bar{t})\tau_\pm(zq,t,\bar{t}+[\beta])=0.
\end{aligned}$$

Proof. Eq. (27) follows from the Hirota-Miwa equation (23), (25), (26) and Lemma 2.1. Noting that we have $\tau(s,t-[\alpha],\bar{t})=\tau(s+1,t,\bar{t}-[\alpha^{-1}])$, we have (28) in the same way. Eq. (29) follows from Eq. (23). □

Remark 2.1. Note that we may write

$$(1-\alpha q^n\varepsilon)\prod_{k=1}^{n}\frac{(1-\alpha a_k)}{(1-\alpha q a_k)}=\exp\left(-\sum_{i=1}^{\infty}M_i\alpha^i\right), \qquad (30)$$

$$(1-\beta/q^n\varepsilon)\prod_{k=1}^{n}\frac{(1-\beta/a_k)}{(1-\beta/q a_k)}=\exp\left(-\sum_{i=1}^{\infty}\overline{M}_i\beta^i\right), \qquad (31)$$

$$M_i=\frac{1-q^i}{i}\left(a_1^i+\cdots+a_n^i+q^{ni}\varepsilon^i+q^{(n+1)i}\varepsilon^i+q^{(n+2)i}\varepsilon^i+\cdots\right), \qquad (32)$$

$$\overline{M}_i=\frac{1-q^{-i}}{i}\left(a_1^{-i}+\cdots+a_n^{-i}+q^{-ni}\varepsilon^{-i}+q^{-(n+1)i}\varepsilon^{-i}+\cdots\right). \qquad (33)$$

Proposition 2.2. *By expanding (27) in α, subtracting some constant multiples or time-differentials of lower order equations, we have the Hirota equations*

$$(D_{t_1}+M_1)\tau_-(z)\cdot\tau_+(z)=\varepsilon\tau_-(z/q)\cdot\tau_+(zq), \qquad (34)$$

$$(D_{t_2}+2M_2)\tau_-(z)\cdot\tau_+(z)=\varepsilon(D_{t_1}+M_1)\tau_-(z/q)\cdot\tau_+(zq), \qquad (35)$$

$$\begin{aligned}
&(D_{t_3}+3M_3)\tau_-(z)\cdot\tau_+(z)+\frac{1}{8}(D_{t_1}+M_1)^3\tau_-(z)\cdot\tau_+(z) \qquad (36)\\
&=\frac{3}{4}\varepsilon(D_{t_2}+2M_2)\tau_-(z/q)\cdot\tau_+(zq)+\frac{3}{8}\varepsilon(D_{t_1}+M_1)^2\tau_-(z/q)\cdot\tau_+(zq).
\end{aligned}$$

Here M_i's are defined in (32).

Thus we found that Eq. (18) coincide with Eq. (34) under the identification $t = t_1, \bar{t} = \bar{t}_1$. Eqs. (19) and (20) coincide with the first nontrivial equation from Eqs. (28) and (29) respectively.

In the next section, we will check that Eqs. (35) and Eq. (36) also agree with the equations derived from the Poisson structure.

3. Poisson algebra and 2D Toda hierarchy

3.1. *elementary and power sum symmetric functions*

We need some facts about the symmetric functions.[10] Let $x = (x_1, x_2, \cdots)$ be an infinite set of independent indeterminates. Let $e_n(x)$ be the n-th elementary symmetric function, and $p_n(x)$ be the n-th power sum function. The generating functions for them are given by $E(y) = \sum_{n=0}^{\infty} e_n(x)y^n = \prod_{i=1}^{\infty}(1 + x_i y)$, and $P(y) = \sum_{n=1}^{\infty} \frac{1}{n} p_n(x)y^n = -\log E(-y)$. Solving the equation $P'(y) = -E'(-y)/E(-y)$, we have

$$p_n = \begin{vmatrix} e_1 & 1 & 0 & \cdots 0 \\ 2e_2 & e_1 & 1 & \cdots \\ 3e_3 & e_2 & e_1 & \cdots \\ \vdots & & & \ddots \vdots \\ ne_n & e_{n-1} & e_{n-2} & \cdots e_1 \end{vmatrix}. \tag{37}$$

3.2. *Integrals of motion from Ding-Iohara algebra*

First we introduce some notations. We denote the constant term f_0 of a series $f(z) = \sum_{n \in \mathbf{Z}} f_n z^n$ by $[f(z)]_1$. We also use the same symbol for the case of a series with several variables. For examples, by $[f(z_1, z_2)]_1$ we denote the constant term $f_{0,0}$ of the series $f(z_1, z_2) = \sum_{n_1, n_2 \in \mathbf{Z}} f_{n_1, n_2} z_1^{n_1} z_2^{n_2}$.

Definition 3.1. Define the integrals of motion by

$$I_k = \left[\prod_{1 \le i < j \le k} \frac{1 - w_j/w_i}{1 - q w_j/w_i} \eta(w_1)\eta(w_2) \cdots \eta(w_k) \right]_1, \tag{38}$$

$$\bar{I}_k = \left[\prod_{1 \le i < j \le k} \frac{1 - w_j/w_i}{1 - q^{-1} w_j/w_i} \xi(w_1)\xi(w_2) \cdots \xi(w_k) \right]_1, \tag{39}$$

where the rational factors in w_i's should be understood in the sense of the series as $(1 - w_j/w_i)/(1 - q^{\pm 1} w_j/w_i) = 1 + (1 - q^{\mp 1}) \sum_{n>0} (q^{\pm 1} w_j/w_i)^n$.

For example, we have $I_1 = \eta_0$ and $I_2 = \eta_0^2 + (1-q^{-1})\sum_{n>0} q^n \eta_{-n}\eta_n$, and so on. Based on the argument given in Ref. 9 with considering the classical limit ($t \to 1$), one can prove the following.

Proposition 3.1. *We have the commutativity* $\{I_k, I_l\} = 0$, $\{\overline{I}_k, \overline{I}_l\} = 0$ *and* $\{I_k, \overline{I}_l\} = 0$.

3.3. *Integrals of motion associated with t and* $\overline{t}$

Some explicit calculations show us that the integrals I_k and $\overline{I}_k$ does not correspond to the Toda times $t_1, t_2, \cdots$ and $\overline{t}_1, \overline{t}_2, \cdots$ in general. Hence our task is to find a suitable set of integrals, which we call M_k and $\overline{M}_k$.

At present, unfortunately, it is not easy to do the task purely within the framework of Poisson algebra. However, with the knowledge of the values of the integrals I_k and $\overline{I}_k$ on the n-soliton solution (25), (26), we can guess the correct formula.

Conjecture 3.1. *Let* $\tau_\pm(z, t, \overline{t})$ *be as in (25) and (26), and set* $\eta(t) = \varepsilon\tau_-(z/q, t, \overline{t})\tau_+(zq, t, \overline{t})/\tau_-(z, t, \overline{t})\tau_+(z, t, \overline{t})$, *namely as in (5). Then one can define the quantities* I_k*'s and* $\overline{I}_k$*'s by (38) and (39) using this* $\eta(z)$ *associated with the n-soliton solution.*

Then the quantities I_k*'s and* $\overline{I}_k$*'s are independent of t and* $\overline{t}$. *The values are given by the follwing specialization of the elementary symmetric functions as*

$$\begin{aligned} I_k &= q^{-k(k-1)/2}(1-q)(1-q^2)\cdots(1-q^k) \\ &\quad\times e_k(a_1, \cdots, a_n, q^n\varepsilon, q^{n+1}\varepsilon, \cdots), \end{aligned} \tag{40}$$

$$\begin{aligned} \overline{I}_k &= q^{k(k-1)/2}(1-q^{-1})(1-q^{-2})\cdots(1-q^{-k}) \\ &\quad\times e_k(a_1^{-1}, \cdots, a_n^{-1}, q^{-n}\varepsilon^{-1}, q^{-n-1}\varepsilon^{-1}, \cdots). \end{aligned} \tag{41}$$

As for the statement about I_k in Eq. (40), see Ref. 2.

Remark 3.1. For small k, we have

$$\begin{aligned} I_1 &= (1-q)(a_1 + \cdots + a_n) + q^n\varepsilon, \\ I_2 &= q^{-1}(1-q)(1-q^2)(a_1a_2 + a_1a_3 + \cdots + a_{n-1}a_n) \\ &\quad + q^{n-1}(1-q^2)(a_1 + \cdots + a_n)\varepsilon + q^{2n}\varepsilon^2. \end{aligned}$$

Definition 3.2. Set $I'_k = q^{k(k-1)/2}((1-q)(1-q^2)\cdots(1-q^k))^{-1} I_k$ and $\overline{I}'_k = q^{-k(k-1)/2}((1-q^{-1})(1-q^{-2})\cdots(1-q^{-k}))^{-1}\overline{I}_k$ for simplicity of display.

Define

$$M_k = \frac{1-q^k}{k}\begin{vmatrix} I'_1 & 1 & 0 & \cdots 0 \\ 2I'_2 & I'_1 & 1 & \cdots \\ 3I'_3 & I'_2 & I'_1 & \cdots \\ \vdots & & & \ddots \vdots \\ kI'_k & I'_{k-1} & I'_{k-2} & \cdots I'_1 \end{vmatrix}, \tag{42}$$

$$\overline{M}_k = \frac{1-q^{-k}}{k}\begin{vmatrix} \overline{I}'_1 & 1 & 0 & \cdots 0 \\ 2\overline{I}'_2 & \overline{I}'_1 & 1 & \cdots \\ 3\overline{I}'_3 & \overline{I}'_2 & \overline{I}'_1 & \cdots \\ \vdots & & & \ddots \vdots \\ k\overline{I}'_k & \overline{I}'_{k-1} & \overline{I}'_{k-2} & \cdots \overline{I}'_1 \end{vmatrix}. \tag{43}$$

Remark 3.2. Conjecture 3.1 implies that if $\tau_\pm(z,t,\overline{t})$ be as in (25) and (26), M_k's and $\overline{M}_k$'s are given by

$$M_k = \frac{1-q^k}{k} p_k(a_1, \cdots, a_n, q^n\varepsilon, q^{n+1}\varepsilon, \cdots),$$
$$\overline{M}_k = \frac{1-q^{-k}}{k} p_k(a_1^{-1}, \cdots, a_n^{-1}, q^{-n}\varepsilon^{-1}, q^{-n-1}\varepsilon^{-1}, \cdots).$$

3.4. *Formulas for M_2 and M_3*

Now we come back to our study of the Poisson algebra to check the higher Hirota equations (35) and (36).

It is desirable to find some reasonably simple expressions for M_k's. At present, however, we only have the following partial results.

Lemma 3.1. *We have $M_1 = [\eta(w)]_1$ and*

$$M_2 = \left[\left(\frac{1}{2} + \frac{qw_2/w_1}{1-qw_2/w_1}\right)\eta(w_1)\eta(w_2)\right]_1, \tag{44}$$

$$M_3 = \left[\left(\frac{1}{3} + \frac{qw_3/w_2}{(1-qw_2/w_1)(1-qw_3/w_2)}\right)\eta(w_1)\eta(w_2)\eta(w_3)\right]_1. \tag{45}$$

Proof. From (42), we have $M_1 = I_1 = [\eta(w)]_1$, and

$$\begin{aligned} M_2 &= \frac{1+q}{2(1-q)} I_1^2 - \frac{q}{1-q} I_2 \\ &= \left[\left(\frac{1+q}{2(1-q)} - \frac{q}{1-q}\frac{1-w_2/w_1}{1-qw_2/w_1}\right)\eta(w_1)\eta(w_2)\right]_1 \\ &= \left[\left(\frac{1}{2} + \frac{qw_2/w_1}{1-qw_2/w_1}\right)\eta(w_1)\eta(w_2)\right]_1. \end{aligned}$$

Next, we have

$$
\begin{aligned}
M_3 &= \frac{1-q^3}{3(1-q)^2} I_1^3 - \frac{q(1-q^3)}{(1-q)^2(1-q^2)} I_1 I_2 + \frac{q^3}{(1-q)(1-q^2)} I_3 \\
&= \Bigg[\Bigg(\frac{1-q^3}{3(1-q)^2} - \frac{q(1-q^3)}{(1-q)^2(1-q^2)} \frac{1-w_3/w_2}{1-qw_3/w_2} \\
&\qquad + \frac{q^3}{(1-q)(1-q^2)} \frac{1-w_2/w_1}{1-qw_2/w_1} \frac{1-w_3/w_1}{1-qw_3/w_1} \frac{1-w_3/w_2}{1-qw_3/w_2} \Bigg) \\
&\quad \times \eta(w_1)\eta(w_1)\eta(w_1) \Bigg]_1 .
\end{aligned}
$$

Since we have the property of the constant term $[f(w_1,w_2,w_3)]_1 = [\mathrm{Sym}_{w_1,w_2,w_3} f(w_1,w_2,w_3)]_1$, and

$$
\begin{aligned}
&\mathrm{Sym}_{w_1,w_2,w_3} \Bigg(\frac{1-q^3}{3(1-q)^2} - \frac{q(1-q^3)}{(1-q)^2(1-q^2)} \frac{1-w_3/w_2}{1-qw_3/w_2} \\
&\qquad + \frac{q^3}{(1-q)(1-q^2)} \frac{1-w_2/w_1}{1-qw_2/w_1} \frac{1-w_3/w_1}{1-qw_3/w_1} \frac{1-w_3/w_2}{1-qw_3/w_2} \Bigg) \\
&= \mathrm{Sym}_{w_1,w_2,w_3} \left(\frac{1}{3} + \frac{qw_3/w_2}{(1-qw_2/w_1)(1-qw_3/w_2)} \right),
\end{aligned}
$$

(45) holds. Here the symbol $\mathrm{Sym}_{w_1,w_2,w_3}$ denotes the symmetrization with respect to w_1, w_2 and w_3. □

Remark 3.3. It is an open problem to find a simple expression as above for $M_4, M_5, \cdots$.

3.5. *Main conjecture and equations with respect to* t_2, t_3

Definition 3.3. Set $\partial_{t_k} * = \{M_k, *\}$ and $\partial_{\bar{t}_k} * = \{*, \overline{M}_k\}$.

Now we are ready to state our conjecture.

Conjecture 3.2. *Calculating ∂_{t_k} and $\partial_{\bar{t}_k}$ by using the Poisson brackets given in Definition 3.3, we recover the same equation derived from the Hirota-Miwa equations (27), (28) and (29) within the Hamiltonian formalism.*

The rest of the paper is devoted to give some evidence of our conjecture.

Proposition 3.2. *Calculating ∂_{t_2} and ∂_{t_1} by using the Poisson brackets given in Definition 3.3, we recover the same equation as in Eq. (35).*

For the sake of simplicity, we introduce a notation. For a series $f(z, w_1, w_2, \cdots, w_n)$ in z and $w_1, w_2, \cdots w_n$, we denote the constant term with respect to w_i's by the symbol $[f(z, w_1, w_2, \cdots, w_n)]_{1,w_1,\cdots,w_n}$. For example, for $f(z, w_1, w_2) = \sum_{i,j,k\in\mathbf{Z}} f_{i,j,k} z^i w_1^k w_2^l$, we have $[f(z, w_1, w_2)]_{1,w_1,w_2} = \sum_{i\in\mathbf{Z}} f_{i,0,0} z^i$.

Proof. From (10), (11) and (44), we have

$$
\begin{aligned}
&\frac{\partial_{t_2}\tau_-(z)}{\tau_-(z)} - \frac{\partial_{t_2}\tau_+(z)}{\tau_+(z)} + 2M_2 \\
&= \left[(\delta(w_1/z) + \delta(w_2/z)) \left(\frac{1}{2} + \frac{q w_2/w_1}{1 - q w_2/w_1} \right) \eta(w_1)\eta(w_2) \right]_{1,w_1,w_2} \\
&= M_1\eta(z) + \eta(z) \left[\frac{q w_2/z}{1 - q w_2/z} \eta(w_2) \right]_{1,w_2} + \eta(z) \left[\frac{q z/w_1}{1 - q z/w_1} \eta(w_1) \right]_{1,w_1} \\
&= M_1\eta(z) + \eta(z)\eta_-(z/q) + \eta(z)\eta_+(zq).
\end{aligned}
$$

Using (5) and (16) (with $t = t_1$), we have the result. □

Finally, we study the Hirota equation involving the third time t_3.

Proposition 3.3. *Calculating ∂_{t_3}, ∂_{t_2} and ∂_{t_1} by using the Poisson brackets given in Definition 3.3, we have*

$$
\begin{aligned}
&(D_{t_3} + 3M_3)\,\tau_-(z)\cdot\tau_+(z) \qquad (46)\\
&= \frac{1}{2}\varepsilon\,(D_{t_2} + 2M_2)\,\tau_-(z/q)\cdot\tau_+(zq) + \frac{1}{2}\varepsilon\,(D_{t_1} + M_1)^2\,\tau_-(z/q)\cdot\tau_+(zq),\\
&(D_{t_1} + M_1)^3\,\tau_-(z)\cdot\tau_+(z) \qquad (47)\\
&= 2\varepsilon\,(D_{t_2} + 2M_2)\,\tau_-(z/q)\cdot\tau_+(zq) - \varepsilon\,(D_{t_1} + M_1)^2\,\tau_-(z/q)\cdot\tau_+(zq).
\end{aligned}
$$

Corollary 3.1. *As a linear combination of the above two, we recover (36) from the the Hamiltonian structure.*

Remark 3.4. For small n, we can check that n-soliton solution (25), (26) satisfy both (46) and (47). At present, however, we do not have a proof for general n.

Proof of Proposition 3.3. They follow from (7), (10), (11) and Lemmas 3.2, 3.3, 3.4 and 3.5 below. □

Proofs of Lemmas 3.2, 3.3,3.4 and 3.5 can be obtained by using Lemma 3.1. All the calculations are tedious but straightforward. Therefore we omit them.

Lemma 3.2. *From (10), (11) and (45), we have*

$$\begin{aligned}
&\frac{(D_{t_3}+3M_3)\,\tau_-(z)\cdot\tau_+(z)}{\tau_-(z)\cdot\tau_+(z)} \\
&=\eta(z)\left(M_2+\frac{1}{2}M_1^2+M_1(\eta_+(zq)+\eta_-(z/q))+\eta_+(zq)\eta_-(z/q)\right) \\
&+\eta(z)\left[\left(\frac{qw_1/w_2}{1-qw_1/w_2}\frac{qw_2/z}{1-qw_2/z}\right.\right. \\
&\qquad\left.\left.+\frac{qw_2/w_1}{1-qw_2/w_1}\frac{qz/w_2}{1-qz/w_2}\right)\eta(w_1)\eta(w_2)\right]_{1,w_1,w_2}.
\end{aligned}$$

Lemma 3.3. *From (7), (10) and (11), we have*

$$\begin{aligned}
&\frac{(D_{t_1}+M_1)^3\,\tau_-(z)\cdot\tau_+(z)}{\tau_-(z)\cdot\tau_+(z)} \\
&=\eta(z)\left(4M_2-M_1^2+M_1(\eta_+(zq)+\eta_-(z/q))-2\eta_+(zq)\eta_-(z/q)\right) \\
&+2\eta(z)\left[\left(\frac{qw_1/w_2}{1-qw_1/w_2}+\frac{qw_2/w_1}{1-qw_2/w_1}\right)\right. \\
&\qquad\left.\times\left(\frac{qw_2/z}{1-qw_2/z}+\frac{qz/w_2}{1-qz/w_2}\right)\eta(w_1)\eta(w_2)\right]_{1,w_1,w_2} \\
&-\eta(z)\left[\left(\frac{qw_1/w_2}{1-qw_1/w_2}-\frac{qw_2/w_1}{1-qw_2/w_1}\right)\right. \\
&\qquad\left.\times\left(\frac{qw_2/z}{1-qw_2/z}-\frac{qz/w_2}{1-qz/w_2}\right)\eta(w_1)\eta(w_2)\right]_{1,w_1,w_2}.
\end{aligned}$$

Lemma 3.4. *From (10), (11) and (44), we have*

$$\begin{aligned}
&\frac{(D_{t_2}+2M_2)\,\tau_-(z/q)\cdot\tau_+(zq)}{\tau_-(z/q)\cdot\tau_+(zq)} \\
&=2M_2+M_1(\eta_+(zq)+\eta_-(z/q)) \\
&+\left[\left(\frac{qw_1/w_2}{1-qw_1/w_2}+\frac{qw_2/w_1}{1-qw_2/w_1}\right)\right. \\
&\qquad\left.\times\left(\frac{qw_2/z}{1-qw_2/z}+\frac{qz/w_2}{1-qz/w_2}\right)\eta(w_1)\eta(w_2)\right]_{1,w_1,w_2}.
\end{aligned}$$

Lemma 3.5. *From (7), (10) and (11), we have*

$$\frac{(D_{t_1}+M_1)^2\,\tau_-(z/q)\cdot\tau_+(zq)}{\tau_-(z/q)\cdot\tau_+(zq)}$$
$$= M_1^2 + M_1(\eta_+(zq)+\eta_-(z/q)) + 2\eta_+(zq)\eta_-(z/q)$$
$$+\left[\left(\frac{qw_1/w_2}{1-qw_1/w_2}-\frac{qw_2/w_1}{1-qw_2/w_1}\right)\right.$$
$$\left.\times\left(\frac{qw_2/z}{1-qw_2/z}-\frac{qz/w_2}{1-qz/w_2}\right)\eta(w_1)\eta(w_2)\right]_{1,w_1,w_2}.$$

Acknowledgments

The authors thank Kenji Kajiwara, Paul Wiegmann and Masatoshi Noumi for stimulating discussions. YT is supprted by The Promotion and Mutual Aid Corporation for Private Schools of Japan.

References

1. J. Shiraishi and Y. Tutiya, Periodic ILW equation with discrete Laplacian, J. Phys. A **42** (2009), no. 40, 404018, 15 pp.
2. Y. Tutiya and J. Shiraishi, On some special solutions to periodic Benjamin-Ono equation with discrete Laplacian, arXiv:0911.5005, to appear in Mathematics and Computers in Simulation.
3. T.B. Benjamin, Internal waves of permanent form in fluids of great depth, J. Fluid. Mech., **29** (1967), 559.
4. H. Ono, Algebraic solitary waves in stratified fluids, J. Phys. Soc. Japan, **30** (1975), 1082.
5. K. Ueno and K. Takasaki, Toda lattice hierarchy, Group Representations and Systems of Differential Equations, K. Okamoto (ed.), Adv. Stud. Pure Math. vol. **4**, pp.1-95 (North-Holland, Amsterdam, Kinokuniya, Tokyo, 1884).
6. P.M. Santini, Inverse Problems **5** (1989) 203.
7. D.R. Lebedev, A.O. Radul, Commun. Math. Phys. **91** (1983) 543.
8. A.G. Abanov, E. Bettelheim and P. Wiegmann, Integrable hydrodynamics of Calogero-Sutherland model: bidirectional Benjamin-Ono equation, J. Phys. A: Math. Theor. **42** (2009) 135201 (24pp).
9. B. Feigin, K. Hashizume, A. Hoshino, J. Shiraishi and S. Yanagida, A commutative algebra on degenerate $\mathbb{CP}^1$ and Macdonald polynomials, J. Math. Phys. **50** (2009), no. 9, 095215, 42 pp.
10. I.G. Macdonald, Symmetric Functions and Hall Polynomials, 2nd edn. Oxford University Press, 1995.
11. R. Hirota, Discrete analogue of a generalized Toda equation, J. Phys. Soc. Jpn. **50** (1981) 3785-3791.
12. T. Miwa, On Hirota's difference equations, Proc. Japan Acad. **58** (1982) 9-12.

KP AND TODA TAU FUNCTIONS IN BETHE ANSATZ

KANEHISA TAKASAKI

Graduate School of Human and Environmental Studies
Kyoto University
Yoshida, Sakyo, Kyoto, 606-8501, Japan
E-mail: takasaki@math.h.kyoto-u.ac.jp

Recent work of Foda and his group on a connection between classical integrable hierarchies (the KP and 2D Toda hierarchies) and some quantum integrable systems (the 6-vertex model with DWBC, the finite XXZ chain of spin 1/2, the phase model on a finite chain, etc.) is reviewed. Some additional information on this issue is also presented.

Keywords: Six-vertex model; XXZ model; domain wall boundary condition; Bethe ansatz; KP hierarchy; Toda hierarchy; tau function.

1. Introduction

Searching for a connection between classical and quantum integrable systems is an old and new subject, occasionally leading to a breakthrough towards a new area of research. One of the landmarks in this sense is the quantum inverse scattering method, also known as the algebraic Bethe ansatz. Stemming from the classical inverse scattering method, the algebraic Bethe ansatz covers a wide class of integrable systems including solvable models of statistical mechanics on the basis of the Yang-Baxter equations.[1] Moreover, remarkably, it was recognized later that a kind of classical integrable systems (discrete Hirota equations) show up in the so called nested Bethe ansatz.[2]

Recently a new connection was found by Foda and his group.[3–7] They observed that special solutions of the classical integrable hierarchies (the KP and 2D Toda hierarchies) are hidden in quantum (or statistical) integrable systems such as the 6-vertex model under the domain wall boundary condition (DWBC),[3] the finite XXZ chain of spin 1/2,[4,5] and some other quantum integrable systems.[6,7] Their results are based on a determinant formula of physical quantities, namely, the Izergin-Korepin formula for the partition function of the 6-vertex model[8–10] and the Slavnov formula for the scalar product of Bethe states in the XXZ spin chain.[11,12] Those formulae contain a set of free variables, and the determinant in the formula is divided by the Vandermonde determinant of these variables. Foda et al. interpreted the quotient of the determinant by the Vandermonde determinants as a tau function of the KP (or 2D Toda) hierarchy expressed in the so called "Miwa variables".

In this paper, we review these results along with some additional information on this issue. We are particularly interested in the relevance of the 2-component KP (2-KP) and 2D Toda hierarchies. Unfortunately, this research is still in an early stage, and we cannot definitely say which direction this research leads us to. A modest goal will be to understand the algebraic Bethe ansatz better in the perspective of classical integrable hierarchies.

This paper is organized as follows. In Section 2, we start with a brief account of the notion of Schur functions that play a fundamental role in the theory of integrable hierarchies, and introduce the tau function of the KP, 2-KP and 2D Toda hierarchies as a function of both the usual time variables and the Miwa variables. Section 3 deals with the partition function of the 6-vertex model with DWBC. Following the procedure of Foda et al., we rewrite the Izergin-Korepin formula into an almost rational form and show that a main part of the partition function can be interpreted as a KP tau function. Actually, the partition function allows two different interpretations that correspond to two choices of the Miwa variables. We examine a unified interpretation of the partition function as a tau function of the 2-KP (or 2D Toda) hierarchy. In Section 4, we turn to the finite XXZ chain of spin 1/2, and present a similar interpretation to the scalar product of Bethe states (one of which depends on free variables) on the basis of the Slavnov formula. Section 5 is devoted to some other models including the phase model,[13] which is also studied by the group of Foda.[6] For those models, a determinant formula is known to hold for the scalar product of Bethe states both of which depend on free parameters.[14] We consider a special case related to enumeration of boxed plane partitions.

2. Tau functions

2.1. *Schur functions*

Let us review the notion of Schur functions. We mostly follow the notations of Macdonald's book.[15]

For N variables $\boldsymbol{x} = (x_1, \ldots, x_N)$ and a partition $\lambda = (\lambda_1, \lambda_2, \ldots, \lambda_N)$

($\lambda_1 \geq \lambda_2 \geq \cdots \geq \lambda_N \geq 0$) of length ($\lambda$) $\leq N$, the Schur function $s_\lambda(\boldsymbol{x})$ can be defined by Weyl's character formula

$$s_\lambda(\boldsymbol{x}) = \frac{\det(x_j^{\lambda_i - i + N})_{i,j=1}^N}{\Delta(\boldsymbol{x})}, \tag{1}$$

where $\Delta(\boldsymbol{x})$ is the Vandermonde determinant

$$\Delta(\boldsymbol{x}) = \det(x_j^{-i+N})_{i,j=1}^N = \prod_{1 \leq i < j \leq N} (x_i - x_j).$$

By one of the Jacobi-Trudi formulae, $s_\lambda(\boldsymbol{x})$ can be expressed as a determinant of the form

$$s_\lambda(\boldsymbol{x}) = \det(h_{\lambda_i - i + j}(\boldsymbol{x}))_{i,j=1}^N, \tag{2}$$

where $h_n(\boldsymbol{x})$, $n = 0, 1, 2, \ldots$, are the completely symmetric functions

$$h_n(\boldsymbol{x}) = \sum_{1 \leq k_1 \leq k_2 \leq \cdots \leq k_n \leq N} x_{k_1} x_{k_2} \cdots x_{k_n} \quad \text{for} \quad n \geq 1, \quad h_0(\boldsymbol{x}) = 1.$$

The complete symmetric functions $h_n(\boldsymbol{x})$ themselves can be identified with the Schur functions for partitions with a single part:

$$h_n(\boldsymbol{x}) = s_{(n)}(\boldsymbol{x}), \quad (n) := (n, 0, \ldots, 0).$$

Another form of the Jacobi-Trudi formulae uses on the elementary symmetric functions $e_n(\boldsymbol{x})$. Since we shall not use it in the following, its details are omitted here.

The complete symmetric functions have the generating function

$$\sum_{n=0}^{\infty} h_n(\boldsymbol{x}) z^n = \prod_{k=1}^{N} (1 - x_k z)^{-1} = \exp\left(-\sum_{k=1}^{N} \log(1 - x_k z)\right).$$

Since $\log(1 - x_k z)$ has a Taylor expansion of the form

$$\log(1 - x_k z) = -\sum_{n=1}^{\infty} \frac{x_k^n}{n} z^n,$$

this generating function can be rewritten as

$$\sum_{n=0}^{\infty} h_n(\boldsymbol{x}) z^n = \exp\left(\sum_{n=1}^{\infty} t_n z\right), \tag{3}$$

where t_n's are defined as

$$t_n = \frac{1}{n} \sum_{k=1}^{N} x_k^n. \tag{4}$$

In view of (2) and (3), one can redefine the complete symmetric functions and the Schur functions as functions $h_n[\boldsymbol{t}]$ and $s_\lambda[\boldsymbol{t}]$ [a] of $\boldsymbol{t} = (t_1, t_2, \ldots)$, namely,

$$\sum_{n=0}^{\infty} h_n[\boldsymbol{t}] z^n = \exp\left(\sum_{n=1}^{\infty} t_n z^n\right) \tag{5}$$

and

$$s_\lambda[\boldsymbol{t}] = \det(h_{\lambda_i - i + j}[\boldsymbol{t}])_{i,j=1}^{N}, \tag{6}$$

where λ is understood to be an infinite decreasing sequence $\lambda = (\lambda_1, \lambda_2, \ldots)$ with $\lambda_i = 0$ for all but a finite number of i's, and N is arbitrary integer greater than or equal to $l(\lambda) = \max\{i \mid \lambda_i \neq 0\}$. The right hand side of (6) is independent of N, because the lower left block of the matrix therein for $i > l(\lambda)$ and $j \leq l(\lambda)$ is zero and the lower right block for $i, j > l(\lambda)$ is an upper triangular matrix with 1 on the diagonal line.

This is a place where a connection with the KP hierarchy[17] shows up. Namely, the variables $\boldsymbol{t} = (t_1, t_2, \ldots)$ are nothing but the "time variables" of the KP hierarchy, and the Schur functions $s_\lambda[\boldsymbol{t}]$ are special tau functions. As first pointed out by Miwa,[18] viewing the tau function as a function of the $\boldsymbol{x}$ variables leads to a discrete (or difference) analogue of the KP hierarchy. For this reason, the $\boldsymbol{x}$ variables are sometimes referred to as "Miwa variables" in the literature of integrable systems.

2.2. *Tau functions of KP hierarchy*

Let us use the notation $\tau[\boldsymbol{t}]$ for the tau function in the usual sense (namely, a function of $\boldsymbol{t}$), and let $\tau(\boldsymbol{x})$ denote the function obtained from $\tau[\boldsymbol{t}]$ by the change of variables (4). It is the latter that plays a central role in this paper.

A general tau function of the KP hierarchy is a linear combination of the Schur functions

$$\tau[\boldsymbol{t}] = \sum_{\lambda} c_\lambda s_\lambda[\boldsymbol{t}], \tag{7}$$

where the coefficients c_λ are Plücker coordinates of a point of an infinite dimensional Grassmann manifold (Sato Grassmannian).[19] Roughly speaking, the Sato Grassmannian consists of linear subspaces $W \simeq \mathbf{C}^{\mathrm{N}}$ of a fixed

[a]These convenient notations are borrowed from Zinn-Justin's paper.[16]

linear space $V \simeq \mathbf{C}^{\mathbf{Z}}$. We shall not pursue those fully general tau functions in the following.

We are interested in a smaller (but yet infinite dimensional) class of tau functions such that $c_\lambda = 0$ for all partitions with $l(\lambda) > N$. This corresponds to a submanifold $\mathrm{Gr}(N, \infty)$ of the full Sato Grassmannian. The Plücker coordinates c_λ are labelled by partitions of the form $\lambda = (\lambda_1, \ldots, \lambda_N)$, and given by finite determinants as

$$c_\lambda = \det(f_{i,l_j})_{i,j=1}^N, \quad l_i := \lambda_i - i + N. \tag{8}$$

Note that the sequences λ_i's and l_i's of non-negative integers are in one-to-one correspondence:

$$\infty > \lambda_1 \geq \cdots \lambda_N \geq 0 \quad \longleftrightarrow \quad \infty > l_1 > \cdots > l_N \geq 0.$$

The $N \times \infty$ matrix

$$F = (f_{ij})_{i=1,\ldots,N,\ j=0,1,\ldots}$$

of parameters represent a point of the Grassmann manifold $\mathrm{Gr}(N, \infty)$.

By the Cauchy-Binet formula, the tau function $\tau(\boldsymbol{x})$ in the $\boldsymbol{x}$-picture can be expressed as

$$\tau(\boldsymbol{x}) = \sum_{\infty > l_1 > \cdots > l_N \geq 0} \frac{\det(f_{i,l_j})_{i,j=1}^N \det(x_i^{l_j})_{i,j=1}^N}{\Delta(\boldsymbol{x})} = \frac{\det(f_i(x_j))_{i,j=1}^N}{\Delta(\boldsymbol{x})}, \tag{9}$$

where $f_i(x)$'s are the power series of the form

$$f_i(x) = \sum_{l=0}^{\infty} f_{il} x^l.$$

In particular, if there is a positive integer M such that

$$f_{ij} = 0 \quad \text{for} \quad i \geq M + N$$

(in other words, $f_i(x)$'s are polynomials of degree less than $M + N$), the Plücker coordinate c_λ vanishes for all Young diagrams not contained in the $N \times M$ rectangular Young diagram, namely,

$$c_\lambda = 0 \quad \text{for} \quad \lambda \not\subseteq (M^N) := (\underbrace{M, \ldots, M}_{N})$$

The tau function $\tau[\boldsymbol{t}]$ thereby becomes a linear combination of a finite number of Schur function, hence a polynomial in $\boldsymbol{t}$. Geometrically, these solutions of the KP hierarchy sit on the finite dimensional Grassmann manifold $Gr(N, N + M)$ of the Sato Grassmannian.

2.3. *Tau functions of 2-KP hierarchy*

The tau function $\tau[\boldsymbol{t},\bar{\boldsymbol{t}}]$ of the 2-component KP (2-KP) hierarchy is a function of two sequences $\boldsymbol{t} = (t_1, t_2, \ldots)$ and $\bar{\boldsymbol{t}} = (\bar{t}_1, \bar{t}_2, \ldots)$ of time variables, and can be expressed as

$$\tau[\boldsymbol{t},\bar{\boldsymbol{t}}] = \sum_{\lambda,\mu} c_{\lambda\mu} s_\lambda[\boldsymbol{t}] s_\mu[\bar{\boldsymbol{t}}], \tag{10}$$

where $c_{\lambda\mu}$'s are Plücker coordinates of a point of a 2-component analogue of the Sato Grassmannnian (which is, actually, isomorphic to the one-component version).[19]

The aforementioned class of tau functions of the KP hierarchy can be generalized to the 2-component case. Such tau functions correspond to points of the submanifold $\mathrm{Gr}(M+N, 2\infty)$ of the 2-component Sato Grassmannian. For those tau functions, the Plücker coordinates $c_{\lambda\mu}$ vanish if $l(\lambda) > M$ or $l(\mu) > N$; the remaining Plücker coordinates are given by finite determinants of a matrix with two rectangular blocks of size $(M+N) \times M$ and $(M+N) \times N$ as

$$c_{\lambda\mu} = \det(f_{i,l_j} \mid g_{i,m_k}), \tag{11}$$

where i is the row index ranging over $i = 1, \ldots, M+N$ and j, k are column indices in the two blocks ranging over $j = 1, \ldots, M$ and $k = 1, \ldots, N$, respectively. l_j's and m_k's are related to the parts of $\lambda = (\lambda_j)_{j=1}^M$ and $\mu = (\mu_i)_{j=1}^N$ as

$$l_j = \lambda_j - j + M, \quad m_k = \mu_k - k + N.$$

By the change of variables from $\boldsymbol{x}$ and $\boldsymbol{y}$ to

$$t_n = \frac{1}{n}\sum_{j=1}^M x_j^n, \quad \bar{t}_n = \frac{1}{n}\sum_{k=1}^N y_k^n, \tag{12}$$

the tau function $\tau[\boldsymbol{t},\bar{\boldsymbol{t}}]$ is converted to the $(\boldsymbol{x},\boldsymbol{y})$-picture $\tau(\boldsymbol{x},\boldsymbol{y})$. Again by the Cauchy-Binet formula, $\tau(\boldsymbol{x},\boldsymbol{y})$ turns out to be a quotient of two determinants as

$$\tau(\boldsymbol{x},\boldsymbol{y}) = \frac{\det(f_i(x_j) \mid g_i(y_k))}{\Delta(\boldsymbol{x})\Delta(\boldsymbol{y})}, \tag{13}$$

where the denominator is the determinant with the same block structure as (11), and $f_i(x)$ and $g_j(y)$ are power series of the form

$$f_i(x) = \sum_{l=0}^\infty f_{il} x^l, \quad g_i(y) = \sum_{l=0}^\infty g_{il} y^l.$$

2.4. *Tau function of 2D Toda hierarchy*

The 2-KP hierarchy is closely related to the 2D Toda hierarchy.[20] The tau function $\tau_s[\boldsymbol{t},\bar{\boldsymbol{t}}]$ of the 2D Toda hierarchy depends on a discrete variable (lattice coordinate) s alongside the two series of time variables $\boldsymbol{t}$ and $\bar{\boldsymbol{t}}$. For each value of s, $\tau_s[\boldsymbol{t},\bar{\boldsymbol{t}}]$ is a tau function of the 2-KP hierarchy, and these 2-KP tau functions are mutually connected by a kind of Bäcklund transformations. Consequently, $\tau_s[\boldsymbol{t},\bar{\boldsymbol{t}}]$ can be expressed as shown in (10) with the coefficients $c_{s\lambda\mu}$ depending on s.

Actually, it is more natural to use $s_\lambda[\boldsymbol{t}]s_\mu[-\bar{\boldsymbol{t}}]$ rather than $s_\lambda[\boldsymbol{t}]s_\mu[\bar{\boldsymbol{t}}]$ for the Schur function expansion of the Toda tau function.[21] (Note that $s_\mu[-\bar{\boldsymbol{t}}]$ can be rewritten as

$$s_\mu[-\bar{\boldsymbol{t}}] = (-1)^{|\mu|}s_{{}^{\mathrm{t}}\mu}[\boldsymbol{t}],$$

where ${}^{\mathrm{t}}\mu$ denotes the transpose of μ.) Expanded in these product of tau functions as

$$\tau_s[\boldsymbol{t},\bar{\boldsymbol{t}}] = \sum_{\lambda,\mu} c_{s\lambda\mu}s_\lambda[\boldsymbol{t}]s_\mu[-\bar{\boldsymbol{t}}], \tag{14}$$

the coefficients $c_{s\lambda\mu}$ become Plücker coordinates of an infinite dimensional flag manifold. Intuitively, they are minor determinants

$$c_{s\lambda\mu} = \det(g_{\lambda_i-i+s,\mu_j-j+s})_{i,j=1}^{\infty} \tag{15}$$

of an infinite matrix $g = (g_{ij})_{i,j\in\mathbf{Z}}$, though this definition requires justification.[21] In particular, if g is a diagonal matrix, the coefficients $c_{s\lambda\mu}$ are also diagonal (namely, $c_{s\lambda\mu} \propto \delta_{\lambda\mu}$) and the Schur function expansion (14) simplifies to the "diagonal" form

$$\tau_s[\boldsymbol{t},\bar{\boldsymbol{t}}] = \sum_{\lambda} c_{s\lambda}s_\lambda[\boldsymbol{t}]s_\lambda[-\bar{\boldsymbol{t}}], \quad c_{s\lambda} = \prod_{i=1}^{\infty} g_{\lambda_i-i+s}. \tag{16}$$

If we reformulate the 2D Toda hierarchy on the semi-infinite lattice $s \geq 0$, the infinite determinants defining c_λ's are replaced by finite determinants, and $\tau_s[\boldsymbol{t},\bar{\boldsymbol{t}}]$ itself becomes a finite determinant. We shall encounter an example of such tau functions in the next section.

3. 6-vertex model with DWBC

3.1. *Setup of model*

We consider the 6-vertex model on an $N \times N$ square lattice with inhomogeneity parameters $\boldsymbol{u} = (u_1,\ldots,u_N)$ and $\boldsymbol{v} = (v_1,\ldots,v_N)$ assigned to

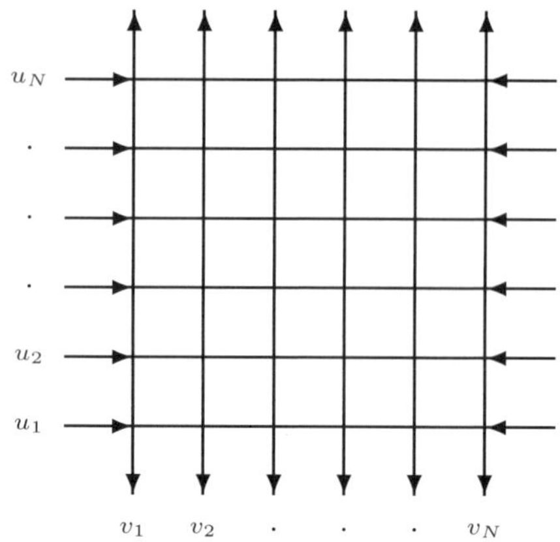

Fig. 1. Square lattice with DWBC

the rows and columns. The boundary of the lattice is supplemented with extra edges pointing outwards, and the domain-wall boundary condition (DWBC) is imposed on these extra edges. Namely, the arrows on the extra edges on top and bottom of the boundary are pointing outwards, and those on the other extra edges are pointing inwards (see figure 1).

The vertex at the intersection of the i-th row and the j-th column is given the following weight w_{ij} determined by the configuration of arrows on the adjacent edges:

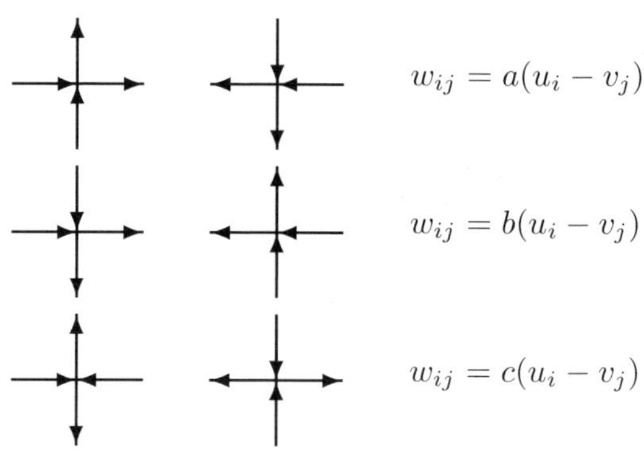

The weight functions $a(u), b(u), c(u)$ are defined as

$$a(u) = \sinh(u + \gamma), \quad b(u) = \sinh u, \quad c(u) = \sinh \gamma, \tag{17}$$

where γ is a parameter. Thus the partition function of this model is defined as a function of the inhomogeneity parameters $\boldsymbol{u}$ and $\boldsymbol{v}$:

$$Z_N = Z_N(\boldsymbol{u}, \boldsymbol{v}) = \sum_{\text{configuration}} \prod_{i,j=1}^{N} w_{ij}.$$

3.2. *Izergin-Korepin formula for Z_N*

According to the result of Korepin[8] and Izergin,[9] the partition function Z_N has the determinant formula

$$Z_N = \frac{\prod_{i,j=1}^{N} \sinh(u_i - v_j + \gamma)\sinh(u_i - v_j)}{\prod_{1\le i<j\le N} \sinh(u_i - u_j)\sinh(v_j - v_i)} \times \det\left(\frac{\sinh\gamma}{\sinh(u_i - v_j + \gamma)\sinh(u_i - v_j)}\right)_{i,j=1}^{N}, \tag{18}$$

which one can rewrite as

$$Z_N = \frac{\sinh^N \gamma}{\prod_{1\le i<j\le N} \sinh(u_i - u_j)\sinh(v_j - v_i)} \times \det\left(\frac{\prod_{k=1}^{N} \sinh(u_i - v_k + \gamma)\sinh(u_i - v_k)}{\sinh(u_i - v_j + \gamma)\sinh(u_i - v_j)}\right)_{i,j=1}^{N} \tag{19}$$

and

$$Z_N = \frac{\sinh^N \gamma}{\prod_{1\le i<j\le N} \sinh(u_i - u_j)\sinh(v_j - v_i)} \times \det\left(\frac{\prod_{k=1}^{N} \sinh(u_k - v_j + \gamma)\sinh(u_k - v_j)}{\sinh(u_i - v_j + \gamma)\sinh(u_i - v_j)}\right)_{i,j=1}^{N}. \tag{20}$$

If we introduce the new variables and parameters[3]

$$x_i := e^{2u_i}, \quad y_i := e^{2v_i}, \quad q := e^{-\gamma},$$

we can rewrite these formulae as

$$Z_N = C_N \prod_{i,j=1}^{N} (x_i q^{-1} - y_j q)(x_i - y_j) \times \frac{1}{\Delta(\boldsymbol{x})\Delta(\boldsymbol{y})} \det\left(\frac{q^{-1} - q}{(x_i q^{-1} - y_j q)(x_i - y_j)}\right)_{i,j=1}^{N}, \tag{21}$$

$$Z_N = \frac{C_N(q^{-1}-q)^N}{\Delta(\boldsymbol{x})\Delta(\boldsymbol{y})} \det\left(\frac{\prod_{k=1}^N (x_i q^{-1} - y_k q)(x_i - y_k)}{(x_i q^{-1} - y_j q)(x_i - y_j)}\right)_{i,j=1}^N, \quad (22)$$

and

$$Z_N = \frac{C_N(q^{-1}-q)^N}{\Delta(\boldsymbol{x})\Delta(\boldsymbol{y})} \det\left(\frac{\prod_{k=1}^N (x_k q^{-1} - y_j q)(x_k - y_j)}{(x_i q^{-1} - y_j q)(x_i - y_j)}\right)_{i,j=1}^N, \quad (23)$$

where $C_N = C_N(\boldsymbol{u}, \boldsymbol{v})$ is an exponential function of a linear combination of u_i's and v_i's. Apart from this simple factor, Z_N thus reduces to a rational function of $\boldsymbol{x} = (x_1, \ldots, x_N)$ and $\boldsymbol{y} = (y_1, \ldots, y_N)$.

3.3. *KP and 2-KP tau functions hidden in Z_N*

As pointed out by Foda et al.,[3] two KP tau functions are hidden in these determinant formulae of Z_N. Firstly, if y_i's are considered to be constants, the $\boldsymbol{x}$-dependent part of (22) give the function

$$\tau_1(\boldsymbol{x}) = \frac{\det(f_j(x_i))_{i,j=1}^N}{\Delta(\boldsymbol{x})}, \quad f_j(x) := \frac{\prod_{k=1}^N (xq^{-1} - y_k q)(x - y_k)}{(xq^{-1} - y_j q)(x - y_j)}. \quad (24)$$

This is a tau function of the KP hierarchy with respect to $t_n = \frac{1}{n}\sum_{k=1}^N x_k^n$. Moreover, since $f_j(x)$'s are polynomials in x, this tau function is as a polynomial in $\boldsymbol{t}$. In the same sense, if x_i's are considered to be constants, the $\boldsymbol{y}$-dependent part of (23) gives the function

$$\tau_2(\boldsymbol{y}) = \frac{\det(g_i(y_j))_{i,j=1}^N}{\Delta(\boldsymbol{y})}, \quad g_i(y) := \frac{\prod_{k=1}^N (x_i q^{-1} - yq)(x_i - y)}{(x_i q^{-1} - yq)(x_i - y)}, \quad (25)$$

which is a polynomial tau function of the KP hierarchy with respect to $\bar{t}_n = \frac{1}{n}\sum_{k=1}^N y_k^n$.

Thus, apart from an irrelevant factor, Z_N is a tau function of the KP hierarchy with respect to $\boldsymbol{x}$ and $\boldsymbol{y}$ separately. It will be natural to ask whether a tau function of the 2-KP hierarchy is hidden in Z_N.

A partial answer can be found in the work of Stroganov[22] and Okada.[23] According to their results, if $q = e^{\pi i/3}$, the partition function coincides, up to a simple factor, with a single Schur function of $(\boldsymbol{x}, \boldsymbol{y})$ as

$$Z_N = (\text{simple factor}) s_\lambda(\boldsymbol{x}, \boldsymbol{y}), \quad (26)$$

where λ is the double staircase partition

$$\lambda = (N-1, N-1, N-2, N-2, \ldots, 1, 1)$$

of length $2N$. By the way, for any partition $\lambda = (\lambda_1, \ldots, \lambda_{2N})$ of length $\leq 2N$, the Weyl character formula for $s_\lambda(\boldsymbol{x}, \boldsymbol{y})$ reads

$$s_\lambda(\boldsymbol{x}, \boldsymbol{y}) = \frac{\det(x_j^{\lambda_i - i + 2N} \mid y_k^{\lambda_i - i + 2N})}{\Delta(\boldsymbol{x}, \boldsymbol{y})},$$

where the row index i ranges over $i = 1, \ldots, 2N$ and the column indices j, k in the two blocks over $j, k = 1, \ldots, N$. Multiplying this function by $\Delta(\boldsymbol{x}, \boldsymbol{y})/\Delta(\boldsymbol{x})\Delta(\boldsymbol{y})$ gives

$$\frac{\Delta(\boldsymbol{x}, \boldsymbol{y})}{\Delta(\boldsymbol{x})\Delta(\boldsymbol{y})} s_\lambda(\boldsymbol{x}, \boldsymbol{y}) = \frac{\det(x_j^{\lambda_i - i + 2N} \mid y_k^{\lambda_i - i + 2N})}{\Delta(\boldsymbol{x})\Delta(\boldsymbol{y})}, \tag{27}$$

which may be thought of as a 2-KP tau function of the form (13).

Another answer, which is valid for arbitrary values of q, was found by Zinn-Justin (private communication). Let us note that (21) can be rewritten as

$$Z_N = C_N(q^{-1} - q)^N \prod_{i,j=1}^{N} (1 - q^{-2} x_i y_j^{-1})(1 - x_i y_j^{-1}) \prod_{1 \leq i < j \leq N} (-y_i y_j)$$
$$\times \frac{1}{\Delta(\boldsymbol{x})\Delta(\boldsymbol{y}^{-1})} \det\left(\frac{1}{(1 - q^{-2} x_i y_j^{-1})(1 - x_i y_j^{-1})}\right)_{i,j=1}^{N}, \tag{28}$$

where

$$\boldsymbol{y}^{-1} = (y_1^{-1}, \ldots, y_N^{-1}).$$

The last part of thins expression, namely the quotient of the determinant by the Vandermonde determinants $\Delta(\boldsymbol{x})\Delta(\boldsymbol{y}^{-1})$, may be thought of as a 2-KP tau function with respect to the time variables

$$t_n = \frac{1}{n} \sum_{k=1}^{N} x_k^n, \quad \bar{t}_n = -\frac{1}{n} \sum_{k=1}^{N} y_k^{-n}. \tag{29}$$

This is a special case of the tau functions

$$\tau(\boldsymbol{x}, \boldsymbol{y}) = \frac{\det(h(x_i y_j^{-1}))_{i,j=1}^{N}}{\Delta(\boldsymbol{x})\Delta(\boldsymbol{y}^{-1})}, \tag{30}$$

considered by Orlov and Shiota,[24] where $h(z)$ is an arbitrary power series of the form

$$h(z) = \sum_{n=0}^{\infty} h_n z^n, \quad h_n \neq 0 \quad \text{for} \quad n \geq 0.$$

By the Cauchy-Binet formula, $\tau(\boldsymbol{x}, \boldsymbol{y})$ can be expanded as

$$\tau(\boldsymbol{x}, \boldsymbol{y}) = \sum_{\lambda=(\lambda_1,\ldots,\lambda_N)} c_\lambda s_\lambda(\boldsymbol{x}) s_\lambda(\boldsymbol{y}^{-1}), \quad c_\lambda := \prod_{i=1}^{N} h_{\lambda_i - i + N}. \tag{31}$$

This is an analogue (for a semi-infinite lattice) of the Toda tau functions of the diagonal form (16). Note that the role of the lattice coordinate s is played by N. Since the number of the Miwa variables in (29) also depends on N, translation to the lanugage of the 2D Toda hierarchy is somewhat tricky, but this tricky situation is rather common in random matrix models.[24] Thus, though not of the type shown in (13), the last part of (28) turns out to be a 2-KP tau function.

Lastly, let us mention that Korepin and Zinn-Justin considered the partition function in the homogeneous limit as $u_i, v_j \to 0$.[25] In that limit, the partition function reduces, up to a simple factor, to a special tau function of the 1D Toda equation, and can be treated as an analogue of random matrix models.

4. Scalar product of states in finite XXZ spin chain

4.1. *L- and T-matrices for spin 1/2 chain*

We consider a finite XXZ spin chain of spin 1/2 and length N with inhomongeneity parameters ξ_l, $l = 1, \ldots, N$. To define local L-matrices, let us introduce the 2×2 matrix $L(u) = (L_{ij}(u))_{i,j=1,2}$ of the 2×2 blocks

$$L_{11}(u) = a(u)\frac{1+\sigma^3}{2} + b(u)\frac{1-\sigma^3}{2}, \quad L_{12}(u) = c(u)\sigma^-,$$
$$L_{21}(u) = c(u)\sigma^+, \quad L_{22}(u) = b(u)\frac{1+\sigma^3}{2} + a(u)\frac{1-\sigma^3}{2},$$

where $\sigma^\pm$ and σ^3 are the Pauli matrices

$$\sigma^+ = \begin{pmatrix} 0 & 1 \\ 0 & 0 \end{pmatrix}, \quad \sigma^- = \begin{pmatrix} 0 & 0 \\ 1 & 0 \end{pmatrix}, \quad \sigma^3 = \begin{pmatrix} 1 & 0 \\ 0 & -1 \end{pmatrix}.$$

These 2×2 blocks are understood to act on the single spin space $\mathbf{C}^2$. The structure functions $a(u), b(u), c(u)$ are the same as the weight functions (17) for the 6-vertex model, and built into the R-matrix

$$R(u-v) = \begin{pmatrix} a(u-v) & 0 & 0 & 0 \\ 0 & b(u-v) & c(u-v) & 0 \\ 0 & c(u-v) & b(u-v) & 0 \\ 0 & 0 & 0 & a(u-v) \end{pmatrix}.$$

Let $L^{(l)}(u-\xi_l) = (L^{(l)}_{ij}(u-\xi))_{i,j=1,2}$ be the local L-matrix at the l-th site,

$$L^{(l)}_{ij}(u-\xi_l) = \cdots \otimes 1 \otimes L_{ij}(u-\xi_l) \otimes 1 \otimes \cdots,$$

and define the T-matrix as

$$T(u) = \begin{pmatrix} A(u) & B(u) \\ C(u) & D(u) \end{pmatrix} = L^{(1)}(u-\xi_1)\cdots L^{(N)}(u-\xi_N).$$

The matrix elements of these matrices, hence the trace of the T-matrix

$$\mathcal{T}(u) = \operatorname{Tr} T(u) = A(u) + D(u)$$

as well, are operators on the full spin space $V = \bigotimes_{l=1}^{N} \mathbf{C}^2$. The L-matrices satisfy the local intertwining relations

$$\begin{aligned} R(u-v)&(L^{(l)}(u)\otimes I)(I\otimes L^{(m)}(v)) \\ &= (I\otimes L^{(m)}(v))(L^{(l)}(u)\otimes I)R(u-v), \qquad (32) \end{aligned}$$

where $R(u-v)$, $L^{(l)}(u)\otimes I$ and $I\otimes K^{(m)}(v)$ are understood to be 4×4 matrices (of scalars or of spin operators on V) acting on the tensor product $\mathbf{C}^2 \otimes \mathbf{C}^2$ of two copies of the auxiliary space $\mathbf{C}^2$. These local intertwining relations lead to the global intertwining relation

$$R(u-v)(T(u)\otimes I)(I\otimes T(v)) = (I\otimes T(v))(T(u)\otimes I)R(u-v) \qquad (33)$$

for the T-matrix. This is a compact expression of many bilinear relations among the matrix elements of $T(u)$ and $T(v)$, such as

$$\begin{aligned} A(u)B(v) &= f(u-v)B(v)A(u) - g(u-v)B(u)A(v), \\ D(u)B(v) &= f(u-v)B(v)D(u) - g(u-v)B(u)D(v), \\ C(u)B(v) &= g(u-v)(A(u)D(v) - A(v)D(u)) \end{aligned} \qquad (34)$$

and

$$\begin{aligned} [A(u),A(v)] = 0, \quad [B(u),B(v)] = 0, \\ [C(u),C(v)] = 0, \quad [D(u),D(v)] = 0, \end{aligned} \qquad (35)$$

where

$$f(u) = \frac{a(u)}{b(u)} = \frac{\sinh(u+\gamma)}{\sinh u}, \quad g(u) = \frac{c(u)}{b(u)} = \frac{\sinh\gamma}{\sinh u}.$$

A consequence of those relations is the fact that $\mathcal{T}(u)$ and $\mathcal{T}(v)$ commute for any values of u, v:

$$[\mathcal{T}(u), \mathcal{T}(v)] = 0. \qquad (36)$$

The algebraic Bethe ansatz is a method for constructing simultaneous eigenstates (called "Bethe states") of $\mathcal{T}(u)$ for all values of u.

4.2. *Algebraic Bethe ansatz*

Let us introduce the pseudo-vacuum $0\rangle$ and its dual $\langle 0|$:

$$\langle 0| = \bigotimes_{l=1}^{N} (\,1\ 0\,) \in V^*, \quad |0\rangle = \bigotimes_{l=1}^{N} \begin{pmatrix} 1 \\ 0 \end{pmatrix} \in V.$$

They indeed satisfy the vacuum conditions

$$\begin{aligned} \langle 0|B(u) &= 0, & C(u)|0\rangle &= 0, \\ A(u)|0\rangle &= \alpha(u)|0\rangle, & D(u)|0\rangle &= \delta(u)|0\rangle, \\ \langle 0|A(u) &= \alpha(u)\langle 0|, & \langle 0|D(u) &= \delta(u)\langle 0|, \end{aligned} \tag{37}$$

where

$$\alpha(u) = \prod_{l=1}^{N} \sinh(u - \xi_l + \gamma), \quad \delta(u) = \prod_{l=1}^{N} \sinh(u - \xi_l).$$

For notational convenience, we introduce the reflection coefficients

$$r(u) = \frac{\alpha(u)}{\delta(u)}.$$

Bethe states are generated from $|0\rangle$ by the action of $B(v_j)$'s. Suppose that v_j's satisfy the Bethe equations

$$r(v_i) \prod_{j \neq i} \frac{\sinh(v_i - v_j - \gamma)}{\sinh(v_i - v_j + \gamma)} = 1, \quad i = 1, \ldots, n. \tag{38}$$

The state $\prod_{i=1}^{n} B(v_i)|0\rangle$ then becomes an eigenstate of $\mathcal{T}(u)$:

$$\begin{aligned} &\mathcal{T}(u) \prod_{i=1}^{n} B(v_i)|0\rangle \\ &\quad = \left(\alpha(u) \prod_{i=1}^{n} f(v_i - u) + \delta(u) \prod_{i=1}^{n} f(u - v_i) \right) \prod_{i=1}^{n} B(v_i)|0\rangle. \end{aligned} \tag{39}$$

Let us remark that the operators $A(u), B(u), C(u), D(u)$ are related to a row-to-row transfer matrix of the 6-vertex model on the square lattice. One can thereby derive the identities[10]

$$\langle 0| \prod_{i=1}^{N} C(u_i) = \langle \bar{0}| Z_N(u_1, \ldots, u_N, \xi_1, \ldots, \xi_N),$$

$$\prod_{i=1}^{N} B(u_i)|0\rangle = Z_N(u_1, \ldots, u_N, \xi_1, \ldots, \xi_N)|\bar{0}\rangle,$$

where u_i's are free (namely, not required to satisfy the Bethe equations) variables, and $|\bar{0}\rangle$ and $\langle\bar{0}|$ denote the "anti-pseudo-vacuum" and its dual:

$$\langle\bar{0}| = \bigotimes_{l=1}^{N} (\,0\;1\,) \in V^*, \quad |\bar{0}\rangle = \bigotimes_{l=1}^{N} \begin{pmatrix} 0 \\ 1 \end{pmatrix} \in V.$$

4.3. *Slavnov formula for scalar product*

Let $\boldsymbol{u} = (u_1, \ldots, u_n)$ be free variables and $\boldsymbol{v} = (v_1, \ldots, v_n)$ satisfy the Bethe equations (38). The Slavnov formula[11,12] for the scalar product

$$S_n(\boldsymbol{u}, \boldsymbol{v}) = \langle 0| \prod_{i=1}^{n} C(u_i) \prod_{i=1}^{n} B(v_i) |0\rangle$$

reads

$$S_n(\boldsymbol{u}, \boldsymbol{v}) = \frac{\prod_{i=1}^{n} \delta(u_i)\delta(v_i) \prod_{i,j=1}^{n} \sinh(u_i - v_j + \gamma)}{\prod_{1\le i<j\le n} \sinh(u_i - u_j)\sinh(v_j - v_i)} \det(H_{ij})_{i,j=1}^{n}, \tag{40}$$

where

$$H_{ij} = \frac{\sinh\gamma}{\sinh(u_i - v_j + \gamma)\sinh(u_i - v_j)} \left(1 - r(u_i) \prod_{k\neq i} \frac{\sinh(u_i - v_k - \gamma)}{\sinh(u_i - v_k + \gamma)}\right).$$

One can rewrite this formula as

$$S_n(\boldsymbol{u}, \boldsymbol{v}) = \frac{\sinh^n\gamma \prod_{i=1}^{n} \delta(v_i)}{\prod_{1\le i<j\le n} \sinh(u_i - u_j)\sinh(v_j - v_i)} \det(K_{ij})_{i,j=1}^{n}, \tag{41}$$

where

$$K_{ij} = \frac{\delta(u_i) \prod_{k\neq j} \sinh(u_i - v_k + \gamma) - \alpha(u_i) \prod_{k\neq j} \sinh(u_i - v_k - \gamma)}{\sinh(u_i - v_j)}.$$

4.4. *KP tau function hidden in $S_n(\boldsymbol{u}, \boldsymbol{v})$*

If we introduce the new variables and parameters[4]

$$x_i := e^{2u_i}, \quad y_i := e^{2v_i}, \quad z_i := e^{2\xi_i}, \quad q := e^{-\gamma},$$

the Slavnov formula can be converted to the almost rational form

$$S_n(\boldsymbol{u}, \boldsymbol{v}) = \frac{C_n \sinh^n\gamma \prod_{i=1}^{n} \delta(v_i)}{\Delta(\boldsymbol{y})} \frac{\det(f_j(x_i))_{i,j=1}^{n}}{\Delta(\boldsymbol{x})}, \tag{42}$$

where $C_n = C_n(\boldsymbol{u}, \boldsymbol{v})$ is an exponential function of a linear combination of u_i's and v_i's, and

$$f_j(x) = \frac{\prod_{l=1}^{N}(x - z_l)\prod_{k\neq j}(q^{-1}x - qy_k) - \prod_{l=1}^{N}(q^{-1}x - qz_l)\prod_{k\neq j}(qx - q^{-1}y_k)}{x - y_j}.$$

Thus, as pointed out by Foda et al.,[4] a KP tau function of the form (9) is hidden in $S(\boldsymbol{u}, \boldsymbol{v})$. Moreover, since the Bethe equations (38) imply the equations

$$\prod_{l=1}^{N}(y_i - z_l)\prod_{k\neq j}(q^{-1}y_i - qy_k) = \prod_{l=1}^{N}(q^{-1}y_i - qz_l)\prod_{k\neq j}(qy_i - q^{-1}y_k) \quad (i = 1, \ldots, n)$$

for y_i's, the numerator of $f_j(x)$ can be factored out by the denominator $x - y_j$. Thus $f_j(x)$'s turn out to be polynomials in x.

5. Scalar product of states in models at $q = 0$

A class of solvable models, such as the phase model[13] and and the totally asymmetric simple exclusion process (TASEP) model,[26] can be formulated by a set of 2×2 L-matrices $L^{(l)}(u)$, $l = 1, 2, \ldots, N$, and an R-matrix of the form

$$R(u - v) = \begin{pmatrix} f(u-v) & 0 & 0 & 0 \\ 0 & 1 & g(u-v) & 0 \\ 0 & g(u-v) & 0 & 0 \\ 0 & 0 & 0 & f(u-v) \end{pmatrix},$$

where

$$f(u - v) = \frac{u^2}{u^2 - v^2}, \quad g(u - v) = \frac{uv}{u^2 - v^2}.$$

This R-matrix is obtained as a "crystal" (namely, $q \to 0$) limit of the R-matrix of the 6-vertex model and the XXZ spin chain. Unlike the XXZ chain of spin 1/2, the L-matrices are not given by 2×2 blocks of the R-matrix and take a model-dependent form. We define the T-matrix as

$$T(u)T(u) = \begin{pmatrix} A(u) & B(u) \\ C(u) & D(u) \end{pmatrix} = L^{(1)}(u) \cdots L^{(N)}(u)$$

and consider the scalar product

$$S_n(\boldsymbol{u},\boldsymbol{v}) = \langle 0|\prod_{i=1}^{n} C(u_i)\prod_{i=1}^{n} B(v_i)|0\rangle.$$

Remarkably, even if both $\boldsymbol{u} = (u_1,\ldots,u_n)$ and $\boldsymbol{v} = (v_1,\ldots,v_n)$ are free variables, the scalar product for those models has a determinant formula[14] of the form

$$S_n(\boldsymbol{u},\boldsymbol{v}) = (\text{simple factor}) \prod_{1\le i<j\le n} \frac{u_i u_j}{u_i^2 - u_j^2}\frac{v_j v_i}{v_j^2 - v_i^2} \det(K_{ij})_{i,j=1}^{n}, \quad (43)$$

where

$$K_{ij} = \frac{\alpha(u_i)\delta(v_j)(v_j/u_i)^{n-1} - \delta(u_i)\alpha(v_j)(u_i/v_j)^{n-1}}{(u_i^2 - v_j^2)/u_i v_j},$$

$\alpha(u)$ and $\delta(u)$ being determined by the action of $A(u)$ and $D(u)$ on the pseudo-vacuum. If $\alpha(u)$ and $\delta(u)$ are given explicitly, we will be able to obtain a KP tau function (and hopefully a 2-KP tau function as well).

Such an interpretation can be found most clearly in the cases of the phase model[27–29] and the 4-vertex model,[30] both of which are related to enumeration of boxed plane partitions. The scalar product of Bethe states in these cases becomes, up to a simple factor, a sum of products of two Schur functions:

$$S(\boldsymbol{u},\boldsymbol{v}) = (\text{simple factor}) \sum_{\lambda\subseteq(N^n)} s_\lambda(u_1^2,\ldots,u_n^2)s_\lambda(v_1^{-2},\ldots,v_n^{-2}). \quad (44)$$

By the same reasoning as the interpretation of (31), one can see that this sum is a tau function of the 2-KP hierarchy (or, rather, the 2D Toda hierarchy as Zuparic argued[6]) with respect to the time variables

$$t_n = \frac{1}{n}\sum_{i=1}^{n} u_i^{2n}, \quad \bar{t}_n = -\frac{1}{n}\sum_{i=1}^{n} v_i^{-2n}.$$

Actually, this case admits yet another interpretation. In the course of deriving (44), the scalar product is shown to be a generating function for counting plane partitions in a box of size $n\times n\times N$. It is more or less well known[16] that the Schur function $s_{(N^n)}$ gives such a generating function. Thus, in terms of this Schur function, the scalar product can be expressed as

$$S(\boldsymbol{u},\boldsymbol{v}) = (\text{simple factor}) s_{(N^n)}(u_1^2,\ldots,u_N^2,v_1^2,\ldots,v_N^2). \quad (45)$$

This is similar to the formula (26) of the partition function of the 6-vertex model for $q = e^{\pi i/3}$; one can thereby find an interpretation as a KP or 2-KP tau function with respect to the time variables

$$t_n = \frac{1}{n}\sum_{i=1}^{n} u_i^{2n}, \quad \bar{t}_n = \frac{1}{n}\sum_{i=1}^{n} v_i^{2n}.$$

Acknowledgments

The author thanks Omar Foda and Paul Zinn-Justin for useful comments and discussion. This work is partly supported by Grant-in-Aid for Scientific Research No. 19104002, No. 19540179 and No. 21540218 from the Japan Society for the Promotion of Science.

References

1. M. Jimbo (ed.), "Yang-Baxter equations in Integrable Systems", World Scientific Publishing, 1990.
2. A. Zabrodin, Discrete Hirota's equation in integrable models, Int. J. Mod. Phys. **B11** (1997), 3125–3158.
3. O. Foda, M. Wheeler and M. Zuparic, Domain wall partition functions and KP, J. Stat. Mech. (2009), P03017.
4. O. Foda, M. Wheeler and M. Zuparic, XXZ scalar products and KP, Nucl. Phys. **B820** [FS] (2009), 649–663.
5. O. Foda and G. Schrader, XXZ scalar products, Miwa variables and discrete KP, arXiv:1003.2524.
6. M. Zuparic, Phase model expectation values and the 2-Toda hierarchy, J. Stat. Mech. (2009), P08010.
7. M. Zuparic, Studies in integrable quantum lattice models and classical hierarchies, arXiv:0908.3936.
8. V. Korepin, Calculation of norms of Bethe wave functions, Comm. Math. Phys. **86** (1982), 391–418.
9. A.G. Izergin, Partition function of the six-vertex model in a finite volume, Soviet Phys. Dokl. **32**(1987), 878–879.
10. A.G. Izergin, D.A. Coker and V.E. Korepin, Determinant formula for the six vertex model, J. Phys. A: Math. Gen **25** (1991), 4315–4334.
11. N. Slavnov, Calculation of scalar products of wave functions and form factors in the framework of the algebraic Bethe ansatz, Theor. Math. Phys. **79** (1989), 502–508.
12. N. Kitanine, J. Maillet, V. Terras, Form Factors of the XXZ Heisenberg spin-1/2 finite chain, Nucl. Phys. **B554** (1999), 647–678.
13. N. Bogoliubov, A. Izergin and N. Kitanine, Correlation functions for a strongly correlated boson system, Nucl. Phys. **B516** [FS] (1998), 501–528.
14. V.E. Korepin, N.M. Bogoliubov and A.G. Izergin, "Quantum inverse scattering method and correlation functions", Cambridge University Press, 1993.

15. I. Macdonald, "Symmetric Functions and Hall Polynomials", Oxford University Press, USA, 1999.
16. P. Zinn-Justin, Six-vertex, loop and tiling models: Integrability and combinatorics, arXiv:0901.0665.
17. T. Miwa, M. Jimbo and E. Date, "Solitons", Cambridge University Press, 2000.
18. T. Miwa, On Hirota's difference equation, Proc. Japan Acad., Ser. A, **58** (1982), 8–11.
19. M. Sato and Y. Sato, Soliton equations as dynamical systems on infinite dimensional Grassmann manifold, in "Nonlinear PDE in Applied Science" (Tokyo, 1992), Lecture Notes in Num. Appl. Anal. vol. 5 (1982), pp. 259–271.
20. K. Ueno and K. Takasaki, Toda Lattice Hierarchy, in "Group Representations and Systems of Differential Equations", Advanced Studies in Pure Mathematics vol. 4 (Kinokuniya, 1984), pp. 1–95.
21. K. Takasaki, Initial value problem for the Toda lattice hierarchy, in "Group Representations and Systems of Differential Equations", Advanced Studies in Pure Mathematics vol. 4 (Kinokuniya, 1984), pp. 139–163.
22. Yu. Stroganov, Izergin-Korepin determinant at a third root of unity, Theor. Math. Phys. **146** (2006), 53–62.
23. S. Okada, Enumeration of symmetry classes of alternating sign matrices and characters of classical groups, J. Algebraic Combin. **23** (2006), 43–69.
24. A.Yu. Orlov and T. Shiota, Schur function expansion for normal matrix model and associated discrete matrix models, Phys. Lett. **A343** (2005), 384–396.
25. V. Korepin and P. Zinn-Justin, Thermodynamic limit of the six-vertex model with domain wall boundary conditions, J. Phys. A: Math. Gen. **33** (2000), 7053–7066.
26. N.M. Bogoliubov, Determinantal representation of the time-dependent stationary correlation function for the totally asymmetric simple exclusion model, SIGMA **5** (2009), 052 (11 pages).
27. N. Bogoliubov, Boxed plane partitions as an exactly solvable boson model, J. Phys. A: Math. Gen. **38** (2005), 9415–9430.
28. K. Shigechi and M. Uchiyama, Boxed skew plane partitions and integrable phase model, J. Phys. A: Math. Gen. **38** (2005), 10287–10306.
29. N. Tsilevich, Quantum inverse scattering method for the q-boson model and symmetric functions, Funct. Anal. Appl. **40** (2006), 207–217.
30. N.M. Bogoliubov, Four-vertex models and random tilings, Theor. Math. Phys. **155** (2008), 523–535.

INTRODUCTION TO MIDDLE CONVOLUTION FOR DIFFERENTIAL EQUATIONS WITH IRREGULAR SINGULARITIES

KOUICHI TAKEMURA*

Department of Mathematical Sciences, Yokohama City University 22-2 Seto, Kanazawa-ku, Yokohama 236-0027, Japan

Dedicated to Professor Tetsuji Miwa on his sixtieth birthday

We introduce middle convolution for systems of linear differential equations with irregular singular points, and we present a tentative definition of the index of rigidity for them. Under some assumption, we show a list of terminal patterns of irreducible systems of linear differential equations by iterated application of middle convolution when the index is positive or zero.

Keywords: Middle convolution; irregular singularity; Euler's integral transformation; index of rigidity.

1. Introduction

Middle convolution was originally introduced by N. Katz in his book "Rigid local systems",[7] and several others studied and reformulated it. Dettweiler and Reiter[3,4] defined middle convolution for systems of Fuchsian differential equations written as

$$\frac{dY}{dz} = \left(\sum_{i=1}^{r} \frac{A^{(i)}}{z - t_i} \right) Y. \tag{1}$$

Note that Eq.(1) has singularities at $\{t_1, \dots, t_r, \infty\}$ and they are all regular. On the theory of middle convolution, the index of rigidity plays important roles and it is preserved by two operation, addition and middle convolution. Addition is related to multiplying a function to solutions of

*Current address: Department of Mathematics, Faculty of Science and Technology, Chuo University, 1-13-27 Kasuga, Bunkyo-ku Tokyo 112-8551, Japan.
E-mail: takemura@math.chuo-u.ac.jp

differential equations, and middle convolution is related to applying Euler's integral transformation to them. Middle convolution may change the size of differential equations. It was essentially established by Katz[7] that every irreducible system of Fuchsian differential equations whose index of rigidity is two is reduced to the system of rank one. The procedure to obtain rank one system is called Katz' algorithm. Subsequently it is shown that the system has integral representations of solutions which are calculated by following Katz' algorithm.

In this paper we study middle convolution for systems of linear differential equations with irregular singularities, which are written as

$$\frac{dY}{dz} = \left(-\sum_{j=1}^{m_0} A_j^{(0)} z^{j-1} + \sum_{i=1}^{r} \sum_{j=0}^{m_i} \frac{A_j^{(i)}}{(z-t_i)^{j+1}} \right) Y. \tag{2}$$

If $m_i = 0$ $(i \neq 0)$ (resp. $m_0 = 0$), then the point $z = t_i$ (resp. the point $z = \infty$) is regular singularity. In particular if $m_0 = \cdots = m_r = 0$, then the system is Fuchsian. There are several important linear differential equations which are not Fuchsian, e.g., Kummer's confluent hypergeometric equation and Bessel's equation. We want to extend the theory of middle convolution to include those equations.

A pioneering work on middle convolution with irregular singularities was carried out by Kawakami.[8] He focused on the differential equations

$$(zI_n - T)\frac{d\Psi}{dz} = A\Psi, \tag{3}$$

which is called Okubo normal form if T is a diagonal matrix. Eq.(3) is called generalized Okubo normal form, if T may not be diagonalizable. Kawakami constructed a map from generalized Okubo normal forms to linear differential equations with irregular singularities with the condition $m_0 = 0$, and he showed that the map is surjective, i.e. every irreducible equation written as Eq.(2) with the condition $m_0 = 0$ corresponds to an irreducible equation of generalized Okubo normal form. He considered middle convolution by using generalized Okubo normal forms, because Euler's integral transformation is on well with (generalized) Okubo normal forms. Yamakawa[13] gave a geometric interpretation to Kawakami's map and investigated middle convolution with irregular singularities with the condition $m_0 \leq 1$ by Harnad duality.

In this paper, we construct middle convolution including the case $m_0 \neq 0$ directly, i.e. without generalized Okubo normal forms nor Harnad duality. Our construction is explicit as we will discuss in sections 2 and 3.

We firstly define convolution matrices, which are compatible with Euler's integral transformation (see Theorem 2.1). The size of convolution matrices is nM, where n is the size of given matrices $A^{(i)}_j$ in Eq.(2) and $M = r + \sum_{i=0}^{r} m_i$. The module defined by convolution matrices may not be irreducible. We consider a quotient of convolution matrices to obtain irreducible modules in section 3, which leads to the definition of middle convolution. We can describe the quotient spaces explicitly, and it is an advantage of our construction. We propose a tentative definition of the index of rigidity, which is expected to be preserved by middle convolution.

We study middle convolution further on the case that $m_i \leq 1$ and $A^{(i)}_1$ is semi-simple for all i in section 4. In particular, we show that the index of rigidity is preserved by middle convolution on this case. By applying middle convolutions and additions appropriately, the size of differential equations may be possibly decreased. Under the assumption of this section, we show that if the index of rigidity is positive and the system of differential equations is irreducible, then the size of differential equations can be decreased to one by iterated application of middle convolution and addition. Moreover we show a list of terminal patterns obtained by iterated application of middle convolution and addition for the case that the index of rigidity is zero.

We give some comments for future reference in section 5.

2. Convolution

Let $\mathbf{A} = (A^{(0)}_{m_0}, \dots, A^{(0)}_1, A^{(1)}_{m_1}, \dots, A^{(r)}_0)$ be a tuple of matrices acting on the finite-dimensional vector space V ($\dim V = n$). The tuple $\mathbf{A}$ is attached with the system of differential equations (2). We denote by $\langle \mathbf{A} \rangle$ the algebra generalted by $A^{(i)}_j$ ($i = 0, \dots, r$, $j = \delta_{i,0}, \dots m_i$). The $\langle \mathbf{A} \rangle$-module V (or $\langle \mathbf{A} \rangle$) is called irreducible if there is no proper subspace $W (\subset V)$ such that $A^{(i)}_j W \subset W$ for any i, j.

Set

$$M = r + \sum_{i=0}^{r} m_i, \quad V' = \bigoplus_{i=0}^{r} V^{(i)} = \mathbb{C}^{nM}, \tag{4}$$
$$V^{(0)} = V^{\oplus m_0}, \quad V^{(i)} = V^{\oplus (m_i+1)}, \ (i \geq 1).$$

We fix $\mu \in \mathbb{C}$ and define convolution matrices $\tilde{A}^{(i)}_j$ ($i = 0, \dots, r$, $j =$

$\delta_{i,0}, \dots m_i$) acting on V' by

$$\begin{pmatrix} u^{(0)}_{m_0} \\ \vdots \\ u^{(0)}_1 \\ u^{(1)}_{m_1} \\ \vdots \\ u^{(r)}_0 \end{pmatrix} = \tilde{A}^{(i)}_j \begin{pmatrix} v^{(0)}_{m_0} \\ \vdots \\ v^{(0)}_1 \\ v^{(1)}_{m_1} \\ \vdots \\ v^{(r)}_0 \end{pmatrix}, \tag{5}$$

where $v^{(i')}_{j'}, u^{(i')}_{j'} \in V$ ($i' = 0, \dots, r$, $j' = \delta_{i',0}, \dots m_{i'}$) and $u^{(i')}_{j'}$ are given by

$$u^{(i')}_{j'} = \begin{cases} \mu v^{(i')}_{j'-j} & i' = i,\ j' > j, \\ \displaystyle\sum_{i''=0}^{r} \sum_{j''=\delta_{i'',0}}^{m_{i''}} A^{(i'')}_{j''} v^{(i'')}_{j''} & i = 0,\ i' = i,\ j' = j, \\ \mu v^{(i')}_0 + \displaystyle\sum_{i''=0}^{r} \sum_{j''=\delta_{i'',0}}^{m_{i''}} A^{(i'')}_{j''} v^{(i'')}_{j''} & i \neq 0, i' = i, j' = j, \\ 0 & \text{otherwise.} \end{cases} \tag{6}$$

Namely

$$\tilde{A}^{(0)}_j = \begin{pmatrix} \left.\begin{matrix} \mu I_n & & \\ & \ddots & \\ & & \mu I_n \end{matrix}\right\} m_0 - j & & & & & & & & \\ A^{(0)}_{m_0} \cdots A^{(0)}_2 \; A^{(0)}_1 \; A^{(1)}_{m_1} \cdots A^{(1)}_0 \; A^{(2)}_{m_2} \cdots A^{(r)}_0 \\ \\ \\ \end{pmatrix}, \tag{7}$$

$$\begin{matrix} \tilde{A}^{(i)}_j = \\ (i \geq 1) \end{matrix} \begin{pmatrix} \left.\begin{matrix} \\ \\ \end{matrix}\right\} m_0 + (m_1+1) + \cdots + (m_{i-1}+1) \\ \left.\begin{matrix} \mu I_n & & \\ & \ddots & \\ & & \mu I_n \end{matrix}\right\} m_i - j \\ A^{(0)}_{m_0} \cdots A^{(i-1)}_{\delta_{i,0}} \; A^{(i)}_{m_i} \cdots A^{(i)}_1 \; A^{(i)}_0 + \mu I_n \; A^{(i+1)}_{m_{i+1}} \cdots A^{(r)}_0 \\ \\ \\ \end{pmatrix}, \tag{8}$$

where I_n is the unit matrix of size n. We denote the tuple of convolution matrices by $\tilde{\mathbf{A}} = (\tilde{A}^{(0)}_{m_0}, \dots, \tilde{A}^{(0)}_1, \tilde{A}^{(1)}_{m_1}, \dots, \tilde{A}^{(r)}_0)$. Note that the definition of convolution matrices is a straightforward generalization of Dettweiler and Reiter.[3] Then we can show that the convolution corresponds to Euler's integral transformation on linear differential equations, which is also analogous to Dettweiler and Reiter.[4]

Theorem 2.1. *Assume that* $Y = \begin{pmatrix} y_1(z) \\ \vdots \\ y_n(z) \end{pmatrix}$ *is a solution of*

$$\frac{dY}{dz} = \left(-\sum_{j=1}^{m_0} A_j^{(0)} z^{j-1} + \sum_{i=1}^{r} \sum_{j=0}^{m_i} \frac{A_j^{(i)}}{(z-t_i)^{j+1}} \right) Y. \tag{9}$$

Let γ *be a cycle such that* $\int_\gamma \frac{d}{dw}(r(w) y_j(w)(z-w)^\mu) dw = 0$ *for any* j *and any rational function* $r(w)$. *Then the function* U *defined by*

$$U = \begin{pmatrix} U^{(0)}_{m_0}(z) \\ \vdots \\ U^{(0)}_1(z) \\ U^{(1)}_{m_1}(z) \\ \vdots \\ U^{(r)}_0(z) \end{pmatrix}, \quad U^{(0)}_j(z) = - \begin{pmatrix} \int_\gamma w^{j-1} y_1(w)(z-w)^\mu dw \\ \vdots \\ \int_\gamma w^{j-1} y_n(w)(z-w)^\mu dw \end{pmatrix}, \tag{10}$$

$$\begin{matrix} U^{(i)}_j(z) = \\ (i \neq 0) \end{matrix} \begin{pmatrix} \int_\gamma (w-t_i)^{-j-1} y_1(w)(z-w)^\mu dw \\ \vdots \\ \int_\gamma (w-t_i)^{-j-1} y_n(w)(z-w)^\mu dw \end{pmatrix} \tag{11}$$

satisfies

$$\frac{dU}{dz} = \left(-\sum_{j=1}^{m_0} \tilde{A}_j^{(0)} z^{j-1} + \sum_{i=1}^{r} \sum_{j=0}^{m_i} \frac{\tilde{A}_j^{(i)}}{(z-t_i)^{j+1}} \right) U. \tag{12}$$

The proof will be given in our forthcoming paper.[12]

Let γ_t is a cycle turning the point $w = t$ anti-clockwise. Then the contour $[\gamma_z, \gamma_{t_i}] = \gamma_z \gamma_{t_i} \gamma_z^{-1} \gamma_{t_i}^{-1}$ for $i = 0, \dots, r$ ($t_0 = \infty$) satisfies the condition of Theorem 2.1. We should also consider cycles which reflect the Stokes phenomena if there exists an irregular singularity.

3. Middle convolution and the index of rigidity

3.1. *Middle convolution*

We have defined convolution in section 2, although convolution may not preserve irreducibility. We now consider a quotient of convolution. We define the subspaces $\mathcal{K}$, $\mathcal{L}'(\mu)$ and $\mathcal{L}(\mu)$ of $V' = V^{\oplus M}$ by

$$\begin{array}{l}\mathcal{K}^{(i)} = \\ (i \geq 1)\end{array} \left\{ \begin{pmatrix} v^{(i)}_{m_i} \\ v^{(i)}_{m_i-1} \\ \vdots \\ v^{(i)}_0 \end{pmatrix} \in V^{(i)} \middle| \begin{pmatrix} A^{(i)}_{m_i} & A^{(i)}_{m_i-1} & \dots & A^{(i)}_0 \\ 0 & A^{(i)}_{m_i} & \dots & A^{(i)}_1 \\ 0 & 0 & \ddots & \vdots \\ 0 & \cdots & 0 & A^{(i)}_{m_i} \end{pmatrix} \begin{pmatrix} v^{(i)}_{m_i} \\ v^{(i)}_{m_i-1} \\ \vdots \\ v^{(i)}_0 \end{pmatrix} = \begin{pmatrix} 0 \\ 0 \\ \vdots \\ 0 \end{pmatrix} \right\}, \tag{13}$$

$$\mathcal{K}^{(0)} = \{0\} (\subset V^{(0)}), \quad \mathcal{K} = \bigoplus_{i=0}^{r} \mathcal{K}^{(i)},$$

$$\mathcal{L}'(\mu) = \left\{ \begin{pmatrix} v^{(0)}_{m_0} \\ \vdots \\ v^{(0)}_1 \\ v^{(1)}_{m_1} \\ \vdots \\ v^{(r)}_0 \end{pmatrix} \middle| \begin{array}{c} v^{(i)}_j = 0, \quad (i \neq 0, j \neq 0), \\ v^{(1)}_0 = \cdots = v^{(r)}_0 = -\ell, \\ \begin{pmatrix} A^{(0)}_{m_0} & \dots & A^{(0)}_1 & A^{(0)}_0 - \mu I_n \\ 0 & A^{(0)}_{m_0} & \dots & A^{(0)}_1 \\ 0 & 0 & \ddots & \vdots \\ 0 & \cdots & 0 & A^{(0)}_{m_0} \end{pmatrix} \begin{pmatrix} v^{(0)}_{m_0} \\ \vdots \\ v^{(0)}_1 \\ \ell \end{pmatrix} = \begin{pmatrix} 0 \\ \vdots \\ 0 \\ 0 \end{pmatrix} \end{array} \right\},$$

$$\begin{array}{l}\mathcal{L}(\mu) = \mathcal{L}'(\mu), \\ (\mu \neq 0)\end{array} \quad \mathcal{L}(0) = \left\{ \begin{pmatrix} v^{(0)}_{m_0} \\ \vdots \\ v^{(0)}_1 \\ v^{(1)}_{m_1} \\ \vdots \\ v^{(r)}_0 \end{pmatrix} \middle| \sum_{i=0}^{r} \sum_{j=\delta_{i,0}}^{m_i} A^{(i)}_j v^{(i)}_j = 0 \right\},$$

where $A^{(0)}_0 = -(A^{(1)}_0 + \cdots + A^{(r)}_0)$.

Proposition 3.1. *We have* $\tilde{A}^{(i)}_j \mathcal{K} \subset \mathcal{K}$, $\tilde{A}^{(i)}_j \mathcal{L}(\mu) \subset \mathcal{L}(\mu)$ *and* $\tilde{A}^{(i)}_j \mathcal{L}'(\mu) \subset \mathcal{L}'(\mu)$ *for all* i, j.

Proof. We show that $\tilde{A}^{(i)}_j \mathcal{L}'(\mu) \subset \mathcal{L}'(\mu)$. Assume that $v = (v^{(0)}_{m_0} \dots v^{(0)}_1 v^{(1)}_{m_1} \dots v^{(r)}_0)^{\mathbf{T}} \in \mathcal{L}'(\mu)$. Then $v^{(i)}_j = 0$ $(i \neq 0, j \neq 0)$, $v^{(1)}_0 = \cdots = v^{(r)}_0 = -\ell$ and

$(v^{(0)}_{m_i} \dots v^{(0)}_1\ \ell)^{\mathbf{T}}$ satisfies the definition of $\mathcal{L}'(\mu)$. In particular we have

$$\sum_{i'=0}^{r}\sum_{j'=\delta_{i',0}}^{m_{i'}} A^{(i')}_{j'}v^{(i')}_{j'} = \sum_{j'=1}^{m_0} A^{(0)}_{j'}v^{(0)}_{j'} - \sum_{i=1}^{r} A^{(i)}_0\ell \qquad (14)$$

$$= \sum_{j'=1}^{m_0} A^{(0)}_{j'}v^{(0)}_{j'} + A^{(0)}_0\ell = \mu\ell.$$

Then $\tilde{A}^{(0)}_j v$ $(j \neq 0)$ is written as

$$\tilde{A}^{(0)}_j v = \begin{pmatrix} \mu v^{(0)}_{m_0-j} \\ \vdots \\ \mu v^{(0)}_1 \\ \sum_{j=1}^{m_0} A^{(0)}_j v^{(0)}_j + \sum_{i=1}^{r} A^{(i)}_0 v^{(i)}_0 \\ 0 \\ \vdots \end{pmatrix} = \mu \begin{pmatrix} v^{(0)}_{m_0-j} \\ \vdots \\ v^{(0)}_1 \\ \ell \\ 0 \\ \vdots \end{pmatrix}, \qquad (15)$$

and $(v^{(0)}_{m_0-j} \cdots v^{(0)}_1\ \ell\ 0 \dots)^{\mathbf{T}}$ satisfies the definition of $\mathcal{L}'(\mu)$. Hence $\tilde{A}^{(0)}_j v \in \mathcal{L}'(\mu)$ for $j = 1, \dots, m_0$. We also have $\tilde{A}^{(i)}_j v = 0$ for $i \neq 0$, because $v^{(i)}_{m_i} = \dots = v^{(i)}_1 = 0$ and $\mu v^{(i)}_0 + \sum_{i'=0}^{r}\sum_{j'=\delta_{i',0}}^{m_{i'}} A^{(i')}_{j'}v^{(i')}_{j'} = -\mu\ell + \mu\ell = 0$.

The other cases are shown similarly. □

We define $mc_\mu(V)$ to be the $\langle mc_\mu(\mathbf{A})\rangle$-module $V^{\oplus M}/(\mathcal{K} + \mathcal{L}(\mu))$ where $mc_\mu(\mathbf{A})$ is the tuple of matrices on $V^{\oplus M}/(\mathcal{K} + \mathcal{L}(\mu))$ whose actions are determined by $\tilde{\mathbf{A}}$, and we call it the middle convolution of V with the parameter μ. The following propositions are analogues to Dettweiler and Reiter,[3] which will be shown in our forthcoming paper.[12]

Proposition 3.2. *(i) If $\mu \neq 0$, then $\mathcal{K} \cap \mathcal{L}(\mu) = \{0\}$.*
(ii) If $\mu = 0$, then $\mathcal{K} + \mathcal{L}'(0) \subset \mathcal{L}(0)$.
(iii) If the $\langle \mathbf{A}\rangle$-module V is irreducible and $\mu = 0$, then $\mathcal{K} \cap \mathcal{L}'(0) = \{0\}$ and $\dim \mathcal{K} + \dim \mathcal{L}'(0) \leq n(M-1)$.

Proposition 3.3. *Assume that the $\langle \mathbf{A}\rangle$-module V is irreducible.*
(i) $mc_0(V) \simeq V$ as $\langle \mathbf{A}\rangle$-modules.
(ii) The $\langle mc_\mu(\mathbf{A})\rangle$-module $mc_\mu(V)$ is irreducible and $V \simeq mc_{-\mu}(mc_\mu(V))$ for any μ.

3.2. *Addition*

Let Y be a solution of Eq.(2). Then the function

$$Y' = \exp\left(-\sum_{j=1}^{m_0} \frac{\mu_j^{(0)}}{j} z^j - \sum_{i=1}^{r}\sum_{j=1}^{m_i} \frac{\mu_j^{(i)}}{j(z-t_i)^{j+1}}\right) \prod_{i=1}^{r} (z-t_i)^{\mu_0^{(i)}} Y \tag{16}$$

satisfies the equation

$$\frac{dY'}{dz} = \left(-\sum_{j=1}^{m_0} (A_j^{(0)} + \mu_j^{(0)} I_n) z^{j-1} + \sum_{i=1}^{r}\sum_{j=0}^{m_i} \frac{A_j^{(i)} + \mu_j^{(i)} I_n}{(z-t_i)^{j+1}}\right) Y'. \tag{17}$$

We now define addition for the tuple $\mathbf{A} = (A_{m_0}^{(0)}, \dots, A_1^{(0)}, A_{m_1}^{(1)}, \dots, A_0^{(r)})$ by

$$M_{\overline{\mu}}(\mathbf{A}) = \mathbf{A} + \overline{\mu} I_n = (A_{m_0}^{(0)} + \mu_{m_0}^{(0)} I_n, \dots, A_0^{(r)} + \mu_0^{(r)} I_n), \tag{18}$$

where $\overline{\mu} = (\mu_{m_0}^{(0)}, \dots, \mu_1^{(0)}, \mu_{m_1}^{(1)}, \dots, \mu_0^{(r)}) \in \mathbb{C}^M$.

3.3. *Index of rigidity*

Let $\mathbf{A} = (A_{m_0}^{(0)}, \dots, A_1^{(0)}, A_{m_1}^{(1)}, \dots, A_0^{(r)})$ be a tuple of matrices acting on V. Set

$$A^{(i)} = \begin{pmatrix} A_{m_i}^{(i)} & A_{m_i-1}^{(i)} & \dots & A_0^{(i)} \\ 0 & A_{m_i}^{(i)} & \dots & A_1^{(i)} \\ 0 & 0 & \ddots & \vdots \\ 0 & \cdots & 0 & A_{m_i}^{(i)} \end{pmatrix} \begin{matrix} \in \mathrm{End}(V^{\oplus(m_i+1)}) \\ (i = 0, \dots, r) \end{matrix}, \tag{19}$$

and

$$\mathcal{C}^{(i)} = \left\{ C^{(i)} = \begin{pmatrix} C_{m_i}^{(i)} & C_{m_i-1}^{(i)} & \dots & C_0^{(i)} \\ 0 & C_{m_i}^{(i)} & \dots & C_1^{(i)} \\ 0 & 0 & \ddots & \vdots \\ 0 & \cdots & 0 & C_{m_i}^{(i)} \end{pmatrix} \middle| A^{(i)} C^{(i)} = C^{(i)} A^{(i)} \right\}. \tag{20}$$

We define the index of rigidity by

$$\mathrm{idx}(\mathbf{A}) = \sum_{i=0}^{r} \dim(\mathcal{C}^{(i)}) - (M-1)(\dim(V))^2, \tag{21}$$

where $M = r + \sum_{i=0}^{r} m_i$. The condition $A^{(i)} C^{(i)} = C^{(i)} A^{(i)}$ is equivalent to

$$\sum_{j=0}^{k} \left(A_{m_i-j}^{(i)} C_{m_i-k+j}^{(i)} - C_{m_i-k+j}^{(i)} A_{m_i-j}^{(i)} \right) = 0, \quad k = 0, \dots, m_i. \tag{22}$$

The following proposition is readily obtained by Eq.(22):

Proposition 3.4. *The index of rigidity is preserved by addition, i.e.* $\mathrm{idx}(M_{\overline{\mu}}(\mathbf{A})) = \mathrm{idx}(\mathbf{A})$.

Conjecture 3.1. *If the $\langle \mathbf{A} \rangle$-module V is irreducible, then the index of rigidity is preserved by middle convolution, i.e.* $\mathrm{idx}(mc_\mu(\mathbf{A})) = \mathrm{idx}(\mathbf{A})$.

We will prove the conjecture for a special case in section 4 (see Proposition 4.2).

We define the local index of rigidity by

$$\mathrm{idx}_i(\mathbf{A}) = \mathrm{idx}_i[A^{(i)}_{m_i}, \dots, A^{(i)}_0] = \dim(\mathcal{C}^{(i)}) - (m_i + 1)\,(\dim(V))^2. \tag{23}$$

Then we have

$$\mathrm{idx}(\mathbf{A}) = \sum_{i=0}^{r} \mathrm{idx}_i(\mathbf{A}) + 2(\dim(V))^2. \tag{24}$$

3.4. *Example*

We consider irreducible systems of differential equations of size two written as

$$\frac{dY}{dz} = \left(-A^{(0)}_1 + \frac{A^{(1)}_0}{z} \right) Y, \;\; Y = \begin{pmatrix} y_1(z) \\ y_2(z) \end{pmatrix}. \tag{25}$$

It has an irregular singularity at $z = \infty$ and a regular singularity at $z = 0$.

First we consider the case that $A^{(0)}_1$ is semi-simple. It follows from irreducibility that $A^{(0)}_1$ is not scalar. By applying addition (i.e. multiplying $e^{\nu' z} z^{\alpha'}$ to the solution Y) and gauge transformation (i.e. multiplying a constant matrix to the solution Y), we may assume that $A^{(0)}_1$ is a diagonal matrix with the eigenvalues 0 and $-\nu (\neq 0)$ and $A^{(1)}_0$ has the eigenvalues 0 and $-\gamma$. Then we may set

$$A^{(0)}_1 = \begin{pmatrix} 0 & 0 \\ 0 & -\nu \end{pmatrix}, \quad A^{(1)}_0 = \begin{pmatrix} -\alpha & k \\ \frac{\alpha(\gamma-\alpha)}{k} & \alpha - \gamma \end{pmatrix}. \tag{26}$$

It follows from irreducibility that $k \neq 0$ and $\alpha \neq 0$. By eliminating $y_2(z)$ in Eq.(25), we have a second order linear differential equation,

$$z\frac{d^2 y_1}{dz^2} + (\gamma - \nu z)\frac{dy_1}{dz} - \alpha\nu y_1 = 0. \tag{27}$$

If $\nu = 1$, then Eq.(27) represents the confluent hypergeometric differential equation. Eq.(27) for the case $\nu \neq 0$ reduces to the confluent hypergeometric differential equation by changing the variable $z' = \nu z$. The index of rigidity for Eq.(25) is two, because $\dim(\mathcal{C}^{(0)}) = 4$ and $\dim(\mathcal{C}^{(1)}) = 2$ which follows from $\nu \neq 0$ and $k \neq 0$.

We investigate middle convolution mc_μ for the matrices in Eq.(26). Convolution matrices are given as

$$\tilde{A}_1^{(0)} = \begin{pmatrix} A_1^{(0)} & A_0^{(1)} \\ 0 & 0 \end{pmatrix} = \begin{pmatrix} 0 & 0 & -\alpha & k \\ 0 & -\nu & \frac{\alpha(\gamma-\alpha)}{k} & \alpha-\gamma \\ 0 & 0 & 0 & 0 \\ 0 & 0 & 0 & 0 \end{pmatrix}, \tag{28}$$

$$\tilde{A}_0^{(1)} = \begin{pmatrix} 0 & 0 \\ A_1^{(0)} & A_0^{(1)} + \mu I_2 \end{pmatrix} = \begin{pmatrix} 0 & 0 & 0 & 0 \\ 0 & 0 & 0 & 0 \\ 0 & 0 & -\alpha+\mu & k \\ 0 & -\nu & \frac{\alpha(\gamma-\alpha)}{k} & \alpha-\gamma+\mu \end{pmatrix}.$$

The dimension of the space $\mathcal{K}(\simeq \mathcal{K}_1)$ is one and the space is described as

$$\mathcal{K} = \begin{pmatrix} 0 \\ \mathrm{Ker}(A_0^{(1)}) \end{pmatrix} = \mathbb{C}\begin{pmatrix} 0 \\ 0 \\ k \\ \alpha \end{pmatrix}. \tag{29}$$

The space $\mathcal{L}(\mu)$ $(\mu \neq 0)$ is described as

$$\begin{aligned} \mathcal{L}(\mu) &= \left\{ \begin{pmatrix} v_1^{(0)} \\ -\ell \end{pmatrix} \middle| \begin{pmatrix} A_1^{(0)} & -A_0^{(1)} - \mu \\ 0 & A_1^{(0)} \end{pmatrix} \begin{pmatrix} v_1^{(0)} \\ \ell \end{pmatrix} = \begin{pmatrix} 0 \\ 0 \end{pmatrix} \right\} \\ &= \mathrm{Ker} \begin{pmatrix} 0 & 0 & -\alpha+\mu & k \\ 0 & -\nu & \frac{\alpha(\gamma-\alpha)}{k} & \alpha-\gamma+\mu \\ 0 & 0 & 0 & 0 \\ 0 & 0 & 0 & -\nu \end{pmatrix}. \end{aligned} \tag{30}$$

Hence the dimension of the space $\mathcal{L}(\alpha)$ for the case $\mu \neq \alpha$ (resp. $\mu = \alpha$) is one (resp. two). We concentrate on the case $\mu = \alpha (\neq 0)$. A basis of the space $\mathcal{L}(\alpha)$ is given by

$$\begin{pmatrix} 1 \\ 0 \\ 0 \\ 0 \end{pmatrix}, \begin{pmatrix} 0 \\ \alpha(\gamma-\alpha) \\ k\nu \\ 0 \end{pmatrix}. \tag{31}$$

Set

$$S = \begin{pmatrix} 0 & 1 & 0 & 0 \\ 1 & 0 & \alpha(\gamma-\alpha) & 0 \\ 0 & 0 & k\nu & k \\ 0 & 0 & 0 & \alpha \end{pmatrix}. \tag{32}$$

Then we have

$$S^{-1}\tilde{A}_1^{(0)}S = \begin{pmatrix} -\nu & 0 & 0 & 0 \\ 0 & 0 & -k\nu\alpha & 0 \\ 0 & 0 & 0 & 0 \\ 0 & 0 & 0 & 0 \end{pmatrix}, \quad S^{-1}\tilde{A}_0^{(1)}S = \begin{pmatrix} \alpha-\gamma & 0 & 0 & 0 \\ 0 & 0 & 0 & 0 \\ 1/\alpha & 0 & 0 & 0 \\ -\nu/\alpha & 0 & 0 & \alpha \end{pmatrix}. \tag{33}$$

Since the second, the third and the fourth columns of the matrix S are divisors of the quotient space $mc_\alpha(\mathbb{C}^2) = (\mathbb{C}^2)^{\oplus 2}/(\mathcal{K}+\mathcal{L}(\alpha))$, the matrix elements of $\tilde{A}_0^{(1)}$ and $\tilde{A}_1^{(0)}$ appear as $(1,1)$-elements of Eq.(33). Hence Eq.(25) is transformed to

$$\frac{dy}{dz} = \left(\nu + \frac{\alpha-\gamma}{z}\right) y \tag{34}$$

by the middle convolution mc_α. The solutions of Eq.(34) is given by $y = c\exp(\nu z)z^{\alpha-\gamma}$ (c: a constant).

We are going to recover Eq.(25) from Eq.(34) and obtain integral representations of solutions of Eq.(25), which arise from the equality $mc_{-\alpha}mc_\alpha = \mathrm{id}$. We apply middle convolution $mc_{-\alpha}$ to Eq.(34). Then we have

$$\frac{dW}{dz} = \left(-\begin{pmatrix} -\nu & \alpha-\gamma \\ 0 & 0 \end{pmatrix} + \frac{1}{z}\begin{pmatrix} 0 & 0 \\ -\nu & -\gamma \end{pmatrix}\right) W. \tag{35}$$

It follows from Theorem 2.1 that the function

$$W = \begin{pmatrix} \int_C \exp(\nu w)w^{\alpha-\gamma}(z-w)^{-\alpha}dw \\ \int_C \exp(\nu w)w^{\alpha-\gamma-1}(z-w)^{-\alpha}dw \end{pmatrix}, \tag{36}$$

is a solution of Eq.(35) by choosing a cycle C appropriately. For simplicity we assume $\nu \in \mathbb{R}_{>0}$. Then we can take cycles C which start from $w=-\infty$, move along a real axis, turn the point $w=z$ or $w=0$ and come back to $w=-\infty$. By setting

$$W = \begin{pmatrix} \alpha-\gamma & -k \\ \nu & 0 \end{pmatrix} \tilde{W}, \tag{37}$$

we recover the matrices in Eq.(26) such as the function $\tilde{W}$ satisfies Eq.(25). Consequently we obtain integral representations of solutions of Eq.(25)

which are expressed as

$$\tilde{W} = \frac{1}{k\nu}\begin{pmatrix} 0 & k \\ -\nu & \alpha-\gamma \end{pmatrix}\begin{pmatrix} \int_C \exp(\nu w)w^{\alpha-\gamma}(z-w)^{-\alpha}dw \\ \int_C \exp(\nu w)w^{\alpha-\gamma-1}(z-w)^{-\alpha}dw \end{pmatrix}. \tag{38}$$

In particular, the function

$$y(z) = \int_C \exp(\nu w)w^{\alpha-\gamma-1}(z-w)^{-\alpha}dw \tag{39}$$

satisfies Eq.(27), and we obtain integral representations of solutions of the confluent hypergeometric differential equation.

Next we consider the case that $A_1^{(0)}$ is nilpotent. Set

$$A_1^{(0)} = \begin{pmatrix} 0 & -1 \\ 0 & 0 \end{pmatrix}, \quad A_0^{(1)} = \begin{pmatrix} a_{1,1} & a_{1,2} \\ a_{2,1} & a_{2,2} \end{pmatrix}. \tag{40}$$

Then it follows from irreducibility that $a_{2,1} \neq 0$. The index of rigidity is also two. But we cannot reduce to rank one case by applying additions $A_1^{(0)} \to A_1^{(0)} + \alpha I_2$, $A_0^{(1)} \to A_0^{(1)} + \beta I_2$ and middle convolution mc_μ, because $\dim(\mathcal{L}'(\mu)) \leq 1$ for any α, β, μ, which follows from $a_{2,1} \neq 0$. Note that solutions of the differential equations determined by Eq.(40) are expressed in terms of Bessel's function by setting $z = x^2$.

4. The case $m_i \leq 1$ for all i

In this section, we investigate the index of rigidity and middle convolution for the case $m_i \leq 1$ for all i (see Eq.(2)). The case $m_i = 0$ is included to the case $m_i = 1$ by setting $A_1^{(i)} = 0$. We assume that $A_1^{(i)}$ is semi-simple for all i and $V(= \mathbb{C}^n)$ is irreducible as $\langle \mathbf{A} \rangle$-module.

4.1. *Index of rigidity*

To study the index of rigidity, we investigate Eq.(22) for the case $m_i = 1$. We ignore the superscript $^{(i)}$. Then Eq.(22) is written as

$$A_1 C_1 = C_1 A_1, \quad A_1 C_0 - C_0 A_1 + A_0 C_1 - C_1 A_0 = 0. \tag{41}$$

By the assumption that A_1 is semi-simple, we diagonalize A_1 as

$$P^{-1}A_1 P = \begin{pmatrix} d_1 I_{n_1} & 0 & \dots & 0 \\ 0 & d_2 I_{n_2} & \dots & 0 \\ 0 & 0 & \ddots & \vdots \\ 0 & \cdots & 0 & d_k I_{n_k} \end{pmatrix}, \quad d_i \neq d_j \ (i \neq j). \tag{42}$$

Write

$$P^{-1}A_0P = \begin{pmatrix} A_0^{[1,1]} & A_0^{[1,2]} & \dots & A_0^{[1,k]} \\ A_0^{[2,1]} & A_0^{[2,2]} & \dots & A_0^{[2,k]} \\ \vdots & \vdots & \ddots & \vdots \\ A_0^{[k,1]} & A_0^{[k,2]} & \dots & A_0^{[k,k]} \end{pmatrix}, \tag{43}$$

where $A_0^{[i,j]}$ is a $n_i \times n_j$ matrix. It follows from $A_1C_1 = C_1A_1$ that C_1 is written as

$$P^{-1}C_1P = \begin{pmatrix} C_1^{[1]} & 0 & \dots & 0 \\ 0 & C_1^{[2]} & \dots & 0 \\ 0 & 0 & \ddots & \vdots \\ 0 & \cdots & 0 & C_1^{[k]} \end{pmatrix}, \tag{44}$$

where $C_1^{[l]}$ is a $n_l \times n_l$ matrix. Then

$$P^{-1}(A_0C_1 - C_1A_0)P = \tag{45}$$

$$\begin{pmatrix} A_0^{[1,1]}C_1^{[1]} - C_1^{[1]}A_0^{[1,1]} & A_0^{[1,2]}C_1^{[2]} - C_1^{[1]}A_0^{[1,2]} & \dots & A_0^{[1,k]}C_1^{[k]} - C_1^{[1]}A_0^{[1,k]} \\ A_0^{[2,1]}C_1^{[1]} - C_1^{[2]}A_0^{[2,1]} & A_0^{[2,2]}C_1^{[2]} - C_1^{[2]}A_0^{[2,2]} & \dots & A_0^{[2,k]}C_1^{[k]} - C_1^{[2]}A_0^{[2,k]} \\ \vdots & \vdots & \ddots & \vdots \\ A_0^{[k,1]}C_1^{[1]} - C_1^{[k]}A_0^{[k,1]} & A_0^{[k,2]}C_1^{[2]} - C_1^{[k]}A_0^{[k,2]} & \dots & A_0^{[k,k]}C_1^{[k]} - C_1^{[k]}A_0^{[k,k]} \end{pmatrix}.$$

By writing

$$P^{-1}C_0P = \begin{pmatrix} C_0^{[1,1]} & C_0^{[1,2]} & \dots & C_0^{[1,k]} \\ C_0^{[2,1]} & C_0^{[2,2]} & \dots & C_0^{[2,k]} \\ \vdots & \vdots & \ddots & \vdots \\ C_0^{[k,1]} & C_0^{[k,2]} & \dots & C_0^{[k,k]} \end{pmatrix}, \tag{46}$$

we have

$$P^{-1}(A_1C_0 - C_0A_1)P = \tag{47}$$

$$\begin{pmatrix} 0 & (d_1 - d_2)C_0^{[1,2]} & \dots & (d_1 - d_k)C_0^{[1,k]} \\ (d_2 - d_1)C_0^{[2,1]} & 0 & \dots & (d_2 - d_k)C_0^{[2,k]} \\ \vdots & \vdots & \ddots & \vdots \\ (d_k - d_1)C_0^{[k,1]} & (d_k - d_2)C_0^{[k,2]} & \dots & 0 \end{pmatrix}.$$

It follows from $A_1C_0 - C_0A_1 + A_0C_1 - C_1A_0 = 0$ that $A_0^{[i,i]}C_1^{[i]} = C_1^{[i]}A_0^{[i,i]}$ and $C_0^{[i,j]}$ $(i \neq j)$ is determined as $C_0^{[i,j]} = -(A_0^{[i,j]}C_1^{[j]} - C_1^{[i]}A_0^{[i,j]})/(d_i - d_j)$.

Elements of $C_0^{[i,i]}$ are not restricted by relations. Hence the dimension of solutions of Eq.(41) is

$$\sum_{l=1}^{k}\{(n_l)^2+\dim Z(A_0^{[l,l]})\}, \tag{48}$$

where $Z(A_0^{[l,l]})=\{X\in\mathbb{C}^{n_l\times n_l}|XA_0^{[l,l]}=A_0^{[l,l]}X\}$. Let $I_{a,b}$ be the $a\times b$ matrix whose (i,j)-element is given by $\delta_{i,j}$, $\mathbf{q}=(q_1,\dots,q_p)\in\mathbb{Z}^p$ $(q_1+\dots+q_p=n,\ q_1\geq\dots\geq q_p\geq 1)$, $\underline{\lambda}=(\lambda_1,\dots,\lambda_p)\in\mathbb{C}^p$. Following Oshima,[10] set

$$L(\mathbf{q};\,\underline{\lambda})=\begin{pmatrix}\lambda_1 I_{q_1} & I_{q_1,q_2} & 0 & \cdots \\ 0 & \lambda_2 I_{q_2} & I_{q_2,q_3} & \ddots \\ 0 & 0 & \lambda_3 I_{q_3} & \ddots \\ \vdots & \ddots & \ddots & \ddots\end{pmatrix}. \tag{49}$$

Every matrix is conjugate to $L(\mathbf{q};\,\underline{\lambda})$ for some $\mathbf{q}$, $\underline{\lambda}$. Note that if $\lambda_i\neq\lambda_j$ $(i\neq j)$ then the matrix $L(\mathbf{q};\,\underline{\lambda})$ is conjugate to the diagonal matrix whose multiplicity of the eigenvalue λ_i is q_i. If the matrix $A_0^{[l,l]}$ is conjugate to $L(n_{l,1},\dots,n_{l,p_l};\,d_{l,1},\dots,d_{l,p_l})$, then the dimension of solutions of Eq.(41) is given by

$$\sum_{l=1}^{k}\left\{(n_l)^2+\sum_{j=1}^{p_l}(n_{l,j})^2\right\}. \tag{50}$$

We denote the type of multiplicities of the matrices (A_1,A_0) which are expressed as Eqs.(42), (43) and $A_0^{[l,l]}$ $(l=1,\dots,k)$ is conjugate to $L(n_{l,1},\dots,n_{l,p_l};\lambda_{l,1},\dots,\lambda_{l,p_l})$ by

$$(n_1,n_2,\dots,n_k)-((n_{1,1},\dots,n_{1,p_1}),(n_{2,1},\dots,n_{2,p_2}),\dots,(n_{k,1},\dots,n_{k,p_k})). \tag{51}$$

Note that $n_l=n_{l,1}+\dots+n_{l,p_l}$ $(l=1,\dots,k)$. Then the local index of rigidity of the matrices (A_1,A_0) is calculated as

$$2n^2-\sum_{l=1}^{k}\left\{(n_l)^2+\sum_{j=1}^{p_l}(n_{l,j})^2\right\}. \tag{52}$$

If $A_1=0$, then $k=1$, $n_1=n$ and we simplify the notation $(n_1)-((n_{1,1},\dots,n_{1,p_1}))$ by $(n_{1,1},\dots,n_{1,p_1})$. Note that the notation $(n_{1,1},\dots,n_{1,p_1})$ was already adapted by Kostov[9] and Oshima[10] for the case of regular singularity.

By combining Eq.(52) with Eq.(24) we have the following proposition:

Proposition 4.1. *We assume that* $m_0 = \cdots = m_r = 1$, $A_1^{(i)}$ *are semi-simple for* $i = 0, \ldots, r$ *and* $\mathbb{C}^n$ *is irreducible as* $\langle \mathbf{A} \rangle$*-module. Let*

$$\begin{aligned}&(n_1^{(i)}, n_2^{(i)}, \ldots, n_{k^{(i)}}^{(i)})- \\ &((n_{1,1}^{(i)}, \ldots, n_{1,p_1^{(i)}}^{(i)}), (n_{2,1}^{(i)}, \ldots, n_{2,p_2^{(i)}}^{(i)}), \ldots, (n_{k^{(i)},1}^{(i)}, \ldots, n_{k^{(i)},p_{k^{(i)}}^{(i)}}^{(i)})),\end{aligned} \tag{53}$$

$(n_1^{(i)} \geq n_2^{(i)} \geq \cdots \geq n_{k^{(i)}}^{(i)},\ n_{j,1}^{(i)} \geq \cdots \geq n_{j,p_j^{(i)}}^{(i)}\ (j = 1, \ldots, k^{(i)}))$ *be the type of multiplicities of* $(A_1^{(i)}, A_0^{(i)})$*. Then the index of rigidity is equal to*

$$\mathrm{idx}(\mathbf{A}) = \sum_{i=0}^{r} \sum_{j=1}^{k^{(i)}} \left((n_j^{(i)})^2 + \sum_{j'=1}^{p_j^{(i)}} (n_{j,j'}^{(i)})^2 \right) - 2rn^2. \tag{54}$$

4.2. *Subspace*

Next we investigate solutions of the equations

$$A_1 v_0 = 0,\ A_1 v_1 + A_0 v_0 = 0, \tag{55}$$

for the case that the matrices A_1 and A_0 are expressed as Eqs.(42), (43) and $d_1 = 0$ to understand the subspaces $\mathcal{K}^{(i)}$ and $\mathcal{L}'(\lambda)$. Write

$$v_1 = P^{-1} \begin{pmatrix} v_1^{[1]} \\ \vdots \\ v_1^{[n]} \end{pmatrix},\ v_0 = P^{-1} \begin{pmatrix} v_0^{[1]} \\ \vdots \\ v_0^{[n]} \end{pmatrix},\ (v_1^{[l]}, v_0^{[l]} \in \mathbb{C}^{n_l}). \tag{56}$$

It follows from $A_1 v_0 = 0$ that $v_0^{[i]} = 0\ (i \geq 2)$. Hence

$$P^{-1}(A_0 v_0 + A_1 v_1) = \begin{pmatrix} A_0^{[1,1]} v_0^{[1]} \\ A_0^{[1,2]} v_0^{[1]} + d_2 v_1^{[2]} \\ \vdots \\ A_0^{[1,k]} v_0^{[1]} + d_k v_1^{[k]} \end{pmatrix} = 0, \tag{57}$$

$v_0^{[1]} \in \mathrm{Ker}(A_0^{[1,1]})$ and $v_1^{[l]}$ $(l \geq 2)$ is determined as $-A_0^{[1,l]} v_0^{[l]}/d_l$. The elements of $v_1^{[1]}$ are independent. Hence the dimension of solutions of Eq.(55) is $n_1 + \dim(\mathrm{Ker}(A_0^{[1,1]}))$. If the matrix $A_0^{[1,1]}$ is conjugate to $L(n_{1,1}, \ldots, n_{1,p_1}; d_{1,1}, \ldots, d_{1,p_1})$ and $d_{1,1} = 0$, then the dimension of solutions of Eq.(55) is $n_1 + n_{1,1}$.

4.3. Middle convolution

We now study the matrices for which the middle convolution is applied. Let $i \in \{1, \dots, r\}$ and $m_i = 1$. We investigate the matrices $\tilde{A}_1^{(i)}$ and $\tilde{A}_0^{(i)}$ for the case that the matrices $A_1^{(i)}$ and $A_0^{(i)}$ are expressed as Eqs.(42), (43), $d_1 = 0$ and $\mathbb{C}^n$ is irreducible as $\langle \mathbf{A} \rangle$-module. By changing the order of the direct sum $V' = (\mathbb{C}^n)^{\oplus M}$, the matrices $\tilde{A}_1^{(i)}$ and $\tilde{A}_0^{(i)}$ are expressed as

$$(QP^{\oplus M})^{-1} \tilde{A}_1^{(i)} QP^{\oplus M} = \begin{pmatrix} P^{-1}A_1P & P^{-1}A_0P + \mu I_n & P^{-1}\overline{A}P^{\oplus(M-2)} \\ 0 & 0 & 0 \\ 0 & 0 & 0 \end{pmatrix},$$
$$(QP^{\oplus M})^{-1} \tilde{A}_0^{(i)} QP^{\oplus M} = \begin{pmatrix} \mu & 0 & 0 \\ P^{-1}A_1P & P^{-1}A_0P + \mu I_n & P^{-1}\overline{A}P^{\oplus(M-2)} \\ 0 & 0 & 0 \end{pmatrix}, \tag{58}$$

where $\overline{A} = \left(A_{m_0}^{(0)} \dots A_{\delta_{i,1}}^{(i-1)} \; A_{m_{i+1}}^{(i+1)} \dots \right)$ and Q represents the change of the order of the direct sum $(\mathbb{C}^n)^{\oplus M}$. Set

$$A_1^{\#} = P \begin{pmatrix} 0 \cdot I_{n_1} & 0 & \dots & 0 \\ 0 & (d_2)^{-1} I_{n_2} & \dots & 0 \\ 0 & 0 & \ddots & \vdots \\ 0 & \dots & 0 & (d_k)^{-1} I_{n_k} \end{pmatrix} P^{-1}, \tag{59}$$
$$X = -P^{-1} A_1^{\#} \left(A_0 + \mu I_n \; \overline{A} \right) P^{\oplus(M-1)},$$
$$R = -P^{-1} A_1^{\#} (A_0 + \mu I_n) P.$$

Since the j-th row blocks of the matrix $P^{-1}(A_1 A_1^{\#} - I_n)P$ zero for $j \geq 2$, the j-th row blocks of the matrix $X' = P^{-1}A_1PX + P^{-1}\left(A_0 + \mu I_n \; \overline{A} \right) P^{\oplus(M-1)}$ and $P^{-1}A_1PR + P^{-1}(A_0 + \mu I_n)P$ are also zero for $j \geq 2$. We denote the size of the matrices $\tilde{A}_1^{(i)}$, $\tilde{A}_0^{(i)}$ on the space $\mathcal{M} - mc_\mu(\mathbb{C}^n)$ by $n (= nM - \sum_{i=1}^{r} \dim \mathcal{K}^{(i)} - \dim \mathcal{L}(\mu))$. Restrictions of the matrices $\tilde{A}_1^{(i)}$ and $\tilde{A}_0^{(i)}$ to the space $\mathcal{K}^{(j)}$ $(j \neq i)$ and $\mathcal{L}(\mu)$ are zero, which follow from the definitions of the spaces. Thus the matrix $\tilde{A}_1^{(i)}$ on $\mathcal{M}$

is diagonalized as

$$\begin{pmatrix} I & X \\ 0 & I \end{pmatrix}^{-1} (QP^{\oplus M})^{-1} \tilde{A}_1^{(i)} QP^{\oplus M} \begin{pmatrix} I & X \\ 0 & I \end{pmatrix}\Bigg|_{\mathcal{M}'} \tag{60}$$

$$= \begin{pmatrix} P^{-1}A_1P & X' \\ 0 & 0 \end{pmatrix}\Bigg|_{\mathcal{M}'} = \begin{pmatrix} d_2 I_{n_2} & 0 & \cdots & 0 \\ 0 & \ddots & \ddots & 0 \\ \vdots & \ddots & d_k I_{n_k} & 0 \\ 0 & \cdots & 0 & 0 \end{pmatrix},$$

where $\mathcal{M}' = \begin{pmatrix} I & X \\ 0 & I \end{pmatrix}^{-1} (QP^{\oplus M})^{-1}\mathcal{M}$, I is a unit matrix of the suitable size and the dimension of the kernel of $\tilde{A}_1^{(i)}|_{\mathcal{M}}$ is $\tilde{n} - n + n_1$. Since the matrix $P^{-1}A_1P$ is diagonal, the $[l,l]$-block of $RP^{-1}A_1P$ coincides with that of $P^{-1}A_1PR$, and it is equal to the $[l,l]$-block of $-P^{-1}(A_0 + \mu I_n)P$, if $l \geq 2$. Hence

$$\begin{pmatrix} I & X \\ 0 & I \end{pmatrix}^{-1} (QP^{\oplus M})^{-1} \tilde{A}_0^{(i)} QP^{\oplus M} \begin{pmatrix} I & X \\ 0 & I \end{pmatrix}\Bigg|_{\mathcal{M}'} \tag{61}$$

$$= \begin{pmatrix} \mu - RP^{-1}A_1P & \mu X - XX' \\ P^{-1}A_1P & X' \end{pmatrix}\Bigg|_{\mathcal{M}'}$$

$$= \begin{pmatrix} A_0^{[2,2]} + 2\mu I_{n_2} & * & \cdots & * \\ * & \ddots & \ddots & * \\ \vdots & * & A_0^{[k,k]} + 2\mu I_{n_k} & * \\ \vdots & \ddots & * & \overline{X}' \end{pmatrix},$$

where $\overline{X}'$ is expressed as

$$\overline{X}' = \begin{pmatrix} \overline{A}_0^{[1,1]} + \mu I & \overline{X}'' \\ 0 & 0 \end{pmatrix},$$

$\overline{A}_0^{[1,1]}$ is the matrix obtained by replacing the domain and the range of $A_0^{[1,1]}$ to $\mathbb{C}^{n_1}/\mathrm{Ker}A_0^{[1,1]}$. It follows from irreducibility that the rank of $\overline{X}'$ is equal to the size of $\overline{A}_0^{[1,1]}$. Thus, if the matrix $A_0^{[1,1]}$ is conjugate to

$$L(n_1^{\langle 0\rangle}, \dots, n_{p\langle 0\rangle}^{\langle 0\rangle}; 0, \dots, 0) \oplus L(n_1^{\langle -\mu\rangle}, \dots, n_{p\langle -\mu\rangle}^{\langle -\mu\rangle}; -\mu, \dots, -\mu) \tag{62}$$
$$\oplus L(m_{1,1}, \dots, m_{1,p_1'}; \lambda_{1,1}, \dots, \lambda_{1,p_1'}) \qquad (\lambda_{1,j} \neq 0, -\mu),$$

then we have $\tilde{n}-n+n_1^{\langle 0\rangle}\geq n_1^{\langle -\mu\rangle}$ and the matrix $\overline{X}'$ is conjugate to

$$L(\tilde{n}-n+n_1^{\langle 0\rangle}, n_1^{\langle -\mu\rangle}, \dots, n_{p\langle -\mu\rangle}^{\langle -\mu\rangle}; 0, \dots, 0) \tag{63}$$
$$\oplus L(n_2^{\langle 0\rangle}, \dots, n_{p\langle 0\rangle}^{\langle 0\rangle}; \mu, \dots, \mu) \oplus L(m_{1,1}, \dots, m_{1,p'_1}; \lambda_{1,1}+\mu, \dots, \lambda_{1,p'_1}+\mu).$$

If the matrix $A_0^{[l,l]}$ $(l\geq 2)$ is conjugate to $L(m_{l,1},\dots,m_{l,p_l};\lambda_{l,1},\dots,\lambda_{l,p_l})$, then the matrix $A_0^{[l,l]}+2\mu I_{n_l}$ is conjugate to $L(m_{l,1},\dots,m_{l,p_l};\lambda_{l,1}+2\mu,\dots,\lambda_{l,p_l}+2\mu)$. Hence

$$\mathrm{idx}_i(mc_\mu(\mathbf{A}))-\mathrm{idx}_i(\mathbf{A})=(\tilde{n}-n+n_1)^2+\sum_{j=2}^{k}(n_j)^2 \tag{64}$$
$$+(\dim Z(\overline{A}_0^{[1,1]}+\mu))^2+(\tilde{n}-n+n_{1,1})^2+\sum_{j=2}^{k}(\dim Z(A_0^{[j,j]}+2\mu))^2-2\tilde{n}^2$$
$$-\left(\sum_{j=1}^{k}(n_j)^2+(\dim Z(\overline{A}_0^{[1,1]}))^2+n_{1,1}^2+\sum_{j=2}^{k}(\dim Z(A_0^{[j,j]}))^2-2n^2\right)$$
$$=2(n-\tilde{n})(2n-n_1-n_{1,1})=2(n-\tilde{n})(2n-\dim\mathcal{K}^{(i)}),$$

where $n_{1,1}=\dim(\mathrm{Ker}A_0^{[1,1]})$.

We investigate the matrices $\tilde{A}_1^{(0)}$ and $\tilde{A}_0^{(0)}$ for the case that $m_0=1$, the matrices $A_1^{(0)}$ and $A_0^{(0)}$ are expressed as Eqs.(42), (43) and $d_1=0$. By changing the order of the direct sum $V'=(\mathbb{C}^n)^{\oplus M}$, the matrices $\tilde{A}_1^{(0)}$ and $\tilde{A}_0^{(0)}+\mu I$ are simultaneously conjugate to

$$\tilde{A}_1^{(0)}\sim\begin{pmatrix} P^{-1}A_1P & P^{-1}A_0P & P^{-1}\overline{A}P^{\oplus(M-2)} \\ 0 & 0 & 0 \\ 0 & 0 & 0 \end{pmatrix}, \tag{65}$$
$$\tilde{A}_0^{(0)}+\mu I\sim\begin{pmatrix} \mu & 0 & 0 \\ P^{-1}A_1P & P^{-1}A_0P & P^{-1}\overline{A}P^{\oplus(M-2)} \\ 0 & 0 & 0 \end{pmatrix},$$

where $\overline{A}=\left(A_{m_1}^{(1)}\dots A_1^{(1)}\ A_{m_2}^{(2)}\dots\right)$. It follows from similar argument to the case $\tilde{A}_1^{(i)}$, $\tilde{A}_0^{(i)}$ $(i \neq 0)$ that $\tilde{A}_1^{(0)}$ and $\tilde{A}_0^{(0)}$ are simultaneously conjugate to

$$\tilde{A}_1^{(0)}\Big|_{\mathcal{M}} \sim \begin{pmatrix} d_2 I_{n_2} & 0 & \cdots & 0 \\ 0 & \ddots & \ddots & 0 \\ \vdots & \ddots & d_k I_{n_k} & 0 \\ 0 & \cdots & 0 & 0 \end{pmatrix}, \tag{66}$$

$$\tilde{A}_0^{(0)}\Big|_{\mathcal{M}} \sim \begin{pmatrix} A_0^{[2,2]} & * & \cdots & * \\ * & \ddots & \ddots & * \\ \vdots & * & A_0^{[k,k]} & * \\ \vdots & \ddots & * & \overline{X}' \end{pmatrix}, \tag{67}$$

where $\overline{X}'$ is expressed as

$$\overline{X}' = \begin{pmatrix} \overline{A_0}^{[1,1]} - \mu I & \overline{X}'' \\ 0 & -\mu I \end{pmatrix},$$

$\overline{A_0}^{[1,1]}$ is the matrix obtained by replacing the domain and the range of $A_0^{[1,1]}$ to $\mathbb{C}^{n_1}/\mathrm{Ker}(A_0^{[1,1]} - \mu I)$. If the matrix $A_0^{[1,1]}$ is conjugate to

$$\begin{aligned} &L(n_1^{\langle\mu\rangle}, \ldots, n_{p_{\langle\mu\rangle}}^{\langle\mu\rangle}; \mu, \ldots, \mu) \oplus L(n_1^{\langle 0\rangle}, \ldots, n_{p_{\langle 0\rangle}}^{\langle 0\rangle}; 0, \ldots, 0) \\ &\oplus L(m_{1,1}, \ldots, m_{1,p_1'}; \lambda_{1,1}, \ldots, \lambda_{1,p_1'}), \qquad (\lambda_{1,j} \neq 0, \mu), \end{aligned} \tag{68}$$

then we have $\tilde{n} - n + n_1^{\langle\mu\rangle} \geq n_1^{\langle 0\rangle}$ and the matrix $\overline{X}'$ is conjugate to

$$\begin{aligned} &L(\tilde{n} - n + n_1^{\langle\mu\rangle}, n_1^{\langle 0\rangle}, \ldots, n_{p_{\langle 0\rangle}}^{\langle 0\rangle}; -\mu, \ldots, -\mu) \\ &\oplus L(n_2^{\langle\mu\rangle}, \ldots, n_{p_{\langle\mu\rangle}}^{\langle\mu\rangle}; 0, \ldots, 0) \oplus L(m_{1,1}, \ldots, m_{1,p_1'}; \lambda_{1,1} - \mu, \ldots, \lambda_{1,p_1'} - \mu). \end{aligned} \tag{69}$$

Hence we also have

$$\mathrm{idx}_0(mc_\mu(\mathbf{A})) - \mathrm{idx}_0(\mathbf{A}) = 2(n - \tilde{n})(2n - \dim \mathcal{L}(\mu)). \tag{70}$$

We include the case $m_i = 0$ to the case $m_i = 1$ by setting $A_1^{(i)} = 0$ and we have $\tilde{n} = n(2r+1) - \sum_{i=1}^{r} \dim \mathcal{K}^{(i)} - \dim \mathcal{L}(\mu)$. It follows from Eqs.(64),

(70) that

$$\begin{aligned}\operatorname{idx}(mc_\mu(\mathbf{A})) - \operatorname{idx}(\mathbf{A}) &= 2\tilde{n}^2 - 2n^2 + \sum_{i=0}^{r} \{\operatorname{idx}_i(mc_\mu(\mathbf{A})) - \operatorname{idx}_i(\mathbf{A})\} \\ &= 2\tilde{n}^2 - 2n^2 + 2(n-\tilde{n})\left(2(r+1)n - \sum_{i=1}^{r} \dim \mathcal{K}^{(i)} - \dim \mathcal{L}(\mu)\right) \\ &= 2\tilde{n}^2 - 2n^2 + 2(n-\tilde{n})(n+\tilde{n}) = 0.\end{aligned} \tag{71}$$

Hence the index of rigidity is preserved by application of middle convolution, i.e. $\operatorname{idx}(mc_\mu(\mathbf{A})) = \operatorname{idx}(\mathbf{A})$. Namely we obtain the following proposition.

Proposition 4.2. *If $m_i \leq 1$, the matrices $A_1^{(i)}$ are semi-simple for all i and $\langle \mathbf{A} \rangle$ is irreducible, then the index of rigidity is preserved by application of middle convolution, i.e.* $\operatorname{idx}(mc_\mu(\mathbf{A})) = \operatorname{idx}(\mathbf{A})$ *for all $\mu \in \mathbb{C}$.*

4.4. *Classification*

Proposition 4.3. *Assume that $m_i \leq 1$ $(i = 0, \dots, r)$, $A_1^{(i)}$ are semi-simple for all i and $\langle \mathbf{A} \rangle$ is irreducible. We identify the case $m_i = 0$ with the case $m_i = 1$ and $A_1^{(i)} = 0$.*
(i) If $\operatorname{idx}(\mathbf{A}) = 2$, *then* $\mathbf{A}$ *is transformed to the rank one matrices by applying addition and middle convolution repeatedly.*
(ii) If $\operatorname{idx}(\mathbf{A}) = 0$, *then* $\mathbf{A}$ *is transformed to one of the following cases by applying middle convolution and addition repeatedly, where $d \in \mathbb{Z}_{\geq 1}$.*

$$\begin{aligned}&\textit{Four singularities}: && \{(d,d),\ (d,d),\ (d,d),\ (d,d)\}, \\ &\textit{Three singularities}: && \{(d,d,d),\ (d,d,d),\ (d,d,d)\}, \\ & && \{(2d,2d),\ (d,d,d,d),\ (d,d,d,d)\}, \\ & && \{(3d,3d),\ (2d,2d,2d),\ (d,d,d,d,d,d)\}, \\ & && \{(d,d)-((d),(d)),\ (d,d),\ (d,d)\},\end{aligned} \tag{72}$$

Two singularities : $\{(d,d)-((d),(d)),\ (d,d)-((d),(d))\}$,
$\{(d,d,d)-((d),(d),(d)),\ (d,d,d)\}$,
$\{(d,d,d,d)-((d),(d),(d),(d)),\ (2d,2d)\}$,
$\{(2d,2d)-((d,d),(d,d)),\ (d,d,d,d)\}$,
$\{(3d,2d)-((d,d,d),(2d)),\ (d,d,d,d,d)\}$,
$\{(2d,2d,2d)-((d,d),(d,d),(d,d)),\ (3d,3d)\}$,
$\{(3d,3d,2d)-((d,d,d),(d,d,d),(2d)),\ (4d,4d)\}$,
$\{(5d,4d,3d)-((d,d,d,d,d),(2d,2d),(3d)),\ (6d,6d)\}$,
$\{(5d,4d)-((d,d,d,d,d),(2d,2d)),\ (3d,3d,3d)\}$,
$\{(3d,3d)-((d,d,d),(d,d,d)),\ (2d,2d,2d)\}$,
$\{(5d,3d)-((d,d,d,d,d),(3d)),\ (2d,2d,2d,2d)\}$,
$\{(4d,3d)-((2d,2d),(3d)),\ (d,d,d,d,d,d,d)\}$.

Proof. (i) It is enough to show that the size of matrices can be decreased by appropriate application of addition and middle convolution, because the size of matrices is reduced to one by applying many times.

Let

$$\begin{aligned}&(n_1^{(i)},n_2^{(i)},\ldots,n_{k^{(i)}}^{(i)})-\\&((n_{1,1}^{(i)},\ldots,n_{1,p_1^{(i)}}^{(i)}),(n_{2,1}^{(i)},\ldots,n_{2,p_2^{(i)}}^{(i)}),\ldots,(n_{k^{(i)},1}^{(i)},\ldots,n_{k^{(i)},p_{k^{(i)}}^{(i)}}^{(i)})),\end{aligned}\tag{73}$$

$(n_1^{(i)}\geq n_2^{(i)}\geq\cdots\geq n_{k^{(i)}}^{(i)},\ n_{j,1}^{(i)}\geq\cdots\geq n_{j,p_j^{(i)}}^{(i)})$ be the type of multiplicities of $(A_1^{(i)},A_0^{(i)})$. Note that $n_{j,1}^{(i)}+\cdots+n_{j,p_j^{(i)}}^{(i)}=n_j^{(i)}$. Then

$$(n_j^{(i)})^2+\sum_{j'}(n_{j,j'}^{(i)})^2\leq(n_j^{(i)})^2+n_{j,1}^{(i)}\sum_{j'}n_{j,j'}^{(i)}=n_j^{(i)}(n_j^{(i)}+n_{j,1}^{(i)}).\tag{74}$$

There exists a number $l^{(i)}$ such that

$$\frac{n}{n_{l^{(i)}}^{(i)}}\left((n_{l^{(i)}}^{(i)})^2+\sum_j(n_{l^{(i)},j}^{(i)})^2\right)\geq\sum_j(n_j^{(i)})^2+\sum_j\sum_{j'}(n_{j,j'}^{(i)})^2,\tag{75}$$

because if not we have contradiction as

$$\sum_l \left((n_l^{(i)})^2 + \sum_{j'} (n_{l,j'}^{(i)})^2 \right) \tag{76}$$

$$< \sum_l \frac{n_l^{(i)}}{n} \left\{ \sum_j (n_j^{(i)})^2 + \sum_j \sum_{j'} (n_{j,j'}^{(i)})^2 \right\} = \sum_j (n_j^{(i)})^2 + \sum_j \sum_{j'} (n_{j,j'}^{(i)})^2.$$

By combining with Eq.(74), we have

$$n(n_{l(i)}^{(i)} + n_{l(i),1}^{(i)}) \geq \sum_j (n_j^{(i)})^2 + \sum_j \sum_{j'} (n_{j,j'}^{(i)})^2. \tag{77}$$

Recall that the index of rigidity is calculated as

$$\mathrm{idx}(\mathbf{A}) = \sum_{i=0}^{r} \left(\sum_j (n_j^{(i)})^2 + \sum_j \sum_{j'} (n_{j,j'}^{(i)})^2 \right) - 2rn^2. \tag{78}$$

If $\sum_{i=0}^{r} n_{l(i)}^{(i)} + n_{l(i),1}^{(i)} \leq 2rn$, then

$$\sum_{i=0}^{r} \left(\sum_j (n_j^{(i)})^2 + \sum_j \sum_{j'} (n_{j,j'}^{(i)})^2 \right) \leq \sum_{i=0}^{r} n(n_{l(i)}^{(i)} + n_{l(i),1}^{(i)}) \leq 2rn^2, \tag{79}$$

which contradicts to $\mathrm{idx}(\mathbf{A}) = 2$. Hence

$$\sum_{i=0}^{r} (n_{l(i)}^{(i)} + n_{l(i),1}^{(i)}) \geq 2rn + 1. \tag{80}$$

We apply addition in order that $\dim \mathcal{K}^{(i)} = n_{l(i)}^{(i)} + n_{l(i),1}^{(i)}$ for $i = 1, \dots, r$ and the dimension of the kernel of $A_1^{(0)}$ is $n_{l(0)}^{(0)}$. Let μ be the value such that $\dim \mathcal{L}(\mu) = n_{l(0)}^{(0)} + n_{l(0),1}^{(0)}$. If $\mu = 0$, then it follows from Eq.(80) that $\dim \mathcal{K} + \dim \mathcal{L}'(0) - n(M-1) = \sum_{i=0}^{r} (n_{l(i)}^{(i)} + n_{l(i),1}^{(i)}) - 2nr < 0$ and it contradicts to Proposition 3.2 (iii). Thus $\mu \neq 0$, the size of matrices obtained by middle convolution is

$$(2r+1)n - \sum_{i=0}^{r} (n_{l(i)}^{(i)} + n_{l(i),1}^{(i)}), \tag{81}$$

and it is no more than $n-1$, which follows from Eq.(80). Hence the size can be decreased by addition and middle convolution.

(ii) It is sufficient to consider the case that the size of matrices cannot be decreased by addition and middle convolution. Let $l^{(i)}$ be the number which satisfies Eq.(75). If $\sum_{i=0}^{r}(n^{(i)}_{l^{(i)}}+n^{(i)}_{l^{(i)},1})>2rn$, then the size is decreased by middle convolution (see Eq.(81)) or the system is reducible (the case $\mu=0$). Hence $\sum_{i=0}^{r}(n^{(i)}_{l^{(i)}}+n^{(i)}_{l^{(i)},1})\le 2rn$. If $\sum_{i=0}^{r}(n^{(i)}_{l^{(i)}}+n^{(i)}_{l^{(i)},1})<2rn$, then we have $\sum_{i=0}^{r}\left(\sum_j(n^{(i)}_j)^2+\sum_j\sum_{j'}(n^{(i)}_{j,j'})^2\right)<2rn^2$ as Eq.(79) and it contradicts to $\mathrm{idx}(\mathbf{A})=0$. Thus $\sum_{i=0}^{r}(n^{(i)}_{l^{(i)}}+n^{(i)}_{l^{(i)},1})=2rn$. Since $\mathrm{idx}(\mathbf{A})=0$, all inequalities in Eqs.(77), (79) are equalities. In particular we have $(n^{(i)}_{l^{(i)}})^2+\sum_{j'}(n^{(i)}_{l^{(i)},j'})^2=n^{(i)}_{l^{(i)}}(n^{(i)}_{l^{(i)}}+n^{(i)}_{l^{(i)},1})$, which leads to $n^{(i)}_{l^{(i)},j'}=n^{(i)}_{l^{(i)},1}$ for $1\le j'\le p^{(i)}_{l^{(i)}}$. Then $((n^{(i)}_{l^{(i)}})^2+\sum_j(n^{(i)}_{l^{(i)},j})^2)n/n^{(i)}_{l^{(i)}}=n(n^{(i)}_{l^{(i)}}+n^{(i)}_{l^{(i)},1})=\sum_j(n^{(i)}_j)^2+\sum_j\sum_{j'}(n^{(i)}_{j,j'})^2$. Hence if Eq.(75) is satisfied, then the inequality in Eq.(75) is replaced by equality. Therefore $n((n^{(i)}_l)^2+\sum_j(n^{(i)}_{l,j})^2)\le n^{(i)}_l(\sum_j(n^{(i)}_j)^2+\sum_j\sum_{j'}(n^{(i)}_{j,j'})^2)$ for all l. It follows from summing up with respect to l that $((n^{(i)}_l)^2+\sum_j(n^{(i)}_{l,j})^2)n/n^{(i)}_l=\sum_j(n^{(i)}_j)^2+\sum_j\sum_{j'}(n^{(i)}_{j,j'})^2$ for all l. By setting $l=l^{(i)}$ and repeating the discussion above, we have

$$n^{(i)}_{l,j}=n^{(i)}_{l,1}=\frac{n^{(i)}_l}{p^{(i)}_l},\quad n^{(i)}_{l,1}(p^{(i)}_l+1)=n^{(i)}_{1,1}(p^{(i)}_1+1),\tag{82}$$

for all $j\in\{1,\dots,p^{(i)}_l\}$ and $l\in\{1,\dots,k^{(i)}\}$. Since $\sum_{i=0}^{r}(n^{(i)}_1+n^{(i)}_{1,1})=2rn$, we have

$$\sum_{i=0}^{r}\left(2-\frac{n^{(i)}_1+n^{(i)}_{1,1}}{n}\right)=2.\tag{83}$$

Hence $r\ge 1$.

If $n^{(i)}_1=n$ and $p^{(i)}_1=1$, then the matrices $A^{(i)}_1$ and $A^{(i)}_0$ are scalar, and we may omit the singularity corresponding to $A^{(i)}_1$ and $A^{(i)}_0$, because they are transformed to $A^{(i)}_1=A^{(i)}_0=0$ by applying addition. If $n^{(i)}_1=n$ and $p^{(i)}_1\ge 2$, then we have $k^{(i)}=1$, $n^{(i)}_{1,1}=\cdots=n^{(i)}_{1,p^{(i)}_1}$ and $2-(n^{(i)}_1+n^{(i)}_{1,1})/n=1-1/p^{(i)}_1$. Thus $1/2\le 2-(n^{(i)}_1+n^{(i)}_{1,1})/n<1$ and the equality holds iff $p^{(i)}_1=2$. Since the matrix $A^{(i)}_1$ is scalar, this case can be regarded as $m_i=0$ by applying addition.

If $n^{(i)}_1\ne n$, then $k^{(i)}\ge 2$ and we have $n=n^{(i)}_1+\cdots+n^{(i)}_{k^{(i)}}\ge k^{(i)}n^{(i)}_{k^{(i)}}$, $n^{(i)}_{k^{(i)},1}\le n^{(i)}_{k^{(i)}}$ and

$$2-\frac{n^{(i)}_1+n^{(i)}_{1,1}}{n}=2-\frac{n^{(i)}_{k^{(i)}}+n^{(i)}_{k^{(i)},1}}{n}\ge 2-\frac{2n^{(i)}_{k^{(i)}}}{n}\ge 2-\frac{2}{k^{(i)}}.\tag{84}$$

Hence we have $2-(n_1^{(i)}+n_{1,1}^{(i)})/n \geq 1$ and the equality holds iff $k^{(i)}=2$, $p_1^{(i)}=p_2^{(i)}=1$ and $n_1^{(i)}=n_2^{(i)}=n/2$.

We consider the case $k^{(i)}=1$ for all i. Then $\sum_{i=0}^{r}(1-1/p_1^{(i)})=2$ and we have solution for the cases $r=3$ and $r=2$. If $r=3$, then $p_1^{(0)}=p_1^{(1)}=p_1^{(2)}=p_1^{(3)}=2$, i.e. the case $\{(d,d),\,(d,d),\,(d,d),\,(d,d)\}$ $(d=n/2)$. If $r=2$, then $(p_1^{(0)},\,p_1^{(1)},\,p_1^{(2)})=(3,3,3)$, $(2,\,4,\,4)$, $(2,\,3,\,6)$ or their permutations, i.e. the cases $\{(d,d,d),\,(d,d,d),\,(d,d,d)\}$ $(d=n/3)$, $\{(2d,2d),\,(d,d,d,d),\,(d,d,d,d)\}$ $(d=n/4)$ or $\{(3d,3d),\,(2d,2d,2d),\,(d,d,d,d,d,d)\}$ $(d=n/6)$.

We consider the case $\#\{i \mid k^{(i)}\geq 2\}\geq 2$. Then it follows from Eq.(84) that $\#\{i \mid k^{(i)}\geq 2\}=2$, $k^{(0)}=k^{(1)}=2$, $p_1^{(i)}=p_2^{(i)}=1$ and $n_1^{(i)}=n_2^{(i)}=n/2$ $(i=0,1)$. Hence we obtain the case $\{(d,d)\text{–}((d),(d)),\ (d,d)\text{–}((d),(d))\}$ $(d=n/2)$.

We consider the case $\#\{i \mid k^{(i)}\geq 2\}=1$ and $r\geq 2$. We set $k^{(0)}\geq 2$, $k^{(1)}=\cdots=k^{(r)}=1$ for simplicity. It follows from Eq.(83), $2-(n_1^{(0)}+n_{1,1}^{(0)})/n\geq 1$ and $2-(n+n_{1,1}^{(i)})/n\geq 1/2$ $(i\geq 1)$ that $r=2$, $n_1^{(0)}=n_2^{(0)}=n/2$, $p_1^{(1)}=p_1^{(2)}=2$ and $n_{1,1}^{(i)}=n_{1,2}^{(i)}=n/2$ $(i=1,2)$. It corresponds to the case $\{(d,d)-((d),(d)),\ (d,d),\ (d,d)\}$ $(d=n/2)$.

The remaining case is $\#\{i \mid k^{(i)}\geq 2\}=1$ and $r=1$. We set $k^{(0)}\geq 2$, $k^{(1)}=1$ for simplicity. By Eq.(83), we have

$$\frac{n_1^{(0)}+n_{1,1}^{(0)}}{n}=1-\frac{1}{p_1^{(1)}}. \tag{85}$$

It follows from $p_1^{(1)}\geq 2$ and Eq.(84) that $2-(n_1^{(0)}+n_{1,1}^{(0)})/n\leq 3/2$ and $k^{(0)}\leq 4$.

If $k^{(0)}=4$, then inequalities in Eq.(84) must be equalities, $p_1^{(0)}=p_2^{(0)}=p_3^{(0)}=p_4^{(0)}=1$ and $n_1^{(0)}=n_2^{(0)}=n_3^{(0)}=n_4^{(0)}=n/4$. We have $p_1^{(1)}=2$ and $n_1^{(1)}=n_2^{(1)}=n/2$. Hence we obtain the case $\{(d,d,d,d)-((d),\,(d),\,(d),\,(d)),\,(2d,2d)\}$ $(d=n/4)$.

If $k^{(0)}=3$, then we have $(n_1^{(0)}+n_{1,1}^{(0)})/n\leq 2/3$. Since $(n_1^{(0)}+n_{1,1}^{(0)})/n+1/p_1^{(1)}=1$, we have $p_1^{(1)}=3$ or 2, i.e. $(n_1^{(0)}+n_{1,1}^{(0)})/n=2/3$ or $1/2$. If $(n_1^{(0)}+n_{1,1}^{(0)})/n=2/3$, then $p_1^{(0)}=p_2^{(0)}=p_3^{(0)}=1$, $n_i^{(0)}=n_{i,1}^{(0)}=n/3$ $(i=1,2,3)$, $p_1^{(1)}=3$ and $n_1^{(1)}=n_2^{(1)}=n_3^{(1)}=n/3$, i.e. the case $\{(d,d,d)-((d),\,(d),\,(d)),\,(d,d,d)\}$ $(d=n/3)$. If $(n_1^{(0)}+n_{1,1}^{(0)})/n=1/2$, then we have

$$\begin{aligned}2n_{1,1}^{(0)}(p_1^{(0)}+1)&=2n_{2,1}^{(0)}(p_2^{(0)}+1)=2n_{3,1}^{(0)}(p_3^{(0)}+1)=n \\ &=n_1^{(0)}+n_2^{(0)}+n_3^{(0)}=n_{1,1}^{(0)}p_1^{(0)}+n_{2,1}^{(0)}p_2^{(0)}+n_{3,1}^{(0)}p_3^{(0)}.\end{aligned} \tag{86}$$

By summing up, we have $2(n^{(0)}_{1,1}+n^{(0)}_{2,1}+n^{(0)}_{3,1}) = n^{(0)}_{1,1}p^{(0)}_1+n^{(0)}_{2,1}p^{(0)}_2+n^{(0)}_{3,1}p^{(0)}_3$. Since $n^{(0)}_1 \geq n^{(0)}_2 \geq n^{(0)}_3$, we have $n^{(0)}_{1,1} \leq n^{(0)}_{2,1} \leq n^{(0)}_{3,1}$, $p^{(0)}_1 \geq p^{(0)}_2 \geq p^{(0)}_3$ and $p^{(0)}_3 \leq 2$. If $p^{(0)}_3 = 2$, then $p^{(0)}_1 = p^{(0)}_2 = 2$. If $p^{(0)}_3 = 1$, then we have $2n^{(0)}_{3,1} = n^{(0)}_{1,1}(p^{(0)}_1+1) = n^{(0)}_{2,1}(p^{(0)}_2+1)$ and $3n^{(0)}_{3,1} = n^{(0)}_{1,1}p^{(0)}_1+n^{(0)}_{2,1}p^{(0)}_2$. Hence $n^{(0)}_{3,1} = n^{(0)}_{1,1}+n^{(0)}_{2,1}$ and $1/(p^{(0)}_1+1)+1/(p^{(0)}_2+1) = 1/2$. Therefore $(p^{(0)}_1+1, p^{(0)}_2+1, p^{(0)}_3+1) = (3,3,3)$, $(6,3,2)$ or $(4,4,2)$. If $(p^{(0)}_1+1, p^{(0)}_2+1, p^{(0)}_3+1) = (3,3,3)$, then $n^{(0)}_{1,1} = n^{(0)}_{2,1} = n^{(0)}_{3,1} = n/3$ and we obtain the case $\{(2d,2d,2d)-((d,d),(d,d),(d,d)),\ (3d,3d)\}$ $(d=n/6)$. If $(p^{(0)}_1+1, p^{(0)}_2+1, p^{(0)}_3+1) = (6,3,2)$, $n^{(0)}_{1,1} = n/12$, $n^{(0)}_{2,1} = n/6$ and $n^{(0)}_{3,1} = n/4$ and we obtain the case $\{(5d,4d,3d)-((d,d,d,d,d),(2d,2d),(3d)),\ (6d,6d)\}$ $(d=n/12)$. If $(p^{(0)}_1+1, p^{(0)}_2+1, p^{(0)}_3+1) = (4,4,2)$, then $n^{(0)}_{1,1} = n^{(0)}_{2,1} = n/8$, $n^{(0)}_{3,1} = n/4$, and we obtain the case $\{(3d,3d,2d)-((d,d,d),(d,d,d),(2d)),\ (4d,4d)\}$ $(d=n/8)$.

We investigate the case $k^{(0)} = 2$. It follows from Eq.(83) and $n = n^{(0)}_{1,1}p^{(0)}_1+n^{(0)}_{2,1}p^{(0)}_2$ that

$$n^{(0)}_{1,1}(p^{(0)}_1+1) = n^{(0)}_{2,1}(p^{(0)}_2+1) = \left(1-\frac{1}{p^{(1)}_1}\right)(n^{(0)}_{1,1}p^{(0)}_1+n^{(0)}_{2,1}p^{(0)}_2). \quad (87)$$

By erasing the term $p^{(0)}_1$, we obtain

$$n^{(0)}_{2,1}+(p^{(1)}_1-1)n^{(0)}_{1,1} = (p^{(1)}_1-2)n^{(0)}_{2,1}p^{(0)}_2. \quad (88)$$

Hence $p^{(1)}_1 \geq 3$. It follows from $n^{(0)}_1 \geq n^{(0)}_2$ that $n^{(0)}_{1,1} \leq n^{(0)}_{2,1}$, $p^{(0)}_1 \geq p^{(0)}_2$, $p^{(1)}_1 \geq 1+(p^{(1)}_1-1)n^{(0)}_{1,1}/n^{(0)}_{2,1} = (p^{(1)}_1-2)p^{(0)}_2$ and $p^{(0)}_2 \leq p^{(1)}_1/(p^{(1)}_1-2)$. We consider the case $p^{(1)}_1 = 3$. Then $p^{(0)}_2 \leq 3$. If $p^{(0)}_2 = 1$, then $n^{(0)}_{1,1} = 0$ and it cannot occur. If $p^{(0)}_2 = 2$, then $n^{(0)}_{2,1} = 2n^{(0)}_{1,1}$ and $p^{(0)}_1 = 5$. It corresponds to the case $\{(5d,4d)-((d,d,d,d,d),(2d,2d)),\ (3d,3d,3d)\}$ $(d = n/9)$. If $p^{(0)}_2 = 3$, then $n^{(0)}_{1,1} = n^{(0)}_{2,1}$ and $p^{(0)}_1 = 3$. It corresponds to the case $\{(3d,3d)-((d,d,d),(d,d,d)),\ (2d,2d,2d)\}$ $(d=n/6)$. We consider the case $p^{(1)}_1 \geq 4$. Then $p^{(0)}_2 \leq p^{(1)}_1/(p^{(1)}_1-2) \leq 2$. If $p^{(0)}_2 = 2$, then $p^{(1)}_1 = 4$, $n^{(0)}_{1,1} = n^{(0)}_{2,1}$ and $p^{(0)}_1 = 2$. It corresponds to the case $\{(2d,2d)-((d,d),(d,d)),\ (d,d,d,d)\}$ $(d=n/4)$. If $p^{(0)}_2 = 1$, then $n^{(0)}_{1,1} = n^{(0)}_{2,1}(p^{(1)}_1-3)/(p^{(1)}_1-1)$ and $p^{(0)}_1 = 1+4/(p^{(1)}_1-3)$. Hence 4 is divisible by $p^{(1)}_1-3$ and we have $p^{(1)}_1 = 4, 5$ or 7. If $p^{(1)}_1 = 4$, then $p^{(0)}_1 = 5$ and $3n^{(0)}_{1,1} = n^{(0)}_{2,1}$. It corresponds to the case $\{(5d,3d)-((d,d,d,d,d),(3d)),\ (2d,2d,2d,2d)\}$ $(d=n/8)$. If $p^{(1)}_1 = 5$, then $p^{(0)}_1 = 3$ and $2n^{(0)}_{1,1} = n^{(0)}_{2,1}$. It corresponds to the case $\{(3d,2d)-((d,d,d),(2d)),\ (d,d,d,d,d)\}$ $(d=n/5)$. If $p^{(1)}_1 = 7$, then $p_1^{(0)} = 2$ and $3n_{1,1}^{(0)} = 2n_{2,1}^{(0)}$. It corresponds to the case $\{(4d, 3d) - ((2d, 2d), (3d)),\ (d, d, d, d, d, d, d)\}$ $(d = n/7)$.

Thus we have exhausted all the cases. □

Remark that some patterns in the list of Proposition 4.3 may not be realized as an irreducible system of differential equations. In fact the patterns corresponding to Fuchsian systems as

$$\begin{aligned}
&\{(d, d),\ (d, d),\ (d, d),\ (d, d)\},\\
&\{(d, d, d),\ (d, d, d),\ (d, d, d)\},\\
&\{(2d, 2d),\ (d, d, d, d),\ (d, d, d, d)\},\\
&\{(3d, 3d),\ (2d, 2d, 2d),\ (d, d, d, d, d, d)\},
\end{aligned} \tag{89}$$

for the case $d \geq 2$ (resp. $d = 1$) cannot be realized (resp. can be realized) as irreducible systems, which was established by Kostov[9] and Crawley-Boevey.[2]

5. Concluding remarks

We give comments for future reference.

In this paper we gave a tentative definition of the index of rigidity for systems of linear differential equations which may have irregular singularities, and we should clarify the correctness (or incorrectness) of our definition. A key point would be Conjecture 3.1, which is compatible with Proposition 4.2, and we should consider the case that the coefficient matrices are not semi-simple.

Crawley-Boevey[2] made a correspondence between systems of Fuchsian differential equations and roots of Kac-Moody root systems, which was applied for solving additive Deligne-Simpson problem. Boalch[1] gave a generalization of Crawley-Boevey's work to the cases which include an irregular singularity. It would be hopeful to develop studies on this direction to understand several properties of differential equations.

Laplace transformation (or Fourier transformation) has been a powerful tool for analysis of differential equations. It is known that Okubo normal form fits well with Laplace transformation. In fact, Okubo normal form

$$(xI_n - T)\frac{d\Psi}{dx} = A\Psi, \tag{90}$$

is transformed to

$$\frac{dV}{dz} = \left(T - \frac{A + I_n}{z}\right) V, \tag{91}$$

which is Birkhoff canonical form of Poincaré rank one by Laplace transformation

$$V(z) = \int_C \exp(zx)\Psi(x)dx. \tag{92}$$

Assume that the matrices T and A are written as

$$T = \begin{pmatrix} t_1 I_{n_1} & 0 & \dots & 0 \\ 0 & t_2 I_{n_2} & \dots & 0 \\ 0 & 0 & \ddots & \vdots \\ 0 & \cdots & 0 & t_k I_{n_k} \end{pmatrix}, \; A = \begin{pmatrix} A^{[1,1]} & A^{[1,2]} & \dots & A^{[1,k]} \\ A^{[2,1]} & A^{[2,2]} & \dots & A^{[2,k]} \\ \vdots & \vdots & \ddots & \vdots \\ A^{[k,1]} & A^{[k,2]} & \dots & A^{[k,k]} \end{pmatrix}, \tag{93}$$

where $t_i \neq t_j$ $(i \neq j)$ and $A^{[i,j]}$ is a $n_i \times n_j$ matrix, and we further assume that A, $A^{[i,i]}$ $(i = 1, \dots, k)$ are semi-simple and Eq.(90) is irreducible. Then the index of rigidity for Okubo normal form (Eq.(90)) is equal to

$$\sum_{j=1}^{k} \left((n_j)^2 + \dim(Z(A^{[j,j]})) \right) + \dim(Z(A)) - n^2, \tag{94}$$

which was described by Haraoka[5] and Yokoyama.[14] On the other hand, the index of rigidity of Eq.(91) can be calculated as a special case of section 4 (see Eq.(54)), and it is equal to Eq.(94). Hence the index of rigidity in this paper fits well with Laplace transformation.

We now observe an example of Laplace transformation. Set

$$T = \begin{pmatrix} 0 & 1 \\ 0 & 0 \end{pmatrix}, \quad A = - \begin{pmatrix} a_{1,1}+1 & a_{1,2} \\ a_{2,1} & a_{2,2}+1 \end{pmatrix}, \tag{95}$$

in Eq.(91), which corresponds to Eq.(40). It follows from irreducibility that $a_{2,1} \neq 0$. By (inverse) Laplace transformation, we obtain Eq.(90), which is rewritten as

$$\frac{d\Psi}{dx} = \left\{ -\frac{1}{x^2} \begin{pmatrix} a_{2,1} & a_{2,2}+1 \\ 0 & 0 \end{pmatrix} - \frac{1}{x} \begin{pmatrix} a_{1,1}+1 & a_{1,2} \\ a_{2,1} & a_{2,2}+1 \end{pmatrix} \right\} \Psi, \tag{96}$$

and it can be reduced to a scalar differential equation by middle convolution as the example in section 3.4. Therefore we should develop a theory of Laplace transformation as well as the theory of middle convolution which is based on Euler's integral transformation (Theorem 2.1).

Several important functions are written as a solution of single differential equations of higher order

$$y^{(n)} + a_1(z)y^{(n-1)} + \dots + a_{n-1}(z)y' + a_n(z)y = 0, \tag{97}$$

where $a_i(z)$ $(i = 1, \dots, n)$ are rational functions which may have poles at prescribed points $\{t_1, \dots, t_r\}$. Note that we need to treat delicately on writing Eq.(97) into the form of systems of differential equations (2) to reflect the depth of the singularities. Oshima[11] formulated middle convolution (Euler's transformation) for single differential equations of higher order, and Hiroe[6] studied it for the case that the differential equation has an irregular singularity at $z = \infty$ and regular singularities. Hiroe's result includes a part of the content of section 4 in this paper with a different situation. Moreover he clarified a structure of Kac-Moody root system, which is based on Boalch's study.[1] Studies on this direction should be developed further.

Acknowledgments

The author would like to thank Hiroshi Kawakami and Daisuke Yamakawa for discussions and comments. He was supported by the Grant-in-Aid for Young Scientists (B) (No. 19740089) from the Japan Society for the Promotion of Science.

References

1. P. Boalch, Irregular connections and Kac-Moody root systems, Preprint, arXiv:0806.1050 [math.DG].
2. W. Crawley-Boevey, *Duke Math. J.* **118** 339 (2003).
3. M. Dettweiler and S. Reiter, *J. Symbolic Comput.* **30** 761 (2000).
4. M. Dettweiler and S. Reiter, *J. Algebra* **318** 1 (2007).
5. Y. Haraoka, *Adv. Math.* **169** 187 (2002).
6. K. Hiroe, Twisted Euler transform of differential equations with an irregular singular point, Preprint, arXiv:0912.5124 [math.CA].
7. N. M. Katz, *Rigid local systems* (Princeton University Press, Princeton, 1996).
8. H. Kawakami, Generalized Okubo systems and the middle convolution, to appear in *Int. Math. Res. Notices* (2010).
9. V. P. Kostov, *J. Algebra* **281** 83 (2004).
10. T. Oshima, Classification of Fuchsian systems and their connection problem, Preprint, arXiv:0811.2916 [math.CA].
11. T. Oshima, Fractional calculas of Weyl algebra and Fuchsian differential equations, Preprint.
12. K. Takemura, Middle convolution for systems of linear differential equations with irregular singularities, in preparation.
13. D. Yamakawa., Middle Convolution and Harnad Duality, Preprint, arXiv:0911.3863 [math.CA]
14. T. Yokoyama, *Math. Nachr.* **279** 327 (2006).

DIFFERENTIAL EQUATIONS COMPATIBLE WITH BOUNDARY RATIONAL qKZ EQUATION

YOSHIHIRO TAKEYAMA

Department of Mathematics
Graduate School of Pure and Applied Sciences
Tsukuba University, Tsukuba, Ibaraki 305-8571, Japan
E-mail: takeyama@math.tsukuba.ac.jp

Dedicated to Professor Tetsuji Miwa on his sixtieth birthday

We give differential equations compatible with the rational qKZ equation with boundary reflection. The total system contains the trigonometric degeneration of the bispectral qKZ equation of type $(C_n^\vee, C_n)$ which in the case of type GL_n was studied by van Meer and Stokman. We construct an integral formula for solutions to our compatible system in a special case.

Keywords: Boundary rational qKZ equation; bispectral qKZ equation; double affine Hecke algebra.

1. Introduction

In this paper we give differential equations compatible with the rational version of the quantum Knizhnik-Zamolodchikov (qKZ) equation[8] with boundary reflection, which we call the *boundary rational qKZ equation.*

Let $V = \mathbb{C}^{2N}$ be a vector space of even dimension. The boundary rational qKZ equation is the following system of difference equations for an unknown function $f(x\,|\,y)$ on $(\mathbb{C}^\times)^N \times \mathbb{C}^n$ taking values in $V^{\otimes n}$:

$$
\begin{aligned}
&f(x\,|\,\ldots, y_m - c, \ldots) \\
&= R_{m,m-1}(y_m - y_{m-1} - c) \cdots R_{m,1}(y_m - y_1 - c) K_m(y_m - c/2\,|\,x, \beta) \\
&\times R_{1,m}(y_1 + y_m) \cdots R_{m-1,m}(y_{m-1} + y_m) \\
&\times R_{m,m+1}(y_m + y_{m+1}) \cdots R_{m,n}(y_m + y_n) \\
&\times K_m(y_m\,|\,\underline{1}, \alpha)\, R_{m,n}(y_m - y_n) \cdots R_{m,m+1}(y_m - y_{m+1}) f(x\,|\,\ldots, y_m, \ldots)
\end{aligned}
$$

for $1 \le m \le n$, where c, α and β are parameters and $\underline{1} = (1, \ldots, 1) \in (\mathbb{C}^\times)^N$. The linear operator $R(\lambda)$ on $V^{\otimes 2}$ is the rational R-matrix, and

$K(\lambda \,|\, x, \beta) \in \mathrm{End}(V)$ is the boundary K-matrix which is a linear sum of the identity and the reflection of the basis of V with exponent $x = (x_1, \ldots, x_N)$ (see Eq. (2) below). The indices of R and K in the right hand side signify the position of the components of $V^{\otimes n}$ on which they act. The boundary rational qKZ equation can be regarded as a slight generalization of an additive degeneration of Cherednik's trigonometric qKZ equation[2] associated with the root system of type C_n. Such equation was also derived as that for correlation functions of the integrable spin chains with a boundary.[3,10] Recently combinatorial aspects of a special polynomial solution have attracted attention.[1,5,9]

In this paper we give commuting differential operators in the form

$$D_a(x \,|\, y) = c x_a \frac{\partial}{\partial x_a} + L_a(x \,|\, y) \qquad (1 \le a \le N),$$

where $L_a(x \,|\, y)\,(1 \le a \le N)$ are commuting linear operators acting on $V^{\otimes n}$, and prove that the system consisting of the boundary rational qKZ equation and the differential equations $D_a(x \,|\, y) f(x \,|\, y) = 0\,(1 \le a \le N)$ is compatible (see Theorem 3.1 below). Such compatible system is obtained for the differential KZ or the qKZ equations without boundary reflection by Etingof, Felder, Markov, Tarasov and Varchenko in more general settings.[6,7,15,16]

In Ref. 11 van Meer and Stokman constructed a consistent system of q-difference equations, which they call the bispectral qKZ equation, using the double affine Hecke algebra (DAHA)[4] of type GL_n. The key ingredients are Cherednik's intertwiners and the dual anti-involution. As mentioned in Ref. 11, their construction can be extended to arbitrary root system. In this paper we consider the case of type $(C_n^\vee, C_n)$. The DAHA of this type also has intertwiners and dual anti-involution,[14] and hence we can construct the bispectral qKZ equation. Now recall that the DAHA has a trigonometric degeneration.[4] In this degeneration the bispectral qKZ equation turns into a system of differential equations called the affine KZ equation (see Section 1.1.3 of Ref. 4) and additive difference equations. We prove that the system is contained in our compatible system of differential equations and the boundary rational qKZ equation with $N = n$ restricted to a subspace of $V^{\otimes n}$ isomorphic to the group algebra of the Weyl group of type C_n. As will be seen in Sec. 4.6 the differential operator $D_a(x \,|\, y)$ does not literally appear in the affine KZ equation because some of its parts act by zero on the subspace. Thus our compatible system gives a non-trivial generalization of the trigonometric degeneration of the bispectral qKZ equation of type $(C_n^\vee, C_n)$.

As previously mentioned the boundary rational qKZ equation can be regarded as an additive degeneration of the qKZ equation on the root system of type C_n. For the qKZ equation of type C_n, Mimachi obtained an integral formula for solutions in a special case.[12] Similar construction works for the boundary rational qKZ equation with $\alpha = \beta = k/2$, where k is a parameter contained in the R-matrix, and the exponent x restricted to the hyperplane $x_2 = \cdots = x_N = 1$. We prove that the solutions obtained in this way satisfy the differential equation $D_1(x_1, 1, \ldots, 1 \,|\, y)f = 0$. Thus we get solutions of our compatible system in a special case.

The rest of this paper is organized as follows. In Sec. 2 we give the definition of the boundary rational qKZ equation. In Sec. 3 we construct the commuting differential operators and prove that they are compatible with the boundary rational qKZ equation. In Sec. 4 we calculate the trigonometric degeneration of the bispectral qKZ equation of type $(C_n^\vee, C_n)$ and prove that it is contained in our compatible system. In Sec. 5 we give an integral formula for solutions of our compatible system with $\alpha = \beta = k/2$ and $x = (x_1, 1, \ldots, 1)$.

2. Boundary rational qKZ equation

Let n and N be positive integers, and c, k, α and β be generic non-zero complex numbers. Let $V = \oplus_{a=1}^{N} (\mathbb{C}v_a \oplus \mathbb{C}v_{\overline{a}})$ be a vector space with the basis $\{v_l\}_{l=1,\ldots,N,\overline{1},\ldots,\overline{N}}$.

The rational R-matrix acting on $V^{\otimes 2}$ is defined by

$$R(\lambda) := \frac{\lambda + kP}{\lambda + k},$$

where P is the transposition $P(u \otimes v) := v \otimes u$. It is a rational solution of the Yang-Baxter equation

$$\begin{aligned} &R_{12}(\lambda_1 - \lambda_2)R_{13}(\lambda_1 - \lambda_3)R_{23}(\lambda_2 - \lambda_3) \\ &\quad = R_{23}(\lambda_2 - \lambda_3)R_{13}(\lambda_1 - \lambda_3)R_{12}(\lambda_1 - \lambda_2) \end{aligned} \tag{1}$$

on $V^{\otimes 3}$, where $R_{ij}(\lambda)$ is the linear operator acting on the tensor product of the i-th and the j-th component of $V^{\otimes 3}$ as $R(\lambda)$.

For $x = (x_1, \ldots, x_N) \in (\mathbb{C}^\times)^N$ we define the reflection operator $T(x) \in \mathrm{End}(V)$ by

$$T(x)(v_a) := x_a^{-1} v_{\overline{a}}, \quad T(x)(v_{\overline{a}}) := x_a v_a \qquad (1 \le a \le N)$$

and set

$$K(\lambda \,|\, x, \beta) := \frac{\lambda T(x) + \beta}{\lambda + \beta}. \tag{2}$$

Then the operator $K(\lambda \,|\, x, \beta)$ satisfies the boundary Yang-Baxter equation:

$$\begin{aligned} &R_{12}(\lambda_1 - \lambda_2)K_1(\lambda_1 \,|\, x, \beta)R_{21}(\lambda_1 + \lambda_2)K_2(\lambda_2 \,|\, x, \beta) \\ &= K_2(\lambda_2 \,|\, x, \beta)R_{21}(\lambda_1 + \lambda_2)K_1(\lambda_1 \,|\, x, \beta)R_{12}(\lambda_1 - \lambda_2) \end{aligned} \tag{3}$$

on $V^{\otimes 2}$, where $K_j(\lambda \,|\, x, \beta)$ is the linear operator acting on the j-th component of $V^{\otimes 2}$ as $K(\lambda \,|\, x, \beta)$.

For $x = (x_1, \ldots, x_N) \in (\mathbb{C}^\times)^N$ and $y = (y_1, \ldots, y_n) \in \mathbb{C}^n$, we define the linear operator $Q_m(x \,|\, y)$ $(1 \le m \le n)$ acting on $V^{\otimes n}$ by

$$\begin{aligned} &Q_m(x \,|\, y) \\ &:= R_{m,m-1}(y_m - y_{m-1} - c) \cdots R_{m,1}(y_m - y_1 - c)K_m(y_m - c/2 \,|\, x, \beta) \\ &\times R_{1,m}(y_1 + y_m) \cdots R_{m-1,m}(y_{m-1} + y_m) \\ &\times R_{m,m+1}(y_m + y_{m+1}) \cdots R_{m,n}(y_m + y_n) \\ &\times K_m(y_m \,|\, \underline{1}, \alpha)\, R_{m,n}(y_m - y_n) \cdots R_{m,m+1}(y_m - y_{m+1}), \end{aligned}$$

where $\underline{1} := (1, \ldots, 1)$ and the lower indices of R and K in the right hand side signify the position of the components in $V^{\otimes n}$ on which they act.

Proposition 2.1. *For $1 \le l, m \le n$ we have*

$$Q_m(x \,|\, \ldots, y_l - c, \ldots)Q_l(x \,|\, y) = Q_l(x \,|\, \ldots, y_m - c, \ldots)Q_m(x \,|\, y).$$

Proof. It follows from Eq. (1) and Eq. (3). □

Let $f(x \,|\, y)$ be a function on $(\mathbb{C}^{\otimes N}) \times \mathbb{C}^n$ taking values in $V^{\otimes n}$. We denote by Δ_m $(1 \le m \le n)$ the shift operator with respect to y_m:

$$\Delta_m f(x \,|\, y) := f(x \,|\, y_1, \ldots, y_m - c, \ldots, y_n).$$

From Proposition 2.1 the following system of difference equations is consistent:

$$\Delta_m f(x \,|\, y) = Q_m(x \,|\, y) f(x \,|\, y) \qquad (m = 1, \ldots, n). \tag{4}$$

Definition 2.1. We call the system (4) of difference equations the *boundary rational qKZ equation.*

3. Compatible differential equations

3.1. *Commuting differential operators*

We denote by $e_{ab} \in \mathrm{End}(V)$ $(a, b \in \{1, \ldots, N, \bar{1}, \ldots, \bar{N}\})$ the matrix unit acting by $e_{ab}v_p = \delta_{bp}v_a$. In this section, for $u \in \mathrm{End}(V)$ and $1 \le j \le$

n, we denote by $u^{(j)} \in \mathrm{End}(V^{\otimes n})$ the linear operator acting on the j-th component of $V^{\otimes n}$ as u. For $1 \le a, b \le N$ we set

$$E_{ab} := e_{ab} + e_{\overline{a}\overline{b}}, \quad \overline{E}_{ab} := e_{a\overline{b}} + e_{\overline{a}b},$$

and define $\mathbf{X}_{ab}, \mathbf{Y}_{ab}, \mathbf{Z}_{ab} \in \mathrm{End}(V^{\otimes n})$ by

$$\mathbf{X}_{ab} := \sum_{1 \le i < j \le n} \left(e_{ab}^{(i)} E_{ba}^{(j)} + e_{\overline{b}\overline{a}}^{(i)} E_{ab}^{(j)} \right),$$

$$\mathbf{Y}_{ab} := \sum_{1 \le i < j \le n} \left(e_{a\overline{b}}^{(i)} \overline{E}_{ba}^{(j)} + e_{b\overline{a}}^{(i)} \overline{E}_{ab}^{(j)} \right),$$

$$\mathbf{Z}_{ab} := \sum_{1 \le i < j \le n} \left(e_{\overline{a}b}^{(i)} \overline{E}_{ba}^{(j)} + e_{\overline{b}a}^{(i)} \overline{E}_{ab}^{(j)} \right).$$

Note that $\mathbf{Y}_{ab} = \mathbf{Y}_{ba}$ and $\mathbf{Z}_{ab} = \mathbf{Z}_{ba}$.

Define the linear operators $A_a(y)$ and $B_a(x)\,(1 \le a \le N)$ on $V^{\otimes n}$ by

$$A_a(y) := \sum_{j=1}^{n} y_j (e_{aa}^{(j)} - e_{\overline{a}\overline{a}}^{(j)}) + 2\alpha \sum_{j=1}^{n} e_{a\overline{a}}^{(j)} + k \left(-\sum_{p=1}^{a-1} \mathbf{X}_{pa} + \sum_{p=a+1}^{N} \mathbf{X}_{ap} + \sum_{p=1}^{N} \mathbf{Y}_{ap} \right)$$

and

$$B_a(x) := 2 \frac{\alpha + \beta x_a}{x_a^2 - 1} \sum_{j=1}^{n} \overline{E}_{aa}^{(j)} + k \left\{ \sum_{p=1}^{a-1} \frac{x_a}{x_a - x_p} (\mathbf{X}_{ap} + \mathbf{X}_{pa}) + \sum_{p=a+1}^{N} \frac{x_p}{x_a - x_p} (\mathbf{X}_{ap} + \mathbf{X}_{pa}) + \sum_{p=1}^{N} \frac{1}{x_a x_p - 1} (\mathbf{Y}_{ap} + \mathbf{Z}_{ap}) \right\}.$$

Set

$$L_a(x \,|\, y) := A_a(y) + B_a(x) \qquad (1 \le a \le N).$$

By direct calculation we can check the commutativity:

Lemma 3.1. *For $1 \le a, b \le N$ we have*

$$[A_a(y), A_b(y)] = 0, \qquad [L_a(x \,|\, y), L_b(x \,|\, y)] = 0.$$

Now define the differential operators $D_a(x \,|\, y)\,(1 \le a \le N)$ by

$$D_a(x \,|\, y) := c x_a \frac{\partial}{\partial x_a} + L_a(x \,|\, y).$$

Proposition 3.1. *The differential operators $D_a(x \,|\, y)\,(a = 1, \ldots, N)$ commute with each other.*

Proof. It follows from Lemma 3.1 and the equality $x_a \frac{\partial L_b}{\partial x_a} = x_b \frac{\partial L_a}{\partial x_b}$ for $1 \le a, b \le N$. □

3.2. *Compatibility*

In this subsection we prove the main theorem:

Theorem 3.1. *For $1 \le a \le N$ and $1 \le m \le n$ we have*

$$[D_a(x\,|\,y), \Delta_m^{-1} Q_m(x\,|\,y)] = 0. \tag{5}$$

Hence the system of equations

$$\begin{cases} \Delta_m f(x\,|\,y) = Q_m(x\,|\,y) f(x\,|\,y) \ (1 \le m \le n), \\ D_a(x\,|\,y) f(x\,|\,y) = 0 \qquad\qquad (1 \le a \le N) \end{cases} \tag{6}$$

is compatible.

To prove Theorem 3.1 we rewrite the linear operators $L_a(x\,|\,y)$ $(1 \le a \le N)$ as follows. For $\lambda \in \mathbb{C}^\times$ and $\gamma \in \mathbb{C}$, we define $I_a(\lambda\,|\,\gamma) \in \mathrm{End}(V)$ by

$$I_a(\lambda\,|\,\gamma) := \gamma(e_{aa} - e_{\overline{a}\overline{a}}) + 2\frac{\alpha + \beta\lambda}{\lambda^2 - 1} e_{\overline{a}a} + 2\frac{\alpha + \beta\lambda^{-1}}{1 - \lambda^{-2}} e_{a\overline{a}} \quad (1 \le a \le N).$$

For $x = (x_1, \ldots, x_N) \in (\mathbb{C}^\times)^N$ we define $M_a(x) \in \mathrm{End}(V^{\otimes 2})$ $(1 \le a \le N)$ by

$$M_a(x) := \frac{2k}{x_a - x_a^{-1}} (x_a e_{a\overline{a}} + x_a^{-1} e_{\overline{a}a}) \otimes (e_{a\overline{a}} + e_{\overline{a}a})$$
$$+ k \sum_{\substack{p=1 \\ p \ne a}}^{N} \left(\frac{x_a}{x_a - x_p} U_{ap} + \frac{x_p}{x_a - x_p} U_{pa} + \frac{x_a x_p}{x_a x_p - 1} J_{ap} + \frac{1}{x_a x_p - 1} K_{ap} \right),$$

where

$$U_{ab} := e_{ab} \otimes E_{ba} + e_{\overline{b}\overline{a}} \otimes E_{ab}, \qquad J_{ab} := e_{a\overline{b}} \otimes \overline{E}_{ba} + e_{b\overline{a}} \otimes \overline{E}_{ab},$$
$$K_{ab} := e_{\overline{a}b} \otimes \overline{E}_{ba} + e_{\overline{b}a} \otimes \overline{E}_{ab}.$$

Note that $J_{ab} = J_{ba}$ and $K_{ab} = K_{ba}$. Then we have

$$L_a(x\,|\,y) = \sum_{j=1}^{n} I_a(x_a\,|\,y_j)^{(j)} + \sum_{1 \le i < j \le n} M_a(x)^{(i,j)}$$

for $1 \le a \le N$, where $M_a(x)^{(i,j)}$ is the linear operator acting on the tensor product of the i-th and the j-th component of $V^{\otimes n}$ as $M_a(x)$.

Lemma 3.2. *For $x = (x_1, \ldots, x_N)$ and $1 \le a \le N$ we have*

$$R_{12}(y_1 - y_2)\left(I_a(x_a|y_1)^{(1)} + I_a(x_a|y_2)^{(2)} + M_a(x)^{(1,2)}\right)R_{12}(y_1 - y_2)^{-1}$$
$$= I_a(x_a|y_1)^{(1)} + I_a(x_a|y_2)^{(2)} + M_a(x)^{(2,1)}$$

on $V^{\otimes 2}$, where $M_a(x)^{(2,1)} = PM_a(x)P$.

Proof. Note that if $h \in \mathrm{End}(V^{\otimes 2})$ is symmetric, i.e. $PhP = h$, then h commutes with the R-matrix. We extract symmetric parts from the operator $I_a(x_a|y_1)^{(1)} + I_a(x_a|y_2)^{(2)} + M_a(x)^{(1,2)}$ as follows. In the following we enclose symmetric parts in a square bracket [ ]. First we have

$$I_a(x_a|y_1)^{(1)} + I_a(x_a|y_2)^{(2)}$$
$$= \left[y_1 \sum_{j=1}^{2} (e_{aa}^{(j)} - e_{\bar{a}\bar{a}}^{(j)}) + 2\frac{\alpha + \beta x_a}{x_a^2 - 1} \sum_{j=1}^{2} e_{\bar{a}a}^{(j)} + 2\frac{\alpha + \beta x_a^{-1}}{1 - x_a^{-2}} \sum_{j=1}^{2} e_{a\bar{a}}^{(j)} \right]$$
$$+ (y_2 - y_1)(e_{aa}^{(2)} - e_{\bar{a}\bar{a}}^{(2)}).$$

Note that $U_{ap} + U_{pa}$ and $J_{ap} + K_{ap}$ are symmetric. Using

$$U_{ap} = \left[e_{\bar{p}\bar{a}}^{(1)} e_{ap}^{(2)} + e_{ap}^{(1)} e_{\bar{p}\bar{a}}^{(2)} \right] + e_{\bar{p}\bar{a}}^{(1)} e_{\bar{a}\bar{p}}^{(2)} + e_{ap}^{(1)} e_{pa}^{(2)},$$
$$J_{ap} = \left[e_{p\bar{a}}^{(1)} e_{a\bar{p}}^{(2)} + e_{a\bar{p}}^{(1)} e_{p\bar{a}}^{(2)} \right] + e_{p\bar{a}}^{(1)} e_{\bar{a}p}^{(2)} + e_{a\bar{p}}^{(1)} e_{\bar{p}a}^{(2)},$$

we divide $M_a(x)^{(1,2)}$ as

$$M_a(x)^{(1,2)} = \frac{2k}{x_a - x_a^{-1}} \left[x_a e_{a\bar{a}}^{(1)} e_{a\bar{a}}^{(2)} + x_a^{-1} e_{\bar{a}a}^{(1)} e_{\bar{a}a}^{(2)} + x_a^{-1} (e_{a\bar{a}}^{(1)} e_{\bar{a}a}^{(2)} + e_{\bar{a}a}^{(1)} e_{a\bar{a}}^{(2)}) \right]$$
$$+ k \sum_{\substack{p=1 \\ p \neq a}}^{N} \left[\frac{x_p}{x_a - x_p} (U_{ap} + U_{pa}) + \frac{1}{x_a x_p - 1} (J_{ap} + K_{ap}) \right]$$
$$+ k \left[\sum_{\substack{p=1 \\ p \neq a}}^{N} (e_{\bar{p}\bar{a}}^{(1)} e_{ap}^{(2)} + e_{ap}^{(1)} e_{\bar{p}\bar{a}}^{(2)} + e_{p\bar{a}}^{(1)} e_{a\bar{p}}^{(2)} + e_{a\bar{p}}^{(1)} e_{p\bar{a}}^{(2)}) - (e_{aa}^{(1)} e_{aa}^{(2)} + e_{\bar{a}\bar{a}}^{(1)} e_{\bar{a}\bar{a}}^{(2)}) \right]$$
$$+ k \sum_{p}{}' (e_{ap}^{(1)} e_{pa}^{(2)} + e_{\bar{p}\bar{a}}^{(1)} e_{\bar{a}\bar{p}}^{(2)}),$$

where $\sum_p'$ is a sum over all indices $p \in \{1, \ldots, N, \bar{1}, \ldots, \bar{N}\}$. Thus we find

$$I_a(x_a|y_1)^{(1)} + I_a(x_a|y_2)^{(2)} + M_a(x)^{(1,2)} = [(\text{symmetric part})]$$
$$+ (y_2 - y_1)(e_{aa}^{(2)} - e_{\bar{a}\bar{a}}^{(2)}) + k \sum_{p}{}' (e_{ap}^{(1)} e_{pa}^{(2)} + e_{\bar{p}\bar{a}}^{(1)} e_{\bar{a}\bar{p}}^{(2)}). \qquad (7)$$

Now we make use of the intertwining property of the R-matrix:

$$R(\lambda)\left(\lambda\cdot 1\otimes e_{lm}+k{\sum_{p}}' e_{pm}\otimes e_{lp}\right)=\left(\lambda\cdot 1\otimes e_{lm}+k{\sum_{p}}' e_{lp}\otimes e_{pm}\right)R(\lambda),$$

$$R(\lambda)\left(\lambda\cdot 1\otimes e_{lm}-k{\sum_{p}}' e_{lp}\otimes e_{pm}\right)=\left(\lambda\cdot 1\otimes e_{lm}-k{\sum_{p}}' e_{pm}\otimes e_{lp}\right)R(\lambda)$$

for any $l,m\in\{1,\ldots,N,\bar{1},\ldots,\bar{N}\}$. For $g\in \mathrm{GL}(V^{\otimes 2})$ we denote by $\mathrm{Ad}(g)$ the adjoint action $h\mapsto ghg^{-1}$ on $\mathrm{End}(V^{\otimes 2})$. Then we see that

$$\begin{aligned}
&\mathrm{Ad}(R_{12}(y_1-y_2))\left((y_2-y_1)(e^{(2)}_{aa}-e^{(2)}_{\bar{a}\bar{a}})+k{\sum_{p}}'(e^{(1)}_{ap}e^{(2)}_{pa}+e^{(1)}_{p\bar{a}}e^{(2)}_{\bar{a}p})\right)\\
&=(y_2-y_1)(e^{(2)}_{aa}-e^{(2)}_{\bar{a}\bar{a}})+k{\sum_{p}}'(e^{(1)}_{pa}e^{(2)}_{ap}+e^{(1)}_{\bar{a}p}e^{(2)}_{p\bar{a}})\\
&=(y_2-y_1)(e^{(2)}_{aa}-e^{(2)}_{\bar{a}\bar{a}})+2k\,e^{(1)}_{\bar{a}a}e^{(2)}_{a\bar{a}}\\
&+k\Bigg\{\sum_{\substack{p=1\\p\neq a}}^{N}(e^{(1)}_{pa}e^{(2)}_{ap}+e^{(1)}_{\bar{p}a}e^{(2)}_{a\bar{p}}+e^{(1)}_{\bar{a}p}e^{(2)}_{p\bar{a}}+e^{(1)}_{\bar{a}\bar{p}}e^{(2)}_{\bar{p}\bar{a}})+(e^{(1)}_{aa}e^{(2)}_{aa}+e^{(1)}_{\bar{a}\bar{a}}e^{(2)}_{\bar{a}\bar{a}})\Bigg\}.
\end{aligned}$$

Adding this to the symmetric part in Eq. (7) we obtain $I_a(x_a|y_1)^{(1)}+I_a(x_a|y_2)^{(2)}+M_a(x)^{(2,1)}$. □

Proof of Theorem 3.1. We split the operator $Q_m(x\,|\,y)$ into three parts:

$$Q_m(x\,|\,y)=Q'_m(y)\,K_m(y_m-c/2\,|\,x,\beta)\,Q''_m(y), \tag{8}$$

where

$$Q'_m(y):=R_{m,m-1}(y_m-y_{m-1}-c)\cdots R_{m,1}(y_m-y_1-c)$$

and $Q''_m(y)$ is determined by Eq. (8). Then the equality (5) is equivalent to

$$\begin{aligned}
&cx_a\left(\frac{\partial}{\partial x_a}K_m(y_m-c/2\,|\,x,\beta)\right)K_m(y_m-c/2\,|\,x,\beta)^{-1}\\
&+\mathrm{Ad}(Q'_m(y))(L_a(x\,|\ldots,y_m-c,\ldots))\\
&-\mathrm{Ad}\left(K_m(y_m-c/2\,|\,x,\beta)Q''_m(y)\right)(L_a(x\,|\ldots,y_m,\ldots))=0.
\end{aligned}\tag{9}$$

The first term of Eq. (9) is equal to

$$\frac{c(y_m - c/2)}{(y_m - c/2)^2 - \beta^2} \left\{ (y_m - c/2)(e_{aa}^{(m)} - e_{\bar{a}\bar{a}}^{(m)}) - \beta(x_a e_{a\bar{a}}^{(m)} - x_a^{-1} e_{\bar{a}a}^{(m)}) \right\}. \quad (10)$$

From Lemma 3.2 we see that the second term is equal to

$$\sum_{\substack{j=1 \\ j\neq m}}^{n} I_a(x_a \,|\, y_j)^{(j)} + I_a(x_a \,|\, y_m - c)^{(m)} + \sum_{\substack{j=1 \\ j\neq m}}^{n} M_a(x)^{(m,j)}$$
$$+ \sum_{\substack{1\leq i<j\leq n \\ i,j\neq m}}^{n} M_a(x)^{(i,j)}. \quad (11)$$

Let us calculate the third term. Using Lemma 3.2 and

$$\mathrm{Ad}(K(\gamma \,|\, \underline{1}, \alpha))(I_a(\lambda \,|\, \gamma)) = I_a(\lambda \,|\, -\gamma),$$
$$\mathrm{Ad}(K_2(\gamma \,|\, \underline{1}, \alpha))(M_a(x)^{(1,2)}) = M_a(x)^{(1,2)},$$

we obtain

$$\mathrm{Ad}(Q''_m(y))(L_a(x \,|\, y)) = \sum_{\substack{j=1 \\ j\neq m}}^{n} I_a(x_a \,|\, y_j)^{(j)} + I_a(x_a \,|\, -y_m)^{(m)}$$
$$+ \sum_{\substack{j=1 \\ j\neq m}}^{n} M_a(x)^{(m,j)} + \sum_{\substack{1\leq i<j\leq n \\ i,j\neq m}}^{n} M_a(x)^{(i,j)}.$$

What remains is to calculate the image of the second and the third term above by the operator $\mathrm{Ad}(K_m(y_m - c/2 \,|\, x, \beta))$. By direct calculation we find

$$\mathrm{Ad}(K(\gamma - c/2 \,|\, x, \beta))(I_a(x_a \,|\, -\gamma))$$
$$= I_a(x_a \,|\, \gamma) + \frac{c\beta}{(\gamma - c/2)^2 - \beta^2} \left\{ \beta(e_{aa} - e_{\bar{a}\bar{a}}) + (\gamma - c/2)(x_a^{-1} e_{\bar{a}a} - x_a e_{a\bar{a}}) \right\}$$

and

$$\mathrm{Ad}(K_1(\gamma \,|\, x, \beta))(M_a(x)^{(1,2)}) = M_a(x)^{(1,2)}.$$

Using these formulas we see that the third term of Eq. (9) is equal to

$$(-1) \times \left(\sum_{j=1}^{n} I_a(x_a \,|\, y_j)^{(j)} + \sum_{\substack{j=1 \\ j\neq m}}^{n} M_a(x)^{(m,j)} + \sum_{\substack{1\leq i<j\leq n \\ i,j\neq m}}^{n} M_a(x)^{(i,j)} \right.$$
$$\left. + \frac{c\beta}{(y_m - c/2)^2 - \beta^2} \left\{ \beta(e_{aa}^{(m)} - e_{\bar{a}\bar{a}}^{(m)}) + (y_m - c/2)(x_a^{-1} e_{\bar{a}a}^{(m)} - x_a e_{a\bar{a}}^{(m)}) \right\} \right). \quad (12)$$

The sum of (10), (11) and (12) is zero and this completes the proof. □

4. The bispectral qKZ equation and its degeneration

4.1. *The double affine Hecke algebra of type* $(C_n^\vee, C_n)$

Here we recall the definition and some properties of the double affine Hecke algebra of type $(C_n^\vee, C_n)$.[14] We denote by $\mathbb{F} := \mathbb{C}(q^{1/2}, t^{1/2}, t_0^{1/2}, t_n^{1/2}, u_0^{1/2}, u_n^{1/2})$ the coefficient field.

Definition 4.1. The *double affine Hecke algebra* $\mathbb{H}$ of type $(C_n^\vee, C_n)$ is the unital associative $\mathbb{F}$-algebra generated by $X_i^{\pm 1}$ $(1 \le i \le n)$ and T_i $(0 \le i \le n)$ satisfying the following relations:

(i) quadratic Hecke relations

$$(T_i - t_i^{1/2})(T_i + t_i^{-1/2}) = 0 \qquad (0 \le i \le n),$$

where $t_i^{1/2} := t^{1/2}$ for $1 \le i < n$.

(ii) braid relations

$$T_iT_{i+1}T_iT_{i+1} = T_{i+1}T_iT_{i+1}T_i \quad (i = 0, n-1),$$
$$T_iT_{i+1}T_i = T_{i+1}T_iT_{i+1} \quad (1 \le i < n), \qquad T_iT_j = T_jT_i \quad (|i-j| \ge 2).$$

(iii) relations between X and T

$$X_iX_j = X_jX_i \quad (\forall i, j), \quad T_iX_j = X_jT_i \quad (|i-j| \ge 2 \text{ or } (i,j) = (n, n-1)),$$
$$T_iX_iT_i = X_{i+1} \quad (1 \le i \le n-1), \quad X_n^{-1}T_n^{-1} = T_nX_n + (u_n^{1/2} - u_n^{-1/2}),$$
$$q^{-1/2}T_0^{-1}X_1 = q^{1/2}X_1^{-1}T_0 + (u_0^{1/2} - u_0^{-1/2}).$$

Noumi found the polynomial representation of $\mathbb{H}$ given as follows.[13] Let $W = \langle s_0, \ldots, s_n \rangle$ be the affine Weyl group of type C_n. The group W acts on the Laurent polynomial ring $\mathbb{F}[X^{\pm 1}] = \mathbb{F}[X_1^{\pm 1}, \ldots, X_n^{\pm 1}]$ by

$$(s_0f)(X) = f(qX_1^{-1}, X_2, \ldots, X_n), \tag{13}$$
$$(s_if)(X) = f(\ldots, X_{i+1}, X_i, \ldots) \qquad (1 \le i < n), \tag{14}$$
$$(s_nf)(X) = f(X_1, \ldots, X_{n-1}, X_n^{-1}). \tag{15}$$

Define the $\mathbb{F}$-linear operators $\widehat{T}_i$ $(0 \le i \le n)$ on $\mathbb{F}[X^{\pm 1}]$:

$$\widehat{T}_0 := t_0^{1/2} + t_0^{-1/2}\frac{(1 - q^{1/2}t_0^{1/2}u_0^{1/2}X_1^{-1})(1 + q^{1/2}t_0^{1/2}u_0^{-1/2}X_1^{-1})}{1 - qX_1^{-2}}(s_0 - 1),$$

$$\widehat{T}_i := t^{1/2} + t^{-1/2}\frac{1 - tX_i/X_{i+1}}{1 - X_i/X_{i+1}}(s_i - 1) \qquad (1 \le i < n),$$

$$\widehat{T}_n := t_n^{1/2} + t_n^{-1/2}\frac{(1 - t_n^{1/2}u_n^{1/2}X_n)(1 + t_n^{1/2}u_n^{-1/2}X_n)}{1 - X_n^2}(s_n - 1).$$

Then the map $T_i \mapsto \widehat{T}_i$ and $X_i \mapsto X_i$ (left multiplication) gives a representation of $\mathbb{H}$ on $\mathbb{F}[X^{\pm 1}]$. It is faithful and hence $\mathbb{H}$ is isomorphic to the $\mathbb{F}$-subalgebra of the smashed product algebra $\mathbb{F}(X)\#W$ generated by the difference operators $\widehat{T}_i\,(0 \le i \le n)$ and the multiplication operators $f(\underline{X}) \in \mathbb{F}[X^{\pm 1}]$ (see, e.g., Sec. 2.1 in Ref. 11 for the definition of the smashed product algebra). Hereafter we identify $\mathbb{H}$ with the subalgebra of $\mathbb{F}(X)\#W$.

The subalgebra H_0 generated by $T_i\,(1 \le i \le n)$ is isomorphic to the Hecke algebra of type C_n. Denote by $W_0 := \langle s_1, \ldots, s_n \rangle$ the finite Weyl group of type C_n. Let $w = s_{j_1} \cdots s_{j_r}$ be a reduced expression of $w \in W_0$. Then the element $T_w := T_{j_1} \cdots T_{j_r}$ is well-defined for $w \in W_0$. The set $\{T_w\}_{w \in W_0}$ gives a basis of H_0.

Set

$$Y_i := T_i \cdots T_{n-1}(T_n \cdots T_0)T_1^{-1} \cdots T_{i-1}^{-1} \quad (1 \le i \le n).$$

They satisfy

$$Y_iY_j = Y_jY_i \quad (\forall i,j), \qquad T_iY_j = Y_jT_i \quad (|i-j| \ge 2 \text{ or } (i,j) = (n, n-1)),$$
$$T_iY_{i+1}T_i = Y_i \quad (1 \le i \le n-1), \qquad T_n^{-1}Y_n = Y_n^{-1}T_n + (t_0^{1/2} - t_0^{-1/2})$$

and

$$q^{-1/2}Y_1^{-1}U_n^{-1} = q^{1/2}U_nY_1 + (u_0^{1/2} - u_0^{-1/2}),$$

where $U_n := X_1^{-1}T_0Y_1^{-1}$. The subalgebra H generated by $T_i\,(1 \le i \le n)$ and $Y_i^{\pm 1}$ $(1 \le i \le n)$ is called the *affine Hecke algebra* of type C_n.

Let $* : \mathbb{F} \to \mathbb{F}$ be the $\mathbb{C}$-algebra involution defined by $(t_0^{1/2})^* = u_n^{1/2}$ and the other parameters $q^{1/2}, t^{1/2}, t_n^{1/2}, u_0^{1/2}$ are fixed. It uniquely extends to the $\mathbb{C}$-algebra anti-involution on $\mathbb{H}$ such that

$$T_0^* = U_n, \quad T_i^* = T_i, \quad X_i^* = Y_i^{-1}, \quad Y_i^* = X_i^{-1} \qquad (1 \le i \le n).$$

The anti-involution $*$ is called duality anti-involution. For a Laurent polynomial f of n-variables with the coefficients in $\mathbb{F}$, we define $f^\diamond$ by the equality $f^\diamond(Y) = (f(X))^*$.

Define the elements $\widetilde{S}_i$ $(0 \le i \le n)$ in $\mathbb{F}(X)\#W$ by

$$\widetilde{S}_0 := t_0^{-1/2}(1 - q^{1/2}t_0^{1/2}u_0^{1/2}X_1^{-1})(1 + q^{1/2}t_0^{1/2}u_0^{-1/2}X_1^{-1})s_0,$$
$$\widetilde{S}_i := t^{-1/2}(1 - tX_i/X_{i+1})s_i \quad (1 \le i < n),$$
$$\widetilde{S}_n := t_n^{-1/2}(1 - t_n^{1/2}u_n^{1/2}X_n)(1 + t_n^{1/2}u_n^{-1/2}X_n)s_n.$$

In fact they belong to the subalgebra $\mathbb{H}$ since we have

$$\begin{aligned}\widetilde{S}_0 &= (1 - qX_1^{-2})T_0 - (t_0^{1/2} - t_0^{-1/2}) - (u_0^{1/2} - u_0^{-1/2})q^{1/2}X_1^{-1},\\ \widetilde{S}_i &= (1 - X_i/X_{i+1})T_i - (t^{1/2} - t^{-1/2}) \quad (1 \le i < n),\\ \widetilde{S}_n &= (1 - X_n^2)T_n - (t_n^{1/2} - t_n^{-1/2}) - (u_n^{1/2} - u_n^{-1/2})X_n.\end{aligned} \tag{16}$$

The elements $\widetilde{S}_i$ and their dual $\widetilde{S}_i^*$ $(0 \le i \le n)$ will play a fundamental role in the construction of the bispectral qKZ equation.

4.2. *The bispectral qKZ equation*

Here we construct the bispectral qKZ equation of type $(C_n^\vee, C_n)$. See Ref. 11 for the details in the case of type GL_n.

Hereafter we set the parameters $q^{1/2}, t^{1/2}, \ldots$ to generic complex values and consider $\mathbb{H}$ as a $\mathbb{C}$-algebra. Then we have the Poincaré-Birkhoff-Witt (PBW) decomposition of H_0 and $\mathbb{H}$ as $\mathbb{C}$-vector spaces:

$$H \simeq H_0 \otimes \mathbb{C}[Y^{\pm 1}], \qquad \mathbb{H} \simeq \mathbb{C}[X^{\pm 1}] \otimes H_0 \otimes \mathbb{C}[Y^{\pm 1}].$$

Set $\mathbb{L} := \mathbb{C}[X^{\pm 1}] \otimes \mathbb{C}[Y^{\pm 1}]$. The DAHA $\mathbb{H}$ has $\mathbb{L}$-module structure defined by

$$(f \otimes g).h = f(x)\, h\, g(y) \qquad (f \otimes g \in \mathbb{L},\ h \in \mathbb{H}). \tag{17}$$

We consider $\mathbb{L}$ as the ring of regular functions on $T \times T$, where T is the n-dimensional torus $T := (\mathbb{C}^\times)^n$. From the PBW decomposition, any element of $\mathbb{H}$ can be regarded as an H_0-valued regular function on $T \times T$. Let $\mathbb{K}$ be the field of meromorphic functions on $T \times T$. Then $H_0^{\mathbb{K}} := \mathbb{K} \otimes_{\mathbb{L}} \mathbb{H}$ is a left $\mathbb{K}$-module of H_0-valued meromorphic functions on $T \times T$. Any element $F \in H_0^{\mathbb{K}}$ is uniquely written in the form $F = \sum_{w \in W_0} f_w.T_w$ $(f_w \in \mathbb{K})$.

Denote the translations in W by

$$\epsilon_i := s_i \cdots s_{n-1}(s_n \cdots s_0)s_1 \cdots s_{i-1} \qquad (1 \le i \le n).$$

Then W is a semi-direct product $W \simeq W_0 \ltimes \Gamma$ of the finite Weyl group $W_0 := \langle s_1, \ldots, s_n \rangle$ and the lattice $\Gamma \simeq \mathbb{Z}^n$ generated by ϵ_i $(1 \le i \le n)$. Define the involution ${}^\diamond : W \to W$ by $w_0^\diamond = w_0$ for $w_0 \in W_0$ and $\epsilon_i^\diamond = \epsilon_i^{-1}$ $(1 \le i \le n)$.

Then $W \times W$ acts on $\mathbb{L}$ by $(w, w')(f(X) \otimes g(Y)) = (wf)(X) \otimes (w'^{\diamond} g)(Y)$. The action naturally extends to that on $\mathbb{K}$. Now we define the action of $W \times W$ on $H_0^{\mathbb{K}}$ by $(w, w')(F) := \sum_{w \in W_0} (w, w')(f_w).T_w$ for $F = \sum_{w \in W_0} f_w.T_w \in H_0^{\mathbb{K}}$.

Let $w = s_{j_1} \cdots s_{j_l}$ be a reduced expression of $w \in W$. Then the element $\widetilde{S}_w := \widetilde{S}_{j_1} \cdots \widetilde{S}_{j_l} \in \mathbb{H}$ is well-defined. We denote by $d_w(X)$ the Laurent polynomial uniquely determined by the equality $\widetilde{S}_w = d_w(X) w$ in $\mathbb{F}(X)\#W$.

For $(w, w') \in W \times W$, consider the $\mathbb{C}$-linear endomorphism $\widetilde{\sigma}_{(w,w')}$ on $\mathbb{H}$ defined by

$$\widetilde{\sigma}_{(w,w')}(h) := \widetilde{S}_w \, h \, \widetilde{S}^*_{w'}.$$

Then we have

$$\widetilde{\sigma}_{(w,w')}(f.h) = (w, w')(f).\widetilde{\sigma}_{(w,w')}(h) \qquad (f \in \mathbb{L},\ h \in \mathbb{H}). \tag{18}$$

The map $\widetilde{\sigma}_{(w,w')}$ extends to the $\mathbb{C}$-linear endomorphism on $H_0^{\mathbb{K}}$ satisfying Eq. (18) for all $f \in \mathbb{K}$ and $h \in \mathbb{H}$.

We define $\tau(w, w') \in \mathrm{End}_{\mathbb{C}}(H_0^{\mathbb{K}})$ $(w, w' \in W)$ by

$$\tau(w, w')(F) := d_w(X)^{-1} d^{\diamond}_{w'}(Y)^{-1}.\widetilde{\sigma}_{(w,w')}(F) \qquad (F \in H_0^{\mathbb{K}}).$$

Using the equality

$$d_{w_1}(X)^{-1}(w_1 d_{w_2})(X)^{-1} \widetilde{S}_{w_1} \widetilde{S}_{w_2} = w_1 w_2 = d_{w_1 w_2}(X)^{-1} \widetilde{S}_{w_1 w_2},$$

we see that τ is a group homomorphism $\tau : W \times W \to \mathrm{GL}_{\mathbb{C}}(H_0^{\mathbb{K}})$. From the definition of τ, the operator

$$C(w, w') := \tau(w, w') \cdot (w, w')^{-1}$$

acting on $H_0^{\mathbb{K}}$ is $\mathbb{K}$-linear.

Now consider the equation

$$\tau(\mu, \nu)(F) = F \qquad (\forall \mu, \nu \in \Gamma)$$

for $F \in H_0^{\mathbb{K}}$. It is rewritten as

$$C(\mu, \nu) F(\mu X \,|\, \nu^{-1} Y) = F(X \,|\, Y) \qquad (\forall \mu, \nu \in \Gamma), \tag{19}$$

where $\mu Z := (q^{\mu_1} Z_1, \ldots, q^{\mu_n} Z_n)$ for $Z = (Z_1, \ldots, Z_n)$ and $\mu = \epsilon_1^{\mu_1} \cdots \epsilon_n^{\mu_n} \in \Gamma$. Thus Eq. (19) is a system of linear q-difference equations. Since τ is a group homomorphism, the system is holonomic.

Definition 4.2. We call the holonomic system of q-difference equations (19) the *bispectral qKZ equation of type* $(C_n^{\vee}, C_n)$.

The bispectral qKZ equation (19) essentially consists of the two systems:

$$C(\epsilon_a,1)F(\ldots,qX_a,\ldots|Y)=F(\ldots,X_a,\ldots|\,Y)\quad(1\le a\le n),\tag{20}$$
$$C(1,\epsilon_m)F(X|\ldots,q^{-1}Y_m,\ldots)=F(X\,|\ldots,Y_m,\ldots)\quad(1\le m\le n).\tag{21}$$

The system (20) is called the *quantum affine Knizhnik-Zamolodchikov* (QAKZ) *equation of type* C_n. See Section 1.3.6 of Ref. 4 for construction of the QAKZ equation associated with arbitrary root system. The system (21) is dual of (20). In the rest of this subsection we compute the operator $C(1,\epsilon_m)\,(1\le m\le n)$ explicitly.

Denote by H^* the subalgebra of $\mathbb{H}$ generated by $X_i^{\pm1}\,(1\le i\le n)$ and $T_i\,(1\le i\le n)$. The duality anti-involution $*$ gives the isomorphism $H\simeq H^*$. We define the anti-algebra homomorphism $\eta_R: H^*\to \mathrm{End}_{\mathbb{K}}(H_0^{\mathbb{K}})$ by

$$\eta_R(A)\left(\sum_{w\in W_0}f_w.T_w\right):=\sum_{w\in W_0}f_w.(T_wA),$$

where $A\in H^*$ and $f_w\in\mathbb{K}\,(w\in W_0)$. Applying the dual anti-involution to (16), we obtain explicit formulas for $\widetilde{S}_i^*$ in terms of $T_i\,(1\le i\le n), U_n$ and $Y_i^{\pm1}\,(1\le i\le n)$. Then we see that $C(1,s_i)\,(0\le i\le n)$ are given by

$$C(1,s_i)=\begin{cases}\mathcal{K}_0(Y_1) & (i=0),\\ \mathcal{R}_i(Y_{i+1}/Y_i) & (1\le i<n),\\ \mathcal{K}_n(Y_n) & (i=n),\end{cases}$$

where $\mathcal{K}_0(Y),\mathcal{R}_i(Y)\,(1\le i<n)$ and $\mathcal{K}_n(Y)$ are defined by

$$\begin{aligned}\mathcal{K}_0(Y)&:=\frac{u_n^{1/2}}{(1-q^{1/2}u_0^{1/2}u_n^{1/2}Y)(1+q^{1/2}u_0^{-1/2}u_n^{1/2}Y)}\\&\quad\times\left\{(1-qY^2)\,\eta_R(U_n)-(u_n^{1/2}-u_n^{-1/2})-(u_0^{1/2}-u_0^{-1/2})q^{1/2}\,Y\right\},\\ \mathcal{R}_i(Y)&:=\frac{t^{1/2}}{1-tY}\left\{(1-Y)\eta_R(T_i)-(t^{1/2}-t^{-1/2})\right\}\quad(1\le i<n),\\ \mathcal{K}_n(Y)&:=\frac{t_n^{1/2}}{(1-t_0^{1/2}t_n^{1/2}Y^{-1})(1+t_0^{-1/2}t_n^{1/2}Y^{-1})}\\&\quad\times\left\{(1-Y^{-2})\,\eta_R(T_n)-(t_n^{1/2}-t_n^{-1/2})-(t_0^{1/2}-t_0^{-1/2})Y^{-1}\right\}.\end{aligned}$$

Then we have

$$\begin{aligned} C(1,\epsilon_m) &= \mathcal{R}_m(Y_{m+1}/Y_m)\cdots\mathcal{R}_{n-1}(Y_n/Y_m)\mathcal{K}_n(Y_m) \\ &\times \mathcal{R}_{n-1}(Y_m^{-1}Y_n^{-1})\cdots\mathcal{R}_m(Y_m^{-1}Y_{m+1}^{-1}) \\ &\times \mathcal{R}_{m-1}(Y_{m-1}^{-1}Y_m^{-1})\cdots\mathcal{R}_1(Y_1^{-1}Y_m^{-1}) \\ &\times \mathcal{K}_0(Y_m^{-1})\mathcal{R}_1(qY_1/Y_m)\cdots\mathcal{R}_{m-1}(qY_{m-1}/Y_m). \end{aligned} \tag{22}$$

4.3. *The degenerate double affine Hecke algebra*

In this subsection we consider the trigonometric degeneration of the DAHA of type $(C_n^\vee, C_n)$. We refer to Ref. 4 for the general theory on degeneration of the DAHA.

In the following we make use of X_i, T_i, Y_i $(1 \le i \le n)$ as generators of $\mathbb{H}$. Note that T_0 is recovered from them by $T_0 = T_1^{-1}\cdots T_{n-1}^{-1}\cdot T_n^{-1}\cdots T_1^{-1}Y_1$. Let $\hbar$ be a small parameter. We set

$$q^{1/2} = e^{\hbar c/2}, \ t^{1/2} = e^{\hbar k/2}, \tag{23}$$

$$t_0^{1/2} = e^{\hbar k_0/2}, \ t_n^{1/2} = e^{\hbar k_n/2}, \ u_0^{1/2} = e^{\hbar k_0^*/2}, \ u_n^{1/2} = e^{\hbar k_n^*/2},$$

$$X_i = x_i, \quad T_i = s_i + \hbar\widetilde{T}_i + o(\hbar), \quad Y_i = e^{\hbar y_i} \qquad (1 \le i \le n) \tag{24}$$

and take the limit $\hbar \to 0$. In Eq. (24) we introduced accessory generators $\widetilde{T}_i$ $(1 \le i \le n)$ to avoid rewriting formulas in the form of Lusztig's relations. For example, substitute Eq. (23) and Eq. (24) into the relations

$$(T_i - t^{1/2})(T_i + t^{-1/2}) = 0, \quad T_iY_{i+1}T_i = Y_i \qquad (1 \le i \le n)$$

and expand them into power series of $\hbar$. Taking the zeroth and the first order terms we obtain

$$s_i^2 = 1, \quad s_i\widetilde{T}_i + \widetilde{T}_is_i = ks_i, \quad \widetilde{T}_is_i + s_iy_{i+1}s_i + s_i\widetilde{T}_i = y_i.$$

Eliminating $\widetilde{T}_i$, we find $y_is_i = s_iy_{i+1} + k$. Thus we get closed relations among x_i, s_i and y_i $(1 \le i \le n)$. The parameters k_0, k_n, k_0^* and k_n^* appear only in the form of $k_0 + k_n$ and $k_0^* + k_n^*$. Setting $\alpha := (k_0 + k_n)/2$ and $\beta := (k_0^* + k_n^*)/2$, we obtain the trigonometric degeneration of $\mathbb{H}$:

Definition 4.3. The *degenerate double affine Hecke algebra* $\overline{\mathbb{H}}$ of type $(C_n^\vee, C_n)$ is the unital associative algebra generated by x_i, s_i and y_i $(1 \le$

$i \leq n$) satisfying the following relations:

$$
\begin{aligned}
&s_i^2 = 1 \quad (1 \leq i \leq n), \qquad s_i s_{i+1} s_i = s_{i+1} s_i s_{i+1} \quad (1 \leq i < n),\\
&s_{n-1} s_n s_{n-1} s_n = s_n s_{n-1} s_n s_{n-1},\\
&s_i x_i s_i = x_{i+1} \quad (1 \leq i < n), \qquad s_n x_n s_n = x_n^{-1},\\
&y_i s_i = s_i y_{i+1} + k \quad (1 \leq i < n), \qquad y_n s_n = -s_n y_n + 2\alpha,\\
&[s_i, x_j] = 0 = [s_i, y_j] \quad (|i-j| > 1 \text{ or } (i,j) = (n, n-1)),\\
&[y_i, x_j] = \begin{cases} k(\tilde{s}_{ji} - s_{ji})x_j & (i > j)\\ c x_i + 2(\alpha x_i^{-1} + \beta) r_i & \\ \quad + k(\sum_{1 \leq l < i} s_{li} x_l + \sum_{i < l \leq n} s_{il} x_i + \sum_{l (\neq i)} \tilde{s}_{il} x_i) & (i = j)\\ k(\tilde{s}_{ij} x_j - s_{ij} x_i) & (i < j), \end{cases}
\end{aligned}
$$

where

$$
\begin{aligned}
&s_{ij} = s_{ji} := (s_i \cdots s_{j-1})(s_{j-2} \cdots s_i) \quad (i < j),\\
&r_i := (s_i \cdots s_n)(s_{n-1} \cdots s_i) \quad (1 \leq i \leq n), \qquad \tilde{s}_{ij} := r_i r_j s_{ij}.
\end{aligned}
$$

The subalgebra generated by $s_i \, (1 \leq i < n)$ is isomorphic to the group algebra $\mathbb{C}[W_0]$. The subalgebra $\overline{H}$ generated by $s_i \, (1 \leq i < n)$ and $y_i \, (1 \leq i \leq n)$ is called the *degenerate affine Hecke algebra* of type C_n. The subalgebra generated by $x_i \, (1 \leq i \leq n)$ and $s_i \, (1 \leq i < n)$ is isomorphic to the group algebra $\mathbb{C}[W]$ of the affine Weyl group through the map $x_1^{-1} r_1 \mapsto s_0$ and $s_i \mapsto s_i \, (1 \leq i \leq n)$. Hereafter we identify them. We have the PBW decomposition:

$$
\begin{aligned}
&\mathbb{C}[W] \simeq \mathbb{C}[x^{\pm 1}] \otimes \mathbb{C}[W_0], \quad \overline{H} \simeq \mathbb{C}[W_0] \otimes \mathbb{C}[y],\\
&\overline{\mathbb{H}} \simeq \mathbb{C}[x^{\pm 1}] \otimes \mathbb{C}[W_0] \otimes \mathbb{C}[y],
\end{aligned}
$$

where $\mathbb{C}[x^{\pm 1}] := \mathbb{C}[x_1^{\pm 1}, \ldots, x_n^{\pm 1}]$ and $\mathbb{C}[y] := \mathbb{C}[y_1, \ldots, y_n]$.

We regard $\overline{\mathbb{L}} := \mathbb{C}[x^{\pm 1}] \otimes \mathbb{C}[y]$ as the ring of regular functions on $(\mathbb{C}^\times)^n \times \mathbb{C}^n$. Denote by $\overline{\mathbb{K}}$ the field of meromorphic functions on $(\mathbb{C}^\times)^n \times \mathbb{C}^n$. We give $\overline{\mathbb{L}}$-module structure to $\overline{\mathbb{H}}$ in the same way as Eq. (17). Then $\mathbb{C}[W_0]^{\overline{\mathbb{K}}} := \overline{\mathbb{K}} \otimes_{\overline{\mathbb{L}}} \overline{\mathbb{H}}$ is the vector space of $\mathbb{C}[W_0]$-valued meromorphic functions. Any element $G \in \mathbb{C}[W_0]^{\overline{\mathbb{K}}}$ is uniquely represented in the form $G = \sum_{w \in W_0} g_w . w$ where $g_w \in \overline{\mathbb{K}}$.

We define two maps

$$
\overline{\eta}_L : \overline{H} \to \mathrm{End}_{\overline{\mathbb{K}}}(\mathbb{C}[W_0]^{\overline{\mathbb{K}}}), \qquad \overline{\eta}_R : \mathbb{C}[W] \to \mathrm{End}_{\overline{\mathbb{K}}}(\mathbb{C}[W_0]^{\overline{\mathbb{K}}}) \tag{25}
$$

by

$$\overline{\eta}_L(h)\left(\sum_{w\in W_0} g_w.w\right) := \sum_{w\in W_0} g_w.(hw),$$
$$\overline{\eta}_R(\xi)\left(\sum_{w\in W_0} g_w.w\right) := \sum_{w\in W_0} g_w.(w\xi).$$

Then the map $\overline{\eta}_L$ (resp. $\overline{\eta}_R$) is an algebra (resp. anti-algebra) homomorphism. Hence they determine $(\overline{H}, \mathbb{C}[W])$-bimodule structure on $\mathbb{C}[W_0]^{\mathbb{K}}$.

4.4. *Degeneration of the bispectral qKZ equation (Y-side)*

In the following two subsections, we consider the trigonometric degeneration of the bispectral qKZ equation (20) and (21).

First we consider Eq. (21). Recall that the operator $C(1, \epsilon_m)$ is explicitly given by (22). Denote by $\overline{\mathcal{K}}_0(y), \overline{\mathcal{R}}_i(y), \overline{\mathcal{K}}_n(y)$ the zeroth order term of the power series expansion of $\mathcal{K}_0(e^{\hbar y}), \mathcal{R}_i(e^{\hbar y}), \mathcal{K}_n(e^{\hbar y})$ as $\hbar \to 0$. We have

$$\begin{aligned}
\overline{\mathcal{K}}_0(y) &= \frac{1}{y+\beta+c/2}\left((y+c/2)\,\overline{\eta}_R(x_1^{-1}r_1)+\beta\right),\\
\overline{\mathcal{R}}_i(y) &= \frac{1}{y+k}\left(y\,\overline{\eta}_R(s_i)+k\right) \quad (1\le i<n),\\
\overline{\mathcal{K}}_n(y) &= \frac{1}{y-\alpha}\left(y\,\overline{\eta}_R(s_n)-\alpha\right).
\end{aligned}$$

Suppose that $F = F(X\,|\,Y) \in H_0^{\mathbb{K}}$ is expanded as

$$F = \hbar^M\left(\overline{G} + o(1)\right) \qquad (\hbar \to 0) \tag{26}$$

for some $M \in \mathbb{Z}$ and $\overline{G} \in \mathbb{C}[W_0]^{\mathbb{K}} \setminus \{0\}$. Taking the lowest degree term with respect to $\hbar$, we obtain the trigonometric degeneration of Eq. (21):

$$\overline{C}(1, \epsilon_m)\Delta_m \overline{G} = \overline{G} \quad (1 \le m \le n), \tag{27}$$

where

$$\begin{aligned}
\overline{C}(1,\epsilon_m) := {}& \overline{\mathcal{R}}_m(y_{m+1}-y_m)\cdots\overline{\mathcal{R}}_{n-1}(y_n-y_m)\overline{\mathcal{K}}_n(y_m)\\
&\times \overline{\mathcal{R}}_{n-1}(-y_m-y_n)\cdots\overline{\mathcal{R}}_m(-y_m-y_{m+1})\\
&\times \overline{\mathcal{R}}_{m-1}(-y_{m-1}-y_m)\cdots\overline{\mathcal{R}}_1(-y_1-y_m)\\
&\times \overline{\mathcal{K}}_0(-y_m)\overline{\mathcal{R}}_1(y_1-y_m+c)\cdots\overline{\mathcal{R}}_{m-1}(y_{m-1}-y_m+c).
\end{aligned} \tag{28}$$

4.5. Degeneration of the bispectral qKZ equation (X-side)

Next we consider the trigonometric degeneration of Eq. (20). One can calculate it by rewriting the operator $C(\epsilon_a, 1)$ into the form of Eq. (22) and taking the limit as $\hbar \to 0$. Here we calculate the limit more directly.

The QAKZ equation (20) is equivalent to $\tau(\epsilon_a, 1)(F) = F$ $(1 \le a \le n)$, which is explicitly given by

$$\sum_{w \in W_0} ((\epsilon_a, 1)(f_w) \cdot d_{\epsilon_a}(X)^{-1}).\widetilde{S}_{\epsilon_a} T_w = \sum_{w \in W_0} f_w.T_w, \tag{29}$$

where $F = \sum_{w \in W_0} f_w.T_w$ $(f_w \in \mathbb{K})$. By direct calculation we have

$$d_{\epsilon_a}(X) = d^{(0)}_{\epsilon_a}(x)\left(1 - \hbar\, d^{\dagger}_{\epsilon_a}(x) + o(\hbar)\right),$$

where

$$d^{(0)}_{\epsilon_a}(x) := (1 - x_a^2)^2 \prod_{p(\ne a)} (1 - x_a/x_p)(1 - x_a x_p)$$

and

$$\begin{aligned} d^{\dagger}_{\epsilon_a}(x) := {} & \alpha \frac{1 + x_a^2}{1 - x_a^2} + 2\beta \frac{x_a}{1 - x_a^2} \\ & + c\left(\frac{x_a^2}{1 - x_a^2} - \sum_{p=1}^{a-1} \frac{x_a}{x_a - x_p} \right) + k \sum_{p(\ne a)} \left(\frac{1}{1 - x_a x_p} - \frac{x_a}{x_a - x_p} \right). \end{aligned}$$

To calculate the limit of $\widetilde{S}_{\epsilon_a}$ as $\hbar \to 0$, we use the accessory generators $\widetilde{T}_i$ $(1 \le i \le n)$ and the relations

$$\begin{aligned} & s_i \widetilde{T}_i + \widetilde{T}_i s_i = k s_i \quad (1 \le i < n), \qquad s_n \widetilde{T}_n + \widetilde{T}_n s_n = k_n s_n, \\ & \widetilde{T}_i x_i = x_{i+1}(\widetilde{T}_i - k), \ x_i \widetilde{T}_i = (\widetilde{T}_i - k) x_{i+1} \quad (1 \le i < n), \\ & \widetilde{T}_n x_n = x_n^{-1}(\widetilde{T}_i - k_n) - k_n^* \end{aligned}$$

derived from (i) and (iii) in Definition 4.3. Then we obtain

$$\widetilde{S}_{\epsilon_a} = d^{(0)}_{\epsilon_a}(x)\left(1 + \hbar\, \widetilde{S}^{\dagger}_{\epsilon_a} + o(\hbar)\right),$$

where

$$\begin{aligned} \widetilde{S}^{\dagger}_{\epsilon_a} &:= \Phi_a - c\left(\frac{x_a^2}{1 - x_a^2} - \sum_{p=1}^{a-1} \frac{x_a}{x_a - x_p} \right), \\ \Phi_a &:= y_a + 2\,\frac{\alpha + \beta x_a}{x_a^2 - 1} r_a \\ & \quad + k\left(\sum_{p=1}^{a-1} \frac{x_a}{x_a - x_p} s_{pa} + \sum_{p=a+1}^{n} \frac{x_p}{x_a - x_p} s_{ap} + \sum_{p(\ne a)} \frac{1}{x_a x_p - 1} \tilde{s}_{ap} \right). \end{aligned}$$

Note that

$$d^{\dagger}_{\epsilon_a}(x) + \widetilde{S}^{\dagger}_{\epsilon_a} = \Phi_a + cx_a \frac{\partial \log U(x)}{\partial x_a},$$

where

$$U(x) := \prod_{p=1}^{n} \left(x_p^{\frac{k(n-1)+\alpha}{c}} (1+x_p)^{\frac{\beta-\alpha}{c}} (1-x_p)^{-\frac{\beta+\alpha}{c}} \right) \times \prod_{1\le p<l\le n} ((1-x_p x_l)(x_p - x_l))^{-\frac{k}{c}}. \tag{30}$$

Suppose that $F = F(X \mid Y) \in H_0^{\mathbb{K}}$ is expanded as in Eq. (26). Then we have

$$(\epsilon_a, 1)F - F = \hbar^{M+1}(cx_a \frac{\partial \overline{G}}{\partial x_a} + o(1)).$$

Hence the trigonometric degeneration of Eq. (29) is given by

$$\left(cx_a \frac{\partial}{\partial x_a} + L_a + cx_a \frac{\partial \log U(x)}{\partial x_a} \right) \overline{G} = 0, \tag{31}$$

where

$$L_a := \overline{\eta}_L(y_a) + 2\,\frac{\alpha + \beta x_a}{x_a^2 - 1} \overline{\eta}_L(r_a)$$

$$+ k \left(\sum_{p=1}^{a-1} \frac{x_a}{x_a - x_p} \overline{\eta}_L(s_{pa}) + \sum_{p=a+1}^{n} \frac{x_p}{x_a - x_p} \overline{\eta}_L(s_{ap}) + \sum_{p(\neq a)} \frac{1}{x_a x_p - 1} \overline{\eta}_L(\tilde{s}_{ap}) \right).$$

Remark 4.1. The equation (31) is the semi-classical limit of the QAKZ equation (20), and hence it could be regarded as the affine KZ equation of type $(C_n^{\vee}, C_n)$. See Section 1.3.2 of Ref. 4 for details of such correspondence in the GL_n case.

4.6. *Embedding into the compatible system*

As was seen in Sec. 4.4 and Sec. 4.5, the trigonometric degeneration of the bispectral qKZ equation is the system of equations (27) and (31) for $\overline{G} \in \mathbb{C}[W_0]^{\mathbb{K}}$. Compatibility of the system formally follows from the relation $\tau(1, \epsilon_m)(\tau(\epsilon_a, 1) - 1) = (\tau(\epsilon_a, 1) - 1)\tau(1, \epsilon_m)$. In this subsection we prove that the system is contained in our compatible one (6).

Define $G \in \mathbb{C}[W_0]^{\mathbb{K}}$ by

$$G(x \mid y) = U(x)\overline{G}(x \mid y), \tag{32}$$

where $U(x)$ is given by Eq. (30). Note that the operator $\overline{C}(1,\epsilon_m)\Delta_m$ commutes with multiplication by any function in x. Hence the modified function G also satisfies Eq. (27). Thus the system of equations (27) and (31) is equivalent to

$$\overline{C}(1,\epsilon_m)\Delta_m G = G \qquad (1 \le m \le n), \tag{33}$$

$$\left(cx_a\frac{\partial}{\partial x_a} + L_a\right) G = 0 \qquad (1 \le a \le n) \tag{34}$$

for $G \in \mathbb{C}[W_0]^{\overline{\mathbb{K}}}$.

We realize the $(\overline{H}, \mathbb{C}[W])$-bimodule $\mathbb{C}[W_0]^{\overline{\mathbb{K}}}$ in the scalar extension $(V^{\otimes n})^{\overline{\mathbb{K}}} := \overline{\mathbb{K}} \otimes_{\mathbb{C}} V^{\otimes n}$ as follows. First we define the left action of W_0 on V by

$$s_i(v_a) = v_{\sigma_i(a)}, \quad s_i(v_{\overline{a}}) = v_{\overline{\sigma_i(a)}} \qquad (1 \le i < n),$$

where $\sigma_i := (i, i+1)$ is the transposition, and

$$s_n(v_a) = \begin{cases} v_{\overline{n}} \ (a = n) \\ v_a \ (a \ne n), \end{cases} \qquad s_n(v_{\overline{a}}) = \begin{cases} v_n \ (a = n) \\ v_{\overline{a}} \ (a \ne n). \end{cases}$$

View $V^{\otimes n}$ as a tensor representation of W_0. We extend it to the $\overline{\mathbb{K}}$-linear action on $(V^{\otimes n})^{\overline{\mathbb{K}}}$. Let $(V^{\otimes n})_0 := \mathbb{C}[W_0](v_1 \otimes \cdots \otimes v_n)$ be the cyclic submodule and $(V^{\otimes n})_0^{\overline{\mathbb{K}}} := \overline{\mathbb{K}} \otimes_{\mathbb{C}} (V^{\otimes n})_0$. Denote the left action on $(V^{\otimes n})_0^{\overline{\mathbb{K}}}$ by $\rho_L : \mathbb{C}[W_0] \to \mathrm{End}_{\overline{\mathbb{K}}}(V^{\otimes n})_0^{\overline{\mathbb{K}}}$.

Note that the operators $A_a(y), B_a(x)$ and $Q_m(x\,|\,y)$ are $\overline{\mathbb{K}}$-linear, and hence belong to $\mathrm{End}_{\overline{\mathbb{K}}}(V^{\otimes n})_0^{\overline{\mathbb{K}}}$.

Lemma 4.1. *The following relations hold on* $(V^{\otimes n})_0^{\overline{\mathbb{K}}}$*:*

$$\begin{aligned} &A_i(y)\rho_L(s_i) = \rho_L(s_i)A_{i+1}(y) + k \quad (1 \le i < n), \\ &A_n(y)\rho_L(s_n) = -\rho_L(s_n)A_n(y) + 2\alpha, \\ &[A_i(y), \rho_L(s_j)] = 0 \quad (|i-j| > 1 \ or\, (i,j) = (n-1, n)). \end{aligned}$$

From Lemma 4.1 the action ρ_L is extended to that of $\overline{H}$ which maps $y_a \mapsto A_a(y)$. We also denote it by ρ_L. Now consider the $\overline{\mathbb{K}}$-linear map

$$\phi : \mathbb{C}[W_0]^{\overline{\mathbb{K}}} \to (V^{\otimes n})_0^{\overline{\mathbb{K}}}, \quad w \mapsto \rho_L(w)(v_1 \otimes \cdots \otimes v_n) \quad (w \in W_0).$$

Proposition 4.1. *The map* ϕ *gives an isomorphism between the left* $\overline{H}$*-modules* $(\overline{\eta}_L, \mathbb{C}[W_0]^{\overline{\mathbb{K}}})$ *and* $(\rho_L, (V^{\otimes n})_0^{\overline{\mathbb{K}}})$*. In particular we have* $\phi\overline{\eta}_L(y_a)\phi^{-1} = A_a(y)$.

Proof. It is clear that ϕ commutes with the left action of $\mathbb{C}[W_0]$. From $A_a(y)(v_1 \otimes \cdots \otimes v_n) = y_a v_1 \otimes \cdots \otimes v_n$ and Lemma 4.1, ϕ commutes also with the action of $y_a \in \overline{H}$ $(1 \leq a \leq n)$. □

Recall that $\mathbb{C}[W_0]^{\overline{\mathbb{K}}}$ is a right $\mathbb{C}[W]$-module with the action $\overline{\eta}_R$ (see Eq. (25)). Now we define a right action of $\mathbb{C}[W]$ on $(V^{\otimes n})_0^{\overline{\mathbb{K}}}$:

Lemma 4.2. *There exists an anti-algebra homomorphism* $\rho_R : \mathbb{C}[W] \to \mathrm{End}_{\overline{\mathbb{K}}}(V^{\otimes n})_0^{\overline{\mathbb{K}}}$ *such that*

$$\rho_R(s_0) = T_1(x), \quad \rho_R(s_i) = P_{i,i+1} \ (1 \leq i < n), \quad \rho_R(s_n) = T_n(\underline{1}).$$

Lemma 4.2 follows from $T(x)^2 = 1$ and the braid relation $P_{i,i+1}P_{i+1,i+2}P_{i,i+1} = P_{i+1,i+2}P_{i,i+1}P_{i+1,i+2}$ $(1 \leq i \leq n-2)$. We denote the right action of $\mathbb{C}[W]$ on $(V^{\otimes n})_0^{\overline{\mathbb{K}}}$ by ρ_R.

Proposition 4.2. *The map ϕ commutes with the right action of $\mathbb{C}[W]$ on $\mathbb{C}[W_0]^{\overline{\mathbb{K}}}$ and $(V^{\otimes n})_0^{\overline{\mathbb{K}}}$. Therefore ϕ is an isomorphism between $(\overline{H}, \mathbb{C}[W])$-bimodules.*

Proof. Set $v^\dagger := v_1 \otimes \cdots \otimes v_n$. For $w \in W_0$ we have

$$\phi(ws_i) = \rho_L(w)\rho_L(s_i)v^\dagger = \rho_L(w)P_{i,i+1}v^\dagger = P_{i,i+1}\rho_L(w)v^\dagger = P_{i,i+1}\phi(w)$$

for $1 \leq i < n$, and

$$\phi(ws_n) = \rho_L(w)\rho_L(s_n)v^\dagger = \rho_L(w)T_n(\underline{1})v^\dagger = T_n(\underline{1})\rho_L(w)v^\dagger = T_n(\underline{1})\phi(w).$$

If wx_1^{-1} is equal to $x_j^{-1}w$ (resp. $x_j w$) in $\mathbb{C}[W]$, the first component of $\rho_L(wr_1)v^\dagger \in V^{\otimes n}$ is $v_{\bar{j}}$ (resp. v_j). Hence we get

$$\begin{aligned}\phi(ws_0) &= \phi(wx_1^{-1}r_1) = \phi(x_j^{-1}wr_1) = x_j^{-1}\phi(wr_1) \\ &= T_1(x)\rho_L(w)v^\dagger = T_1(x)\phi(w).\end{aligned}$$

□

Now we send the equations (33) and (34) on $\mathbb{C}[W_0]$ by ϕ. First consider Eq. (33). From Eq. (28) and Proposition 4.2 we find

$$\begin{aligned}&\phi\overline{C}(1,\epsilon_m)\phi^{-1} \\ &= p_m(y)^{-1}\big((y_m - y_{m+1})P_{m,m+1} - k\big)\cdots\big((y_m - y_n)P_{n-1,n} - k\big) \\ &\times \big(y_m T_n(\underline{1}) - \alpha\big)\big((y_m + y_n)P_{n-1,n} - k\big)\cdots\big((y_m + y_{m+1})P_{m,m+1} - k\big) \\ &\times \big((y_m + y_{m-1})P_{m-1,m} - k\big)\cdots\big((y_m + y_1)P_{1,2} - k\big)\big((y_m - c/2)T_1(x) - \beta\big) \\ &\times \big((y_m - y_1 - c)P_{1,2} - k\big)\cdots\big((y_m - y_{m-1} - c)P_{m-1,m} - k\big),\end{aligned}$$

where

$$p_m(y) := (y_m - \alpha)(y_m - \beta - c/2) \times \prod_{j=1}^{m-1}(y_m - y_j - c - k) \prod_{j=m+1}^{n}(y_m - y_j - k) \prod_{j(\neq m)}(y_m + y_j - k).$$

Using

$$\lambda P - k = (\lambda - k)PR(\lambda)^{-1}, \quad \lambda T(x) - \beta = (\lambda - \beta)K(\lambda \,|\, x, \beta)^{-1},$$

we obtain

$$\phi \overline{C}(1, \epsilon_m)\phi^{-1} = Q_m(x \,|\, y)^{-1}.$$

Hence, sending Eq. (33) by ϕ, we get the boundary rational qKZ equation

$$\Delta_m \phi(G) = Q_m(x \,|\, y)\phi(G) \qquad (1 \le m \le n)$$

on $(V^{\otimes n})_0$.

Next let us consider Eq. (34). Note that the following equalities hold on $(V^{\otimes n})_0$:

$$\sum_{j=1}^{n} \overline{E}_{aa}^{(j)} = \rho_L(r_a), \quad \mathbf{Y}_{aa} + \mathbf{Z}_{aa} = 0,$$
$$\mathbf{X}_{ab} + \mathbf{X}_{ba} = \rho_L(s_{ab}), \quad \mathbf{Y}_{ab} + \mathbf{Z}_{ab} = \rho_L(\tilde{s}_{ab}) \qquad (a \neq b).$$

Therefore we have

$$\phi L_a \phi^{-1} = A_a(y) + B_a(x) = L_a(x \,|\, y).$$

The Euler operator $x_a \frac{\partial}{\partial x_a}$ commutes with ϕ. Consequently, sending Eq. (34) by ϕ, we obtain the differential equation

$$D_a(x \,|\, y)\phi(G) = 0 \qquad (1 \le a \le n).$$

As a result we find

Proposition 4.3. *The trigonometric degeneration (27) and (31) of the bispectral qKZ equation of type $(C_n^\vee, C_n)$ is equivalent to the compatible system (6) with $N = n$ restricted to $(V^{\otimes n})_0$ through the gauge transform (32).*

5. Integral formula for solutions in a special case

We construct an integral formula for solutions to the compatible system (6) with

$$\alpha = \beta = k/2$$

and the variable x restricted to the hyperplane

$$x = (x_1, 1, \dots, 1).$$

Hence only one differential operator $D_1(x\,|\,y)$ enters in the system. In the following we set

$$x_1 = e^{2\pi i\lambda}$$

and assume that

$$\operatorname{Im} c > 0, \qquad \operatorname{Im} k > 0$$

for simplicity's sake.

Our solutions take values in the $2n$-dimensional subspace of $V^{\otimes n}$ determined as follows. Denote by $\tilde{V}$ the subspace of V spanned by v_a and $v_{\bar{a}}$ $(2 \le a \le N)$. We fix a non-zero vector $\tilde{v} \in \tilde{V}^{\otimes(n-1)}$ satisfying

$$P_{i,i+1}\tilde{v} = \tilde{v} \quad (1 \le \forall i \le n-2), \qquad T_{n-1}(\underline{1})\,\tilde{v} = \tilde{v}.$$

Now define the vectors $u_j \in V^{\otimes n}$ $(1 \le j \le 2n)$ by

$$u_j := P_{1,j}(v_1 \otimes \tilde{v}), \quad u_{2n+1-j} := P_{j,n}(\tilde{v} \otimes v_{\bar{1}}) \qquad (1 \le j \le n),$$

where $P_{1,1} = P_{n,n} = id$. Our solutions take values in the subspace $\mathcal{V} := \oplus_{j=1}^{2n}\mathbb{C}u_j$.

Define the rational functions $g_j(t) = g_j(t\,|y)$ $(1 \le j \le 2n)$ by

$$g_j(t) := \frac{1}{t - y_j}\prod_{p=1}^{j-1}\frac{t - y_p - k}{t - y_p},$$

$$g_{2n+1-j}(t) := \frac{1}{t + y_j}\prod_{p=j+1}^{n}\frac{t + y_p - k}{t + y_p}\prod_{p=1}^{n}\frac{t - y_p - k}{t - y_p},$$

for $1 \le j \le n$. Set $\mathcal{G} := \sum_{j=1}^{2n}\mathbb{C}g_j(t)$.

Denote by $\mathcal{W}$ the $\mathbb{C}$-vector space spanned by the functions in the form

$$\frac{P(e^{2\pi i t/c})}{\prod_{p=1}^{n}(1 - e^{2\pi i(t-y_p)/c})(1 - e^{2\pi i(t+y_p)/c})}, \tag{35}$$

where P is a polynomial whose coefficients are entire and periodic functions in $y_1, \dots, y_n$ with period c.

Now we define a pairing I between $\mathcal{G}$ and $\mathcal{W}$ by

$$I(g, W) := \int_{C(y)} \varphi(t \,|\, y) g(t) W(e^{2\pi i t/c})\, dt \qquad (g \in \mathcal{G},\, W \in \mathcal{W}), \tag{36}$$

where the kernel function φ is defined by

$$\varphi(t \,|\, y) := e^{-\frac{2\pi i\lambda}{c} t} \prod_{p=1}^{n} \frac{\Gamma\left(\frac{t-y_p-k}{-c}\right)\Gamma\left(\frac{t+y_p-k}{-c}\right)}{\Gamma\left(\frac{t-y_p}{-c}\right)\Gamma\left(\frac{t+y_p}{-c}\right)}.$$

The contour $C(y)$ is a deformation of the real line $(-\infty, +\infty)$ such that the poles at $\pm y_p + k + c\mathbb{Z}_{\geq 0}$ $(1 \leq p \leq n)$ are above $C(y)$ and the poles at $\pm y_p + c\mathbb{Z}_{\leq 0}$ $(1 \leq p \leq n)$ are below $C(y)$. Suppose that $W(e^{2\pi i t/c})$ is given by Eq. (35). From the Stirling formula we see that the integral (36) converges if

$$\operatorname{Re}\lambda < \deg P < \operatorname{Re}\lambda + 2n. \tag{37}$$

For $W \in \mathcal{W}$ satisfying the degree condition (37) we set

$$f(\lambda \,|\, y) := \sum_{j=1}^{2n} I(g_j, W) u_j. \tag{38}$$

Proposition 5.1. *The function f satisfies the boundary rational qKZ equation (4) with $\alpha = \beta = k/2$ and $x = (e^{2\pi i\lambda}, 1, \ldots, 1)$.*

Proof. In the proof below we need to signify the dependence of $W \in \mathcal{W}$ on y. For that purpose we set

$$\tilde{g}(t \,|\, y') = \sum_{j=1}^{2n} g_j(t \,|\, y') u_j,$$

$$\tilde{f}(\lambda \,|\, y' \,|\, y) = \int_{C(y')} \varphi(t \,|\, y') \tilde{g}(t \,|\, y') W(e^{2\pi i t/c} \,|\, y) dt,$$

for $y' = (y'_1, \ldots, y'_n)$ such that each coordinate y'_j belongs to the set $\{\pm y_a + cl \,|\, 1 \leq a \leq n,\, l \in \mathbb{Z}\}$.

By direct calculation we get

$$P_{l,l+1} R_{l,l+1}(y_l - y_{l+1})\, \tilde{g}(t \,|\, \ldots, y_l, y_{l+1}, \ldots) = \tilde{g}(t \,|\, \ldots, y_{l+1}, y_l, \ldots)$$

for $1 \leq l \leq n-1$, and

$$K_n(y_n \,|\, 1, k/2)\, \tilde{g}(t \,|\, y_1, \ldots, y_{n-1}, y_n) = \tilde{g}(t \,|\, y_1, \ldots, y_{n-1}, -y_n).$$

Since $\varphi(t \,|\, y)$ and $C(y)$ are invariant under the transposition $y_l \leftrightarrow y_{l+1}$ and the reflection $y_n \to -y_n$, the above equalities where $\tilde{g}(t \,|\, \ldots)$ is replaced by

$\tilde{f}(\lambda\,|\ldots|\,y)$ also hold. From this fact and the periodicity of W in y, it is enough to prove that

$$K_1(y_1 - c/2\,|\,x, k/2)\tilde{f}(\lambda\,|\,-y_1, y_2, \ldots, y_n\,|\,y) = \tilde{f}(\lambda\,|\,y_1 - c, y_2, \ldots, y_n\,|\,y) \tag{39}$$

with $x = (e^{2\pi i\lambda}, 1, \ldots, 1)$. We abbreviate $y^{(1)} = (-y_1, y_2, \ldots, y_n)$ and $y^{(2)} = (y_1 - c, y_2, \ldots, y_n)$. Taking the coefficients of u_j $(1 \le j \le 2n)$ in Eq. (39) we have the equalities to prove:

$$\int_{C(y^{(1)})} \varphi(t\,|\,y^{(1)})g_j(t\,|\,y^{(1)})W(e^{2\pi it/c})dt$$
$$= \int_{C(y^{(2)})} \varphi(t\,|\,y^{(2)})g_j(t\,|\,y^{(2)})W(e^{2\pi it/c})dt \qquad (j \neq 1, 2n), \tag{40}$$

$$\int_{C(y^{(1)})} \varphi(t\,|\,y^{(1)})\,\frac{e^{2\pi i\lambda}(y_1 - c/2)g_{2n}(t\,|\,y^{(1)}) + (k/2)g_1(t\,|\,y^{(1)})}{y_1 - c/2 + k/2}\,W(e^{2\pi it/c})dt$$
$$= \int_{C(y^{(2)})} \varphi(t\,|\,y^{(2)})g_1(t\,|\,y^{(2)})W(e^{2\pi it/c})dt, \tag{41}$$

$$\int_{C(y^{(1)})} \varphi(t\,|\,y^{(1)})\,\frac{e^{-2\pi i\lambda}(y_1 - c/2)g_1(t\,|\,y^{(1)}) + (k/2)g_{2n}(t\,|\,y^{(1)})}{y_1 - c/2 + k/2}\,W(e^{2\pi it/c})dt$$
$$= \int_{C(y^{(2)})} \varphi(t\,|\,y^{(2)})g_{2n}(t\,|\,y^{(2)})W(e^{2\pi it/c})dt. \tag{42}$$

First we prove Eq. (40). If $j \neq 1, 2n$, we have

$$\frac{g_j(t\,|\,y^{(1)})}{g_j(t\,|\,y^{(2)})} = \frac{t + y_1 - k}{t + y_1}\,\frac{t - y_1 + c}{t - y_1 - k + c}. \tag{43}$$

Note that $\varphi(t\,|\,y^{(1)}) = \varphi(t\,|\,y)$ and

$$\frac{\varphi(t\,|\,y^{(2)})}{\varphi(t\,|\,y)} = \frac{t + y_1 - k}{t + y_1}\,\frac{t - y_1 + c}{t - y_1 - k + c}. \tag{44}$$

It is equal to the right hand side of Eq. (43), and hence

$$\varphi(t\,|\,y^{(1)})g_j(t\,|\,y^{(1)}) = \varphi(t\,|\,y^{(2)})g_j(t\,|\,y^{(2)}) \qquad (j \neq 1, 2n).$$

Thus the integrands in the both hand sides of Eq. (40) are the same. Since the integrand has no poles at $-y_1 + k, -y_1 + c, y_1$ and $y_1 + k - c$, we can deform $C(y^{(1)})$ to $C(y^{(2)})$ without crossing poles. Thus we obtain Eq. (40).

In the rest we only prove Eq. (41). The proof of Eq. (42) is similar. We separate the integrand in the left hand side and first consider

$$\frac{y_1 - c/2}{y_1 - c/2 + k/2}\int_{C(y^{(1)})} \varphi(t\,|\,y^{(1)})\,e^{2\pi i\lambda}g_{2n}(t\,|\,y^{(1)})W(e^{2\pi it/c})dt. \tag{45}$$

We have

$$\varphi(t\,|\,y^{(1)})\,e^{2\pi i\lambda}g_{2n}(t\,|\,y^{(1)})=\varphi(t-c\,|\,y)\,\frac{1}{t-y_1-k}.$$

Changing $t\to t+c$ we see that the integral (45) is equal to

$$\frac{y_1-c/2}{y_1-c/2+k/2}\int_{C(y^{(1)})-c}\varphi(t\,|\,y)\,\frac{1}{t+c-y_1-k}\,W(e^{2\pi it/c})dt.$$

Since the integrand has no poles at $\pm y_j\,(1\le j\le n)$, the contour $C(y^{(1)})-c$ can be deformed to the contour C', which is a deformation of the real line such that the points $-y_1+k+c\mathbb{Z}_{\ge 0}, y_1+k+c\mathbb{Z}_{\ge -1}$ and $\pm y_j+k+c\mathbb{Z}_{\ge 0}\,(2\le j\le n)$ are above C', and the points $-y_1+c\mathbb{Z}_{\le 0}, y_1+c\mathbb{Z}_{\le -1}$ and $\pm y_j+c\mathbb{Z}_{\le 0}\,(2\le j\le n)$ are below C'. The integrand of the rest part of the left hand side of Eq. (41) has no pole at y_1, hence the contour $C(y^{(1)})$ can be deformed to C'. As a result the left hand side becomes

$$\begin{aligned}&\int_{C'}\varphi(t\,|\,y)\left(\frac{y_1-c/2}{y_1-c/2+k/2}\frac{1}{t+c-y_1-k}+\frac{k/2}{y_1-c/2+k/2}\frac{1}{t+y_1}\right)\\&\qquad\times W(e^{2\pi it/c})dt\\&=\int_{C'}\varphi(t\,|\,y)\frac{1}{t+c-y_1-k}\frac{t+y_1-k}{t+y_1}W(e^{2\pi it/c})dt.\end{aligned}$$

On the other hand, using Eq. (44) we see that the right hand side of Eq. (41) is equal to

$$\int_{C(y^{(2)})}\varphi(t\,|\,y)\,\frac{1}{t+c-y_1-k}\,\frac{t+y_1-k}{t+y_1}\,W(e^{2\pi it/c})dt.$$

Since the integrand is regular at $t=-y_1+c$, the contour $C(y^{(2)})$ can be deformed to C' without crossing poles. Thus we get the equality (41). □

Proposition 5.2. *If $\alpha=\beta=k/2$ and $x=(e^{2\pi i\lambda},1,\ldots,1)$, we have*

$$D_1(x\,|\,y)f(\lambda\,|\,y)=-\frac{ke^{2\pi i\lambda}}{e^{2\pi i\lambda}-1}f(\lambda\,|\,y).$$

Proof. Since $\mathbf{Y}_{11}$ and $\mathbf{Z}_{11}$ act as zero on $\mathcal{V}$, we find

$$D_1(x\,|\,y)|_{\mathcal{V}} = \frac{c}{2\pi i}\frac{\partial}{\partial\lambda} + \sum_{j=1}^{n} y_j(e_{11}^{(j)} - e_{\bar{1}\bar{1}}^{(j)}) + \frac{k}{e^{2\pi i\lambda}-1}\left(\sum_{j=1}^{n} e_{\bar{1}1}^{(j)} + \sum_{p=2}^{N}(\mathbf{X}_{p1} + \mathbf{Z}_{1p})\right) + \frac{ke^{2\pi i\lambda}}{e^{2\pi i\lambda}-1}\left(\sum_{j=1}^{n} e_{1\bar{1}}^{(j)} + \sum_{p=2}^{N}(\mathbf{X}_{1p} + \mathbf{Y}_{1p})\right)$$

if $x = (e^{2\pi i\lambda}, 1, \ldots, 1)$ and $\alpha = \beta = k/2$. For $1 \le j \le n$ we have

$$\sum_{p=2}^{N}(\mathbf{X}_{p1} + \mathbf{Z}_{1p})v_j = \sum_{l=j+1}^{2n-j} v_l + \sum_{l=2n+2-j}^{2n} v_l, \qquad \sum_{p=2}^{N}(\mathbf{X}_{1p} + \mathbf{Y}_{1p})v_j = \sum_{l=1}^{j-1} v_l$$

and

$$\sum_{p=2}^{N}(\mathbf{X}_{p1} + \mathbf{Z}_{1p})v_{2n+1-j} = \sum_{l=2n+2-j}^{2n} v_l,$$

$$\sum_{p=2}^{N}(\mathbf{X}_{1p} + \mathbf{Y}_{1p})v_{2n+1-j} = \sum_{l=1}^{j-1} v_l + \sum_{l=j+1}^{2n-j} v_l.$$

From the calculation above we obtain

$$D_1(x\,|\,y)f(\lambda\,|\,y) = \sum_{j=1}^{2n} I(h_j, W)u_j,$$

where the rational functions h_j $(1 \le j \le 2n)$ are given by

$$h_j(t\,|\,y) = -(t - y_j)g_j(t\,|\,y) + \frac{k}{e^{2\pi i\lambda}-1}\sum_{l=1}^{j-1} g_l(t\,|\,y) + \frac{ke^{2\pi i\lambda}}{e^{2\pi i\lambda}-1}\sum_{l=j+1}^{2n} g_l(t\,|\,y),$$

$$h_{2n+1-j}(t\,|\,y) = -(t + y_j)g_j(t\,|\,y) + \frac{k}{e^{2\pi i\lambda}-1}\sum_{l=1}^{2n-j} g_l(t\,|\,y) + \frac{ke^{2\pi i\lambda}}{e^{2\pi i\lambda}-1}\sum_{l=2n+2-j}^{2n} g_l(t\,|\,y)$$

for $1 \le j \le n$.

Note that

$$g_j(t\,|\,y) = \frac{1}{k}\left(\prod_{l=1}^{j-1}\frac{t-y_l-k}{t-y_l} - \prod_{l=1}^{j}\frac{t-y_l-k}{t-y_l}\right),$$

$$g_{2n+1-j}(t\,|\,y) = \frac{1}{k}\left(\prod_{l=j+1}^{n}\frac{t+y_l-k}{t+y_l} - \prod_{l=j}^{n}\frac{t+y_l-k}{t+y_l}\right)\prod_{l=1}^{n}\frac{t-y_l-k}{t-y_l}$$

for $1 \le j \le n$. Using these formulas we get

$$\begin{aligned} h_j(t\,|\,y) &= -\frac{ke^{2\pi i\lambda}}{e^{2\pi i\lambda}-1}g_j(t\,|\,y) \\ &\quad + \frac{1}{1-e^{2\pi i\lambda}}\left(1 - e^{2\pi i\lambda}\prod_{l=1}^{n}\frac{t-y_l-k}{t-y_l}\frac{t+y_l-k}{t+y_l}\right) \end{aligned}$$

for $1 \le j \le 2n$. The last term in the right hand side is related to the kernel function φ by

$$e^{2\pi i\lambda}\prod_{l=1}^{n}\frac{t-y_l-k}{t-y_l}\frac{t+y_l-k}{t+y_l} = \frac{\varphi(t-c\,|\,y)}{\varphi(t\,|\,y)}.$$

Therefore we have

$$\begin{aligned} &\int_{C(y)}\varphi(t\,|\,y)\left(1 - e^{2\pi i\lambda}\prod_{l=1}^{n}\frac{t-y_l-k}{t-y_l}\frac{t+y_l-k}{t+y_l}\right)W(e^{2\pi it/c})dt \\ &= \int_{C(y)}\left(\varphi(t\,|\,y) - \varphi(t-c\,|\,y)\right)W(e^{2\pi it/c})dt \\ &= \left(\int_{C(y)} - \int_{C(y)-c}\right)\varphi(t\,|\,y)W(e^{2\pi it/c})dt. \end{aligned} \tag{46}$$

Since the integrand $\varphi(t\,|\,y)W(e^{2\pi it/c})$ is regular at $t = \pm y_j$ $(1 \le j \le n)$, the integral (46) is equal to zero. Therefore we obtain

$$\begin{aligned} D_1(x\,|\,y)f(e^{2\pi i\lambda}\,|\,y) &= \sum_{j=1}^{2n} I(h_j, W)u_j \\ &= -\frac{ke^{2\pi i\lambda}}{e^{2\pi i\lambda}-1}\sum_{j-1}^{2n} I(g_j, W)u_j = -\frac{ke^{2\pi i\lambda}}{e^{2\pi i\lambda}-1}f(\lambda\,|\,y). \end{aligned}$$

□

The linear operator $L_1(x\,|\,y)$ commutes with multiplication by any function in x. Hence, from Proposition 5.1 and Proposition 5.2, we finally get

Theorem 5.1. *Set*

$$\tilde{f}(\lambda \,|\, y) := (e^{2\pi i\lambda} - 1)^{k/c} f(\lambda \,|\, y),$$

where f is defined by Eq. (38)*. Then the function $\tilde{f}$ is a solution to the compatible system* (6) *with $\alpha = \beta = k/2$ and $x = (e^{2\pi i\lambda}, 1, \ldots, 1)$ for any $W \in \mathcal{W}$ satisfying the degree condition* (37)*.*

Acknowledgments

The research of the author is supported by Grant-in-Aid for Young Scientists (B) No. 20740088. The author is deeply grateful to Saburo Kakei, Masahiro Kasatani, Michitomo Nishizawa and Yoshihisa Saito for valuable discussions. He also thanks the referee for critical comments that improved the paper considerably.

References

1. Cantini, L., qKZ equations and ground state of the O(1) loop model with open boundary conditions, arXiv:0903.5050.
2. Cherednik, I., Quantum Knizhnik-Zamolodchikov equations and affine root systems, *Comm. Math. Phys.* **150** (1992), no. 1, 109–136.
3. Cherednik, I., Factorizing particles on a half line, and root systems, *Theoret. and Math. Phys.* **61** (1984), no. 1, 977–983.
4. Cherednik, I., *Double affine Hecke algebras*, London Mathematical Society Lecture Note Series, 319, Cambridge University Press, Cambridge, 2005.
5. Di Francesco, P. and Zinn-Justin, P., Quantum Knizhnik-Zamolodchikov equation: reflecting boundary conditions and combinatorics, *J. Stat. Mech. Theory Exp.* **2007**, no. 12, P12009.
6. Etingof, P. and Varchenko, A., Dynamical Weyl groups and applications, *Adv. Math.* **167** (2002), no. 1, 74–127.
7. Felder, G., Markov, Y., Tarasov, V. and Varchenko, A., Differential equations compatible with KZ equations, *Math. Phys. Anal. Geom.* **3** (2000), no. 2, 139–177.
8. Frenkel, I. B. and Reshetikhin, N. Yu., Quantum affine algebras and holonomic difference equations, *Comm. Math. Phys.* **146** (1992), no. 1, 1–60.
9. de Gier, J., Ponsaing A. and Shigechi K., Exact finite size groundstate of the O($n = 1$) loop model with open boundaries, *J. Stat. Mech. Theory Exp.* **2009**, no. 4, P04010.
10. Jimbo, M., Kedem, R., Konno, H., Miwa, T. and Weston, R., Difference equations in spin chains with a boundary, *Nuclear Phys. B* **448** (1995), no. 3, 429–456.
11. van Meer, M. and Stokman, J., Double affine Hecke algebras and bispectral quantum Knizhnik-Zamolodchikov equations, arXiv:0812.1005.

12. Mimachi, K., A solution of the quantum Knizhnik Zamolodchikov equation of type C_n, *Comm. Math. Phys.* **197** (1998), no. 1, 229–246.
13. Noumi, M., Macdonald-Koornwinder polynomials and affine Hecke rings, Sūrikaisekikenkyūsho kōkyūroku No. 919 (1995), 44–55 (Japanese).
14. Sahi, S., Nonsymmetric Koornwinder polynomials and duality, *Ann. of Math. (2)* **150** (1999), no. 1, 267–282.
15. Tarasov, V. and Varchenko, A., Difference equations compatible with trigonometric KZ differential equations, *Internat. Math. Res. Notices* **2000**, no. 15, 801–829.
16. Tarasov, V. and Varchenko, A., Dynamical differential equations compatible with rational qKZ equations, *Lett. Math. Phys.* **71** (2005), no. 2, 101–108.

CLASSIFICATION OF SOLUTIONS TO THE REFLECTION EQUATION FOR THE CRITICAL $\mathbf{Z}_N$-SYMMETRIC VERTEX MODEL I

YUJI YAMADA

Department of Mathematics, Rikkyo University
3-34-1 Nishi Ikebukuro, Toshima-ku, Tokyo 171-8501, Japan
E-mail: yamaday@rikkyo.ac.jp

We classify and list up all the meromorphic solutions $K(z)$ to the reflection equation associated to the critical $\mathbf{Z}_N$-symmetric vertex model under two assumptions that none of the diagonal elements is constantly zero and that there is at least a pair of elements $K_b^a(z)K_a^b(z) \neq 0$ We make explicit the matrix elements of $K(z)$, parameters they have and the relations among parameters.

Keywords: Yang-Baxter equation; reflection equation.

1. Introduction

The Yang-Baxter equation

$$R^{01}(z_1)R^{02}(z_1z_2)R^{12}(z_2)$$
$$= R^{12}(z_2)R^{02}(z_1z_2)R^{01}(z_1) \in \mathrm{End}(\mathbf{C}^N \otimes \mathbf{C}^N \otimes \mathbf{C}^N),$$

guarantees the commutativity among the transfer matrices $\mathcal{T}(z)$

$$\mathcal{T}(z_1)\mathcal{T}(z_2) = \mathcal{T}(z_2)\mathcal{T}(z_1)$$

under the cyclic boundary condition

$$\mathcal{T}(z) := \mathrm{tr}_0\left(R^{01}(z)R^{02}(z)\cdots R^{0l}(z)\right)$$
$$\in \mathrm{End}(\mathbf{C}^N \otimes \overbrace{\mathbf{C}^N \otimes \cdots \otimes \mathbf{C}^N}^{l \text{ times}}).$$

Sklyanin[1] proposed the reflection equation

$$R^{12}(z_1z_2^{-1})K^1(z_1)R^{21}(z_1z_2)K^2(z_2)$$
$$= K^2(z_2)R^{12}(z_1z_2)K^1(z_1)R^{21}(z_1z_2^{-1}) \quad \in \mathrm{End}(\mathbf{C}^N \otimes \mathbf{C}^N), \qquad (1)$$

which in turn guarantees the commutativity among the transfer matrices $\mathcal{T}_K(z)$ with a fixed boundary condition specified by $K(z)$, a solution to the reflection equation (1),

$$\begin{aligned}\mathcal{T}_K(z) &:= \mathrm{tr}_0\left(K_+(z)\mathcal{T}(z^{-1})^{-1}K(z)\mathcal{T}(z)\right),\\ K_+(z) &= K\left(z^{-1}q^{-N/2}\right),\\ \mathcal{T}_K(z_1)\mathcal{T}_K(z_2) &= \mathcal{T}_K(z_2)\mathcal{T}_K(z_1).\end{aligned}$$

They constitutes another commutative family of transfer matrices $\mathcal{T}_K(z)$ corresponding to a solution $K(z)$ to (1) beside $\mathcal{T}(z)$.

We classify meromorphic solutions $K(z)$ to the reflection equation (1) in this paper under two assumptions,

(∗) none of the diagonal elements of the K-matrix $K(z)$ is zero, (2)

(∗∗) the existence of a pair (a, b) such that $K^a_b(z)K^b_a(z) \neq 0$ and $a \neq b$, where $K(z) = \left(K^i_j(z)\right)_{ij}$. (3)

All the discussions are done under these assumptions even when not mentioned explicitly throughout except examples in Section 10. The solutions are completely classified in four classes in Theorem 10.1 according to the numbers of the elements of the sets P, S and T specified in Definition 7.1 and Proposition 7.3. There are $\frac{1}{2}n(n+1)$ solutions for each $N \geq 2$ ($N = 2n$ or $N = 2n+1$), and we make explicit all the elements of solutions, parameters they have and the relations among these parameters in Theorem 10.1. The investigations on more degenerate cases, which do not satisfy the assumptions (2) and (3), are left to the subsequent paper.[10]

The organization of this paper is as follows. In Section 2 and Section 3, we describe the R-matrix $R(z)$ and the reflection equation we deal with, and define the notion of similarities among solutions to the reflection equation. We review the necessary results from our previous paper[8] in Section 4. The notations we employ in this paper are also fixed there. The components of the reflection equations are divided into fifteen groups. Important is Proposition 4.1 which states that the reflection equation is equivalent to a part of them, the components of type I, VI, X, XIII and XIV, and Lemma 4.4. We prove some consequences of Lemma 4.4 in Section 5. We analyze the components of type I in Section 6, and derive Proposition 6.1, the equivalent conditions to the type I components. The diagonal elements of $K(z)$ are determined in Section 7 upto similarities in Definition 3.2. This part may be the core part of the classifying procedure. Proposition 6.1 is invoked to pull out the information of the off-diagonal elements from the diagonal ones in Section 8. In Section 9, we write down the off-diagonal elements explicitly, and obtain the relations among parameters in $K(z)$. The classification is written down in Theorem 10.1, and we end up Section 10 with some examples.

2. Critical $\mathbf{Z}_N$-symmetric vertex model

We fix the standard orthonormal basis $\{e_0, e_1, \cdots, e_{N-1}\}$ of the vector space $\mathbf{C}^N$, and extend their indices to all integers by defining $e_{j+N} = e_j$. We define the matrix elements M^i_j of a matrix $M \in \mathrm{End}(\mathbf{C}^N)$ with respect to this basis by

$$Me_j = \sum_{i=0}^{N-1} e_i M^i_j,$$

and two matrices g and h by

$$ge_j = \omega^j e_j, \qquad he_j = e_{j+1},$$

where $\omega = e^{2\sqrt{-1}\pi/N}$. They satisfy $gh = \omega hg$.
The R-matrix of the critical $\mathbf{Z}_N$-symmetric vertex model of Belavin

$$R(z) \in \mathrm{End}(\mathbf{C}^N \otimes \mathbf{C}^N), \quad R(z)e_k \otimes e_l = \sum_{i,j=0}^{N-1} e_i \otimes e_j R^{ij}_{kl}(z)$$

is defined by

Definition 2.1 (critical Z_n-symmetric R-matrices).

$$R^{ij}_{kl}(z) =: \delta_{i+j,k+l} \cdot S^{i-k,j-k}(z), \tag{4}$$

where $q \neq \pm 1$ is a parameter in $\mathbf{C}$ and

$$S^{ab}(z) = \begin{cases} \dfrac{q^2z^2-1}{q^2-1} & \text{for } a \equiv b \equiv 0 \\ z^{(N-a^*)} & \text{for } a \equiv 1, 2, \cdots, N-1, \ b \equiv 0 \\ q^{(b^*-N)} \cdot \dfrac{z^N-1}{q^2-1} & \text{for } a \equiv 0, \ b \equiv 1, 2, \cdots, N-1 \\ 0 & \text{for } a, b \equiv 1, 2, \cdots, N-1 \end{cases} \tag{5}$$

with $a^* := a - N\left[\dfrac{a}{N}\right]$, a^* being the integer a in the interval $[0, N)$ congruent to a modulo N.

The following symmetries of critical $\mathbf{Z}_N$-symmetric R-matrices are immediate from the definition above.

Proposition 2.1. *The critical $\mathbf{Z}_N$-symmetric vertex model of Belavin $R(z)$ has two symmetries,*

$$\begin{aligned} \mathbf{Z}_N - \textit{symmetry} \ &: R^{ij}_{kl}(z) = R^{i+\beta,j+\beta}_{k+\beta,l+\beta}(z) \ \textit{for} \ \beta \in \mathbf{Z}/N\mathbf{Z} \\ \textit{conservation law} \ &: R^{ij}_{kl}(z) = \delta_{i+j,k+l} R^{ij}_{kl}(z), \end{aligned}$$

which are equivalent to

$$(h \otimes h)^{-1} R(z)(h \otimes h) = (g \otimes g)^{-1} R(z)(g \otimes g) = R(z).$$

It satisfies the Yang-Baxter equation.

Theorem 2.1 (Refs. 4–6). *The critical $\mathbf{Z}_N$-symmetric vertex model of Belavin $R(u)$ satisfies the Yang-Baxter equation*

$$\begin{aligned} &R^{01}(z_1)R^{02}(z_1z_2)R^{12}(z_2) \\ &= R^{12}(z_2)R^{02}(z_1z_2)R^{01}(z_1) \in \textit{End}(\mathbf{C}^N \otimes \mathbf{C}^N \otimes \mathbf{C}^N), \end{aligned}$$

and the unitary relation

$$\begin{aligned} PR^{01}(z)PR^{01}(z^{-1}) &= \rho_R(z) Id_{\mathbf{C}^N \otimes \mathbf{C}^N} \in \textit{End}(\mathbf{C}^N \otimes \mathbf{C}^N), \\ \rho_R(z) &= S^{00}(z)S^{00}(z^{-1}), \end{aligned}$$

where, for example, $R^{02}(z)$ acts as $R(z)$ on the 0th and the 2nd component of $\mathbf{C}^N \otimes \mathbf{C}^N \otimes \mathbf{C}^N$ and as an identity on the 1st component (We are now counting from zero), and $P \in \textit{End}(\mathbf{C}^N \otimes \mathbf{C}^N)$ is the permutation, $P(x \otimes y) = y \otimes x$.

3. Reflection equation

In this paper, we will consider solutions $K(z) \in \mathrm{End}(\mathbf{C}^N)$ to the reflection equation associated to the critical $\mathbf{Z}_N$-symmetric vertex model $R(z)$ defined in (4) and (5).

Definition 3.1 (reflection equation). The reflection equation is

$$\begin{aligned} &R^{12}(z_1z_2^{-1})K^1(z_1)R^{21}(z_1z_2)K^2(z_2) \\ &= K^2(z_2)R^{12}(z_1z_2)K^1(z_1)R^{21}(z_1z_2^{-1}) \in \mathrm{End}(\mathbf{C}^N \otimes \mathbf{C}^N), \end{aligned} \tag{6}$$

where $R(z)$ is the critical $\mathbf{Z}_N$-symmetric vertex model defined in Definition 2.1 and

$$\begin{aligned} R^{12}(z) = R(z), \quad &R^{21}(z) = PR(z)P, \\ K^1(z) = K(z) \otimes Id, \quad &K^2(z) = Id \otimes K(z). \end{aligned}$$

The K-matrices inherits the symmetries of the R-matrices in Proposition 2.1 in the following sense.

Proposition 3.1. *If $K(z)$ is a solution to the reflection equation (6), then*

$$K_{\alpha\beta}(z) = g^{-\alpha}h^{-\beta}K(z)h^{\beta}g^{\alpha} \quad (\alpha,\beta \in \{0,1,2,\cdots,N-1\})$$

is also a solution.

The proof is obtained by Proposition 2.1 and Definition 3.1.

Definition 3.2. If two solutions $K(z)$ and $K'(z)$ to the reflection equation satisfy

$$K'(z) = f(z)g^{-\alpha}h^{-\beta}K(z)h^{\beta}g^{\alpha}$$

with some not identically zero function $f(z)$, then we say that $K'(z)$ is similar to $K(z)$.

4. Summary of results in the previous paper[8] and notations

In this section, we summarize several results in the previous paper.[8] The reflection equation is a matrix equation in $\mathrm{End}(\mathbf{C}^N \otimes \mathbf{C}^N)$, whose components are represented by four indices in $\{0,1,2,\cdots,N-1\}$.

Definition 4.1 (matrix elements for the reflection equation). We denote the $(ab|cd)$-element of the reflection equation associated to the critical $\mathbf{Z}_N$-symmetric vertex model $R(z)$ by $(ab|cd)$.

$$(ab|cd) := \begin{pmatrix} R^{12}(z_1z_2^{-1})K^1(z_1)R^{21}(z_1z_2)K^2(z_2) \\ \qquad -K^2(z_2)R^{12}(z_1z_2)K^1(z_1)R^{21}(z_1z_2^{-1}) \end{pmatrix}^{ab}_{cd}, \\ (a,b,c,d \in \{0,1,2,\cdots,N-1\}).$$

The reflection equation consists of N^4 scaler equations. We divide them into 15 groups according to the situation of the coincidence of these four indices $a,b,c,d \in \mathbf{Z}/N\mathbf{Z}$.

Lemma 4.1. *If we suppose that a,b,c,d denote mutually different numbers in $\{0,1,2,\cdots,N-1\}$, the N^4 components of the reflection equation are divided into non-intersecting 15 groups as follows*

I	$(ab\|cd)$	IX	$(ab\|ab)$
II	$(ab\|cc)$	X	$(ab\|ba)$
III	$(aa\|bc)$	XI	$(ab\|aa)$
IV	$(ab\|ac)$	XII	$(aa\|ab)$
V	$(ab\|cb)$	XIII	$(aa\|ba)$
VI	$(ab\|bc)$	XIV	$(ab\|bb)$
VII	$(ab\|ba)$	XV	$(aa\|aa)$
VIII	$(aa\|bb)$		

For example, the equations of type I above consists of $N(N-1)(N-2)(N-3)$ equations. The explicit forms of type I, VI, X, XIII and XIV are written in Proposition 4.1. We prepare some notations.

Definition 4.2 (matrix elements of K-matrix).
Let $K(z)$ be a solution to the reflection equation (6). We define $\kappa^a_b(w)$ *for* $a,b \in \{0,1,2,\cdots,N-1\}$ *by*

$$w := z^N \quad \textit{and} \quad \kappa^a_b(w) := z^{a+b} K^a_b(z).$$

We can write the components $(ab|cd)$ of the reflection equations only in w's and $\kappa^a_b(w)$'s. Because the reflection equation is homogeneous quadratic equation in $\kappa^{a_1}_{b_1}(w_1)$'s and $\kappa^{a_2}_{b_2}(w_2)$'s, $(w_1 = z_1^N, w_2 = z_2^N)$, we often meet terms of the type $\kappa^a_b(w_1)\kappa^c_d(w_2)$.

Definition 4.3. We abbreviate $\kappa^a_b(w_1)\kappa^c_d(w_2)$ simply to $\kappa^{a\,c}_{b\,d}$.

$$\kappa^{a\,c}_{b\,d} := \kappa^a_b(w_1)\kappa^c_d(w_2).$$

We have defined α^* by $\alpha - N\,[\alpha/N]$, which is the integer in $\{0,1,2,\cdots,N-1\}$ equivalent to α modulo N in Definition 2.1. If $K'(z) = h^{-1}K(z)h$, then

$$\begin{aligned} K'(z) &= \left(K^{(a-1)^*}_{(b-1)^*}(z)\right)_{a,b=0,1,\cdots,N-1} \\ &= z^2\left(z^{-(a-1)^*-(b-1)^*-2}\,\kappa^{(a-1)^*}_{(b-1)^*}(w)\right)_{a,b=0,1,\cdots,N-1}. \end{aligned}$$

Remarking that

$$(a-1)^* + (b-1)^* + 2 = \begin{cases} 2N & a=0,\ b=0 \\ N+b & a=0,\ 1 \le b < N \\ N+a & 1 \le a < N,\ b=0 \\ a+b & 1 \le a < N,\ 1 \le b < N \end{cases},$$

we have

Lemma 4.2. *If $K'(z) = \left(z^{-a-b}\kappa'^a_b(w)\right)_{a,b=0,1,\cdots,N-1}$ is defined by*

$$\kappa'^a_b(w) = \begin{cases} w^{-2}\kappa^{N-1}_{N-1}(w) & a=0,\ b=0 \\ w^{-1}\kappa^{N-1}_{b-1}(w) & a=0,\ 1\le b<N \\ w^{-1}\kappa^{a-1}_{N-1}(w) & 1\le a<N,\ b=0 \\ \kappa^{a-1}_{b-1}(w) & 1\le a<N,\ 1\le b<N \end{cases},$$

where $K(z) = \left(z^{-a-b}\kappa^a_b(w)\right)_{a,b=0,1,\cdots,N-1}$, then $K'(z)$ is similar to $K(z)$ in the sense in Definition 3.2.

Definition 4.4. For two functions $f(w)$ and $g(w)$, we define $\langle f,g\rangle_{w,w'}$ by

$$\langle f,g\rangle_{w,w'} := f(w)g(w') - f(w')g(w).$$

When the variables are explicit, we omit them and simply write $\langle f,g\rangle$.

Definition 4.5. Three mutually different integers a_1, a_2 and a_3 uniquely determine the element σ in the symmetric group $\mathcal{S}_3$ such that $a_{\sigma(1)} < a_{\sigma(2)} < a_{\sigma(3)}$. We define $P(a_1,a_2,a_3)$ by

$$P(a_1,a_2,a_3) = \begin{cases} 1 \text{ when } \mathrm{sgn}(\sigma) = 1 \\ 0 \text{ when } \mathrm{sgn}(\sigma) = -1 \end{cases}.$$

Definition 4.6. We define $\delta(Q)$ for a proposition Q by

$$\delta(Q) = \begin{cases} 1 \text{ when a proposition } Q \text{ is true} \\ 0 \text{ when a proposition } Q \text{ is false} \end{cases}.$$

Definition 4.7. Let f_j and g_j $(j = 1,2,\cdots,l)$ be meromorphic functions in w. If two functions f_1 and g_1 satisfy

$$f_1(w_1)g_1(w_2) = f_1(w_2)g_1(w_1)$$

for arbitrary w_1 and w_2, then we write

$$f_1 \sim g_1.$$

If $f_j(w_1)g_j(w_2) = f_j(w_2)g_j(w_1)$ hold for $j = 1,2,\cdots,l$ and arbitrary w_1, w_2, then we write

$$(f_1, f_2, \cdots, f_l) \sim (g_1, g_2, \cdots, g_l).$$

We remark that the relation "$\sim$" is not an equivalence one.

Example 4.1.

(i) If f is the identically zero, then $f \sim g$ holds for any function g; $0 \sim g$.

(ii) $(1,w,w^2) \sim (0,w,w^2)$, $(1,w,w^2) \sim (1,0,w^2)$, $(1,w,w^2) \sim (1,w,0)$, $(1,w,w^2) \sim (1,0,0)$, $(1,w,w^2) \sim (0,w,0)$, $(1,w,w^2) \sim (0,0,w^2)$, $(1,w,w^2) \sim (0,0,0)$.

We often invoke the following lemma.

Lemma 4.3. *If two meromorphic functions $f(w)$ and $g(w)$ satisfy $\langle f,g\rangle_{w,w'} = f(w)g(w') - f(w')g(w) = 0$ for arbitrary w_1 and w_2 in $\mathbf{C}$, namely $f \sim g$, then there exist two constants C and C' such that $Cf(w) = C'g(w)$. If $f(w)g(w) \neq 0$, then two constants can be taken as $CC' \neq 0$.*

The following Lemma 4.4 and Proposition 4.1 constitute the main part in our previous paper.[8]

Definition 4.8. Let $a,b,c \in \{0,1,2,\cdots,N-1\}$ be $a<b<c$. Then there are six elements of $K(z)$ whose indices are a,b or c. We define a set C_{abc} by $C_{abc} := \left\{\kappa^a_b(w), \kappa^a_c(w), \kappa^b_c(w), \kappa^b_a(w), \kappa^c_a(w), \kappa^c_b(w)\right\}$.

Lemma 4.4. *There are only three possibilities for maximal subsets in C_{abc} consisting of only non-zero elements, $\left\{\kappa^a_b(w), \kappa^a_c(w), \kappa^b_c(w), \kappa^b_a(w)\right\}$, $\left\{\kappa^a_b(w), \kappa^a_c(w), \kappa^c_a(w), \kappa^c_b(w)\right\}$, $\left\{\kappa^b_c(w), \kappa^b_a(w), \kappa^c_a(w), \kappa^c_b(w)\right\}$, and they satisfy*

$$\begin{aligned}
\left(\kappa^a_b(w), \kappa^a_c(w), \kappa^b_c(w), \kappa^b_a(w)\right) &\sim \left(f(w), f(w), wf(w), f(w)\right),\\
\left(\kappa^a_b(w), \kappa^a_c(w), \kappa^c_a(w), \kappa^c_b(w)\right) &\sim \left(f(w), f(w), f(w), f(w)\right),\\
\left(\kappa^b_c(w), \kappa^b_a(w), \kappa^c_a(w), \kappa^c_b(w)\right) &\sim \left(f(w), w^{-1}f(w), f(w), f(w)\right)
\end{aligned}$$

for some non-zero function $f(w)$.

Proposition 4.1. *The reflection equation (6), which consists of N^4 scalar equations with the N^2 unknown functions $\kappa^a_b(w)$ is equivalent to Lemma 4.4 and five components* I, VI, X, XIII *and* XIV,

$$\mathrm{I}(a,b,c,d) : \begin{bmatrix} \left(P(a,c,b) - P(a,c,d)\right)\left(1 - w_1^{-1}w_2\right)\kappa^{ab}_{cd} \\ +(w_1w_2^{-1})^{-\delta(a<b)}\kappa^{ba}_{cd} - (w_1w_2^{-1})^{-\delta(c<d)}\kappa^{ab}_{dc} \end{bmatrix} = 0, \quad (7)$$

$$\mathrm{VI}(a,b,c) : \begin{bmatrix} (w_1 - w_2^{-1})\begin{pmatrix} (w_1w_2^{-1})^{-\delta(a<b)}\kappa^{ba}_{bc} \\ -(w_1w_2^{-1})^{-\delta(b<c)}\kappa^{ab}_{cb} \end{pmatrix} \\ -P(b,c,a)(w_1 - w_2)(1 - w_1^{-1}w_2^{-1})\kappa^{ab}_{bc} \\ +(w_1 - w_2)\sum_{j=0}^{N-1}(w_1w_2)^{-\delta(b<j)}\kappa^{aj}_{jc} \end{bmatrix} = 0, \quad (8)$$

$$\mathrm{X}(a,b):\left[\begin{array}{l}(w_1-w_2^{-1})\begin{pmatrix}(w_1w_2^{-1})^{-\delta(b<a)}\kappa_{ab}^{ab}\\-(w_1w_2^{-1})^{-\delta(a<b)}\kappa_{ba}^{ba}\end{pmatrix}\\+(w_1-w_2)\begin{pmatrix}\displaystyle\sum_{j=0}^{N-1}(w_1w_2)^{-\delta(a<j)}\kappa_{jb}^{bj}\\-\displaystyle\sum_{j=0}^{N-1}(w_1w_2)^{-\delta(b<j)}\kappa_{aj}^{ja}\end{pmatrix}\end{array}\right]=0, \tag{9}$$

$$\mathrm{XIII}(a,b):\left[\begin{array}{l}(w_1-w_2^{-1})\left(\kappa_{ba}^{aa}-(w_1w_2^{-1})^{-\delta(b<a)}\kappa_{ab}^{aa}\right)\\-(w_1-w_2)\displaystyle\sum_{j=0}^{N-1}(w_1w_2)^{-\delta(a<j)}\kappa_{bj}^{ja}\end{array}\right]=0, \tag{10}$$

$$\mathrm{XIV}(a,b):\left[\begin{array}{l}(w_1-w_2^{-1})\left(\kappa_{bb}^{ab}-(w_1w_2^{-1})^{-\delta(a<b)}\kappa_{bb}^{ba}\right)\\-(w_1-w_2)\displaystyle\sum_{j=0}^{N-1}(w_1w_2)^{-\delta(b<j)}\kappa_{jb}^{aj}\end{array}\right]=0, \tag{11}$$

where roman letter number I, VI, X, XIII and XIV are corresponding to the grouping in Lemma 4.1. We are assuming that $a,b,c,d\in\{0,1,2,\cdots,N-1\}$ *are all different. We consider* $\mathrm{X}(a,b)$ *only for* $a<b$ *because* $\mathrm{X}(a,b)$ *and* $\mathrm{X}(b,a)$ *are equivalent to each other.*

5. Consequences of Lemma 4.4

We prove some consequences of Lemma 4.4, which are important in the latter discussions.

Lemma 5.1. *Except the diagonal elements, the number of nonzero elements in one column of* $K(z)$ *is at most two.*

This Lemma is directly derived from the following Lemma.

Lemma 5.2. *One of* $(\kappa_\alpha^\beta(w),\kappa_\alpha^\gamma(w),\kappa_\alpha^\delta(w))$ *is zero for mutually different* $\alpha,\beta,\gamma,\delta\in\{0,1,2,\cdots,N-1\}$.

Proof. By Lemma 4.4, we know that, for $a<b<c<d$,

	κ_b^a	κ_c^a	κ_c^b	κ_a^b	κ_a^c	κ_b^c
(i)	1	1	w	1	0	0
(ii)	1	1	0	0	1	1
(iii)	0	0	1	w^{-1}	1	1

	κ_b^a	κ_d^a	κ_d^b	κ_a^b	κ_a^d	κ_b^d
(iv)	1	1	w	1	0	0
(v)	1	1	0	0	1	1
(vi)	0	0	1	w^{-1}	1	1

	κ_c^a	κ_d^a	κ_d^c	κ_a^c	κ_a^d	κ_c^d
(vii)	1	1	w	1	0	0
(viii)	1	1	0	0	1	1
(ix)	0	0	1	w^{-1}	1	1

	κ_c^b	κ_d^b	κ_d^c	κ_b^c	κ_b^d	κ_c^d
(x)	1	1	w	1	0	0
(xi)	1	1	0	0	1	1
(xii)	0	0	1	w^{-1}	1	1

upto a common constant and a common factor.

Concerning the elements in the a-th column, we have, where the asterisk $*$

	κ_a^b	κ_a^c	κ_a^d
(i)	1	0	$*$
(ii)	0	1	$*$
(iii)	w^{-1}	1	$*$
(iv)	1	$*$	0
(v)	0	$*$	1
(vi)	w^{-1}	$*$	1
(vii)	$*$	1	0
(viii)	$*$	0	1
(ix)	$*$	w^{-1}	1

means that we do not know whether the element there is zero or not. We will first show that one of $(\kappa_a^b, \kappa_a^c, \kappa_a^d)$ has to be zero. We remark that at least one of (i), (ii) and (iii) always holds. Similarly, one of (iv), (v) and (vi), and one of (vii), (viii) and (ix) always hold, respectively.

- If (i) holds, Lemma 5.2 is true because $\kappa_a^c = 0$.
- If (ii) holds, it is true because $\kappa_a^b = 0$.
- If (iii) and (iv) hold, it is true because $\kappa_a^d = 0$.
- If (iii) and (v) hold, it is true because $\kappa_a^b = 0$.
- If (iii), (vi) and (vii) hold, it is true because $\kappa_a^d = 0$.
- If (iii), (vi) and (viii) hold, it is true because $\kappa_a^c = 0$.

The case left to investigate here is the case when (iii), (vi) and (ix) hold. If (iii), (vi) and (ix) hold, then

	κ_a^b	κ_a^c	κ_a^d
(iii)	w^{-1}	1	$*$
(vi)	w^{-1}	$*$	1
(ix)	$*$	w^{-1}	1

.

There are only three possible cases of nonzero elements which satisfy (iii), (vi) and (ix) simultaneously. They are

$$(\kappa_a^b, \kappa_a^c, \kappa_a^d) = \begin{cases} (w^{-1}, 1, 0) \\ (w^{-1}, 0, 1) \\ (0, w^{-1}, 1) \end{cases}$$

up to a common factor and a common constant. We can find that Lemma 5.2 is true about the a-th column. The proofs for the b-th, c-th and d-th column are all the same. We omit them. □

Lemma 5.3. *If there are mutually different* $p, q \in \{0, 1, 2, \cdots, N-1\}$ *such that* $\kappa_{qp}^{pq} \neq 0$*, then*

$$\kappa_p^j = 0 \text{ for any } j \neq p, \text{ and } \kappa_q^k = 0 \text{ for any } k \neq q.$$

Proof. By Lemma 4.4, we know that, for mutually different $a < b < c$, that there are only three possibilities among the six elements $\kappa_b^a, \kappa_c^a, \kappa_c^b, \kappa_a^b, \kappa_a^c, \kappa_b^c$, that are

	κ_b^a	κ_c^a	κ_c^b	κ_a^b	κ_a^c	κ_b^c
(i)	1	1	w	1	0	0
(ii)	1	1	0	0	1	1
(iii)	0	0	1	w^{-1}	1	1

,

upto a common constant and a common factor.

We will study the cases when a set $\{a, b, c\} \subset \{0, 1, 2, \cdots, N-1\}$ (a, b, c are mutually different) contains p and q, where $\kappa_q^p \kappa_p^q \neq 0$ and $p < q$.

- If $a < p < q$, namely the case (iii) above

	κ_p^a	κ_q^a	κ_q^p	κ_a^p	κ_a^q	κ_p^q
(iii)	0	0	1	w^{-1}	1	1

,

then $\kappa_p^a(w) = 0$.

- If $p < b < q$, namely the case (ii) above,

	κ_b^p	κ_q^p	κ_q^b	κ_p^b	κ_p^q	κ_b^q
(ii)	1	1	0	0	1	1

,

then $\kappa_p^b(w) = 0$.
- If $p < q < c$, namely the case (i) above,

	κ_q^p	κ_c^p	κ_c^q	κ_p^q	κ_p^c	κ_q^c
(i)	1	1	w	1	0	0

,

then $\kappa_p^c(w) = 0$.

We proved that $\kappa_p^j = \kappa_q^j = 0$ for $j \neq p, q$ if $\kappa_q^p \kappa_p^q \neq 0$. □

6. Type I components of the reflection equation

We consider the type I components (7) of the reflection equation in this section. When we fix $a < b < c < d$ in $\{0, 1, 2, \cdots, N-1\}$, there are 24 type I components. We divide them into 6 subgroups;

$$\begin{array}{ll}
\text{(i)} & \mathrm{I}(a,b,c,d), \mathrm{I}(a,b,d,c), \mathrm{I}(b,a,c,d), \mathrm{I}(b,a,d,c), \\
\text{(ii)} & \mathrm{I}(c,d,a,b), \mathrm{I}(c,d,b,a), \mathrm{I}(d,c,a,b), \mathrm{I}(d,c,b,a), \\
\text{(iii)} & \mathrm{I}(a,c,b,d), \mathrm{I}(a,c,d,b), \mathrm{I}(c,a,b,d), \mathrm{I}(c,a,d,b), \\
\text{(iv)} & \mathrm{I}(b,d,a,c), \mathrm{I}(b,d,c,a), \mathrm{I}(d,b,a,c), \mathrm{I}(d,b,c,a), \\
\text{(v)} & \mathrm{I}(a,d,b,c), \mathrm{I}(a,d,c,b), \mathrm{I}(d,a,b,c), \mathrm{I}(d,a,c,b), \\
\text{(vi)} & \mathrm{I}(b,c,a,d), \mathrm{I}(b,c,d,a), \mathrm{I}(c,b,a,d), \mathrm{I}(c,b,d,a).
\end{array}$$

In each subgroup, only the two equations are independent of the four. We write them down explicitly.

$$\begin{array}{ll}
\text{(i)} & \langle w\kappa_c^a(w), \kappa_d^b(w)\rangle = 0, \\
 & (w_1 - w_2)w_1^{-1}\kappa_{dc}^{ba} = \langle \kappa_c^b(w), \kappa_d^a(w)\rangle, \\
\text{(ii)} & \langle w\kappa_a^c(w), \kappa_b^d(w)\rangle = 0, \\
 & (w_1 - w_2)w_1^{-1}\kappa_{b\ a}^{d\ c} = \langle \kappa_a^d(w), \kappa_b^c(w)\rangle, \\
\text{(iii)} & \langle w\kappa_b^a(w), \kappa_d^c(w)\rangle = 0, \\
 & \langle \kappa_d^a(w), \kappa_b^c(w)\rangle = 0, \\
\text{(iv)} & \langle w\kappa_a^b(w), \kappa_c^d(w)\rangle = 0, \\
 & \langle \kappa_c^b(w), \kappa_a^d(w)\rangle = 0, \\
\text{(v)} & \langle \kappa_c^a(w), \kappa_b^d(w)\rangle = 0, \\
 & (w_1 - w_2)\kappa_{c\ b}^{a\ d} = \langle w\kappa_b^a(w), \kappa_c^d(w)\rangle, \\
\text{(vi)} & \langle \kappa_d^b(w), \kappa_a^c(w)\rangle = 0, \\
 & (w_1 - w_2)\kappa_{a\ d}^{c\ b} = \langle \kappa_d^c(w), w\kappa_a^b(w)\rangle.
\end{array}$$

For example, the relation $w_1\kappa_{c\ d}^{a\ b} - w_2\kappa_{d\ c}^{b\ a} = 0$ can be written in $w\kappa_c^a(w) \sim \kappa_d^b(w)$ in the notation in Definition 4.7. We have proved the following Lemma.

Lemma 6.1. *The type I components of the reflection equation in Lemma 4.1, which consists of $N(N-1)(N-2)(N-3)$ scalar equations, is equivalent to that the following 12 relations hold for any $a < b < c < d$ in $\{0, 1, 2, \cdots, N-1\}$;*

$$w\kappa_c^a(w) \sim \kappa_d^b(w) \tag{12}$$
$$w\kappa_a^c(w) \sim \kappa_b^d(w) \tag{13}$$
$$w\kappa_b^a(w) \sim \kappa_d^c(w) \tag{14}$$
$$w\kappa_a^b(w) \sim \kappa_c^d(w) \tag{15}$$
$$\kappa_d^a(w) \sim \kappa_b^c(w) \tag{16}$$
$$\kappa_c^b(w) \sim \kappa_a^d(w) \tag{17}$$
$$\kappa_c^a(w) \sim \kappa_b^d(w) \tag{18}$$
$$\kappa_d^b(w) \sim \kappa_a^c(w) \tag{19}$$
$$(w_1 - w_2)w_1^{-1}\kappa_{d\ c}^{b\ a} = \langle \kappa_c^b(w), \kappa_d^a(w) \rangle \tag{20}$$
$$(w_1 - w_2)w_1^{-1}\kappa_{b\ a}^{d\ c} = \langle \kappa_a^d(w), \kappa_b^c(w) \rangle \tag{21}$$
$$(w_1 - w_2)\kappa_{c\ b}^{a\ d} = \langle w\kappa_b^a(w), \kappa_c^d(w) \rangle \tag{22}$$
$$(w_1 - w_2)\kappa_{a\ d}^{c\ b} = \langle \kappa_d^c(w), w\kappa_a^b(w) \rangle \tag{23}$$

For example, the equation (20) above gives us the information about the relation between $\kappa_d^a(w)$ and $\kappa_c^b(w)$. If $\kappa_{dc}^{ba} = 0$, we have $\kappa_d^a(w) \sim \kappa_c^b(w)$. If $\kappa_{dc}^{ba} \neq 0$, then $\kappa_{dc}^{ab} \neq 0$. Because $\kappa_{dd}^{ab} \neq 0$, the first case of Lemma 4.4 implies the existence of the nonzero constants C_1 and C_2 such that $C_1 w\kappa_d^a(w) = C_2\kappa_d^b(w)$. Because $\kappa_{cc}^{ab} \neq 0$, also the first case of Lemma 4.4 implies the existence of the nonzero constants C_3 and C_4 such that $C_3 w\kappa_c^a(w) = C_4\kappa_c^b(w)$. Because $\kappa_{cd}^{bb} \neq 0$, also the first case of Lemma 4.4 implies the existence of the nonzero constants C_5 and C_6 such that $C_5\kappa_c^b(w) = C_6\kappa_d^b(w)$. So we have $w\kappa_d^a(w) \sim \kappa_d^b(w)$, $\kappa_d^a(w) \sim \kappa_c^a(w)$ and $w\kappa_c^a(w) \sim \kappa_c^b(w)$. The similar arguments for (21), (22) and (23) lead to

Lemma 6.2. *For any $a < b < c < d \in \{0, 1, 2, \cdots, N-1\}$, we obtain*

(i) If $\kappa_{dc}^{ba} = 0$, then $\kappa_c^b(w) \sim \kappa_d^a(w)$.
If $\kappa_{dc}^{ba} \neq 0$, then $\kappa_{dc}^{ab} \neq 0$, $\kappa_c^b(w) \sim w\kappa_d^a(w)$, $\kappa_c^a(w) \sim \kappa_d^a(w)$ and $\kappa_d^b(w) \sim w\kappa_d^a(w)$.

(ii) *If* $\kappa^{dc}_{ba} = 0$, *then* $\kappa^{c}_{b}(w) \sim \kappa^{d}_{a}(w)$.
If $\kappa^{dc}_{ba} \neq 0$, *then* $\kappa^{cd}_{ba} \neq 0$, $\kappa^{d}_{a}(w) \sim w\kappa^{c}_{b}(w)$, $\kappa^{c}_{a}(w) \sim \kappa^{c}_{b}(w)$ *and* $\kappa^{d}_{b}(w) \sim w\kappa^{c}_{b}(w)$.

(iii) *If* $\kappa^{ad}_{cb} = 0$, *then* $\kappa^{d}_{c}(w) \sim w\kappa^{a}_{b}(w)$.
If $\kappa^{ad}_{cb} \neq 0$, *then* $\kappa^{da}_{cb} \neq 0$, $\kappa^{d}_{c}(w) \sim \kappa^{a}_{b}(w)$, $\kappa^{a}_{c}(w) \sim \kappa^{a}_{b}(w)$ *and* $\kappa^{d}_{c}(w) \sim \kappa^{a}_{b}(w)$.

(iv) *If* $\kappa^{cb}_{ad} = 0$, *then* $w\kappa^{b}_{a}(w) \sim \kappa^{c}_{d}(w)$.
If $\kappa^{cb}_{ad} \neq 0$, *then* $\kappa^{bc}_{ad} \neq 0$, $\kappa^{c}_{d}(w) \sim w^2\kappa^{b}_{a}(w)$, $\kappa^{c}_{a}(w) \sim w\kappa^{b}_{a}(w)$ *and* $\kappa^{b}_{d}(w) \sim w\kappa^{b}_{a}(w)$.

6.1. *Relations between diagonal and non-diagonal elements of K-matrix*

In this subsection, we derive formulas that express

$$\kappa^{ab}_{bc} \quad (a, b \text{ and } c \text{ are mutually different})$$

by

$$\kappa^{ab'}_{cb'} \quad \text{and} \quad \kappa^{b'a}_{b'c} \quad (b' = 0, 1, 2, \cdots, N-1)$$

from type VI, XIII and XIV components in Lemma 4.1.
By interchanging w_1 and w_2 in (10), we have

$$\mathrm{XIII}'(a,b) : \begin{bmatrix} (w_2 - w_1^{-1})\left((w_1 w_2^{-1})^{\delta(c<a)}\kappa^{aa}_{ca} - \kappa^{aa}_{ac}\right) \\ -(w_1 - w_2)\displaystyle\sum_{j=0}^{N-1}(w_1 w_2)^{-\delta(a<j)}\kappa^{aj}_{jc} \end{bmatrix} = 0.$$

When we fix two different integers $a, c \in \{0, 1, 2, \cdots, N-1\}$, we can consider that N equations; $\mathrm{XIII}'(a,c)$, $\mathrm{VI}(a,b,c)$ $(b \neq a, c)$ and $\mathrm{XIV}(a,c)$ are the simultaneous equations for κ^{ab}_{bc} $(b = 0, 1, 2, \cdots, N-1)$.
We write down these equations and solve them after defining some notation.

Definition 6.1.

$$\begin{aligned} A^{c}_{a} &:= w_2(w_1 w_2^{-1})^{\delta(c<a)}\kappa^{aa}_{ca} - w_2\kappa^{aa}_{ac} \\ B^{ac}_{b} &:= w_1(w_1 w_2^{-1})^{\delta(b<c)}\kappa^{ab}_{cb} - w_1(w_1 w_2^{-1})^{-\delta(a<b)}\kappa^{ba}_{bc} \\ &\quad + P(a,b,c)(w_1 - w_2)\kappa^{ab}_{bc} \\ C^{a}_{c} &:= w_1\kappa^{ac}_{cc} - w_1(w_1 w_2^{-1})^{-\delta(a<c)}\kappa^{ca}_{cc} \end{aligned}$$

With the notation above, XIII$'(a,c)$, VI(a,b,c) $(b \neq a,c)$ and XIV(a,c) are written as

$$\mathrm{XIII}'(a,c) : (w_1 - w_2)\sum_{j=0}^{N-1}(w_1w_2)^{-\delta(a<j)}\kappa_{j\,c}^{a\,j} = (1 - w_1^{-1}w_2^{-1})A_a^c, \quad (24)$$

$$\mathrm{VI}(a,b,c) : (w_1 - w_2)\sum_{j=0}^{N-1}(w_1w_2)^{-\delta(b<j)}\kappa_{j\,c}^{a\,j} = (1 - w_1^{-1}w_2^{-1})B_b^{ac}, \quad (25)$$

$$\mathrm{XIV}(a,c) : (w_1 - w_2)\sum_{j=0}^{N-1}(w_1w_2)^{-\delta(c<j)}\kappa_{j\,c}^{a\,j} = (1 - w_1^{-1}w_2^{-1})C_c^a. \quad (26)$$

We will write (24), (25) and (26) in a matrix form.

Definition 6.2.

$$M := \left((w_1w_2)^{-\delta(i<j)}\right)_{i,j=0,1,2\cdots,N-1}$$
$$= \begin{pmatrix} 1 & w_1^{-1}w_2^{-1} & w_1^{-1}w_2^{-1} & \cdots & w_1^{-1}w_2^{-1} \\ 1 & 1 & w_1^{-1}w_2^{-1} & \cdots & w_1^{-1}w_2^{-1} \\ \vdots & \vdots & \ddots & & \vdots \\ 1 & 1 & \cdots & 1 & w_1^{-1}w_2^{-1} \\ 1 & 1 & \cdots & 1 & 1 \end{pmatrix},$$

$$\mathbf{k}(a,c) := {}^t\left(\kappa_{0\,c}^{a\,0}, \kappa_{1\,c}^{a\,1}, \cdots, \kappa_{N-1\,c}^{a\;N-1}\right),$$

$$\mathbf{v}(a,c) := \begin{cases} {}^t\left(B_0^{ac}, \cdots, B_{a-1}^{ac}, A_a^c, B_{a+1}^{ac}, \cdots, B_{c-1}^{ac}, C_c^a, B_{c+1}^{ac}, \cdots, B_{N-1}^{ac}\right) & (a<c) \\ {}^t\left(B_0^{ac}, \cdots, B_{c-1}^{ac}, C_c^a, B_{c+1}^{ac}, \cdots, B_{a-1}^{ac}, A_a^c, B_{a+1}^{ac}, \cdots, B_{N-1}^{ac}\right) & (c<a) \end{cases}.$$

Then the simultaneous equations, (24), (25) and (26), turn into one N-by-N matrix equation

$$(w_1 - w_2)M\mathbf{k}(a,c) = (1 - w_1^{-1}w_2^{-1})\mathbf{v}(a,c). \quad (27)$$

Because the inverse of the matrix M is

$$M^{-1} = = \frac{1}{1-w_1^{-1}w_2^{-1}} \begin{pmatrix} 1 & 0 & 0 & \cdots & 0 & -w_1^{-1}w_2^{-1} \\ -1 & 1 & 0 & 0 & \cdots & 0 \\ 0 & -1 & 1 & 0 & & \vdots \\ \vdots & & \ddots & \ddots & \ddots & 0 \\ 0 & \cdots & & -1 & 1 & 0 \\ 0 & 0 & \cdots & 0 & -1 & 1 \end{pmatrix},$$

the equation (27) is solved as

$$\begin{aligned} (w_1-w_2)\mathbf{k}(a,c)_0 &= (w_1-w_2)\kappa_{0\,c}^{a\,0} = \mathbf{v}(a,c)_0 - w_1^{-1}w_2^{-1}\mathbf{v}(a,c)_{N-1}, \\ (w_1-w_2)\mathbf{k}(a,c)_b &= (w_1-w_2)\kappa_{b\,c}^{a\,b} = -\mathbf{v}(a,c)_{b-1} + \mathbf{v}(a,c)_b \\ & \qquad\qquad \text{for} \quad b = 1,2,\cdots,N-1. \end{aligned}$$

With some care about the order of a, b, and c involved in A_c^a, B_b^{ac} and C_c^a, we have the equivalent conditions to all of the components of types VI, XIII and XIV.

Proposition 6.1. *The following conditions are equivalent to the $N^2(N-1)$ components VI (8), XIII (10) and XIV (11) of the reflection equation (3.1).*

(i) If $0 < a < c \leq N-1$, *then*

$$\begin{aligned} 0 &= \langle \kappa_c^a, w(\kappa_a^a - \kappa_{a-1}^{a-1})\rangle_{w_1w_2}, \\ 0 &= \langle \kappa_c^a, \kappa_{a+1}^{a+1} - \kappa_a^a\rangle_{w_1w_2}, \\ (w_1-w_2)\kappa_{0\,c}^{a\,0} &= \langle \kappa_c^a, w\kappa_0^0 - w^{-1}\kappa_{N-1}^{N-1}\rangle_{w_1w_2}, \\ (w_1-w_2)\kappa_{b\,c}^{a\,b} &= \langle \kappa_c^a, w(\kappa_b^b - \kappa_{b-1}^{b-1})\rangle_{w_1w_2} \quad \textit{for} \quad 1 \leq b \leq a-1, \\ (w_1-w_2)\kappa_{b\,c}^{a\,b} &= w_2\langle \kappa_c^a, \kappa_{b+1}^{b+1} - \kappa_b^b)\rangle_{w_1w_2} \quad \textit{for} \quad a+1 \leq b \leq c-1, \\ (w_1-w_2)\kappa_{b\,c}^{a\,b} &= \langle w\kappa_c^a, \kappa_b^b - \kappa_{b-1}^{b-1})\rangle_{w_1w_2} \quad \textit{for} \quad c+1 \leq b \leq N-1. \end{aligned}$$

(ii) If $a = 0 < c \leq N-1$, *then*

$$\begin{aligned} 0 &= \langle \kappa_c^0, w\kappa_0^0 - w^{-1}\kappa_{N-1}^{N-1}\rangle_{w_1w_2}, \\ 0 &= \langle \kappa_c^0, \kappa_1^1 - \kappa_0^0\rangle_{w_1w_2} \\ (w_1-w_2)\kappa_{b\,c}^{0\,b} &= w_2\langle \kappa_c^0, \kappa_{b+1}^{b+1} - \kappa_b^b)\rangle_{w_1w_2} \quad \textit{for} \quad 1 \leq b \leq c-1 \\ (w_1-w_2)\kappa_{b\,c}^{0\,b} &= \langle w\kappa_c^0, \kappa_b^b - \kappa_{b-1}^{b-1})\rangle_{w_1w_2} \quad \textit{for} \quad c+1 \leq b \leq N-1 \end{aligned}$$

(iii) If $0 \le c < a < N-1$, *then*

$$\begin{aligned}
0 &= \langle \kappa_c^a, \kappa_a^a - \kappa_{a-1}^{a-1} \rangle_{w_1 w_2},\\
0 &= \langle w\kappa_c^a, \kappa_{a+1}^{a+1} - \kappa_a^a \rangle_{w_1 w_2},\\
(w_1 - w_2)\kappa_{b\,c}^{a\,b} &= \langle \kappa_c^a, w(\kappa_{b+1}^{b+1} - \kappa_b^b)\rangle_{w_1 w_2} \quad \textit{for} \quad 0 \le b \le c-1,\\
(w_1 - w_2)\kappa_{b\,c}^{a\,b} &= w_1 \langle \kappa_c^a, \kappa_b^b - \kappa_{b-1}^{b-1})\rangle_{w_1 w_2} \quad \textit{for} \quad c+1 \le b \le a-1,\\
(w_1 - w_2)\kappa_{b\,c}^{a\,b} &= \langle w\kappa_c^a, \kappa_{b+1}^{b+1} - \kappa_b^b)\rangle_{w_1 w_2} \quad \textit{for} \quad a+1 \le b \le N-2,\\
(w_1 - w_2)\kappa_{N-1\,c}^{a\quad N-1} &= \langle w\kappa_c^a, w(w\kappa_0^0 - w^{-1}\kappa_{N-1}^{N-1})\rangle_{w_1 w_2}.
\end{aligned}$$

(iv) If $0 \le c < N-1 = a$, *then*

$$\begin{aligned}
0 &= \langle \kappa_c^{N-1}, w\kappa_0^0 - w^{-1}\kappa_{N-1}^{N-1} \rangle_{w_1 w_2},\\
0 &= \langle \kappa_c^{N-1}, \kappa_{N-1}^{N-1} - \kappa_{N-2}^{N-2} \rangle_{w_1 w_2},\\
(w_1 - w_2)\kappa_{b\quad c}^{N-1\,b} &= \langle \kappa_c^{N-1}, w(\kappa_{b+1}^{b+1} - \kappa_b^b)\rangle_{w_1 w_2} \quad \textit{for} \quad 0 \le b \le c-1,\\
(w_1 - w_2)\kappa_{b\quad c}^{N-1\,b} &= w_1 \langle \kappa_c^{N-1}, \kappa_b^b - \kappa_{b-1}^{b-1})\rangle_{w_1 w_2} \quad \textit{for} \quad c+1 \le b \le N-2.
\end{aligned}$$

7. Determination of the diagonal elements

In this section, we determine the diagonal elements of the $K(z)$ under the assumptions (2) and (3), in the case that there exist integers $p < q$ such that $\kappa_{qp}^{pq} \neq 0$. We remark that, when we specify an integer $p \in \{0, 1, 2, \cdots, N-1\}$, the integer q such that $\kappa_{qp}^{pq} \neq 0$ is uniquely determined by Lemma 5.3.

Definition 7.1. For a solution to the reflection equation $K(z) := \left(z^{-a-b}\kappa_b^a(w)\right)_{a,b=0,1,2,\cdots}$, we define three sets $P, Q, R \subset \{0, 1, 2\cdots, N-1\}$ by

$$\begin{aligned}
P &:= \left\{p \,\middle|\, \text{there exists an integer } q \text{ such that } \kappa_{qp}^{pq} \neq 0 \text{ and } p < q\right\},\\
Q &:= \left\{q \,\middle|\, \text{there exists an integer } p \text{ such that } \kappa_{qp}^{pq} \neq 0 \text{ and } p < q\right\},\\
R &= S \cup T := \{0, 1, 2, \cdots, N-1\} \setminus (P \cup Q),\\
S &:= \left\{s \,\middle|\, \begin{array}{l} \kappa_{rs}^{sr} = 0 \text{ for any } r \neq s, \\ \text{and there exists an element } q \in Q \text{ such that } s < q \end{array}\right\},\\
T &:= \left\{t \,\middle|\, \kappa_{rs}^{tt} = 0 \text{ for any } r \neq t, \text{ and } q < t \text{ for any } q \in Q\right\}.
\end{aligned}$$

The elements of P are denoted by $p_1 < p_2 < \cdots < p_l$, and $q_j \in Q$ denotes the corresponding element $p_j \in P$, which satisfies $\kappa_{q_j p_j}^{p_j q_j} \neq 0$. We call such an element $q \in Q$ the corresponding element to $p \in P$. We remark that p_1 is the least element in $P \cup Q$. The elements of S and T are also denoted by $s_1 < s_2 < \cdots < s_m$ and $t_1 < t_2 < \cdots < t_{m'}$ ($m' = N - 2l - m$), respectively.

The goal of this section is to prove the following theorem.

Theorem 7.1. *If a solution $K(z)$ to the reflection equation (3.1) satisfies the assumptions (2) and (3), then there are integers L and m such that $1 < l < [N/2]$ and $0 \leq m < N - 2l$, and $K(z)$ is similar to the solution $K'(z) = \left(z^{-a-b}\kappa(w)\right)_{a,b=0,1,2,\cdots,N-1}$ which has the following properties*

(i) The subsets P, S, Q and T defined in Definition 7.1 for $K'(z)$ are

$$P = \{0, 1, 2, \cdots, l-1\}, \quad S = \{l, l+1, \cdots, l+m-1\},$$
$$Q = \{l+m, l+m+1, \cdots, 2l+m-1\}, \quad T = \{2l+m, \cdots, N-1\},$$

and the corresponding element to $p \in P$ is $q = 2l + m - p - 1 \in Q$.

(ii) The diagonal elements of $K'(z)$ are described by at most four parameters in $\mathbf{C}$,

$$A_0, \; A_1 \; (A_1 \text{ exists if } \sharp(S) = m > 0),$$
$$A_2 \; (A_2 \text{ exists if } \sharp(T) = N - 2l - m > 0), \; A_3,$$

and some not constantly zero function $f(w)$ as follows,

$$\kappa^p_p(w) = f(w)\left[(A_0 - A_1)w^{-1} - A_2 + A_3\right] \text{ for } 0 \leq p \leq l-1,$$
$$\kappa^s_s(w) = f(w)\left[A_0 w^{-1} - A_2 + A_3 - A_1 w\right] \text{ for } l \leq s \leq l+m-1,$$
$$\kappa^q_q(w) = f(w)\left[-A_2 + A_3 + (A_0 - A_1)w\right] \text{ for } l+m \leq q \leq 2l+m-1,$$
$$\kappa^t_t(w) = f(w)\left[A_3 + (A_0 - A_1)w - A_2 w^2\right] \text{ for } 2l+m \leq t \leq N-1.$$

(iii) For $0 \leq p \leq l-1$ and the corresponding $q \in Q$, $\kappa^p_q(w)$ and $\kappa^q_p(w)$ has the forms

$$\kappa^p_q(w) = u_p f(w)(w - w^{-1}), \quad \kappa^q_p(w) = v_p f(w)(w - w^{-1}),$$

where u_p and v_p are parameters in $\mathbf{C}$.

(iv) Among the parameters there are relations,

$$u_p v_p = u_{p'} v_{p'} \quad (p, p' = 0, 1, \cdots, l-1)$$
$$A_0 A_1 = A_2 A_3 \quad (\text{if } m > 0)$$
$$A_0 A_1 = u_p v_p \quad (p = 0, 1, \cdots, l-1 \text{ and if } m > 0)$$
$$A_2 A_3 = u_p v_p \quad (p = 0, 1, \cdots, l-1 \text{ and if } N - 2l - m > 0)$$

for any $p \in P$. None of u_p, v_p, A_0, A_1, A_2 and A_3 is zero if it exists because of the assumption (3); $\kappa^{p_j\, q_j}_{q_j\, p_j} \neq 0$ for $j = 1, 2, \cdots l$.

And the solution $K'(z)$ which has the properties specified above satisfies all the components of type X *(9) of the reflection equation.*

We will prove Theorem 7.1 step by step.

Proposition 7.1. *There exist four parameters B_p, D_p, u_p, v_p and a not constantly zero function $f_p(w)$ such that*

$$\kappa_p^p(w) = f_p(w)(B_p w^{-1} - D_p), \quad \kappa_q^q(w) = f_p(w)(B_p w - D_p),$$
$$\kappa_q^p(w) = u_p f_p(w)(w - w^{-1}), \quad \kappa_p^q(w) = v_p f_p(w)(w - w^{-1}).$$

Proof. Because of Lemma 5.3, (9) is equivalent to

$$\mathrm{X}(p,q): \begin{bmatrix} (w_1 - w_2^{-1})\left((w_1 w_2^{-1})^{-\delta(q<p)}\kappa_{pq}^{pq} - (w_1 w_2^{-1})^{-\delta(p<q)}\kappa_{qp}^{qp}\right) \\ +(w_1 - w_2)\left(\sum_{j=0}^{N-1}(w_1 w_2)^{-\delta(p<j)}\kappa_{jq}^{qj} - \sum_{j=0}^{N-1}(w_1 w_2)^{-\delta(q<j)} d_{pj}^{jp}\right) \end{bmatrix}$$
$$= \begin{bmatrix} (w_1 - w_2^{-1})\left(\kappa_{pq}^{pq} - (w_1 w_2^{-1})^{-1}\kappa_{qp}^{qp}\right) \\ +(w_1 - w_2)\left(\kappa_{pq}^{qp} + (w_1 w_2)^{-1}\kappa_{qq}^{qq} - d_{pp}^{pp} - \kappa_{pq}^{qp}\right) \end{bmatrix}$$
$$= \begin{bmatrix} (w_1 - w_2^{-1})\left(\kappa_{pq}^{pq} - (w_1 w_2^{-1})^{-1}\kappa_{qp}^{qp}\right) \\ +(w_1 - w_2)\left((w_1 w_2)^{-1}\kappa_{qq}^{qq} - \kappa_{pp}^{pp}\right) \end{bmatrix}$$
$$= \begin{bmatrix} \left(\kappa_p^p(w_1) - \kappa_q^q(w_1)\right)\left(w_2\kappa_p^p(w_2) - w_2^{-1}\kappa_q^q(w_2)\right) \\ -\left(w_1\kappa_p^p(w_1) - w_1^{-1}\kappa_q^q(z_1)\right)\left(\kappa_p^p(w_2) - \kappa_q^q(w_2)\right) \end{bmatrix} = 0$$

in this case. By Lemma 4.3 we have

$$\kappa_p^p(w) = f_p(w)(B_p w^{-1} - D_p), \quad \kappa_q^q(w) = f_p(w)(B_p w - D_p),$$

where B_p, D_p and $f_p(w)$ are constants and a not constantly zero function by the assumption (2). We remark $q \in Q$ is uniquely determined by $p \in P$. We substitute above into $\mathrm{XIII}(p,q)$ and $\mathrm{XIII}(q,p)$,

$$\mathrm{XIII}(p,q): \begin{bmatrix} (w_1 - w_2^{-1})\left(\kappa_{qp}^{pp} - (w_1 w_2^{-1})^{-\delta(q<p)}\kappa_{pq}^{pp}\right) \\ -(w_1 - w_2)\sum_{j=0}^{N-1}(w_1 w_2)^{-\delta(p<j)}\kappa_{qj}^{jp} \end{bmatrix}$$
$$= \begin{bmatrix} (w_1 - w_2^{-1})\left(\kappa_{qp}^{pp} - \kappa_{pq}^{pp}\right) \\ -(w_1 - w_2)\left(\kappa_{qp}^{pp} + (w_1 w_2)^{-1}\kappa_{qq}^{qp}\right) \end{bmatrix} = 0,$$

$$\mathrm{XIII}(q,p): \begin{bmatrix} (w_1 - w_2^{-1})\left(\kappa_{pq}^{qq} - (w_1 w_2^{-1})^{-\delta(p<q)}\kappa_{qp}^{qq}\right) \\ -(w_1 - w_2)\sum_{j=0}^{N-1}(w_1 w_2)^{-\delta(q<j)}\kappa_{pj}^{jq} \end{bmatrix}$$
$$= \begin{bmatrix} (w_1 - w_2^{-1})\left(\kappa_{pq}^{qq} - (w_1 w_2^{-1})^{-1}\kappa_{qp}^{qq}\right) \\ -(w_1 - w_2)\left(\kappa_{pp}^{pq} + \kappa_{pq}^{qq}\right) \end{bmatrix} = 0,$$

then we obtain

$$\kappa_p^p(w_2)\left[\kappa_q^p(w_1)\cdot f_p(w_2)(w_2-w_2^{-1})-f_p(w_1)(w_1-w_1^{-1})\cdot\kappa_q^p(w_2)\right]=0,$$
$$\kappa_q^q(w_2)\left[\kappa_p^q(w_1)\cdot f_p(w_2)(w_2-w_2^{-1})-f_p(w_1)(w_1-w_1^{-1})\cdot\kappa_p^q(w_2)\right]=0,$$

respectively. Because $\kappa_p^p(w)$ and $\kappa_q^q(w)$ are not constantly zero by the assumption (2), we can show the existence of the constants $u_p, v_p \in \mathbf{C}$ which satisfy

$$\kappa_q^p(w)=u_p f_p(w)(w-w^{-1}),\quad \kappa_p^q(w)=v_p f_p(w)(w-w^{-1}).\qquad\square$$

By invoking Lemma 6.1, we will obtain a kind of "exclusion rule" for elements in P and Q. It helps to decrease the number of cases to study in the subsequent discussion. We remark that, if $p\neq p'$, $\kappa_{q\,p}^{p\,q}\neq 0$ $(p<q)$ and $\kappa_{q'\,p'}^{p'\,q'}\neq 0$ $(p'<q')$, then p,q,p' and q' are different from each other owing to Lemma 5.3.

Lemma 7.1. *If $i<j$, then $p_i<p_j<q_j<q_i$ or $p_i<q_i<p_j<q_j$.*

Proof. It is enough for us to prove that the case of $p_i<p_j<q_i<q_j$ cannot happen. If $a_1<a_2<b_1<b_2$, (12) and (13) in Lemma 6.1 say that $C_1w_1w_2\kappa_{b_1\,a_1}^{a_1\,b_1}(w)=C_2\kappa_{b_2\,a_2}^{a_2\,b_2}$ for some constants C_1 and C_2, and (18) and (19) say that $C_3\kappa_{b_1\,a_1}^{a_1\,b_1}=C_4\kappa_{b_2\,a_2}^{a_2\,b_2}$ for some constants C_3 and C_4. One of $\kappa_{b_1\,a_1}^{a_1\,b_1}$ or $\kappa_{b_2\,a_2}^{a_2\,b_2}$ must be zero. $\square$

Lemma 7.2. *If $i<j<k$, then $p_i<p_j<q_j<q_i<p_k<q_k$, $p_i<q_i<p_j<p_k<q_k<q_j$, or $p_i<p_j<p_k<q_k<q_j<q_i$.*

Proof. According to Lemma 7.1, it is enough for us to show that the cases $p_i<q_i<p_j<q_j<p_k<q_k$ and $p_i<p_j<q_j<p_k<q_k<q_i$ are impossible. If $a<b<a'<b'$, then $C_1w_1w_2\kappa_{b\,a}^{a\,b}(w)=C_2\kappa_{b'\,a'}^{a'\,b'}$ for some constants C_1 and C_2. by (14) and (15). If $a<a'<b'<b$, then $C_3\kappa_{b\,a}^{a\,b}=C_4\kappa_{b'\,a'}^{a'\,b'}$ for some constants C_3 and C_4. by (16) and (17). We therefore have, for $a_1<a_2<b_2<a_3<b_3<b_1$,

$$\begin{aligned}C_5\kappa_{b_1\,a_1}^{a_1\,b_1}&=C_6\kappa_{b_2\,a_2}^{a_2\,b_2},\\ C_7\kappa_{b_1\,a_1}^{a_1\,b_1}&=C_8\kappa_{b_3\,a_3}^{a_3\,b_3},\\ C_9w_1w_2\kappa_{b_2\,a_2}^{a_2\,b_2}&=C_{10}\kappa_{b_3\,a_3}^{a_3\,b_3}.\end{aligned}$$

We similarly have, for $a_1 < b_1 < a_2 < b_2 < a_3 < b_3$,

$$\begin{aligned} C_{11} w_1 w_2 \kappa^{a_1 b_1}_{b_1 a_1} &= C_{12} \kappa^{a_2 b_2}_{b_2 a_2}, \\ C_{13} w_1 w_2 \kappa^{a_1 b_1}_{b_1 a_1} &= C_{14} \kappa^{a_3 b_3}_{b_3 a_3}, \\ C_{15} w_1 w_2 \kappa^{a_2 b_2}_{b_2 a_2} &= C_{16} \kappa^{a_3 b_3}_{b_3 a_3}. \end{aligned}$$

In both cases above, at least one of the three; $\kappa^{a_1 b_1}_{b_1 a_1}$, $\kappa^{a_2 b_2}_{b_2 a_2}$, or $\kappa^{a_3 b_3}_{b_3 a_3}$, must be zero. □

Lemma 7.3. *For a solution $K(z)$ to the reflection equation (6), there is an integer l' ($1 \le l' \le l$) such that*

$$\begin{aligned} &p_1 < p_2 < \cdots < p_{l'} < q_{l'} < q_{l-1} < \cdots < q_1 \\ &\quad < p_{l'+1} < p_{l'+2} < \cdots < p_l < q_l < q_{l-1} < \cdots < q_{l'+1}, \end{aligned}$$

where p_j and q_j are elements in P and Q in Definition 7.1, respectively.

Proof. The proof is done by induction on l.

(i) When $l = 1$, it is trivial because we are assuming the existence of $p < q$ such that $\kappa^{pq}_{qp} \neq 0$.
(ii) When $l = 2$, it is true because of Lemma 7.1.
(iii) When the Lemma is true for $l \le k$, then there is an integer m such that

$$\begin{aligned} &p_1 < p_2 < \cdots < p_m < q_m < q_{m-1} < \cdots < q_1 \\ &\quad < p_{m+1} < p_{m+2} < \cdots < p_k < q_k < q_{k-1} < \cdots < q_m. \end{aligned}$$

By Lemma 7.2, we also have that $p_m < p_{k+1} < q_{k+1} < q_m$ or $p_{l''} < q_{l''} < p_{k+1} < q_{k+1}$. In the former this Lemma is true with $l' = m + 1$. In the latter case, we know that $p_1 < q_1 < p_k < q_k < p_{k+1} < q_{k+1}$ is impossible by Lemma 7.2. This Lemma is true with $l' = m$ because $p_k < p_{k+1} < q_{k+1} < q_k$ by Lemma 7.1. □

Corollary 7.1. *For a solution $K(z)$ to the reflection equation (6) there is a similar solution $K'(z)$ to $K(z)$, whose P and Q defined in Definition 7.1 become*

$$\begin{aligned} &P = \{p_j \,|1 = p_1 < p_2 < \cdots < p_l\}, \\ &Q = \{q_j \,|q_l < q_{l-1} < \cdots < q_1\}, \\ &p_1 < p_2 < \cdots < p_l < q_l < q_{l-1} < \cdots < q_1, \\ &q_l - p_l \le (N-1) - q_1. \end{aligned}$$

Proof. For any solution $K(z)$, there is an integer l such that

$$\begin{aligned} &p_1 < p_2 < \cdots < p_{l'} < q_{l'} < q_{l'-1} < \cdots < q_1 \\ &\quad < p_{l'+1} < p_{l'+2} < \cdots < p_l < q_l < q_{l-1} < \cdots < q_{l'+1} \end{aligned}$$

by Lemma 7.3. We define

$$\alpha = \begin{cases} p_1 & \text{when} \quad q_{l'} - p_{i'} \leq N - 1 - q_1 + p_1, \\ p_1 + l + q_{l'} - p_{l'} & \text{when} \quad q_{l'} - p_{i'} \leq N - 1 - q_1 + p_1, \end{cases}.$$

Then $h^{-\alpha}K(z)h^{\alpha}$ has the desired property. □

Remark 7.1. If $p < p'$, $\kappa_{qp}^{pq} \neq 0$ and $\kappa_{q'p'}^{p'q'} \neq 0$, then it is enough for us to consider only the case $p < p' < q' < q$ owing to Corollary 7.1.

From now on, we only deal with $K(z)$, satisfying Corollary 7.1.

We will show that the function $f_p(w)$ and the constants B_p and D_p are independent of $p \in P$.

Lemma 7.4. *If $p < p' < q' < q$, $\kappa_{qp}^{pq} \neq 0$ and $\kappa_{q'p'}^{p'q'} \neq 0$, then*

$$\begin{aligned} &\kappa_p^p(w) = \kappa_{p'}^{p'}(w) = f(w)(Bw^{-1} - D), \quad \kappa_q^q(w) = \kappa_{q'}^{q'}(w) = f(w)(Bw - D). \\ &\kappa_q^p(w) = u_p f(w)(w - w^{-1}), \quad \kappa_p^q(w) = v_p f(w)(w - w^{-1}), \\ &\kappa_{q'}^{p'}(w) = u_{p'} f(w)(w - w^{-1}), \quad \kappa_{p'}^{q'}(w) = v_{p'} f(w)(w - w^{-1}). \end{aligned}$$

Proof. In Proposition 7.1 we obtained, for $p, p' \in P$ $(p < p')$

$$\begin{aligned} &\kappa_q^p(w) = u_p f_p(w)(w - w^{-1}), \quad \kappa_p^q(w) = v_p f_p(w)(w - w^{-1}), \\ &\kappa_{q'}^{p'}(w) = u_{p'} f_{p'}(w)(w - w^{-1}), \quad \kappa_{p'}^{q'}(w) = v_{p'} f_{p'}(w)(w - w^{-1}), \end{aligned}$$

where $q, q' \in Q$ are the corresponding elements to p and p', respectively. Because $p < p'$, we only have to consider the case of $p < p' < q' < q$. The equations (16) and (17) in Lemma 6.1 implies the existence of nonzero constants C and C' such that

$$\begin{aligned} C\kappa_q^p(w_1)\kappa_p^q(w_2) &= C'\kappa_{q'}^{p'}(w_1)\kappa_{p'}^{q'}(w_2). \\ Cu_p v_p f_p(w_1) f_p(w_2) &= C' u_{p'} v_{p'} f_{p'}(w_1) f_{p'}(w_2). \end{aligned}$$

Without any loss of generalities we can retake the constants $u_p C v_p$, $u_{p'}$ and $v_{p'}$ to satisfy $Cu_p v_p = C' u_{p'} v_{p'}$, then we obtain $f(w) := f_p(w) = f_{p'}(w)$ for any $p, p' \in P$. When we substitute $\kappa_p^p(w)$, $\kappa_q^q(w)$, $\kappa_{p'}^{p'}(w)$, $\kappa_{q'}^{q'}(w)$ and $\kappa_{q'}^{p'}(w)$ in Proposition 7.1 into $\mathrm{VI}(p', p, q')$

$$\begin{aligned}
\mathrm{VI}(p',p,q') \ : \ & \left[\begin{array}{l} (w_1 - w_2^{-1}) \left((w_1 w_2^{-1})^{-\delta(p<p')} \kappa_{pq'}^{pp'} - (w_1 w_2^{-1})^{-\delta(p<q')} \kappa_{q'p}^{p'p} \right) \\ +(w_1 - w_2) \displaystyle\sum_{j=0}^{N-1} (w_1 w_2)^{-\delta(p<j)} \kappa_{j\ q'}^{p'\,j} \end{array}\right] \\
&= \left[\begin{array}{l} (w_1 - w_2^{-1}) \left(w_1^{-1} w_2 \kappa_{pq'}^{pp'} - w_1^{-1} w_2 \kappa_{q'p}^{p'p} \right) \\ +(w_1 - w_2) \left((w_1 w_2)^{-1} \kappa_{p'q'}^{p'p'} + (w_1 w_2)^{-1} \kappa_{q'q'}^{p'q'} \right) \end{array}\right] \\
&= \frac{u_{p'} f_p(w_1) f_p(w_2)(1 - w_1 w_2)(w_1 - w_2)}{w_1^2 w_2^2} \\
&\quad \times [(B_p - B_{p'})(w_1 w_2 + 1) - (D_p - D_{p'})(w_1 + w_2)] = 0,
\end{aligned}$$

we have

$$B := B_p = B_{p'}, \quad D := D_p = D_{p'}.$$

because we are assuming that $\kappa_{q'\,p'}^{p'\,q'} \neq 0$, namely $u_{p'} v_{p'} \neq 0$. □

We determine $\kappa_r^r(w)$ for $r \in R = S \cup T$ in Definition 7.1 by $\mathrm{VI}(p, r, q)$ (8). Since $\kappa_q^a(w) = 0$ for $a \neq q \in Q$ by Lemma 5.3, we obtain

$$\mathrm{VI}(p,r,q) : \left[\begin{array}{l} (w_1 - w_2^{-1}) \begin{pmatrix} (w_1 w_2^{-1})^{-\delta(p<r)} \kappa_{rq}^{rp} \\ \quad -(w_1 w_2^{-1})^{-\delta(r<q)} \kappa_{qr}^{pr} \end{pmatrix} \\ +(w_1 - w_2) \begin{pmatrix} (w_1 w_2)^{-\delta(r<p)} \kappa_{pq}^{pp} \\ \quad +(w_1 w_2)^{-\delta(r<q)} \kappa_{qq}^{pq} \end{pmatrix} \end{array}\right] = 0. \tag{28}$$

When we substitute the results of Lemma 7.4 into (28), we have the following Lemma.

Lemma 7.5. *If $\kappa_{qp}^{pq} \neq 0$ and $r \in R$, then*

$$\kappa_r^r(w) = \begin{cases} f(w) \left(Bw - D + E_r w(w - w^{-1}) \right) & (p < q < r), \\ f(w) \left(Bw^{-1} - D + E_r (w - w^{-1}) \right) & (p < r < q), \\ f(w) \left(Bw^{-1} - Dw^{-2} + E_r w^{-1}(w - w^{-1}) \right) & (r < p < q), \end{cases} \tag{29}$$

where B, D and $f(w)$ are specified in Lemma 7.4, and where E_r ($r \in R$) is a constant.

Proof. The proof will be done by case-by-case study according to the order among p, q and r.

(proof for the case $p < q < r$) After substituting the results in Lemma 7.4 and $\kappa_r^r(w) = f(w)e_r(w)$ with $e_r(w)$ meromorphic into (28), we have

$$u_p w_1^{-1} w_2^{-1} f(w_1) f(w_2) \times \begin{bmatrix} (w_1^2-1)\begin{bmatrix} -w_1\left(w_2 e_r(w_2) - B + Dw_2\right) \\ +\left(e_r(w_2) - Bw_2 + Dw_2^2\right) \end{bmatrix} \\ +\begin{bmatrix} w_2\left(w_1 e_r(w_1) - B + Dw_1\right) \\ -\left(e_r(w_1) - Bw_1 + Dw_1^2\right) \end{bmatrix}(w_2^2-1). \end{bmatrix} = 0 \qquad (30)$$

Because $\kappa_{qp}^{pq} \neq 0$, we have $u_p f(w) \neq 0$. The equation (30) becomes

$$(w_1^2-1)\begin{bmatrix} w_1\left(w_2 e_r(w_2) - B + Dw_2\right) \\ -\left(e_r(w_2) - Bw_2 + Dw_2^2\right) \end{bmatrix} = \begin{bmatrix} w_2\left(w_1 e_r(w_1) - B + Dw_1\right) \\ -\left(e_r(w_1) - Bw_1 + Dw_1^2\right) \end{bmatrix}(w_2^2-1). \qquad (31)$$

Since the right-hand side of (31) is holomorphic in w_2, the left-hand side of (31) is holomorphic in w_2. Furthermore, because the right-hand side of (31) is a polynomial of degree 3 in w_2, the left-hand side of (31) is also a polynomial of degree 3 in w_2. We especially know that $w_2 e_r(w_2)$ and $e_r(w_2)$ contained in the left-hand side of (31) are polynomials of degree 3 in w_2. This in turn implies that $e_r(w)$ has the form

$$e_r(w) = E_r w^2 + F_r w + G_r, \qquad E_r, F_r, G_r \in \mathbf{C}.$$

By (31), both of $w_2 e_r(w_2) - B + Dw_2$ and $e_r(w_2) - Bw_2 + Dw_2^2$ in the left-hand side must have the factor $w_2^2 - 1$ in the right-hand side. It leads to that $F_r - B = 0$ and $E_r + G_r + D = 0$. Then we obtain

$$\begin{aligned} \kappa_r^r(w) &= f(w)e_r(w) = f(w)\left(E_r w^2 + Bw - D - E_r\right) \\ &= E_r f(w) w(w - w^{-1}) + \kappa_q^q(w). \end{aligned}$$

It is easy to check that this $e_r(w)$ satisfies (31).

(proof for the case $p < r < q$) The equation VI(p, r, q) (28) turns into

$$(w_1^2-1)\begin{bmatrix} w_1(w_2^2 e_r(w) - Bw_2 + D) \\ -(w_2 e_r(w_2) - Bw_2^2 + Dw_2) \end{bmatrix} = \begin{bmatrix} w_2(w_1 e_r(w_1) - Bw_1 + Dw_1^2) \\ -(w_1 e_r(w_1) - Bw_1^2 + Dw_1^3) \end{bmatrix}(w_2^2-1) \qquad (32)$$

Because the right-hand side is a polynomial in w_2 of degree 4, we know that both of $w_2^2 e_r(w_2)$ and $w_2 e_r(w_2)$ in the left-hand side are also polynomials of degree 3. It implies that $e_r(w)$ must have the form

$$e_r(w) = E_r w + F_r + G_r w^{-1}, \qquad E_r, F_r, G_r \in \mathbf{C}.$$

Since both of $w_2^2 e_r(w_2)$ and $w_2 e_r(w_2)$ in the left-hand side have the factor $w_2^2 - 1$ contained in the right-hand side, we finally have the result that

$$\begin{aligned}\kappa_r^r(w) &= f(w)e_r(w) = f(w)\left(E_r w - D + (B - E_r)w^{-1}\right)\\ &= E_r f(w)(w - w^{-1}) + \kappa_p^p(w).\end{aligned}$$

It is easy to check that this $e_r(w)$ satisfies (32).

(proof for the case $r < p < q$) The discussion is similar to above two cases. The equation VI(p, r, q) (28) turns into

$$\begin{aligned}&(w_1^2 - 1)\begin{bmatrix} w_1(w_2^3 e_r(w) - Bw_2^2 + Dw_2) \\ -(w_2^2 e_r(w_2) - Bw_2^3 + Dw_2^2) \end{bmatrix}\\ &= \begin{bmatrix} w_2(w_1^3 e_r(w_1) - B + Dw_1) \\ -(w_1^2 e_r(w_1) - Bw_1 + Dw_1^2) \end{bmatrix}(w_2^2 - 1) \qquad (33)\end{aligned}$$

Because the right-hand side of (33) is a polynomial in w_2 of degree 3, both of $w_2^3 e_r(w)$ and $w_2^2 e_r(w)$ in the left-hand side are also polynomials in w_2 of degree 3. It implies that $e_r(w)$ must have the form

$$e_r(w) = E_r + F_r w^{-1} + G_r w^{-2}, \qquad E_r, F_r, G_r \in \mathbf{C}.$$

Since $w_2^3 e_r(w) - Bw_2^2 + Dw_2$ and $w_2^2 e_r(w_2) - Bw_2^3 + Dw_2^2$ in the left-hand side of (33) have the factor $w_2^2 - 1$, we have

$$\begin{aligned}\kappa_r^r(z) &= f(w)\left(E_r + Bw^{-1} - (D + E_r)w^{-2}\right)\\ &= E_r f(w)w^{-1}(w - w^{-1}) + w^{-2}\kappa_q^q(w).\end{aligned}$$

It is easy to check that this $e_r(w)$ satisfies (33). □

We have used X(p, q) ($p \in P$ and $q \in Q$ is the corresponding element to p) to prove Proposition 7.1. Other components of type X (9) yield relations among the parameters B, D, u_p, v_p ($p \in P$) and E_r ($r \in R = S \cup T$).

Proposition 7.2. *For $p \in P$, corresponding $q \in Q$ and $r \in R = S \cup T$, X(p, r), X(q, r), X(r, p) and X(r, q) are equivalent to the following conditions.*

(i) If $p < q < r$ or $r < p < q$, then $(D + E_r)E_r - u_p v_p = 0$.
(ii) If $p < r < q$, then $(E_r - B)E_r - u_p v_p = 0$.

Proof. The proof is done separately in each three cases $p < q < r$, $p < r < q$ and $r < p < q$. We remark again that we only have to consider X(a, b) with $a < b$ (Proposition 4.1).

(i) In the case of $p < q < r$, substituting

$$\kappa_p^p(w) = f(w)(Bw^{-1} - D), \quad \kappa_p^p(w) = f(w)(Bw^{-1} - D),$$
$$\kappa_r^r(w) = f(w)\left(Bw - D + E_r w(w - w^{-1})\right),$$
$$\kappa_q^p(w) = u_p f(w)(w - w^{-1}), \quad \kappa_p^q(w) = v_p f(w)(w - w^{-1}),$$

into X(p, r) and X(q, r)

$$\begin{aligned} \mathrm{X}(p,r) : &\begin{bmatrix} (w_1 - w_2^{-1})\left[\kappa_{pr}^{pr} - w_{12}^{-1}\kappa_{rp}^{rp}\right] \\ +(w_1 - w_2)\left[w_1^{-1}w_2^{-1}\kappa_{rr}^{rr} - \kappa_{pp}^{pp} - \kappa_{pq}^{qp}\right] \end{bmatrix} \\ &= f(w_1)f(w_2)(w_1 - w_2)(w_1 - w_1^{-1})(w_2 - w_2^{-1}) \\ &\qquad \times [(D + E)E - u_p v_p] = 0, \\ \mathrm{X}(q,r) : &\begin{bmatrix} (w_1 - w_2^{-1})\left[\kappa_{qr}^{qr} - w_{12}^{-1}\kappa_{rq}^{rq}\right] \\ +(w_1 - w_2)\left[w_1^{-1}w_2^{-1}\kappa_{rr}^{rr} - \kappa_{qq}^{qq} - \kappa_{qp}^{pq}\right] \end{bmatrix} \\ &= f(w_1)f(w_2)(w_1 - w_2)(w_1 - w_1^{-1})(w_2 - w_2^{-1}) \\ &\qquad \times [(D + E_r)E_r - u_p v_p] = 0, \end{aligned}$$

we obtain $(D + E_r)E_r - u_p v_p = 0$.

(ii) In the case of $p < r < q$, substituting

$$\kappa_p^p(w) = f(w)(Bw^{-1} - D), \quad \kappa_p^p(w) = f(w)(Bw^{-1} - D),$$
$$\kappa_r^r(w) = f(w)\left(Bw^{-1} - D + E(w - w^{-1})\right),$$
$$\kappa_q^p(w) = x_{pq} f(w)(w - w^{-1}), \quad \kappa_p^q(w) = x_{qp} f(w)(w - w^{-1}),$$

into

$$\begin{aligned} \mathrm{X}(p,r) : &\begin{bmatrix} (w_1 - w_2^{-1})\left[\kappa_{pr}^{pr} - w_{12}^{-1}\kappa_{rp}^{rp}\right] \\ +(w_1 - w_2)\left[w_1^{-1}w_2^{-1}\kappa_{rr}^{rr} - \kappa_{pp}^{pp} - w_1^{-1}w_2^{-1}\kappa_{pq}^{qp}\right] \end{bmatrix} \\ &= f(w_1)f(w_2)(w_1 - w_2)(w_1 - w_1^{-1})(w_2 - w_2^{-1}) \\ &\qquad \times [(E_r - B)E_r - u_p v_p] = 0, \\ \mathrm{X}(r,q) : &\begin{bmatrix} (w_1 - w_2^{-1})\left[\kappa_{rq}^{rq} - w_{12}^{-1}\kappa_{qr}^{qr}\right] \\ +(w_1 - w_2)\left[\kappa_{pq}^{qp} + w_1^{-1}w_2^{-1}\kappa_{qq}^{qq} - \kappa_{rr}^{rr}\right] \end{bmatrix} \\ &= -f(w_1)f(w_2)(w_1 - w_2)(w_1 - w_1^{-1})(w_2 - w_2^{-1}) \\ &\qquad \times [(E_r - B)E_r - u_p v_p] = 0, \end{aligned}$$

we obtain $(E_r - B)E_r - u_p v_p = 0$.

(iii) In the case of $r < p < q$, substituting

$$\kappa_p^p(w) = f(w)(Bw^{-1} - D), \quad \kappa_p^p(w) = f(w)(Bw^{-1} - D),$$
$$\kappa_r^r(w) = f(w)\left(w^{-2}(Bw - D) + E_r w^{-1}(w - w^{-1})\right),$$
$$\kappa_q^p(w) = u_p f(w)(w - w^{-1}), \quad \kappa_p^q(w) = v_p f(w)(w - w^{-1}),$$

into

$$\begin{aligned} X(r,p) : & \begin{bmatrix} (w_1 - w_2^{-1}) \left[\kappa_{rp}^{rp} - w_{12}^{-1} \kappa_{pr}^{pr}\right] \\ +(w_1 - w_2) \left[w_1^{-1} w_2^{-1} \kappa_{pp}^{pp} + w_1^{-1} w_2^{-1} \kappa_{qp}^{pq} - \kappa_{rr}^{rr}\right] \end{bmatrix} \\ & = -f(w_1) f(w_2)(w_1 - w_2)(w_1 - w_1^{-1})(w_2 - w_2^{-1}) \\ & \qquad \times \left[(D + E_r)E_r - u_p v_p\right] = 0, \\ X(r,q) : & \begin{bmatrix} (w_1 - w_2^{-1}) \left[\kappa_{rq}^{rq} - w_{12}^{-1} \kappa_{qr}^{qr}\right] \\ +(w_1 - w_2) \left[w_1^{-1} w_2^{-1} \kappa_{qq}^{qq} + w_1^{-1} w_2^{-1} \kappa_{qp}^{pq} - \kappa_{rr}^{rr}\right] \end{bmatrix} \\ & = -f(w_1) f(w_2)(w_1 - w_2)(w_1 - w_1^{-1})(w_2 - w_2^{-1}) \\ & \qquad \times \left[(D + E_r)E_r - u_p v_p\right] = 0, \end{aligned}$$

we obtain $(D + E_r)E_r - u_p v_p = 0$. □

We will have another "exclusion rule" for elements in P, Q and $R = S \cup T$ by Lemma 7.5 and Proposition 7.2.

Lemma 7.6. *If $1 \leq i < j \leq l$ and $r \in R$, then $p_i < p_j < r < q_j < q_i$ or $p_i < p_j < q_j < q_i < r$, where $p_i, p_j \in P$, $q_i, q_j \in Q$ and R are specified in Corollary 7.1.*

Proof. It is enough for us to show that neither $p_i < r < p_j < q_j < q_i$ nor $p_i < p_j < q_j < r < q_i$ occurs. If $p_i < r < p_j < q_j < q_i$, then both $p_i < r < q_i$ and $r < p_j < q_j$ hold. Lemma 7.5 tells us that

$$\begin{aligned} \kappa_r^r(w) &= f(w) \left[Bw^{-1} - D + E_r(w - w^{-1})\right] \\ &= f(w) \left[Bw^{-1} - Dw^{-2} + E_r w^{-1}(w - w^{-1})\right], \end{aligned}$$

namely that $D = 0$ and $E_r = 0$ because $f(w) \neq 0$ by the assumption (2). By Proposition 7.2, $E_r = 0$ implies $u_{p_i} v_{p_i} = 0$, but it is in contradiction with $\kappa_{q_i\, p_i}^{p_i\, q_i} = 0$. The proof that $p_i < p_j < q_j < r < q_i$ is impossible is all the same. □

We can now have the elaborate version of Corollary 7.1.

Proposition 7.3. *For any solution $K(z)$ to the reflection equation (6) there is a similar solution $K'(z)$ to $K(z)$, whose P, Q, S and T in Definition*

7.1 are characterized by two integers l and m such that

$$
\begin{aligned}
&1 \le l \le [N/2], \qquad 0 \le m \le N-2l-m,\\
&P = \{p_j, 1 \le j \le l \,|\, p_j = j-1\},\\
&S = \{s_j, 1 \le j \le m \,|\, s_j = l+j-1\},\\
&Q = \{q_j, 1 \le j \le l \,|\, q_j = 2l+m-j\},\\
&T = \{t_j, 1 \le j \le N-2l-m \,|\, t_j = 2l+m+j-1\}.
\end{aligned}
$$

We remark that $0 \le m \le N-2l-m$ implies $\sharp(S) \le \sharp(T)$.

Proof. The proof is directly obtained from Definition 7.1, Lemma 7.3 and Lemma 7.6. □

By invoking Lemma 7.5, Proposition 7.3 restricts the forms of $\kappa_r^r(w)$ for $r \in R = S \cup T$.

Lemma 7.7.

(i) $\kappa_s^s(w) = f(w)\left[Bw^{-1} - D + E_s(w - w^{-1})\right]$ *for* $s \in S$.
(ii) $\kappa_t^t(w) = f(w)\left[Bw - D + E_t w(w - w^{-1})\right]$ *for* $t \in T$.

Concerning the type X components (9) of the reflection equation (6), we have used

- $\mathrm{X}(p,q)$ ($q \in Q$ is the corresponding element to $p \in P$) in Proposition 7.1.
- $\mathrm{X}(p,r)$, $\mathrm{X}(q,r)$, $\mathrm{X}(r,q)$ ($p \in P$, $q \in Q$ and $r \in R$) in Proposition 7.2.

Other type X components, $\mathrm{X}(r,r')$, $\mathrm{X}(p,p')$, $\mathrm{X}(p,q')$ and $\mathrm{X}(q,q')$ for $p < p' \in P$, $q' < q \in Q$ and $r < r' \in R$. yield relations among parameters B, D u_p, v_p, ($p \in P$) and E_r ($r \in R = S \cup T$).

Lemma 7.8.

(i) If $m = \sharp(S) \ge 1$, *then* $F := E_s = E_{s'}$ *for* $s, s' \in S$.
(ii) If $N - 2l - m = \sharp(T) \ge 1$, *then* $G := E_t = E_{t'}$ *for* $t, t' \in T$.
(iii) If $m = \sharp(S) \ge 1$ *and* $N - 2l - m = \sharp(T) \ge 1$,
then $(F - B)F - (G + D)G = 0$ *for* $s \in S$ *and* $t \in T$.
(iv) If $l = \sharp(P) \ge 2$, *then* $u_p v_p - u_{p'} v_{p'} = 0$ *for* $p, p' \in P$.

Proof. We first consider $\mathrm{X}(r, r')$ (9) for $r, r' \in R = S \cup T$ $(r < r')$. If $r < r'$, then $\mathrm{X}(r, r')$ (9) becomes

$$\begin{bmatrix} (w_1 - w_2^{-1}) \left(\kappa_{r\,r'}^{r\,r'} - w_{12}^{-1} \kappa_{r'\,r}^{r'\,r}\right) \\ +(w_1 - w_2) \left(w_1^{-1} w_2^{-1} \kappa_{r'\,r'}^{r'\,r'} - \kappa_{r\,r}^{r\,r}\right) \end{bmatrix} = 0 \tag{34}$$

because $\kappa_{a\,s}^{s\,a} = \kappa_{s\,a}^{a\,s} = 0$ for $a \neq a$.

(i) If $s < s' \in S$, then

$$\kappa_s^s(w) = f(w) \left[Bw^{-1} - D + E_s(w - w^{-1})\right],$$
$$\kappa_s^s(w) = f(w) \left[Bw^{-1} - D + E_s(w - w^{-1})\right],$$

by Lemma 7.7. When we substitute them into (34), we obtain

$$f(w_1)f(w_2)\frac{(w_1 - w_2)(w_1 - w_1^{-1})(w_2 - w_2^{-1})}{w_1 w_2}$$
$$\times (E_s - E_{s'})(B - E_{s'} - E_s w_1 w_2) = 0.$$

Because the function $f(w)$ is not constantly zero by the assumption (2), we have $(E_s - E_{s'})(B - E_{s'} - E_s w_1 w_2) = 0$. If $B - E_{s'} - E_s w_1 w_2 = 0$, namely $B - E_s = E_s = 0$, then $u_p v_p = 0$ by Proposition 7.2. But it is in contradiction with $\kappa_{qp}^{pq} \neq 0$ for $p \in P$. We conclude that $F := E_s = E_{s'}$ for any $s, s' \in S$.

(ii) If $t < t' \in S$, then

$$\kappa_t^t(w) = f(w) \left[Bw - D + E_t w(w - w^{-1})\right],$$
$$\kappa_{t'}^{t'}(w) = f(w) \left[Bw - D + E_{t'} w(w - w^{-1})\right],$$

by Lemma 7.7. When we substitute them into (34), we obtain

$$f(w_1)f(w_2)\frac{(w_1 - w_2)(w_1 - w_1^{-1})(w_2 - w_2^{-1})}{w_1 w_2}$$
$$\times (E_t - E_{t'})(E_{t'} w_1 w_2 + E_t + D) = 0.$$

The similar arguments just above leads us to that $G := E_t = E_{t'}$ for any $t, t' \in T$.

(iii) If $s \in S$ and $t \in T$, then (34) becomes

$$f(w_1)f(w_2)(w_1 - w_2)(w_1 - w_1^{-1})(w_2 - w_2^{-1})((G + D)G - (F - B)F) = 0$$

(iv) When we substitute

$$\kappa^p_p(w) = \kappa^{p'}_{p'} = f(w)(Bw^{-1} - D)$$
$$\kappa^q_q(w) = \kappa^{q'}_{q'} = f(w)(Bw - D)$$
$$\kappa^p_q(w) = u_p f(w)(w - w^{-1}), \quad \kappa^q_p(w) = v_p f(w)(w - w^{-1}),$$
$$\kappa^{p'}_{q'}(w) = u_{p'} f(w)(w - w^{-1}), \quad \kappa^{q'}_{p'}(w) = v_{p'} f(w)(w - w^{-1})$$

into $\mathrm{X}(p,p')$, $\mathrm{X}(p,q')$, $\mathrm{X}(q,q')$, the terms contained the diagonal elements all vanish. We obtain $u_p v_p - u_{p'} v_{p'} = 0$. □

Proof of Theorem 7.1.

(i) Proposition 7.3.
(ii) We define $A_0 := B - F$, $A_1 := -F$, $A_2 := -G$, and $A_3 := -D - G$. Lemma 7.4, Lemma 7.7 and Lemma 7.8 give the results.
(iii) Lemma 7.4.
(iv) Proposition 7.2 and Lemma 7.8.

And we have checked at each step, the diagonal elements together with $\kappa^p_q(w)$ and $\kappa^q_p(w)$ satisfies $\mathrm{X}(a,b)$ ($a, b \in \{0, 1, 2, \cdots, N-1\} = P \cup S \cup Q \cup T$, $a < b$);

- for $a = p \in P$ and $b = q \in Q$, where q is the corresponding element to p, at Proposition 7.1,
- for $a = p \in P$ and $b = q' \in Q$ at Lemma 7.8,
- for ($a = p \in P$ and $b = p' \in P$) and ($a = q \in Q$ and $b = q' \in Q$) at Lemma 7.8,
- for ($a = p \in P$ and $b = r \in R = S \cup T$), ($a = q \in P$ and $b = t \in T$) and ($a = s \in S$ and $b = r \in Q$) at Proposition 7.2,
- for ($a = r \in R = S \cup T$ and $b = r' \in R = S \cup T$) at Lemma 7.8. □

8. Determination of the off diagonal elements

We prove the following theorem in this section.

Theorem 8.1. *Let $K(z)$ be a solution to the reflection equation (6) whose diagonal elements are specified in Theorem 7.1. If $a, b, c \in \{0, 1, 2, \cdots, N-1\}$ be three different integers, then*

$$\kappa^{a\,b}_{b\,c} \equiv 0$$

except the cases

(i) $(a, b, c) = (p_1, q_1, t)$ *for* $t \in T$,
(ii) $(a, b, c) = (p_l, q_l, s)$ *for* $s \in S$,
(iii) $(a, b, c) = (q_1, p_1, t)$ *for* $t \in T$,
(iv) $(a, b, c) = (q_l, p_l, s)$ *for* $s \in S$,

where the integer $l = \sharp(P)$ *is specified in Theorem 7.1.*

Lemma 5.3 restrict the range for us to study.

Lemma 8.1. *If* $a, b, c \in \{0, 1, 2, \cdots, N-1\}$ *be three different integers, then*

$$\kappa^{a\,b}_{b\,c} \equiv 0$$

except the cases

(i) $(a, b, c) = (p_j, q_j, r'')$ *for* $r'' \in S \cup T$.
(ii) $(a, b, c) = (q_j, p_j, r'')$ *for* $r'' \in S \cup T$,
(iii) $(a, b, c) = (r, r', r'')$ *for* $r, r', r'' \in T$,
(iv) $(a, b, c) = (p, r', r'')$ *for* $p \in P$ *and* $r', r'' \in S \cup T$,
(v) $(a, b, c) = (q, r', r'')$ *for* $q \in Q$ *and* $r', r'' \in S \cup T$,

where $q_j \in Q$ *is the corresponding to* $p_j \in P$.

Proof.

(i) We will show that $\kappa^{a\,b}_{b\,c} \equiv 0$ when $c \in P \cup Q$. If $c = p_j \in P$, then $\kappa^{b}_{p_j}(w) \neq 0$ $(b \neq p_j)$ only when $b = q_j$ by Lemma 5.3. If $a \neq p_j, q_j$, then $\kappa^{a}_{q_j}(w) \equiv 0$ also by Lemma 5.3, namely, that $\kappa^{a\,b}_{b\,p} \equiv 0$ for $p \in P$. We can prove that $\kappa^{a\,b}_{b\,q} \equiv 0$ for $q \in Q$ in a similar way.
(ii) The cases left to consider are

$$\kappa^{r\,p'}_{p'\,r''}, \kappa^{r\,q'}_{q'\,r''}, \kappa^{p\,p'}_{p'\,r''}, \kappa^{q\,q'}_{q'\,r''}, \kappa^{p\,q'}_{q'\,r''}, \kappa^{q\,p'}_{p'\,r''}, \kappa^{r\,r'}_{r'\,r''}, \kappa^{p\,r'}_{r'\,r''}, \kappa^{q\,r'}_{r'\,r''},$$

for $p, p' \in P$ $(p \neq p')$, $q \neq q' \in Q$ $(q \neq q')$ and mutually different $r, r', r'' \in S \cup T$. The first four cases are zero because $\kappa^{r}_{p'}(w)$, $\kappa^{r}_{q'}$, $\kappa^{p}_{p'}(w)$ and $\kappa^{q}_{q'}(w)$ are zero by Lemma 5.3. The fifth and the sixth cases are not zero only when p and q', or p' and q' are corresponding to each other. $\square$

We prepare one Lemma which is derived from Lemmas 6.1 and 6.2

Lemma 8.2. *If* $\kappa^{p\,q}_{q\,p} \neq 0$ *for* $p < q$, *then* $\kappa^{a}_{b}(w) \equiv 0$ *for any* $a < p < b < q$, $p < a < q < b$, $b < p < a < q$ *and* $p < b < q < a$.

Proof. We first recall that $\kappa^{p}_{q} \sim \kappa^{q}_{p}(w)$ by Theorem 7.1.

(i) If $a < p < b < q$ and $\kappa_b^a(w) \neq 0$, then $\kappa_q^p(w) \sim w\kappa_b^a(w)$ by (12). By Lemma 6.2-iii, that $\kappa_{b\,p}^{a\,q} \neq 0$ implies that $\kappa_p^q(w) \sim \kappa_b^a(w)$. Because this is in contradiction to $\kappa_q^p \sim \kappa_p^q(w)$, $\kappa_b^a(w)$ must be zero.
(ii) If $p < a < q < b$ and $\kappa_b^a(w) \neq 0$, then $w\kappa_q^p(w) \sim \kappa_b^a(w)$ by (12). By Lemma 6.2-iv, that $\kappa_{b\,p}^{a\,q} \neq 0$ implies that $\kappa_p^q(w) \sim \kappa_b^a(w)$. Because this is in contradiction to $\kappa_q^p \sim \kappa_p^q(w)$, $\kappa_b^a(w)$ must be zero.
(iii) If $b < p < a < q$ and $\kappa_b^a(w) \neq 0$, then $\kappa_p^q(w) \sim w\kappa_b^a(w)$ by (13). By Lemma 6.2-iv, that $\kappa_{b\,q}^{a\,p} \neq 0$ implies that $\kappa_q^p(w) \sim \kappa_b^a(w)$. Because this is in contradiction to $\kappa_q^p \sim \kappa_p^q(w)$, $\kappa_b^a(w)$ must be zero.
(iv) If $p < b < q < a$ and $\kappa_b^a(w) \neq 0$, then $w\kappa_p^q(w) \sim \kappa_b^a(w)$ by (13). By Lemma 6.2-iii, that $\kappa_{b\,q}^{a\,p} \neq 0$ implies that $\kappa_q^p(w) \sim \kappa_b^a(w)$. Because this is in contradiction to $\kappa_q^p \sim \kappa_p^q(w)$, $\kappa_b^a(w)$ must be zero. □

We can directly show that some of non-diagonal elements of $K(z)$ are constantly zero by Lemma 8.2 above.

Proposition 8.1.

$$\kappa_t^s(w) \equiv 0, \quad \kappa_s^t(w) \equiv 0 \quad \textit{for } s \in S \textit{ and } t \in T, \tag{35}$$
$$\kappa_t^p(w) \equiv 0, \quad \kappa_p^t(w) \equiv 0 \quad \textit{for } p \in P\ (p \neq p_1) \textit{ and } t \in T, \tag{36}$$
$$\kappa_s^p(w) \equiv 0, \quad \kappa_p^s(w) \equiv 0 \quad \textit{for } p \in P\ (p \neq p_l) \textit{ and } s \in S, \tag{37}$$
$$\kappa_t^q(w) \equiv 0, \quad \kappa_q^t(w) \equiv 0 \quad \textit{for } q \in Q\ (q \neq q_1) \textit{ and } t \in T, \tag{38}$$
$$\kappa_s^q(w) \equiv 0, \quad \kappa_q^s(w) \equiv 0 \quad \textit{for } q \in Q\ (q \neq q_l) \textit{ and } s \in S, \tag{39}$$

where $l = \sharp(P)$.

Proof. Because $\kappa_{q_1\,p_1}^{p_1\,q_1} \neq 0$ and

$$p_1 < s < q_1 < t \quad \text{for } s \in S \text{ and } t \in T,$$
$$p_1 < p_j < q_1 < t \quad \text{for } 1 < j \leq l \text{ and } t \in T,$$
$$p_1 < q_j < q_1 < t \quad \text{for } 1 < j \leq l \text{ and } t \in T,$$

we obtain (35), (36) and (38) by Lemma 8.2. Also Because $\kappa_{q_l\,p_l}^{p_l\,q_l} \neq 0$ and

$$p_j < p_l < s < q_l \quad \text{for } 1 \leq j < l \text{ and } s \in S,$$
$$p_l < s < q_l < q_j \quad \text{for } 1 \leq j < l \text{ and } s \in S,$$

we obtain (37) and (39). □

We will determine the nonzero cases among specified in Lemma 8.1.

Lemma 8.3. $\kappa_{r'\,r''}^{r\,r'} \equiv 0$ *for mutually different* $r, r', r'' \in S \cup T$.

Proof. If $\{r, r', r''\}$ contains both elements in S and T, then $\kappa^{r\,r'}_{r'\,r''} \equiv 0$ because $\kappa^s_t(w)$ and $\kappa^t_s(w)$ are zero for $s \in S$ and $t \in T$ by Proposition 8.1. We only have to determine that $\kappa^{s\,s'}_{s'\,s''} = 0$ for mutually different $s, s', s'' \in S$ and that $\kappa^{t\,t'}_{t'\,t''} = 0$ for mutually different $t, t', t'' \in T$. We first show that $\kappa^{s\,s'}_{s'\,s''} = 0$ by dividing cases according to the order among s, s' and s''.

(i) If $s < s' < s''$, then

$$(w_1 - w_2)\kappa^{s\,s'}_{s'\,s''} = w_2\langle \kappa^s_{s''}(w), \kappa^{s'+1}_{s'+1}(w) - \kappa^{s'}_{s'}(w)\rangle \tag{40}$$

by Proposition 6.1. We know $s'+1 \in S$ because $s' < s'' \in S$, and that $\kappa^{s'+1}_{s'+1}(w) - \kappa^{s'}_{s'}(w) = 0$ by Theorem 7.1. We obtain $\kappa^{s\,s'}_{s'\,s''} = 0$ by (40).

(ii) If $s < s'' < s'$, then

$$(w_1 - w_2)\kappa^{s\,s'}_{s'\,s''} = \langle \kappa^s_{s''}(w), \kappa^{s'}_{s'}(w) - \kappa^{s'-1}_{s'-1}(w)\rangle \tag{41}$$

by Proposition 6.1. We know $s'-1 \in S$ because $s' > s'' \in S$, and that $\kappa^{s'}_{s'}(w) - \kappa^{s'-1}_{s'-1}(w) = 0$ by Theorem 7.1. We obtain $\kappa^{s\,s'}_{s'\,s''} = 0$ by (41).

(iii) If $s' < s < s''$, then $w\kappa^s_{s''}(w) \sim \kappa^s_{s''}(w)$ by Lemma 4.4, and

$$(w_1 - w_2)\kappa^{s\,s'}_{s'\,s''} = \langle \kappa^s_{s''}(w), w(\kappa^{s'}_{s'}(w) - \kappa^{s'-1}_{s'-1}(w))\rangle \tag{42}$$

by Proposition 6.1. We also have that $\kappa^{s'}_{s''}(w) \sim \kappa^{q_1}_{p_1}(w)$ and $\kappa^s_{s''}(w) \sim \kappa^{q_1}_{p_1}(w)$ by (17) because $p_1 < s' < s'' < q_1$ and $p_1 < s < s'' < q_1$, respectively. Because $\kappa^{q_1}_{p_1}(w) \neq 0$ by Theorem 7.1, we have that $\kappa^{s'}_{s''}(w) = 0$ or $\kappa^s_{s''}(w) = 0$. We obtain $\kappa^{s\,s'}_{s'\,s''} = 0$ by (42).

(iv) If $s' < s'' < s$, then

$$(w_1 - w_2)\kappa^{s\,s'}_{s'\,s''} = \langle \kappa^s_{s''}(w), w(\kappa^{s'+1}_{s'+1}(w) - \kappa^{s'}_{s'}(w))\rangle \tag{43}$$

by Proposition 6.1. We know $s'+1 \in S$ because $s' < s'' \in S$, and that $\kappa^{s'+1}_{s'+1}(w) - \kappa^{s'}_{s'}(w) = 0$ by Theorem 7.1. We obtain $\kappa^{s\,s'}_{s'\,s''} = 0$ by (43).

(v) If $s'' < s < s'$, then $w\kappa^{s'}_{s''}(w) \sim \kappa^s_{s'}(w)$ by Lemma 4.4. We also have that $\kappa^s_{s'}(w) \sim \kappa^{q_1}_{p_1}(w)$ and $\kappa^{s'}_{s''}(w) \sim \kappa^{p_1}_{q_1}(w)$ by (17) and (16) because $p_1 < s < s' < q_1$ and $p_1 < s'' < s' < q_1$, respectively. Because $\kappa^{p_1\,q_1}_{q_1\,p_1} \neq 0$ and $\kappa^{p_1}_{q_1}(w) \sim \kappa^{q_1}_{p_1}(w)$ by Theorem 7.1, we have that $\kappa^{s'}_{s''}(w) = 0$ or $\kappa^s_{s'}(w) = 0$. We obtain $\kappa^{s\,s'}_{s'\,s''} = 0$.

(vi) If $s'' < s' < s$, then

$$(w_1 - w_2)\kappa^{s\,s'}_{s'\,s''} = w_1\langle \kappa^s_{s''}(w), \kappa^{s'}_{s'}(w) - \kappa^{s'-1}_{s'-1}(w)\rangle \tag{44}$$

by Proposition 6.1. We know $s'-1 \in S$ because $s' > s'' \in S$, and that $\kappa^{s'}_{s'}(w) - \kappa^{s'-1}_{s'-1}(w) = 0$ by Theorem 7.1. We obtain $\kappa^{s\,s'}_{s'\,s''} = 0$ by (44).

The proof that $\kappa^{t\,t'}_{t'\,t''} = 0$ $(t, t', t'' \in T)$ is done in a similar way. In the cases $t < t' < t''$, $t < t'' < t'$, $t' < t'' < t$ and $t'' < t' < t$, we obtain $\kappa^{t\,t'}_{t'\,t''} = 0$ because $\kappa^{t'+1}_{t'+1}(w) - \kappa^{t'}_{t'}(w) = 0$ or $\kappa^{t'}_{t'}(w) - \kappa^{t'-1}_{t'-1}(w) = 0$.
If $t' < t < t''$, then $w\kappa^{t'}_{t''}(w) \sim \kappa^{t}_{t''}(w)$ by Lemma 4.4, and

$$(w_1 - w_2)\kappa^{t\,t'}_{t'\,t''} = \langle \kappa^{t}_{t''}(w), w(\kappa^{t'}_{t'}(w) - \kappa^{t'-1}_{t'-1}(w))\rangle \tag{45}$$

by Proposition 6.1. We also have that $\kappa^{s'}_{s''}(w) \sim w\kappa^{p_1}_{q_1}(w)$ and $\kappa^{s}_{s''}(w) \sim w\kappa^{q_1}_{p_1}(w)$ by (14) because $p_1 < q_1 < t' < t''$ and $p_1 < q_1 < s < s''$, respectively. Because $\kappa^{p_1}_{q_1}(w) \neq 0$ by Theorem 7.1, we have $\kappa^{t'}_{t''}(w) = 0$ or $\kappa^{t}_{t''}(w) = 0$. We obtain $\kappa^{t\,t'}_{t'\,t''} = 0$ by (45).
If $t'' < t' < t$, then $\kappa^{t}_{t'}(w) \sim w\kappa^{t'}_{t''}(w)$ by Lemma 4.4. $\kappa^{t}_{t'}(w) \sim w\kappa^{q_1}_{p_1}(w)$ and $\kappa^{t'}_{t''}(w) \sim w\kappa^{q_1}_{p_1}(w)$ by (15) because $p_1 < q_1 < t' < t$ and $p_1 < q_1 < t'' < t'$, respectively. Because $\kappa^{q_1}_{p_1}(w) \neq 0$ by Theorem 7.1, we have $\kappa^{t}_{t'}(w) = 0$ or $\kappa^{t'}_{t''}(w) = 0$. We obtain $\kappa^{t\,t'}_{t'\,t''} = 0$. □

Lemma 8.4. *$\kappa^{p\,r'}_{r'\,r''} \equiv 0$ for $p \in P$ and $r', r'' \in S \cup T$ $(r' \neq r'')$.*

Proof. Because $\kappa^{p}_{s}(w) \equiv 0$ for $p_l \neq p \in P$ and $\kappa^{p}_{t}(w) \equiv 0$ for $p_1 \neq p \in P$ by Proposition 8.1 we only have to consider $\kappa^{p_1\,t'}_{t'\,r''}$ $(t' \in T\ ,\ r'' \in S \cup T)$ and $\kappa^{p_l\,s'}_{s'\,r''}$ $(s' \in S\ ,\ r'' \in S \cup T)$. Furthermore since $\kappa^{s}_{t}(w) \equiv 0$ and $\kappa^{t}_{s}(w) \equiv 0$ for $s \in S$ and $t \in T$ by Proposition 8.1, it is enough for us to study $\kappa^{p_1\,t'}_{t'\,t''}$ $(t', t'' \in T\ t' \neq t'')$ and $\kappa^{p_l\,s'}_{s'\,s''}$ $(s', s'' \in S\ s' \neq s'')$.
If $p_1 < s' < s''$, then

$$(w_1 - w_2)\kappa^{p_l\,s'}_{s'\,s''} = w_2\langle \kappa^{p_l}_{s''}(w), \kappa^{s'+1}_{s'+1}(w) - \kappa^{s'}_{s'}(w)\rangle. \tag{46}$$

by Proposition 6.1. Because $s' < s'' \in S$, $s' + 1 \in S$. We have $\kappa^{s'+1}_{s'+1}(w) - \kappa^{s'}_{s'}(w) = 0$ by Theorem 7.1 and $\kappa^{p_l\,s'}_{s'\,s''} \equiv 0$ by (46).
If $p_l < s'' < s'$, then

$$(w_1 - w_2)\kappa^{p_l\,s'}_{s'\,s''} = \langle w\kappa^{p_l}_{s''}(w), \kappa^{s'}_{s'}(w) - \kappa^{s'-1}_{s'-1}(w)\rangle. \tag{47}$$

Because $s' > s'' \in S$, $s' - 1 \in S$. We have $\kappa^{s'}_{s'}(w) - \kappa^{s'-1}_{s'-1}(w) = 0$ by Theorem 7.1 and $\kappa^{p_l\,s'}_{s'\,s''} \equiv 0$ by (47).
The proof that $\kappa^{p_1\,t'}_{t'\,t''} \equiv 0$ $(t', t'' \in T\ t' \neq t'')$ is all the same. □

Lemma 8.5. *$\kappa^{q\,r'}_{r'\,r''} \equiv 0$ for $q \in Q$ and $r', r'' \in S \cup T$ $(r' \neq r'')$.*

Proof. Because $\kappa^{q}_{s}(w) \equiv 0$ for $q_l \neq q \in Q$ and $\kappa^{q}_{t}(w) \equiv 0$ for $q_1 \neq q \in P$ by Proposition 8.1 we only have to consider $\kappa^{q_1\,t'}_{t'\,r''}$ $(t' \in T\ ,\ r'' \in S \cup T)$ and $\kappa^{q_l\,s'}_{s'\,r''}$ $(s' \in S\ ,\ r'' \in S \cup T)$. Furthermore since $\kappa^{s}_{t}(w) \equiv 0$ and $\kappa^{t}_{s}(w) \equiv 0$

for $s \in S$ and $t \in T$ by Proposition 8.1, it is enough for us to study $\kappa^{q_1\,t'}_{t'\,t''}$ ($t', t'' \in T$ $t' \neq t''$) and $\kappa^{q_l\,s'}_{s'\,s''}$ ($s', s'' \in S$ $s' \neq s''$).
If $q_1 < t' < t''$, then

$$(w_1 - w_2)\kappa^{q_1\,t'}_{t'\,t''} = w_2\langle \kappa^{q_1}_{t''}(w), \kappa^{t'+1}_{t'+1}(w) - \kappa^{t'}_{t'}(w)\rangle, \tag{48}$$

by Proposition 6.1. Because $t' < t'' \in T$, $t'+1 \in T$. We have $\kappa^{t'+1}_{t'+1}(w) - \kappa^{t'}_{t'}(w) = 0$ by Theorem 7.1 and $\kappa^{q_1\,t'}_{t'\,t''} \equiv 0$ by (48).
If $q_1 < t'' < t'$, then

$$(w_1 - w_2)\kappa^{q_1\,t'}_{t'\,t''} = \langle w\kappa^{q_1}_{t''}(w), \kappa^{t'}_{t'}(w) - \kappa^{t'-1}_{t'-1}(w)\rangle, \tag{49}$$

by Proposition 6.1. Because $t' > t'' \in T$, $t'-1 \in T$. We have $\kappa^{t'}_{t'}(w) - \kappa^{t'-1}_{t'-1}(w) = 0$ by Theorem 7.1 and $\kappa^{q_1\,t'}_{t'\,t''} \equiv 0$ by (49).
If $s' < s'' < q_l$, then

$$(w_1 - w_2)\kappa^{q_l\,s'}_{s'\,s''} = \langle \kappa^{q_l}_{s''}(w), w(\kappa^{s'+1}_{s'+1}(w) - \kappa^{s'}_{s'}(w))\rangle, \tag{50}$$

by Proposition 6.1. Because $s' < s'' \in T$, $s'+1 \in S$. We have $\kappa^{s'+1}_{s'+1}(w) - \kappa^{s'}_{s'}(w) = 0$ by Theorem 7.1 and $\kappa^{q_l\,s'}_{s'\,s''} \equiv 0$ by (50).
If $s'' < s' < q_l$, then

$$(w_1 - w_2)\kappa^{q_l\,s'}_{s'\,s''} = w_1\langle \kappa^{q_l}_{s''}(w), \kappa^{s'}_{s'}(w) - \kappa^{s'-1}_{s'-1}(w)\rangle, \tag{51}$$

by Proposition 6.1. Because $s' > s'' \in T$, $s'-1 \in S$. We have $\kappa^{s'}_{s'}(w) - \kappa^{s'-1}_{s'-1}(w) = 0$ by Theorem 7.1 and $\kappa^{q_l\,s'}_{s'\,s''} \equiv 0$ by (51). □

Lemma 8.6. *$\kappa^{p_j\,q_j}_{q_j\,r} \equiv 0$ ($r \in S \cup T$) except $\kappa^{p_1\,q_1}_{q_1\,t}$ ($t \in T$) and $\kappa^{p_l\,q_l}_{q_l\,s}$ ($s \in S$), and $\kappa^{q_j\,p_j}_{p_j\,r} \equiv 0$ ($r \in S \cup T$) except $\kappa^{q_1\,p_1}_{p_1\,t}$ ($t \in T$) and $\kappa^{q_l\,p_l}_{p_l\,s}$ ($s \in S$).*

Proof. Because $\kappa^{q}_{r}(w) \equiv 0$ except $\kappa^{q_1}_{s}(w)$ ($s \in S$) and $\kappa^{q_l}_{t}(w)$ ($t \in T$) by Proposition 8.1, we have $\kappa^{p_j\,q_j}_{q_j\,r} \equiv 0$ ($r \in S \cup T$) except $\kappa^{p_1\,q_1}_{q_1\,t}$ ($t \in T$) and $\kappa^{p_l\,q_l}_{q_l\,s}$ ($s \in S$). Because $\kappa^{p}_{r}(w) \equiv 0$ except $\kappa^{p_1}_{t}(w)$ ($t \in T$) and $\kappa^{p_l}_{s}(w)$ ($s \in S$) by Proposition 8.1, we have $\kappa^{q_j\,p_j}_{p_j\,r} \equiv 0$ ($r \in S \cup T$) except $\kappa^{q_1\,p_1}_{p_1\,t}$ ($t \in T$) and $\kappa^{q_l\,p_l}_{p_l\,s}$ ($s \in S$). □

Proof of Theorem 8.1. Theorem 8.1 is proved by Lemma 8.1, Lemma 8.3, Lemma 8.4, Lemma 8.5 and Lemma 8.6. □

9. Relations among parameters

We first prove the following theorem in this section.

Theorem 9.1. *Let $K(z)$ be a solution to the reflection equation (6) whose diagonal elements are specified in Theorem 7.1. If $a, b \in \{0, 1, 2, \cdots, N-1\}$ be two different integers, then*

$$\kappa^a_b(w) \equiv 0$$

except the cases

(i) $(a,b) = (p_j, q_j), (q_j, p_j)$ *for* $j = 1, 2, \cdots, l$
(ii) $(a,b) = (p_1, t)$ *for* $t \in T$,
(iii) $(a,b) = (p_l, s)$ *for* $s \in S$,
(iv) $(a,b) = (q_1, t)$ *for* $t \in T$,
(v) $(a,b) = (s, q_l)$ *for* $s \in S$,

where the integer $l = \sharp(P)$ *is specified in Theorem 7.1.*

Proof. Except the cases specified above, we will show

(i) $\kappa^p_a \equiv 0$ for $p \in P$ and $a \neq p$,
(ii) $\kappa^s_a \equiv 0$ for $s \in S$ and $a \neq s$,
(iii) $\kappa^q_a \equiv 0$ for $q \in Q$ and $a \neq q$,
(iv) $\kappa^t_a \equiv 0$ for $t \in T$ and $a \neq t$,

in the followings.

(i)(a) $\kappa^{p_i}_{p_j}(w) \equiv 0$ and $\kappa^{p_i}_{q_j}(w) \equiv 0$ $(i \neq j)$ by Lemma 5.3.
(b) If $\kappa^{p_j}_s(w) \neq$ $(j \neq l,\ s \in S)$, then $\kappa^{p_j}_{p_l}(w) \neq 0$ by Lemma 6.2 because that $\kappa^{p_j q_l}_{s\ p_l} \neq 0$ and that $p_j < p_l < s < q_l$. This is in contradiction to $\kappa^{p_j}_{p_l}(w) \equiv 0$.
(c) If $\kappa^{p_j}_t(w) \neq$ $(j \neq 1,\ t \in T)$, then $\kappa^{p_j}_{p_1}(w) \neq 0$ by Lemma 6.2 because that $\kappa^{p_j q_1}_{t\ p_1} \neq 0$ and that $p_1 < p_j < q_1 < t$. This is in contradiction to $\kappa^{p_j}_{p_1}(w) \equiv 0$.

(ii) (a) $\kappa^s_p(w) \equiv 0$ $(s \in S,\ p \in P)$ by Lemma 5.3.
(b) If $\kappa^s_{s'}(w) \neq 0$ $(s, s' \in S\ s < s')$, then

$$\begin{aligned} 0 = (w_1 - w_2)\kappa^{s0}_{0s'} &= \langle \kappa^s_{s'}(w), w\kappa^0_0(w) - w^{-1}\kappa^{N-1}_{N-1}(w) \rangle \\ &= A_3 f(w) \langle \kappa^s_{s'}(w), w - w^{-1} \rangle \end{aligned} \tag{52}$$

because of that $0 < s < s'$ and of Proposition 6.1 and Theorem 7.1. We remark that the set T in Theorem 7.1 in not empty in this case because $1 \leq \sharp(S) \leq \sharp(T)$. It implies that $N - 1 \in T$ and that $w\kappa_0^0(w) - w^{-1}\kappa_{N-1}^{N-1}(w) = A_3 f(w)(w - w^{-1})$. Moreover we also have

$$0 = (w_1 - w_2)\kappa_{q_l\,s'}^{s\;q_l} = \langle w\kappa_{s'}^{s}(w), \kappa_{q_l}^{q_l}(w) - \kappa_{q_l-1}^{q_l-1}(w)\rangle$$
$$= A_0 f(w)\langle w\kappa_{s'}^{s}(w), w - w^{-1}\rangle \qquad (53)$$

because of that $s < s' < q_l$, Proposition 6.1 and Theorem 7.1. We remark that $\kappa_{q_l}^{q_l}(w) - \kappa_{q_l-1}^{q_l-1}(w) = A_0 f(w)(w - w^{-1})$ because $q_l - 1 \in S$. We obtain $\kappa_{s'}^{s}(w) \equiv 0$ by (52) and (53). If $\kappa_{s'}^{s}(w) \neq 0$ $(s, s' \in S\ s > s')$, then

$$0 = (w_1 - w_2)\kappa_{N-1\,s'}^{s\;\;N-1} = \langle w\kappa_{s'}^{s}(w), w(w\kappa_0^0(w) - w^{-1}\kappa_{N-1}^{N-1}(w))\rangle$$
$$= A_3 f(w)\langle w\kappa_{s'}^{s}(w), w(w - w^{-1})\rangle \qquad (54)$$
$$0 = (w_1 - w_2)\kappa_{p_l\,s'}^{s\;p_l} = \langle \kappa_{s'}^{s}(w), w(\kappa_{p_l+1}^{p_l+1}(w) - \kappa_{p_l}^{p_l}(w))\rangle$$
$$= -A_1 f(w)\langle \kappa_{s'}^{s}(w), w(w - w^{-1})\rangle \qquad (55)$$

because of that $s' < s < N-1$ and $p_l < s' < s$, respectively, Proposition 6.1 and Theorem 7.1. We obtain $\kappa_{s'}^{s}(w) \equiv 0$ by (54) and (55).

(c) $\kappa_q^s(w) \equiv 0$ $(s \in S,\ q \in Q)$ by Lemma 5.3.
(d) $\kappa_t^s(w) \equiv 0$ $(s \in S,\ t \in T)$ by Proposition 8.1.

(iii) (a) $\kappa_{p_j}^{q_i}(w) \equiv 0$ and $\kappa_{q_j}^{q_i}(w) \equiv 0$ $(i \neq j)$ by Lemma 5.3.
(b) If $\kappa_s^{q_j}(w) \neq$ $(j \neq l,\ s \in S)$, then $\kappa_{q_j}^{q_l}(w) \neq 0$ by Lemma 6.2 because that $\kappa_{q_l\,s}^{p_l\,q_j} \neq 0$ and that $p_l < s < q_l < q_j$. This is in contradiction to $\kappa_{q_j}^{q_l}(w) \equiv 0$.
(c) If $\kappa_t^{q_j}(w) \neq$ $(j \neq 1,\ t \in T)$, then $\kappa_{q_1}^{q_j}(w) \neq 0$ by Lemma 6.2 because that $\kappa_{q_1\,t}^{p_1\,q_j} \neq 0$ and that $p_1 < q_j < q_1 < t$. This is in contradiction to $\kappa_{q_1}^{q_j}(w) \equiv 0$.

(iv) (a) $\kappa_p^t(w) \equiv 0$ $(t \in T,\ p \in P)$ by Lemma 5.3.
(b) $\kappa_s^t(w) \equiv 0$ $(t \in T,\ s \in S)$ by Proposition 8.1.
(c) $\kappa_q^t(w) \equiv 0$ $(t \in T,\ q \in Q)$ by Lemma 5.3.
(d) If $\kappa_{t'}^t(w) \neq 0$ $(t, t' \in T\ t < t')$, then

$$0 = (w_1 - w_2)\kappa_{0\,t'}^{t\,0} = \langle \kappa_{t'}^{t}(w), w\kappa_0^0(w) - w^{-1}\kappa_{N-1}^{N-1}(w)\rangle$$
$$= A_3 f(w)\langle \kappa_{t'}^{t}(w), w - w^{-1}\rangle$$

because of that $0 < t < t'$ and $N-1 \in T$, Proposition 6.1 and Theorem 7.1. It implies $\kappa_{t'}^t(w) \sim (w - w^{-1})$. On the other hand, we have $\kappa_{t'}^t(w) \sim w\kappa_{q_1}^{p_1}(w) = u_p w f(w)(w - w^{-1})$ because of (14) in Lemma 6.1. We obtain $\kappa_{t'}^t(w) \equiv 0$. If $\kappa_{t'}^t(w) \neq 0$ $(t, t' \in T\ t > t')$, then

$$0 = (w_1 - w_2)\kappa_{q_1\, t'}^{t\ \, q_1} = \langle \kappa_{s'}^{s}(w), w(\kappa_{q_1+1}^{q_1+1}(w) - \kappa_{q_1}^{q_1}(w))\rangle$$
$$= A_2 f(w)\langle w\kappa_{s'}^{s}(w), w^2(w - w^{-1})\rangle$$

because of that $q_1 < t' < t$, Proposition 6.1 and Theorem 7.1. On the other hand, we have $\kappa_{t'}^{t}(w) \sim w\kappa_{p_1}^{q_1}(w) = v_p w f(w)(w - w^{-1})$ because of (15) in Lemma 6.1. We obtain $\kappa_{t'}^{t}(w) \equiv 0$. □

Lemma 4.4 determine the possible form of the non-diagonal elements.

Proposition 9.1. *Let $K(z)$ be a solution to the reflection equation (6) whose diagonal elements are specified in Theorem 7.1. If $a, b \in \{0, 1, 2, \cdots, N-1\}$ be two different integers, then*

(i) $\kappa_{q_j}^{p_j}(w) = u_{p_j} f(w)(w - w^{-1})$ *and* $\kappa_{p_j}^{q_j}(w) = u_{p_j} f(w)(w - w^{-1})$ *for* $j = 1, 2, \cdots, l$,
(ii) $\kappa_{s}^{p_l}(w) = x_s f(w)(w - w^{-1})$ *and* $\kappa_{q_l}^{s}(w) = \overline{x}_s f(w)(w - w^{-1})$ *for* $s \in S$,
(iii) $\kappa_{t}^{p_1}(w) = y_t f(w)(w - w^{-1})$ *and* $\kappa_{t}^{q_1}(w) = \overline{y}_t w f(w)(w - w^{-1})$ *for* $t \in T$,

where the integer $l = \sharp(P) = \sharp(Q)$ and $m = \sharp(S)$ are specified in Theorem 7.1 and where x_s, $\overline{x}_s$ $(s \in S)$ and y_t, $\overline{y}_t$ $(t \in T)$ are parameters in **C**. *Other off-diagonal elements $K(z)$ not specified above are constantly zero. These off-diagonal elements satisfy Lemma 4.4.*

Proof. We have to consider C_{abc} in Definition 4.8 only when one of a, b and c, say α, has two integers β and γ in $\{0, 1, 2, \cdots, N-1\}$ such that $\kappa_{\beta\gamma}^{\alpha\alpha} \neq 0$, $\kappa_{\alpha\gamma}^{\beta\alpha} \neq 0$ or $\kappa_{\alpha\alpha}^{\beta\gamma} \neq 0$, and $-beta$ and γ are also members of $\{a, b, c\}$. If it is not the case, C_{abc} has at most one element, and Lemma 4.4 is trivially fulfilled.
The elements α in $\{0, 1, 2, \cdots, N-1\}$ which has two integers β and γ such that $\kappa_{\beta\gamma}^{\alpha\alpha} \neq 0$, $\kappa_{\alpha\gamma}^{\beta\alpha} \neq 0$ or $\kappa_{\alpha\alpha}^{\beta\gamma} \neq 0$ are p_1, $s \in S$, p_l, q_l, q_1 and $t \in T$. The relevant C_{abc} are only $C_{p_1 q_1 t}$ $(t \in T)$ and $C_{p_l s q_l}$ $(s \in S)$.
In $C_{p_1 q_1 t}$ $(t \in T)$,

$$\kappa_{q_1}^{p_1}(w) = u_{p_1} f(w)(w - w^{-1}) \neq 0,$$
$$\kappa_{p_1}^{q_1}(w) = v_{p_1} f(w)(w - w^{-1}) \neq 0,$$

and $p_1 < q_1 < t$, we obtain

$$\kappa_{t}^{p_1}(w) = y_t f(w)(w - w^{-1}),$$
$$\kappa_{t}^{q_1}(w) = \overline{y}_t f(w)(w - w^{-1}),$$

for some parameters y_t, $\overline{y}_t$.

In $C_{p_l s q_l}$ $(s \in S)$,

$$\kappa_{q_l}^{p_l}(w) = u_{p_l} f(w)(w - w^{-1}) \neq 0,$$
$$\kappa_{p_l}^{q_l}(w) = v_{p_l} f(w)(w - w^{-1}) \neq 0,$$

and $p_l < s < q_l$, we obtain

$$\kappa_s^{p_l}(w) = x_s f(w)(w - w^{-1}),$$
$$\kappa_{q_l}^{s}(w) = \overline{x}_s f(w)(w - w^{-1}),$$

for some parameters x_s, $\overline{x}_s$. □

These parameters have to satisfy the conditions in Proposition 6.1 and Lemma 6.1. The proof of the following is straightforward from Theorem 7.1.

Lemma 9.1.

$$w\kappa_0^0(w) - w^{-1}\kappa_{N-1}^{N-1}(w) = A_3(w - w^{-1}),$$

$$\kappa_1^1(w) - \kappa_0^0(w) = \begin{cases} 0 & \textit{if } l = \sharp(P) > 1 \\ A_0(w - w^{-1}) & \textit{if } l = \sharp(P) = 1 \textit{ and } m = \sharp(S) = 0 \\ -A_1(w - w^{-1}) & \textit{if } l = \sharp(P) = 1 \textit{ and } m = \sharp(S) > 0 \end{cases},$$

$$\kappa_{p_l+1}^{p_l+1}(w) - \kappa_{p_l}^{p_l}(w) = \begin{cases} (A_0 - A_1)(w - w^{-1}) & \textit{if } m = \sharp(S) = 0 \\ -A_1(w - w^{-1}) & \textit{if } m = \sharp(S) > 0 \end{cases}$$

$$\kappa_{q_l}^{q_l}(w) - \kappa_{q_l-1}^{q_l-1}(w) = A_0(w - w^{-1}),$$

$$\kappa_{q_1+1}^{q_1+1}(w) - \kappa_{q_1}^{q_1}(w) = -A_2 w(w - w^{-1}) \quad \textit{if } \sharp(T) > 0$$

$$\kappa_{N-1}^{N-1}(w) - \kappa_{N-2}^{N-2}(w) = \begin{cases} 0 & \textit{if } \sharp(T) = 0 \\ -A_2 w(w - w^{-1}) & \textit{if } \sharp(T) = 1 \\ 0 & \textit{if } \sharp(T) > 1 \end{cases}$$

Corollary 9.1.

$$w\kappa_0^0(w) - w^{-1}\kappa_{N-1}^{N-1}(w) \sim w - w^{-1},$$
$$\kappa_1^1(w) - \kappa_0^0(w) \sim w - w^{-1},$$
$$\kappa_{p_l+1}^{p_l+1}(w) - \kappa_{p_l}^{p_l}(w) \sim w - w^{-1},$$
$$\kappa_{q_l}^{q_l}(w) - \kappa_{q_l-1}^{q_l-1}(w) \sim w - w^{-1},$$
$$\kappa_{N-1}^{N-1}(w) - \kappa_{N-2}^{N-2}(w) \sim w(w - w^{-1}).$$

Proposition 9.2. *Let $K(z)$ be a solution to the reflection equation (6), whose elements are specified in Theorem 7.1 and Proposition 9.1. The necessary and sufficient conditions for $K(z)$ to satisfy the conditions in Proposition 6.1 is that the parameters A_0, A_1, A_2, A_3, u_{p_1}, v_{p_1}, u_{p_l}, v_{p_l} satisfy*

$$\begin{cases} u_{p_l}\overline{x}_s = x_s A_0 \\ v_{p_l} x_s = \overline{x}_s A_1 \end{cases} \quad (s \in S),$$

$$\begin{cases} u_{p_1}\overline{y}_t = y_t A_2 \\ v_{p_1} y_t = \overline{y}_t A_3 \end{cases} \quad (t \in T).$$

Proof. If $\kappa^{ab}_{bc} = 0$ and $\kappa^a_c(w) = 0$, then the conditions in Proposition 6.1 are trivially satisfied as $0 = \langle 0, g\rangle$, where g is a function of w. If $\kappa^{ab}_{bc} = 0$ and $\kappa^a_c(w) \neq 0$, then the conditions in Proposition 6.1 are satisfied as $0 = \langle g, g\rangle$ or $0 = \langle g, 0\rangle$ by Corollary 9.1. The relation among the parameters arise when $\kappa^{ab}_{bc} \neq 0$ and $\kappa^a_c(w) \neq 0$. Such case are specified in Theorem 8.1. We will study them.

(i) In the case of $(a, b, c) = (p_1, q_1, t)$ $(t \in T)$, we have

$$(w_1 - w_2)\kappa^{p_1 q_1}_{q_1 t} = w_2\langle \kappa^{p_1}_t(w), \kappa^{q_1+1}_{q_1+1}(w) - \kappa^{q_1}_{q_1}(w)\rangle$$

by Proposition 6.1 because $p_1 < q_1 < t$. Substituting the results for $\kappa^{p_1}_{q_1}(w)$, $\kappa^{q_1}_t(w)$ and $\kappa^{p_1}_t(w)$, in Proposition 9.1, we obtain

$$\begin{aligned} &(w_1 - w_2)\cdot u_{p_1}(w_1 - w_1^{-1})\cdot \overline{y}_t w_2(w_2 - w_2^{-1}) \\ &= w_2\langle y_t(w - w^{-1}), -A_2 w(w - w^{-1})\rangle, \\ &= -y_t A_2 w_2(w_2 - w_1)(w_1 - w_1^{-1})(w_2 - w_2^{-1}), \\ &u_{p_1}\overline{y}_t = y_t A_2. \end{aligned}$$

(ii) In the case of $(a, b, c) = (p_l, q_l, s)$ $(s \in S)$, we have

$$(w_1 - w_2)\kappa^{p_l q_l}_{q_l s} = \langle w\kappa^{p_l}_s(w), \kappa^{q_l}_{q_l}(w) - \kappa^{q_1-1}_{q_1-1}(w)\rangle$$

by Proposition 6.1 because $p_l < s < q_l$. Substituting the results for $\kappa^{p_l}_{q_l}(w)$, $\kappa^{q_l}_s(w)$ and $\kappa^{p_l}_s(w)$, in Proposition 9.1, we obtain

$$\begin{aligned} &(w_1 - w_2)\cdot u_{p_l}(w_1 - w_1^{-1})\cdot \overline{x}_s(w_2 - w_2^{-1}) \\ &= \langle x_s w(w - w^{-1}), A_0(w - w^{-1})\rangle, \\ &= x_s A_0(w_1 - w_2)(w_1 - w_1^{-1})(w_2 - w_2^{-1}), \\ &u_{p_l}\overline{x}_s = x_s A_0. \end{aligned}$$

(iii) In the case of $(a, b, c) = (q_1, p_1, t)$ $(t \in T)$, we have

$$(w_1 - w_2)\kappa^{q_1 p_1}_{p_1 t} = (w_1 - w_2)\kappa^{q_1 0}_{0\ t} = \langle \kappa^{q_1}_t(w), w\kappa^0_0(w) - w^{-1}\kappa^{N-1}_{N-1}(w)\rangle$$

by Proposition 6.1 because $0 = p_1 < q_1 < t$. Substituting the results for $\kappa^{q_1}_{p_1}(w)$, $\kappa^{p_1}_t(w)$ and $\kappa^{q_1}_t(w)$, in Proposition 9.1, we obtain

$$\begin{aligned}
&(w_1 - w_2) \cdot v_{p_1}(w_1 - w_1^{-1}) \cdot y_t(w_2 - w_2^{-1}) \\
&= \langle \overline{y}_t w(w - w^{-1}), A_3(w - w^{-1})\rangle, \\
&= \overline{y}_t A_3(w_1 - w_2)(w_1 - w_1^{-1})(w_2 - w_2^{-1}), \\
&v_{p_1} y_t = \overline{y}_t A_3.
\end{aligned}$$

(iv) In the case of $(a, b, c) = (q_l, p_l, s)$ $(s \in S)$, we have

$$(w_1 - w_2)\kappa^{q_l\,p_l}_{p_l\,s} = \langle \kappa^{q_l}_s(w), w(\kappa^{p_l+1}_{p_l+1}(w) - \kappa^{p_l}_{p_l}(w))\rangle$$

by Proposition 6.1 because $0 = p_1 < s < q_1$. Substituting the results for $\kappa^{q_l}_{p_l}(w)$, $\kappa^{p_l}_s(w)$ and $\kappa^{q_l}_s(w)$, in Proposition 9.1, we obtain

$$\begin{aligned}
&(w_1 - w_2) \cdot v_{p_l}(w_1 - w_1^{-1}) \cdot x_s(w_2 - w_2^{-1}) \\
&= \langle \overline{x}_s w(w - w^{-1}), -A_1 w(w - w^{-1})\rangle, \\
&= -\overline{x}_s A_1(w_2 - w_1)(w_1 - w_1^{-1})(w_2 - w_2^{-1}), \\
&v_{p_l} x_s = \overline{x}_s A_1.
\end{aligned}$$

□

Proposition 9.3. *Let $K(z)$ be a solution to the reflection equation (6) whose elements are specified in Theorem 7.1 and Proposition 9.1. The necessary and sufficient conditions for $K(z)$ to satisfy the conditions in Lemma 6.1 is that the parameters x_s, $\overline{x}_s$ $(s \in S)$, y_t, $\overline{y}_t$ $(t \in T)$, satisfy*

$$\begin{aligned}
x_s \overline{x}_{s'} &= \overline{x_s} x_{s'} \quad (s, s' \in S), \\
y_t \overline{y}_{t'} &= \overline{y_t} y_{t'} \quad (t, t' \in T).
\end{aligned}$$

Proof. The conditions in Lemma 6.1 we have to consider are

(i) (20) for $(a, b, c, d) = (p_1, q_1, t, t')$ $(t, t' \in T)$,

$$(w_1 - w_2)w_1^{-1}\kappa^{q_1\,p_1}_{t'\,t} = \langle \kappa^{q_1}_t(w), \kappa^{p_1}_{t'}(w)\rangle,$$

(ii) (22) for $(a, b, c, d) = (p_l, s, s', q_l)$ $(s, s' \in S)$,

$$(w_1 - w_2)\kappa^{p_l\,q_l}_{s'\,s} = \langle w\kappa^{p_l}_s(w), \kappa^{q_l}_{s'}(w)\rangle.$$

We can check that the other conditions are fulfilled with ease. Substituting the results of Proposition 9.1 into above, we obtain the results in the statement.

□

10. Main results

We summarized briefly what we have proved so far.

- Proposition 4.1, which states that, that

$$K(z) = \left(z^{-a-b}\kappa_b^a(w)\right)_{a,b=0,1,\cdots,N-1} \qquad (w = z^N)$$

  is a solution to the reflection equation is equivalent to that $K(z)$ satisfy the conditions in Lemma 4.4 and type I, type VI, type X, type XIII and type XIV components of the reflection equation,
- Theorem 7.1, which determines the diagonal elements of $K(z)$ by Lemma 4.4 and type X components of the reflection equation, and the results inevitably satisfy the type X components.
- Theorem 9.1, which determines the non-zero off-diagonal elements of $K(z)$,
- Proposition 9.1, Proposition 9.2 and Proposition 9.3, which describe the necessary and sufficient conditions for $K(z)$ to satisfy Lemma 4.4, type VI, VI, VI components, and type I components, respectively.

We already have proved the main theorem for us. We write it down explicitly below according to the cases

(i) $S = \emptyset$ and $T = \emptyset$,
(ii) $S = \emptyset$ and $T \neq \emptyset$,
(iii) $\sharp(P) = 1$, $S \neq \emptyset$ and $T \neq \emptyset$,
(iv) $\sharp(P) > 1$, $S \neq \emptyset$ and $T \neq \emptyset$,

where $P, Q, R \subset \{0, 1, 2, \cdots\}$ are in Definition 7.1 and Proposition 7.3. We remark that the case $S \neq \emptyset$ and $T = \emptyset$ does not occur because $\sharp(S) = m \leq N - 2l - m = \sharp(T)$.

Theorem 10.1. *If* $K(z) = \left(z^{-a-b}\kappa_b^a(w)\right)_{a,b=0,1,\cdots,N-1}$ *is a solution to the reflection equation (6), which satisfies the assumptions (2) and (3), then there exist two integers* $1 \leq l \leq [N/2]$ *and* $0 \leq m \leq [(N-2l)/2]$ *and* $K(z)$ *is similar to only one of (i), (ii), (iii) and (iv) below.*

(i) When $S = \emptyset$ *and* $T = \emptyset$*, namely when* N *is even and* $N = 2l$*,* $K(z)$ *has two parameters*

$$A_0, A_3$$

for the diagonal elements, and N *parameters*

$$u_l, v_l \quad (l = 0, 1, \cdots, N/2 - 1)$$

for the off-diagonal elements, which satisfy the relations

$$u_p v_p = u_{p'} v_{p'} \quad (p, p' = 0, 1, 2, \cdots N/2 - 1).$$

The non-zero elements of $K(z)$ *are*

$$\begin{aligned}
\kappa_p^p(w) &= A_0 w^{-1} + A_3 \quad (p = 0, 1, 2, \cdots N/2 - 1), \\
\kappa_q^q(w) &= A_0 w + A_3 \quad (q = N/2, N/2 + 1, \cdots, N - 1), \\
\kappa_{N-p-1}^p(w) &= u_p(w - w^{-1}) \quad (p = 0, 1, 2, \cdots N/2 - 1), \\
\kappa_{N-q-1}^q(w) &= v_p(w - w^{-1}) \quad (q = N/2, N/2 + 1, \cdots, N - 1).
\end{aligned}$$

(ii) *When* $S = \emptyset$ *and* $T \neq \emptyset$*, namely when* $N > 2$*,* $1 \leq l \leq [(N-1)/2]$ *and* $m = 0$*,* $K(z)$ *has three parameters*

$$A_0, A_2, A_3$$

for the diagonal elements, and $2N - 2l$ *parameters*

$$\begin{aligned}
&u_p, v_p \quad (p = 0, 1, \cdots l - 1) \\
&y_t, \overline{y}_t \quad (t = 2l, 2l + 1, \cdots, N - 1)
\end{aligned}$$

for the off-diagonal elements, which satisfy the relations

$$\begin{aligned}
u_p v_p &= A_2 A_3 \quad (p = 0, 1, \cdots, l - 1) \\
y_t \overline{y}_{t'} &= \overline{y}_t y_{t'} \quad (t, t' = 2l, 2l + 1, \cdots, N - 1), \\
u_0 \overline{y}_t &= y_t A_2 \quad (t = 2l, 2l + 1, \cdots, N - 1), \\
v_0 y_t &= \overline{y}_t A_3 \quad (t = 2l, 2l + 1, \cdots, N - 1).
\end{aligned}$$

The non-zero elements of $K(z)$ *are*

$$\begin{aligned}
\kappa_p^p(w) &= A_0 w^{-1} - A_2 + A_3 \quad (p = 0, 1, 2, \cdots l - 1), \\
\kappa_q^q(w) &= A_0 w - A_2 + A_3 \quad (q = l, l + 1, \cdots, 2l - 1), \\
\kappa_t^t(w) &= A_3 + A_0 w - A_2 w^2 \quad (t = 2l, 2l + 1, \cdots, N - 1), \\
\kappa_{2l-p-1}^p(w) &= u_p(w - w^{-1}) \quad (p = 0, 1, 2, \cdots l - 1), \\
\kappa_{2l-q-1}^q(w) &= v_p(w - w^{-1}) \quad (q = l, l + 1, \cdots, 2l - 1), \\
\kappa_t^0(w) &= y_t(w - w^{-1}) \quad (t = 2l, 2l + 1, \cdots, N - 1), \\
\kappa_t^{2l-1}(w) &= \overline{y}_t w(w - w^{-1}) \quad (t = 2l, 2l + 1, \cdots, N - 1),
\end{aligned}$$

(iii) *When* $\sharp(P) = 1$*,* $S \neq \emptyset$ *and* $T \neq \emptyset$*, namely when* $N > 2$*,* $l = 1$ *and* $0 < m \leq [N/2 - 1]$*,* $K(z)$ *has four parameters*

$$A_0, A_1, A_2, A_3$$

for the diagonal elements, and $2N-2$ *parameters*

$$\begin{aligned}
&u_0, v_0\\
&x_s, \overline{x}_s \quad (s=1,2,\cdots,m)\\
&y_t, \overline{y}_t \quad (t=m+2,m+3,\cdots,N-1)
\end{aligned}$$

for the off-diagonal elements, which satisfy the relations

$$\begin{aligned}
u_0v_0 &= A_0A_1\\
u_0v_0 &= A_2A_3\\
x_s\overline{x}_{s'} &= \overline{x}_s x_{s'} \quad (s,s'=1,2,\cdots,m),\\
u_0\overline{x}_s &= x_sA_0 \quad (s=1,2,\cdots,m),\\
v_0x_s &= \overline{x}_sA_1 \quad (s=1,2,\cdots,m).\\
y_t\overline{y}_{t'} &= \overline{y}_t y_{t'} \quad (t,t'=m+2,m+3,\cdots,N-1),\\
u_0\overline{y}_t &= y_tA_2 \quad (t=m+2,m+3,\cdots,N-1),\\
v_0y_t &= \overline{y}_tA_3 \quad (t=m+2,m+3,\cdots,N-1).
\end{aligned}$$

The non-zero elements of $K(z)$ *are*

$$\begin{aligned}
\kappa_0^0(w) &= (A_0-A_1)w^{-1}-A_2+A_3,\\
\kappa_s^s(w) &= A_0w^{-1}-A_2+A_3-A_1w \quad (s=1,2,\cdots,m),\\
\kappa_{m+1}^{m+1}(w) &= (A_0-A_1)w-A_2+A_3,\\
\kappa_t^t(w) &= A_3+(A_0-A_1)w-A_2w^2 \quad (t=m+2,m+3,\cdots,N-1),\\
\kappa_{m+1}^0(w) &= u_0(w-w^{-1}),\\
\kappa_0^{m+1}(w) &= v_0(w-w^{-1}),\\
\kappa_s^0(w) &= x_s(w-w^{-1}), \quad (s=1,2,\cdots,m),\\
\kappa_s^{m+1}(w) &= \overline{x}_s(w-w^{-1}) \quad (s=1,2,\cdots,m),\\
\kappa_t^0(w) &= y_t(w-w^{-1}) \quad (t=m+2,m+3,\cdots,N-1),\\
\kappa_t^{m+1}(w) &= \overline{y}_t w(w-w^{-1}) \quad (t=m+2,m+3,\cdots,N-1),
\end{aligned}$$

(iv) *When* $\sharp(P)>1$, $S\neq\emptyset$ *and* $T\neq\emptyset$, *namely when* $N>2$, $1<l<[N/2]$ *and* $0<m\leq[(N-2l)/2]$, $K(z)$ *has four parameters*

$$A_0, A_1, A_2, A_3$$

for the diagonal elements, and $2N-2l$ *parameters*

$$\begin{aligned}
&u_p, v_p \quad (p=0,1,\cdots,l-1)\\
&x_s, \overline{x}_s \quad (s=l,l+1,\cdots,l+m-1)\\
&y_t, \overline{y}_t \quad (t=2l+m,2l+m+1,\cdots,N-1)
\end{aligned}$$

for the off-diagonal elements, which satisfy the relations

$$\begin{aligned}
u_p v_p &= A_0 A_1 \quad (p = 0, 1, \cdots, l-1) \\
u_p v_p &= A_2 A_3 \quad (p = 0, 1, \cdots, l-1) \\
x_s \overline{x}_{s'} &= \overline{x}_s x_{s'} \quad (s, s' = l, l+1, \cdots, l+m-1), \\
u_{l-1} \overline{x}_s &= x_s A_0 \quad (s = l, l+1, \cdots, l+m-1), \\
v_{l-1} x_s &= \overline{x}_s A_1 \quad (s = l, l+1, \cdots, l+m-1). \\
y_t \overline{y}_{t'} &= \overline{y}_t y_{t'} \quad (t, t' = 2l+m, 2l+m+1, \cdots, N-1), \\
u_0 \overline{y}_t &= y_t A_2 \quad (t = 2l+m, 2l+m+1, \cdots, N-1), \\
v_0 y_t &= \overline{y}_t A_3 \quad (t = 2l+m, 2l+m+1, \cdots, N-1).
\end{aligned}$$

The non-zero elements of $K(z)$ are

$$\begin{aligned}
\kappa_p^p(w) &= (A_0 - A_1) w^{-1} - A_2 + A_3 \\
&\qquad (p = 0, 1, \cdots, l-1), \\
\kappa_s^s(w) &= A_0 w^{-1} - A_2 + A_3 - A_1 w, \\
&\qquad (s = l, l+1, \cdots, l+m-1), \\
\kappa_q^q(w) &= (A_0 - A_1) w - A_2 + A_3, \\
&\qquad (q = l+m, l+m+1, \cdots, 2l+m-1), \\
\kappa_t^t(w) &= A_3 + (A_0 - A_1) w - A_2 w^2, \\
&\qquad (t = 2l+m, 2l+m+1, \cdots, N-1), \\
\kappa_{2l+m-p-1}^p(w) &= u_p (w - w^{-1}) \quad (p = 0, 1, \cdots, l-1), \\
\kappa_p^{2l+m-p+1}(w) &= v_p (w - w^{-1}) \quad (p = 0, 1, \cdots, l-1), \\
\kappa_s^{l-1}(w) &= x_s (w - w^{-1}), \quad (s = l, l+1, \cdots, l+m-1), \\
\kappa_s^{2l+m-1}(w) &= \overline{x}_s (w - w^{-1}) \quad (s = l, l+1, \cdots, l+m-1), \\
\kappa_t^0(w) &= y_t (w - w^{-1}) \quad (t = 2l+m, 2l+m+1, \cdots, N-1), \\
\kappa_t^{2l+m-1}(w) &= \overline{y}_t w (w - w^{-1}) \quad (t = 2l+m, 2l+m+1, \cdots, N-1),
\end{aligned}$$

We can check with the direct calculation that $K(z)$ in Theorem 10.1 satisfies the initial condition $K(1) = \rho_0 Id_{\mathbf{C}^N}$ and the unitarity $K(z)K(z^{-1}) = \rho_K(z) Id_{\mathbf{C}^N \otimes \mathbf{C}^N}$

Proposition 10.1. *The K-matrix $K(z)$ in Theorem 10.1 satisfies the initial condition and the unitarity*

$$\begin{aligned}
K(1) &= \rho_0 Id_{\mathbf{C}^N}, \\
K(z)K(z^{-1}) &= \rho_K(z) Id_{\mathbf{C}^N \otimes \mathbf{C}^N},
\end{aligned}$$

where ρ_0 and $\rho_k(z)$ are

(1) When $S = \emptyset$ and $T = \emptyset$, namely when N is even and $N = 2l$,

$$\rho_0 = A_0 + A_3,$$
$$\rho_K(z) = (A_0 w^{-1} + A_3)(A_0 w + A_3) - u_0 v_0 (w - w^{-1})^2.$$

(2) When $S = \emptyset$ and $T \neq \emptyset$, namely when $N > 2$, $1 \leq l \leq [(N-1)/2]$ and $m = 0$,

$$\rho_0 = A_0 - A_2 + A_3,$$
$$\rho_K(z) = (A_0 w^{-1} - A_2 w^{-2} + A_3)(A_0 w - A_2 w^2 + A_3).$$

(3) When $\sharp(P) = 1$, $S \neq \emptyset$ and $T \neq \emptyset$, namely when $N > 2$, $l = 1$ and $0 < m \leq [N/2 - 1]$,

$$\rho_0 = A_0 - A_1 - A_2 + A_3$$
$$\rho_K(z) = (A_0 w^{-1} - A_1 w^{-1} - A_2 w^{-2} + A_3)(A_0 w - A_1 w - A_2 w^2 + A_3).$$

(4) When $\sharp(P) > 1$, $S \neq \emptyset$ and $T \neq \emptyset$, namely when $N > 2$, $1 < l < [N/2]$ and $0 < m \leq [(N-2l)/2]$,

$$\rho_0 = A_0 - A_1 - A_2 + A_3$$
$$\rho_K(z) = (A_0 w^{-1} - A_1 w^{-1} - A_2 w^{-2} + A_3)(A_0 w - A_1 w - A_2 w^2 + A_3).$$

We can count the number of solutions.

Corollary 10.1. *The number of solutions $\mathcal{N}_N$ for $N \geq 2$ to the reflection equation (6) classified in Theorem10.1 is*

$$\mathcal{N}_N = \frac{1}{2} n(n+1) \quad (N = 2n,\ N = 2n+1).$$

Proof.

(i) If $N = 2n$, $S = \emptyset$ and $T = \emptyset$, the number of solution of this case is 1.

(ii) If $N = 2n$, $S = \emptyset$ and $T \neq \emptyset$, the number of solution of this case is $n-1$ because $l = 1, 2, \cdots, n-1$.

(iii) If $N = 2n$, $\sharp(P) = 1$, $S \neq \emptyset$ and $T \neq \emptyset$, the number of solution of this case is $n-1$ because $m = 1, 2, \cdots, n-1$.

(iv) If $N = 2n$, $\sharp(P) > 1$, $S \neq \emptyset$ and $T \neq \emptyset$, the number of solution of this case is

$$\sum_{l=2}^{n-1} \sum_{m=1}^{n-l} 1 = \frac{1}{2} n^2 - \frac{3}{2} n + 1,$$

because $l = 2, 3 \cdots, n-1$ and $m = 1, 2, \cdots, n-l$.

We have

$$\mathcal{N}_{2n} = 1 + (n-1) + (n-1) + \left(\frac{1}{2}n^2 - \frac{3}{2}n + 1\right) = \frac{1}{2}n(n+1).$$

We can count the number of each kind of solutions.

(i) If $N = 2n+1$, $S = \emptyset$ and $T = \emptyset$, the number of solution of this case is 0.

(ii) If $N = 2n+1$, $S = \emptyset$ and $T \neq \emptyset$, the number of solution of this case is n because $l = 1, 2, \cdots, n$.

(iii) If $N = 2n$, $\sharp(P) = 1$, $S \neq \emptyset$ and $T \neq \emptyset$, the number of solution of this case is $n-1$ because $m = 1, 2, \cdots, n-1$.

(iv) If $N = 2n$, $\sharp(P) > 1$, $S \neq \emptyset$ and $T \neq \emptyset$, the number of solution of this case is

$$\sum_{l=2}^{n-1}\sum_{m=1}^{n-l} 1 = \frac{1}{2}n^2 - \frac{3}{2}n + 1,$$

because $l = 1, 2, \cdots, n$ and $m = 1, 2, \cdots, n-l$.

We have

$$\mathcal{N}_{2n+1} = 0 + n + (n-1) + \left(\frac{1}{2}n^2 - \frac{3}{2}n + 1\right) = \frac{1}{2}n(n+1). \qquad \square$$

We define an N-by-N diagonal matrix $D(z)$ to express some example.

Definition 10.1.

$$D(z) := \mathrm{diag}(1, z, z^2, \cdots, z^{N-1}).$$

Example 10.1. When $N = 2$, the solution in Theorem 10.1 is well-known[1].[3] That is

$$D(z)K(z)D(z) = \begin{pmatrix} A_0 w^{-1} + A_3 & u_0(w - w^{-1}) \\ v_0(w - w^{-1}) & A_0 w + A_3 \end{pmatrix},$$

where there is no relation among parameters.

Example 10.2.[9] When $N = 3$, there is only one solution upto the similarity, which is corresponding to the case ($N = 3$, $l = 1$, $m = 0$) in Theorem 10.1. That is

$$D(z)K(z)D(z)$$
$$= \begin{pmatrix} A_0 w^{-1} - A_2 + A_3 & u_0(w - w^{-1}) & y_2(w - w^{-1}) \\ v_0(w - w^{-1}) & A_0 w - A_2 + A_3 & \overline{y}_2 w(w - w^{-1}) \\ 0 & 0 & A_3 - A_1 w - A_2 w^2 \end{pmatrix},$$

where there are relations among parameters,

$$\begin{aligned} u_0 v_0 &= A_2 A_3, \\ u_0 \overline{y}_2 &= y_2 A_2, \\ v_0 y_2 &= \overline{y}_2 A_3 \end{aligned}$$

The relations above can be rewritten in the determinant form,

$$\text{rank}\begin{bmatrix} u_0 & A_3 & y_2 \\ A_2 & v_0 & \overline{y}_2 \end{bmatrix} = 1,$$

by which the parameter space is isomorphic to $\mathbf{C} \oplus (\mathbf{P}^1 \otimes \mathbf{P}^2)$, and birationally to $\mathbf{C} \oplus \mathbf{P}^3$. (The space $\mathbf{C}$ corresponds to the parameter A_0.)

Example 10.3. In the case of $l = 1$ and $m = 0$, the relations are rewritten

$$\text{rank}\begin{bmatrix} u_0 & A_3 & y_2 & y_3 & \cdots & y_{N-1} \\ A_2 & v_0 & \overline{y}_2 & \overline{y}_3 & \cdots & \overline{y}_{N-1} \end{bmatrix} = 1,$$

which implies that the parameter space is isomorphic to $\mathbf{C} \oplus (\mathbf{P}^1 \otimes \mathbf{P}^{N-1})$, and birationally to $\mathbf{C} \oplus \mathbf{P}^N$.

References

1. E. K. Sklyanin, "Boundary conditions for integrable quantum systems", J. Phys. A: Math. Gen. 21, pp2375-2389, (1988)
2. A.A. Belavin, "Dynamical symmetries of integrable quantum system", Nucl. Phys. B180 [FS21], pp189-200, (1981)
3. P.P. Kulish and E. K. Sklyanin, "The general $U_q(sl(2))$ invariant XXZ integrable quantum spin chain", J. Phys. A: Math. Gen. 24, L435-L439", (1991)
4. I.V. Cherednik, "On the properties of factorized S-matrices in elliptic functions", Sov. J. Nucl. Phys. 36, pp320-324, (1982)
5. A. Bovier, "Factorized S-matrices and generalized Baxter model", J. Math. Phys. 24, pp631-641, (1983)
6. C.A. Tracy, "Embedded elliptic curve and Yang-Baxter equations", Phisica 16D, pp203-222, (1985)
7. M.P. Richey and C.A. Tracy, "$\mathbf{Z}_n$ Baxter model: Symmetries and the Belavin's parameterization", J. Stat. Phys. 42, pp311-348, (1986)
8. Y. Yamada, "A remark on solutions of reflection equation for the critical $\mathbf{Z}_N$-symmetric vertex model", J. Phys. A: Math. Gen. 37, pp521-535, (2004)
9. Y. Yamada, "Segre threefold and $N = 3$ reflection equation", Phys. Lett. A. 298, pp350-360, (2002)
10. Y.Yamada, in preparation.